Basic
Mathematics

FOR SCIENCE AND ENGINEERING

Basic
Mathematics

FOR SCIENCE AND ENGINEERING

The Late Paul G. Andres
Illinois Institute of Technology

Hugh J. Miser
Operations Analyst, United States Air Force

Haim Reingold
Illinois Institute of Technology

JOHN WILEY & SONS, INC.

New York · London · Sydney

BOOKS BY ANDRES, P. G., MISER, H. J., and
REINGOLD, H.

Basic Mathematics for Engineers
Basic Mathematics for Science and Engineering

Library of Congress Catalog Card Number: 55-8369 √

PRINTED IN THE UNITED STATES OF AMERICA

Preface ─────────────────────

This book is a revision of an earlier volume by the same authors published in 1944 under the title *Basic Mathematics for Engineers*. The principal changes effected in the revision are these:

1. The text has been thoroughly reviewed and revised to insure that it meets the needs of scientific as well as engineering curricula. To this end, the applications treated, as well as many exercises and illustrative examples, have been drawn from varied fields of science as well as engineering.

2. The portion of the book devoted to analytic geometry has been expanded.

3. The exercises have been replaced.

Many lesser changes have been made in order to strengthen and improve the book. However, the organization and objectives of the book remain the same, and all of the principal features that distinguished the earlier volume from other texts in this field have been retained.

This book presents the mathematics required for the intelligent pursuit of elementary science and engineering courses and serves as a preparation for a course in the calculus. It contains the topics from algebra, trigonometry, and analytic geometry that are needed to meet this dual objective.

The book is intended for use with students who have had at least two years of high school mathematics. However, it is completely self-contained except that it relies on the usual simple facts from arithmetic and geometry. Since the necessary facts from geometry are collected in the appendix, it is possible to use the book successfully with intelligent students who have had very little mathematical training beyond ordinary arithmetic, or who have largely forgotten their high school mathematics.

An attempt has been made to render the explanations and discussions unusually clear and complete so that the book can also be used for home study, or in courses where greater reliance than usual is placed on the student's reading.

Many illustrative examples are worked out in full so that the student can see in operation the principles discussed in the text and so that he has a clear procedure after which to pattern his own work.

One purpose of the book is to show the student how elementary mathematics is used in scientific and engineering applications. Therefore, the book contains many uses of scientific and engineering symbols and terminology, and numerous illustrations, graphs, examples, and exercises from science and engineering. Thus, without sacrificing any of the mathematical content and rigor expected of a textbook at this level, we have attempted to utilize the practical sophistication of the modern student.

Certain special features of the book are designed to make it especially adapted to scientific and engineering curricula:

1. Principles of accuracy in numerical computations are discussed in the first chapter, and the student is reminded of them throughout the text.

2. The slide rule is discussed in the first chapter, and the student can use it to solve problems throughout the book.

3. Graphical methods are stressed.

4. Since the trigonometric functions are needed early in many science and engineering courses, they are introduced early in the book.

5. A brief introductory chapter on vectors is placed early in the text.

6. The graphs of the trigonometric functions are discussed fully.

7. The Doolittle method of solving simultaneous linear equations is treated.

8. Two introductory chapters on differential and integral calculus are supplied for use in courses where the fundamental ideas of calculus are needed.

9. Each chapter begins with a brief discussion that places the material of the chapter in a scientific and engineering setting and concludes with a summary of the material treated in the chapter.

10. There is an abundance of exercises for the student, many being chosen from engineering and the physical sciences.

11. The arrangement of chapters allows considerable flexibility in the order in which they are used. For example, the book is particularly suited for use where parallel science or engineering courses require the early introduction of such topics as trigonometric functions, vectors, and the fundamental concepts of calculus.

Since Professor Andres completed his work on the manuscript of this book shortly before his death on August 26, 1953, it appears now in a form that accords with his wishes. Thus this book, the last fruit of a cooperation begun with training courses for the Signal Corps in 1942, is Professor Andres' final contribution to education for science and engineering, a field he served notably during his lifetime.

HUGH J. MISER
HAIM REINGOLD

May 1955

Contents

1

Numerical Computations

A great many modern scientific and technological problems demand for their solution great care and skill in computation. Therefore scientists and engineers must have the ability to compute easily, quickly, and accurately. This ability depends on a knowledge of the fundamental properties of numbers, on skill developed from much practice, and on the use of various mechanical devices to speed and simplify the computational procedures. Further, especially if the numbers entering the computation were obtained experimentally, it is important to be able to estimate the accuracy of the result.

In this chapter we shall study the fundamental properties of numbers and the operations with them; we shall learn to use the slide rule, a device for simplifying and speeding computations; and finally we shall give certain rules which can be used to estimate the accuracy of our results.

We assume, of course, that the reader is already familiar to some extent with the ordinary processes of arithmetic. However, in order to establish a firm foundation for later work, in the early part of this chapter we shall reconsider some fundamental arithmetical ideas.

1–1. The Natural Numbers

The number system that we use today is the result of a long development. The necessity of counting objects forced man to create the simplest numbers, called the **natural numbers,** which we denote by the symbols 1, 2, 3, 4, etc. Problems of everyday life suggested certain operations with these numbers, and finally the four operations of addition, subtraction, multiplication, and division were developed.

Various properties of the natural numbers and of the operations with them were discovered quite intuitively and were used almost instinctively. As a basis for the more complicated problems and methods which are to be studied in this book, a knowledge of the following laws and properties is essential.

1

1. *The process of counting starts with the natural number one and can be carried on indefinitely.*

2. *Of two given natural numbers, one occurs earlier than the other in counting;* the natural number encountered earlier is said to be **smaller** than the later one, and the later one is said to be **greater** than the earlier one. For brevity, we use the sign $<$ as an abbreviation for the phrase **is smaller than** and the sign $>$ for **is greater than.** For example, $3 < 5$ and $5 > 3$. Note that the *narrowing end* of the symbol is nearer the smaller natural number, and the *widening end* is nearer the larger natural number. In this connection, the phrase *is less than* is frequently used in place of the phrase *is smaller than.*

3. *If a first natural number is greater than a second, and the second is greater than a third, then the first natural number is greater than the third.* For example, if $7 > 4$ and $4 > 2$, then $7 > 2$.

4. *The result of an addition or multiplication does not depend on the arrangement of the natural numbers.* This statement is called the **commutative law.**

Example 1.
$$3 + 5 + 7 = 7 + 5 + 3 = 5 + 7 + 3 = 15.$$
Example 2.
$$3 \times 7 = 7 \times 3 = 21.$$

5. *In the addition or multiplication of more than two natural numbers, the result is the same, no matter how the natural numbers are grouped in performing the computation.* This statement is called the **associative law** or the **law of grouping**.

Example 3.
$$17 + 13 + 26 + 14 = 70;$$
$$17 + 13 = 30, \ 26 + 14 = 40,$$
$$30 + 40 = 70.$$

Example 4.
$$3 \times 7 \times 2 \times 5 = 210;$$
$$3 \times 7 = 21, \ 2 \times 5 = 10,$$
$$21 \times 10 = 210.$$

6. *If the sum of two or more natural numbers is to be multiplied by a natural number, the result can be found by multiplying each of the first two or more natural numbers by the latter and adding the products.* This statement is called the **distributive law.**

Example 5. The sum of 13 and 18 is to be multiplied by 7.

(*a*) Add: $13 + 18 = 31.$
   Multiply: $31 \times 7 = 217.$

(*b*) Multiply each: $13 \times 7 = 91, \ 18 \times 7 = 126.$
   Add: $91 + 126 = 217.$

1-2. The Positive and Negative Integers

If we add any two natural numbers, the sum is a natural number; if we multiply any two natural numbers, the product is a natural number. However, we cannot always subtract one natural number from another and obtain a natural number as an answer. For example, $5 - 3 = 2$, but there is no natural number representing the difference $3 - 3$ or the difference $3 - 5$. Generally, we obtain a natural number when we subtract a smaller natural number from a larger one, but there is no natural number representing either a natural number minus itself or a smaller natural number minus a larger one. On the other hand, it is desirable to have answers for all subtraction problems, *so desirable, in fact, that we shall invent some new numbers to provide answers to the subtraction problems whose answers are not natural numbers.*

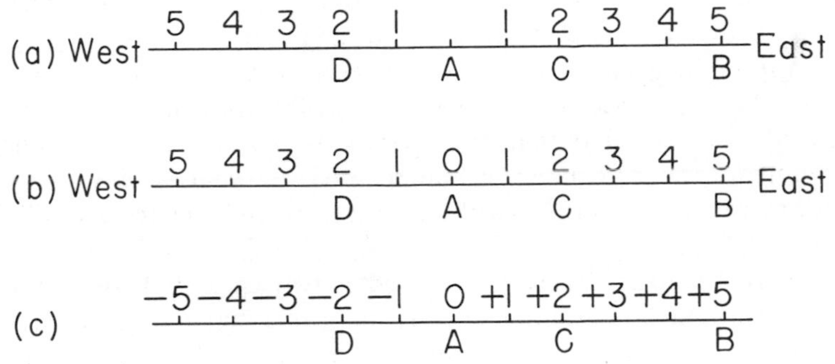

FIG. 1–1

Let us assume that the line in Fig. 1–1*a* represents an east-west highway, that the natural numbers and their corresponding points to the right of A represent distances in miles east from A, and that the natural numbers and their corresponding points to the left of A represent distances in miles west from A. Suppose that a car starts at A, travels 5 miles east to B, and then returns 3 miles to C. We may think of its journey as corresponding to the subtraction $5 - 3 = 2$. On the other hand, suppose that the car returns 5 miles from B to the starting point A. This journey corresponds to the subtraction $5 - 5$, for which no natural number provides an answer. In order to provide an answer to this subtraction, *we invent a number* 0, *called* **zero,** which we attach to A, as shown in Fig. 1–1*b*. Then $5 - 5 = 0$, $7 - 7 = 0$, $10 - 10 = 0$, etc.

Now let us suppose that, after traveling from A to B, the car travels west 7 miles to D. This journey corresponds to the subtraction $5 - 7$, for which the natural numbers provide no answer. Thus we cannot use the number 2 at D as our answer, as we did when we wrote the number 2 at C as the answer for the subtraction $5 - 3$. But here we recognize that each natural number in Fig. 1–1a has two points corresponding to it, one east and one west of A. Thus we may go one step further and think of the numbers as being east and west, with "east 2" at C and "west 2" at D, etc. However, we prefer to choose terms not associated with any particular example, and so we call the point at C "plus 2" and the point at D "minus 2", abbreviating these terms by the symbols $+2$ and -2, respectively. *In this way we invent the* **positive integers** $+1$, $+2$, $+3$, $+4$, *etc., and the* **negative integers** -1, -2, -3, -4, *etc., and agree that they correspond to the points on a line as shown in Fig.* 1–1c. Finally, of course, we agree that the positive integer $+2$ is the same as the natural number 2, $+5$ is the same as 5, etc. Having made these agreements, we find that the subtraction $5 - 7$ that we have been thinking about has the answer -2, the negative integer corresponding to the point D.

The positive integers, the negative integers, and zero compose the set of **integers.** Later (see Sec. 1–6), when the rules for operating with integers are stated, we shall find that the operations of addition, subtraction, and multiplication with integers lead to integers; that is, any of these operations can be performed with any integers and the results will be integers.

The signs $+$ and $-$ are each now used in two senses. First, the processes of addition and subtraction are indicated by these signs; and, second, the signs are used to distinguish the positive and negative integers. In the first sense they are termed *signs of operation*, and in the second they are termed *signs of quality*. When used as signs of operation they are always expressed, but $+$ is usually omitted as a sign of quality. Thus the signs in $20 - 5 + 7$ are signs of operation, but the signs of $+6$ and -8 are signs of quality. Here 6 is usually written instead of $+6$.

Positive and negative signs are used to indicate direction in many scientific and engineering problems. If a spring is fastened to a rigid support at one end, and the other is acted on by several forces, some pushing and some pulling, we may designate those forces that tend to compress the spring as positive and those that tend to elongate the spring as negative. Similarly, the ammeter on the dashboard of an automobile has positive numbers to indicate that the battery is being charged, whereas negative numbers show discharge. In this case the sign of the number indicates the direction of flow of electric current.

A convenient standard representation of the integers is suggested by the car problem above. Draw a horizontal straight line and mark an arbitrary point 0 on it. After choosing a convenient unit of measurement, place the integer $+1$ one unit to the right of 0, place the integer -1 one unit to the left of 0, place the integer $+2$ two units to the right of 0, etc., as shown in Fig. 1–2.   An arrow is usually placed on such a line to indicate the direction in which distances are measured from 0 for the positive integers.

In connection with this representation of the integers, we note that, if one positive integer is greater than a second, the first lies to the right of the second on the line.   This remark can be used to extend the ideas of *greater than* and *less than* to the negative integers and zero, as follows:

1. *If one integer lies to the right of a second integer on the line, the first is greater than the second.*   Thus, $5 > 3$, $2 > 0$, $5 > -3$, $-5 > -7$, and $0 > -6$

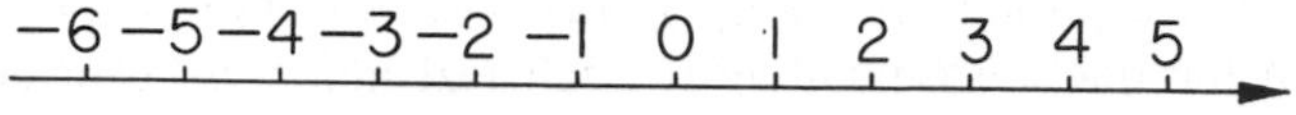

$$-6 \; -5 \; -4 \; -3 \; -2 \; -1 \quad 0 \quad 1 \quad 2 \quad 3 \quad 4 \quad 5$$

Fig. 1–2

2. *If one integer lies to the left of a second integer on the line, the first is less than the second.*   Thus $3 < 5$, $0 < 3$, $-2 < 1$, $-3 < 0$, and $-5 < -3$.

Exercises

Place the proper sign, $<$ or $>$, between the integers of the following pairs.

1. 4, 2.	**2.** 4, 9.	**3.** $-4, 0$.	**4.** $-2, 3$.
5. $-4, -2$.	**6.** $-6, -10$.	**7.** 5, -5.	**8.** 4, -1.
9. 6, -2.	**10.** $-4, 2$.	**11.** 1, -4.	**12.** 6, 1.
13. $-3, -10$.	**14.** 3, -3.	**15.** $-5, 6$.	**16.** $-7, -2$.

1–3. Rational and Irrational Numbers: The System of Real Numbers

The product of any two natural numbers is a natural number.   However, the quotient of two natural numbers is not necessarily a natural number.   To provide answers for all quotients consisting of a natural number divided by a natural number, the positive fractions with which the reader is familiar were introduced.   By introducing also the corresponding negative fractions, it is possible (see Sec. 1–6) to provide a result for the quotient of any two integers (with one exception discussed in Sec. 1–7).

The integers and the fractions form the set of **rational numbers.**   Since a fraction is the quotient of two integers, and an integer can be thought of as itself divided by the integer one, we may characterize rational numbers as follows: *A rational number is one that can be written as the quotient of two integers.*

There are processes that lead from rational numbers to the introduction of numbers that are not rational; these numbers are called **irrational.** For instance, it can be shown that it is not possible to find a rational number which when multiplied by itself gives 2. In other words, the square root of 2 is not rational. Other operations, beyond the scope of this book, lead also to irrational numbers.

Rational and irrational numbers form the set (or system) of **real numbers.** The term **number** is frequently used to mean real number.

In Chapter 12 we shall find that the operation of taking square roots leads to the introduction of other new elements, the **imaginary numbers.**

The point on the line of Fig. 1–2 corresponding to any real number can now be located as follows. *Measure from* 0 *a distance equal to the given number times the unit of measurement, measuring to the right if the given number is positive, to the left if it is negative.* In this way there is a point on the line corresponding to each real number. Conversely, *the real number corresponding to a given point on the line can be determined by measuring the number of units the point lies to the right or left of* 0. In this way there is a real number corresponding to each point on the line. We shall assume that these rules for determining the correspondence between the real numbers and the points on the line yield the following situation: *To each real number there corresponds one and only one point on the line and to each point on the line there corresponds one and only one real number.*

When a correspondence of this kind has been set up between the real numbers and the points on a line, we shall speak of the line as a **number line.** The point on the number line corresponding to 0 is called the **origin.** The direction on the number line which is the direction the positive numbers are plotted from the origin is called the **positive direction;** the other direction is called the **negative direction.** Although it is frequently convenient to draw a number line horizontally with its positive direction to the right as in Fig. 1–2, this arrangement is not necessary; number lines are frequently drawn in other orientations.

Using the representation of numbers as points on a number line drawn horizontally with its positive direction to the right, we can extend in a very simple way the ideas of *greater than* and *less than* to all real numbers by restating the definitions given at the end of the preceding section with *real number* replacing *integer:*

1. *If one real number lies to the right of a second real number on the number line, the first is greater than the second.*

2. *If one real number lies to the left of a second real number on the number line, the first is less than the second.*

By these definitions, every positive number is greater than 0 and every negative number is less than 0.

*The **absolute value** of a positive number is the number itself; the absolute value of a negative number is the positive number obtained by changing its sign of quality from − to +.*

The absolute value is denoted by parallel vertical bars on either side of the number; the absolute values of $+5$ and -5 are denoted by $|+5|$ and $|-5|$, respectively. By the definition, $|+5| = |-5| = 5$.

Exercises

Locate the given numbers on a number line, and insert the proper inequality sign, $>$ or $<$, between the numbers of each pair.

1. 2, 5.	**2.** 7, 1.	**3.** -3, 4.	**4.** -2, -4.
5. -8, 6.	**6.** $\frac{3}{4}$, $\frac{7}{8}$.	**7.** $\frac{9}{2}$, $-\frac{1}{2}$.	**8.** 1.4, -2.3.
9. -6.2, -3.1.	**10.** 0, -3.	**11.** $-4\frac{1}{2}$, -4.	**12.** -3, 3.
13. -3, 5.	**14.** $-\frac{6}{7}$, $-\frac{9}{10}$.	**15.** -100, 50.	**16.** -2π, π.

Locate the numbers in each exercise on a number line and then arrange the numbers in each exercise in increasing order.

17. 3, -2, π, -6.	**18.** $\frac{1}{2}$, -6.5, 0, 10.
19. 1.2, -3.5, 6.1, 0.	**20.** -18, -100, 60, -50.
21. 8, -16, 0, -10.	**22.** π, -2π, 4, -6.
23. 1, -10, 0.5, -4.5.	**24.** $\frac{5}{2}$, $-\frac{7}{2}$, $-\frac{1}{2}$, 4.

State the absolute value of each of the following numbers.

25. 0, -2, 3, -5.	**26.** 14, -10, $-\frac{3}{2}$, $\frac{5}{2}$.
27. π, $-\pi$, $\frac{3}{2}$, $-\frac{3}{2}$.	**28.** 8.5, -6.2, -3.6, 0.1.
29. -6, -10, 12, -8.	**30.** 4, 6, -1, 1.5.

1–4. The Basic Generalization of Algebra

In arithmetic, operations are performed with specific numbers. However, when it is desired to make general arguments about numbers, letters are used, such as a, b, C, D, x, y, z. These letters, in any particular application, represent numbers and may be replaced by numbers at the will of the reader. Thus by using letters we may designate, in effect, an entire group of numbers at once instead of being limited to one at a time. This generalization of the idea of the number, in addition to furnishing an extremely economical means of expressing ideas about physical quantities, makes possible the achievement of many general results which apply to all or large groups of numbers. It is the basic characteristic of algebra (see Sec. 2–1) and furnishes the mathematician, the scientist, and the engineer with a practical tool of great power. Many uses of this tool in practical applications will be seen in this book.

1–5. The Signs of Arithmetic and Algebra

Concept	Sign	Example
Addition	$+$	$5 + 2$
Subtraction	$-$	$3 - 2$
Multiplication	$\cdot$ or $\times$	$3 \cdot 4$ or 3×4
Division	$\div$	$6 \div 2$ or $\frac{6}{2}$
Signs of grouping — Parentheses	$(\ )$	$(2 + 3)$
Brackets	$[\ ]$	$[2 - (3 - 6)]$
Braces	$\{\ \}$	$\{1 - [3 + (3 - 2)]\}$
Equality	$=$	$3 = 3$
Not equal	$\neq$	$7 \neq 5$
Less than	$<$	$2 < 5$
Greater than	$>$	$2 > -3$
Equal to or less than	$\leq$	$a \leq b$
Equal to or greater than	$\geq$	$c \geq d$
Plus or minus	$\pm$	$x = \pm 2$
Minus or plus	$\mp$	$x = \mp 4$
Positive square root	$\sqrt{\ }$	$\sqrt{4} = +2$
Negative square root	$-\sqrt{\ }$	$-\sqrt{4} = -2$
Cube root	$\sqrt[3]{\ }$	$\sqrt[3]{8} = 2$

1–6. The Laws of Signs

The following rules govern the operations of addition, subtraction, multiplication, and division in the system of real numbers:

1. *Addition and subtraction.*

(a) *To add two numbers with like signs of quality, add their absolute values and attach their common sign of quality to the result.*

Example 1.

$$+5 + (+4) = |+5| + |+4| = + |5 + 4| = +9 = 9.$$

Example 2.

$$(-5) + (-4) = -(|-5| + |-4|) = -(5 + 4) = -9.$$

(b) *To add two numbers with unlike signs of quality, subtract the smaller absolute value from the larger, and attach to the result the sign of quality of the number with the larger absolute value.*

Example 3. $6 + (-2) = + |6 - 2| = +4 = 4.$

Example 4. $7 + (-9) = - |9 - 7| = -2.$

(c) *To subtract one number from another, change the sign of quality of the number to be subtracted (the subtrahend), and proceed as in addition.*

By this rule addition and subtraction may be regarded as essentially the same process.

Example 5.
$$12 - (+20) = - \mid 20 - 12 \mid = -8.$$

2. *Multiplication and division.*

(a) *The product of two numbers of like signs of quality is the product of their absolute values; the product of two numbers of unlike signs of quality is the negative of the product of their absolute values.*

Example 6.
$$12 \times 4 = \mid 12 \mid \times \mid 4 \mid = 48.$$

Example 7.
$$(12) \times (-4) = -(\mid 12 \mid \times \mid -4 \mid) = -(12 \times 4) = -48.$$

(b) *The quotient of two numbers of like signs of quality is the quotient of their absolute values; the quotient of two numbers of unlike signs of quality is the negative of the quotient of their absolute values.*

Example 8.
$$\frac{-6}{-2} = \frac{\mid -6 \mid}{\mid -2 \mid} = \frac{6}{2} = 3.$$

Example 9.
$$\frac{10}{-5} = - \frac{\mid 10 \mid}{\mid -5 \mid} = - \frac{10}{5} = -2.$$

The **reciprocal** of a number is defined as one divided by the number. The reciprocal of 6 is $\frac{1}{6}$; the reciprocal of -2 is $-\frac{1}{2}$. We will see in Sec. 1–10 that *division by a given number is the same as multiplication by the reciprocal of that number, so that multiplication and division may be regarded as essentially the same process.*

1-7. Operations with Zero

Where a is any number, the operations with zero are defined as follows.
$$a \pm 0 = a.$$
$$a \cdot 0 = 0.$$
$$\frac{0}{a} = 0, \quad \text{if} \quad a \neq 0.$$

Note that division by zero is not included. To see why this is so we need only examine the definition of the quotient q of a and b, $q = a/b$ if $a = b \cdot q$. Let $a = 5$ and $b = 0$. Then the quotient $5/0$ should yield the number q such that $5 = 0 \cdot q$. But there is no number q such that $0 \cdot q = 5$, since by definition 0 times any number is zero. Hence, neither 5 nor any other nonzero number can be divided by zero without contradiction. By the definition, the quotient $0/0$ must be a number q such that $0 = q \cdot 0$. Now any value of q satisfies this relation, and hence no unique value of q is determined. We therefore assign no particular value to $0/0$. Since it is desirable that the number system be free of contradiction and that the process of division lead to a definite result, *we exclude the process of division by zero from arithmetic.*

1–8. Fundamental Laws of Arithmetic

It is important to note that *the general properties of positive integers and of the operations with them,* as described in Sec. 1–1, *hold for all real numbers,* with the exception of statements 1 and 2, which obviously apply only to natural numbers. However, statement 2 can be reworded as follows. *If two real numbers are unequal, then either the first is greater than the second or the second is greater than the first.*

For completeness, we shall restate these properties here more briefly with letters.

1. **If $a \neq b$, then either $a > b$ or $b > a$.**
2. **If $a > b$ and $b > c$, then $a > c$.**
3. **Commutative law:**
   $a + b = b + a, a \times b = b \times a.$
4. **Associative law:**
   $(a + b) + c = a + (b + c), (a \times b) \times c = a \times (b \times c).$
5. **Distributive law:**
   $(a + b) \times c = a \times c + b \times c.$

Exercises

1. Add:
   - (*a*) 4 and −10,
   - (*b*) −2 and −14,
   - (*c*) 7 and 3,
   - (*d*) 7 and −10,
   - (*e*) 32 and −15.

2. Add:
   - (*a*) 8 and −34,
   - (*b*) 5 and −10,
   - (*c*) −6 and 40,
   - (*d*) −13 and −9,
   - (*e*) 8 and −5.

3. Add:
   - (*a*) 13, −30, and 6,
   - (*b*) −12, −6, and 20,
   - (*c*) 18, 19, and −20,
   - (*d*) 11, −37, and 26,
   - (*e*) 8, −14, −6, and 13.

4. Add:
   - (*a*) 10, −19, 30, and 70,
   - (*b*) −77, 29, −16, and 37,
   - (*c*) −56, −14, 69, and 13,
   - (*d*) 19, 36, 0, and −101,
   - (*e*) 12, 0, −43, and 29.

5. Subtract:
   - (*a*) 6 from −9,
   - (*b*) −17 from 32,
   - (*c*) −9 from −17,
   - (*d*) 6 from −18,
   - (*e*) 14 from −6.

6. Subtract:
   - (*a*) 11 from 6,
   - (*b*) −9 from 11,
   - (*c*) 13 from −8,
   - (*d*) −19 from −6,
   - (*e*) 10 from −36.

7. Subtract:
   (a) −6 from −2,
   (b) 13 from −16,
   (c) 17 from −42,
   (d) 49 from 0,
   (e) 23 from −23.

8. Subtract:
   (a) −23 from 47,
   (b) −19 from −62,
   (c) −32 from 19,
   (d) −38 from 0,
   (e) 17 from 0.

9. Multiply:
   (a) 4 and −6,
   (b) −6 and 14,
   (c) 19 and −23,
   (d) −28 and −9,
   (e) 21 and −4.

10. Multiply:
    (a) −7 and 13,
    (b) 7 and 0,
    (c) −13 and 0,
    (d) −15 and −9,
    (e) −9 and −42.

11. Multiply.
    (a) 4, 6, and −8,
    (b) −12, −3, and 25,
    (c) −24, 6, and −8,
    (d) 12, −8, and −23,
    (e) 0, −23, and 111.

12. Multiply:
    (a) 2, −7, and 5,
    (b) 8, 9, and −21,
    (c) −6, −4, and −9,
    (d) −7, −5, and 0,
    (e) −21, −6, and −1.

13. Divide:
    (a) 0 by −17,
    (b) 32 by −8,
    (c) −12 by −12,
    (d) 46 by −23,
    (e) −36 by 18.

14. Divide:
    (a) −18 by −9,
    (b) −28 by −7,
    (c) −32 by 8,
    (d) −96 by −12,
    (e) −45 by −15.

15. Divide:
    (a) −12 by −24,
    (b) −18 by 36,
    (c) 24 by 0,
    (d) 0 by −46,
    (e) 36 by −108.

16. Divide:
    (a) 23 by 0,
    (b) −144 by 48,
    (c) 0 by −19,
    (d) −31 by 62,
    (e) 42 by −14.

17. Multiply:
    (a) 4 by the sum of 6 and −9,
    (b) −6 by the sum of 14 and −11,
    (c) 0 by the sum of 23 and −19,
    (d) −9 by the sum of −9 and −13,
    (e) 12 by the sum of −8 and 23.

18. Multiply:
    (a) 4 by the sum of −9 and −5,
    (b) −37 by the sum of −12 and −8,
    (c) 0 by the sum of 34 and −45,
    (d) −6 by the sum of 12 and −4,
    (e) −8 by the sum of 14 and −2.

1–9. Signs of Grouping

A table of the signs of grouping was given in Sec. 1–5. When a given number multiplies the quantity within signs of grouping, according to the distributive law, each number forming the sum or difference in the signs of grouping is to be multiplied by the given number. In such cases the sign of multiplication appearing in front of the bracket is often omitted. Other properties of operations with signs of grouping will be evident from the following illustrative examples.

Example 1. $8 \times (5 - 6) = 8(5 - 6) = 40 - 48 = -8$. Also $8 \cdot (5 - 6) = 8(-1) = -8$.

Example 2. $(-8) \times (-5) \cdot (+6) = -8(-5)(+6) = -(-40)(6) = -(-240) = 240$.

Example 3. $-1(8 - 6 + 4) = -8 + 6 - 4 = -2 - 4 = -6$, or $-1(8 - 6 + 4) = -1(6) = -6$.

Example 4. $+[8 + (-6)] = +8 + (-6) = 8 - 6 = 2$, or $+ [8 - 6] = 8 - 6 = 2$.

Example 5. $-(8 - 6) = -8 - (-6) = -8 + 6 = -2$.

Example 6. $5 - 3 + 6 = 5 + (-3 + 6)$, or $5 - 3 + 6 = 5 - (+3 - 6) = 8$.

As already indicated in the preceding examples, we can state the following rules concerning the removal and insertion of signs of grouping preceded by a + or − sign. A **term** is a number of an expression that is set off from the other numbers by + or − signs.

Rule 1. Signs of grouping preceded by a plus sign may be removed or inserted by rewriting each enclosed term or group of terms with its original sign.

Rule 2. Signs of grouping preceded by a minus sign may be removed or inserted if the sign of each of the enclosed terms or groups of terms is changed.

Example 7. $8 - \{16 + [(4 - 3) + 5] - 2\} = 8 - \{16 + [4 - 3 + 5] - 2\}$
$= 8 - \{16 + 4 - 3 + 5 - 2\} = 8 - 16 - 4 + 3 - 5 + 2 = 13 - 25 = -12$.
Also $8 - \{16 + [(4 - 3) + 5] - 2\} = 8 - \{16 + [1 + 5] - 2\} = 8 - \{16 + 6 - 2\}$
$= 8 - \{22 - 2\} = 8 - \{20\} = 8 - 20 = -12$.

Exercises

Simplify the following expressions by removing the signs of grouping:

1. $4 + (5 - 7)$.

2. $(3 - 9) - 13$.

3. $6 - (4 - 6)$.

4. $7 - 2(14 - 8)$.

5. $15 - 0(4 - 7)$.

6. $9 - [5 + 3(6 - 7)]$.

7. $-8 + [6 - 2(4 + 5)]$.

8. $8 - [6 - 0(4 - 2) + 4]$.

9. $(3 - 2) - 6(4 - 8) + 12$.

10. $-[8 + 6(2 - 7) - 8(7)]$.

11. $-[8 - 0(3 - 1) + 14]$.

12. $6[9 - 3(4 - 8) + 2(10 - 4)]$.

13. $-2[(3 + 9) - (7 - 9) + 24]$.

14. $13 - 2\{9 - [4(6 - 7) + 4(1 + 6)] + 9\}$.

15. $-8 + 6\{14 - [3(4 - 7) - 3(2 - 4)] + 9\}$.

16. $\{12 + [21 - 13 + (14 - 8)] + 13\} - 20$.

17. $\{13 - [15 - (3 + 8 - 17)] - 19\}$.

18. $\{21 + [12 - 0(2 - 8)]\} - [21 + 8(4 - 4)]$.

19. $\{13 - [12 + 4(6 - 8)]\} + 2\{6 - 3(4 - 2) + 2[4 - (6 - 7)]\}$.

20. $\{21 - 3[9(4 - 6) - 3]\} - 26(14 - 8)$.

1–10. Fractions

The term fraction was used in Sec. 1–3 to refer to rational numbers that are not integers. More generally the term refers to the quotient of any two numbers. The **dividend,** or number to be **divided,** is the **numerator;** the **divisor,** or the number by which the numerator is divided, is the **denominator.**

1. *The sum of two fractions with a common denominator is the sum of their numerators divided by their common denominator.* Symbolically,

$$\frac{a}{c} + \frac{b}{c} = \frac{a+b}{c}.$$

2. *The product of two fractions is the product of their numerators divided by the product of their denominators.* Symbolically,

$$\frac{a}{b} \cdot \frac{c}{d} = \frac{a \cdot c}{b \cdot d}.$$

Since it is obvious that $-a = (-1)a$, $-b = (-1)b$,

$$\frac{-a}{b} = (-1)\frac{a}{b}, \quad \text{and} \quad \frac{-1}{1} = \frac{1}{-1} = -1,$$

it is easy to see that

$$\frac{-a}{b} = \frac{a}{-b} = -\frac{a}{b},$$

and

$$\frac{a}{c} - \frac{b}{c} = \frac{a-b}{c}.$$

The last formula gives the *rule for subtraction of fractions.*

Since $\frac{c}{c} = 1$, we have

$$\frac{a}{b} = \frac{a \cdot c}{b \cdot c}.$$

Hence, *the numerator and denominator of a fraction can each be multiplied (or divided) by the same number without changing the value of the fraction.*

A fraction is said to be inverted by the interchange of its numerator and denominator. For example, inverting $\frac{3}{5}$ we get $\frac{5}{3}$; inverting $\frac{a}{b}$ we get $\frac{b}{a}$.

Since

$$\frac{\dfrac{a}{b}}{\dfrac{c}{d}} = \frac{\dfrac{a}{b}\cdot b\cdot d}{\dfrac{c}{d}\cdot b\cdot d} = \frac{a\cdot d}{b\cdot c},$$

we have

$$\frac{\dfrac{a}{b}}{\dfrac{c}{d}} = \frac{a}{b}\cdot\frac{d}{c},$$

or, *the quotient of two fractions is equal to the fraction of the dividend multiplied by the inverted fraction of the divisor.* Note in the above example that $\frac{a}{b}$ is the dividend and $\frac{c}{d}$ the divisor.

Since the reciprocal of a number is one divided by that number, it follows that the *reciprocal of a fraction is equal to that fraction inverted.*

An integer is said to be a **multiple** of a second integer if the quotient of the first by the second is an integer. In this case the second integer is said to **divide evenly** into the first integer.

An integer is a **prime integer** *if it is not* 0, $+1$, *or* -1 *and if it can be divided evenly only by itself, its negative, and* ±1. Thus 3, 5, -11 are primes, whereas $-4, 0, 12$ are not. Every integer different from $0, \pm1$ can be resolved into a product of positive prime integers (called **factors**) with the sign of quality of the given integer prefixed. Thus $-12 = -2\cdot2\cdot3$, $+39 = +3\cdot13$, etc.

If the numerator and denominator of a fraction are integers, and if we resolve the numerator and denominator of a fraction each into the products of their positive prime factors, it is possible to eliminate factors common to numerator and denominator by dividing both by these factors. When all such common factors have been eliminated from numerator and denominator, the fraction is in its **lowest terms.**

Example 1.
$$\frac{156}{72} = \frac{2\cdot2\cdot3\cdot13}{2\cdot2\cdot2\cdot3\cdot3} = \frac{13}{2\cdot3} = \frac{13}{6}.$$

Example 2.
$$\frac{-96}{108} = -\frac{2\cdot2\cdot2\cdot2\cdot2\cdot3}{2\cdot2\cdot3\cdot3\cdot3} = -\frac{2\cdot2\cdot2}{3\cdot3} = -\frac{8}{9}.$$

Example 3.
$$\frac{5}{18}\cdot\frac{6}{10} = \frac{5\cdot6}{18\cdot10} = \frac{5\cdot3\cdot2}{2\cdot3\cdot3\cdot2\cdot5} = \frac{1}{6}.$$

Example 4.

$$\frac{\frac{3}{8}}{\frac{5}{6}} = \frac{3}{8} \cdot \frac{6}{5} = \frac{3 \cdot \cancel{2} \cdot 3}{\cancel{2} \cdot 2 \cdot 2 \cdot 5} = \frac{9}{20}.$$

Example 5.

$$\frac{5}{8} + \frac{6}{8} - \frac{3}{8} = \frac{5 + 6 - 3}{8} = \frac{8}{8} = 1.$$

The least common multiple (*designated as* **LCM**) *of a group of integers is the smallest positive integer that is a multiple of each of the given integers.* Thus the least common multiple is the smallest positive integer into which each of the given integers will divide evenly.

The greatest common factor (*designated as* **GCF**) *of a group of integers is the largest positive integer that divides evenly into each integer of the group.*

We observe that, if the quotient of two integers is an integer, every positive prime factor of the denominator greater than 1 occurs at least as many times as a prime factor of the numerator. Using this observation, we arrive at a simple method for finding the least common multiple and the greatest common factor of a set of integers.

1. *Factor each integer of the set into the product of positive primes.*

2. *To find the least common multiple, select each prime factor the greatest number of times it occurs in any one number of the set. The product of these factors is the least common multiple.*

3. *To find the greatest common factor, select each prime factor the greatest number of times it occurs as a factor of every number of the set. The product of these numbers is the greatest common factor.*

Example 6. Find the greatest common factor and least common multiple of 56, −32, and 96.

Factoring each expression, we obtain

$$56 = 2 \cdot 2 \cdot 2 \cdot 7,$$
$$-32 = -2 \cdot 2 \cdot 2 \cdot 2 \cdot 2,$$
$$96 = 2 \cdot 2 \cdot 2 \cdot 2 \cdot 2 \cdot 3.$$

By the rule above, the GCF = $2 \cdot 2 \cdot 2 = 8$, and the LCM = $2 \cdot 2 \cdot 2 \cdot 2 \cdot 2 \cdot 3 \cdot 7 = 672$. As a check on the GCF, $56/8 = 7$, $-32/8 = -4$, and $96/8 = 12$. Since 7, −4, 12 have no common factors other than +1 or −1, 8 is the correct GCF. As a check on the LCM, $672/56 = 12$, $672/(-32) = -21$, and $672/96 = 7$; since 12, −21, and 7 have no common factors, 672 is the correct LCM.

When fractions with different denominators are to be added, a common denominator can be obtained by multiplying numerator and denominator of the fractions by certain integers. The most convenient denominator is the least common multiple of the given denominators, termed the **least common denominator.** Once the common denominator has been obtained, the previous rule of addition applies.

Example 7.   Add $\frac{5}{8} - \frac{7}{12} + \frac{13}{36}$.

We may write this sum as

$$\frac{5}{2 \cdot 2 \cdot 2} - \frac{7}{2 \cdot 2 \cdot 3} + \frac{13}{2 \cdot 2 \cdot 3 \cdot 3}.$$

Thus the LCM of the denominators is $2 \cdot 2 \cdot 2 \cdot 3 \cdot 3 = 72$, so that the sum becomes

$$\frac{5(3 \cdot 3)}{2 \cdot 2 \cdot 2 \cdot 3 \cdot 3} + \frac{-7(2 \cdot 3)}{2 \cdot 2 \cdot 2 \cdot 3 \cdot 3} + \frac{13(2)}{2 \cdot 2 \cdot 2 \cdot 3 \cdot 3} = \frac{45}{72} + \frac{-42}{72} + \frac{26}{72}$$

$$= \frac{45 - 42 + 26}{72} = \frac{29}{72}.$$

Example 8.   Simplify

$$\frac{\frac{5}{8} - \frac{5}{6} + \frac{1}{12}}{1 - \frac{2}{3} - \frac{1}{15}}.$$

The expression can be simplified as follows:

$$\frac{\dfrac{5}{2 \cdot 2 \cdot 2} - \dfrac{5}{2 \cdot 3} + \dfrac{1}{2 \cdot 2 \cdot 3}}{1 - \dfrac{2}{3} - \dfrac{1}{3 \cdot 5}} = \frac{\dfrac{15}{24} - \dfrac{20}{24} + \dfrac{2}{24}}{\dfrac{15}{15} - \dfrac{10}{15} - \dfrac{1}{15}}$$

$$= \frac{\dfrac{15 - 20 + 2}{24}}{\dfrac{15 - 10 - 1}{15}} = \frac{\dfrac{-3}{24}}{\dfrac{4}{15}} = -\frac{3}{24} \cdot \frac{15}{4} = -\frac{15}{32}.$$

Exercises

Write the following fractions in lowest terms:

1. (*a*) $\dfrac{32}{144}$,

(*b*) $\dfrac{24}{18}$,

(*c*) $\dfrac{357}{42}$.

2. (*a*) $\dfrac{-84}{124}$,

(*b*) $\dfrac{-144}{728}$,

(*c*) $\dfrac{81}{288}$.

3. (*a*) $\dfrac{256}{500}$,

(*b*) $\dfrac{800}{728}$,

(*c*) $\dfrac{-74}{111}$.

4. (*a*) $\dfrac{-58}{87}$,

(*b*) $-\dfrac{144}{256}$,

(*c*) $\dfrac{242}{396}$.

Find the LCM of

5. 5, 25, 10, and 4.

6. 3, 5, −10, and 15.

7. 4, 6, −10, and 20.

8. 12, 15, −30, and −50.

9. 96, 36, −18, and 144.

10. −81, 108, 90, and 135.

11. 98, 56, 63, and 24.

12. 48, 108, −60, and 144.

Find the GCF of

13. 48, 56, and 96.

14. −26, 39, and 78.

15. 500, −125, and 75.

16. −84, 28, and 98.

17. −66, 242, and 44.

18. −100, 144, and 28.

19. 81, 108, and −36.

20. 256, −144, and 64.

Simplify the following expressions, stating your result as a fraction in lowest terms:

21. $\frac{1}{8} + \frac{2}{3} - \frac{5}{12}$.

22. $\frac{5}{12} - \frac{7}{16} + \frac{5}{24}$.

23. $\frac{11}{15} - \frac{9}{20} + \frac{17}{30}$.

24. $\frac{13}{22} - \frac{5}{4} - \frac{1}{11}$.

25. $\frac{7}{10} - \frac{5}{6} + \frac{2}{15}$.

26. $-\frac{9}{10} + \frac{3}{25} - \frac{1}{20}$.

27. $\frac{3}{14} - \frac{5}{21} + \frac{3}{49}$.

28. $2 - \frac{11}{12} - \frac{11}{15}$.

29. $\frac{7}{12} \cdot \frac{36}{49}$.

30. $\frac{6}{7} \cdot \left(-\frac{14}{9}\right) \cdot \frac{24}{36}$.

31. $-\frac{5}{8} \cdot \frac{4}{10} \cdot \frac{12}{25}$.

32. $\left(-\frac{7}{12}\right)\left(-\frac{3}{5}\right) \cdot \frac{4}{7}$.

33. $\dfrac{\frac{27}{49}}{\frac{63}{14}}$.

34. $\dfrac{\frac{14}{13}}{\frac{49}{39}}$.

35. $\dfrac{\dfrac{-144}{110}}{\dfrac{96}{132}}$.

36. $-\dfrac{\dfrac{-81}{42}}{-\dfrac{18}{39}}$.

37. $\dfrac{2 + \frac{1}{2}}{\frac{5}{3}}$.

38. $\dfrac{3 - \frac{4}{5}}{\frac{1}{2} + \frac{3}{5}}$.

39. $\dfrac{\frac{1}{9} + \frac{5}{6} - \frac{1}{12}}{\frac{7}{8} + \frac{3}{16} - \frac{5}{12}}$.

40. $\dfrac{\frac{2}{3} - \frac{7}{6} + \frac{1}{2}}{\frac{5}{8} + \frac{7}{12} + \frac{1}{18}}$.

41. $\dfrac{\frac{23}{24} - \frac{7}{8} + \frac{1}{12}}{1 - \frac{1}{2} - \frac{2}{3}}$.

42. $\dfrac{\frac{1}{5} - \frac{3}{25} + \frac{7}{10}}{\frac{1}{8} + \frac{1}{40} - \frac{3}{50}}$.

43. $\dfrac{\frac{6}{5} - \frac{7}{30} - \frac{7}{12}}{1 + \frac{3}{4} + \frac{1}{3}}$.

44. $\dfrac{\frac{2}{3} + \frac{1}{2} \cdot \frac{11}{14}}{2 + \frac{1}{6} \cdot \frac{22}{39}}$.

45. $\dfrac{\frac{6}{5} - \frac{7}{10}}{\frac{1}{2} + \frac{1}{4} - \frac{1}{6}} \cdot \dfrac{\frac{13}{36}}{\frac{5}{9}}$.

46. $\dfrac{\frac{5}{3} + \frac{1}{4} - \frac{7}{15}}{\frac{7}{9} + \frac{1}{4} + \frac{1}{6}} \cdot \left(-\dfrac{\frac{25}{32}}{\frac{5}{24}}\right)$.

47. $\dfrac{\frac{1}{2} + \frac{1}{3}}{\frac{1}{6} + \frac{1}{9}} \cdot \dfrac{\frac{1}{2} - \frac{1}{8}}{\frac{1}{3} - \frac{1}{9}}$.

48. $\dfrac{\frac{7}{12} - \frac{5}{6}}{\frac{1}{5} - \frac{1}{20}} \cdot \dfrac{\frac{7}{20} - \frac{2}{5}}{\frac{5}{16} - \frac{3}{8}}$.

49. A truck that weighs $3\frac{5}{8}$ tons when empty is carrying $5\frac{1}{2}$ tons of cargo. If the truck has 6 wheels and each wheel carries the same weight, find the weight carried by a single wheel.

50. If the cost of electric energy is 4 cents for each kilowatt used throughout 1 hr, or 4 cents per kilowatt-hour, and if an electric iron uses $\frac{9}{20}$ kilowatt and a lamp uses $\frac{3}{100}$ kilowatt, find the cost of operating the iron and 6 lamps for 2 hr and $13\frac{1}{3}$ min.

51. If cast iron weighs $\frac{1}{4}$ lb per cu in., what is the weight of a rectangular cast-iron block whose dimensions are $3\frac{1}{2}$ by 8 by $3\frac{5}{7}$ in.?

52. If water weighs $62\frac{1}{2}$ lb per cu ft, what is the weight of water in a rectangular can whose dimensions are $2\frac{2}{5}$ by 12 by $4\frac{4}{5}$ in.?

53. A triangle whose base is $1\frac{1}{3}$ in. and whose altitude is $2\frac{2}{5}$ in. lies in a rectangle $3\frac{3}{5}$ by $2\frac{2}{3}$ in.   What portion of the square's area lies in the triangle?

54. Five water pipes come together at one place.   If water flows into the junction from two pipes at the rate of $3\frac{1}{3}$ and $1\frac{3}{5}$ gal per min, respectively, and if water flows away from the junction in two pipes at the rate of $2\frac{2}{9}$ and $\frac{5}{6}$ gal per min, respectively, at what rate is water flowing in the fifth pipe?   Is the water in the fifth pipe flowing to or from the junction?

55. One thousand board-feet of red oak weigh $3666\frac{2}{3}$ lb after air drying.   If the green lumber weighs $1\frac{6}{11}$ times the weight of the dry lumber, how many board-feet of lumber must be cut in order to obtain 1380 lb of green lumber?

56. A streamlined train is scheduled to run between La Junta and Dodge City in $2\frac{7}{12}$ hr.   On three successive days the train made the run in $158\frac{1}{2}$, 157, and $156\frac{1}{12}$ min. What was the average deviation from schedule for the three trips?

57. The stress on the wire in a cable is equal to the load on the cable divided by the area of the cross section of the wire in the cable.   If a force of $687\frac{1}{2}$ lb is carried by a steel cable $\frac{5}{16}$ in. in diameter and the wire occupies $\frac{7}{8}$ of the cross section of the cable, compute the stress in pounds per square inch.   (Let $\pi = \frac{22}{7}$.)

58. The reservoir back of Grand Coulee Dam contains 3,160,755 gal of water.   If $35\frac{31}{25}$ cu ft of water weigh 2240 lb and $13\frac{11}{25}$ gal weigh 112 lb, how many cubic feet of water does the reservoir contain?

59. If there are $5\frac{1}{6}$ grams of sulfuric acid in 62 kilograms of solution, how much sulfuric acid is there in 12 kilograms of solution?

60. A water-culture solution in which plants may be grown to the flowering and fruiting stages weighs 250 grams and consists of calcium nitrate $\frac{1}{4}$ gram, potassium nitrate, magnesium sulfate, and acid potassium phosphate each $\frac{1}{16}$ gram, ferric chloride $\frac{1}{50}$ gram, and the remainder distilled water.   How much of each component is required to make a solution weighing 100 grams?

1–11. Exponents

Exponents are introduced as an abbreviation for the product of several like factors.   For example, in place of $2 \cdot 2 \cdot 2 \cdot 2 \cdot 2$, we write 2^5.   The reader will see, by the time he has finished this book, that the exponent idea has become a tool of great usefulness, both in theory and in applications.

If a is any number and n is a positive integer, the product of n of the quantities a is denoted by a^n.   The following are the definitions needed here:

$$\underbrace{a \cdot a \cdot a \cdots a}_{n \text{ factors}} = a^n,$$

where n is called the exponent,

a is called the base.

a^n **is read** $\begin{cases} \text{"}a \text{ to the exponent } n\text{,"} \\ \text{"}a \text{ to the } n\text{th power,"} \\ \text{"the } n\text{th power of } a\text{."} \end{cases}$

When a number with its sign of quality is written with an exponent, it is necessary to include both the number and its sign of quality in parentheses; otherwise the exponent is not understood to affect the sign of quality. For example, $+2^4 = +(2)^4 = +(+2)^4 = +16$, and $-2^4 = -(2)^4 = -(+2)^4 = -16$, while $(-2)^4 = +16$.

If m and n are positive integers, we have, from our definition,

$$a^n \cdot a^m = \underbrace{(a \cdot a \cdots a)}_{n \text{ factors}} \cdot \underbrace{(a \cdot a \cdots a)}_{m \text{ factors}} = \underbrace{a \cdot a \cdots a}_{m+n \text{ factors}},$$

whence

$$a^n \cdot a^m = a^{m+n}.$$

Now, if $n > m$, we have

$$\frac{a^n}{a^m} = \frac{(a \cdot a \cdots a)}{(a \cdot a \cdots a)} = \frac{(n \text{ factors})}{(m \text{ factors})} = a \cdot a \cdots a, \ n - m \text{ factors},$$

and

$$\frac{a^n}{a^m} = a^{n-m}.$$

The reader may also show very readily that

$$(a^n)^m = a^{m \cdot n}.$$

Thus the addition of exponents takes the place of multiplication, subtraction of exponents takes the place of division, and multiplication of exponents takes the place of raising a quantity to a power.

As a consequence of the definition, if n is a positive integer,

$$(a \cdot b)^n = (a \cdot b) \cdot (a \cdot b) \cdots (a \cdot b), \ n \text{ factors},$$

$$= \underbrace{(a \cdot a \cdots a)}_{n \text{ factors}} \cdot \underbrace{(b \cdot b \cdots b)}_{n \text{ factors}},$$

which gives

$$(a \cdot b)^n = a^n \cdot b^n.$$

Consider now

$$\frac{a^3}{a^5} = \frac{\not a \cdot \not a \cdot \not a}{\not a \cdot \not a \cdot \not a \cdot a \cdot a} = \frac{1}{a^2}.$$

If we make use formally of the rule of exponents for division, disregarding the requirement that the exponent of the numerator be larger than the exponent of the denominator, we get

$$\frac{a^3}{a^5} = a^{3-5} = a^{-2}.$$

Now a^{-2} has not yet been defined, but, since $a^3/a^5 = 1/a^2$, we are led naturally to define $a^{-2} = 1/a^2$. More generally, if n is a positive integer, we define

$$a^{-n} = \frac{1}{a^n}.$$

Thus changing the sign of the exponent converts any number to its reciprocal.

Similarly, since on one hand $a^4/a^4 = 1$ and on the other $a^4/a^4 = a^{4-4} = a^0$, we are led to the general definition,

$$a^0 = 1, \quad \text{if} \quad a \neq 0.$$

Having a meaning now for negative integral and zero exponents, we may abandon the requirement that $n > m$ in the quotient rule and state simply that

$$\frac{a^n}{a^m} = a^{n-m}.$$

Similarly, the reader will see readily that the requirement that m and n be positive may now be omitted in all the rules.

We have now arrived at an extension of the idea of a positive integral exponent which includes negative integral exponents and zero exponents, and the processes of multiplication, division, and raising to positive and negative integral and zero powers. For present considerations, these concepts and principles will suffice, but a further extension will be made in Chapter 8.

To summarize:

If a and b are any numbers, and m and n are integers, we have

Definitions: $\underbrace{a \cdot a \cdots a}_{n \text{ factors}} = a^n$, **if n is positive.**

$$a^{-n} = \frac{1}{a^n}.$$

$$a^0 = 1, \qquad a \neq 0.$$

Rules: $a^n \cdot a^m = a^{m+n}.$

$$\frac{a^n}{a^m} = a^{n-m}.$$

$$(a^n)^m = a^{m \cdot n}.$$

$$(a \cdot b)^n = a^n \cdot b^n.$$

Five other important consequences of these rules are worth stating.

$$\left(\frac{a}{b}\right)^n = \frac{a^n}{b^n}.$$

$$\frac{a^n}{a^m} = \frac{1}{a^{m-n}} = a^{n-m}.$$

$$(a^n \cdot b^m)^q = a^{n \cdot q} \cdot b^{m \cdot q}.$$

$$\frac{a^{-n} \cdot b}{c} = \frac{b}{a^n \cdot c}.$$

$$\frac{a}{b^{-n} \cdot c} = \frac{a \cdot b^n}{c}.$$

Example 1. $2^3 \cdot 3^2 \cdot 2^8 \cdot 3^5 = 2^3 \cdot 2^8 \cdot 3^2 \cdot 3^5 = 2^{3+8} \cdot 3^{2+5} = 2^{11} \cdot 3^7.$

Example 2. $\dfrac{3^4 \cdot 2^3}{3^7 \cdot 2^2} = \left(\dfrac{3^4}{3^7}\right) \cdot \left(\dfrac{2^3}{2^2}\right) = 3^{4-7} \cdot 2^{3-2} = 3^{-3} \cdot 2 = \dfrac{2}{3^3}.$

Example 3. $\left(\dfrac{6^2 \cdot 2}{3^4 \cdot 2}\right)^3 = \left(\dfrac{[2 \cdot 3]^2 \cdot 2}{3^4 \cdot 2}\right)^3 = \left(\dfrac{2^2 \cdot 3^2 \cdot 2}{3^4 \cdot 2}\right)^3 = \left(\dfrac{2^3 \cdot 3^2}{2 \cdot 3^4}\right)^3$

$$= [(2^{3-1}) \cdot (3^{2-4})]^3 = (2^2 \cdot 3^{-2})^3 = \left(\frac{2^2}{3^2}\right)^3 = \frac{2^6}{3^6}.$$

It is often of considerable help in simplifying an expression involving exponents to reduce all the integers involved to the products of their prime factors.   The previous example is an illustration, as well as the following one.

Example 4.

$$\frac{12^2 \cdot 5^2 \cdot 60}{3^5 \cdot (-2)^4 \cdot (-5)^3} = \frac{(2^2 \cdot 3)^2 \cdot 5^2 \cdot (2^2 \cdot 5 \cdot 3)}{3^5 \cdot 2^4 \cdot (-1)^4 \cdot 5^3 \cdot (-1)^3}$$

$$= \frac{2^4 \cdot 3^2 \cdot 5^2 \cdot 2^2 \cdot 5 \cdot 3}{3^5 \cdot 2^4 \cdot 5^3 \cdot (-1)} = \frac{2^6 \cdot 3^3 \cdot 5^3}{-2^4 \cdot 3^5 \cdot 5^3}$$

$$= -2^{6-4} \cdot 3^{3-5} \cdot 5^{3-3} = -2^2 \cdot 3^{-2} \cdot 5^0$$

$$= -\frac{2^2}{3^2}.$$

Exercises

1. Write the following numbers as powers of 10: 100,000,000; 10,000; 100; 1; 0.1; 0.001; 0.000001.

2. Write the following numbers without exponents: 10^9, 10^6, 10^3, 10^0, 10^{-2}, 10^{-6}, 10^{-9}.

Simplify, using the laws of exponents. State all of your results in terms of prime integers and positive exponents.

3. $2^3 \cdot 2^5$.

4. $3^6 \cdot 3^3 \cdot 3^{-8}$.

5. $5^6 \cdot 5^7 \cdot 5^{-10}$.

6. $2^3 \cdot 2^5 \cdot 2^{-13}$.

7. $2^3 \cdot 3^5 \cdot 4^{-2}$.

8. $2^3 \cdot 5^{-2} \cdot 3^4 \cdot 4^{-3} \cdot 5$.

9. $\dfrac{6^3 \cdot 5^2 \cdot 4}{10^3 \cdot 3^4 \cdot 2}$.

10. $\dfrac{12^3 \cdot 5^6 \cdot 7^3}{100^3 \cdot 28^4 \cdot 3^3}$.

11. $\dfrac{(-3)^4 \cdot 16 \cdot 36 \cdot 3^{-2}}{12^4 \cdot 8 \cdot 2^{-5}}$.

12. $\dfrac{12 \cdot 15 \cdot 10^3}{9 \cdot 100^2 \cdot 2^{-1}}$.

13. $\dfrac{(-16)^{-2} \cdot (-8)^3 \cdot 2^{-1}}{10^0}$.

14. $\dfrac{-49 \cdot 128 \cdot 3^4}{(-14)^2 \cdot (-12)^5 \cdot 2^{-3}}$.

15. $\dfrac{128 \cdot 10^{-6}}{25^{-4}}$.

16. $\dfrac{10^{10} \cdot 10^{-6} \cdot (-25)^{-3}}{(-50)^{-4} \cdot 10^8 \cdot 5^{-2}}$.

17. $\left(\dfrac{260 \cdot 39^3}{13^4 \cdot 60^2} \right)^{-1}$.

18. $\left[\dfrac{-72 \cdot 256}{(-2)^{10}} \right]^{-2}$.

19. $\left(\dfrac{144 \cdot 5^3}{120^3} \right) \cdot \left(\dfrac{3^2}{5^3} \right)^0$.

20. $\dfrac{15^3 \cdot 144 \cdot 3^{-3}}{-2^6 \cdot 5 \cdot 9}$.

21. $\dfrac{(-28)^{-3} \cdot (-49)^2}{7^0 \cdot 32^{-1} \cdot 10}$.

22. $\dfrac{(-20)^3 \cdot (-3)^3}{250 \cdot 27 \cdot 16 \cdot 6^0}$.

23. $\dfrac{36^{-4} \cdot 144^2 \cdot 9^{-2}}{81^{-2} \cdot (-2)^{15}}$.

24. $\dfrac{46^3 \cdot 14^3 \cdot 10^{-2}}{(-2)^5 \cdot (-7)^3 \cdot 23^3}$.

25. $\left(\dfrac{3^2 \cdot 15}{25 \cdot 27} \right)^{-2} \cdot \left(\dfrac{6^2 \cdot 25}{150^2} \right)^{-1}$.

26. $\left(\dfrac{13^2 \cdot 14^2}{91^2} \right)^3 \cdot \left(\dfrac{6^3 \cdot 8^2}{48^2} \right)^0$.

27. $\dfrac{56^3 \cdot (-108)^3 \cdot 2^{-10}}{-32 \cdot 49^2 \cdot 3^4 \cdot 7^{-1}}$.

28. $\left(\dfrac{14 \cdot 16 \cdot 36^{-2}}{49 \cdot 128 \cdot 9} \right)^0$.

1–12. Scientific and Engineering Notation of Numbers

Very large or very small numbers are frequently encountered in scientific work. The distance of the earth from the sun is 93,000,000 miles.

Light travels at a speed of about 300,000,000 meters per second. The number of molecules in a cubic centimeter of gas at a temperature of $0°$ C and a pressure of 76 cm of mercury is about 27,050,000,000,000,000,000. The thickness of an oil film on water is 0.000,000,2 in.   Scientific notation provides a convenient scheme for representation of such quantities.

A positive number is expressed in scientific notation when it is written as the product of an integral power of 10 and a number between 1 and 10.   A number in scientific notation has the form

$$M \times 10^n$$

where *M is a number between 1 and 10, and n is an integer.*

The quantities already mentioned, which are in ordinary (or **positional**) notation, may be written in scientific notation as follows:

$$93{,}000{,}000 = 9.3 \times 10{,}000{,}000 = 9.3 \times 10^7,$$

$$300{,}000{,}000 = 3 \times 100{,}000{,}000 = 3 \times 10^8,$$

$$27{,}050{,}000{,}000{,}000{,}000{,}000 = 2.705 \times 10{,}000{,}000{,}000{,}000{,}000{,}000$$

$$= 2.705 \times 10^{19},$$

$$0.000{,}000{,}2 = 2 \times 0.000{,}000{,}1 = 2 \times 10^{-7}.$$

The student may verify the following practical rule.

To change the form of a number from positional to scientific notation:

(a) Move the decimal point to a position such that only one nonzero digit appears to its left, thus obtaining a number between 1 and 10.

(b) Multiply this number by 10^n, where $|n|$ equals the number of places the decimal point has been moved, n being positive if the decimal point has been moved to the left, and negative if the decimal point has been moved to the right.

A second rule for determining the value of the exponent n is often used:

(a) If the number is greater than 1, then n is 1 less than the number of digits to the left of the decimal point.

(b) If the number is less than 1, then n is negative, and its absolute value is 1 greater than the number of zeros between the decimal point and the leftmost nonzero digit.

The rule for writing in positional notation a number expressed in scientific notation can be readily inferred from the rules given above.

Multiplication and division of very large or very small numbers can be simplified by the use of scientific notation and the laws of exponents. Scientific notation can also be used to advantage in slide-rule computations.

Table 8 in the appendix lists some of the units that are used in scientific and engineering practice.

Exercises

Express each of the quantities in the following statements in scientific notation, dropping all the zeros after the rightmost nonzero digit.

1. A common unit of power, the horsepower, is the work done at the rate of 33,000 ft-lb per min.

2. During a normal lifetime the human heart beats approximately 2,210,000,000 times.

3. The sun possesses 3,085,000 times the mass of Mars.

4. The distance from the earth to the moon varies between 217,000 and 248,000 miles.

5. The diameter of the earth is slightly more than 7,910 miles.

6. All forms of radiant energy are propagated at a velocity of 30,000,000,000 cm per sec.

7. A light-year, the distance light travels in a year, is approximately 5,870,000,000,000 miles.

8. The wavelength of X rays used in making industrial radiographs is about 0.000,000,014 cm.

9. The probable value of the mass of a hydrogen atom is
0.000,000,000,000,000,000,000,001,661,7 gram.

10. An oil film has a thickness of about 0.000,000,5 cm.

11. The frequency of rotation of the earth on its axis is 0.000,011,6 cycles per second.

12. The maximum rate of plant growth is of the order of 0.03 mm per sec.

Express each of the quantities in the following statements in positional notation.

13. The number of molecules in 1 cc of gas at $0°C$ and normal pressure is about 3×10^{19}.

14. Each square foot of the earth's surface receives enough heat energy annually from the sun to raise a weight of 10.56×10^8 lb to a height of 1 ft.

15. Microscopic analysis disclosed that the diameter of an average red blood corpuscle is 3.1×10^{-5} in.

16. Colloidal silver has a particle diameter of approximately 6.1×10^{-6} cm.

17. The diameter of the universe according to the theory of relativity is estimated to be 2×10^9 light-years.

18. One ampere of current is produced when 6.3×10^{18} electrons move past every point in an electric circuit in 1 sec.

19. The electric charge on an electron is 1.591×10^{-19} coulomb.

20. Soldier's Field Stadium in Chicago has a maximum seating capacity of 2×10^5.

21. The range of light radiation visible to the human eye covers the wavelengths between 4×10^3 and 8×10^3 angstrom units.

22. The frequency of rotation of the Andromeda Nebula is estimated to be 1.7×10^{-15} cycle per second.

23. An airliner travels at a scheduled speed of 2.2×10^4 ft per min.

24. The average power in connected speech is about 32×10^{-6} watt.

1-13. Significant Figures

The **digits** are the ten symbols 0, 1, 2, 3, 4, 5, 6, 7, 8, 9. **Figure** is synonymous with digit.

The voltmeter dial shown in Fig. 1–3 has primary and secondary division marks on its scale; the distance between the two adjacent primary marks corresponds to a difference of 10 volts; and the distance between two adjacent secondary marks corresponds to a difference of 5 volts.

Let us suppose that a reading is taken from the dial as shown in Fig. 1–3 by simply noting the number corresponding to the primary division mark nearest the needle; in this way we obtain a reading of 220 volts. By this rough reading we are assured only that the actual value is closer to 220 volts than 230 volts or to 210 volts. We know that, no matter how the reading is refined, the digit 2 in the hundred's place is accurate. In

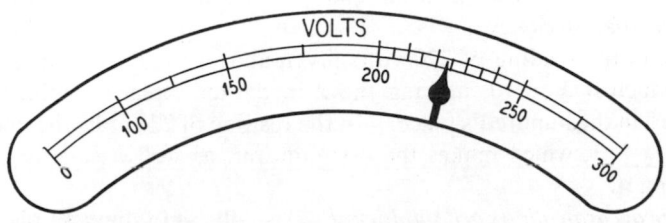

Fig. 1–3

general, *a digit in an observed quantity which will not be changed by a refinement of the observation is called an* **accurate digit.** Since by the reading taken on the meter we know only that the correct value is closer to 220 than to 210 or to 230, a refinement of the reading might give any number from 215 to 225. Thus the digit in the ten's place might become a 1 instead of a 2, and the digit in the unit's place might be any of the digits. We say that these digits are **inaccurate.** In general, *a digit in an observed quantity which may be changed by a refinement of the observation is called an* **inaccurate digit.**

As we have made the reading, the only digits in the number 220 of any significance for computations are the first two. The zero merely serves as an occupant of the unit's place.

Now suppose we inspect the meter scale more carefully, estimating to the nearest unit where the needle lies between secondary scale marks, and obtain a reading of 224 volts. Here all three digits have significance, but the reading means only that the actual value is closer to 224 volts than to 223 volts or to 225 volts. In the reading 224, the digits in the ten's and hundred's places are accurate, and the unit's place digit is inaccurate.

Observational data in this way are never absolutely accurate, many possibilities of error entering into almost any given experiment. Numbers obtained from experiments thus contain accurate digits followed by inaccurate digits. Also, as we shall see later, computations with numbers of limited accuracy contain errors which limit the accuracy of results. Since it is important to know which digits of a result are accurate and which are inaccurate, we shall investigate the influence of the accuracy of given data on results computed from this data.

In the paragraphs above we have called certain digits significant. In order to be a little more precise, we shall agree on the following criterion, which will be used throughout this discussion to distinguish between the significant and nonsignificant digits of quantities which are of limited accuracy.

A digit is significant if the maximum error in the number in which it is contained is less than or at most equal to one half of one unit in the place which the digit occupies.

Thus, in the reading of 220 volts given above, the maximum error is 5 volts, which is $\frac{1}{2} \times 10$, making the 2 in the ten's place significant, and also the 2 in the hundred's place. In the reading of 224 volts the maximum error is $\frac{1}{2} \times 1$, which makes the 4 significant, as well as all the integers preceding it.

All the accurate digits are significant. Usually, as in the examples above, the *rightmost significant digit is inaccurate.* Consequently, the rightmost significant digit will be called the *inaccurate significant digit*, and sometimes for the sake of simplicity, when only significant digits are under consideration, the *inaccurate digit.*

It is worth while to establish notational conventions which make clear how many digits of a given number are significant. Two such agreements are in common use.

1. *The significant figures in a number in positional notation consist of*

   (a) *All nonzero digits.*

   (b) *Zero digits that:*

   (1) *Lie between significant digits.*

   (2) *Lie to the right of the decimal point, and at the same time lie to the right of a nonzero digit.*

   (3) *Are specifically indicated by the context to be significant.*

2. *The significant figures in a number written in scientific notation ($M \times 10^n$) consist of all the digits expressed explicitly in M.*

*Significant figures are counted from left to right, starting with the leftmost
nonzero digit.*

Example.

Number	Significant Figures	Number of Significant Figures
35.62	3, 5, 6, 2	4
5.600	5, 6, 0, 0	4
3020	3, 0, 2	3
0.00046	4, 6	2
0.000850	8, 5, 0	3
3.0080	3, 0, 0, 8, 0	5
5.00	5, 0, 0	3
9.3×10^7	9, 3	2
5.00×10^8	5, 0, 0	3
2.705×10^{19}	2, 7, 0, 5	4
2×10^{-7}	2	1

Exercises

State the number of significant figures, according to the agreements of this section,
in the numbers given in the exercises following Sec. 1–12.

1–14. Accuracy in Computations

In computations involving observed data accurate only to a certain
number of significant figures, it is important to be able to judge how many
figures of the result are significant.   Simple considerations form the
basis of such a judgment.   These will be pointed out here through
examples, and specific rules of thumb will be given.   However, these rules
of thumb do not cover all cases; in fact, it is easy to invent examples that
violate them.   Therefore, the student should regard them not as inflexible,
but simply as a convenient expression of the common sense that should
govern one's computations in some simple situations.   In more complicated cases, arguments like those given below can be made to arrive at
conclusions about the accuracy of results.

Given three numbers 2.88×10^3, 4.346×10, and 1.376×10^{-1} which
are assumed to have been obtained by measurement or experimental means
and are therefore accurate only to the number of figures indicated in
accordance with the agreements of the preceding section.   The sum of
these numbers is obtained as follows:

$$2.88 \times 10^3 = 2880$$
$$4.346 \times 10 -\quad 43.46$$
$$1.376 \times 10^{-1} =\quad\ \ 0.1376$$
$$\overline{\qquad\qquad 2923.5976}$$

At greatest, the error in 2880 is 5, the error in 43.46 is 0.005, and the error
in 0.1376 is 0.00005.   Thus the greatest possible total error in the sum

would be 5.00505.   However, the second two errors are so small compared with the first that they may well be ignored.   Hence the greatest error in the sum is about 5.

Since the error in the sum may be as much as 5, the sum may be as much as 2928 or as little as 2918.   Hence the last six digits of the sum are inaccurate.   The last five digits are not significant and may be discarded by *rounding off* the sum to its three significant figures, expressing the result as a number of three significant figures nearest to 2923.5976.   This number is 2920, or 2.92×10^3.

For addition and subtraction we may state the following rule.

The rightmost significant figure in a sum (or difference) occurs in the leftmost place at which an inaccurate digit occurs in any of the numbers entering into the sum (or difference).

Given two numbers 3.71×10^2 and 8.216×10^3 which are accurate only to the number of significant figures indicated in accordance with the previous agreements.   The product is $3.71 \times 10^2 \times 8.216 \times 10^3$ $= 3.71 \times 8.216 \times 10^5$. This multiplication is carried out by the ordinary process of arithmetic as follows.

$$
\begin{array}{rl}
8.216 & \\
3.71 & \\
\hline
8216 & \quad (1)\\
57512 & \quad (2)\\
24648 & \quad (3)\\
\hline
30.48136 & \quad (4)
\end{array}
$$

Since line 1 is really the product of 8.216 and 0.01, line 2 is the product of 8.216 and 0.7, and line 3 is the product of 8.216 and 3, this multiplication can be written

$$
\begin{array}{rl}
8.216 & \\
3.71 & \\
\hline
0.08216 & \quad (5)\\
5.7512 & \quad (6)\\
24.648 & \quad (7)\\
\hline
30.48136 & \quad (8)
\end{array}
$$

Now in line 5 the inaccuracy of 3.71 has been multiplied by 8.216.   At worst, if the inaccuracy of 3.71 is 0.005, the inaccuracy of the product in line 5 is about $8.216 \times 0.005 = 0.041080$, whence the number in line 5 could be as little as 0.04108 or as much as 0.12324.   Thus the digit in the tenth's place in line 5 is inaccurate, and all the boldface digits in line 5 are inaccurate. Similarly, at worst the inaccuracy of 8.216 is 0.0005, making the inaccuracy in line 6 about $0.0005 \times 0.7 = 0.00035$, and the inaccuracy in line 7 about $0.0005 \times 3 = 0.0015$. Arguing as above, we can show that the boldface digits in lines 6 and 7 are inaccurate. Since line 8 is the sum of lines 5, 6, and 7, its inaccuracy may be as much as 0.04108 $+ 0.00035 + 0.0015 = 0.04293$, or slightly less than 0.05. It follows then that 30.48136 is inaccurate in the first place after the decimal point, and that only the first three digits of the product are significant. If we round off to three significant figures, the result is 30.5. The result of the original problem is then

$$3.71 \times 10^2 \times 8.126 \times 10^3 = 3.71 \times 8.126 \times 10^5$$
$$= 30.5 \times 10^5 = 3.05 \times 10^6.$$

The reader may study what happens in division in like manner. For the operations of multiplication and division we may state the following rule:

The product (or quotient) of numbers is accurate at most to the number of significant figures contained in the least accurate factor. The least accurate factor is the number entering into the computation that has the least number of significant figures.

In the preceding example the result was accurate at most to three figures. The product of 6.8×10^2 and 4.5186×10 will be accurate at most to two figures, since 6.8×10^2 is accurate only to two figures.

It should be emphasized again that examples can be constructed easily that deviate from the rules given above, but for the usual numerical computations met in science and engineering these rules are valid, or nearly enough valid for practical purposes. It is convenient to adopt these rules as a practical working criterion for the accuracy of results, in order that one may avoid a tedious analysis of every computation. In any doubtful case one should, of course, make an analysis of the computation rather than rely on the rules of thumb given above.

1–15. Rounding off Numbers

In rounding off a number after a computation, the number is chosen that has the required number of significant figures and is closest to the number to be rounded off. *When either of two numbers equally close to the number to be rounded off can be chosen, we shall adopt the convention of choosing the one ending in an even digit.*

Example.

| | Rounded off to | | |
Number	3 Figures	4 Figures	5 Figures
0.666666	0.667	0.6667	0.66667
0.312341	0.312	0.3123	0.31234
51.2155	51.2	51.22	51.216
26.5455	26.5	26.55	26.546
18.3545	18.4	18.35	18.354

1–16. Computation with Numbers of Limited Accuracy

In order to achieve in numerical computations an economy of effort and a uniform accuracy commensurable with the accuracy of the numbers given, the directions below should be followed.

1. **Addition and subtraction.** Perform all the operations with the complete numbers given, and then round off the final result. Round off the result so that its rightmost significant figure appears in the leftmost place at which an inaccurate figure occurs in any number entering into the sum or difference.

2. **Multiplication.** Perform all the operations with the complete numbers given, and then round off the final result to the number of significant digits in the least accurate factor.

3. **Division.** Perform the operations with the complete quantities given, obtaining for each division an answer to one more digit than the number of significant digits in the least accurate given quantity. Then round off the final result to the number of significant figures in the least accurate given quantity.

4. **Combined operations.** When multiplications and divisions both occur in an expression, perform the multiplications first.

Exercises

In the following exercises assume that the numbers are accurate only to the number of digits specified in the agreements of Sec. 1–13. Perform the computations, and round off the answers in accordance with the directions given in Sec. 1–15 and summarized in Sec. 1–16, and state your results in scientific notation.

Add:

1. 693, 7.42, and 32.6.
2. 89, 2.479, and 436.
3. 46, 79.1, and −376.
4. 967, 36.1, and 27.2.
5. 0.00235, 0.9317, and 0.006729.
6. 0.03958, 0.0567, and 23,456.
7. 78.9, −44.88, and −32.44.
8. 59.61, −32.1, and −670.
9. 3800, 469, and −456.1.
10. 0.00800, 0.32160, and 0.2135.

Multiply:

11. 7.2 by 28.6.
12. 4.8 by 36.2.
13. 3.33 by 167.2.
14. −4.93 by −0.000334.
15. 0.00230 by 0.987.
16. 7300 by 0.00231.
17. 23,000 by 0.234.
18. 567 by 0.04500.
19. 987 by 0.569.
20. 23.4 by 6.7.

Divide:

21. 7.9 by 2.34.
22. 4.6 by 34.67.
23. 56.9 by 6.7.
24. 678 by 0.00234.
25. 37,000 by 3459.
26. 456 by 0.34.
27. 3478 by 7600.
28. 36.0 by 0.023.
29. 0.03060 by 0.4162.
30. 967 by 0.180.

Add:

31. 232, 3.56×10^3, and 1.367×10^2.

32. 2.68×10^3, 8.56×10^2, and 1.892×10^4.

33. -6.498×10^{-2}, 1.34×10^{-1}, and 9.62×10^{-3}.

34. 5.567×10^2, 5.782, and -1.34×10.

35. -4.43×10^4, 6.932×10^3, 1.239×10^4.

36. 6.732×10^{-5}, 4.679×10^{-4}, and 9.34×10^{-4}.

Multiply:

37. 4.349×10^7 by 3.45×10^{-6}.   **38.** 6.72×10^2 by 1.567×10^{-4}.

39. 9.452×10^{-9} by 3.21×10^{-2}.   **40.** -3.649×10^5 by 3.45×10^{-3}.

41. 8.286×10^2 by -4.762×10^{-5}.   **42.** 9.23×10^{-8} by -3.49×10.

Perform the indicated operations:

43. $\dfrac{1.23 \times 10^{-2}}{3.6 \times 10^{-3}}$.   **44.** $\dfrac{6.782 \times 10^3}{1.46 \times 10^2}$.

45. $\dfrac{(8.62 \times 10^3) \cdot (7.93 \times 10^2)}{3.467 \times 10^4}$.   **46.** $\dfrac{(7.32 \times 10^{-3}) \cdot (6.912 \times 10^{-2})}{3.67 \times 10^2}$.

47. $\dfrac{6.732 \times 10^3}{(6.231 \times 10^2) \cdot (2.612 \times 10^2)}$.   **48.** $\dfrac{4.6241 \times 10^5}{(6.21 \times 10^{-2}) \cdot (7.82 \times 10^8)}$.

49. $\dfrac{(6.231 \times 10^3) \cdot (3.27 \times 10)}{(9.413 \times 10^4) \cdot (6.728 \times 10^{-1})}$.   **50.** $\dfrac{(9.672 \times 10^{-6}) \cdot (6.793 \times 10^{-3})}{(8.1234 \times 10^{-4}) \cdot (2.134 \times 10^{-2})}$.

1–17. Description of the Slide Rule

The slide rule is an instrument on which the processes of multiplication, division, finding squares, and finding square roots can be performed to three-figure accuracy simply and quickly.   Other computations may also be performed, as will be seen later, but in this chapter we will be concerned only with the four processes mentioned.   However, addition and subtraction cannot be performed on the slide rule.   The theory governing the construction of a slide rule will be found in Chapter 9; only its operation will be discussed here.

We shall discuss the 10-in. Mannheim-type slide rule, named after its inventor Lieutenant Mannheim of the French artillery, who devised it in 1859.   However, the techniques discussed here may be applied to any rule that the reader has in his possession.

The central sliding part of the rule is the **slide;** the part surrounding it is the **body.**   The glass runner is the **indicator,** and the fine line on the indicator is the **hairline.**   The mark associated with the primary number 1 on any scale is the **index** of the scale.   Two positions on different scales are **opposite** for a given position of the slide if the hairline can be brought to cover both positions simultaneously, without moving the slide.

The Mannheim rule has the four scales A, B, C, and D, on one side.

The A and B scales are identical, and the C and D scales are identical. The B and C scales are on a slide which can be moved back and forth (see Fig. 1–4). The distances between successive digits are unequal. The distance between successive integers on the C and D scales is double the distance between the same integers on the A and B scales.

Since wide variations in scales exist (even among different makes of Mannheim rules), the following remarks are general suggestions which each user may follow in interpreting his own slide rule.

1. Every C or D scale is divided into nine principal divisions by **primary marks** numbered 1, 2, 3, · · · , 9, 1. The space between any two primary marks is divided into ten parts by nine **secondary marks.** These are not numbered except between the primary marks 1 and 2. Between the secondary marks appear unnumbered **tertiary marks;** some spaces between secondary marks are divided into ten parts, some into five parts, and others into two.

2. Every A or B scale has two portions, identical with each other, which are each divided into nine principal divisions by primary marks numbered as on the C and D scales. Unnumbered secondary and tertiary marks also appear on these scales.

3. Every K scale has three portions, identical with each other, divided much like the two portions of the A and B scales.

On none of these scales is the distance between consecutive numbers equal, the distances being greater on the left end of the scales. Thus there are more subdivisions toward the left. The following points should be carefully noted.

(*a*) Where there are ten subdivisions of a given division, each subdivision counts 0.1 of the value of the given division.

(*b*) Where there are five subdivisions of a given division, each subdivision counts 0.2 of the value of the given division.

(*c*) Where there are two subdivisions of a given division, each subdivision counts 0.5 of the value of the given division.

Many errors are made in reading the slide rule because of failure to observe these facts. Practice, which leads to the ability to recognize at a glance the values of the different markings, is the only answer. As we shall see, the reading between any successive tertiary marks must be based on an estimate. Hence in making these estimates it is necessary to have firmly in mind the number associated with the space under consideration.

1–18. Locating Numbers on the Slide Rule

It is important to observe that the decimal point has no bearing on the position associated with a number on the C and D scales, and, in certain operations, on the A and B scales.

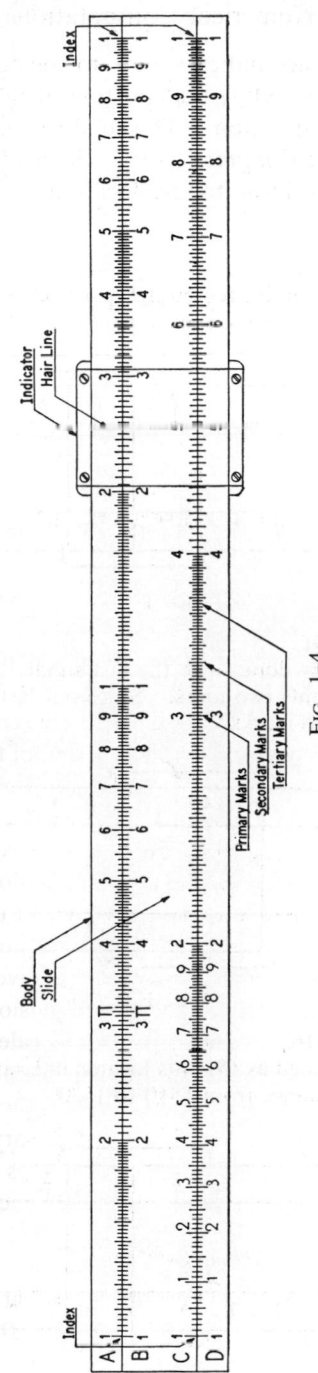

FIG. 1-4

The following examples indicate how to locate three- and four-digit numbers on the *D* scale.   Only that portion of the scale is shown which contains the number in question.   The subdivisions are shown as usually found on a 10-in. rule *at that place on it*.   The reader will see immediately the application of these ideas to the location of numbers on the other scales.

Example 1.   Locate 151.

This is easily done since in this region the space between 1 and 2 is divided into ten parts and each of these into ten more parts.

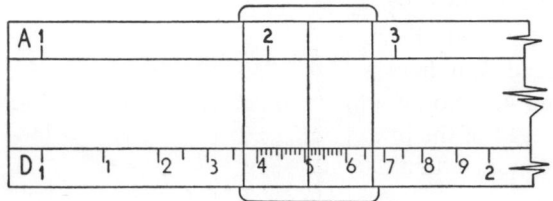

FIG. 1–5

Example 2.   Locate 251.

This is a little less easily done since the last subdivision consists of only five marks, each of which counts two units.   Since our last digit is 1, we must estimate halfway between two marks.

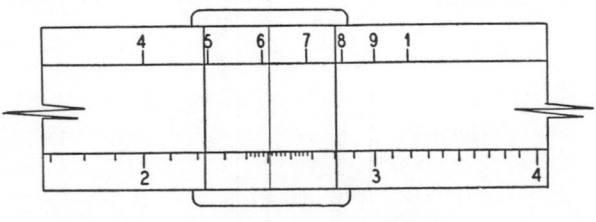

FIG. 1–6

Example 3.   Locate 1516.

The number 1510 is located as 151 was located in Example 1.   To locate 1516 we estimate 0.6 of the distance from 1510 to 1520.

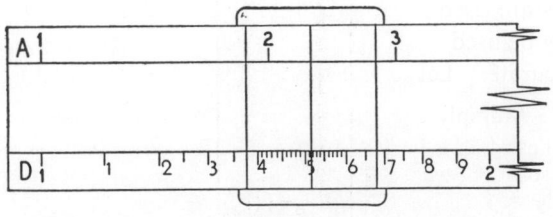

FIG. 1–7

Exercises

Locate the following numbers on the *D* scale. Then locate each one on the *A* scale.

1. 30.	**2.** 35.	**3.** 350.	**4.** 136.
5. 14.	**6.** 376.	**7.** 925.	**8.** 762.
9. 424.	**10.** 348.	**11.** 3.48.	**12.** 0.986.
13. 5.48.	**14.** 13.46.	**15.** 1288.	**16.** 1176.
17. 16.80.	**18.** 776.	**19.** 523.	**20.** 15.44.
21. 67.2.	**22.** 996.	**23.** 378.	**24.** 1146.

1–19. Accuracy of the Slide Rule

Between the primary marks 1 and 4 on the *D* scale of a 10-in. rule, it is possible to read four figures, although the last figure is very inaccurate between 2 and 4. No attempt should be made to read more than three digits to the right of the primary mark numbered 4. This means roughly that the maximum attainable accuracy is one part in 1000. Hence the

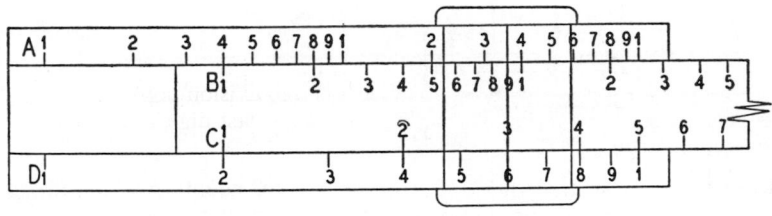

FIG. 1–8

right-hand end of the scale gives computational accuracy to three significant figures, the left end to four significant figures. But, in general, computations usually involve both ends of the scale, and we are limited to an accuracy of three significant figures.

These remarks must be modified by the reader at his own judgment to fit the rule in his possession, depending on its length, the care used in its manufacture, and other factors.

1–20. Multiplication by the Slide Rule

Two scales are used in conjunction. Either the *C* and *D* scales *or* the *A* and *B* may be used. The *C* and *D* scales, having larger scale divisions, are more accurate. Let us use them on a simple example.

Example 1. Multiply 2 × 3 on the slide rule.

The multiplicand, 2, is located on the *D* scale and the left 1 (the left index) of the *C* scale is brought even with it. The hairline may be used to assist in this alignment if desired. Then the hairline is moved along the *C* scale to the multiplier, 3. The product, 6, is directly below the 3 on the *D* scale. See Fig. 1–8.

While the rule is in this adjustment it might be noted that the products of 2 by the integers 2, 4, and 5 may also be obtained by looking on the D scale below each of these on the C scale, without changing the adjustment of the slide.   But the multiplication of 2 by the integers 6, 7, 8, and 9 is impossible, as the slide is now set, because the D scale does not extend far enough.   Multiplication by these numbers is accomplished by putting the right index, rather than the left, even with the 2 as shown in Fig. 1–9.

Example 2.   Multiply 2 × 6 on the slide rule.

The product, 12, is found to be two tenths of the way between the large 1 and 2, which *in this case*, is interpreted as being two tenths of the way between 10 and 20, that is, 12.   Had we been using the left half of the A and B scales for the multiplication, the use of the right index could have been avoided, since the A scale (used in the same manner as the D scale) is continuous and can take care of products greater than 10.   And, for these simple numbers, the A and B scales would have been sufficiently accurate.

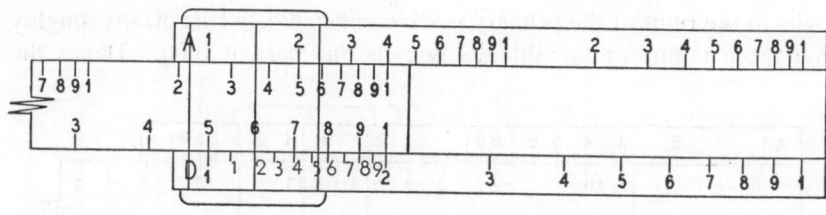

Fig. 1–9

In order to avoid the trial-and-error process of deciding which index to use when multiplying numbers with 2 or more digits on the C and D scales, the following suggestion will be found useful: *When the product of the first digits of each of the two factors is less than 10, use the left index; when it is greater than 10 use the right index.*

While you are learning to multiply on the slide rule, it is well to check the rules by applying them to a number of simple problems for which the correct answer is known.   For products of more than two numbers, consider the product of the first two (which you will mark with the hairline without reading it) as the multiplicand for the next multiplication.

The slide rule merely gives the significant digits of the product; the position of the decimal point must be determined by other methods.

Example 3.   Multiply 12.6 × 35.9 × 187 on the slide rule.

The significant digits for 12.6 × 35.9 are 452, and for the entire product they are 846.   Now rounding off the numbers, the product must be in the neighborhood of 10 × 35 × 200 = 70,000.   Hence the proper result is 84,600 = 8.46 × 10^4.

The position of the decimal point in a slide-rule computation can be determined by a mental computation with the simple numbers obtained from the given numbers by rounding them off to one or two significant figures.

An alternative method to be used in complicated computations will be given later, but many persons prefer to use the method shown above in all their work. The authors suggest that the student adopt whatever method he finds to be the easier and faster.

We have, then, this general procedure.

To multiply two numbers:

1. *Find the multiplicand on the D (or A) scale.*
2. *Bring the proper index of the C (or B) scale even with it.*
3. *Locate the multiplier on the C (or B) scale.*
4. *With the aid of the hairline find the number on the D (or A) scale opposite the multiplier. This number gives the significant digits of the product.*
5. *Determine the decimal point by means of a mental computation with the given numbers rounded off.*

Exercises

Perform the following multiplications, obtaining results accurate to as many places as your slide rule will allow, and express these results in scientific notation.

1. 4.00×17.00.   2. 13.00×5.00.

3. 19.00×4.45.   4. 16.00×4.67.

5. 22.4×35.6.   6. 21.7×33.7.

7. 1.545×45.6.   8. 17.83×5.08.

9. 7.77×21.6.   10. 5.62×13.84.

11. 0.00672×1.364.   12. 96.8×3.16.

13. 3.67×48.1.   14. 0.00612×8.88.

15. 0.619×3.76.   16. 76.2×0.00814.

17. 0.00318×0.0616.   18. 88.2×0.0216.

19. $61.2 \times 4.34 \times 0.489$.   20. $3.23 \times 6.76 \times 0.00871$.

21. $6.66 \times 8.19 \times 0.00433$.   22. $31.8 \times 4.67 \times 8.92$.

23. $13.66 \times 4.89 \times 41.2$.   24. $0.00819 \times 6.72 \times 8.82$.

1–21. Division by the Slide Rule

Division is easily understood after multiplication has been studied. The following example, with Fig. 1–10, will suffice to explain the procedure.

Example. Divide 8 by 4 on the slide rule. (For the solution, see Fig. 1–10.)

To divide one number by another:

1. *Locate the dividend on the D (or A) scale with the aid of the hairline.*

2. *Locate the divisor on the C (or B) scale and bring it opposite the dividend at the hairline.*

3. *With the aid of the hairline find the number on the D (or A) scale opposite the index of the slide. This number gives the significant digits of the quotient.*

4. *Determine the position of the decimal point by a mental computation with the given figures rounded off.*

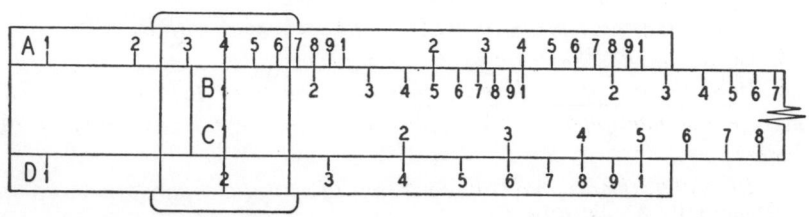

Fig. 1–10

Exercises

Perform the following computations, obtaining results accurate to as many places as your slide rule will allow, and express these results in scientific notation.

1. $65.0 \div 5.00$.

2. $144.0 \div 120.0$.

3. $65.0 \div 104.0$.

4. $48.0 \div 63.0$.

5. $29.2 \div 5.66$.

6. $932 \div 69.1$.

7. $0.00832 \div 98.1$.

8. $0.0818 \div 0.0632$.

9. $14.89 \div 0.00981$.

10. $986 \div 13.89$.

11. $4880 \div 3.98$.

12. $16.82 \div 957$.

13. $0.00662 \div 0.781$.

14. $7.63 \div 0.0002145$.

15. $69.3 \div 786$.

16. $12.46 \div 698$.

17. $69,800 \div 9.39$.

18. $367 \div 4890$.

19. $\dfrac{72.3 \times 89.1}{124.6}$.

20. $\dfrac{89.3 \times 13.82}{66,900.}$.

21. $\dfrac{0.00932 \times 0.0816}{6.39}$.

22. $\dfrac{6.98 \times 0.00716}{0.00811}$.

23. $\dfrac{1482 \times 6790}{6.87}$.

24. $\dfrac{(6.81 \times 10^2) \cdot (4.19 \times 10^3)}{6.97 \times 10^{-4}}$.

1–22. Finding Squares by the Slide Rule

To find the square of a number:

1. *Locate the number to be squared on the D scale.*

2. *By the aid of the hairline find the number opposite it on the A scale. This gives the significant figures of the square of the original number.*

3. *Determine the position of the decimal point by a mental computation with the given number rounded off.*

Example.   Square 5 on the slide rule.

The slide is not used in this operation and therefore is not shown in the figure.

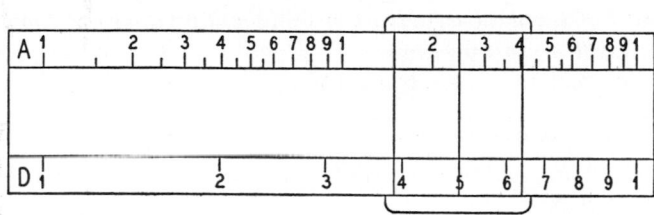

FIG. 1–11

Occasionally a rule, other than a Mannheim, may be found that does not have the *A* and *D* scales on the same side.   However, these rules have the *A* and *D* scales matched and a hairline on both sides so that the process is the same even though the rule has to be turned over.

For accurate readings, a hairline that is exactly perpendicular to the scales is essential.   This is true for any operation, but it is more important here because the *A* and *D* scales are widely separated.   A good check on the perpendicularity of the hairline is to see if it gives the exact square of some simple number such as 2 or 3.   If found in error, the hairline should be adjusted or replaced.

Exercises

Find the squares of the following numbers, obtaining results accurate to as many places as your slide rule will allow, and express these results in scientific notation.

1. 13.00.	**2.** 47.0.
3. 760.	**4.** 34.5.
5. 13.46.	**6.** 48.6.
7. 0.789.	**8.** 0.0843.
9. 5670.	**10.** 57,700.
11. 8.84.	**12.** 19.84.
13. 0.00318.	**14.** 82.8.
15. 438.	**16.** 112.6.
17. 1824.	**18.** 14.38.
19. 0.000672.	**20.** 34,800.
21. 8.24.	**22.** 39.6.
23. 45.8.	**24.** 6.73.

1–23. Finding Square Roots by the Slide Rule

This process is just the reverse of that described in the previous section.   However, certain care must be taken since the *A* scale has two identical

halves.   The question is which to use in a given case.   The process of
taking the square root in arithmetic requires that the number whose root
is to be taken be marked off in periods of two digits beginning at the decimal
point, as for example, 4 56, 35 69.32, 4 59.63 18, or 0.08 17 1.   The first
digit of the root is different when the leftmost period contains one digit
from what it is when the leftmost period contains two digits, even though
the sequence of digits is the same.   The result by slide rule must, of course,
differ correspondingly.   We frame the procedure as follows.

To find the square root of a number:

1. *Mark off the number into periods of two digits so that the decimal point
is a point of division.*

2. (*a*) *If the number whose square root is sought has one digit in the
leftmost period, locate it on the left half of the A scale.*

   (*b*) *If the number whose square root is sought has two digits in the
leftmost period, locate it on the right half of the A scale.*

*Statements a and b apply to decimal fractions less than 1 if we ignore
zeros between the decimal point and the leftmost nonzero digit.*

3. *The significant digits of the square root of the number will be found
opposite it on the D scale.*

4. *Determine the position of the decimal point by a mental computation
with the given number rounded off.*

Example 1.   Find the square root of 156.2.
Dividing the number into periods, we obtain 1 56.2.   By 2*a*, we locate 1562 on
the left *A* scale, and opposite on the *D* scale we read 125.   Since the required
root, by mental computation, is between 12 and 13, the result is $12.5 = 1.25 \times 10$.

Example 2.   Find the square root of 0.00869.
Dividing the number into periods, we obtain 0.00 86 9.   By 2*b* we locate 869
on the right-hand *A* scale, and opposite on the *D* scale we read 932.   Since the
required root, by mental computation, is between 0.09 and 0.10, the result is
$0.0932 = 9.32 \times 10^{-2}$.

Exercises

Find the square roots of the following numbers, obtaining results accurate to as many
places as your slide rule will allow, and express these results in scientific notation:

1. 36.0.	**2.** 3.60.	**3.** 360.
4. 3600.	**5.** 256.	**6.** 1440.
7. 25.6.	**8.** 50.4.	**9.** 389.
10. 4750.	**11.** 16,900.	**12.** 18,450.
13. 0.0469.	**14.** 0.000462.	**15.** 0.0000962.
16. 0.00395.	**17.** 698,000.	**18.** 76,200.
19. 5.89×10^2.	**20.** 3.46×10^3.	**21.** 6.43×10^{-4}.
22. 8.98×10^{-3}.	**23.** 6.46×10^{10}.	**24.** 5.88×10^{-7}.

1–24. Finding Cubes and Cube Roots by the Slide Rule

If the reader possesses a rule with a K scale, he will discover quickly how to find cubes, using the K and D scales. The rule for finding cube roots is similar to that for finding square roots, but the given numbers must be marked off into periods of three digits instead of two.

1–25. The Use of the Slide Rule in Extended Computations

Several useful hints on the use of the slide rule will appear in connection with the following computation:

$$\frac{346 \times 17.1}{21\ 2}.$$

We may multiply 346 by 17.1 and divide the result by 221, or we may divide 346 by 221 and then multiply the result by 17.1. The latter is the shorter method since one setting of the slide will work the entire problem. To carry out the second process, first find 346 on the D scale and bring 221 on the C scale opposite it by the aid of the hairline. The result of this division appears on the D scale directly below the left index (at the digits 1566), but *we need not read this because whatever it reads is merely the point at which the index should be set for the ensuing multiplication.* So, *without moving the slide or even reading what is under the index, move the hairline to the next multiplier* (17.1) on the C scale. The result is the digits 268 on the D scale.

Although the first method is not quite so easy, let us follow it through as a check. First multiply 346 $\times$ 17.1. This gives the digits 592. Then divide this by 221. Again we obtain the digits 268. In general, *where several multiplications and divisions are to be performed, the easiest and most accurate method is to alternate between multiplication and division* instead of doing all the multiplying and then all the dividing. The reason is that, in general, fewer moves of the slide are required, and therefore there is less chance for error. It is very good practice to work first by one method and then by the other, as a check. Since

$$\frac{350 \times 20}{200} = 35,$$

the result of the computation is 26.8.

It should be observed that this method may fail, because the final reading may be beyond the limits of scale D. If we replace, for instance, the factor 17.1 by 7.60 using the first method, we shall find that the multiplier 7.60 is outside the body of the slide rule. In order to read the product, the slide must be reset so that the right index of it takes the place of the left index.

which has been marked with the hairline. Then, after the resetting of the slide, we move the hairline to the multiplier 7.60 on C and read 11.9 on scale D.

The necessity of resetting the slide can be avoided by the use of the scales A and B, if the accuracy of these scales is sufficient.

Scientific notation is sometimes useful in determining the position of the decimal point, especially if the numbers are very large or very small. For example,

$$\frac{3{,}460{,}000 \times 0.000171}{22{,}100}$$

can be written as

$$\frac{3.46 \times 10^6 \times 1.71 \times 10^{-4}}{2.21 \times 10^4} = \frac{3.46 \times 1.71}{2.21} \times 10^{6-4-4}$$

$$= \frac{3.46 \times 1.71}{2.21} \times 10^{-2}.$$

Performing the slide-rule computations on

$$\frac{34.46 \times 1.71}{2.21},$$

we obtain significant digits 268, and the decimal point is easily determined from $\dfrac{3.5 \times 2}{2} = 3.5$.   Thus the result of the computation is

$$2.68 \times 10^{-2} = 0.0268.$$

Exercises

Perform the following computations on the slide rule, obtaining results accurate to as many places as your slide rule will allow, and express these results in scientific notation:

1. $\dfrac{21.6 \times 893}{465}$.

2. $\dfrac{0.0608 \times 91.7}{48.6}$.

3. $\dfrac{64{,}600 \times 989}{0.606}$.

4. $\dfrac{6.81 \times 1.822}{6190}$.

5. $\dfrac{3240}{9620 \times 0.0844}$.

6. $\dfrac{41.2}{3180 \times 4.61}$.

7. $\left(\dfrac{41.6}{816}\right)^2$.

8. $\left(\dfrac{329}{627}\right)^2$.

9. $\left(\dfrac{31.3 \times 42.8}{664}\right)^2.$

10. $\left(\dfrac{418 \times 327}{67{,}600}\right)^2.$

11. $\sqrt{\dfrac{48.2}{31.6}}.$

12. $\sqrt{\dfrac{639}{27.2}}.$

13. $\sqrt{\dfrac{4.89 \times 69.7}{73.9}}.$

14. $\sqrt{\dfrac{84.3 \times 956}{693{,}000}}.$

15. $\dfrac{\sqrt{6.86} \times 91.2}{329}.$

16. $\dfrac{\sqrt{469} \times \sqrt{39.6}}{61.8}.$

17. $\left(\sqrt{\dfrac{69.8}{44.7}}\right) \cdot 3.29.$

18. $\left(\dfrac{46.8}{7.36}\right)^2 \cdot 0.0762.$

19. $\left(\dfrac{81.2 \times 31.9}{4180}\right)^2 \cdot 89.6.$

20. $47.8 \cdot \sqrt{\dfrac{36.2 \times 9.87}{0.00418}}.$

Progress Report

This chapter can be divided into three parts:

1. A review of some simple ideas from arithmetic.

   (a) How operations lead from the natural numbers to the integers, the rational numbers, and the irrational numbers, and thus to the system of real numbers.

   (b) The laws of signs.

   (c) Operations with fractions.

   (d) Exponents.

   (e) Scientific notation.

2. A discussion of computations with numbers of limited accuracy.

   (a) Significant digits.

   (b) Rules for determining the accuracy of results computed from numbers of limited accuracy.

3. The use of the slide rule in multiplying, dividing, squaring, taking square roots, and in combinations of these operations.

The student not only should be thoroughly familiar with the principles introduced in this chapter but also should acquire skill in computation by much practice.

2

Simple Algebraic Operations

The basic generalization of algebra, as we have seen, consists of using a letter to represent any one of a set of numbers. Algebra, which is founded on this generalization, provides the scientist and engineer with a powerful, practical tool.

An algebraic formula related to a scientific or engineering investigation describes in compact form the relations between the quantities under study; algebriac transformations of this formula provide new perspectives on these relations. Skill in the derivation and manipulation of algebraic formulas is essential for the scientist and engineer; in this chapter we shall consider some of the simpler algebraic operations.

The formulas encountered in engineering and science usually give relations between physical quantities. These quantities can be designated by letters of the English alphabet, by letters of the Greek alphabet, or by other symbols. Frequently the scientist or engineer uses a notation that has become standard in the field in which he is working. For example, a force which may act on a structural member is frequently designated by F, and a voltage in a circuit is frequently designated by E or e. On the other hand, mathematics texts such as this one, which are concerned to a large extent with algebraic expressions in which the symbols do not refer to specific physical quantities, use a wide variety of letters.

2–1. The Nature of Algebra

Algebra is an extension of arithmetic. Each statement of arithmetic deals with particular numbers: the statement $(20 + 4)^2 = 20^2 + 2 \cdot 20 \cdot 4 + 4^2 = 576$ explains how the square of the sum of the two numbers, 20 and 4, may be computed. It can be shown that the same procedure applies if the numbers 20 and 4 are replaced by any two other numbers. In order to state the general rule, we write symbols, ordinarily letters, instead of particular numbers. Let the number 20 be replaced by the symbol a, which may denote any number, and the number 4 by the symbol b. Then the statement is true that the square of the sum of any two numbers a and b can be computed by the rule

$$(a + b)^2 = a^2 + 2 \cdot a \cdot b + b^2.$$

This is a general rule which remains true no matter what particular numbers may replace the symbols a and b. A rule of this kind is often called **a formula.**

Algebra is the system of rules concerning the operations with numbers. These rules can be most easily stated as formulas in terms of letters, like the rule given above for squaring the sum of two numbers.

Since the letters in algebra are used to represent numbers, all of the laws of arithmetic given in Chapter 1 hold for operations with these letters.

In the same way, all the signs that have been introduced to denote relations between numbers and the operations with them are likewise used with letters.

For convenience the operation of multiplication is generally denoted by a dot or by the letters being placed adjacent to each other. For example $a \cdot b$ is written simply as ab.

The operations of addition, subtraction, multiplication, division, and extracting roots are called **algebraic operations.** Any symbol or combination of symbols obtained by employing algebraic operations is called an **algebraic expression.** Examples of algebraic expressions are given by

$$2x + 7y - \sqrt{x + y}, \qquad 2a^3b + 7ab.$$

When an algebraic expression consists of several parts connected by $+$ or $-$ signs, each part with its preceding sign is called a **term.** In the above examples $2x$, $7y$, and $-\sqrt{x + y}$ are terms of the expression $2x + 7y - \sqrt{x + y}$, and $2a^3b$ and $7ab$ are terms of the expression $2a^3b + 7ab$.

In order to represent different quantities, different letters are used. Although these letters may be picked at random in any given problem, it is usually found convenient in scientific work to set aside a certain letter which is always used to represent physical quantities of a certain class. If there are several quantities in the same class, these may be denoted by subscripts. Thus E refers to voltage in electrical problems, and when several voltages occur in a given problem they are denoted by

$$E_1, E_2, E_3, \cdots, E_n,$$

read E sub-one, E sub-two, $\cdots$, E sub-n, or, if no misunderstanding is possible,

$$E\text{-one}, E\text{-two}, \cdots, E\text{-}n.$$

A similar type of notation employs accents for the same purpose. Thus:

$$I', I'', I''', \cdots$$

is read I prime, I two-prime (or double prime), I three-prime, etc.

In general, any type of symbol may be attached to a letter if it forms a convenient representation of the quantities involved. Such forms as

$$E_0, \; E^*, \; E, \; V_R, \; F_x, \; V_{AB},$$

will be frequently encountered.

2–2. Addition and Subtraction

In adding and subtracting algebraic expressions, we use the commutative and associative laws of addition (Sec. 1–8) and, through the rules for removing and inserting signs of grouping (Sec. 1–9), the distributive law.

Example 1. Add $4E_1 + 7E_2$, $5E_1 - 3E_2$, $2E_2 - E_1$.

The sum can be written as:

$$(4E_1 + 7E_2) + (5E_1 - 3E_2) + (2E_2 - E_1)$$

$$= 4E_1 + 7E_2 + 5E_1 - 3E_2 + 2E_2 - E_1 = 8E_1 + 6E_2.$$

Thus, *algebraic expressions may be given a simpler form by combining similar terms.* Two terms are called **similar,** if they differ only in their numerical factor (called a **coefficient**).

Example 2. Simplify the expression

$$2a + \{5a - [8a - 3b - (2a - b) + 3c] + 10c\}.$$

Removing successively the parentheses, brackets, and braces, we obtain

$$2a + \{5a - [8a - 3b - 2a + b + 3c] + 10c\}$$

$$= 2a + \{5a - 8a + 3b + 2a - b - 3c + 10c\}$$

$$= 2a + 5a - 8a + 3b + 2a - b - 3c + 10c$$

$$= a + 2b + 7c.$$

The last example shows that, *in removing several signs of grouping, it is more convenient to remove the innermost first.*

Exercises

Add:

1. $3a, -5a, 12a, -4a.$

2. $2x + 3y, 4x - 3y, 6y - 5x.$

3. $6xy, -3xy, 11xy.$

4. $3a^2b + 4ab^2, 7a^2b - 2ab^2.$

5. $5E_1 + 2E_2 - 9E_3, 3E_1 - 5E_2 - E_3.$

6. $7ab + 2ac - 4bc, 3bc - ab + 4ac.$

7. $2V_1 - V_2 + 3V_3, -3V_1 + V_2 + 5V_3.$

8. $I_1 + 2I_2 \quad 5I_3, 3I_2 - 2I_1 + I_3.$

9. $3x + 2y, x - y, 2y - 3x, 5x - 7y.$

10. $5E - 11X, -7E + 9IX, 3E - 4IX.$

11. $5ab - 7cd, 3ab + cd, 4cd - ab, 6cd - ab.$

12. $3x^4 - 2x^2y^2 + 5y^4, -x^4 + 4x^2y^2 - y^4, 2x^4 - 3x^2y^2 + 5y^4.$

13. $3R_1 - 2R_2 + R_3, -R_1 + 5R_2 - 3R_3, 4R_1 - 2R_2 - R_3.$

14. $3i_1 + 4i_2 - i_3, -5i_1 + 2i_2 + 4i_3, i_1 - i_2 + 2i_3.$

15. $2E - 5RI - 13ZI, 6RI + 15ZI, -12E + RI - 7ZI.$

16. $3W - 2EI - 7I^2R, 2W + EI + 3I^2R, -EI + 3I^2R, -W - EI.$

17. $R_1 + 2R_2 - 3R_3 + 5R_4, -R_1 - R_2 + 2R_3, 2R_1 - R_3 + 2R_4, R_2 + 4R_3 + 7R_4.$

18. $3I_1 - 5I_2 + 2I_3 - I_4, -I_1 + I_2 - I_4, I_2 - I_3 + 5I_4, I_1 - 2I_2 + I_4.$

19. $2x + 4y - z + 5w, 2y - w, 3x + 2w, -x + 2y - 5z + 7w.$

20. $12a + 2b - 3c + 7d, -b + c - d, 5d + b - 2c, 7b - 2a + d - 3c.$

Subtract the second expression from the first:

21. $10V, 3V.$

22. $10ax, -3ax.$

23. $5a + 3b, 4a + b.$

24. $7x - 9y, 4x - 3y.$

25. $7E + 3IX, 9E - 4IX.$

26. $8E + 5RI + 6XI, 3E + 2RI - XI.$

27. $13ax - 15by - 8cz, \quad 7ax - 2by + cz.$

28. $5W - 3EI + 15I^2R, 4EI - 2W - 3I^2R.$

29. $5x^2 + 7xy - y^2, 2x^2 - xy + y^2.$

30. $4I_3 - 3I_1 + 7I_2, 5I_3 - I_1 - 2I_2.$

31. $6I_1 + 2I_2 + I_3, 12I_3 + 2I_1 - I_2.$

32. $13x^2 - 15xy + 3y^2, 6x^2 + 8xy - 4y^2.$

33. $4Z_1 - 2Z_2 + 12, 3Z_1 - 7Z_2 + 1.$

34. $12x + 8y - 5z - 11w, 16w - 3y - 4z + 2x.$

35. $2R_1 + 7R_2 - 5R_3 - 2R_4, R_3 - 3R_2 - 5R_4.$

Simplify:

36. $(a - 4) - (2a - 5) + (a - 1)$. **37.** $(3x - 2y) - (2x - y) + (x + y)$.

38. $2E - 3IR - (E + IR)$. **39.** $5ac - \{2by + (2ac - by)\}$.

40. $5p - 3q - [(2p + 3q) - (2p - 3q)]$.

41. $x + y - \{x - y + [x + y - (x - y)]\}$.

42. $2e - (ir + e) - \{e - [2ir - (e + ir)] + e\}$.

43. $3E - (5I_1R_1 - I_2R_2) - \{2E - (I_1R_1 + 3I_2R_2)\}$.

44. $2W - \{W + EI - [I^2R - (-2EI)] + 4I^2R\}$.

45. $5x - \{8x - y + [3x - y + 23] - (z + 3y)\}$.

46. $2I_1 - \{3I_1 - [4I_2 + (I_1 - I_3)] + 5I_3\}$.

47. $8y - \{4x - [3x - (y - 2x) + y]\} - 7x$.

48. $-\{-[-(2a + b - c) + 19] - 49 + b - 2c\}$.

49. $5E - (3RI + 2IZ) - [3E - (2RI + IZ)]$.

50. $-7R_1 - \{R_2 + [2R_1 - (R_3 + 3R_2)] - R_2\}$.

2–3. Multiplication

In multiplying algebraic expressions consisting of one term, we use the commutative and associative laws (Sec. 1–8) and the laws of exponents (Sec. 1–11).   An algebraic expression of one term is a **monomial.**

Example 1.
$$(5a^2x^3)(3a^4x^2) = 5 \cdot 3 \cdot a^2 \cdot a^4 \cdot x^3 \cdot x^2 = 15a^{2+4}x^{3+2} = 15a^6x^5.$$

Example 2.
$$(2ab^2c)(3a^2bc)(5abc^2) = 2 \cdot 3 \cdot 5 \cdot a \cdot a^2 \cdot a \cdot b^2 \cdot b \cdot b \cdot c \cdot c \cdot c^2$$
$$= 30a^{1+2+1}b^{2+1+1}c^{1+1+2} = 30a^4b^4c^4.$$

Example 3.
$$(2x^2y^3)^2 = 2^2(x^2)^2(y^3)^2 = 4x^4y^6.$$

Example 4.
$$(3a^4b^3c)^3(2abc^3)^2 = 3^3a^{12}b^9c^3 \cdot 2^2a^2b^2c^6 = 108a^{14}b^{11}c^9.$$

Exercises

Perform the indicated operations:

1. $2xy \cdot 4xy \cdot 5x$. **2.** $3m^2 \cdot 2m^3 \cdot 4m$.

3. $(2a^4)(-5a^3)(-3a^2)$. **4.** $(-5ab^2)(-2a^2b)$.

5. $4IR \cdot 3R$. **6.** $(4xyz)(3xz)$.

7. $7a^2bc^2d \cdot 8ab^3cd^2$. **8.** $(2IR^2)(3I^2X)$.

9. $6I^2R^3 \cdot 5EI$. **10.** $10x^2yz \cdot 4xy^3z^2$.

11. $(-2W^3) \cdot 3WXY^2 \cdot X^2Y$.

12. $2I^2Z^2(-5I)(-3Z^2R^2)$.

13. $2R_1I^2 \cdot 3R_2I(-5R_3I^2)$.

14. $5IR^2 \cdot 7IRZ^2 \cdot 3I^3R$.

15. $5i^2t^3 \cdot 3irs \cdot 6rts^2$.

16. $4RL^2 \cdot 3WR^2 \cdot 2RC^2$.

17. $5xyz(-6yzw) \cdot 2xyzw$.

18. $8E^2I \cdot 6IR^2 \cdot 5ER^2$.

19. $12Irt^2 \cdot 2I^2 \cdot 5I^3r^2t$.

20. $18p^2qr \cdot 2pr^2t \cdot 5q^2rs^3$.

21. $(-7x^3yz) \cdot 3xyz^2 \cdot 5xy^3z$.

22. $15mnp^4 \cdot 2npr^2(-4m^2prs)$.

23. $5a^2bc \cdot 2abc^2d \cdot 4abcd^2 \cdot abcd$.

24. $(2xyzw)(3yzwv)(5xzvu)(xyzwu)$.

25. $5x^4y(-6x^2y^3)(-x^3y^2)(x^2y^2)$.

26. $7W^2L^3 \cdot (-3WL^3)(-5W^3L^4)$.

27. $2a^3b^4 \cdot 8a^2b^3 \cdot 5a^2b^2$.

28. $3x^2yz^2(-2xy^3z) \cdot 6x^3y^2z$.

29. $R_a{}^2R_b{}^3 \cdot R_a{}^3R_c{}^2 \cdot R_b{}^2R_c{}^3$.

30. $(-Z_1{}^2Z_2{}^3) \cdot Z_1{}^3Z_3(-Z_2Z_3{}^3)$.

2–4. Multiplication of Multinomials

Algebraic expressions consisting of more than one term are called **multinomials.** In particular, an expression of two terms is a **binomial;** an expression of three terms is a **trinomial.** In finding the product of multinomials, we make use of the distributive law (Sec. 1–8).

Rule 1. To multiply a multinomial by a monomial, multiply each term of the multinomial by the monomial and add the resulting products.

Example 1.

$$5a^3b(7a^2b^2 - 3ab^3 + 2a^4b) = 35a^5b^3 - 15a^4b^4 + 10a^7b^2.$$

Rule 2. To find the product of two multinomials, take the sum of the products which result from multiplying each term of one multinomial by each term of the other.

Example 2.

$$
\begin{aligned}
(x + y)(3x^2 - 2xy + 5y^2) &= x(3x^2 - 2xy + 5y^2) + y(3x^2 - 2xy + 5y^2) \\
&= (3x^3 - 2x^2y + 5xy^2) + (3x^2y - 2xy^2 + 5y^3) \\
&= 3x^3 - 2x^2y + 5xy^2 + 3x^2y - 2xy^2 + 5y^3 \\
&= 3x^3 + x^2y + 3xy^2 + 5y^3.
\end{aligned}
$$

Exercises

Perform the indicated operations:

1. $x(x + 3)$.

2. $(2x + y)(x - 3y)$.

3. $2I^2R(X + Z)$.

4. $3ab(ab + cd)$.

5. $(R_1 + 4)(2R_2 - 1)$.

6. $5a^2b^3(a^2 + 2ab^2 - b)$.

7. $(x + y)(x - y)$.

8. $(p + q)(p - q)$.

9. $(4I + 1)(3I - 2)$.

10. $(ei + 2)(3ei - 5)$.

11. $(2r^2 + 3z)(r^2 - z)$.

12. $5IZ(2IR - 9IX + XZ)$.

13. $5(a^2 + b^2)(a^2 - b^2)$.

14. $7E(3ER_1 - 2IR_2 + 6IR_3)$.

15. $4ac(3ab - 5ad + 2bc + 7bd)$.

16. $9ei(2t_1 - 4t_2 + 5t_3 + 6t_4)$.

17. $12RI_1(3RI_1 + RI_2 - E_0I_1 + E_0)$.   **18.** $(a + b)(a^2 - ab + b^2)$.

19. $(2P + 3)(P^2 + P - 1)$.   **20.** $(x + y)(x^2 + 2xy + y^2)$.

21. $(2I - 3)(I^2R + R^2 - 2)$.   **22.** $(a^2 - b^2)(a^4 + a^2b^2 + b^4)$.

23. $(x - y - z)(x - y + z)$.   **24.** $(Z + 3R - 5)(2Z - R + 4)$.

25. $(2x^2 + x + 5)(x^2 - 3x + 2)$.   **26.** $(EI - 3IR + 2)(2EI + IR - 5)$.

27. $(2mn - 3np + 5mp)(4mn + 5np - 6mp)$.

28. $(3I_1R_1 - 2E + 4I_2R_2)(ER_1 + I_1 - I_2)$.

29. $(3x + 2y)(5x^2 - 2x^2y + 4xy^2 + 6y^2)$.

30. $(2T + 3)(5RI^2 - 2EI + 3\dfrac{E^2}{T} + 1)$.

2–5. Division of Monomials

In dividing monomials the laws of exponents (Sec. 1–11) are used.

Example 1.

$$5a^2b \div a = \frac{5a^2b}{a} = 5ab.$$

Example 2.

$$\frac{3m^3lt^{-5}}{5m^2l^3t^{-4}} = \frac{3m}{5l^2t}.$$

Exercises

Perform the indicated operations:

1. $4a^8 \div a^3$.   **2.** $-12b^6c^4 \div b^2c$.

3. $\dfrac{(T^2 + 8Y^4)a}{2Z(T^2 + 8Y^4)}$.   **4.** $\dfrac{(3X^2 - 1)^2PL}{RL^2(3X^2 - 1)^2}$.

5. $\dfrac{b^{m+1}}{b^m}$.   **6.** $\dfrac{L^{2n}}{RL^n}$.

7. $24a^8b^3c \div 6a^5bc^3$.   **8.** $3pT^2(E_1 + E_2)^2 \div pT(E_1 + E_2)$.

9. $\dfrac{(a + b)^4bxt}{x(a + b)^3}$.   **10.** $\dfrac{16X_c^3 \cdot 2aX_c \cdot 3Z}{48aX_cZ}$.

11. $M^{x+3}(x + y)^3 \div 2M^{x+1}(x + y)^2$.   **12.** $(a^5 + 5b^2)xy^2z^3 \div yz^2$.

13. $\dfrac{15a^3 \cdot 2b^4c^2 \cdot 7d(a + 1)^3}{35abc(a + 1)}$.   **14.** $\dfrac{10P^mK^2(Z_1 - Z_2)^{m-1}}{2P^{m-1}K(Z_1 - Z_2)^m}$.

15. $\dfrac{(L^2 - 2Z + 5)^4T^{n+1}}{T^2(L^2 - 2Z + 5)^2}$.   **16.** $\dfrac{(p + q)(p - q)(k^2 - pq)}{k^2L(p + q)}$.

17. $\dfrac{N^{p+1}M^qQ^{3-r}}{N^{p-1}M^{q-1}Q^r}$.   **18.** $\dfrac{(l^2 + g)kL^3 \cdot 5(l^2 - g)^2}{(l^2 - g)(l + g)(l^2 + g)L^2}$.

19. $\dfrac{S^{m+1}(RV)^5}{R^{5-m}(SV)^2}$.   **20.** $\dfrac{(AB^3)^2(C + D)^3}{(C + D)^4B^5(A + B)^3}$.

2–6. Division of a Multinomial by a Monomial

According to the laws for division (Sec. 1–10), a multinomial is divided by a monomial by dividing each term separately.

Example 1.

$$\frac{7a^2 + 14ab - 21a^3}{7a} = \frac{7a^2}{7a} + \frac{14ab}{7a} - \frac{21a^3}{7a} = a + 2b - 3a^2.$$

Example 2.

$$\frac{4Z(P + W) - Z^3(P + W)^2 + XZ^2(P + W)^3}{Z^2(P + W)}$$

$$= \frac{4Z(P + W)}{Z^2(P + W)} - \frac{Z^3(P + W)^2}{Z^2(P + W)} + \frac{XZ^2(P + W)^3}{Z^2(P + W)}$$

$$= \frac{4}{Z} - Z(P + W) + X(P + W)^2.$$

Exercises

Perform the indicated operations:

1. $\dfrac{xy^2 + 3xyz}{y}.$

2. $\dfrac{a^3b^2c - 12ab^2c^3}{abc}.$

3. $\dfrac{IR_1^2 + I^2R_1^2 - I^3R_1}{IR_1}.$

4. $\dfrac{X^2Y + XY^2 - XYZ}{XY}.$

5. $\dfrac{\pi fL^2 - \pi fR}{\pi fR}.$

6. $\dfrac{8(x + y) - 3(x + y)^2}{(x + y)}.$

7. $\dfrac{3a^2b^n + 5a^{2n+1} - 2a^nb}{a^2}.$

8. $\dfrac{I^2R - IR^3 + I^4RL}{I^2R}.$

9. $\dfrac{k(a + b)^3 + 2(a + b)^2}{2k(a + b)}.$

10. $\dfrac{(s + 1)^2a - (s + 1)b}{(s + 1)ab}.$

11. $\dfrac{15AB^{n\,|\,2} + 30B^2C^n - 5A^2C^2}{15AB^2C}.$

12. $\dfrac{8L_x(E_p - e_s) + (E_p - e_s)}{E_p - e_s}.$

13. $\dfrac{8py^8 - 6p^2y^5 + 4p^3y^4}{(2py)(4y^2)}.$

14. $\dfrac{5QU + Q(p + 1)^2 - 7(p + 1)}{QU(p + 1)}.$

Divide $18y^4Z^2(p + 1)^m - 9x^2yZ^3 + x^m(p + 1)$ by:

15. $3y^2Zx^3.$ 16. $9mZ(p + 1)^3.$ 17. $18yx^{-2}(p + 1)^{1-m}.$

Divide $12R^6Z^5L^4 + 16RZ^2 - 8(RZL)^3$ by:

18. $R^5Z^{-3}.$ 19. $2(RZL)^4.$ 20. $2Z^{-6}RL^2.$

2–7. Special Products

Certain special products occur so frequently in algebra that it is well to become familiar with them. They are given below and should be verified by the student.

1. $(a + b)(a - b) = a^2 - b^2.$
2. $(a + b)(a + b) = (a + b)^2 = a^2 + 2ab + b^2.$
3. $(a - b)(a - b) = (a - b)^2 = a^2 - 2ab + b^2.$
4. $a(x + y - z) = ax + ay - az.$
5. $(a + b)(x + y) = x(a + b) + y(a + b) = ax + bx + ay + by.$
6. $(x + a)(x + b) = x^2 + (a + b)x + ab.$

Formula 1 states that *the product of the sum and the difference of two numbers is equal to the difference of their squares.*

Formula 2 states that *the square of the sum of two numbers equals the square of the first, plus twice the product of the numbers, plus the square of the second number.*

Formula 3 states that *the square of the difference of two numbers equals the square of the first, minus twice the product of the numbers, plus the square of the second number.*

Similar statements may be supplied by the reader for formulas 4, 5, and 6.

Algebraic computations may be shortened by the use of the above formulas, as explained by the following examples:

Example 1. Find the product $(x - 2)(x + 2)$.

Substituting x for a and 2 for b in formula 1, we obtain
$$(x - 2)(x + 2) = x^2 - 2^2 = x^2 - 4.$$

Example 2. Find the product $(4y + c)^2$.

Substituting $4y$ for a and c for b in formula 2, we obtain
$$(4y + c)^2 = 16y^2 + 8yc + c^2.$$

Example 3. Find the product $(R_1 + 5)(R_1 - 2)$.

Substituting R_1 for x, 5 for a and -2 for b in formula 6, we obtain
$$(R_1 + 5)(R_1 - 2) = R_1^2 + (5 - 2)R_1 + 5(-2) = R_1^2 + 3R_1 - 10.$$

Exercises

Perform the indicated operations:

1. $2(x + 5).$
2. $-5x(2y + 7x).$
3. $e(e^2 + e - 3).$
4. $(p + 3)^2.$
5. $(\lambda f + u)^2.$
6. $(R^2 + z)^2.$
7. $(yb - 3x)^2.$
8. $(r^2 - t^2)^2.$
9. $(R - \frac{2}{3})^2.$
10. $(D + 2a)(D - 2a).$

11. $(IR + 1)(IR - 1)$.

12. $(7x + 4)(7x - 1)$.

13. $(r + 3)(r - 12)$.

14. $(E + 10)(E - 5)$.

15. $(a - 12)(a + 5)$.

16. $(x + 3y)(n - p)$.

17. $(y - 1)^2$.

18. $(\alpha + 5)(\alpha - 12)$.

19. $(u + 8)^2$.

20. $(Z - z)^2$.

21. $(L^2 - x)(L^2 + x)$.

22. $(3 + b)(x + v)$.

23. $(R - \frac{2}{5})^2$.

24. $(6M - 2)(6M + 2)$.

25. $(v + 2a)(u - b)$.

26. $(5x^2 + 8)^2$.

27. $(E_1 + 5)^2$.

28. $(t - \frac{1}{3})^9$.

29. $3a^2(x^2 + 2xz + z^2)$.

30. $3a^2bc(8ab^2c - 9a^2bc^3 + c^3)$.

31. $\left(\dfrac{1}{R_1} + \dfrac{1}{R_2}\right)\left(\dfrac{1}{R_1} - \dfrac{1}{R_2}\right)$.

32. $\left(\dfrac{E}{R} - 1\right)^2$.

33. $(\theta - \frac{1}{2})(\theta + \frac{1}{2})$.

34. $(2T + 3t)(2T - 3t)$.

35. $[3(a + b)]^2$.

36. $(2r - 5pq^2)^2$.

37. $[7(2a - b)]^2$.

38. $(3a - A)^2 - 6aA - A^2$.

39. $\left(\dfrac{a}{b} + \dfrac{p}{q}\right)\left(\dfrac{a}{b} - \dfrac{p}{q}\right)$.

40. $\left(\dfrac{E_1}{E_2} - i_p\right)^2$.

41. $\left(\dfrac{L}{R} + \dfrac{2}{i}\right)\left(\dfrac{L}{R} - \dfrac{2}{i}\right)$.

42. $\left(\dfrac{f^2}{F} + I^2R\right)\left(\dfrac{f^2}{F} - I^2R\right)$.

43. $[(x + 2) + 5][(x + 2) - 5]$.

44. $[a + (b + c)][a - (b + c)]$.

45. $(I + 2)(2 - I)$.

46. $(p + 3q)(3q - p)$.

47. $(I^2R - LM)(LM + I^2R)$.

48. $(l - L)(L - l)$.

49. $(R_1R_2 - 1)(1 - R_1R_2)$.

50. $(E_1E_2 - I_1I_2)(I_1I_2 - E_1E_2)$.

2–8. Simple Factoring

In using the previous six formulas going from the left side of the equations to the right side, the process known as **expansion** was performed. If this process is reversed, going from the right side of the equations to the left, the operation is called **factoring.** In arithmetic the analogous process is that of resolving integers into the products of prime factors.

Typical procedures in simple factoring are indicated in the following examples.

Example 1. Factor $4R^2 - Z^2$.

Using formula 1 of Sec. 2–7, we have

$$(4R^2 - Z^2) = (2R)^2 - Z^2 = (2R + Z)(2R - Z).$$

Example 2.   Find the factors of $9x^2 - 12x + 4$.

Using formula 3 of Sec. 2–7, we have

$$9x^2 - 12x + 4 = (3x)^2 - 2 \cdot (2 \cdot 3x) + 2^2 = (3x - 2)^2.$$

Example 3.   Factor $2r^2s - 4rs^2t + 8r^2s^2t$.

Using formula 4 of Sec. 2–7, we have

$$2r^2s - 4rs^2t + 8r^2s^2t = 2rs(r - 2st + 4rst).$$

The last example illustrates the fact that a *monomial factor of each term of a multinomial is a factor of the multinomial.*

Grouping of terms as in formula 5 of Sec. 2–7 sometimes allows a common factor to be found for all the groups, as illustrated by the following example.

Example 4.   Factor $d^3 - 3d^2 + 6d - 18$.

Grouping terms, we have

$$d^3 - 3d^2 + 6d - 18 = (d^3 - 3d^2) + (6d - 18)$$
$$= d^2(d - 3) + 6(d - 3)$$
$$= (d - 3)(d^2 + 6).$$

Formula 6 of Sec. 2–7 enables us to factor trinomials like $x^2 + px + q$. Such expressions can sometimes be factored by inspection into two binomials $(x + a)(x + b)$ where $a + b = p$ and $ab = q$.

Example 5.   Factor $x^2 - 7x + 10$.

Following the explanation just given, we search for two numbers a and b whose sum is -7 and whose product is 10.   By inspection, these numbers are found to be -5 and -2.   Hence

$$x^2 - 7x + 10 = (x - 5)(x - 2),$$

where $x - 5$ and $x - 2$ are the factors of the given expression.

Example 6.   Factor $5 + 6Z + Z^2$.

As in the previous example, we have

$$5 + 6Z + Z^2 = Z^2 + 6Z + 5 = (Z + 5)(Z + 1).$$

A few suggestions will help the student in factoring:

1. Remove *all* monomial factors first.

2. Factor, if possible, the remaining expression; be sure that the factors are simple.

3. Check by multiplying the factors.

Exercises

Factor the following expressions by removing a monomial factor:

1. $ay^2 + ax^2 + av^2$.   **2.** $5a - 15ab$.

3. $17x^2 - 289x^3$.   **4.** $12y^2 + 6xy^3$.

5. $4L^3 - 8L^2R$.   **6.** $10m^4n^2 - 15m^3n^4$.

7. $2ab^2 + 4a^2b - 6ab$.   **8.** $4r^3st - r^2s^2t^2 + 8r^3s^3t^3$.

9. $a^3l^2 + 3a^2l^3 - 7a^3l^3$.   **10.** $16x^2 - 2abx$.

11. $\dfrac{WL}{R} + WC + W^3 P$.   **12.** $\dfrac{2\pi}{T} + 8\pi^3 C - 4\pi^3 C$.

13. $x^3y^2Z^3 - x^2y^3Z^2 + x^2y^2Z^2$.   **14.** $\pi^2rh + \pi r^2 l - \pi rh^2$.

15. $2a^3 - 4a^2b^2 + 8a^4b$.   **16.** $4n^4 + 12n^3m^3 - 8n^2m$.

17. $7R_1{}^2R_2 - 14R_1R_2{}^2 + 7R_1R_2$.   **18.** $x(a + b) + y(a + b)$.

19. $a^2(x - y)^2 - 3(x - y)^2$.   **20.** $15X(R - L)^3 + 5Y(R - L)^3$.

Factor the following expressions by grouping terms:

21. $x^2 + x + x + 1$.   **22.** $ay + a + y + 1$.

23. $3x + 3a - cx - ca$.   **24.** $2ab - 6b + 2ac - 6c$.

25. $3p + 3prs + pq + pqrs$.   **26.** $w^2k - 2w^2n - u^2k + 2u^2n$.

27. $x^3y^3 + x^4y^2 + x^3y + x^2y^2$.   **28.** $e\pi + fh + f\pi + eh$.

29. $L^2 + RL - ML - MR$.   **30.** $r_1r_2 + r_1r_3r_4 - r_2{}^2 - r_2r_3r_4$.

31. $\alpha_1{}^2\theta^2 + \alpha_1{}^2\beta + \alpha_2{}^2\theta^2 + \alpha_2{}^2\beta$.   **32.** $18G + 21 - 6EG - 7E$.

33. $3CE_p - E_pI_p - 3CI_p + I_p{}^2$.   **34.** $\pi x^2L^2 + \pi x^2 - 2\pi L^2 - 2\pi$.

35. $a^2 + ab + ax + bx - a - b$.   **36.** $C^2RL + C^2EL + C^2RI + C^2EI$.

Factor the following expressions:

37. $x^2 + 6x + 9$.   **38.** $b^2 + 8b + 16$.

39. $M^2 - 2MR + R^2$.   **40.** $r^4s^2 + 2r^2s + 1$.

41. $16R^4 - 1$.   **42.** $x^4 - 9y^2$.

43. $\dfrac{C_1{}^2}{4} - \dfrac{C_2{}^2}{4}$.   **44.** $\pi^2 - \dfrac{r^2}{9}$.

45. $i^4t^2 - r^2s^4$.   **46.** $R_1{}^2 - 26R_1 + 169$.

47. $9f^2 + 6f + 1$.   **48.** $X_L{}^2 - 25X_c{}^2$.

49. $9e^2 - 42e + 49$.   **50.** $4E_s{}^2 - 12E_s + 9$.

51. $Z^2 y^4 - Z^2$.

52. $K^2 L^2 E^2 - 9R^2$.

53. $9R_1{}^2 - 24R_1 R_2 + 16R_2{}^2$.

54. $49X_L{}^2 + 28X_L X_c + 4X_c{}^2$.

55. $x^2 - 8x + 12$.

56. $a^2 - 4a + 3$.

57. $y^2 + 5y - 6$.

58. $b^2 + 7b + 12$.

59. $q^2 + 16q + 28$.

60. $r^2 - 3r - 4$.

61. $K^2 - 13K - 68$.

62. $2W - 24 + W^2$.

63. $\pi \alpha^2 + 5\pi \alpha - 14\pi$.

64. $R^2 - 2R - 35$.

65. $p^2 q^2 - 9p^2 q + 8p^2$.

66. $x^2 + 11ax - 26a^2$.

67. $3y^2 - 9y - 84$.

68. $9P^2 + 6PQ + Q^2$.

69. $2R_1{}^2 + 6R_1 R_2 + 4R_2{}^2$.

70. $2f_s{}^4 + 32f_s{}^2 + 110$.

71. $m^2 k + 2\pi mnk + \pi^2 n^2 k$.

72. $X_C{}^4 - 15X_C{}^2 - 100$.

73. $a^2 x^4 + 12amx^2 + 35m^2$.

74. $5g^2 - 25gb + 30b^2$.

75. $C_1{}^4 C_2{}^2 + 4C_1{}^2 C_2 + 3$.

76. $30p^5 r^7 - 25p^4 r^6 + 5p^3 r^5$.

77. $S^2 + 3SV^2 L^2 + 2V^4 L^4$.

78. $\pi a^2 x^2 + 17\pi ax + 66\pi$.

79. $cui_p{}^2 + 5cui_p e + 6cue^2$.

80. $\alpha^6 L^4 - 5\alpha^3 L^2 R - 14R^2$.

2–9. Lowest Common Multiple

In adding algebraic fractions a common denominator must first be obtained. *The lowest common multiple (designated as LCM) of two or more expressions is the product of all their different factors, each taken the greatest number of times that it occurs as a factor in any one of the expressions.*

The method of finding the LCM is illustrated by the following examples.

Example 1. Find the LCM of

$$ax + 3a, \quad x^3 + 2x^2, \quad x^2 + 5x + 6.$$

Factoring each, we have

$$a(x + 3), \quad x^2(x + 2), \quad (x + 2)(x + 3).$$

The factors a, $(x + 2)$, $(x + 3)$ each occur at most once in any expression; x occurs at most twice. Hence the LCM is $ax^2(x + 2)(x + 3)$.

Example 2. Find the LCM of

$$15a^3 b^2 c, \quad 12a^2 b^3 c^2, \quad 18b^4 c^3, \quad \text{and} \quad 20a^2 b^4.$$

The required LCM is $180a^3 b^4 c^3$, because 180 is the least common multiple of the numbers 15, 12, 18, 20, and no lower powers than a^3, b^4, or c^3 will contain as factors the powers of a, b, c in *all* the given monomials.

Exercises

Find the LCM of the expressions in the following examples:

1. $5ab^3, 7a^2b^4$.

2. $12x^2y^3, 54yZ^2$.

3. $12R^3, 45L^2$.

4. $72x^3y, 96y^2Z^4$.

5. $x^2 + ax, ax - a^2$.

6. $ab - b^2, a^2 - b^2$.

7. $x^2 + x - 6, x^2 + 4x + 3$.

8. $R^2 - 4, R^2 + 9R + 14$.

9. $y^{10} + y^9, y^{12} + y^{11}$.

10. $ab(x + y), a^2b(x^2 - y^2)$.

11. $r^2 - s^2, r - s, rs + s^2$.

12. $Z^2 - 3Z - 10, Z^2 - 25$.

13. $6e - 18, e^2 - 9, e^2 - 6e + 9$.

14. $x(y - z), y(z - x), z(x - y)$.

15. $x^2 - ax - bx + ab, x^2 - b^2$.

16. $ax - ay, bx + by, x^2 - y^2$.

17. $L^2 + 2L - 3, L^2 - 1$.

18. $8m^2 - 4mn, 2mn + n^2, 4m^2 - n^2$.

19. $x^2 + 2xy + y^2, x^2 - y^2, x + y$.

20. $(a + b)^3, a^2 - b^2, (a - b)^2$.

2–10. Fractions: Addition and Subtraction

In arithmetic, fractions are added and subtracted after they have been reduced to the same denominator. In numerical fractions, the common denominator is always taken to be the least common multiple (see Sec. 1–10) of the denominators of the given fractions and is called the **least common denominator,** abbreviated **LCD.** A similar procedure is used to obtain a common denominator for fractions involving algebraic expressions: *The common denominator chosen for fractions involving algebraic expressions is the lowest common multiple of their denominators;* this denominator is called the **lowest common denominator** (abbreviated **LCD,** as with fractions in arithmetic). Thus we see that when dealing with numerical fractions we use the term **least,** whereas in algebraic expressions this is replaced by the term **lowest.** Let the student give the reason for this change.

Before finding the lowest common multiple of the denominators, it is often convenient to reduce each fraction to its lowest terms. A fraction is in its **lowest terms** when all factors common to numerator and denominator have been eliminated by dividing numerator and denominator by these factors.

The method for the addition or subtraction of fractions involving algebraic expressions will become apparent from the following illustrations.

Example 1.

$$\frac{2a}{(x + 3)a} + \frac{3x^2}{(x + 2)x^2} - \frac{13}{x^2 + 5x + 6}.$$

By factoring the denominators of each fraction, we obtain

$$\frac{2a}{(x+3)a} + \frac{3x^2}{(x+2)x^2} - \frac{13}{(x+3)(x+2)}.$$

When each fraction is reduced to its lowest terms, the expression becomes

$$\frac{2}{x+3} + \frac{3}{x+2} - \frac{13}{(x+3)(x+2)}.$$

Finding the LCD and expressing each fraction as one having the LCD, we have

$$\frac{2(x+2)}{(x+3)(x+2)} + \frac{3(x+3)}{(x+2)(x+3)} - \frac{13}{(x+3)(x+2)}.$$

Combining the new numerators and placing over the LCD, we obtain

$$\frac{2(x+2) + 3(x+3) - 13}{(x+2)(x+3)}.$$

Removing the parentheses in the numerator and simplifying, we have

$$\frac{2x+4+3x+9-13}{(x+2)(x+3)} = \frac{5x}{(x+2)(x+3)}.$$

This illustration suggests the following steps in the addition or subtraction of fractions.

1. *Factor the numerator and denominator of each fraction.*
2. *Reduce each fraction to its lowest terms.*
3. *Find the lowest common denominator (LCD).*
4. *Change each fraction to one having for its denominator the LCD (obtained in step 3) by multiplying numerator and denominator of each fraction by the quotient of the LCD by the denominator of the fraction.*
5. *Combine the new numerators, and place over the LCD.*
6. *Simplify the numerator, and reduce the result to lowest terms (sometimes further simplification is possible if the numerator can be factored).*

Example 2.

$$\frac{r+1}{r^2-4r+4} + \frac{r^2+5r+6}{r^2+r-6} - \frac{r^3}{r^2(r-1)}$$

$$= \frac{r+1}{(r-2)(r-2)} + \frac{(r+3)(r+2)}{(r+3)(r-2)} - \frac{r^2 \cdot r}{r^2(r-1)} \qquad \text{Step 1}$$

$$= \frac{r+1}{(r-2)(r-2)} + \frac{r+2}{r-2} - \frac{r}{r-1} \qquad \text{Step 2}$$

$$= \frac{(r+1)(r-1)}{(r-2)(r-2)(r-1)} + \frac{(r+2)(r-2)(r-1)}{(r-2)(r-2)(r-1)} - \frac{r(r-2)(r-2)}{(r-2)(r-2)(r-1)} \qquad \begin{array}{l} \text{Step 3} \\ \text{Step 4} \end{array}$$

$$= \frac{(r+1)(r-1) + (r+2)(r-2)(r-1) - r(r-2)(r-2)}{(r-2)(r-2)(r-1)} \qquad \text{Step 5}$$

$$= \frac{(r^2-1) + (r^3 - r^2 - 4r + 4) - (r^3 - 4r^2 + 4r)}{(r-2)(r-2)(r-1)} \quad \Bigg\}$$

$$= \frac{r^2 - 1 + r^3 - r^2 - 4r + 4 - r^3 + 4r^2 - 4r}{(r-2)(r-2)(r-1)}$$

$$= \frac{4r^2 - 8r + 3}{(r-2)(r-2)(r-1)} \qquad\qquad\qquad\qquad \Bigg\} \qquad \text{Step 6}$$

$$= \frac{(2r-1)(2r-3)}{(r-2)(r-2)(r-1)}.$$

Exercises

Perform the indicated operations:

1. $\dfrac{x}{xy^2} + \dfrac{y}{x^2 y}.$

2. $\dfrac{a}{a^2 bc} - \dfrac{ab}{ab^2 c}.$

3. $\dfrac{t-3}{2t} + \dfrac{3t+7}{2t}.$

4. $\dfrac{x}{y-1} - \dfrac{y}{y^2(y-1)}.$

5. $\dfrac{1}{x-4} + \dfrac{4}{4-x}.$

6. $\dfrac{2}{x-2} + \dfrac{1}{2-x}.$

7. $\dfrac{r}{r+1} - \dfrac{r}{r-1}.$

8. $\dfrac{x-y}{x+y} - \dfrac{a-b}{a+b}.$

9. $\dfrac{2}{W-1} + \dfrac{2}{1-W}.$

10. $\dfrac{a}{a^2-1} - \dfrac{2-a}{1+a}.$

11. $\dfrac{a-c}{b+d} - \dfrac{a}{b}.$

12. $\dfrac{1}{c^n} + \dfrac{1}{c^{n+1}}.$

13. $\dfrac{3-y}{y^n} + \dfrac{1}{y^{n-1}}.$

14. $\dfrac{2}{3} + \dfrac{1}{a+b} + \dfrac{1}{a-b}.$

15. $\dfrac{y-(5x+y)}{5x+y} + 1.$

16. $\dfrac{m+n}{m-n} - \dfrac{m-n}{m+n} - \dfrac{4mn}{m^2-n^2}.$

17. $\dfrac{5}{2x-1} - \dfrac{7}{2x+1} + \dfrac{x}{4x^2-1}.$

18. $\dfrac{R+X}{R-X} - \dfrac{R+2X}{R+X} + \dfrac{2RX}{R^2-X^2}.$

19. $\dfrac{9y^n}{14b^6 c^4} - \dfrac{5b^{n-4}}{21yc^2} + \dfrac{2c^{n-4}}{15yb^5}.$

20. $\dfrac{a-b}{a+b} - \left(\dfrac{a+b}{a-b} + \dfrac{1}{a^2-b^2} \right).$

21. $\dfrac{a}{a^n+1} + \dfrac{a}{a^n-1}.$

22. $\dfrac{WL}{W^2-4y^2} - \dfrac{W^2}{WL+2yL}.$

23. $\dfrac{p+1}{1-p} - \left(\dfrac{1-p}{p+1} + \dfrac{1}{p^2-1} \right).$

24. $\dfrac{z(1-z)}{z^2-9} + \dfrac{z}{3-z} - \dfrac{1-2z}{z+3}.$

25. $\dfrac{x^m + y^m}{x^m - y^m} - \dfrac{x^m - y^m}{x^m + y^m}.$

26. $\dfrac{7}{y-1} - \dfrac{5}{y-2} - \dfrac{7}{y+1}.$

27. $\dfrac{4Z^2 - 2Z}{4Z^2 + 4Z + 1} + \dfrac{5}{2Z+1}.$

28. $\dfrac{R^2}{R^2 + 2RX + X^2} + \dfrac{R-X}{R+X}.$

29. $\dfrac{2}{x} + \dfrac{2}{x^2 + 2x + 1} - \dfrac{1}{x+1}.$

30. $\dfrac{x+y}{x-y} + \dfrac{1}{x^2 - 2xy + y^2} + 1.$

31. $3x - 4 - \dfrac{9x^2 - 16}{3x + 4}.$

32. $2 - \dfrac{x^2}{x^2 + 2ax + a^2} - \dfrac{x-a}{x+a}.$

33. $\dfrac{3}{1-x} + \dfrac{x}{x^2 - 1} - \dfrac{x+1}{(x-1)^2}.$

34. $\dfrac{1}{x^2 - y^2} - \dfrac{x}{y-x} + \dfrac{2}{(x+y)^2}.$

35. $\dfrac{20x^2 + 7x - 5}{9x^2 - 1} - \dfrac{3x+2}{3x-1}.$

36. $c^2 - \dfrac{c^4 + d^4}{c^2 + cd + d^2} - d^2 + 1.$

37. $\dfrac{3}{4z^2 - 4z + 1} + \dfrac{1-z}{1-2z}.$

38. $\dfrac{7}{x^2 + 7x - 18} - \dfrac{8}{x^2 + 6x - 16}.$

39. $\dfrac{2}{x^2 + x - 6} + \dfrac{1}{x^2 - 4} - \dfrac{3}{x^2 - 9}.$

40. $\dfrac{r_1 + r_2}{(r_2 - r_3)(r_1 - r_3)} + \dfrac{r_2 + r_3}{(r_1 - r_3)(r_1 - r_2)} + \dfrac{r_1 + r_3}{(r_1 - r_2)(r_2 - r_3)}.$

41. $\dfrac{1}{(e_1 - e_2)(e_2 - e_3)} - \dfrac{1}{(e_2 - e_1)(e_3 - e_4)} + \dfrac{1}{(e_2 - e_3)(e_3 - e_4)}.$

42. $\dfrac{3x+1}{x-9} - \dfrac{4x-1}{x+2} - \dfrac{x-1}{x^2 - 7x - 18}.$

43. $\dfrac{4 - x^2}{x^2 - 5x - 14} - \dfrac{3x+6}{x+2} + \dfrac{2x+3}{x-7}.$

44. $\dfrac{1}{z+a} + \dfrac{3z}{(z+a)^2} - \dfrac{2z-a}{z^2 - 2az - 3a^2}.$

45. $\dfrac{\alpha^2 x}{(\alpha - \beta)(\alpha - \gamma)} + \dfrac{\beta^2 x}{(\beta - \alpha)(\beta - \gamma)} - \dfrac{\gamma^2 x}{(\alpha - \gamma)(\gamma - \beta)}.$

46. $\dfrac{I_1}{(I_1 - I_2)(I_1 - I_3)} + \dfrac{I_2}{(I_2 - I_1)(I_2 - I_3)} + \dfrac{I_3}{(I_3 - I_1)(I_3 - I_2)}.$

47. $\dfrac{L^2 + 2L - 15}{L - 3} - \dfrac{L^2 - 8L + 7}{L - 7} - \dfrac{L(L-3)}{L^2(L+5)}.$

48. $\dfrac{C+Q}{C-Q} - \dfrac{C^2 + 2QC + Q^2}{C+Q} + \dfrac{C^3 - QC^2 - 2Q^2 C}{C(C^2 - Q^2)}.$

49. $\dfrac{c + 2d}{c^2 - 2cd + d^2} - \dfrac{2}{d-c} - \dfrac{1}{c+2d}.$

50. $\dfrac{a+2}{a^2 - 7a + 12} + \dfrac{a+2}{6 + a^2 - 5a} - \dfrac{a-1}{6a - a^2 - 8}.$

2–11. Fractions: Multiplication and Division

Multiplication and division of fractions involving algebraic expressions follow the same methods given for arithmetic fractions in Secs. 1–10 and 1–11.

To multiply fractions:

1. *Reduce each fraction to its lowest terms.*

2. *Multiply the numerators to form the numerator of the product; multiply the denominators to form the denominator of the product.*

3. *Reduce the resulting fraction to lowest terms.*

Example 1. Simplify:

$$\frac{ax + x^2}{2b - cx} \cdot \frac{2bx - cx^2}{(a + x)^2} \cdot \frac{a^2 + 2ax + x^2}{ax^2 + x^3}$$

$$= \frac{x(a + x)}{2b - cx} \cdot \frac{x(2b - cx)}{(a + x)(a + x)} \cdot \frac{(a + x)(a + x)}{x^2(a + x)} \qquad \textit{Step 1}$$

$$= \frac{x(a + x)}{2b - cx} \cdot \frac{x(2b - cx)}{(a + x)(a + x)} \cdot \frac{a + x}{x^2}$$

$$= \frac{x(a + x)x(2b - cx)(a + x)}{(2b - cx)(a + x)(a + x)x^2} \qquad \textit{Step 2}$$

$$= 1. \qquad \textit{Step 3}$$

Note that it is convenient in step 2 to keep the expressions in factored form.

To divide one fraction by another invert the divisor and multiply it by the dividend.

Example 2. Simplify:

$$\frac{x^2 + 7x + 12}{x^2 + 2x - 15} \div \frac{x + 4}{x + 5}.$$

Factoring the dividend and inverting the divisor, we obtain

$$\frac{(x + 4)(x + 3)}{(x + 5)(x - 3)} \cdot \frac{x + 5}{x + 4}.$$

Multiplying and reducing the result to lowest terms, we have

$$\frac{(x + 4)(x + 3)(x + 5)}{(x + 5)(x - 3)(x + 4)} = \frac{x + 3}{x - 3}.$$

Exercises

Perform the indicated operations and simplify:

1. $\dfrac{5a}{x^3} \cdot \dfrac{14bx^2}{20a^2}$.

2. $\dfrac{18p^2q^2}{35x^2y^4} \cdot \dfrac{7xy^3}{45pq^3}$.

3. $\dfrac{mu^2g^3}{ns^2t^3} \cdot \dfrac{n^3s^2t}{m^3u^2g}$.

4. $\dfrac{LM^2Z^5}{L^5M^4Z^3} \cdot \dfrac{L^2M^3Z}{LMZ^3}$.

5. $\dfrac{e_1e_2e_3}{e_4e_5e_6} \cdot \dfrac{e_2e_3e_4}{e_5e_6e_1} \cdot \dfrac{e_3e_4e_5}{e_6e_1e_2}$.

6. $\dfrac{6a^3}{7b^4x^n} \cdot \dfrac{14x^{n+3}}{3a^{n-1}}$.

7. $\dfrac{R-L}{R+L} \cdot \dfrac{R+L}{R-L}$.

8. $\dfrac{m-n}{x+m} \cdot \dfrac{y+n}{m-n}$.

9. $\dfrac{5t-15}{7} \cdot \dfrac{14}{2t-6}$.

10. $\dfrac{4b}{14x-21y} \cdot (4x-12y)$.

11. $\dfrac{9E_1{}^3}{E_1{}^2 - E_2{}^2} \cdot \dfrac{E_1 - E_2}{3E_1{}^2}$.

12. $\dfrac{F_1{}^2 - F_2{}^2}{\pi F_1 - \pi F_2} \cdot \dfrac{F_1{}^2 + F_2{}^2}{F_1 + F_2}$.

13. $\dfrac{HK^2 - K^3}{H^2 + HK} \cdot \dfrac{H^3 - HK^2}{K^2}$.

14. $\dfrac{C_1{}^2 - C_2{}^2}{C_1{}^2C_2} \cdot \dfrac{C_1C_2{}^2}{C_1 + C_2}$.

15. $\dfrac{ax + x^2}{b + cx} \cdot \dfrac{a + x}{x^3} \cdot \dfrac{bx + cx^2}{(a+x)^2}$.

16. $\dfrac{x^2 + 3x - 10}{x^2 - 6x - 7} \cdot \dfrac{x-7}{x+5}$.

17. $\dfrac{ax + ay + x^2 + xy}{xy + ax + y^2 + ay} \cdot \dfrac{a+y}{a+x}$.

18. $\dfrac{ms - ns + mt - nt}{ms + ns - mt - nt} \cdot \dfrac{m+n}{s+t}$.

19. $\dfrac{x^2 + 13x + 42}{x^2 + 2x - 15} \cdot \dfrac{x^2 + 6x + 5}{x^2 + 4x - 12}$.

20. $\dfrac{y^2 + (a+b)y + ab}{y^2 - (a+c)y + ac} \cdot \dfrac{y^2 - c^2}{y^2 - a^2}$.

21. $\dfrac{i_1{}^4i_2{}^5}{e_1{}^6e_2{}^7} \div \dfrac{i_1{}^2i_2{}^3}{e_1{}^4e_2{}^5}$.

22. $\dfrac{au^n}{4bv^2} \div \dfrac{a^2u^{n-1}}{8b^2v}$.

23. $\dfrac{15X_c{}^6X_l{}^8}{28\pi^3r^5} \div \dfrac{45X_c{}^4X_l{}^6}{7\pi^2r^4}$.

24. $\dfrac{t_1{}^{m+1}t_2{}^m}{s_1{}^ns_2{}^{n-1}} \div \dfrac{t_1{}^mt_2{}^{m-1}}{s_1{}^{n-1}s_2{}^{n-2}}$.

25. $\dfrac{E_1{}^2 - E_2{}^2}{E_1} \div \dfrac{E_1 - E_2}{E_2}$.

26. $\dfrac{I^2 - i^2}{2I + 3i} \div \dfrac{I - i}{4I + 6i}$.

27. $\dfrac{u^4 - v^4}{u^2 - v^2} \div \dfrac{(u + v)^2}{u^2 + v^2}$.

28. $\dfrac{12(ab - b^2)}{a(a + b)^2} \div \dfrac{3(a^2 - b^2)}{a^2(a + b)}$.

29. $\left(M^2 - \dfrac{1}{M^2}\right) \div \left(M + \dfrac{1}{M}\right)$.

30. $(ax + ay) \div \dfrac{y^2 - x^2}{a}$.

31. $\dfrac{p^4 - q^4}{(p + q)^2} \div \dfrac{qp - q^2}{(p + q)}$.

32. $\dfrac{14a}{x - 4} \div \dfrac{7a^2}{x^2 - 11x + 28}$.

33. $\left(u + v - \dfrac{uv}{u + v}\right) \div \left(u + \dfrac{v^2}{u + v}\right)$.

34. $\left(\dfrac{e + 1}{e - 1}\right)^3 \div \left(\dfrac{e + 1}{e - 1}\right)^2$.

35. $\dfrac{x^2 - y^2}{2x + y} \div \dfrac{3x^2 y - 3y^3}{10x + 5y}$.

36. $\left(\dfrac{J_m - 1}{J_m + 2}\right)^2 \div \dfrac{J_m^{\,2} - 1}{J_m^{\,2} - 4}$.

37. $\dfrac{12ax + 40ay}{28cx - 70cy} \div \dfrac{12bx + 40by}{14dx - 35dy}$.

38. $\dfrac{h^2 - 6h + 8}{h^2 + 6h + 9} \div \dfrac{h - 2}{h + 3}$.

39. $\dfrac{g^3 + g^2 - g - 1}{1 - p^2} \div \dfrac{g^2 - 1}{p^2 - 1}$.

40. $\dfrac{r^2 - 12r + 35}{r^2 + 3r + 2} \div \dfrac{r^2 - 25}{r^2 - 4}$.

41. $\left(\dfrac{1}{r_1} + \dfrac{1}{r_2}\right) \div \left(\dfrac{1}{r_1} - \dfrac{1}{r_2}\right)$.

42. $\dfrac{u^3 + u^2 + u + 1}{u^3 - 2u^2 + u - 2} \cdot \dfrac{u^2 + 2u - 3}{u^2 - 1}$.

43. $\left(\dfrac{x^2}{y^2} + \dfrac{2x}{y} + 1\right) \cdot \dfrac{x^2 - y^2}{(x + y)^2}$.

44. $\left(\dfrac{a^2 + b^2}{ab} - 2\right) \div \left(\dfrac{a^2 + b^2}{ab} + 2\right)$.

45. $\dfrac{x^2 + 3x - 4}{x^2 - 5x + 6} \div \dfrac{x^2 + 6x + 8}{x^2 + x - 6}$.

46. $\dfrac{r^2 + 5r + 6}{r^2 + 7r + 12} \cdot \dfrac{r^2 + 9r + 20}{r^2 + 11r + 30}$.

47. $\left(\dfrac{a + b}{2a + b}\right)^n \left(\dfrac{2a + b}{a + b}\right)^{n-1} \div \left(\dfrac{a + b}{2a + b}\right)^{-2}$.

48. $\dfrac{4x^2 + y^2 - z^2 + 4xy}{4x^2 - y^2 - z^2 - 2yz} \div \dfrac{2x + y + z}{2x - y - z}$.

49. $\left(2x - \dfrac{x^2}{x + y}\right)\left(x + \dfrac{y^2 - 2xy}{x + 2y}\right) \div \left(x - y + \dfrac{2y^2}{x + y}\right)$.

50. $\left(\dfrac{a + 8}{a - 1} - a\right)\left(\dfrac{a}{7a - 4} - \dfrac{1}{a + 2}\right) \div \left(a - 6 + \dfrac{4}{a - 2}\right)$.

2–12. Equations

An equation is a statement of equality. Thus

$$a^2 - x^2 = (a + x)(a - x), \qquad (1)$$

$$4x + 2 = 10, \qquad (2)$$

$$x + 1 = x \qquad (3)$$

are equations. In making a statement of equality, we make no assertion concerning whether there are any values of the letters for which the statement is true. The statement may be true for all values of the letters, as equation 1 above; the statement may limit the values of the letters for which it is true, as equation 2 above which it is true only for $x = 2$; or the statement may not be true for any values of the letters, as equation 3 above. A more complete discussion of this matter will be found in Section 7–5.

In this section we shall consider only equations that are true for a single value of a single letter. *The value of the letter for which the equation is true is the* **root** *or* **solution** *of the equation.*

When a statement of equality of this kind is given, our interest is in finding the value of the letter for which it is true. The following rules will aid in finding the root.

1. *The roots of an equation remain the same if the same expression is added to or subtracted from both sides of the equation.*

2. *The roots of an equation remain the same if both sides of an equation are multiplied or divided by the same expression other than zero and not involving the letter whose value is in question.*

Thus the equation

$$2x = 4, \qquad (4)$$

where x is unknown, is true for $x = 2$. To illustrate the first of the above two rules, add $5x$ to both sides of equation 4 and get

$$2x + 5x = 4 + 5x$$

which, like equation 4, is true for only $x = 2$. To illustrate the importance of the restriction in the second of the above two rules, multiply both sides of equation 4 by x and get

$$(2x)x = (4)x,$$

which is true not only for $x = 2$ but also for $x = 0$.

Example 1. Find the value of x which satisfies

$$7x + 13 = 3x + 41.$$

In finding the **solution** of this equation both sides may be decreased by 13 (rule 1), yielding

$$7x + 13 - 13 = 3x + 41 - 13,$$

$$7x = 3x + 28.$$

If rule 1 is applied again, both sides may be decreased by $3x$, thus giving

$$4x = 28.$$

Dividing both sides by 4 (rule 2), we get

$$\frac{4x}{4} = \frac{28}{4}, \qquad x = 7.$$

Although the above rules assure us that $x = 7$ satisfies the given equation, we verify the solution $x = 7$:

$$7 \cdot 7 + 13 = 3 \cdot 7 + 41,$$

$$49 + 13 = 21 + 41,$$

$$62 = 62.$$

It is always a good plan to check the accuracy of one's work by substituting the result in the **original** equation to see whether the equation is true for this value.

Rule 1 is applied very frequently. It is, therefore, desirable to state it in a way that mechanizes its application. Observe that the effect of decreasing both sides of the equation in example 1 by 13 is equivalent to the deletion of the term $+13$ on the left side of the equation and the addition of the term -13 on the right side.

If the equation

$$4x = 28 - 3x$$

is given, then, by applying rule 1, $3x$ may be added to both sides of the equation, yielding

$$4x + 3x = 28 - 3x + 3x$$

$$= 28.$$

The result of the operation consists in omitting the term $-3x$ from the right side and adding the term $+3x$ to the left side. We call this operation **transposition** of the term $3x$. This operation is an application of rule 1 and may be explained in the following way:

Any term of one side of an equation may be transposed to the other side if its sign is changed.

Example 2. Find the value of x that satisfies

$$3x + 7(4 - x) + 6x = 15. \tag{5}$$

Clearing parentheses and combining terms, we obtain

$$3x + 28 - 7x + 6x = 15,$$
$$2x + 28 = 15.$$

Transposing $+28$ from the left to the right side, we get

$$2x = 15 - 28,$$
$$2x = -13.$$

Dividing each side by 2, according to rule 2, gives

$$\frac{2x}{2} = -\frac{13}{2},$$

$$x = -\frac{13}{2}.$$

To check, substitute this solution in equation 5:

$$3(-\tfrac{13}{2}) + 7(4 + \tfrac{13}{2}) + 6(-\tfrac{13}{2}) = 15,$$
$$-\tfrac{39}{2} + 28 + \tfrac{91}{2} - \tfrac{78}{2} = 15,$$
$$-\tfrac{117}{2} + \tfrac{147}{2} = 15,$$
$$\tfrac{30}{2} = 15,$$
$$15 = 15.$$

An equation that can be reduced to the form

$$ax + b = 0, \qquad a \neq 0,$$

is called a **linear equation in x.** All equations considered in this section are of this type.

To solve an equation containing fractions, first reduce each fraction to its lowest terms. Then multiply each side of the equation by the least common denominator of all the denominators. This process is called **clearing of fractions.**

Example 3. Find the value of x that satisfies

$$\frac{5x}{6} - \frac{x - 1}{2} = \frac{3x + 1}{3} + \frac{3}{8}(x + 1).$$

The common denominator is 24. Then

$$\frac{20x}{24} - \frac{12(x - 1)}{24} = \frac{8(3x + 1)}{24} + \frac{9(x + 1)}{24},$$

whence

$$\frac{20x - 12(x - 1)}{24} = \frac{8(3x + 1) + 9(x + 1)}{24}.$$

By rule 2, we multiply both sides of the equation by 24, obtaining

$$20x - 12(x - 1) = 8(3x + 1) + 9(x + 1).$$

Clearing parentheses gives

$$20x - 12x + 12 = 24x + 8 + 9x + 9.$$

By rule 1,

$$20x - 12x - 24x - 9x = 8 + 9 - 12,$$

$$-25x = 5,$$

$$x = \frac{5}{-25},$$

$$x = -\frac{1}{5}.$$

The validity of this result may be verified by substituting it in the original equation.

Exercises

Solve the following equations (a, b, c are to be treated as known quantities):

1. $4x + 11 = 31.$

2. $5y + 10 = 3y + 24.$

3. $3e - 2 = e + 24.$

4. $-3s + 5 = -17.$

5. $5l + 6 = l + 24.$

6. $7r + 3 = r + 21.$

7. $27v + 40 = 22v + 90.$

8. $5s - 45 = 18 - 4s - 9.$

9. $bx + b = ab.$

10. $ax + c = bx + 2c.$

11. $4X_L + \dfrac{X_L}{2} = 36.$

12. $5R - \tfrac{9}{2}R = 15 - 2R.$

13. $6K - 19 = -K + 30.$

14. $245T - 15 = 15T + 445.$

15. $\dfrac{3Q}{2} + 4 = 85 - 3Q.$

16. $12L - 22 = \tfrac{3L}{2} - 3L.$

17. $2(5E - 25) + 6E - 15 = 19E.$

18. $3Z + 2(Z - 3) - 7Z - 3(Z + 1).$

19. $17W + 4(5 - 2W) = 44 + 3W.$

20. $6 - (2 - 4x) = 5x - (1 - 2x).$

21. $3I - 5 = 7(I + 2) + \tfrac{3}{4}I.$

22. $\dfrac{7F}{2} + 3F = \dfrac{F - 1}{3} + \dfrac{38}{3}.$

23. $7(4M - 15) = 5(2M + 5) + 14.$

24. $\tfrac{1}{3}i + 2\tfrac{2}{3} = \tfrac{2}{3}i + 6.$

25. $\dfrac{4z}{3} - 5z + 2 = \dfrac{z}{2} - 1.$

26. $4(t - 5) + 3t = 3(2t - \tfrac{1}{2}).$

27. $\dfrac{u}{2} - 3u + 7 = \dfrac{1}{4} - \dfrac{7u}{2}.$

28. $\dfrac{f + 1}{3} + \dfrac{f + 2}{7} = 5.$

29. $\dfrac{x}{a+1} + \dfrac{x}{a-1} = 2a.$

30. $x^2 + a^2 = (x - b)^2.$

31. $0.11y - 0.21y = -0.53.$

32. $m - 0.02m - 4 = 118.5.$

33. $\dfrac{x+2a}{b} - \dfrac{b-x}{2a} = 0.$

34. $\dfrac{y-a}{b} = \dfrac{y-b}{a}.$

35. $13.75v + 5.25 = 8.25v + 16.25.$

36. $(2t - 3)^2 - 4t^2 = 2(7 - t).$

37. $\dfrac{x}{a-2} + \dfrac{2x}{a+2} - \dfrac{x}{a^2-4} = 3.$

38. $\dfrac{x+3}{a+b} - \dfrac{x-3}{a-b} = 0.$

39. $\dfrac{1}{6}(4w - 2) - \dfrac{2}{3} = \dfrac{5}{2}\left(1 + \dfrac{w}{5}\right).$

40. $\dfrac{1}{2}\left(H - \dfrac{5}{2}\right) - \dfrac{5}{3}\left(\dfrac{H}{2} + \dfrac{1}{2}\right) = \dfrac{H}{2}.$

41. $(x + a)(x + b) - (a - x)(c - x) = 0.$

42. $(t - 1)(t - 2) - (t - 3)(t + 1) - 29 = 0.$

43. $(x - 7)(2x + 1) - (3x - 1)(2x - 5) + (4x - 3)(x + 7) - 25 = 0.$

44. $\dfrac{I-a}{2a} + \dfrac{I+6b}{3b} + \dfrac{I-2c}{4c} = 1.$

45. $\dfrac{e+3}{4} + \dfrac{7e-2}{5} = \dfrac{1}{2}\left(\dfrac{5e-1}{2} + \dfrac{10e+8}{9}\right).$

46. $\dfrac{7k+6}{13} + \dfrac{5k+4}{9} = 3 - \dfrac{4k-1}{3}.$

47. $(q + 1)(q^2 - 4q + 7) + 3q^2 = q^3 + 4.$

48. $(x - 2)(x + 2) - (2x + 3)^2 + (x + 5)(3x - 4) + 31 = 0.$

49. $(2y + a)(a + 2y) + (2y - a)(a - 2y) = 16a^2.$

50. $(x - a)(x - c) = 2(a - c)(c + a) + (x + a)(x + c).$

2–13. Factorable Quadratic Equations

A quadratic equation is one that can be reduced to the form

$$ax^2 + bx + c = 0, \qquad a \neq 0, \tag{1}$$

where a, b, and c are known and x is unknown.

Occasionally the left side of equation 1 can be factored by the type forms of Sec. 2–8. Then the left is the product of two factors. *Now, if* $A \cdot B = 0$, *either* $A = 0$ *or* $B = 0$, *or* $A = 0$ *and* $B = 0$. Hence equation 1 is satisfied if at least one factor is zero. The numbers that satisfy equation 1 are the **roots** or solutions of the quadratic equation.

Two examples will illustrate the process.

Example 1. Find the values of x that satisfy the quadratic equation

$$x^2 - 4x + 3 = 0. \tag{2}$$

Factoring the left side, we obtain

$$(x - 3)(x - 1) = 0.$$

This product is zero if

$$x - 3 = 0 \quad \text{or} \quad x - 1 = 0;$$

hence,

$$x = 3 \quad \text{or} \quad x = 1.$$

Substituting $x = 3$ in equation 2, we obtain

$$(3)^2 - 4(3) + 3 = 0,$$
$$9 - 12 + 3 = 0,$$
$$0 = 0.$$

Substituting $x = 1$, we obtain

$$(1)^2 - 4(1) + 3 = 0,$$
$$1 - 4 + 3 = 0,$$
$$0 = 0.$$

Therefore both solutions satisfy the original equation.

Example 2. Solve

$$-\frac{x}{4} + \frac{3}{2} = \left(\frac{x}{2}\right)^2. \tag{3}$$

Rewriting equation 3 in the form of equation 1, we obtain

$$x^2 + x - 6 = 0. \tag{4}$$

Factoring equation 4 yields

$$(x + 3)(x - 2) = 0.$$

Equating each factor to 0, we obtain

$$x + 3 = 0,$$

First solution:

$$x = -3.$$

$$x - 2 = 0,$$

Second solution:

$$x = 2.$$

Substituting the first solution into equation 3, we obtain

$$-\frac{-3}{4} + \frac{3}{2} = \left(\frac{-3}{2}\right)^2,$$
$$\frac{9}{4} = \frac{9}{4}.$$

Substituting the second solution into equation 3, we obtain

$$-\tfrac{2}{4} + \tfrac{3}{2} = (\tfrac{2}{2})^2,$$

$$1 = 1.$$

The solutions obtained are therefore correct.

Exercises

Solve the following equations by factoring and check the solutions obtained:

1. $x^2 - 9 = 0$. 2. $y^2 - 49 = 0$.

3. $4t^2 - 81 = 0$. 4. $9r^2 - 64 = 0$.

5. $z^2 - 6z + 9 = 0$. 6. $u^2 + 4u + 4 = 0$.

7. $4v^2 + 12v + 9 = 0$. 8. $9R^2 - 6R + 1 = 0$.

9. $4e^2 + 25 = 20e$. 10. $12F = -9F^2 - 4$.

11. $x^2 - 2x - 3 = 0$. 12. $s^2 + 3s - 10 = 0$.

13. $V^2 - 3V - 18 = 0$. 14. $Z^2 + 9Z - 10 = 0$.

15. $t^2 = 6t - 8$. 16. $21 - L_c{}^2 = -4L_c$.

17. $3E^2 - 75 = 0$. 18. $5M_x{}^2 - 20 = 0$.

19. $25Q^2 - \tfrac{9}{49} = 0$. 20. $9I^2 - \tfrac{4}{25} = 0$.

21. $x^2 - 5x = 0$. 22. $y^2 + 7y = 0$.

23. $3u^2 + 2u = 0$. 24. $5t^2 - t = 0$.

25. $5p^2 + 15p = 50$. 26. $3Q^2 + 36 = 21Q$.

27. $\dfrac{i^2}{6} - \dfrac{5i}{6} = 1$. 28. $\dfrac{X^2}{2} - \dfrac{X}{2} = 1$.

29. $(x + 2)^2 + (x - 2)^2 - 40 = 0$. 30. $(y + 3)^2 + (y - 3)^2 = 20$.

31. $(x + 1)^2 + (x + 5)^2 - 10 = 0$. 32. $(x - 3)^2 + (x + 7)^2 = 68$.

2–14. Applications to Scientific and Engineering Formulas

Many experimental or theoretical facts in science are stated by equations. For example, the heat in calories produced by a steady direct current is equal to 0.24 multiplied by the square of the current in amperes, by the resistance in ohms, and by the time in seconds during which the current flows. Symbolically, this law may be expressed by the formula

$$H = 0.24I^2RT, \tag{1}$$

where H is the heat in calories, I the current in amperes, R the resistance in ohms, and T the time in seconds. In the form given, equation 1 gives

H when I, R, and T are known.   Suppose that H, I, and R are given, and T is to be determined.   We state this problem by saying that equation 1 has to be **solved for** T.   In view of the preceding section, we may regard T as unknown and the other letters as known, and by rule 2

$$T = \frac{H}{0.24 I^2 R}.$$

In like manner, if H, I, and T are given and the equation has to be solved for R, we obtain from equation 1 the result

$$R = \frac{H}{0.24 I^2 T}.$$

Many formulas encountered in practical work, like the illustration above, have the form

$$ax + b = 0, \tag{2}$$

where a and b are given and x is to be determined.   The methods of the previous section therefore apply.

Exercises

In the formulas below, consider all the quantities as given except the indicated quantity to be solved for.   A hint concerning the meaning of the formulas is given.   If the reader is interested in a description of the exact physical meaning of the formula and the units involved, an appropriate textbook should be consulted.

Given	Solve for	Physical Application
1. $R_T = R_1 + R_2 + R_3$	R_3	Total resistance in a series circuit
2. $L_T = L_1 + L_2 + 2M$	M	Total inductance of two coils with in-
3. $L_T = L_1 + L_2 - 2M$	L_1	ductive coupling
4. $L_T = L_1 + L_2$	L_2	Total inductance of uncoupled coils
5. $I_x = I_{xo} + Ad^2$	I_{xo}, A	Rotational inertia
6. $T = \dfrac{rhdg}{2}$	h, d	Surface tension
7. $I = \dfrac{E}{R}$	E	Ohm's law
8. $P = IE$	I	Electric power
9. $P = I^2 R$	R	Electric power
10. $p = ei$	i	Instantaneous power in a-c circuits
11. $Fd = Wh$	d, W	Work done on a falling body

	Given	Solve for	Physical Application
12.	$\dfrac{P_2 V_2}{T_2} = \dfrac{P_1 V_1}{T_1}$	$P_1, \dfrac{V_1}{V_2}$	Equation for perfect gas
13.	$A = \dfrac{Rt}{pV}$	R, t	Specific inlet area of jet engine
14.	$C = \tfrac{5}{9}(F - 32)$	F	Centigrade-Fahrenheit relation
15.	$v^2 = v_0{}^2 + 2gh$	$g, v_0{}^2$	Initial-final speed relation
16.	$H = \dfrac{tws}{33,000}$	w	Horsepower
17.	$F = \dfrac{w}{g}\, a$	a, w	One of Newton's laws
18.	$M = \dfrac{mgl^3}{12\pi r^4 s}$	m, l^3	Young's modulus by bending for a cylindrical bar
19.	$d = vt$	t	Distance traveled
20.	$a = \dfrac{v - v_0}{t}$	v_0	Average acceleration
21.	$fd = \tfrac{1}{2}mv^2$	m, d	Work–kinetic-energy relation
22.	$R = \dfrac{kl}{d^2}$	l	Resistance of wire
23.	$R = \dfrac{kl}{A}$	l	Magnetic reluctance
24.	$Ft = mv_2 - mv_1$	m, F	Mechanical impulse
25.	$Q = \dfrac{K(t_2 - t_1)aT}{d}$	T, a	Conduction of heat
26.	$Q = CE$	E	Quantity of charge on a capacitor
27.	$l_t = l_0(1 + \alpha t + \beta t^2)$	α, β	Linear expansion of solids
28.	$L = \pi(R + r) + 2d$	d, r	Length of a belt around two pulleys
29.	$B = \dfrac{\phi}{A}$	ϕ	Magnetic field intensity
30.	$f = \dfrac{8.94 B^2 A}{10^8}$	A	Tractive force of electromagnets
31.	$V = \dfrac{2ga^2(d_1 - d_2)}{9\eta}$	a^2	Stokes's law
32.	$P = \dfrac{K B_{\max}{}^{1.6}}{10^7}$	$B_{\max}{}^{1.6}$	Steinmetz equation for hysteresis loss

Given	Solve for	Physical Application
33. $\dfrac{r_1}{r_2} = \dfrac{r_3}{r_4}$	r_1	Wheatstone bridge formula
34. $E = \dfrac{N\phi}{10^8 t}$	ϕ, N	Average electromotive force generated by cutting magnetic field
35. $F = \dfrac{22.5 B l I}{10^8}$	l, B	Force of magnetic field on a coil carrying a current
36. $Q = 0.4 S_c L D^2$	L	Torque transmitted by a splined shaft
37. $P_n = \dfrac{V_a(V_j - V_a)}{gJ}$	V_j	Net power of turbojet engine
38. $L_{\text{avg}} = \dfrac{1.26 N^2 A \mu}{10^8 l}$	A	Self-inductance of long coils
39. $M = \dfrac{-1.26 N_1 N_2 A \mu}{10^8 l}$	N_2	Mutual inductance
40. $S = \dfrac{d^2 w}{8p}$	w	Stretch of trolley wires
41. $W = \rho V_a \dfrac{\pi}{4} D^2 (e^2 - h^2)$	V_a	Mass flow in centrifugal compressor
42. $E = k \dfrac{M}{N V \rho_i}$	M	Volumetric efficiency for internal-combustion engine
43. $C = \dfrac{8.84 K A}{10^8 d}$	A	Calculation of capacitance
44. $J = \frac{1}{32} \pi (d^4 - d_1^4)$	d_1^4	Polar moment of inertia for hollow circular shaft
45. $X_L = 2\pi f L$	f, L	Inductive reactance
46. $f = \dfrac{v}{\lambda}$	v	Frequency of radio wave in terms of velocity and wavelength
47. $L = 0.0251 d^2 n^2 l$	d^2, l	Inductance of single-layer solenoid
48. $E = K(T^4 - T_0^4)$	T^4	Energy radiated in unit time by a black body
49. $i_{\text{avg}} = C \dfrac{e_2 - e_1}{t_2 - t_1}$	e_2, C	Current flowing in a capacitor
50. $e_{\text{avg}} = L \dfrac{i_2 - i_1}{t_2 - t_1}$	L, i_2	Average induced voltage

2–15. Substitution of Particular Values in Formulas

The formulas given in the preceding section show how to solve for certain quantities if other quantities connected with them are given. By the statement **a quantity is given,** it is understood that the following is true:

1. A unit is known by which the particular quantity is measured. It must be explained, for instance, whether a certain length is measured in feet or in inches.

2. The number of units of measurement in the given quantity is known. This number of units is often called the numerical value of the quantity.

For instance, if the radiant flux of a certain light source is to be given, a unit of measurement has to be chosen first. In this case the unit is the lumen. Thus, if an experiment shows that the radiant flux of this light source is 780 lumens, then 780 is the numerical value of the radiant flux when measured in lumens.

In almost all practical applications it is necessary to compute the numerical value of quantities that are given by mathematical formulas. For example, the heat produced in an electric circuit is computed by the formula

$$H = 0.24I^2RT. \tag{1}$$

In every formula we suppose that the quantities involved in it are measured in certain definite units. As a rule the formula will change if the units of measurements on which it is based are replaced by other units. For example, in formula 1, it is supposed that H is measured in calories, I in amperes, R in ohms, and T in seconds.

The numerical value of H can be computed if the numerical values of I, R, and T are known. To find H it is necessary only to replace the letter symbols I, R, and T by their numerical values and to perform the operations indicated by the formula. Thus, if it is known that $I = 4.0$ amperes, $R = 50.0$ ohms, and $T = 5.000$ min $= 300.0$ sec, then

$$H = 0.24 \times (4.0)^2 \times 50.0 \times 300.0 \text{ calories} = 5.8 \times 10^4 \text{ calories}.$$

In this case we say that *the numerical values $I = 4.0$, $R = 50.0$, $T = 300.0$ have been substituted in formula 1* or that *H has been computed from formula 1 for the particular values $I = 4.0$, $R = 50.0$, $T = 300.0$.*

It should be observed that the particular values that are substituted in a formula are, as a rule, results of measurements and are, therefore, numbers of limited accuracy, given only with a certain number of significant digits. It is therefore important to observe the rules developed in Chapter 1 for dealing with inaccurate numbers. The slide rule is often very useful for the evaluation of formulas and should be used whenever its precision is sufficient.

Example 1. The resistance R of a wire, measured in ohms, is given by the formula

$$R = \frac{kl}{d^2},$$

where l is the length of the wire measured in feet, d its diameter measured in mils (1 mil = 0.001 in.), and k the resistance in ohms per mil-foot of a wire of the same material, that is, a wire of length 1 ft and diameter 1 mil. Compute the resistance of a wire of length 5.00 miles and diameter 80.8 mils when $k = 10.4$ ohms per mil-foot.

Since the formula requires that the length be expressed in feet, the value $l = 5.00$ miles has to be changed to feet, giving

$$l = 5.00 \times 5280 = 26,400 \text{ ft},$$

The substitution of the given data in the formula yields

$$R = \frac{10.4 \times 26,400}{80.8^2} = \frac{10.4 \times 26,400}{6530} = 42.0 \text{ ohms},$$

the computations being performed on a slide rule.

Example 2. The fundamental vibration frequency n in cycles per second of a string is given by the formula

$$n = \frac{1}{2rl}\sqrt{\frac{T}{\pi d}},$$

where r is the radius and l the length in centimeters, T is the tension in dynes, and d is the density in grams per cubic centimeter. Find the value of n for $r = 0.026$, $l = 50.0$, $d = 7.80$, and $T = 9.8 \times 10^6$.

Substitution of the given values in the formula for n gives

$$n = \frac{1}{2 \times 0.026 \times 50.0}\sqrt{\frac{9.80 \times 10^6}{\pi \times 7.80}} = \frac{10^3}{2 \times 0.026 \times 50.0}\sqrt{0.40}$$

$$= \frac{0.63 \times 10^3}{2 \times 2.60 \times 10^{-2} \times 50.0} = \frac{0.63 \times 10^3}{2.60} = 240 \text{ cycles per second},$$

where the computations were performed on a slide rule.

Exercises

In the following problems observe the rules about significant figures given in Sec. 1–13.

1. Given $P = I^2R$, find the value of P in watts when $I = 15.1$ amperes and $R = 25.7$ ohms.

2. Given $H = 0.24I^2R$, find the value of H in calories when $I = 0.52$ amperes and $R = 195$ ohms.

3. Given $P = E^2/R$, find the value of P in watts when $E = 115$ volts and $R = 78$ ohms.

4. Given $v = \sqrt{v_0^2 + 2gh}$, find v in feet per second, when $v_0 = 20$ ft per sec, $g = 32$ ft per sec per sec, and $h = 700$ ft.

5. Given $s = v_0 t - \frac{g}{2}t^2$, find s in feet when $v_0 = 850$ ft per sec, $t = 29$ sec, and $g = 32$ ft per sec per sec.

6. Given $a = \dfrac{v - v_0}{t}$, find a in feet per second per second when $v = 85$ ft per sec, $v_0 = 14$ ft per sec, and $t = 7.6$ sec.

7. The moment of inertia I in biquadratic inches of a rectangular section is given by

$$I = \frac{bd^3}{12},$$

where b is the breadth and d is the depth in inches of the beam. Compute I for a beam having a breadth of 10.6 in. and a depth of 6.5 in.

8. The volume of a sphere of radius r is $\frac{4}{3}\pi r^3$, where $\pi = 3.1416$. What is the volume of a sphere 4 ft in diameter?

9. How many cubic inches of rubber are required to make a hollow rubber ball having an outer radius of $3\frac{1}{4}$ in. and an inner radius of $2\frac{3}{4}$ in.?

10. The distance s in feet through which an object falls owing to the action of gravity is $s = \frac{1}{2}gt^2$, where g is the acceleration due to gravity and t is the time of fall in seconds. If the acceleration due to gravity is 32 ft per sec per sec, how far will an object fall in 7 sec?

11. If two point electric charges of magnitude q and q' measured in electrostatic units are separated by a distance r cm, Coulomb's law states that the force F in dynes that one charge exerts on the other is given by

$$F = \frac{qq'}{kr^2},$$

where k is a constant. Given two point charges of 6 and 10 electrostatic units respectively, separated by a distance of 5 cm, what is the force acting between them, if it is assumed that $k = 1$?

12. The tractive force f in pounds of an electromagnet is given by

$$f = \frac{8.94B^2A}{10^8},$$

where B is measured in gausses and A in square centimeters. Compute f when $B = 1.87 \times 10^4$, $A = 31.3$ sq cm.

13. The centrifugal force in dynes for a mass moving in uniform circular motion is given by

$$F = \frac{4\pi^2 mr}{T^2},$$

where m is measured in grams, r in centimeters, and T in seconds. Compute F when $m = 3.2 \times 10^8$, $r = 2.4 \times 10^2$, and $T = 0.80$.

14. The self-inductance L in henrys of a certain type of coil is given by

$$L = \frac{1.26N^2A\mu}{10^8 l}.$$

Compute L when $N = 450$, $A = 420$ sq cm, $l = 210$ cm, and $\mu = 2200$.

15. The acceleration a in feet per second per second which the gravitational attraction of the earth would produce at the distance of the moon is given by

$$a = \frac{4\pi^2 R}{T^2},$$

where R is measured in miles and T in seconds. Find the value of a when $R = 3.1 \times 10^5$ miles, $T = 27.3$ days.

16. The elastic resilience K in inch-pounds of a bar under tension is given by

$$K = \frac{S^2}{2E} Al,$$

where the stress S and the modulus of elasticity E are measured in pounds per square inch, the area A of the bar in square inches, and its length l in inches. Find the value for K when $S = 1.20 \times 10^4$ lb per sq in., $E = 3.15 \times 10^7$ lb per sq in., $A = 4.46$ sq in , and $l = 18$ ft.

17. Torricelli's formula for the velocity v in centimeters per second of efflux of water from a small orifice in the side of a container is given by

$$v = \sqrt{2gh},$$

where h is measured in centimeters and g, the acceleration of gravity, in centimeters per second per second. Find the velocity of water flowing out of an orifice if $g = 980$ and $h = 610$.

18. The reactance X_c in ohms of a capacitor is given by

$$X_c = \frac{1}{2\pi f C},$$

where f is the frequency of the impressed voltage in cycles per second and C is the capacitance of the capacitor in farads. What is the reactance of a 112-micromicrofarad capacitor at a frequency of 3.289 megacycles?

19. An automobile traveling at a speed of s mph can be stopped in x ft after the brakes are applied according to the empirical formula

$$x = 0.094s^2.$$

What may the maximum speed of the car be if it must be stopped in 42 ft?

20. The velocity V in meters per second of sound in solids is expressed by the formula

$$V = \sqrt{\frac{e}{d}},$$

where e is Young's modulus of elasticity and d is the density of the material. Compute V when $e = 2.0 \times 10^{12}$ and $d = 7.8$.

21. The impedance Z, measured in ohms, of a circuit is given by

$$Z = \sqrt{R^2 + \left(2\pi f L - \frac{1}{2\pi f C}\right)^2}.$$

Compute Z when $R = 145$ ohms, $f = 5.4 \times 10^6$ cycles per second, $L = 3.2 \times 10^{-3}$ henry, and $C = 2.5 \times 10^{-10}$ farad. These units are the ones for which the right member of the formula gives Z in ohms.

2–16. Systems of Two Linear Equations in Two Unknowns

Consider the equation

$$x - 2y = 5, \tag{1}$$

where x and y are both unknown. This equation is satisfied when $x = 7$ and $y = 1$, also when $x = 5$ and $y = 0$. There are many such pairs of values that satisfy equation 1. To find pairs other than those given, choose a value of one letter, say y, arbitrarily, and then from equation 1 find the corresponding value of x. For example, let $y = 3$. Then from equation 1

$$x = 5 + 2y, \tag{2}$$

whence

$$x = 5 + 2 \cdot 3,$$

$$x = 5 + 6 = 11,$$

and the pair of values $x = 11$, $y = 3$ satisfies equation 1.

Consider now a second equation in two unknowns:

$$x + y = 2. \tag{3}$$

As before, this equation is satisfied by many pairs of values.

Is there a pair of values, one for x and one for y, that satisfies both equations 1 and 3? The x value satisfying equation 3 must also satisfy equation 2. Substituting equation 2 in equation 3 gives

$$(5 + 2y) + y = 2,$$

$$5 + 3y = 2,$$

$$3y = -3,$$

$$y = -1.$$

Using this value in equation 2, we have

$$x = 5 + 2(-1),$$

$$x = 5 - 2,$$

$$x = 3.$$

The reader may verify that the pair $x = 3$, $y = -1$ satisfies both equations 1 and 3.

The method for finding the pair of values satisfying both equations indicated above usually applies to pairs of equations of the form

$$a_1 x + b_1 y = c_1,$$
$$\tag{4}$$
$$a_2 x + b_2 y = c_2,$$

where a_1, a_2, b_1, b_2, c_1, c_2, are known and x and y are unknown. A more complete discussion is given in Chapter 13. Equations 4 are termed **linear** because the unknowns x and y enter to the first power only.

To solve a system of two linear equations in two unknowns, solve for one unknown in one equation, and substitute this result in the other equation, thus obtaining one equation in one unknown.

An alternative way of solving a system of two linear equations, which is usually more convenient, is given by the following rule.

Multiply the two equations with numerical factors which are chosen so that the coefficients of one of the two unknowns have the same numerical values in both equations. By adding or subtracting the two equations, a new equation with only one unknown quantity is obtained. Solve this equation. In order to find the second unknown quantity, substitute the value that has been found, and solve for the remaining unknown quantity. An alternative method for finding the second unknown is to repeat the above process of forming equal coefficients for this unknown.

From the following example the reader will see that this method works in a very convenient way.

Example. Find the values of I_1 and I_2 that satisfy

$$7I_1 - 3I_2 = 26,$$
$$2I_1 + 11I_2 = 43.$$

Multiply the first equation by 2, the second by 7:

$$14I_1 - 6I_2 = 52,$$
$$14I_1 + 77I_2 = 301.$$

Subtract the first of these equations from the second:

$$83I_2 = 249,$$
$$I_2 = 3.$$

Substituting this value for I_2 in the first equation, we obtain

$$7I_1 - 9 = 26,$$
$$7I_1 = 35,$$
$$I_1 = 5.$$

The two values $I_1 = 5$ and $I_2 = 3$ constitute the desired solution.

Check: Substituting 5 for I_1 and 3 for I_2 in the given equations, we obtain

$$35 - 9 = 26,$$
$$26 = 26,$$
$$10 + 33 = 43,$$
$$43 = 43.$$

Exercises

Solve the following systems of equations by either of the methods described above and check your answers:

1. $x + y = 5,$
$x + 2y = 8.$

2. $2a + b = 4,$
$a + b = 3.$

3. $u + 2v = 7,$
$5u + 3v = 14.$

4. $2r + s = 10,$
$4r + 3s = 24.$

5. $m - n = 2,$
$m + n = 8.$

6. $2e_1 - e_2 = 5,$
$3e_1 + 5e_2 = 14.$

7. $r_1 + r_2 = 3,$
$3r_1 + 2r_2 = 1.$

8. $p + q = 7,$
$2p + 5q = 20.$

9. $3E_1 - 2E_2 = 28,$
$E_1 - 3E_2 = 7.$

10. $x + 2y = 0,$
$5x - 3y = 13.$

11. $4c + 6d = 0,$
$5c + 2d = -11.$

12. $2i_1 - i_2 = 0,$
$7i_1 + 3i_2 = -13.$

13. $3T_1 + 4T_2 = 17,$
$T_1 + 2T_2 = 7.$

14. $4L - 2C = 2,$
$3L + C = 14.$

15. $7M_1 - 5M_2 = 1,$
$5M_1 + M_2 = 19.$

16. $6x - 5y = -3,$
$4x + 2y = 14.$

17. $2u - 3v + 4 = 0,$
$u + 2v - 5 = 0.$

18. $2Z_1 - 5Z_2 + 7 = 0,$
$Z_1 + 2Z_2 - 1 = 0.$

19. $2s = t - 7,$
$4t + 3s = 6.$

20. $3F_1 = -9 - 7F_2,$
$5F_2 = 1 - 4F_1.$

21. $x = 3y,$
$5x - 11 = 4y.$

22. $3C_1 - 19 = -2C_2,$
$2(C_1 - 10) = 6C_2.$

23. $2(2E_1 - 13) = -7E_2,$
$1 + 4E_2 - 3E_1 = 0.$

24. $5x = 2(3y + 1),$
$4x + 7y + 5 = 0.$

25. $M - 2N - 3 = 0,$
$2(M - 2) + N = 0.$

26. $2L + 5(C - 1) = 0,$
$5L + 2(C - 10) = 0.$

27. $3(2u + v) = 1 - u,$

$4(u - 3) = 1 + v.$

28. $3I_2 + 4 = 2(I_1 - 1),$

$2(I_1 + 2) + 4(I_2 + 1) = 0.$

29. $4(x + y) - 6(x - y) = 3,$

$5(3x + 2y) + 6(2y - x) = 20.$

30. $2(4m - n) - (3m + 5n) = 12,$

$\frac{1}{2}(3m + 5n) + \frac{1}{3}(m + n) = -1.$

31. $2(2r + 3s) - 10r = 9s - 4,$

$5(4r - 3s + 1) = 6(r + s) + 5.$

32. $-2(b - a) = 10 + b,$

$2(a - 5) - 6(b + 1) = 12 - 3a.$

33. $E_1 - E_2 = 1,$

$\frac{2E_1}{5} + \frac{3E_2}{4} = 5.$

34. $\frac{R_1}{3} - \frac{R_2}{4} = 5,$

$\frac{R_1}{8} + \frac{R_2}{3} = 7.$

35. $\frac{L + C}{8} + \frac{L - C}{6} = 5,$

$\frac{L + C}{4} - \frac{L - C}{3} = 10.$

36. $2s + \frac{t - 2}{5} = 21,$

$4s + \frac{t - 4}{6} = 29.$

37. $3p = 2(q - 6),$

$3(2p - 3) + 2(6q + 1) - 17 = 0.$

38. $3(3r - s) - (7r - 2s) = 1,$

$5(r + s) - (3r + s) = 13 + s.$

39. $\frac{2x + y}{3} - \frac{x - 2y}{4} + \frac{5}{72} = 0,$

$\frac{x + y}{2} + \frac{x - y}{5} = \frac{1}{4}.$

40. $\frac{x - 2y}{6} + \frac{x + y}{5} = \frac{13}{30},$

$\frac{3x + 2y}{2} - \frac{x - 2y}{3} = \frac{1}{3}.$

2–17. Applications of Linear Equations in One Unknown to Solution of Problems

In arithmetic, problems are solved by direct computation with the given numbers. In algebra, the method of solving a problem is quite different.

The fundamental idea of algebra is to denote numbers by letter symbols. This idea can be used even when the actual numbers represented by the letter are not known; in this way we can denote by letter symbols quantities that have to be computed in a problem. Thus in solving a problem by the algebraic method, a letter is chosen to represent the unknown quantity; then by using this letter, the statements in the problem are translated into algebraic expressions and equations to which the various processes of algebra can be applied.

The algebraic method of solving problems is much more powerful than the arithmetic method. To appreciate this statement, the reader need only try to solve arithmetically almost any problem in this section.

Example 1.   A city consumes 28,000,000 kilowatts of electric power which is supplied by four sources.   The first source supplies twice as much as the second, and each of the third and fourth sources supplies 1,000,000 kilowatts less than the second.   How much power comes from each source?

To form an algebraic solution, we let x represent an unknown quantity:

x = number of kilowatts supplied by the second source.

Then according to the statement of the problem, the first source supplies twice as much as the second source, whence

$2x$ = number of kilowatts supplied by the first source.

Since each of the third and fourth sources supplies 1,000,000 kilowatts less than the second, we have

$x - 1,000,000$ = number of kilowatts supplied by the third source,

and

$x - 1,000,000$ = number of kilowatts supplied by the fourth source.

The sum of all these is 28,000,000:

$$x + 2x + (x - 1,000,000) + (x - 1,000,000) = 28,000,000.$$

We may now dismiss the requirements of the problem and apply the formal processes of algebra to the equation which has been obtained.   Then it becomes

$$x + 2x + x - 1,000,000 + x - 1,000,000 = 28,000,000,$$

$$5x - 2,000,000 = 28,000,000,$$

$$5x = 30,000,000,$$

$$x = 6,000,000.$$

Recalling the question asked in the problem and using the expressions previously stated, we have the following:

$6,000,000$ = number of kilowatts supplied by the second source,

$12,000,000$ = number of kilowatts supplied by the first source,

$5,000,000$ = number of kilowatts supplied by the third source,

$5,000,000$ = number of kilowatts supplied by the fourth source.

Checking the correctness of the result, we obtain

$6,000,000 + 12,000,000 + 5,000,000 + 5,000,000 = 28,000,000.$

The importance of a clear statement of the relation between the conditions of the problem and the algebraic expressions used in its solution cannot be overemphasized.   Such a statement, which may be regarded as the *English-algebra dictionary* for the problem under consideration, is essential for a clear understanding of the solution and will aid very materially in the construction of this solution.

In connection with the algebraic solution of problems the student should note the following two facts.

1. Any one of the unknown quantities may be denoted by the letter.

2. Any letter symbol may be used to denote the unknown.

Example 2.   Find a number such that the sum of $\frac{1}{3}$, $\frac{1}{4}$, and $\frac{1}{5}$ of the number is equal to 47.

Let x be the number sought.   Therefore

$$\frac{x}{3} = \text{third part of the number,}$$

$$\frac{x}{4} = \text{fourth part of the number,}$$

$$\frac{x}{5} = \text{fifth part of the number.}$$

Thus the sum of the third, fourth, and fifth parts of the number is equal to

$$\frac{x}{3} + \frac{x}{4} + \frac{x}{5},$$

and we have the equation

$$\frac{x}{3} + \frac{x}{4} + \frac{x}{5} - 47,$$

which when solved gives us

$$20x + 15x + 12x = 2820,$$

$$47x = 2820,$$

$$x = 60,$$

the number sought.

To check the correctness of the result we see that the third, fourth, and fifth parts of 60 are 20, 15, and 12, which when added give 47.

In applying algebra to the solution of problems the student should proceed as follows:

1. *Denote the unknown quantity by any symbol. Generally x or y is used. If there are several quantities whose values are not known, denote any one of them by a symbol. If the equation can be solved by the methods of this section, the other unknown quantities can be expressed in terms of this symbol.*

2. *Write down in algebraic language every statement of the given problem or any inference that can possibly be made from it. This should be done by short steps requiring only a few words at a time.*

3. *Using the results of the preceding step, frame an equation by expressing the conditions of the problem in algebraic language.*

4. *Solve the equation resulting from the preceding step.*

5. *Check whether the values found are the true solutions of the problem.*

Example 3. A number is composed of two digits whose sum is 6. If the order of the digits is reversed, we obtain a number that is 36 greater than the first number. Find the number.

Both digits of the number are unknown. We decide to denote one of them, for instance, the digit in the ten's place, by the symbol x. Then

$$6 - x = \text{digit in unit's place.}$$

Hence the number is

$$(6 - x) + 10x.$$

Reversing the order of the digits, we have for the new number

$$10(6 - x) + x.$$

Hence we have the equation

$$10(6 - x) + x = (6 - x) + 10x + 36,$$

$$60 - 10x + x = 6 - x + 10x + 36,$$

$$x = 1,$$

$$6 - x = 5.$$

The number is 15.

Checking, we have

$$51 - 15 = 36.$$

Example 4. A milling machine M can face a certain number of castings in 8 days. Machine N can face them in 10 days. In how many days can the castings be faced by both machines working together?

Let

$$x = \text{number of days required if } M \text{ and } N \text{ work together.}$$

We note that

$$\tfrac{1}{8} = \text{part of work done by } M \text{ in one day.}$$

Hence

$$x \cdot \tfrac{1}{8} = \text{part of the work done by } M \text{ in } x \text{ days.}$$

Likewise

$$\tfrac{1}{10} = \text{part of the work done by } N \text{ in one day,}$$

and

$$x \cdot \tfrac{1}{10} = \text{part of the work done by } N \text{ in } x \text{ days.}$$

Therefore

$$x \cdot \tfrac{1}{8} + x \cdot \tfrac{1}{10} = \text{part of the work done by } M \text{ and } N \text{ in } x \text{ days.}$$

Now, the part of the work done by M in x days is a positive fraction less than 1 and the part of the work done by N in x days is also a fraction less than 1. Then, since the work can be completed in x days by M and N working together, the sum of these two fractions is 1. Therefore,

$$x \cdot \tfrac{1}{8} + x \cdot \tfrac{1}{10} = 1.$$

Solving this equation, we obtain

$$x = 4\tfrac{4}{9}.$$

Therefore M and N working together will do the piece of work in $4\tfrac{4}{9}$ days. The student should check this solution by substitution.

Exercises

1. The sum of two numbers is 32 and their difference is 4. Find the numbers.

2. The sum of the seventh and eighth parts of a number is 30. Find the number.

3. Two men turn out 96 castings in a day. If one man turns out twice as many as the other, how many castings does each one make?

4. A tungsten steel drill will last 3 times as long as a carbon steel drill, and a high-speed steel drill will last 5 times as long as a tungsten steel drill. The average number of hours of operation for the three drills together is 703 hr. What is the life of each?

5. If a number is multiplied by 4, the product is equal to twice the sum of the number and 9. What is the number?

6. Find the two acute angles of a right triangle, if one of them is 4 times the other.

7. Find the angles of a triangle if one of them is twice the second, and the third angle is 20° larger than the first.

8. Find the numbers whose sum is 25 if 3 times the smaller number is equal to twice the larger.

9. Find three consecutive integers such that four times the first, plus five times the second, plus twice the third, is equal to 119.

10. Manganin, an alloy that has an extremely low temperature coefficient of resistance and that is used in the manufacture of standard resistance boxes, contains five times as much manganese as nickel, and $27\frac{1}{3}$ times as much copper as nickel. How much of each element is required to make 1 kg of manganin wire?

11. Three machine guns fire 51 kg of bullets in 3 min. Machine gun *B* fires twice the weight of bullets that machine gun *A* fires, and 5 times the weight of bullets that *C* fires. What is the weight of bullets fired by each machine gun? What is the rate per minute for each gun?

12. Find two consecutive integers, the difference of whose squares is 31.

13. Find two consecutive odd integers, the difference of whose squares is 48.

14. *A* is four times as old as *B*, and in 16 years *A* will be twice as old as *B* will then be. How old is *A*?

15. *A* and *B* have equal amounts of money. *A* gives *B* $18 and then has one fourth as much as *B*. How much did each have at first?

16. A truss is loaded 6 times as much as another; together they carry a load of 217 lb. How much load is there on each truss?

17. The current from a battery is divided among three circuits. The current in circuit 1 is 20 milliamperes greater than the current flowing in circuit 2, and the current in circuit 2 is 20 milliamperes greater than the current flowing in circuit 3. If the total current supplied by the battery is 240 milliamperes, what is the current in each circuit?

18. A square court has the same area as a rectangular court whose length is 12 yards greater and whose width is 8 yards less than the side of the square. What is the area of the court?

19. A student has an average of 70 for 8 grades. How many grades of 80 must he receive in order to bring his average up to 76?

20. A number is composed of two digits whose sum is 8. Reversing the order of the digits, we obtain a number that is 18 greater than the number in its original form. What is the number?

21. A number is composed of two digits whose sum is 12. Reversing the order of the digits, we obtain a number that is 36 smaller than the number in its original form. What is the number?

22. The unit's digit of a two-digit number is 6 greater than the ten's digit. The number is 9 less than four times the sum of the digits. Find the number.

23. A number is composed of two digits. The digit in the ten's place is 6 greater than the digit in the unit's place. The number is 2 more than 8 times the sum of the digits. What is the number?

24. A number is composed of two digits, the digit in the unit's place being 4 less than the digit in the ten's place. If the number is increased by two and the result multiplied by two, a number equal to 15 times the sum of the digits is obtained. What is the number?

25. The digit in the unit's place of a number is 8 less than 3 times the digit in the ten's place. The number has the same value if the order of the digits is reversed. What is the number?

26. A number is composed of three digits. The digit in the unit's place is 2 greater than the digit in the ten's place, which in turn is 3 greater than the digit in the hundred's place. The number is 11 greater than 21 times the sum of the digits. What is the number?

27. One man can do a job in 8 days; a second can do it in 6 days. How long will it take them working together to do the job?

28. A tank is filled by two pipes; the first can fill it in 12 hr and the second in 15 hr. How long will it take both together to fill it?

29. A can do a piece of work in 20 days which B can do in 25 days. In how many days can it be done by both working together?

30. A does an amount of work in 3 hr; B does the same task in 4 hr; and C does the same amount in 5 hr. In how many hours can it be done by all working together?

31. A code operator can send 1000 messages in 8 days. A machine can send 1500 messages in 2 days. How long will it take both working together to send 600 messages?

32. Of four pipes, the first fills a tank in 1 day, the second in 3 days, the third in 4 days, and the fourth in 5 days. How long will it take all the pipes together to fill the tank?

33. A battery is charged by three chargers; the first can charge it in 1 hr and 20 min, the second in 200 min, and the third in 3 hr. How long will it take all three together to charge it?

34. A can do a piece of work in 15 hr which B can do in 25 hr. After A has worked a certain time, B completes the job, working 9 hours longer than A. How many hours did A work?

35. How many gallons of a mixture containing 70 per cent alcohol should be added to 6 gal of a 20 per cent solution to give a 30 per cent solution?

36. What percentage of a 30 per cent solution of hydrochloric acid should be drawn off and replaced by water to give a 15 per cent solution?

2–18. Applications of Linear Equations to the Solution of Engineering Problems

Very often the relations in a problem that are to be translated into algebraic notation are those of physical laws. A few examples of such physical situations are given below.

The speed v of a body which travels a distance d in a time t is given by

$$v = \frac{d}{t}, \tag{1}$$

and is expressed in units of distance per unit time. Thus a man driving 60 miles in 2 hr has a speed of $\frac{60}{2} = 30$ mph. Also from formula 1, the distance traveled is the speed multiplied by the time of travel, or

$$d = vt. \tag{2}$$

The speed of a body is **uniform** if v is the same, regardless of the interval of time during which the distance traveled is measured.

Example 1. A freight and passenger train travel toward each other from towns 390 miles apart, the freight at 30 mph, the passenger train at 45 mph. If the freight train starts half an hour sooner than the passenger train, how many hours will the passenger train travel until it meets the freight train?
Let

$$x = \text{hours the passenger train travels to meet the freight train,}$$

$$x + \tfrac{1}{2} = \text{hours the freight train travels to meet the passenger train,}$$

$$45x = \text{miles the passenger train travels,}$$

$$30(x + \tfrac{1}{2}) = \text{miles the freight train travels.}$$

Hence, since together they traverse the entire 390 miles,

$$45x + 30(x + \tfrac{1}{2}) = 390,$$

$$45x + 30x + 15 = 390,$$

$$75x = 390 - 15,$$

$$75x = 375,$$

$$x = 5 \text{ hr the passenger train travels.}$$

The lever is a bar on which force may be applied to overcome a resistance. The bar is pivoted at some point (called the fulcrum) around which it turns. The teeter-totter, the crowbar, the wheelbarrow, and pliers are examples of levers.

When a lever that is assumed to be weightless is in equilibrium (or balanced) on a fulcrum by weights W_1 and W_2 at distances a_1 and a_2 from the fulcrum, respectively (see Fig. 2–1), it is known that

$$W_1a_1 = W_2a_2. \tag{3}$$

The product of a force and the distance of a point from the line in which the force is acting is called the **moment** or **torque** of the force with respect

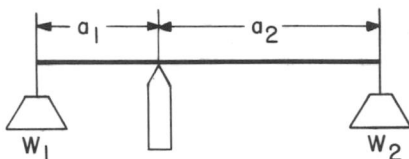

FIG. 2–1

to the point.   Equation 3 then states that the moments with respect to the fulcrum are equal.   More generally, if several weights are placed on either side of the fulcrum, *the lever is in equilibrium if the sum of the moments with respect to the fulcrum on one side of the fulcrum is equal to the sum of the moments on the other.*

Example 2.   A weight of 180 lb is placed 6 ft from a fulcrum, and a weight of 73 lb 5 ft from the fulcrum on the other side.   Considering the lever to be weightless, where should a weight of 55 lb be added to balance the lever?

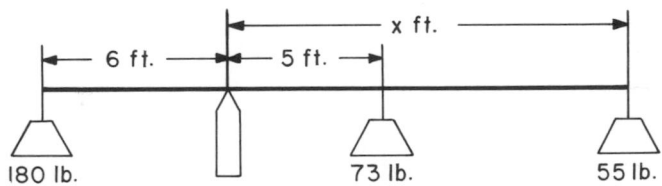

FIG. 2–2

The moment on the left in Fig. 2–2 is greater than that on the right, so that the 55-lb weight must be placed at an unknown distance to the right, which we will call x.

Then

$$180 \cdot 6 = 73 \cdot 5 + 55x,$$
$$1080 = 365 + 55x,$$
$$55x = 1080 - 365,$$
$$55x = 715,$$
$$x = 13 \text{ ft.}$$

Thus the weight should be placed 13 ft to the right of the fulcrum and 8 ft to the right of the 73-lb weight.

If E is the voltage across an electric device D of resistance R ohms, the following formulas are valid when a direct current of I amperes is flowing.

$$E = IR, \quad \text{(Ohm's law)}, \tag{4}$$

$$P = IE = I^2R = \frac{E^2}{R}, \tag{5}$$

$$H = 0.24I^2RT = 0.24PT, \tag{6}$$

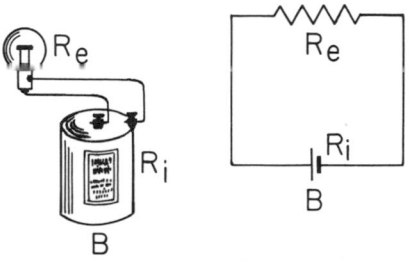

FIG. 2–3

where $E =$ voltage (potential drop) across D,

$I =$ current in amperes,

$H =$ heat in calories developed in D,

$R =$ resistance in ohms,

$P =$ power in watts developed in D,

$T =$ time in seconds,

and 0.24 is accurate to two significant figures.

If these formulas are to be applied to a closed circuit like Fig. 2–3 the total resistance of the circuit consists of the external resistance R_e and the internal resistance R_i of the battery B. If E is the electromotive force developed by the battery, the following formulas are valid.

$$E - I(R_i + R_e),$$

$$P = I^2R_e,$$

$$H = 0.24I^2R_eT = 0.24PT,$$

where P is the power developed in R_e and H is the heat developed in R_e.

Example 3. A 6.0-volt battery supplies power to a 30-watt lamp. What are the resistance of the lamp and the current flowing, and how much heat is generated in 10 min, if the internal resistance of the battery is neglected?
From formula 5,

$$30 = 6 \cdot I,$$

$$I = 5 \text{ amperes.}$$

Then, from formula 4,

$$6 = 5R,$$

$$R = 1.2 \text{ ohms.}$$

Finally, from formula 6,

$$H = 0.24(30)60 \times 10 = 0.24(18,000) = 4.3 \times 10^3 \text{ calories.}$$

Exercises

1. Two columns of troops 72 miles apart, setting out at the same time, travel toward each other at the rates of 5 and 4 mph, respectively. After how many hours will they meet? How far will each have traveled when they meet?

2. An airplane leaves the deck of a carrier and travels north at the rate of 178 mph. The carrier travels north at the rate of 28 mph. When will the airplane pass out of range of wireless communication with the ship if the plane's radio has a range of 600 miles?

3. Jones drove 3 hr and 30 min at a uniform speed. Engine trouble forced him to drive at two fifths of this former speed for an hour. Upon reaching a garage at the end of this time he found that he had driven a distance of 156 miles. What was Jones's original speed?

4. A freight train with an average speed of 40 mph leaves a certain point 5 hr before an express train traveling in the same direction with a speed of 85 mph. How long will it be before the express train overtakes the freight train?

5. The speed of a freight train is two thirds that of an express train. The express train covers 200 miles in 2 hr less time than the slower train. Find the speed of each train.

6. An airplane can cover a distance in 4 hr with the wind, but can travel only two thirds of the way back in the same time. If the airplane travels 150 mph in still air, what is the wind velocity?

7. A traveler sets out to walk at the rate of 4 mph. Twenty hours later a second traveler sets out on a bicycle at 16 mph. How long has the marathon walker been walking when the bicyclist passes him if they travel the same route?

8. From a place *A* a messenger goes to a place *B*, 21 miles distant from *A*, and immediately returns, going at a rate of 4 mph. Simultaneously with the messenger's departure from *A*, another messenger starts from *B* at the rate of 3 mph, goes to *A*, and immediately returns. Find the distance between the two points at which they meet.

9. What voltage is required for a current of 0.8 ampere to flow through a resistance of 10 ohms?

10. An electric heater takes 8 amperes when the voltage across its terminals is 120 volts. What is the resistance?

11. The voltage across a streetcar heater is 220 volts. If the resistance is 6 ohms, what current will flow through the heater?

12. A flashlight operates from a battery of two cells in series, each having an electromotive force of 1.5 volts. If a current of 0.2 ampere flows through the bulb, what is the resistance of the entire circuit? Neglect internal resistance of batteries. (*Note:* If batteries are in series, the electromotive forces are additive.)

13. What is the resistance of an electric iron that takes a current of 8 amperes when used on a 120-volt circuit?

14. What is the internal resistance of a battery of electromotive force 6.4 volts that sends a current of 8 amperes through a resistance of 0.7 ohm?

15. What is the resistance in ohms of a lamp that uses 75 watts if operated at 115 volts?

16. What is the amount of heat developed in 1 min in the streetcar heater described in exercise 11?

17. What is the current in amperes in an electric-power line which transmits 100,000 kilowatts at 50,000 volts?

18. If weights of 160 lb, 100 lb, and 120 lb are placed 8 ft, 6 ft, and 5 ft, respectively, from the fulcrum and on the same side, where must a weight of 496 lb be placed to balance them?

19. How much force can be exerted on a body 4 ft from the fulcrum by a force of 300 lb acting 6 ft from the fulcrum on the opposite side?

20. *A* and *B* together weigh 312 lb. They balance when seated 5 ft and 8 ft, respectively, from the fulcrum on opposite sides. What is the weight of each?

21. If weights of 400 lb and 160 lb are placed on a lever 28 ft apart, where would the fulcrum have to be in order to balance them?

22. Two weights of 250 lb and 150 lb are placed 8 ft and 5 ft, respectively, from the fulcrum of a lever and on the same side; a third weight of 200 lb is placed 4 ft on the other side of the fulcrum, and a fourth weight 5 ft from the fulcrum on the same side as the third weight. What is the amount of the fourth weight if the lever is in equilibrium?

23. A load is placed 8 ft from the fulcrum of a lever and a second load, weighing 600 lb more, is placed 4 ft from the fulcrum on the same side. What is the weight of the two loads if they are balanced by a weight of 6000 lb, placed 7 ft from the fulcrum?

Progress Report

The first chapter of this book dealt with the fundamental operations of arithmetic; this chapter deals with the fundamental operations of algebra, an extension of arithmetic based on the use of a letter symbol to represent any one of a set of numbers.

The chapter is devoted to the following fundamental operations with algebraic expressions:

1. Addition and subtraction.
2. Multiplication.
3. Division (including operations with fractions).
4. Factoring.
5. Solution of three types of algebraic equations:
    (*a*) Linear equations.
    (*b*) Factorable quadratic equations.
    (*c*) Two simultaneous linear equations.

These methods were applied to some problems of the types met in science and engineering.

3 ─────────────────────────

Functions and Their Graphs

Much scientific and engineering literature is concerned with the problem of how one quantity is related to another. For example: For a given car, how is the gas consumption in miles per gallon related to the speed at which the car is driven? How is the distance traveled by a body falling from rest related to the time during which the body has been falling?

In this chapter we shall formulate some mathematical definitions and agree on some notations in connection with relations of this sort. Then we shall discuss the pictorial representation, termed a graph, of such a relation.

A graph furnishes a convenient summary of the relation between quantities, a summary that shows at a glance many of the essential features of the relation. Since graphs make much information available in compact, easily read form, they are universally used by scientists and engineers. For this reason it is important to gain skill in constructing and skill in reading graphs; in this chapter we shall lay the foundation necessary for achieving both of these skills.

3–1. Constants and Variables

In the discussions and formulas of mathematics, physics, engineering, and other branches of science, some of the symbols are intended to represent fixed numbers, while others may be assigned various values in the course of a single discussion.

A symbol that represents a single number throughout a discussion is called a **constant.**

A symbol that represents more than one number in the course of a discussion is called a **variable.** These values may be assigned to the symbol arbitrarily, or they may be determined by some law.

Example. In Sec. 2–18 it was stated that the heat H in calories generated in a light bulb of resistance R ohms, by a current I amperes, during a time T sec is $H = 0.24I^2RT$. Suppose we ask how much heat is generated in a bulb having a resistance of 1.2 ohms in 10, 20, and 30 sec for 5, 10, and 15 amperes of current.

During this discussion 0.24 and R do not change and therefore are constants. I and T change arbitrarily, and therefore I and T are variables. H also changes as I and T vary, and therefore H is also a variable.

An **absolute constant** is a symbol that represents the same number in all discussions. The number 0.24 in the example is an absolute constant; $3, \pi, 6\frac{7}{8}$ are absolute constants.

An **arbitrary constant** is a symbol that represents the same number throughout a given discussion, but may represent another number in another discussion. R is an arbitrary constant in the example; its value may differ in the next discussion, for instance, if a lamp with a different resistance is substituted.

In mathematics the first letters of the alphabet are often used to represent constants, whereas the letters at the end of the alphabet are used to represent variables, especially the letters x, y, and z. However, this practice is not uniform; in many branches of science the symbol for a quantity is frequently chosen as the first letter of the name of the quantity, this choice making it easy to remember the meaning of the symbol.

3-2. Functions

The formula for the volume V of a sphere in terms of its radius r is $V = \frac{4}{3}\pi r^3$. In this formula r and V are variables, since each may assume different values. Further, the formula specifies a definite relation between V and r, since, for any specified value of r, the value of V is determined. For instance, if $r = 3$ in., then $V = 36\pi$ cu in. Thus the value of V may be thought of as depending on r. To express this idea, we say that *V is a function of r.*

If two variables are so related that for each value that may be assigned to one of the variables, there correspond one or more values of the other variable, then the second variable is said to be a **function** *of the first.*

The first variable to which values are assigned arbitrarily is called the **independent variable**; the second variable, whose values are determined, is called the **dependent variable.**

If a specific formula is known that expresses the relation between one variable and another the formula is spoken of as the function. However, a functional relation need not be given by means of an algebraic formula; it may be given by a table of corresponding values or by other information. The set of values that may be assigned to the independent variable, called its **range** of values, may include all real numbers, or it may be limited in some way by physical or mathematical considerations. The following examples will illustrate.

Example 1. Given $y = x^2 - 5x + 6$. Here x may be construed as the independent variable, and y is then a function of x, making y the dependent variable. We speak of $x^2 - 5x + 6$ as the function of x. In this example we may assign any real number to x, and the corresponding value of y is determined by the given function.

Example 2. Given $y = \dfrac{x^2 - 2}{(x + 1)(x - 1)}$. If we construe x as the independent variable, the function of x is $\dfrac{x^2 - 2}{(x + 1)(x - 1)}$. The value of the dependent variable y is specified by this function for any real number x except $x = 1$ and $x = -1$. In these two cases the denominator is zero, leaving the value of the function undefined for these two values of x.

Example 3. A grid bias of -2.8 volts is placed on a 6J5 triode radio tube, and while this bias is held fixed the plate voltage E_p is changed, and the resulting changes in plate current I_p are observed. The following data are obtained.

E_p, volts	2.5	65	70	90	100	118	127	136	142	155
I_p, milliamperes	0.2	1	2	3	4	6	7	8	9	10

For each value of E_p a value of I_p can be found, and hence I_p may be considered a function of E_p. However, the knowledge of the values of I_p is limited to the ten values of E_p given in the table. Hence this function is defined for only ten values of the independent variable E_p. Some knowledge of the behavior of the triode may enable us to predict the approximate values that I_p may have for values of E_p other than those given, but from a mathematical point of view I_p simply remains undefined for values of E_p other than those given in the table. We may, on the other hand, consider I_p as the independent variable, for the table may be construed as giving the values of E_p for ten values of I_p.

Example 4. Given $y = x^2$. Here for any value of x, which we consider the independent variable, the value of the dependent variable y is determined. However, it is possible to consider y as the independent variable and x as the dependent variable. Then, if y is given, the values of x are given by the formula $x = \pm\sqrt{y}$. We note that two values of x are possible, $+\sqrt{y}$ and $-\sqrt{y}$. This function is thus called a **double-valued function.** If the reader will re-examine the definition of function, he will find this case included in the definition. The range of values of y is limited, for there is no real value of x for negative values of y since there is no real square root of a negative number.

The preceding example shows a double-valued function. In general functions that have three, four, and more values of the dependent variable for each value of the independent variable are possible, but we shall deal most frequently with single-valued functions in this book.

The last two examples show also that, if only the formula or table of values relating the two variables is given, we may often choose the dependent and independent variables at our pleasure, keeping in mind only the requirements of the definitions.

3–3. Functional Notation

Since the volume V of a sphere in terms of its radius is $\frac{4}{3}\pi r^3$, we may regard V as a function of r. This statement is conveniently abbreviated by the **functional notation** $V = f(r)$. This equation is read *V equals the f function of r*, or more briefly, *V equals f of r*. It states simply that V depends on r. The reader should note carefully that $f(r)$ does *not* mean f times r.

Thus, in general *the statement that y is a function of x is abbreviated by* $y = f(x)$.

When, in a given discussion, an algebraic formula for the dependent variable in terms of the independent variable is known, the functional notation is used to represent this formula. Thus, in the example above, since $V = f(r)$ and $V = \frac{4}{3}\pi r^3$, we may use $f(r)$ interchangeably with $\frac{4}{3}\pi r^3$. In discussions where complicated expressions are involved this convention is very convenient.

The notation $y = f(x)$ can be used conveniently to indicate that the values of y are to be considered that correspond to a particular value of x. For example, suppose that $y = x^2 - 5x + 6$. Then we may write

$$y = f(x) = x^2 - 5x + 6.$$

If we wish to find the value of y corresponding to $x = 2$, we may replace x by 2. Then we have

$$y = f(2) = 4 - 10 + 6 = 0.$$

In the same way, we obtain

$$f(6) = 12, \qquad f(0) = 6,$$

$$f\left(\frac{a}{2}\right) = \left(\frac{a}{2}\right)^2 - 5 \cdot \frac{a}{2} + 6 = \frac{a^2}{4} - \frac{5a}{2} + 6,$$

$$f(t^2) = (t^2)^2 - 5(t^2) + 6 = t^4 - 5t^2 + 6.$$

In a single discussion, a functional symbol such as $f(x)$ refers to the same function or law of dependence throughout the discussion. If other functional relationships occur in the same discussion, different prefixed letters are used to distinguish the functions. Other letters often used in this way are: g, F, G, ϕ (phi), θ (theta), ψ (psi), μ (mu). However, any letter may be used. In different discussions $f(x)$ may represent different functions of x. Further, the functional notation is used whether an algebraic formula is known for the law of dependence or not. Thus in example 3 of the preceding section, we may write $I_p = f(E_p)$, or $E_p = g(I_p)$, even though no algebraic formula is known for the law of dependence. This functional notation is used for multiple-valued as well as single-valued functions. Its use is as general as the definition of the functional relation itself.

Example 1. The area A of the surface of a sphere as a function of its radius r is $A = g(r) = 4\pi r^2$.

Example 2. Given $f(x) = x^3 - 8x + 10$.

Then

$$f(2) \quad = 8 - 8 \cdot 2 + 10 = 8 - 16 + 10 = 2,$$

$$f(-3) \quad = (-3)^3 - 8(-3) + 10 = -27 + 24 + 10 = 7,$$

$$f(0) \quad = 0 - 8 \cdot 0 + 10 = 10,$$

$$f(a + b) = (a + b)^3 - 8(a + b) + 10,$$

$$f(ab) \quad = (ab)^3 - 8(ab) + 10 = a^3 b^3 - 8ab + 10.$$

3–4. Functions of Several Variables

A dependent variable may be a function of more than one independent variable. Thus in the formula for the volume V of a right circular cylinder, $V = \pi r^2 h$, where h is the altitude, and r is the radius of the base. The volume V is a function of the two independent variables r and h. In functional notation we write $V = f(r, h)$, which is read *V equals the f function of r and h*, or *V equals f of r and h*. The formula for the area A of a trapezoid is $A = \frac{1}{2}h\,(b_1 + b_2)$, where h is the height and b_1 and b_2 are the lengths of the bases. Here A is a function of three independent variables, h, b_1, and b_2. We may state this fact by writing $A = g(h, b_1, b_2)$.

Example. Given $f(x, y) = xy^2 - x^2 + y$.

Then

$$f(0, 1) \quad = 0 \cdot 1^2 - 0^2 + 1 = 1,$$

$$f(3, -2) \quad = 3(-2)^2 - 3^2 + (-2)$$

$$= 12 - 9 - 2 = 1,$$

$$f(a, b) \quad = ab^2 - a^2 + b,$$

$$f(0, a + b) = 0(a + b)^2 - 0^2 + (a + b)$$

$$= a + b.$$

Exercises

Express each of the following statements in functional notation, and then give the exact formula for the function.

1. The area A of a triangle with base 4 is a function of its height h.

2. The volume V of a cube is a function of the length e of one edge of the cube.

3. The area A of the surface of a cube is a function of the length e of one edge of the cube.

4. The circumference C of a circle is a function of its radius r.

5. The area A of a circle is a function of its radius r.

6. The area A of a square is a function of the length d of one side.

7. The area A of the surface of a sphere is a function of its radius r.

8. The area A of an equilateral triangle is a function of the length s of one side.

9. The simple interest I on \$100 at 2 per cent per year is a function of the time t in years.

10. The cost C of x oranges at 40 cents per dozen is a function of x.

11. The area A of a rectangle is a function of its width W and its length L.

12. The volume V of a circular cone is a function of its height h and the radius r of its base.

13. The volume V of a pyramid is a function of its height h and the area A of its base.

14. The volume V of a rectangular solid is a function of its width W, its height H, and its length L.

15. The area A of a triangle is a function of its base b and altitude h.

16. The perimeter P of a right triangle is a function of the legs x and y.

17. The volume V of a right circular cylinder is a function of its altitude h and the radius r of its base.

18. The total area A of the surface of a right circular cylinder is a function of its altitude h and the radius r of its base.

19. The distance d traveled in a time t at a uniform speed v is a function of t and v.

20. The total area A of the surface of a right circular cone is a function of the slant height s and the radius r of the base.

21. Given $f(x) = 2x - 3$, find $f(6)$, $f(0)$, and $f(-2)$.

22. Given $f(x) = x^2 - 3x + 2$, find $f(2)$, $f(-1)$, and $f(4)$.

23. Given $f(x) = \dfrac{x-2}{x+3}$, find $f(1)$, $f(4)$, and $f(-4)$.

24. Given $g(z) = \dfrac{z^2-1}{z^2+1}$, find $g(1)$, $g(0)$, $g(a)$, and $g(-2)$.

25. Given $F(y) = y^3 - 4y + 3$, find $F(1)$, $F(-\tfrac{1}{2})$, and $F(0)$.

26. Given $\theta(x) = x(x-1)(x+2)$, find $\theta(1)$, $\theta(\;1)$, $\theta(2)$, and $\theta(-\tfrac{1}{2})$.

27. Given $\mu(z) = (z-a)(z-b)(z-c)$, find $\mu(a)$, $\mu(b)$, $\mu(0)$, and $\mu(d)$.

28. Given $G(y) = y + \dfrac{1}{y}$, find $G(2)$, $G\left(\dfrac{1}{2}\right)$, $G\left(\dfrac{1}{y}\right)$.

29. Given $\phi(x) = x(x-a)(x-1)$, find $\phi(0)$, $\phi(1)$, $\phi(a)$, and $\phi(a^2)$.

30. Given $f(x) = x^2 - 5x + 6$, find $f(-2) \cdot f(3)$.

31. Given $f(x) = x - \dfrac{1}{x}$, find $\dfrac{f(a^2)}{f(a)}$ and $\dfrac{f(a^4)}{f(a)}$.

32. Given $Q(x) = \dfrac{1}{x-a}$, find $[Q(c) + Q(-c)] \cdot Q(2a)$.

33. Given $\phi(t) = \dfrac{1-t}{1+t}$, find $\phi(1)$, $\phi(t^2)$, $[\phi(t)]^2$, $\phi(1/t)$, $\dfrac{1}{\phi(t)}$, and $\phi[\phi(t)]$.

34. Given $g(z) = a^2 - z^2$, find $g(a)$, $g(a^2)$, $g(z^2)$, $g(a^2 - z^2)$, and $g(a^2) - g(z^2)$.

35. Given $H(t) = t^2 - 1$, find $\dfrac{H(b) - H(a)}{b - a}$.

36. Given $f(x) = x^2$, find $\dfrac{f(x + h) - f(x)}{h}$.

37. Given $g(x) = x^2 - 4x$, find $\dfrac{g(x + h) - g(x)}{h}$.

38. Given $\theta(y) = y^3 - 8$, find $\dfrac{\theta(y + k) - \theta(y)}{k}$.

39. Given $F(x) = x^{\frac{1}{2}}$, find $\dfrac{F(x + \alpha) - F(x)}{\alpha}$.

40. Given $G(x) = \dfrac{1}{x + 1}$, find $\dfrac{G(x + h) - G(x)}{h}$.

41. Given $\psi(y) = y^2 + 1$, find $\psi[\psi(1)]$ and $\psi[\psi(0)]$.

42. Given $f(x) = x - a$, find $f[f(2a)]$ and $f[f(a^2)]$.

43. Given $\theta(y) = y^2 - 3y$, find $\theta[\theta(4)]$ and $\theta[\theta(2)]$.

44. Given $f(x) = x^3 - 1$, find $f\{f[f(1)]\}$.

45. Given $g(x) = x^2 - x$, find $g\{g[g(2)]\}$ and $g\{g[g(-2)]\}$.

46. Given $f(x, y) = 2x + 3y$, find $f(0, 1)$, $f(0, 0)$, $f(1, 1)$, $f(2, -3)$, and $f(-3, 2)$.

47. Given $f(x, y) = xy^2$, find $f(1, 2)$, $f(0, 3)$, $f(0, 0)$, $f(-1, -1)$, and $f(3, -2)$.

48. Given $g(x, z) = x^2 - z^2$, find $g(1, 2)$, $g(-1, -2)$, and $g(0, 0)$.

49. Given $F(w, z) = wz - 2$, find $F(1, 2)$, $F(3, -1)$, $F(0, 1)$, and $F(-1, -1)$.

50. Given $G(x, y) = x^2 + xy + y^2$, find $G(1, -1)$, $G(2, 0)$, $G(0, -2)$, and $G(2, 2)$.

51. Given $f(x, y) = \dfrac{x - y}{xy}$, find $f(1, 1)$, $f(1, -1)$, $f(0, 3)$, and $f(4, 1)$.

52. Given $f(r, h) = \frac{1}{3}\pi r^2 h$, find $f(2, 4)$, $f(3, 2)$, $f(6, 6)$, and $f(0, 1)$.

53. Given $f(h, b) = \frac{1}{2}hb$, find $f(6, 2)$, $f(6, 3)$, $f(4, 4)$, and $f(0, 6)$.

54. Given $F(r, \theta) = \frac{1}{2}r^2\theta$, find $F(2, 2)$, $F(1, 2\pi)$, $F(10, \pi)$, and $F(3, 3\pi/2)$.

55. Given $f(x, y, z) = xyz$, find $f(2, 1, 1)$, $f(-3, -1, 1)$, $f(-1, -3, -2)$, and $f(-1, -3, 2)$.

56. Given $g(x, y, z) = x^2 + 2y^2 + 4z^2$, find $g(0, 0, -1)$, $g(1, -1, -1)$, $g(2, 1, 1)$, and $g(-1, -2, 0)$.

57. Given $F(x, y, z) = xz/y$. find $F(0, 1, 9)$, $F(2, 0, 1)$, $F(9, 2, -1)$, and $F(-1, 1, -1)$.

58. Given $G(r, h, n) = \pi r^2 h n$, find $G(1, 2, 3)$, $G(3, 0, 10)$, $G(4, 1, 3)$, and $G(a, a, n)$.

59. Given $g(p, q, r) = \dfrac{p^2 + q^2}{r^2}$, find $g(1, 1, 1)$, $g(-1, -1, -2)$, $g(2, -1, 3)$, and $g(-1, -3, -5)$.

60. Find the length d of a diagonal of a square as a function of the perimeter p of the square.

61. A right triangle with a hypotenuse of 6 in. has one leg of length x. Find the area A of the triangle as a function of x.

62. A rectangle has an area of 4 sq in. and one side of length x. Find the perimeter p as a function of x.

63. The altitude h of a triangle is twice the length of its base. Find the area A of the triangle as a function of h.

64. An isosceles triangle whose equal sides are 6 in. long has a third side of length x. Find the area A of the triangle as a function of x.

65. A rectangular box having a volume of 36 cu in. is open at the top and stands on a square base with sides of length x. Find the total area A of the four sides and the base as a function of x.

66. A box open at the top is to be made from a square of cardboard 16 in. on a side by cutting out equal squares of side x in. from each of the four corners and turning up the sides. Express the volume V of the box and its surface S as functions of x.

67. A can which is a right circular cylinder has a volume of 36 cu in. If the radius of the base is r, express the height h and the total surface area S of the can as functions of r.

68. A rectangle of sides x and y is inscribed in a circle of radius 6 in. Express y, the area A of the rectangle, and the perimeter P of the rectangle as functions of x.

69. A theater is reserved under the agreement that each of the first 40 persons will pay 50 cents as an admission charge and that each person after the first 40 will pay 45 cents. Find the amount A paid in admissions as a function of the number x of persons who attend the show (*a*) if $x \leq 40$ and (*b*) if $x > 40$.

70. A rectangular garden of 225 square rods is to be laid out so that one side of the rectangle, of length x, is along a neighbor's lot. Fencing costs \$10 a rod, and the neighbor agrees to pay half the cost of the fence along his lot. Express the cost C of fencing in the garden as a function of x.

71. A piece of wire 36 in. long is cut into two pieces, one of which is x in. long. The piece of length x is bent into the form of a square, and the other piece is bent into the form of an equilateral triangle. Express the sum A of the areas of the two figures as a function of x.

72. A piece of wire 48 in. long is cut into two pieces, one of which is x in. long. The piece of length x is bent into the form of a circle, and the other piece is bent into the form of a square. Express the sum A of the areas of the two figures as a function of x.

73. A right circular cylinder is inscribed in a sphere of radius 8 in. If the altitude of the cylinder is y and the radius of the base is x, find y, the volume V, and the total area S of the cylinder as functions of x.

74. A right circular cone with base radius r and height h is inscribed in a sphere of volume 36π cu in. Express r and the volume V of the cone as functions of h.

75. John and James are on a level plain. They leave a given point at the same time on foot, John traveling north and James east. If John walks at a speed of 3 mph and James at a speed of 4 mph, find the distance d between them after t hr as a function of t.

76. A plane flies over a beacon at an altitude of 500 ft and with a speed of 150 mph. If the plane is flying in a horizontal straight line, find the distance d of the plane from the beacon t sec after flying over it as a function of t.

3–5. Inverse Functions

As we have seen, the functional relation expressed in the equation $y = x^2$ can also be expressed in the equation $x = \pm\sqrt{y}$. In general, when an equation of the form $y = f(x)$, which defines y as a function of x, is solved for x in terms of y, taking the form $x = g(y)$, the function $g(y)$ is called the **inverse function** of $f(x)$. Thus in the inverse function $g(y)$, the independent variable y was formerly the dependent variable.

Example 1. Given $y = f(x) = 3x - 5$. To find the inverse function:

$$3x = y + 5, \qquad x = g(y) = \frac{y + 5}{3}.$$

Example 2. Express the volume V of a sphere as a function of its surface area S.

We know that $V = \frac{4}{3}\pi r^3$ and $S = 4\pi r^2$ where r is the radius of the sphere. From the formula $S = 4\pi r^2$ we obtain $r^2 = S/4\pi$ and $r = \pm\sqrt{S/4\pi}$. Since the radius of a sphere is considered to be positive, we choose $r = \sqrt{S/4\pi}$. Substituting this value in the formula for volume, we have

$$V = \frac{4}{3}\pi\left(\sqrt{\frac{S}{4\pi}}\right)^3 = \frac{4}{3}\pi\left(\frac{S}{4\pi}\right)\sqrt{\frac{S}{4\pi}} = \frac{S}{3}\sqrt{\frac{S}{4\pi}} = \frac{S}{6}\sqrt{\frac{S}{\pi}}.$$

Exercises

1. Given $y = f(x) = 4x + 12$. Find the formula for the inverse function $x = g(y)$.
2. Given $y = f(x) = 3x - 12$. Find the formula for the inverse function $x = g(y)$.
3. Given $y = f(x) = x^2 + 4$. Find the formula for the inverse function $x = g(y)$.
4. Given $y = F(x) = 16 - x^2$. Find the formula for the inverse function $x = G(y)$.
5. Given $p = H(q) = q^3 + 1$. Find the formula for the inverse function $q = J(p)$.
6. Given $w = \phi(z) = z^3 - 8$. Find the formula for the inverse function $z = \theta(w)$.
7. Given $x = f(r) = r^2 - a^2$. Find the formula for the inverse function $r = g(x)$.
8. Given $z = p(x) = a^2 - x^2$. Find the formula for the inverse function $x = q(z)$.
9. Express the diameter d of a circle as a function of the circumference C.
10. Express the radius r of a circle as a function of the circumference C.
11. Express the diameter d of a circle as a function of the area A.
12. Express the radius r of a circle as a function of the area A.
13. Express the radius r of a sphere as a function of the volume V.
14. Express the radius r of a sphere as a function of its surface area S.
15. Express the edge e of a cube as a function of its volume V.
16. Express the edge e of a cube as a function of its surface area A.
17. Express the surface area A of a cube as a function of the volume V of the cube.
18. Express the volume V of a cube as a function of its surface area A.
19. Express the surface area S of a sphere as a function of the volume V.
20. Express the circumference C of a circle as a function of the area A.

3–6. The Rectangular Coordinate System

In Sec. 1–3 we saw how a correspondence could be established between the points on a line and the real numbers. In this section we shall see how a correspondence can be established between the points on a plane and pairs of real numbers.

Let XX' be a number line drawn horizontally with its positive direction to the right. Let YY' be another number line drawn perpendicular to XX' so that the point of intersection is the origin on both lines. As a rule, we use the same unit on both lines, but sometimes the choice of different units is advantageous. Let the positive direction of YY' be upward (see Fig. 3–1). The line XX' is called the **x axis** and the line YY' is called the

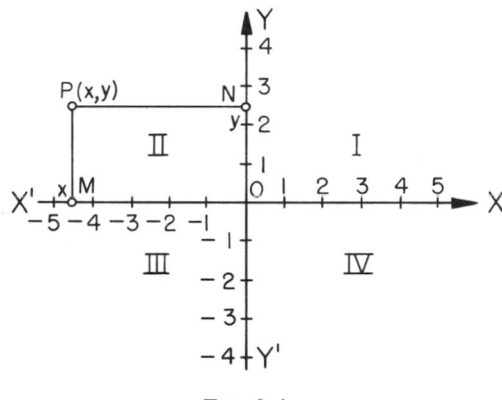

FIG. 3–1

y axis; the two lines together are called the **coordinate axes.** The point of intersection of the two lines is called the **origin.** The coordinate axes may be labeled either as shown in Fig. 3–1 or as shown in Fig. 3–2.

Let P be any point in the plane. From P drop a perpendicular to the x axis and another to the y axis. The perpendicular to the x axis cuts this axis in the point M which corresponds to the number x, called the **abscissa** of P. The perpendicular to the y axis cuts this axis in a point N which corresponds to the number y, called the **ordinate** of P. With P we associate the numbers x and y, called the **rectangular coordinates** of P. Then to indicate that P has coordinates x and y, we write $P(x, y)$, and we refer to the point P as the **point P (x, y)** or, briefly, the **point (x, y).**

Given a pair of numbers (x, y), the point corresponding to this pair of numbers can be found as follows. Erect a perpendicular to the x axis at the point corresponding to the number x. Erect a perpendicular to the y axis at the point corresponding to y. The point of intersection of these two perpendiculars is $P(x, y)$. The process of locating the point with given coordinates is called **plotting the point.**

Instead of erecting the two perpendiculars as just described, it is often more convenient to erect only the perpendicular to the x axis and to measure the distance y directly on this line instead of on the y axis.

The coordinate axes divide the plane into four parts called **quadrants,** numbered I, II, III, and IV as shown in Fig. 3–1.

The system of coordinates with equal units on both axes, as described above, is called the **Cartesian system of rectangular coordinates,** after the

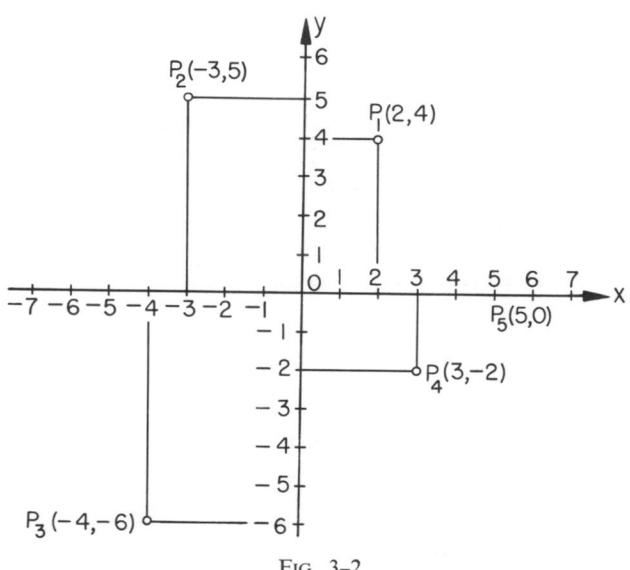

Fig. 3–2

distinguished French mathematician and philosopher Descartes (1596–1650), who was the first man to make extensive use of this system of coordinates. Although the rectangular system is not the only coordinate system in the plane, it is the most convenient one for most purposes in this book and the one most frequently used in practical work. Other systems may be constructed, for instance, by reversing the positive direction on the axes, by using different units on the two axes, or by allowing them to intersect at an angle not a right angle. Later we shall make use of a system of polar coordinates.

Example.   Given $P_1(2, 4)$, $P_2(-3, 5)$, $P_3(-4, -6)$, $P_4(3, -2)$, $P_5(5, 0)$.

These points are plotted in Fig. 3–2. Points are customarily labeled as shown in Fig. 3· 2.

Exercises

Plot each of the following points on rectangular Cartesian coordinates, and if the point does not lie on an axis state the quadrant in which it lies.

1. $(4, 2)$.
2. $(3, 5)$.
3. $(-2, 4)$.
4. $(-3, -5)$.
5. $(4, -7)$.
6. $(6, 0)$.
7. $(0, -8)$.
8. $(-3, -6)$.
9. $(-7, 0)$.
10. $(4\frac{1}{2}, -3\frac{1}{2})$.
11. $(-4.6, -3.2)$.
12. $(10, -5.5)$.

13. In what two quadrants do the points have positive abscissas? negative ordinates?
14. In what two quadrants do the points have positive ordinates? negative abscissas?
15. In what quadrant are the abscissas and ordinates both positive?
16. In what quadrant are the abscissas and ordinates both negative?
17. In what two quadrants do the abscissas and ordinates have opposite signs?
18. What is the abscissa of all points on the y axis?
19. What is the ordinate of all points on the x axis?
20. In what quadrants is the ratio y/x positive? negative?
21. Where do all the points for which $x = 2$ lie?
22. Where do all the points for which $x = -3$ lie?
23. Where do all the points for which $y = 6$ lie?
24. Where do all the points for which $y = -5$ lie?
25. Where do all the points for which $x = 0$ lie?
26. Where do all the points for which $y = 0$ lie?
27. Where do all the points for which $x = -7$ lie?
28. Where do all the points for which $y = -\pi$ lie?
29. Draw the triangle whose vertices are $(5, 6)$, $(3, -1)$, and $(-4, -4)$.
30. Draw the triangle whose vertices are $(0, 7)$, $(-4, 0)$, and $(6, -4)$.
31. Draw the quadrilateral whose vertices, connected in the order given, are $(-8, -2)$, $(-2, -6)$, $(0, 3)$, and $(-5, 4)$.
32. Draw the quadrilateral whose vertices, connected in the order given, are $(8, 0)$, $(-2, 6)$, $(-8, -1)$, and $(0, -6)$.

33. Plot the points in the following table, and connect successive points by straight segments in the order of increasing values of x.

x	-4	-3	-2	-1	0	1	2	3
y	13	3	-3	-5	-3	3	13	27

34. Plot the points in the table of exercise 33, and sketch a smooth curve passing through them in the order of increasing values of x.

35. Plot the points in the following table, and sketch a smooth curve passing through them in the order of increasing values of x.

x	-2	-1	0	1	2	3	4
y	-4	1	0	-1	4	14	56

3–7. Graphs of Functions Given by Tables of Data

The single-valued function $y = f(x)$ gives a number y for each number x in a given range of values for x. The function $f(x)$ may therefore be regarded as specifying a set of number pairs (x, y), one for each x in the given range. The Cartesian coordinate system provides a geometrical

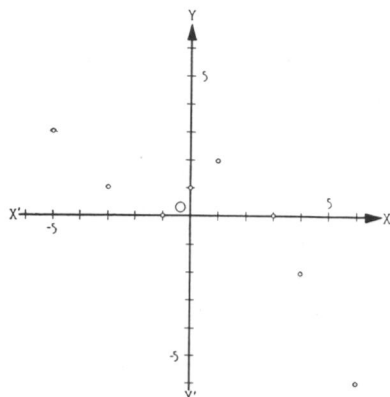

FIG. 3–3

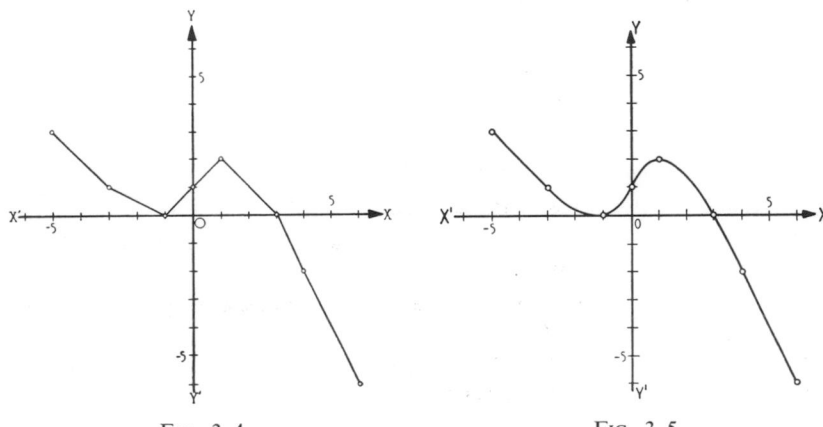

FIG. 3–4   FIG. 3–5

interpretation of a number pair as a point. The set of points in the plane corresponding to the number pairs given by the function $y = f(x)$ is called the **graph of the function** $y = f(x)$. We thus arrive at a very useful geometrical interpretation of the idea of a function.

Example 1. Plot the graph of the function defined by the following table.

x	-5	-3	-1	0	1	3	4	6
y	3	1	0	1	2	0	-2	-6

The graph is shown in Fig. 3–3.   As an aid to the eye, the points are sometimes connected by straight segments as shown in Fig. 3–4 or by some convenient smooth curve as shown in Fig. 3–5.   A **smooth curve** is one without corners, drawn in such a way that its curving is as even as possible.   Most of the quantities that are observed in physics and engineering change in such a manner that the corresponding graph is a smooth curve.

This procedure does *not* imply that the function is defined for other values of x, nor is it intended to extend the definition; it simply serves as a helpful, artistic aid in reading the graph.

Many practical considerations in science, engineering, and business involve relations between two variables which fall under the definition of a function given in Sec. 3–2.   One of the most forceful ways to present such data is to plot them on a Cartesian coordinate system.   *The x axis is replaced by one on which the units of the independent variable are measured; the y axis is replaced by an axis on which the units of the dependent variable are measured.*   It is often convenient to use a different scale on the horizontal axis from that used on the vertical axis.   Sometimes data obtained in connection with practical work involve only positive numbers; it is then sufficient to use only the first quadrant in plotting such data.

Example 2.   A body falling in a vacuum under the attraction of gravity will fall a distance s ft in a time t sec, as given by the table.

t	0	0.5	1.0	1.5	2.0	2.5	3.0
s	0	4.0	16.1	36.2	64.4	100.6	144.9

We may interpret s as a function of t and plot the function on Cartesian coordinates, replacing the x axis by a t axis, the y axis by an s axis.   This procedure is often called, *plotting the curve of the distance s against the time t*, or simply, *plotting s against t*.   It is also convenient to use a much smaller unit on the s axis than on the t axis.   In this example, our intuition suggests that the value of s could be found by experiment for any positive value of t not given in the tables, and we may surmise that the proper values would be approximately represented by the smooth curve drawn in Fig. 3 6 through the points given by the table.

Example 3.   The hourly temperatures, in degrees Fahrenheit, in Chicago, Ill., on a certain date are given in the table.

	A.M.							P.M.		
H	5	6	7	8	9	10	11	12 Noon	1	2
T	3.5	3.5	3.5	4.0	5.0	7.5	12.5	17.5	22.0	25.0

	P.M.									12
H	3	4	5	6	7	8	9	10	11	Midnight
T	27.0	28.0	27.0	23.5	22.5	24.0	27.5	30.0	30.5	30.5

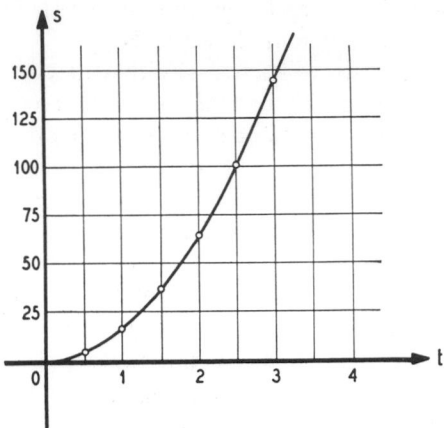

Fig. 3–6

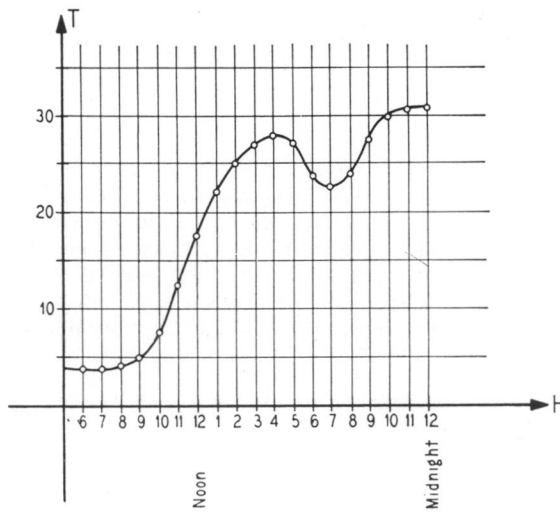

Fig. 3–7

The function is plotted in Fig. 3–7. As in the preceding example, we may surmise that the temperatures for times not given in the table can be approximated fairly closely by drawing a smooth curve through the points given in the table.

Example 4.   Mr. Sam Jones had in his bank account during 1942 the amounts designated in the table.

Date	Jan. 15	Feb. 15	Mar. 15	Apr. 15	May 15	June 15
Amount A	$187	$125	$100	$205	$186	$142

Date	July 15	Aug.15	Sept. 15	Oct. 15	Nov. 15	Dec. 15
Amount A	$217	$108	$146	$150	$50	$19

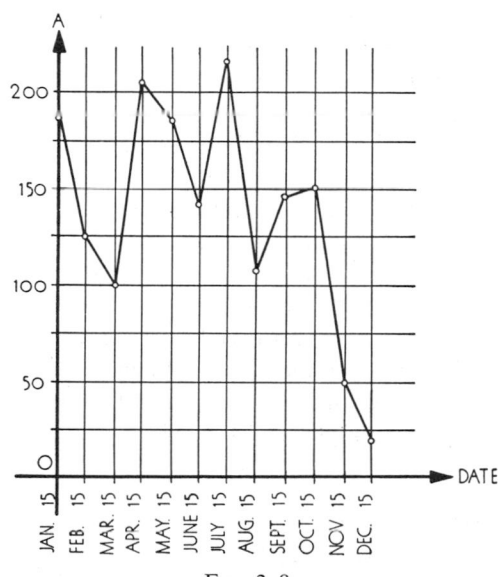

FIG. 3–8

A graph of this function is given in Fig. 3–8.   The x axis is replaced by one on which the dates are given, and the y axis by one on which the amount A is given. The table of values given does not indicate in any way what Mr. Jones had in his account on other days than the fifteenth of each month.   Since this is so, we connect the points plotted on the graph by straight lines, to aid the eye in reading the graph.   However, we note carefully that we cannot infer from them what Mr. Jones's bank account was on days other than those given.

Strictly speaking, the functions in the preceding examples are defined only for the values of the independent variable given in the tables. However, example 4 is different from examples 2 and 3 in one important respect: in example 4 the given data yield no information about what values the function might have if its definition were extended, whereas in examples 2 and 3 we can guess rather accurately what the other values of the function might be in an extended definition.   In example 4 we join the points with straight lines merely as an aid in reading the graph; in examples

2 and 3 we draw a smooth curve through the points as a sign of our conviction that the points plotted are indications of a general trend which might be discovered if the function were defined more fully.

Certain suggestions about the details of plotting a graph will be useful.

1. *Use graph paper (cross-section paper) whenever possible.*

2. *The coordinate axes should be made heavier than the other lines of the network*, especially if cross-section paper is used.

3. *The axes should be properly labeled by the quantities measured on them.* If these labels are omitted, the graph is meaningless.

4. *The scales on the axes should be as large as possible while at the same time the parts of the graph in which we are interested are kept within the space available.* It may be convenient to choose a unit length for the ordinates different from the unit length for the abscissas. Before choosing the units on the axes, it is well to examine the table for the maximum and minimum values of the variables and then select units so as to fit these values into the space available for the graph. The scales should be indicated by numbering points at uniform intervals along the entire length of each axis. Without these labels the graph is meaningless.

5. *In joining the plotted points, proceed from one point to another in order of increasing (or decreasing) values of x.* When a smooth curve through the plotted points is called for by the nature of the problem, first sketch a curve through the points lightly, and then trace it heavily when it appears satisfactory.

3–8. Reading a Graph

Graphs are widely used in engineering and science to indicate the behavior of related physical quantities. Many properties of these functions can then be inferred directly from the graph. The process of finding properties of a function by inspection of the graph representing it is called **reading the graph.**

Example 1. An object is thrown upward with an initial speed of 100 ft per sec. It is known that its distance s from the starting point after t sec is approximately given by this table.

t	0	1	2	3	4	5	6	7
s	0	84	136	156	144	105	34	-74

The graph corresponding to this table is shown in Fig. 3–9. From this graph, we can read the following information. The object moves up for approximately 3 sec and then begins to move down. The highest point reached by the object is about 156 ft. The object returns to its starting point after approximately 6.25 sec.

Studying the graph in Fig. 3–9, we can now answer some questions about the motion of this object.

(a) What height is reached by the object after 4.50 sec? To answer this question, draw a perpendicular to the *t* axis at the point *t* = 4.50. Its length, measured with the unit of the *s* axis, is approximately 130. Thus, 130 ft is roughly the height of the object after 4.50 sec.

(b) After how many seconds is the object 100 ft above the starting point? To answer this question, draw a line perpendicular to the *s* axis through the point *s* = 100. It intersects the graph in two points, whose abscissas are measured to be 1.30 and 5.05. We can state, therefore, that the object is 100 ft above the starting point after 1.30 sec and again after 5.05 sec.

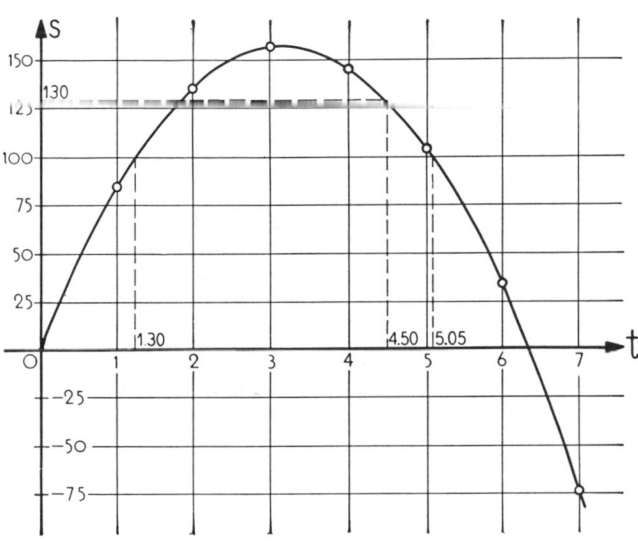

FIG. 3–9

Example 2. When two coils are arranged so that a change in current in one coil causes a voltage to be induced in the other, the two coils are said to possess mutual inductance. The table gives the mutual inductance *M* in henrys of two coils for several distances of separation *s* in centimeters. Plot the curve of the mutual inductance against the distance of separation between the coils. Determine the mutual inductance when the two coils are separated a distance of 7 cm.

s	0	2	4	6	8	10	12	14
M	0.051	0.049	0.041	0.033	0.025	0.017	0.011	0.007

The corresponding graph is plotted in Fig. 3–10. If the mutual inductance for the separation of 7 cm is to be found, draw a vertical line from point 7 to the curve. We note that the corresponding mutual inductance is 0.028 henry.

From the curve we can also read the following information. The mutual inductance does not change rapidly for the first 2 cm of separation. From 2 to 10 cm the mutual inductance decreases more rapidly with separation, after which the mutual inductance changes more gradually as the distance is increased.

From the developments in the preceding paragraphs it may be seen that a function $y = f(x)$ can be given in at least three different ways:

1. *By a formula*, from which the values of y corresponding to given values of x may be computed. For example, $y = x^2 - 3x + 5$.

2. *By a table*, which gives values of y for certain values of x. The functions of Sec. 3–7 are examples of this kind.

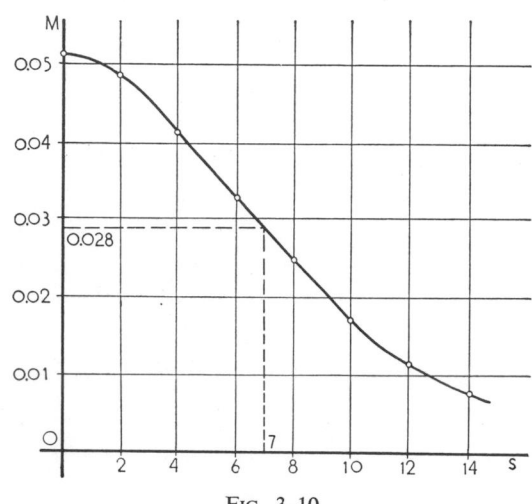

FIG. 3–10

3. *By a graph.* If in example 1 only the graph in Fig. 3–9 were given, it would describe the distance s traveled by the object as a function of time t, for we can find from the graph the value of s corresponding to a given value of t.

Exercises

Plot each set of data on a Cartesian coordinate system, and connect the successive points by straight segments. Then plot the data again and sketch a smooth curve through the points.

1.

x	-6	-4	-2	0	2	4	6	8	10
y	-5	-4	-3	-2	-1	0	1	2	3

2.

x	-4	-3	-2	-1	0	1	2	3	4	5
y	14	12	10	8	6	4	2	0	-2	-4

3.

x	-6	-4	-2	0	2	4	6
y	9	4	1	0	1	4	9

4.

x	-4	-3	-2	-1	0	1	2	3	4
y	-8	-1	4	7	8	7	4	-1	-8

5.

x	-6	-4	-2	0	2	4	6
y	-25	-6	1	2	3	10	29

6.

x	-9	-6	-3	0	3	6	9
y	-27	-8	-1	0	1	8	27

7.

x	-4	-3	-2	-1	0	1	2	3	4
y	-16	-5	0	1	0	-1	0	5	16

8.

x	-4	-3	-2	-1	0	1	2	3	4	5
y	32	0	-8	-5	0	2	0	-3	0	20

9.

x	0	$\frac{1}{2}$	1	$\frac{3}{2}$	2	$\frac{5}{2}$	3
y	-4	$-2\frac{1}{16}$	-1	$-1\frac{3}{16}$	0	$1\frac{1}{16}$	-1

10.

x	0	0.31	0.63	0.94	1.26	1.57	1.88	2.20	2.51	2.83	3.14
y	0	0.31	0.59	0.81	0.95	1.00	0.95	0.81	0.59	0.31	0.00 *

The data in the exercises below were obtained by various experiments. In each case plot the graph of the data on a Cartesian coordinate system, connecting the points by a smooth curve.

11. The hourly temperatures in Chicago, Ill., on a certain day in January are given in the following table. Plot T against H.

	A.M.							P.M.		
Hour H	5	6	7	8	9	10	11	Noon	1	2
Temp. T	6	6	6	5	5	7	10	14	15	17

	P.M.									
Hour H	3	4	5	6	7	8	9	10	11	Midnight
Temp. T	20	21	21	20	19	19	18	19	19	19

12. In testing the brake horsepower P of an engine, a number of readings are taken at various speeds. One set of experimental data is given below, the speed S being given in revolutions per minute. Plot P against S. What is the range of speed for which the horsepower is above 55? Describe in general terms the results of this experiment which you can read from the graph.

S	400	600	800	1000	1200	1400	1600	1800	2000
P	14	22	31	39	47	52	56	57	54

13. As zinc is added to aluminum, the tensile strength of the alloy is increased. For various percentages P of zinc alloyed with aluminum, the corresponding tensile strengths S in pounds per square inch were found to be as follows:

P	0	1	2	3	4	5	6	7	8
S	12,600	13,000	13,500	14,000	14,800	15,700	16,600	18,000	19,200

Plot S against P.

14. In the design of power lines, the vertical sag S in the cable depends on the temperature T. The following set of data was obtained for a 400-ft span, S being measured in feet and T in degrees Fahrenheit. Plot S against T. If the sag is not to exceed 8.1 ft, what is the maximum permissible temperature?

T	−40	−20	0	20	40	60	80	100
S	6.8	7.0	7.2	7.4	7.6	7.8	8.0	8.2

15. An important factor in the treatment of carbon steel is the rate at which the furnace heats and cools. In a typical case, the temperatures T for various times t were found to be as follows:

t	1	2	3	4	5	6	7	8	9	10
T	550	900	1280	1460	1300	1250	1150	1000	880	780

The time t is measured in minutes, and the temperature T in degrees Fahrenheit. Plot T against t. For how long is the temperature of the sample above 1000°?

16. The output voltage e_o in volts of a superheterodyne radio receiver varies with the frequency setting F in kilocycles per second as shown in the following table. Plot the curve, using F as the independent variable. The curve is called a **selectivity curve.**

F	1410	1408	1406	1404	1402	1400	1398	1396	1394	1392	1390
e_o	1.2	1.2	1.6	2.5	4.5	5.5	4.5	2.5	1.6	1.2	1.2

17. A Venturi meter is an apparatus designed to measure the flow of water. For various heights H of water in a certain tank, the rates of discharge D were measured by a Venturi meter. For H in feet and D in cubic feet per second, the following data were obtained. Plot D against H. What is the effect of the height of the water on the discharge?

H	0	0.3	1.0	2.0	3.5	5.5	8.0	11.0
D	0	5	10	15	20	25	30	35

18. When a load is placed on a timber beam, the beam is compressed. The following data were obtained by applying various loads L to a beam and measuring the compression d parallel to the grain of the timber, L being measured in thousands of pounds and d in inches. Plot d against L. What will be the compression for a load of 50 tons? This curve is called a stress-deformation diagram.

L	0	30	67	90	114	124	126
d	0	0.01	0.02	0.03	0.04	0.05	0.055

19. In determining the relative merit of timber, it is necessary to dry or season the wood. The weights W in pounds of a red tie were measured after various times T in months during which the tie was being seasoned, and the following data were obtained. Plot W against T. After what length of seasoning will the loss in weight be 5 per cent of the original weight?

T	0	2	4	6	8	10	12	14	16
W	215	199	184	174	170	168	168	168	167

20. The valve lift L of a tangential cam was found to vary with the rotation θ of the shaft as shown in the table, L being measured in thousandths of an inch and θ in degrees. Plot L against θ. At what values of θ is the valve lift one half of the maximum value?

θ	0	10	20	30	40	50	60	70	80	90	100	110
L	0	20	70	170	260	295	295	260	170	70	20	0

21. In the expansion of a gas at constant temperature, the pressure p times the volume v is equal to a constant. To verify this relation, known as Boyle's law, the following experimental data were obtained, p being measured in pounds per square inch and v in cubic inches. Plot v against p.

p	14.7	16	17.5	21	25	30	37	50	76
v	10	9	8	7	6	5	4	3	2

22. In determining the volt-ampere characteristic of a tungsten lamp, the following data were obtained, the current I being measured in milliamperes and the voltage E in volts. Plot I against E. What is the change in current in milliamperes as the voltage varies between 45 and 75 volts?

E	10	20	30	40	50	60	70	80	90	100	110
I	150	195	235	270	305	340	365	410	425	450	485

23. The deflection on the screen of a cathode-ray tube is a function of the voltage applied to the plates. From the following data, plot a curve of deflection d in inches against the applied voltage E. When the voltage changes from 15 to 25 volts, what is the corresponding change in deflection?

E	52	40	26	13	0	−13	−26	−40	−52
d	0.8	0.6	0.4	0.2	0	−0.2	−0.4	−0.6	−0.8

24. An experiment was made to determine the elongation e of machine steel due to a tensile stress S. Measuring S in thousands of pounds per square inch and e in 0.001 in. per in., the following data were obtained. Plot e againt S. The graph is called a stress-strain diagram.

S	0	30	35	35	36	37	37.5
e	0	1	2	3	4	5	6

25. With a grid bias of -6 volts on a 6J5 triode tube, the values of plate current I_p in milliamperes for various values of plate voltage E_p in volts are found by experiment to be as follows:

E_p	120	140	158	170	180	190	200	210	218	225	235
I_p	0.5	1	2	3	4	5	6	7	8	9	10

Plot I_p against E_p. Find by reading the graph approximately what plate voltage is required if the plate current is to be (a) 1.5 milliamperes, (b) 8.6 milliamperes.

26. The experiment of exercise 25 is performed again with a grid bias of -10 volts, and the following data are obtained.

E_p	215	250	275	295	310	320	325
I_p	0.5	1	2	3	4	5	5.5

Plot I_p against E_p on the same sheet used for the graph of exercise 25. Compare the two graphs. What plate voltage is required if the plate current is to be (a) 1.5 milliamperes, (b) 4.5 milliamperes?

3–9. Graphs of Functions Given by Formulas

If we wish to plot the graph of a function $y = f(x)$, where $f(x)$ is given by an algebraic formula, we must first construct from the given formula a table of number pairs from which the points on the plane may be plotted. However, most functions defined by formulas are defined for many more values of the independent variable than it would be feasible to use in finding number pairs from which to plot points. In fact, most of the functions considered in this chapter which are given by formulas are defined for all real numbers. What we do, therefore, is to select a few convenient number pairs and plot the corresponding points. Then the smooth curve through these points is approximately the graph of the function. This curve is sketched by proceeding from one point to another so that the values of x are taken in order of increasing magnitude. How accurately this curve represents the function depends on how closely together the points are chosen: The closer the points, the more accurately the curve can be drawn.

There is an essential difference between the graphs of the functions of this section and those of the functions in the preceding section: In the preceding section, the curve is drawn simply for convenience in reading the graph, whereas here all the points on the curve are points on the graph of the function.

Since most of the functions we consider in this section are defined for all real numbers, it is convenient to plot only that portion of the graph required for the problem to be investigated by the use of the graph. If nothing is known about the range within which the graph will be used, usually a portion of the graph near the origin of the coordinate system is plotted. It is then understood that the graph extends much further than is shown.

Example 1. Plot the graph of $3x - 4$.

We construct a table of sample values of the function as follows: Let $y = f(x)$ $= 3x - 4$. Then $f(0) = -4$, $f(3) = 5$, etc. Values found in this way, by giving x values and finding the corresponding values of y, fill out the table from which the graph of Fig. 3–11 is plotted.

Joining the points by a smooth curve in order of increasing or decreasing values of x, we see that the graph is a straight line. Although the line extends indefinitely in both directions, only a portion of it is shown in Fig. 3–11.

Example 2. Plot the graph of $2x^2 - 5x + 7$.

Setting $y = f(x) = 2x^2 - 5x + 7$, we find for various values of x the corresponding values of the function as given in the table.

We plot as many of these points as can be located in the space available. Connecting them in the order of increasing values of x by a smooth curve, we have the graph of Fig. 3–12.

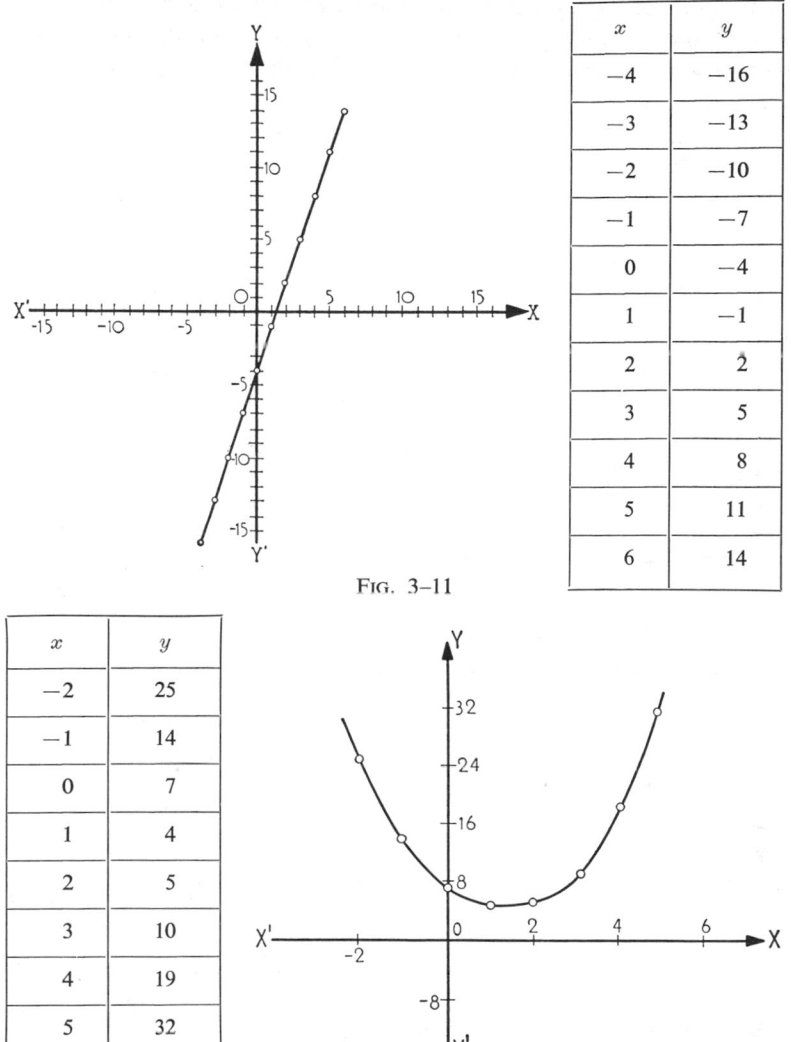

x	y
−4	−16
−3	−13
−2	−10
−1	−7
0	−4
1	−1
2	2
3	5
4	8
5	11
6	14

FIG. 3–11

x	y
−2	25
−1	14
0	7
1	4
2	5
3	10
4	19
5	32

FIG. 3–12

If the graph of the function $y = f(x)$ is plotted, its intersections with the y axis give the values of $f(x)$ for $x = 0$. These values, where the graph meets the y axis, are called **y intercepts.**

The intersections of the graph with the x axis define x values for which the corresponding y value is zero. These values of x are called the **x intercepts.** In order to find x values so that $y = f(x) = 0$, look for the points where the graph of $f(x)$ intersects the x axis.

The values of x for which an equation $f(x) = 0$ is satisfied are called the **roots** of the equation $f(x) = 0$. The inspection of the graph of the function $y = f(x)$ gives a very convenient method for estimating the real roots of $f(x) = 0$.

Example 3. Estimate the roots of $x^2 - 5x + 3 = 0$ by plotting the corresponding graph.

Setting $y = x^2 - 5x + 3$, we obtain the following table of values:

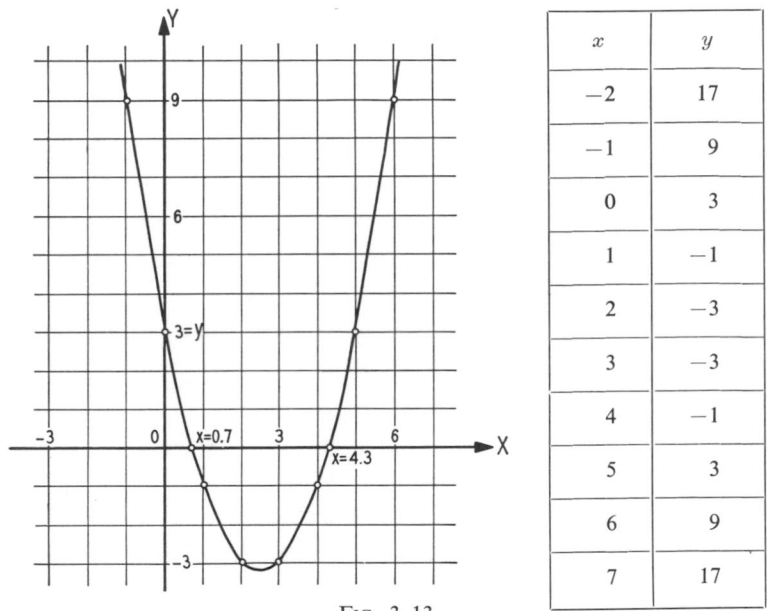

x	y
-2	17
-1	9
0	3
1	-1
2	-3
3	-3
4	-1
5	3
6	9
7	17

FIG. 3–13

The function is plotted in Fig. 3–13, and from where the graph crosses the x axis we infer that the real roots of the function are approximately 4.3 and 0.7. We also observe that the y intercept is 3.

In addition to the suggestions about the details of plotting functions given in Sec. 3–7, the following suggestions will be useful in plotting the graphs of functions given by algebraic formulas.

1. *Intercepts.* The intercepts on the axes are usually of special interest and should be found and plotted whenever possible.

2. *Choosing points to plot.* We can usually obtain a satisfactory graph of the function by choosing the integral values of x near the origin from which to construct the table of values. Any sharp turn in the curve is usually important and may require that points be plotted more thickly. When there is doubt about the nature of a curve between two points, other points in the neighborhood should be found until its behavior is well established. In these matters, however, experience is the best teacher.

Exercises

Plot the graph of each of the following functions:

1. $2x$.

2. $-2x$.

3. $x - 2$.

4. $x + 3$.

5. $2x + 3$.

6. $3x - 4$.

7. x^2.

8. $\frac{1}{4}x^2 - 2$.

9. $1 + x^2$.

10. $4 - x^2$.

11. $x^2 - 2x - 4$.

12. $x^2 - 5x + 6$.

13. $3 - 2x - x^2$.

14. $10 + 3x - 2x^2$.

15. $x^2 + 2x + 2$.

16. $x^2 - 6x + 9$.

17. x^3.

18. $\frac{1}{8}x^3$.

19. $\frac{1}{16}x^4$.

20. $\frac{1}{100}x^5$.

21. $x^3 - 9x$.

22. $6x - x^3$.

23. $2 + 9x - x^3$.

24. $2x^3 - 3x^2 + 2$.

25. $x^3 - 4x^2 - 7x + 10$.

26. $x^3 - 6x^2 + 12x - 8$.

27. $x^4 - 20x + 64$.

28. $-x^4 + 6x^3 - 12x^2 + 10x - 2$.

29. $\dfrac{8}{x^2 + 1}$.

30. $\dfrac{x}{4}$.

31. $\dfrac{4x}{x^2 + 4}$.

32. $x + \dfrac{9}{x}$.

33. $\sqrt{25 - x^2}$.

34. $\pm\sqrt{25 - x^2}$.

35. $\sqrt{9 + x^2}$.

36. $\pm\sqrt{9 + x^2}$.

37. $\pm\sqrt{x + 4}$.

38. $\pm\sqrt{x - 2}$.

39. $\pm\sqrt{x^2 - 9}$.

40. $\dfrac{x - 3}{x}$.

41. $\dfrac{1}{x^2 - 3x - 4}$.

42. $\dfrac{1}{(x - 2)^2}$.

Find the real roots of the following equations to the nearest tenth of a unit by plotting the corresponding graphs.

43. $x^2 - x - 6 = 0$.

44. $x^2 - 2x - 8 = 0$.

45. $x^2 + x - 7 = 0$.

46. $x^2 - 4x + 1 = 0$.

47. $x^2 - x - 8 = 0$.

48. $2x^2 - 2x - 5 = 0$.

49. $x^3 - 3x - 1 = 0$.

50. $x^3 - 5x^2 - 9 = 0$.

51. $x^3 + 4x^2 - 7 = 0$.

52. $x^4 - 15x^2 + 32 = 0$.

By plotting the corresponding graphs show that the following equations have no real roots.

53. $x^2 + x + 2 = 0$.

54. $x^2 - x + 2 = 0$.

55. $x^2 - 2x + 4 = 0$.

56. $x^2 + 6x + 12 = 0$.

57. $x - x^2 - 1 = 0$.

58. $2x - x^2 - 2 = 0$

59. $x^4 + 5x^2 + 4 = 0$.

60. $-x^4 + 6x^3 - 12x^2 + 10x - 6 = 0$.

3–10. Parametric Equations

In plotting the graph of a function $y = f(x)$ we consider pairs of numbers (x, y) for convenient values of x, the corresponding values of y being specified by the function $y = f(x)$. In this way we can think of the pairs (x, y) that we plot as being specified by the function $y = f(x)$. For example,

$$y = x + 1 \tag{1}$$

is a function specifying pairs of numbers (x, y).

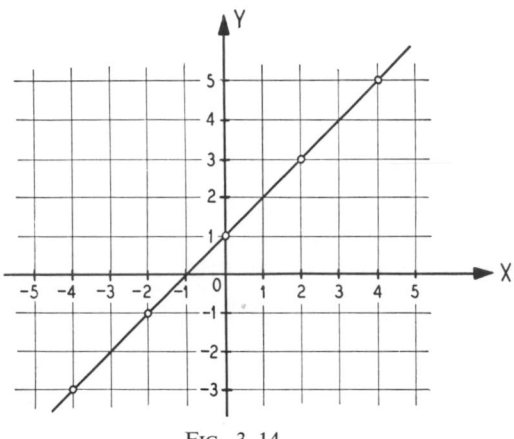

FIG. 3–14

It is also possible, and frequently convenient, to specify pairs of numbers (x, y) by giving both x and y as functions of an auxiliary variable. For example, the relations

$$x = 2t, \qquad y = 2t + 1 \tag{2}$$

specify a pair of real numbers (x, y) for each real number that is assigned to the auxiliary variable t. The graph representing the pairs (x, y) obtained from equations 2 is plotted in example 1.

Example 1. Plot the graph representing equations 2.

Substituting various values of t in equations 2, we obtain the following table:

t	x	y
0	0	1
1	2	3
2	4	5
-1	-2	-1
-2	-4	-3

The graph is shown in Fig. 3–14.

When the pairs (x, y) are given in terms of a third variable t, as

$$x = f(t), \qquad y = g(t), \qquad (3)$$

the auxiliary variable t is called a **parameter,** and equations 3 are called **parametric equations.**

Often it is possible to eliminate the parameter t, obtaining thus a relation between x and y. For example, it is easily seen that eliminating the parameter from equations 2, we obtain equation 1, which verifies that Fig. 3–14 is the correct graph of equations 2. However, if a relation between two variables is given, many sets of parametric equations representing the same relation can be found. For instance, the parametric equations

$$x = t^3, \qquad y = t^3 + 1$$

also represent relation 1.

Many times the parameter chosen has no especial significance. However, in applied problems in physics and engineering a parameter is often chosen that has a geometrical or physical meaning. Suppose, for example, that a body is subjected to various forces and as a result moves about in a plane, and that the position of the body is known at every instant. If we then place a rectangular coordinate system in the plane, we know at every instant the coordinates (x, y) of the position of the body. Thus, if we choose t as a parameter and let t measure time, then for every value of t we know the corresponding values of x and y. In short, x is a function of the time t, and y is a function of the time t, and we have a relation of the form 3 which describes the path of the moving body.

Example 2. A particle moves in the plane in such a manner that at any time t its position is given by the equations $x = 2t$, $y = -4t^2 + 8t + 5$. Plot the path of the particle between $t = -1$ and $t = 3$. At what times is the particle on the x axis?

The graph is shown in Fig. 3–15.

We may verify the correctness of the graph by plotting the graph of the equation obtained when the parameter is eliminated. In this case the resulting equation is $y = -x^2 + 4x + 5$.

To find the values of t for which the particle is on the x axis, we find the values of t for which $y = 0$, in other words, the values of t for which

$$-4t^2 + 8t + 5 = 0.$$

We obtain

$$4t^2 - 8t - 5 = 0,$$

$$(2t + 1)(2t - 5) = 0,$$

$$t = -\tfrac{1}{2}, \tfrac{5}{2}.$$

t	x	y
-1	-2	-7
$-\frac{1}{2}$	-1	0
0	0	5
$\frac{1}{2}$	1	8
1	2	9
$\frac{3}{2}$	3	8
2	4	5
$\frac{5}{2}$	5	0
3	6	-7

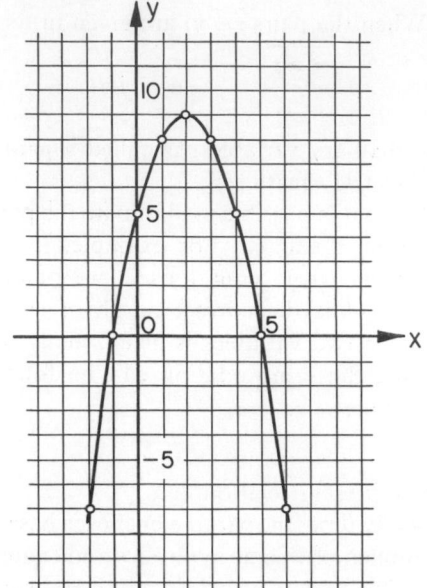

Fig. 3–15

Exercises

Plot the curve corresponding to each set of parametric equations.   In exercises 1–10 find y as a function of x by eliminating the parameter.

1. $x = t, \quad y = 3 + 4t.$

2. $x = t^2 - 1, \quad y = t^2 + 1.$

3. $x = (v - 1)^2, \quad y = v^2 - 2v.$

4. $x = 2w, \quad y = 4w + 3.$

5. $x = 1 - \dfrac{t}{4}, \quad y = t^2.$

6. $x = u - \frac{1}{2}, \quad y = 4u^2 - 4u + 1.$

7. $x = 2t, \quad y = t^2 - 4.$

8. $x = 3z, \quad y = 6z - z^2.$

9. $x = \dfrac{t^2}{4}, \quad y = \dfrac{t^3}{8}.$

10. $x = \dfrac{t}{2} - 6, \quad y = \dfrac{8}{t - 12}.$

11. $x = 2u, \quad y = \dfrac{4}{u}.$

12. $x = v^2 + 2v, \quad y = \dfrac{v^3}{8}.$

13. $x = \dfrac{4}{w}, \quad y = w^2.$

14. $x = 2t, \quad y = \dfrac{2}{t^2 + 1}.$

15. $x = \dfrac{10t}{1 + t^2}, \quad y = \dfrac{5(1 - t^2)}{1 + t^2}.$

16. $x = \dfrac{10t}{1 + t^2}, \quad y = \dfrac{5(t^2 - 1)}{1 + t^2}.$

17. $x = 2t, \quad y = 8t^3 - 6t.$

18. $x = \dfrac{15t}{1 + t^3}, \quad y = \dfrac{15t^2}{1 + t^3}.$

In exercises 19–26 the parametric equations give the motion of a particle in the plane, x and y being given in feet and t in seconds.   Plot a graph of the path of the particle, and answer the questions.

19. $x = 2t$, $y = 6t - t^2$. When is $y = 5$?   At what points on the path of the particle is $y = x$?

20. $x = 2t + 3$, $y = 4t - t^2$.   When does the particle cross the axes?   When is the particle 5 ft from the x axis?

21. $x = t - 1$, $y = t^2 - 4t$.   How long is the particle below the x axis?   the line $y = 5$?   At what times do the coordinates of the particle's position satisfy the relation $y - x = 1$?

22. $x = t^2 - 4t + 2$, $y = 2(t + 1)$.   When are the x and y coordinates of the position of the particle equal?   At what time does the particle cross the x axis?

23. $x = 2t$, $y = t^4/16$ 1. How long is the particle below the x axis?   When is the particle 15 ft above the x axis?

24. $x = t^3/8 + 2$, $y = 2t$.   When does the curve cross the x axis?   the y axis?

25. $x = 2t + 2$, $y = \frac{1}{2}(t^3 + 3t^2 - 9t - 11)$.   By inspecting the graph find the times at which the x and y coordinates of the particle are equal.

26. $x = 3t + 1$, $y = t^3 - 3t^2 + 8$.   At what times is the particle 8 ft above the x axis?

In the following exercises assume that the x axis is parallel and that the y axis is perpendicular to a level surface of the earth.

27. A baseball thrown when $t = 0$ occupies a position t sec after being thrown given by $x = 50t$, $y = 50t - 16t^2$, where x and y are in feet and the x axis is 5 ft above the level of the diamond.   Sketch a graph of the path of the baseball.   Estimate from your graph the maximum height above the diamond attained by the baseball.   How long will the ball be in the air before it is caught 5 ft above the diamond?   At what times is the ball 11 ft above the diamond?

28. A baseball hit by a batter when $t = 0$ occupies a position t sec after being hit given by $x = 64t$, $y = 4 + 64t - 16t^2$, where x and y are in feet and the x axis lies in the surface of the diamond.   Sketch the path of the baseball between the time it is hit and the time that it strikes a signboard in right center field 4 ft above the ground.   When does the ball hit the signboard?   How far is the signboard from home plate?   Estimate from your graph the maximum height above the diamond reached by the ball.

29. A projectile fired horizontally when $t = 0$ from a cliff 64 ft above a lake occupies a position t sec after being fired given by $x = 1000t$, $y = -16t^2$, where x and y are in feet and the x axis passes through the muzzle of the gun from which the projectile is fired. Plot the path of the projectile between the time that it is fired and the time that it hits the lake.   When does it hit the lake?   Where does it hit the lake?

30. A shell from a large naval gun occupies a position t sec after being fired given by $x = 1000t$, $y = 1024t - 16t^2$, where the x axis is 50 ft above the water level.   If the shell hits an enemy vessel at a point 50 ft above the water level, when does it hit the vessel?   How far is the enemy vessel from the gun?   Sketch the path of the projectile. Estimate from your graph the maximum height above the water reached by the projectile.

3–11. Variation

Certain simple functional relations occur so frequently in science and engineering that special terminology is in common use in connection with these relations.   In what follows we shall define some of these terms.

A variable y is said **to vary as** x or **to be proportional to** x if

$$y = kx, \tag{1}$$

where k is a constant.   The constant k is called **the factor** (or **constant**) **of proportionality.**   The quotient y/x has the constant value k when x and y vary so as to satisfy equation 1.   The word **ratio** is used as a synonym for quotient.   The statement that two ratios are equal is sometimes called a **proportion.**

From equation 1 it follows that $x = \dfrac{1}{k}\, y$.   Since $\dfrac{1}{k}$ is also a constant, it follows that x is proportional to y if y is proportional to x.   To emphasize this fact, it is often said that the variables x and y are proportional to one another.

Example 1.   If a force F, measured in newtons, is applied to a free mass of 8 kilograms, the acceleration measured in meters per second per second imparted to the mass is $a = F/8$.   The acceleration a varies as F or is proportional to F, the factor of proportionality being $\tfrac{1}{8}$.   If F changes, a varies so that the ratio F/a has always the same value 8.

A variable y is said **to vary inversely as** another variable x or **to be inversely proportional to** x if

$$y = \frac{k}{x}, \tag{2}$$

where k is a constant.   As before, the constant k is called the factor of proportionality.   From equation 2 it follows that

$$xy = k.$$

Thus, the statement that y varies inversely as x, or x varies inversely as y, is equivalent to the statement that the variables x and y vary in such a way that their product is constant.

Example 2.   If a constant electromotive force of 6 volts is applied across a resistance R, the current produced is inversely proportional to the resistance and is given by the formula $I = 6/R$. I varies inversely as R; the product IR is constant and is equal to 6.

A variable u is said **to vary jointly as the variables** x_1, x_2, x_3 and **to vary inversely as the variables** y_1, y_2, y_3, if

$$u = k\,\frac{x_1 x_2 x_3}{y_1 y_2 y_3}, \tag{3}$$

k being the factor of proportionality.

Example 3. The force F between two small conductors varies jointly as their electric charges Q_1 and Q_2 and inversely as the square of their distance r. This statement is equivalent to the formula

$$F = k \frac{Q_1 Q_2}{r^2},$$

where k is a constant.

In applications, it sometimes happens that quantities are connected by a relation like equation 1, 2, or 3, but the factor of proportionality is unknown. In order to find this unknown value, it is sufficient to know the value of the dependent variable for one set of values of the independent variables. Such a value, if not known, can often be found by an experiment.

Example 4. The electrical resistance R of a wire varies directly as its length l and inversely as the square of its diameter d.
    This statement is equivalent to the formula

$$R = k \frac{l}{d^2}.$$

If k were known, this formula could be used to find the resistance R for given values of l and d. The value of k depends on the units that are used to measure the quantities R, l, d. We suppose that R is measured in ohms, l in feet, and d in mils (1 mil = 0.001 in.). In order to find k, the formula $R = k \dfrac{l}{d^2}$ is applied to a particular case in which the values of R, l, d are known. Suppose that an experiment with a copper wire of length $l = 500.0$ ft and diameter $d = 40.00$ mils yields a resistance of 3.300 ohms. By substituting these values, we obtain

$$3.300 = k \frac{500.0}{1600}, \qquad k = \frac{3.300 \times 1600}{500.0} = 10.56.$$

Hence

$$R = 10.56 \frac{l}{d^2}.$$

This formula can be used to find the resistance for any given values of l and d.

Example 5. The time t required for an elevator to lift a weight varies jointly as the weight W and the distance d through which it is to be lifted and inversely as the power P of the motor. If it requires 20.0 sec for a 5.00-horsepower motor to lift 500.0 lb through 40.0 ft what power is necessary to lift 1250 lb a distance of 124 ft in 32.0 sec?
    The initial sentence of the problem states that

$$t = k \frac{Wd}{P}.$$

In order to find k we use the data that $t = 20.0$ when $W = 500.0$, $d = 40.0$, and $P = 5.00$, obtaining

$$20.0 = k \frac{500.0 \times 40.0}{5.00}, \qquad k = 0.00500 = 5.00 \times 10^{-3}.$$

The complete formula, therefore, is

$$t = 0.00500 \frac{Wd}{P}.$$

In order to answer the question of the problem, we make the substitution $W = 1250$, $d = 124$, $t = 32.0$, obtaining

$$32.0 = 0.00500 \frac{1250 \times 124}{P},$$

$$P = \frac{0.00500 \times 1250 \times 124}{32.0} = 24.2 \text{ hp.}$$

Example 6. If the rate of flow of water through a pipe varies as the square of the radius of the pipe, by how much would the rate of flow be increased if the diameter of the pipe were multiplied by $\frac{3}{2}$?

If the rate of flow is denoted by V and the radius of the pipe by r, then

$$V = kr^2.$$

If the new radius $r_1 = \frac{3}{2}r$, then the new rate of flow is

$$V_1 = kr_1{}^2 = k \cdot \tfrac{9}{4}r = \tfrac{9}{4}kr = \tfrac{9}{4}V.$$

The rate of flow, therefore, is multiplied by $\frac{9}{4}$.

Exercises

In the following exercises make use of the conventions about significant figures treated in Sec. 1–13.

1. The elongation E of a spring balance varies as the applied weight W. If $E = 3.00$ when $W = 20.0$, find E when $W = 15.0$.

2. The deflection D of a beam varies as the cube of the length L. If $D = 0.01250$ when $L = 15.0$, find D when $L = 18.0$.

3. The acceleration g due to gravity varies inversely as the square of the distance d from the center of the earth. Write the formula for g if it is known that $g = 32.0$ ft per sec per sec when $d = 4000$ miles. Find the value of g when d is 8000 miles. Assume that the two measurements of d are accurate to within 5 miles.

4. The distance a body falls, starting from rest in a vacuum near the earth's surface, is proportional to the square of the time during which the body falls. If a body falls 441 ft in 5.25 sec, how far will it fall in 9.50 sec?

5. If the illumination from a light varies inversely as the square of the distance from the light, how much farther from the light must a book which is now 21.0 in. from the light be moved in order to receive exactly one third as much light?

6. The number N of revolutions per minute of a ball governor required to keep the balls h in. below the point of suspension varies inversely as the square root of h. If the balls are 8.0 in. below the point of suspension for 36 rpm, at what speed will they be 2.0 in. below the point of suspension?

7. The rate of vibration of a string under constant tension varies inversely as the length of the string. If a string 48.0 in. long vibrates 256 times per second, what is the length of a string that vibrates 576 times per second?

8. The absolute temperature of a certain quantity of gas varies jointly as the pressure and the volume. If the temperature of the gas is 325° and its pressure is 20.0, what will the temperature become if the volume is exactly doubled and the pressure is changed to 8.0?

9. The force exerted by water on the bottom of a containing vessel varies jointly as the area of the bottom of the vessel and the depth of the water. When the water is 2.50 ft deep, the force on 1.00 sq ft of bottom is 156 lb. Find the force on the bottom of a tank that is a right circular cylinder of diameter 12.5 ft and that has water in it to a depth of 11.2 ft.

10. The velocity that a stream of water must have in order to move a round object varies jointly as the object's specific gravity and the square root of its diameter. If a velocity of 15.0 ft per sec is needed to move a stone with a diameter of 2.25 ft and a specific gravity of 4.00, how large a stone with a specific gravity of 3.00 can be moved by a stream of water with a velocity of 24.5 ft per sec?

11. The force exerted by the wind on a plane surface varies jointly as the area of the surface and the square of the wind's velocity. If the force on 20.0 sq ft is 11.0 lb when the wind velocity is 11.0 mph, find the force on a surface 7.25 ft by 6.50 ft when the wind velocity is 36.5 mph.

12. The horsepower that a shaft can transmit safely varies jointly as its speed and the cube of its diameter. If a shaft of a certain material 3.0 in. in diameter can transmit 36 hp at 75 rpm, what diameter must a shaft of the same material have in order to transmit 45 hp at 125 rpm?

13. The quantity Q of electricity that will flow into a capacitor varies jointly as the capacitance C and the voltage E. If Q is given in coulombs, C in microfarads, and E in volts, the factor of proportionality is 10^{-6}. Find Q when $C = 25$ microfarads and $E = 110$ volts.

14. The stress in the material of a pipe subject to internal pressure varies jointly as the pressure and the internal diameter of the pipe and inversely as the thickness of the pipe. If the stress is 100.0 lb per sq in. when the diameter is 5.00 in., the thickness is 0.75 in, and the internal pressure is 25 lb per sq in., find the stress when the pressure is 42 lb per sq in., if the diameter is increased to 8.50 in. and the thickness is reduced to 0.50 in.

15. The weight that a beam of uniform cross section, supported at both ends, will sustain varies jointly as the breadth of the cross section and the square of the depth of the cross section and inversely as the distance between supports. If a beam 36.0 ft long, 12.0 in. deep, and 12.0 in. broad will support a weight of 1728 lb, find the breadth of a beam of the same material, of the same depth, and 40.0 ft long that will support a weight of 2250 lb.

16. The electrical resistance of a wire varies directly as the length and inversely as the square of the diameter of the wire. If a wire 432 ft long and 4.00 mm in diameter has a resistance of 1.24 ohms, find the length of a wire of the same material whose resistance is 1.44 ohms and whose diameter is 3.00 mm.

17. The energy stored by a capacitor varies jointly as the capacitance of the capacitor and the square of the voltage across the capacitor. If the energy is measured in joules, the capacitance in microfarads, and the voltage in volts, then the constant of proportionality is 5.00×10^{-7}. What is the energy stored in a capacitor of 2.50-microfarad capacitance with 1250 volts applied across it?

18. The weight of a body above the surface of the earth varies inversely as the square of the distance of the body from the center of the earth. If a certain body weighs 55 lb when it is 4.000×10^3 miles from the center of the earth, how much will it weigh when it is 4.400×10^3 miles from the center?

19. A body falling freely from rest in a vacuum falls a distance proportional to the square of the time during which it has been falling. If the body began its fall at the beginning of a certain interval of time, and if this interval is divided into two equal halves, how much farther did the body fall in the second half of the interval than in the first?

20. The electrical resistance of a cable varies directly as its length and inversely as the square of its diameter. A large cable is twice as long and has a cross section whose area is twice as large as that of a smaller cable. What is the relation between the resistances of the two cables?

21. The current in an electric circuit varies directly as the electromotive force and inversely as the resistance. If the resistance is doubled, what proportion of change must occur in the electromotive force in order for the current to be quadrupled?

22. The electrical resistance of a bar of uniform cross section varies directly as the length and inversely as the area of the cross section of the bar. If the resistance of a given bar of annealed aluminum is 1.144×10^{-6} ohm, find the resistance of a bar of the same substance that is 24.0 times as long as the given bar and has a cross section whose area is 2.88 times that of the given bar.

23. The illumination received from a source of light varies directly as the candle-power of the source and inversely as the square of the distance from the source. At what distance will the illumination from a 75-candlepower light be equal to exactly three fifths of the illumination from a 45-candlepower light at 25 ft?

24. Newton's law of gravitation states that the force of attraction between two bodies varies directly as the mass of the two bodies and inversely as the square of the distance between the two bodies. By what factor is the force of attraction multiplied when the distance between them is multiplied by c? By what factor is the force of attraction multiplied when the masses of both bodies are multiplied by c? By what factor is the force of attraction multiplied when the distance between the bodies is multiplied by c and the masses of both bodies are multiplied by c?

25. An electric-power company figures that the cost C per kilowatt-hour for supplying electric current to small consumers is equal to a fixed amount a plus an amount that varies inversely as the number m of kilowatt-hours supplied. If the cost of supplying 64 kilowatt-hours is 4.00 cents and the cost of supplying 128 kilowatt-hours is 3.00 cents what is the cost of supplying 256 kilowatt-hours? If the company charges its customers, at a flat rate of 4.25 cents per kilowatt-hour, how many kilowatt-hours must the company sell a customer in order to make a profit?

3–12. The Straight Line

Consider the equation

$$Ax + By + C = 0, \tag{1}$$

where A, B, and C are constants, and A and B are not both zero, and x and y are variables. *An equation is said to be* **linear** *if it can be reduced to the form 1 by these operations:*

1. Addition or subtraction of the same expression from both sides of the equation.

2. Multiplication or division of both sides of the equation by the same nonzero expression not involving the variables.

Example 1. Given $x - 8y + 32 = 18 - (2x + y)$.

From this equation we obtain

$$x - 8y + 32 = 18 - 2x - y,$$
$$x + 2x + y - 8y + 32 - 18 = 0,$$
$$3x - 7y + 14 = 0.$$

This last equation is of the form 1, where $A = 3$, $B = -7$, and $C = 14$. Thus the given equation is linear.

If the equation

$$3x - 7y + 14 = 0 \tag{2}$$

is solved for y, the result is in the formula

$$y = \tfrac{3}{7}x + 2 \tag{3}$$

by which y is defined as a function of x. Equations 2 and 3 are *equivalent;* that is, each pair of values (x, y) satisfying equation 3 also satisfies equation 2. If, for instance, $x = 7$, then we have from equation 3, $y = 5$. These two values also satisfy equation 2, for $3 \cdot 7 - 7 \cdot 5 + 14 = 21 - 35 + 14 = 0$.

The graph corresponding to $y = \tfrac{3}{7}x + 2$ consists of all points whose coordinates satisfy the relation $y = \tfrac{3}{7}x + 2$. According to the former statement, the coordinates of all those points also satisfy the equation $3x - 7y + 14 = 0$, and the graph belonging to $y = \tfrac{3}{7}x + 2$ consists of all points whose coordinates satisfy the equation $3x - 7y + 14 = 0$.

We shall, therefore, say that the graph corresponding to the function $y = \tfrac{3}{7}x + 2$ corresponds also to the equation $3x - 7y + 14 = 0$.

In this way we are able to construct a graph of each equation of the form 1. The graph consists of all points whose coordinates satisfy the equation $Ax + By + C = 0$, and we call it the graph of this equation.

It will be shown in Chapter 15 that the following statements are true. *The graph of every equation of the form 1 is a straight line. Conversely, every straight line has an equation of the form 1.*

Equation 1 is therefore called the **general linear equation.** Since the variables enter only to the first power, it is said to be of **first degree.**

Since two points determine a straight line, we need to plot only two points to plot a linear equation. Practically, it is a good idea to plot one or two more to insure accuracy. When the line does not pass through the origin or is not parallel to either axis, the two simplest points to plot are the intercepts on the axes.

Example 2. Plot the line $3x - 2y = 10$.

Setting $y = 0$, we see that the x intercept is $\tfrac{10}{3}$, and, setting $x = 0$, we see that the y intercept is -5. When $y = 1$, $3x - 2 = 10$, or $3x = 12$, and $x = 4$.

The line plotted in Fig. 3–16 by drawing the straight line through the intercepts is seen to pass through $(4, 1)$, and we are insured against error.

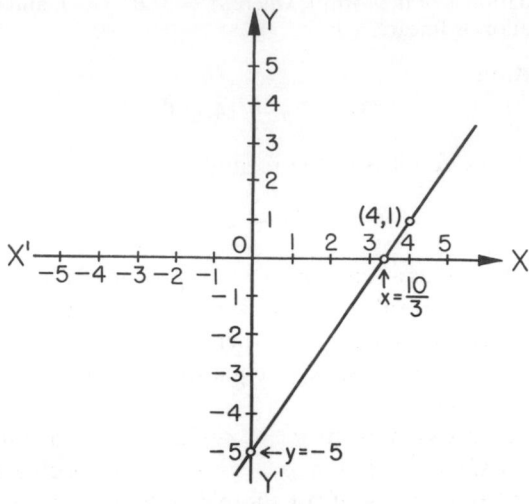

Fig. 3–16

Exercises

Find the intercepts of the following straight lines and plot the graphs of the lines.

1. $x + y - 3 = 0.$

2. $x - 2y + 6 = 0.$

3. $3x - 2y - 12 = 0.$

4. $2x + 5y = 10.$

5. $3x - 4y - 15 = 0.$

6. $3x - 2y = 0.$

7. $6x - 2y - 15 = 0.$

8. $12x - 7y = 30.$

9. $2x = 7.$

10. $3y = -10.$

11. $6L - 7J = 42.$

12. $2X_1 - 3X_2 + 15 = 0.$

13. $E = 10R.$

14. $3E_1 + 5E_2 + 30 = 0.$

15. $R_1 - R_2 = 8.$

16. $3X - 7Y - 35 = 0.$

17. $\frac{1}{2}Z_1 - \frac{1}{4}Z_2 = 4.$

18. $6L - 4C = 13.$

19. $0.25R_1 - 0.50R_2 = 1.75.$

20. $0.24I_1 - 0.48I_2 = 1.44.$

21. During his Arctic expeditions Greely made a number of observations on the velocity of sound at low temperatures that yielded the formula $v - 0.6t = 333$, where v is the velocity in meters per second and t is the temperature in degrees centigrade. Plot the graph of the line and find the intercepts.

22. The hydrometer readings for hydrochloric acid on the Baumé scale B are related to a measure R which is the reciprocal of the readings on the specific-gravity scale by the formula $B + 145R = 145.$ Plot the graph of this relation in the first quadrant.

23. Temperature readings in degrees Fahrenheit F are related to those in degrees centigrade C by the formula $F - \frac{9}{5}C = 32.$ Plot a graph of this relation for values of F between $-100°$ and $+400°$. Find the intercepts.

24. The hydrometer readings for aqua ammonia on the Baumé scale B are related to a measure R which is the reciprocal of the readings on the specific-gravity scale by the formula $B - 140R = -130$.   Plot a graph of this relation, and find the intercepts.

25. If the air outside an empty railroad passenger coach is at a temperature of 32° F and if it is proposed to air-condition the coach by heating the outside air and pumping it through the coach at a rate V between 0 and 25 cu ft per min, then the total heat H in British thermal units per hour per degree of temperature difference from the outside to the inside of the coach is given by the formula $H = 75V + 300$.   Plot the graph of this relation for the indicated range of values of V.

26. In exercises 22 and 24, are the relations between the hydrometer and specific-gravity readings linear? Why? Show that your answer is correct by plotting graphs of the relations between these readings.

3–13. Graphical Solution of Simultaneous Linear Equations in Two Unknowns

In Sec. 2–16 we considered pairs of simultaneous linear equations of the form

$$a_1x + b_1y = c_1,$$

$$\tag{1}$$

$$a_2x + b_2y = c_2,$$

where a_1, b_1, c_1, a_2, b_2, c_2 are constants and x, y are variables. We observed that each equation was satisfied by many pairs (x, y); when there was a pair (x, y) that satisfied both equations simultaneously we learned how to find this pair.

Now, however, we shall ask this question: Does *every* pair of linear equations of the form 1 have a pair of numbers (x, y) satisfying both equations?

In answering this question, we can make use of the fact set forth in Sec. 3–12 that the graph of a linear equation is a straight line. Thus the graph of each equation in (1) is a straight line. For these two lines in the same plane there are three possibilities:

1. *They may intersect in one point.*   Then the coordinates (x, y) of the point of intersection give the numbers that satisfy the two equations simultaneously.   This pair (x, y) is, of course, the same pair that would be obtained by the algebraic method of Sec. 2–16.   Thus it is possible to find a pair of numbers satisfying equations 1 approximately by plotting the two lines and reading their point of intersection from the graph.

2. *They may be parallel.*   Since parallel lines do not have a point in common, there is no pair of numbers (x, y) that satisfies both equations in this case.

3. *They may coincide.*   Then every number pair that satisfies one equation satisfies the other.

Thus our original question is answered in the negative; that is, not every pair of linear equations of the form 1 has a number pair (x, y) that satisfies both equations.   This matter will be discussed further in Chapter 13.

Example 1.   Find algebraically and graphically the values of x and y satisfying

$$2x - y = 3,$$
$$x + y = 3.$$

The graph is shown in Fig. 3-17.   From the graph it appears that the solution is (2, 1).

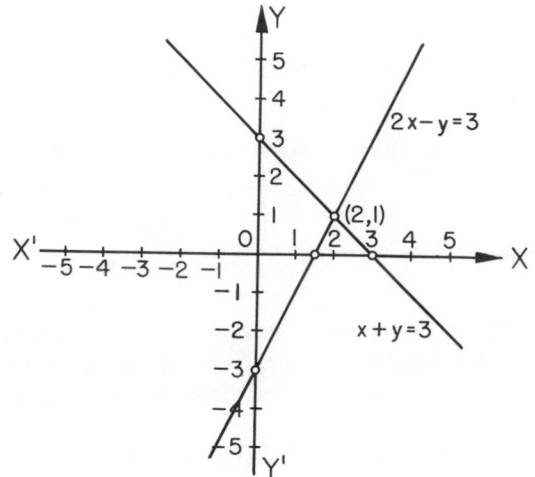

FIG. 3-17

To verify this result algebraically, substituting $y = 3 - x$ from the second equation into the first, we obtain

$$2x - (3 - x) = 3,$$
$$2x - 3 + x = 3,$$
$$3x = 6,$$
$$x = 2,$$

and

$$y = 3 - 2 = 1.$$

Thus the correct solution is (2, 1).

Example 2.   Given:

$$2x + y = 4,$$
$$2x + y = 8.$$

From Fig. 3-18 on which these two lines are plotted, we see that they are parallel, and hence there is no pair of numbers satisfying both.

Subtracting the first equation from the second, we obtain $0 = 4$, which is absurd. When this algebraic process is performed, it is assumed that there is a solution, and, since this assumption leads to a contradiction, the assumption is incorrect, and there is no solution.

Example 3.   Given:

$$x + 2y = 6,$$

$$2x + 4y = 12.$$

The student may verify that these two equations give the same line; consequently every pair of values satisfying one satisfies the other.

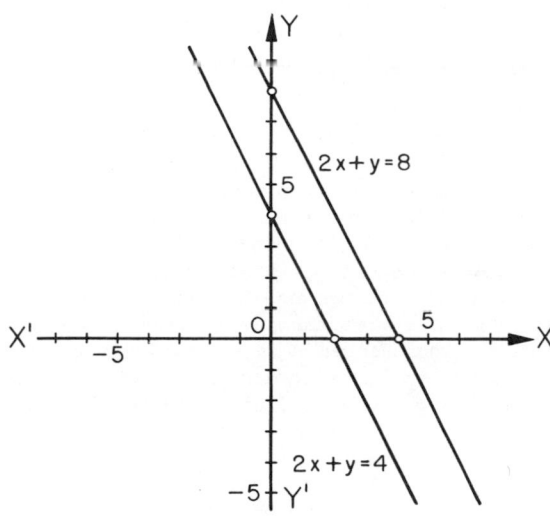

FIG. 3–18

Exercises

Plot the lines in the following pairs of equations, and from the graph state whether the lines intersect, are parallel, or coincide.   If they intersect, estimate the coordinates of the point of intersection from the graph, and then verify your estimate by solving the equations algebraically.

1. $2x + y = 4,$
   $x - y = 5.$

2. $x + y = 6,$
   $3x - 2y = 3.$

3. $2x + y = 5,$
   $2x + y = 10.$

4. $3x - 2y = 3,$
   $6x - 4y = 13.$

5. $2x + y = 4,$
   $6x - 3y - 12 = 0.$

6. $3x - 2y = 3,$
   $9x - 6y - 9 = 0.$

7. $5x + 3y = 11,$
   $2x - y = 11.$

8. $x + y = 5,$
   $2y = 10 - 2x.$

9. $5x + 3y = 11,$
   $5x + 3y + 15 = 0.$

10. $x + y = 8,$
   $3y = 12 - 3x.$

11. $2x - y = 7,$
   $2y = 4x - 14.$

12. $x + y = 6,$
   $5x + 6y + 14 = 0.$

13. $R_1 + 2R_2 = 5,$ 14. $6L - 5C = 12,$ 15. $3R_1 + 5R_2 = 15,$
    $3R_1 + 5R_2 = 30.$ $L + C = 2.$ $5R_2 = 30 - 3R_1.$

16. $E_1 - 3E_2 - 6 = 0,$ 17. $Z_1 - Z_2 = 3,$ 18. $3A + B + 2 = 0,$
    $2E_1 - 6E_2 = 13.$ $3Z_1 - 3Z_2 = 4.$ $5B = A + 22.$

19. $14R + 3S = 21,$ 20. $14R + 3S = 21,$
    $7R + 2S = 10.$ $28R + 6S - 42 = 0.$

21. Between the cetane numbers of 25 and 75, the per cent of paraffin y in straight-run Diesel fuels is related to the cetane number x by the equation $1.15x - y = 8$, and for alkylated benzenes the similar equation is $0.35x + y = 46$. In the given range of cetane numbers is there a cetane number for which the Diesel fuels and alkylated benzenes have the same paraffin content? If so, what is the cetane number?

22. The engine oil pressures y in pounds per square inch in an airplane engine equipped with a standard hopper tank at an altitude x ft are given approximately for ordinary oil and for antifoam oil by the equations $0.00265x + 3y = 270$ and $0.00105x + 2y = 180$, respectively. Is there an altitude at which the engine oil pressures for the two oils are equal? If so, what is the altitude and the pressure?

23. The dimensions above the nominal size y in thousandths of an inch for the diameter x in inches of splined shafts for an external class X fit and for an external class Y fit are given by the equations $0.11x - y = -3.50$ and $0.22x - 2y = -1.60$, respectively. Is there a diameter for which the dimensions above the nominal size are the same for both classes of fit? If so, what are the diameter and the dimension above normal size?

24. The potentials E in volts required by an automobile generator to send a current of I amperes through a discharged storage battery and through a charged storage battery are given by the equations $E - 0.1I = 6.2$ and $E - 0.1I = 7.8$, respectively, where both batteries have a resistance of 0.1 ohm. Is there a potential E for which the same current is sent through both batteries? If so, what is the current and the potential?

Progress Report

This chapter first discussed the definitions of a constant and variable. Then the concept of a functional relation between variables was examined. The remainder of the chapter was devoted to expanding the understanding of the idea of a function by discussing:

1. The functional notation.

2. The pictorial representation of functions by means of graphs.

3. Functions specified by means of parametric equations and their graphs.

4. Functions specified in the language of variation.

5. Linear equations, simultaneous linear equations, and their graphs.

4

Trigonometric Functions

Trigonometry is the study of certain ratios associated with right triangles; these ratios are called the trigonometric functions. These functions were developed originally in connection with the solution of problems involving triangles. However, they are now also used in a wide variety of other ways in many branches of science and engineering. For example, trigonometric functions are used by physicists analyzing the structure of sound waves, by structural engineers determining the

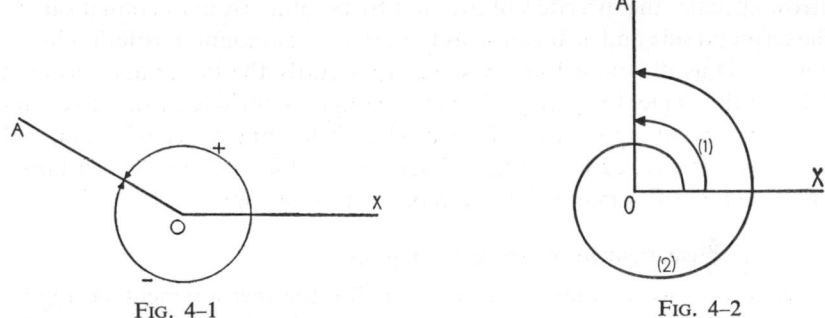

FIG. 4–1   FIG. 4–2

dimensions of steel beams for skyscrapers, by civil engineers engaging in surveying, by mechanical engineers designing internal-combustion engines, by electrical engineers studying electrical phenomena, by meteorologists studying the theory of weather prediction, and by mathematicians studying the properties of functions. Thus scientists and engineers must have a thorough working knowledge of trigonometry.

4–1. Angles

For the purposes of trigonometry, an angle is defined as follows: If a segment in the plane rotates about the point O from a position OX to a position OA, we say that the segment **generates an angle** XOA (see Fig. 4–1). The original position OX of the generating segment is called the

initial side of the angle, the final position OA is called the **terminal side** of the angle. The point O is called the **vertex** of the angle. If the generating segment is rotated in a **counterclockwise** direction to the position OA, the angle XOA is called **positive,** and, if the generating segment is rotated in a **clockwise** direction to the position OA, the angle XOA is called **negative.** The positive angle in Fig. 4–1 is indicated by a $+$ sign, the negative angle by a $-$ sign.

The size of an angle defined in this way is determined by the amount and direction of the rotation used in constructing the angle. Thus two angles can easily be compared with respect to their size. Angle 1 of Fig. 4–2, formed by rotating a segment from the position OX through one quarter of a complete revolution about O in a counterclockwise direction to the position OA, is a different and smaller angle than angle 2, formed by rotating a segment from the position OX through one and one-quarter revolutions about O in a counterclockwise direction to the position OA.

In view of the way an angle is defined in trigonometry, a drawing of an angle must show not only two segments with a common end point at the vertex but also a curved arrow, as in Figs. 4–1 and 4–2. The curved arrow specifies the direction of rotation by pointing from the initial side to the terminal side and is drawn so as to specify the amount of rotation in the angle. Thus the curved arrow serves to identify the initial and terminal sides of the angle, to specify whether the angle is positive or negative, and to aid in specifying the size of the angle. The importance of the curved arrow is emphasized by the fact, illustrated in Fig. 4–2, that given initial and terminal sides may belong to more than one angle.

4–2. The Measurement of Angles: Degrees

There are two common systems used for the measurement of angles. In one the **degree** is the unit of measure; in the other, the **radian** is the unit of measure.

The angle of **one degree** *is the angle that requires* $\frac{1}{360}$ *of the rotation needed to obtain one complete revolution.* Thus a complete revolution is divided into 360 equal parts called degrees. Each degree is divided into 60 equal parts called **minutes,** and each minute into 60 equal parts called **seconds.** The symbols, $°$, $'$, $''$ are used to denote degrees, minutes, and seconds, respectively. Thus an angle of 31 degrees, 15 minutes, and 10 seconds may be written $31°\ 15'\ 10''$.

A frequent practice in science and engineering is to use decimal parts of degrees instead of minutes and seconds, replacing, for example, $17°\ 30'$ by $17.50°$. Therefore the computations in this book will frequently make use of the decimal parts of degrees instead of minutes and seconds.

A positive angle is measured by a positive number of degrees; a negative angle is measured by a negative number of degrees. Thus the positive angle in Fig. 4–1 is 150°, and the negative angle is −210°.

The trigonometric definition of an angle not only gives a meaning to both positive and negative angles, but also gives a meaning to angles larger than 360°, that is, larger than one complete revolution. For example, angle 2 in Fig. 4–2 is an angle of 450°. In fact, *the trigonometric definition of angle gives a meaning to angles of any size, positive or negative.*

Occasionally it is necessary to change from minutes and seconds to decimal parts of a degree, or vice versa. *To express an angle in decimal parts of a degree when it is given in minutes and seconds, first convert the seconds into a decimal part of a minute, and then the minutes into a decimal part of a degree.*

Example 1.

$$31° \, 45' \, 54'' = 31° \, 45' + \left(\frac{54}{60}\right)'$$

$$= 31° \, 45' + 0.90'$$

$$= 31° \, 45.90'$$

$$= 31° + \left(\frac{45.90}{60}\right)°$$

$$= 31.7650°.$$

To perform the inverse operation, multiply the decimal part of the given angle by 60 to obtain the number of minutes, and multiply the decimal part of the minutes so obtained by 60 to get the seconds.

Example 2.

$$31.7650° = 31° + 0.7650°$$

$$= 31° \, 45.90' \quad \text{(since } 0.7650° = 0.7650 \times 60' = 45.90')$$

$$= 31° \, 45' + 0.90'$$

$$= 31° \, 45' \, 54'' \quad \text{(since } 0.90' = 0.90 \times 60'' = 54'').$$

When an angle is specified to a limited accuracy in accordance with the conventions about significant figures developed in Chapter 1, it is important to know how many significant figures should be used when it is converted from one form of notation to the other. For example, an angle measured to the nearest one hundredth of a degree is in error by at most $\frac{1}{200}$ of a degree, that is, by at most 0.3'. Therefore, it may be concluded that *an angle accurate in decimal notation to the nearest one hundredth of a degree is accurate in minutes and seconds to the nearest minute.* In like manner the following table can be compiled.

Angle Given in Decimal Notation Accurate to Nearest	Maximum Error in Measurement of Angle	Angle Can Be Written in Minutes and Seconds Accurate to Nearest
0.1°	0.05° = 3′	10′
0.01°	0.005° = 0.3′	1′
0.001°	0.0005° = 1.8″	10″
0.0001°	0.00005° = 0.18″	1″

As another example, suppose that an angle is measured accurate to the nearest minute. Then the angle is in error by at most one half-minute, that is, by at most 0.0083°. Since the error would have to be at most 0.005° for the angle to be accurate to the nearest hundredth of a degree, in this example the angle is not quite accurate to the nearest hundredth of a degree. However, in most similar practical cases the error in measurement can be expected to be less than the maximum by a sufficient amount to justify writing the angle in decimal notation to the nearest hundredth of a degree. In like manner the following table can be compiled, with the recognition, of course, that the accuracy specified in the last column is not quite justified if the maximum error can be expected to be present.

Angle Given in Minutes and Seconds Accurate to Nearest	Maximum Error in Measurement of Angle	Angle Can Be Written in Decimal Notation Accurate to Nearest
10′	5′ = 0.083°	0.1°
1′	0.5′ = 0.0083°	0.01°
10″	5″ = 0.0014°	0.001°
1″	0.5″ = 0.00014°	0.0001°

In any practical situation where some knowledge may be available of the magnitude of the possible errors in a measurement, this knowledge will form the basis for a decision on how accurately to specify an angle in decimal or minute and second notation. Since such knowledge will never be available in the exercises in this book, it will be convenient to adopt the following convention:

For converting angles in degrees from decimal to minute and second notation and vice versa:

Accuracy to the nearest   *accuracy to the nearest*

10′		0.1°
1′	*is equivalent to*	0.01°
10″		0.001°
1″		0.0001°

Examples 1 and 2 above observe this convention.

Example 3. Express 11° 10′ 13″ in decimal parts of a degree to the number of decimal places permitted by the given data.

This angle has evidently been read to the nearest second. According to the convention just adopted, the result should be expressed to the nearest ten thousandth of a degree:

$$11° \, 10' \, 13'' = 11° \, 10' + \left(\frac{13}{60}\right)'$$

$$= 11° \, 10.2167'$$

$$= 11° + \left(\frac{10.2167}{60}\right)°$$

$$= 11.1703°.$$

Exercises

Convert to decimal parts of a degree, carrying out computations to the number of places permitted by the data given.

1. 17° 36′.	**2.** 17° 48′.	**3.** 161° 30′.	**4.** 531° 31′.
5. 141° 35′.	**6.** −35° 0′ 4″.	**7.** 42′.	**8.** 32″.
9. 18° 19′ 23″.	**10.** 74° 41′ 20″.	**11.** 212° 0′ 30″.	**12.** 625° 19′ 22″.
13. 1′ 18″.	**14.** 179° 40′ 9″.	**15.** −43° 27′ 35″.	**16.** −721° 11′ 43″.
17. 34° 30′.	**18.** 34° 31′ 0″.	**19.** 21° 0′.	**20.** 21° 1′.
21. 112° 13′ 22″.	**22.** −44° 18′ 18″.	**23.** 56° 48′ 5″.	**24.** 18° 0′ 1″.

Change to degrees, minutes, and seconds, writing the results as accurately as the given data will permit:

25. 27.10°.	**26.** 27.11°.	**27.** 63.87°.	**28.** 4.6°.
29. 612.384°.	**30.** 57.326°.	**31.** 7.5007°.	**32.** 322.698°.
33. −38.6855°.	**34.** −87.01°.	**35.** 26.20506°.	**36.** 20.26°.
37. 20.29°.	**38.** 162.59604°.	**39.** −0.01°.	

Using a protractor, construct the following angles and indicate the corresponding rotation with an arrow (as in Sec. 4–1):

40. 30°.	**41.** 600°.	**42.** −900°.	**43.** 120°.
44. 170°.	**45.** 235°.	**46.** −310°.	**47.** −50°.

4–3. The Measurement of Angles: Radians

In the second system used for the measurement of angles, the radian is the unit of measure.

A **radian** *is the measure of an angle which, placed with its vertex at the center of any circle, subtends on the circumference an arc equal in length to the radius of the circle.*

Thus, if we take a circle with center at O and radius r (Fig. 4–3), and from a point A on the circumference measure an arc AB of length r, the angle AOB is by definition an angle of 1 radian.

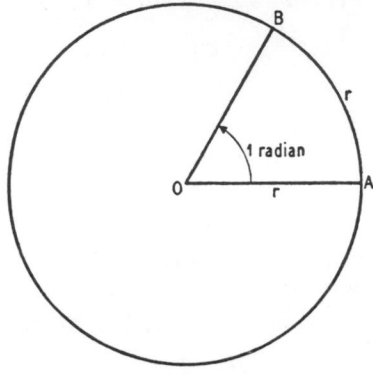

FIG. 4–3

We are now faced with the problem of expressing the measure of any given angle, say θ, in radians. If we place θ with its vertex at the center O of the circle, and its initial side on the initial side OA of angle AOB (Fig. 4–4), then the angle θ will intercept on the circumference of the circle

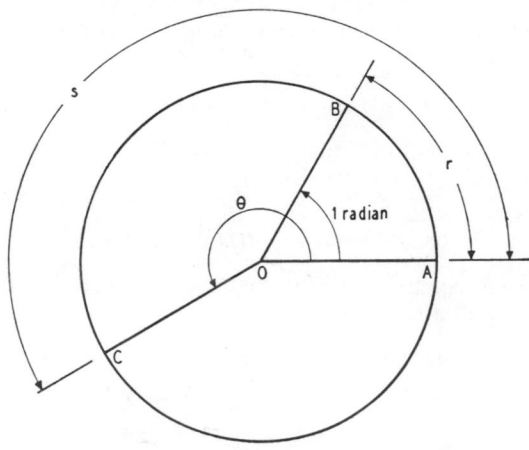

FIG. 4–4

an arc whose length we shall call s. Then from the theorem in geometry which states that, *on the same circle, central angles are proportional to their intercepted arcs*, we may write

$$\frac{\text{Angle } \theta \quad (\text{measured in radians})}{\text{Angle } AOB \quad (\text{measured in radians})} = \frac{\text{arc } AC}{\text{arc } AB}.$$

However, since angle AOB equals 1 radian and since the arcs AC and AB are of lengths s and r, respectively, we obtain

$$\frac{\theta \text{ (in radians)}}{1} = \frac{s}{r} \quad \text{or} \quad \theta = \frac{s}{r}.$$

Usually this is written

$$s = r\theta. \tag{1}$$

In words, formula 1 states that *the length of an arc of a circle is equal to the radius of the circle multiplied by the measure in radians of the angle subtended by the arc at the center of the circle.*

Example 1. In a circle having a radius of 5.0 ft, how long is the arc intercepted by a central angle of 1.5 radians?

Since the angle is given in radians, formula 1 may be applied directly. Since $r = 5.0$, $\theta = 1.5$,

$$s = r\theta = 5.0 \times 1.5 = 7.5 \text{ ft.}$$

Since r is measured in feet, s is measured in feet also.

Example 2. If a point on the circumference of a rotating wheel is moving at the rate of 400 ft per sec and the radius of the wheel is 2 ft, at what angular velocity is the wheel rotating?

By the **angular velocity** of the wheel is meant the rate, in this case measured in radians per second, at which a spoke of the wheel is rotating. Taking one second as a basis, $s = 400$, $r = 2$, and formula 1 gives $400 = 2\theta$, or $\theta = \frac{400}{2} = 200$ radians per second.

It is often necessary to convert an angle from degrees to radians or from radians to degrees. Since the circumference of a circle is equal to $2\pi r$ (where $\pi = 3.1416$ approximately), one complete revolution considered as an angle measured in radians is found by formula 1 to be

$$\theta \quad (1 \text{ revolution}) = \frac{2r\pi}{r} = 2\pi \text{ radians.}$$

Similarly, one complete revolution as an angle measured in degrees is given by

$$\theta \quad (1 \text{ revolution}) = 360°.$$

Hence

$$2\pi \text{ radians} = 360°.$$

From this equation

$$\textbf{1 radian} = \frac{180°}{\pi} = 57.2958 \cdots \textbf{ degrees.} \tag{2}$$

$$\textbf{1}° = \frac{\pi}{180} \textbf{ radians} = 0.0174533 \cdots \textbf{ radians.} \tag{3}$$

These expressions may be used for converting an angle from one system of measurement to the other, as stated in the following rules.

1. *To convert degrees to radians, divide the number of degrees by* $180/\pi$, *or multiply by* $\pi/180$.

2. *To convert radians to degrees, multiply the number of radians by* $180/\pi$.

Note that when the approximations 57.2958 and 0.0174533 are used, the desired accuracy determines the number of significant figures to employ.

Example 3.   Convert to radians: (*a*) 60°, (*b*) 76.24°.

Using rule 1 in both cases, we have

(*a*) $60° \div \dfrac{180°}{\pi} = 60 \times \dfrac{\pi}{180} = \dfrac{\pi}{3}$ radians.

(*b*) $76.24° \div 57.30° = 1.331$ radians.

Example 4.   Convert to degrees: (*a*) $\dfrac{\pi}{2}$ radians, (*b*) 5.716 radians.

Using rule 2 in both cases, we get

(*a*) $\dfrac{\pi}{2} \times \dfrac{180°}{\pi} = 90°$.

(*b*) $5.716 \times 57.30° = 327.5°$.

It is frequently desirable to maintain a certain accuracy of measurement in converting from degrees to radians and vice versa.   Since the conversion from degrees to radians when the angles are given in decimal instead of minute and second notation is performed by carrying out either a multiplication by a standard multiplier or a division by a standard divisor, the rules about significant figures developed in Chapter 1 can be used. If the number of significant figures in the standard multiplier or divisor is chosen to be not less than the number of significant figures in the given angle, the angle converted to radians can be written with the same number of significant figures as the angle in degrees.   Similar remarks apply to the conversion from radians to the decimal-degree notation.   If the angle is given in degrees, minutes, and seconds, one need only convert it to decimal notation in accordance with the convention developed in Sec. 4–2 and then apply the procedure described above in order to convert the angle to radians.

Converting angles from degrees to radians or from radians to degrees is greatly simplified by the use of *conversion tables* which are frequently printed in handbooks and various other collections of tables.   Table 6 in the appendix is such a table.

Example 5.   Convert 69.317° to radians, giving the result to as many significant figures as warranted by the given angle.

Since the given angle has 5 significant figures, we must choose at least 5 significant figures in the multiplier.   The multiplication of 69.317 by 0.017453 gives 1.209,789,601, which is rounded off to 5 significant figures to yield 69.317° = 1.2098 radians.

This result may be checked with Table 6. Since this table gives entries only for 69.3° and 69.4°, we must estimate the amount to be added to the entry for 69.3° to obtain the value for 69.317° by means of a process called *interpolation*:

$$0.100 \left[0.017 \left[\begin{array}{l} 69.300° = 1.2095 \text{ radians} \\ \\ 69.317° = \\ \\ 69.400° = 1.2113 \text{ radians} \end{array} \right] d \right] 0.0018$$

From the differences indicated by the brackets on the left and right, respectively, we note that, as the degrees increase by 0.100 the radians increase by 0.0018. Therefore, as the degrees increase by 0.017, the radians increase by $d = \dfrac{0.017}{0.100} (0.0018) = 0.000306$, which is rounded to 0.0003. Then 69.317° = 1.2095 + 0.0003 = 1.2098 radians.

Example 6. Convert 0.8328 radian to degrees, minutes, and seconds, giving the results to as many significant figures as warranted by the given angle.

Since the angle is given to four significant figures, we obtain $0.8328 \times 57.30° = 47.72° = 47° \, 43'$, where the angle is given to the nearest minute in accordance with the convention of Sec. 4–2. The result 47.72° should be checked in Table 6 by the student; in this check a process of interpolation similar to that used in the previous example will be required.

Exercises

Convert the following angles to radians, assuming the given numbers to be exact. Give the results in terms of π.

1. 15°, 30°.

2. 45°, 60°.

3. 90°, 180°, 270°.

4. 360°, 720°.

5. 135°, 210°.

6. 300°, 330°.

7. 70°, 150°.

8. 540°.

9. 600°.

10. 1620°.

11. −150°.

12. −280°.

13. −310°.

14. 25°, −25°.

15. 198°, −198°.

Convert from radians to degrees; give answers to the nearest hundredth of a degree.

16. $\dfrac{\pi}{12}, \dfrac{\pi}{6}$.

17. $\dfrac{\pi}{4}, \dfrac{3\pi}{4}$.

18. $\dfrac{\pi}{2}, \dfrac{3\pi}{2}$.

19. $2\pi, 4\pi$.

20. $\dfrac{5\pi}{4}, \dfrac{9\pi}{4}$.

21. $\dfrac{2\pi}{3}, \dfrac{7\pi}{3}$.

22. $\dfrac{19\pi}{3}$.

23. $\dfrac{\pi}{9}$.

24. $\dfrac{23\pi}{12}$.

25. $\dfrac{117\pi}{36}$.

26. $\dfrac{-\pi}{12}$.

27. $\dfrac{-5\pi}{9}$.

28. $\dfrac{7\pi}{360}$.

29. $-3\pi, 3\pi$.

30. $\dfrac{-12\pi}{50}, \dfrac{12\pi}{50}$.

Express the following angles in terms of radians.   Make the conversion to radians by means of a multiplier or divisor, and express the results as accurately as the given data permit.   Use Table 6 in the appendix to check your results.

31. 31.0°.   **32.** 97.8°.   **33.** 256.2°.   **34.** 395.7°.

35. 15.47°.   **36.** 87.39°.   **37.** 257.65°.   **38.** 576.70°.

39. 12.609°.   **40.** 39.814°.   **41.** 195.619°.   **42.** 45.0045°.

43. 59.6175°.   **44.** 573.006°.   **45.** 95.5165°.

Change the following angles from radians to degrees and decimal fractions of degrees. Make the conversion by means of a multiplier or divisor, and express the result as accurately as the given angle permits.   Use Table 6 in the appendix to check your results.

46. 0.673.   **47.** 1.219.   **48.** 0.315.   **49.** 2.416.

50. 3.525.   **51.** 4.639.   **52.** 5.472.   **53.** 0.0319.

54. 0.4364.   **55.** 5.6721.   **56.** 2.3455.   **57.** 3.1065.

58. 0.12352.   **59.** 5.67293.   **60.** 2.67931.

Express the following angles in radians as accurately as the given data permit.

61. 31° 14′.   **62.** 16° 57′.   **63.** 98° 26′.

64. 142° 53′.   **65.** 256° 30′.   **66.** 356° 20′.

67. 49° 40′.   **68.** 157° 56′.   **69.** 108° 53′.

70. 218° 43′.   **71.** 220° 27′ 20″.   **72.** 323° 18′ 30″.

73. 198° 37′ 43″.   **74.** 18° 46′ 39″.   **75.** 370° 18′ 46″.

In solving the following problems give the results as accurately as the given data permit.

76. In a circle having a radius of 5.00 in., what is the length of the arc intercepted by a central angle of 5.48 radians?

77. If a central angle intercepts an arc of 21 in. on the circumference of a circle 7.0 in. in diameter, what is the angle in radians?   in degrees?

78. A steam gage has a scale 8.2 in. long which is an arc of a circle with a radius of 3.5 in.   Through how many degrees must a pointer pivoted at the centre of the circle be free to move in order to cover all parts of the scale?

79. An electric current of 10.0 amperes causes an ammeter needle to deflect 82° from the position it takes when no current flows.   If the needle is 4.5 in. long and a circular scale is to be placed at the tip of the needle, how long must the scale be in order to read a maximum current of 10.0 amperes?

80. A racing car has front and rear wheels of 26.5 and 27.5 in. diameters, respectively. In traveling a lap of 2.50 miles how many complete revolutions does each wheel make? What angle in radians does a scar on a front-wheel tire tread generate in that distance? a mark on the tread of one of the rear-wheel tires?   If the car travels a distance (measured to the nearest foot) of 830 ft in 5.00 sec, what is the angular velocity of each size of wheel?

In exercises 81–85 assume that the earth is a perfect sphere with diameter equal to 7918 miles, the mean diameter of the earth.

81. City Hall, New York, has a latitude of 40.72° north. What is the distance from this point to the equator?

82. London has a latitude of 51.48° north. What is its distance to the equator?   to the geographic North Pole?

83. Chicago has a latitude of 41.83° north. What is its distance from the geographic South Pole?

84. How far apart are two points on the same meridian of longitude, one at 42.16° north latitude, the other at 68.37° south latitude?

85. The earth rotates about its axis once each 24 hr. Find the velocity in miles per hour and feet per second of a point on the Equator. What is the angular velocity of a radius connecting a point on the Equator with the center of the earth?

86. The angular velocity of the rotor of a steam turbine is 342.5 radians per second. What is the number of revolutions per minute? If the radius of the rotor is 17.8 in., what is the velocity in feet per minute of a point on the circumference?

87. If the cutting tool on a lathe is designed to cut material at a rate no greater than 600.0 ft per min, find the maximum diameter of a cylindrical piece that may be turned at 180.0 rpm.

88. Two pulleys of diameters 6.4 and 10.2 in. are belted together. If the smaller pulley is driven with an angular velocity of 72 radians per second, find the linear velocity of the belt and the angular velocity of the larger pulley. How many revolutions per minute does each pulley make?

89. If S is the speed of an automobile in miles per hour and d is the outer diameter of the tires in inches, find a formula for expressing R, the revolutions per second of the wheels, in terms of S and d. Express A, the angular velocity in radians per second, in terms of S and d. Find R and A for an automobile with wheels 26.5 in. in diameter, traveling at 65 mph.

90. Prove that $\frac{1}{2}r^2\theta$ is the formula for the area of the sector of a circle of radius r if the sector is formed by a central angle θ measured in radians. Use the geometric theorem that the area of a sector is half the product of the radius of the circle and the length of the arc of the sector.

Exercises 91–93 make use of the formula developed in exercise 90.

91. In a circle of radius 8.00 in., what is the area of a sector formed by an angle of exactly $\pi/4$ radians?   by an angle of exactly $2\pi/5$ radians?   by an angle of 146°?   by an angle of 31.9°?

92. A beam of light from a flashlight sweeps through an angle of 24°. If the range of the light is 60.0 ft, how large an area can be illuminated by the flashlight?

93. A sector whose central angle is 185° is cut from a circular piece of aluminum of diameter 10.0 in. The radii bounding the sector are then brought together to make the sector into a right circular cone. What is the total area of this cone?

94. A tachometer indicates that an airplane propeller is rotating at the rate of 46.52 revolutions per second. Find the angular velocity of the propeller in radians per second. Through how many degrees does the propeller turn in exactly 1 sec? What is the period, that is, the number of seconds required for exactly one revolution?

95. A certain type of camera shutter consists of a circular disk with a sector cut out; the shutter rotates back of the lens and allows light to pass only when the open part of the disk passes by the lens. An engineer wishes to design a motion-picture camera that will make exactly 24 exposures per second, each exposure lasting exactly 0.02 sec. What will be the size in radians of the angle of the sector cut from the disk?   in degrees?

96. The diameter of the earth at the Equator is 7926.68 miles.   Find the distance between two points on the Equator whose longitude differs by 1°, measured to the nearest second.

97. An automobile flywheel 18.24 in. in diameter rotates at an angular velocity of 248.20 radians per second.   Find the velocity in feet per second of a point on the circumference.

98. Unless acted on by a force other than gravity, a pendulum can be expected to swing back and forth in the same place as long as it is in motion.   Therefore, if the earth rotates, it can be expected that a pendulum suspended above the earth and free to continue motion in the same plane will have the earth rotate underneath it.   In 1851 Foucault proved that the earth rotated by suspending from the dome of the Pantheon in Paris an iron ball 1 ft in diameter with a wire 200 ft long and observing that the earth rotated under this pendulum at a rate of 11.3° per hr.   Through how many degrees did the pendulum swing in 2 hr. and 45 min?   Through how many radians?

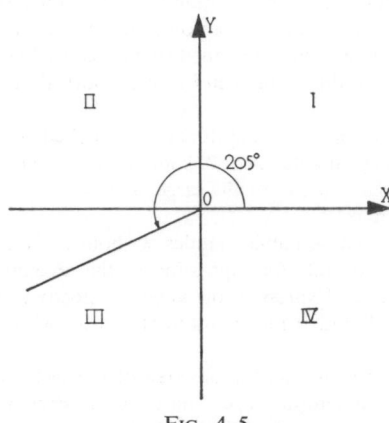

Fig. 4–5

4–4. The Definitions of the Trigonometric Functions

The concepts of the Cartesian coordinate system, of the function, and of the angle have been discussed as separate topics earlier in this text. Combining these three ideas in a convenient manner, we now introduce the **trigonometric functions** on which the whole subject of trigonometry is based.

Any angle may be superimposed on a rectangular coordinate system with its vertex at the origin and its initial side lying along the positive x axis.   An angle so located is said to be in **standard position.**   Angles in standard position, with certain exceptions, may be classified according to the quadrants in which their terminal sides lie.   For example, a positive angle of 205° is said to lie in the third quadrant, for in standard position its terminal side lies in the third quadrant (Fig. 4–5).   We cannot classify in this way angles such as 0°, 90°, 180°, whose terminal sides lie on one of the coordinate axes.   These angles are called **quadrantal** angles.

Now consider any angle θ. This angle, in standard position, might lie in any one of the four quadrants (Fig. 4–6). Regardless of the location of the angle, we select P, *any* point on the terminal side of the angle, and denote the coordinates of P by (x, y). From P draw a line perpendicular to the x axis. In each case this gives us what we call a **right triangle of reference** associated with θ. This triangle does not necessarily include the angle θ with which it is associated. The hypotenuse (denoted by r) of this right triangle of reference is the *distance* of P from

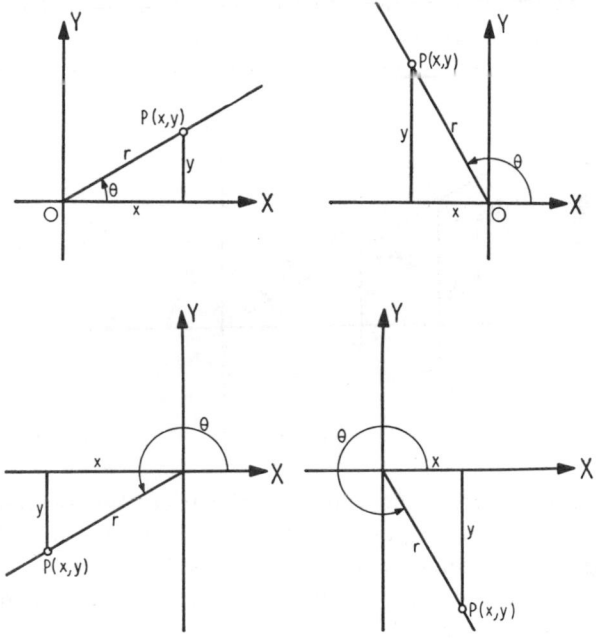

FIG. 4–6

the origin, and this distance is always considered *positive*. The other two sides of the triangle are in every case the coordinates of the point P, and these sides are positive or negative according as x and y are positive or negative. It should be noted that, for the quadrantal angles, a point P can still be chosen, but the right triangle of reference will degenerate into a triangle which has one side of length zero and the other two sides equal in length but possibly opposite in sign.

From the three quantities or lengths, x, y, and r which we have introduced, six different ratios may be defined. Regardless of the quadrant in which the angle lies, and even if it is a quadrantal angle, these definitions are the same.

Given any angle θ in standard position, we define

$$\text{sine } \theta = \frac{y}{r}, \qquad \text{cosecant } \theta = \frac{r}{y},$$

$$\text{cosine } \theta = \frac{x}{r}, \qquad \text{secant } \theta = \frac{r}{x},$$

$$\text{tangent } \theta = \frac{y}{x}, \qquad \text{cotangent } \theta = \frac{x}{y},$$

where x, y, and r are the sides of any right triangle of reference associated with θ. These six names are usually abbreviated to sin θ, cos θ, tan θ, csc θ, sec θ, cot θ. These are the trigonometric functions of θ.

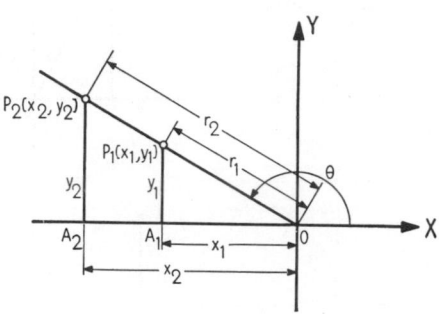

FIG. 4–7

For any given angle θ the values of the trigonometric functions are the same, regardless of the position of P on the terminal side of the reference triangle, for different choices of P result in similar triangles of reference, and, from plane geometry, the ratios of corresponding sides of similar triangles are equal.

For example, let θ be an angle in the second quadrant (Fig. 4–7). Choose any two points, P_1 and P_2, on the terminal side of θ, and form the right triangles of reference, OP_1A_1 and OP_2A_2, associated with these points. Since these triangles are similar

$$\frac{y_1}{r_1} = \frac{y_2}{r_2}, \quad \frac{x_1}{r_1} = \frac{x_2}{r_2}, \quad \text{etc.} \tag{1}$$

Now,

$$\sin \theta \quad \text{(determined from } \Delta\ OP_1A_1) \quad = \frac{y_1}{r_1},$$

and

$$\sin \theta \quad \text{(determined from } \Delta\ OP_2A_2) \quad = \frac{y_2}{r_2}.$$

Therefore, substituting in relation 1, we obtain

$$\sin \theta \quad (\text{from } \Delta\, OP_1A_1) \quad = \sin \theta \quad (\text{from } \Delta\, OP_2A_2).$$

In like manner for the other functions,

$$\cos \theta \quad (\text{from } \Delta\, OP_1A_1) \quad = \frac{x_1}{r_1} = \frac{x_2}{r_2} = \cos \theta \quad (\text{from } \Delta\, OP_2A_2),$$

$$\tan \theta \quad (\text{from } \Delta\, OP_1A_1) \quad = \frac{y_1}{x_1} = \frac{y_2}{x_2} = \tan \theta \quad (\text{from } \Delta\, OP_2A_2), \text{ etc.}$$

Thus, *for a given angle θ, the value of the trigonometric function is independent of the triangle of reference which is used.* However, for different angles, it is easy to see that, except for certain special cases which will be discussed later, the triangles of reference are not similar, and hence the trigonometric ratios are not the same. Since for every value of the angle θ there corresponds one value for each of the trigonometric ratios, these ratios are called the **trigonometric functions of the angle.**

4–5. Signs of the Trigonometric Functions in the Various Quadrants

On the basis of the definitions of the trigonometric functions it is easy to determine whether a given function of an angle is positive or negative. For example, if θ is any angle in the second quadrant, $\cos \theta = x/r$ is a negative quantity because r is always positive and x, in this case the abscissa of a point in the second quadrant, is negative. In like manner the reader may verify the information summarized in the following table.

	Quadrant I	Quadrant II	Quadrant III	Quadrant IV
$\sin \theta = \dfrac{y}{r}$	+	+	−	−
$\cos \theta = \dfrac{x}{r}$	+	−	−	+
$\tan \theta = \dfrac{y}{x}$	+	−	+	−
$\cot \theta = \dfrac{x}{y}$	+	−	+	−
$\sec \theta = \dfrac{r}{x}$	+	−	−	+
$\csc \theta = \dfrac{r}{y}$	+	+	−	−

4–6. Evaluation of the Trigonometric Functions When the Angle is Known

Method 1.   Construction. The six trigonometric functions of a given angle may be found approximately by the accurate construction of a right triangle of reference. Suppose it is desired to find one of or all the functions of 72°. Constructing the angle by means of a protractor and dropping a perpendicular to the x axis from any point on the terminal side, we have a right triangle of reference (Fig. 4–8). Then after measuring the sides of this triangle and obtaining $x = 1.25$ in., $y = 3.8$ in., $r = 4.0$ in., we can evaluate the ratios:

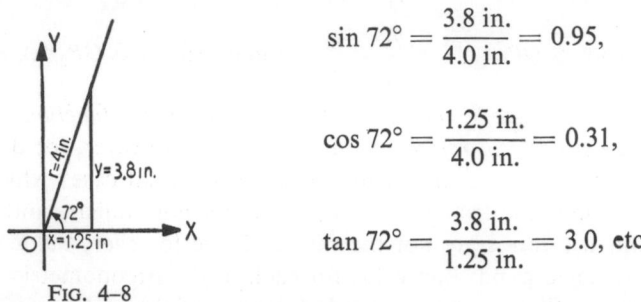

FIG. 4–8

$$\sin 72° = \frac{3.8 \text{ in.}}{4.0 \text{ in.}} = 0.95,$$

$$\cos 72° = \frac{1.25 \text{ in.}}{4.0 \text{ in.}} = 0.31,$$

$$\tan 72° = \frac{3.8 \text{ in.}}{1.25 \text{ in.}} = 3.0, \text{ etc.}$$

Since the lengths measured vary with the accuracy of construction and the type of scale used, the values found by this method are very rough approximations.

The above example furnishes an illustration of the fact that the trigonometric functions are independent of the unit of measurement. No matter what unit of measurement is used, the results are the same.

Method 2. Tables. By methods that are beyond the scope of this book, the trigonometric functions for any angle may be found to any desired degree of accuracy. These results are usually compiled in the form of tables. The use of these tables is discussed later in the chapter.

Exercises

By construction, find the six trigonometric functions of the following angles. Carry out all work to two significant figures.

1. 60°.   **2.** 45°.   **3.** 75°.   **4.** 137°.

5. 221°.   **6.** 325°.   **7.** −30°.   **8.** 410°.

9. 36°.   **10.** −240°.   **11.** 240°.   **12.** 164°.

4–7. Evaluation of the Trigonometric Functions When One Function Is Known

This process can best be shown by the use of examples.

Example 1. Given $\sin \theta = \frac{3}{5}$, where θ is an angle in the first quadrant, find the other trigonometric functions of θ.

Since $\sin \theta = y/r$, then $y/r = \frac{3}{5}$ for the particular angle in which we are interested. Then for some particular triangle of reference, the lengths of two sides will be $y = 3$, $r = 5$ (Fig. 4–9). By the Pythagorean theorem, which states that the square of the hypotenuse of a right triangle equals the sum of the squares of the remaining two sides, we may find the third side x, for

$$3^2 + x^2 = 5^2, \qquad x^2 = 25 - 9,$$

whence

$$x = \pm\sqrt{16}, \quad \text{or} \quad x = \pm 4.$$

Since θ is in the first quadrant, we select the value $x = +4$. Now, knowing the three sides of a triangle of reference, we can find the other five trigonometric ratios: $\cos \theta = \frac{4}{5}$, $\tan \theta = \frac{3}{4}$, $\cot \theta = \frac{4}{3}$, $\sec \theta = \frac{5}{4}$, $\csc \theta = \frac{5}{3}$.

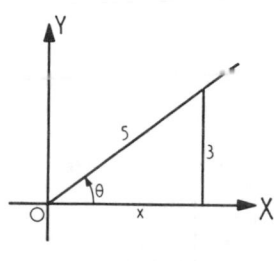

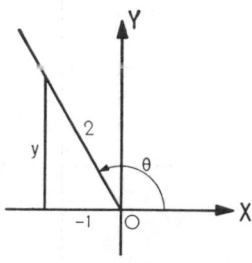

Fig. 4–9   Fig. 4–10

In this process it is not necessary to know the exact size of the angle θ, nor is it necessary that the right triangle of reference be very accurately drawn, for the length of the third side is found algebraically rather than geometrically.

Example 2. Given $\sec \theta = -2$, and $\sin \theta$ positive, find the other trigonometric functions of θ.

The angle θ must be in the second quadrant, for that is the only quadrant in which the secant is negative and the sine positive. Since $\sec \theta = \dfrac{r}{x}$, then $\dfrac{r}{x} = \dfrac{2}{-1}$, for r is always positive. Thus we take θ in the second quadrant, and for a triangle of reference we select one with $r = 2$ and $x = -1$ (Fig. 4–10). By the Pythagorean theorem:

$$2^2 = y^2 + (-1)^2 \quad \text{or} \quad y^2 = 4 - 1,$$

whence

$$y = \pm\sqrt{3}.$$

The positive root must be selected since y is positive in the second quadrant. Therefore the other five functions are: $\sin \theta = \dfrac{\sqrt{3}}{2}$, $\cos \theta = -\dfrac{1}{2}$, $\tan \theta = \dfrac{\sqrt{3}}{-1}$ $= -\sqrt{3}$, $\cot \theta = -\dfrac{1}{\sqrt{3}} = -\dfrac{\sqrt{3}}{3}$, $\csc \theta = \dfrac{2}{\sqrt{3}} = \dfrac{2\sqrt{3}}{3}$.

Example 3. Given $\cot \theta = -\frac{1}{3}$, where θ is between $0°$ and $360°$, find the other trigonometric functions of θ.

Since the cotangent is negative in the second and fourth quadrants, there are two possible angles θ. Since $\cot \theta = \dfrac{x}{y}$, then in the second quadrant $\dfrac{x}{y} = \dfrac{-1}{3}$ and $x = -1$, $y = 3$, making $r = \sqrt{(3)^2 + (-1)^2} = \sqrt{9 + 1} = \sqrt{10}$; whereas in the fourth quadrant $\dfrac{x}{y} = \dfrac{1}{-3}$ and therefore $x = 1$, $y = -3$, making $r = \sqrt{(-3)^2 + 1^2} = \sqrt{10}$. Thus,

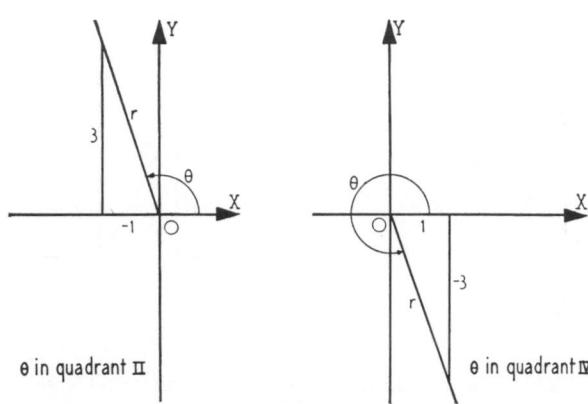

θ in quadrant II θ in quadrant IV

FIG. 4–11

For θ in the Second Quadrant

$$\sin \theta = \frac{3}{\sqrt{10}} \quad \text{or} \quad \frac{3\sqrt{10}}{10},$$

$$\cos \theta = \frac{-1}{\sqrt{10}} \quad \text{or} \quad -\frac{\sqrt{10}}{10},$$

$$\tan \theta = \frac{3}{-1} \quad \text{or} \quad -3,$$

$$\sec \theta = \frac{\sqrt{10}}{-1} \quad \text{or} \quad -\sqrt{10},$$

$$\csc \theta = \frac{\sqrt{10}}{3}.$$

For θ in the Fourth Quadrant

$$\sin \theta = \frac{-3}{\sqrt{10}} \quad \text{or} \quad \frac{-3\sqrt{10}}{10},$$

$$\cos \theta = \frac{1}{\sqrt{10}} \quad \text{or} \quad \frac{\sqrt{10}}{10},$$

$$\tan \theta = \frac{-3}{1} \quad \text{or} \quad -3,$$

$$\sec \theta = \frac{\sqrt{10}}{1} \quad \text{or} \quad \sqrt{10},$$

$$\csc \theta = \frac{\sqrt{10}}{-3} \quad \text{or} \quad -\frac{\sqrt{10}}{3}.$$

The work of this section further illustrates that the trigonometric functions are numbers. Hence we may treat them as numbers. That is, the trigonometric functions may be added, subtracted, multiplied, divided, etc., as numbers. For example, $(\sin \theta) \cdot (\sin \theta) = (\sin \theta)^2$, which is usually written $\sin^2 \theta$. If $\sin \theta = \frac{3}{4}$, then $\sin^2 \theta = (\frac{3}{4})^2 = \frac{9}{16}$.

Example 4. For the angle of example 2, evaluate $\dfrac{\sin \theta \cos \theta}{\tan^2 \theta} - 1$.

$$\frac{\sin \theta \cos \theta}{\tan^2 \theta} - 1 = \frac{\dfrac{\sqrt{3}}{2} \cdot \left(-\dfrac{1}{2}\right)}{(-\sqrt{3})^2} - 1 = \frac{-\dfrac{\sqrt{3}}{4}}{3} - 1$$

$$= -\frac{\sqrt{3}}{12} - 1 = -\left(\frac{\sqrt{3} + 12}{12}\right).$$

Exercises

Find the six trigonometric functions of θ; θ is in the standard position, and its terminal side passes through the given point.

1. (3, 4).   **2.** (5, 2).   **3.** (−1, 3).   **4.** (−2, −3).

5. (9, 4).   **6.** (3, −5).   **7.** (4, 1).   **8.** (−4, −1).

9. (3, −3.)   **10.** (7, 5).   **11.** (6, 0).   **12.** (8, 0).

13. (0, −3).   **14.** (−7, 0).   **15.** (−45, 25).

List the six trigonometric functions of an angle θ if:

16. $\sin \theta = \frac{2}{3}$ and θ lies in the first quadrant.

17. $\cos \theta = \frac{1}{4}$ and θ lies in the third quadrant.

18. $\sec \theta = -\frac{7}{5}$ and θ lies in the second quadrant.

19. $\tan \theta = \frac{3}{11}$ and θ lies in the first quadrant.

20. $\csc \theta = -\frac{18}{5}$ and θ lies in the fourth quadrant.

21. $\cot \theta = \frac{2}{3}$, $\sin \theta$ positive.

22. $\csc \theta = \frac{5}{3}$, $\cos \theta$ negative.

23. $\sin \theta = \frac{7}{12}$, $\cos \theta$ negative.

24. $\tan \theta = 3$, $\cos \theta$ negative.

25. $\csc \theta = \frac{9 \cdot 1}{5}$, $\tan \theta$ negative.

26. $\sec \theta = -7$, $\tan \theta$ positive.

27. $\sin \theta = \frac{4}{5}$, $\sec \theta$ negative.

28. $\cos \theta = \frac{4}{5}$, $\sin \theta$ positive.

List the six trigonometric functions for each angle θ greater than or equal to $0°$ and less than $360°$ if:

29. $\sin \theta = \frac{3}{7}$.   **30.** $\cos \theta = \frac{4}{9}$.   **31.** $\cos \theta = -\frac{4}{9}$.

32. $\csc \theta = \frac{15}{4}$.   **33.** $\sec \theta = \pm 2$.   **34.** $\csc \theta = \pm \frac{5}{2}$.

35. $\cot \theta = \pm \frac{11}{4}$ **36.** $\sec \theta = 1.2$.   **37.** $\tan \theta = 0.7$.

38. $\sec \theta = 6$.   **39.** $\sin \theta = -0.4$.   **40.** $\tan \theta = -2.1$.

41. $\tan \theta = -2$.   **42.** $\cos \theta = 0$.   **43.** $\sin \theta = -1$.

44. $\cos \theta = -\dfrac{\sqrt{5}}{5}$.   **45.** $\sec \theta = \sqrt{2}$.   **46.** $\sin \theta = \dfrac{n}{m}$.

47. $\tan \theta = m.$ **48.** $\cot \theta = -\dfrac{\sqrt{1-a^2}}{a}.$ **49.** $\cos \vartheta = -\dfrac{4a}{3}.$

50. $\tan \theta = 2.0.$ **51.** $\csc \theta = -7.$ **52.** $\sec \theta = 3212.$

53. $\sec \theta = -\tfrac{9}{7}.$ **54.** $\tan \theta = 2.0002.$ **55.** $\sin \theta = -0.1875.$

Compute the following expressions, assuming that θ is an angle in the first or second quadrant:

56. $\dfrac{\sin \theta + \cos \theta}{1 + \sec \theta}$ if $\tan \theta = -\dfrac{3}{2}.$ **57.** $\dfrac{\sin^2 \theta + \cos^2 \theta}{1 + \tan^2 \theta}$ if $\sec \theta = -\dfrac{9}{5}.$

58. $\cos \theta + \sin \theta \sec \theta$ if $\sec \theta = -\tfrac{11}{3}.$

59. $\dfrac{\sin \theta \cos \theta}{\sec^2 \theta}$ if $\tan \theta = \dfrac{7}{4}.$

60. $\dfrac{(1 + \tan^2 \theta)(\sin^2 \theta)}{\cos \theta (1 + \cot^2 \theta)}$ if $\sin \theta = \dfrac{4}{5}.$

61. $\dfrac{\tan \theta \sec \theta}{1 + \sin \theta} + 2 \cos \theta$ where $\sec \theta = \dfrac{8}{3}.$

62. $\dfrac{\sin \theta \cos^2 \theta}{2 + 2 \tan^2 \theta + \sec^2 \theta}$ where $\cot \theta = \dfrac{1}{6}.$

63. $\dfrac{\cos \theta \sec \theta}{\tan \theta \cot \theta} \div \dfrac{\sin \theta}{1 - \cos^2 \theta}$ where $\csc \theta = \dfrac{18}{5}.$

64. $\dfrac{1 + \sin \theta + \cos \theta}{1 + \sin \theta - \cos \theta} \cdot \cos \theta$ where $\sin \theta = \dfrac{\sqrt{3}}{3}.$

Compute the following expressions, assuming that θ is an angle in the third or fourth quadrant:

65. $\tfrac{3}{2} \sin \theta - \cos^2 \theta$ where $\tan \theta = \tfrac{4}{5}.$

66. $\dfrac{A \sin \theta + A \cos \theta}{\tfrac{2}{3} \tan \theta}$ where $\cot \theta = \dfrac{3}{8}.$

67. $\dfrac{2F \cos \theta \tan \theta}{7} + 4F \cos \theta \csc \theta$ where $\sin \theta = -\dfrac{4}{9}.$

68. $\dfrac{5}{\sin^2 \theta} + \dfrac{7}{\cos^2 \theta} - \dfrac{2}{\tan^2 \theta}$ where $\cos \theta = \dfrac{6}{7}.$

69. $\dfrac{1 + \sin \theta}{1 - \sin \theta} - \dfrac{1 - \sin \theta}{1 + \sin \theta}$ where $\sec \theta = -3.$

70. $\dfrac{\tan \theta - \sin \theta}{\sin^3 \theta} - \dfrac{\sec \theta}{1 + \cos \theta}$ where $\cot \theta = -2.1.$

71. $\dfrac{1.874E \cot^2 \theta}{1 - \sin^2 \theta \cos \theta} + E$ where $\cos \theta = 0.$

72. $\dfrac{F_1 + F_2 + 7 \sin \theta}{5 \cos^2 \theta}$ where $\cot \theta = 0.$

4–8. Determination of the Trigonometric Functions of Special Angles

There are certain angles whose trigonometric functions can be found very easily with the aid of a few theorems from plane geometry.

The Functions of 30°. Locating an angle of 30° in standard position on the coordinate axes and forming a triangle of reference, we find that we have a 30°, 60° right triangle. In plane geometry, it is shown that a 30°, 60° right triangle has a hypotenuse that is twice as long as the side opposite the 30° angle. Since the triangle of reference is used only in forming ratios, we may conveniently assume that $r = 2$, $y = 1$. Then,

$$x = \sqrt{r^2 - y^2} = \sqrt{4 - 1} = \sqrt{3}.$$

Thus the three sides of the triangle of reference are $r = 2$, $y = 1$, $x = \sqrt{3}$, and the six trigonometric functions are

$$\sin 30° = \tfrac{1}{2}, \qquad\qquad \csc 30° = \tfrac{2}{1} \ \text{ or } \ 2;$$

$$\cos 30° = \frac{\sqrt{3}}{2}, \qquad\qquad \sec 30° = \frac{2}{\sqrt{3}} \ \text{ or } \ \frac{2\sqrt{3}}{3};$$

$$\tan 30° = \frac{1}{\sqrt{3}} \ \text{ or } \ \frac{\sqrt{3}}{3}, \qquad \cot 30° = \frac{\sqrt{3}}{1} \ \text{ or } \ \sqrt{3}.$$

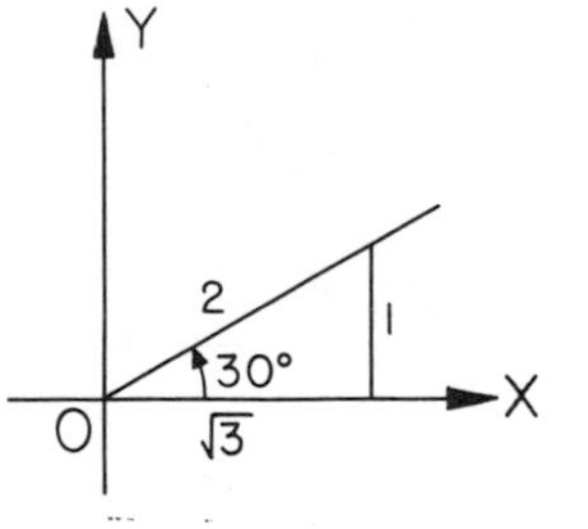

FIG. 4–12

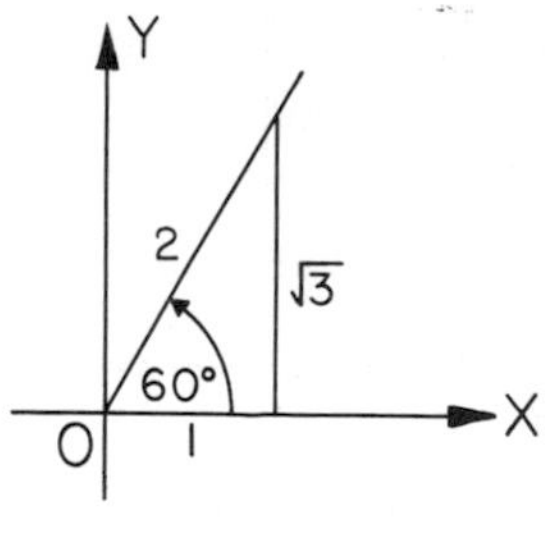

FIG. 4–13

The Functions of 60°. Locating an angle of 60° in standard position on the coordinate axes and forming a triangle of reference, we have a 30°, 60° right triangle. However, now the side x is opposite the 30° angle, and hence it may be assumed that $r = 2$, $x = 1$. From the Pythagorean theorem we obtain $y = \sqrt{3}$. The six trigonometric functions are

$$\sin 60° = \frac{\sqrt{3}}{2}, \qquad\qquad \csc 60° = \frac{2}{\sqrt{3}} \ \text{ or } \ \frac{2\sqrt{3}}{3};$$

$$\cos 60° = \tfrac{1}{2}, \qquad\qquad \sec 60° = \tfrac{2}{1} \ \text{ or } \ 2;$$

$$\tan 60° = \frac{\sqrt{3}}{1} \ \text{ or } \ \sqrt{3}, \qquad \cot 60° = \frac{1}{\sqrt{3}} \ \text{ or } \ \frac{\sqrt{3}}{3}.$$

The Functions of 120°, etc.　Any angle in standard position whose terminal side forms an angle of 30° or 60° with the x axis may be treated in a similar fashion.　For example, 120°, in standard position, forms an angle of 60° with the negative x axis.　Hence, the triangle of reference is a 30°, 60° right triangle.　Since the side x in this triangle is opposite the 30° angle, we may take $r = 2$, $x = -1$, x being negative because the angle lies in the second quadrant.　From the Pythagorean theorem $y = \sqrt{3}$, making

$$\sin 120° = \frac{\sqrt{3}}{2}, \qquad\qquad \csc 120° = \frac{2}{\sqrt{3}} \text{ or } \frac{2\sqrt{3}}{3};$$

$$\cos 120° = \frac{-1}{2}, \text{ or } -\frac{1}{2} \qquad \sec 120° = \frac{2}{-1} \text{ or } -2;$$

$$\tan 120° = \frac{\sqrt{3}}{-1} \text{ or } -\sqrt{3}, \qquad \cot 120° = \frac{-1}{\sqrt{3}} \text{ or } \frac{-\sqrt{3}}{3}.$$

It is easily seen that we can treat in this way any other angle that gives rise to a 30°, 60° right triangle of reference.

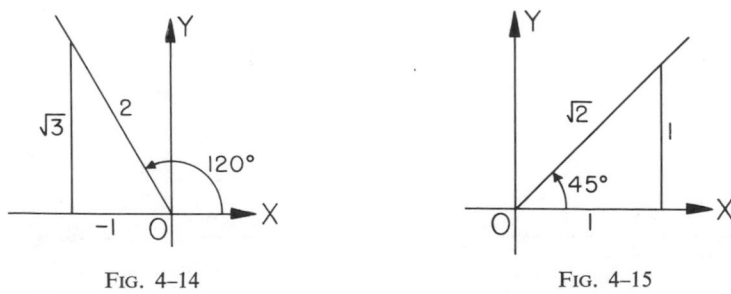

Fig. 4–14　　　　　　　　　　Fig. 4–15

The Functions of 45°.　Locate an angle of 45° in standard position, and form a triangle of reference.　The acute angles of this triangle of reference are each 45° in size, for the acute angles of right triangles are complementary.　From plane geometry, then, the sides opposite these equal angles are of equal length.　Thus for convenience, we may select $x = 1$, $y = 1$, $r = \sqrt{2}$.　The functions are

$$\sin 45° = \frac{1}{\sqrt{2}} \text{ or } \frac{\sqrt{2}}{2}, \qquad \csc 45° = \frac{\sqrt{2}}{1} \text{ or } \sqrt{2};$$

$$\cos 45° = \frac{1}{\sqrt{2}} \text{ or } \frac{\sqrt{2}}{2}, \qquad \sec 45° = \frac{\sqrt{2}}{1} \text{ or } \sqrt{2};$$

$$\tan 45° = \tfrac{1}{1} \text{ or } 1, \qquad\qquad \cot 45° = \tfrac{1}{1} \text{ or } 1.$$

It is easily seen that the functions of any other angle whose terminal side forms an angle of 45° with the x axis may be evaluated in this way.

The Functions of 0°. When an angle of 0° is placed in standard position, the terminal side of the angle coincides with the positive x axis (Fig. 4–16). Thus any triangle of reference chosen has its side $y = 0$, since the perpendicular distance from any point P on the terminal side to the x axis is zero. Such a triangle is often called a **degenerate triangle.** The other two sides of this degenerate triangle are of equal length. For convenience we may select them as $r = 1$, $x = 1$. Thus

$$\sin 0° = \tfrac{0}{1} \text{ or } 0, \qquad \csc 0° = \tfrac{1}{0} \text{ (which is undefined)};$$

$$\cos 0° = \tfrac{1}{1} \text{ or } 1, \qquad \sec 0° = \tfrac{1}{1} \text{ or } 1;$$

$$\tan 0° = \tfrac{0}{1} \text{ or } 0, \qquad \cot 0° = \tfrac{1}{0} \text{ (which is undefined).}$$

The definitions of csc 0° and cot 0° will be discussed further in Chapter 5.

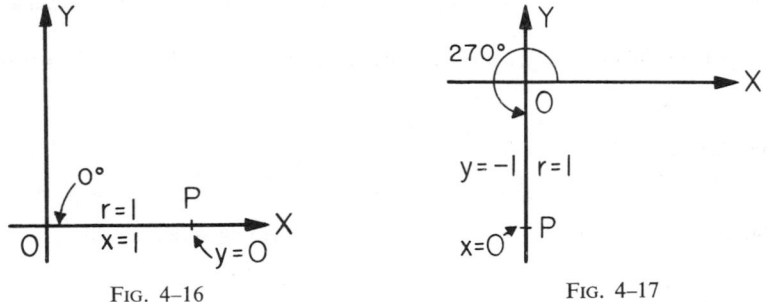

FIG. 4–16   FIG. 4–17

The Functions of 270°. When an angle of 270° is placed in standard position, the terminal side of the angle coincides with the negative y axis. Thus any triangle of reference is degenerate, with the side $x = 0$. We may conveniently assume that the other two sides, being equal in length, are both of length 1, that is, $y = -1$, $r = 1$ (Fig. 4–17). Thus

$$\sin 270° = \frac{-1}{1} \text{ or } -1, \qquad \csc 270° = \frac{1}{-1} \text{ or } -1;$$

$$\cos 270° = \tfrac{0}{1} \text{ or } 0, \qquad \sec 270° = \tfrac{1}{0} \text{ (which is undefined)};$$

$$\tan 270° = \frac{-1}{0} \text{ (which is undefined),} \qquad \cot 270° = \frac{0}{-1} \text{ or } 0.$$

The definitions of sec 270° and tan 270° will be discussed further in Chapter 5. It is evident that the trigonometric functions of any other quadrantal angle may be evaluated in a similar manner.

The quadrantal angles and the angles that have triangles of reference that are 30°, 60° right triangles or 45° right triangles are called the **principal angles.**

Exercises

Find the six trigonometric functions of each of the following angles:

1. $120°$.
2. $240°$.
3. $300°$.
4. $450°$.

5. $750°$.
6. $-45°$.
7. $-120°$.
8. $-240°$.

9. $135°$.
10. $315°$.
11. $765°$.
12. $-765°$.

13. $180°$.
14. $270°$.
15. $360°$.
16. $1080°$.

17. $-180°$.
18. $300°$.
19. $-60°$.
20. $690°$.

21. $\dfrac{5\pi}{3}$ radians.
22. $\dfrac{5\pi}{4}$ radians.
23. $\dfrac{\pi}{2}$ radians.
24. $\dfrac{3\pi}{2}$ radians.

25. $\dfrac{-3\pi}{2}$ radians.
26. $\dfrac{5\pi}{2}$ radians.
27. $\dfrac{\pi}{6}$ radians.
28. $\dfrac{-\pi}{3}$ radians.

29. 3π radians.
30. -2π radians.
31. $\dfrac{-\pi}{6}$ radians.
32. $\dfrac{15\pi}{4}$ radians.

Simplify the following trigonometric expressions as much as possible. For example:

33. $\dfrac{Z \sin 30° \cos 30°}{1 - \cos 30°} = \dfrac{Z \cdot \dfrac{1}{2} \cdot \dfrac{\sqrt{3}}{2}}{1 - \dfrac{1}{2}} = \dfrac{Z \cdot \dfrac{\sqrt{3}}{4}}{\dfrac{1}{2}} = Z \dfrac{\sqrt{3}}{2} \,.$

34. $\sin^2 60° + \cos^2 60°$.

35. $E \sin^2 120° + E \cos^2 120°$.

36. $\dfrac{\sin 210°}{\tan 210°} - \dfrac{\csc 45°}{\sin 30° \sec 60°}\,.$

37. $\dfrac{\csc \dfrac{\pi}{2}}{2 - \sec^2 \pi}\,.$

38. $\dfrac{\tan \dfrac{5\pi}{4} - \sec 0°}{\cos \pi} \cdot \dfrac{\cos \dfrac{5\pi}{6}}{\sin \dfrac{5\pi}{6}}\,.$

39. $3 \csc \dfrac{\pi}{6} \left(\sec \dfrac{3\pi}{2} - 4 \cot \dfrac{11\pi}{4} \right)\,.$

40. $\dfrac{\dfrac{2}{3} \tan \dfrac{7\pi}{4}}{\left(\sqrt{2} - 3 \cos \dfrac{7\pi}{4} \right)}\,.$

41. $\dfrac{4 \cot 45°}{2 \sin 120° \cos 30°} - \dfrac{1}{2}\,.$

42. $I_1 \sin 150° + I_2 \cos 180°$.

43. $\dfrac{\tan 0° \sec x}{1 - \sin^2 90°} + 2 \cot 315°$.

44. $\dfrac{\sin \theta \sec 0°}{2 + \sin 90°} - \dfrac{2 \cos \theta}{5}\,.$

45. $Z_1 \cos \dfrac{3\pi}{4} + 2Z_2 \tan \dfrac{\pi}{4} + 3Z_2 \sin \dfrac{5\pi}{6}$.

46. $\left[\dfrac{\sin \dfrac{\pi}{3} \sin wt}{3 \tan \dfrac{2\pi}{3}} \right] \div \left[\dfrac{\sin \dfrac{3\pi}{2} \cos wt}{\dfrac{4}{5} \sec 3\pi} \right]\,.$

4–9. Functions of Complementary Angles

In trigonometry it is occasionally convenient to speak of the cosine, the cotangent, and the cosecant as the complementary functions or the **cofunctions** of the sine, the tangent, and the secant, respectively. Conversely, the sine, the tangent, and the secant are called the cofunctions of the cosine, the cotangent, and the cosecant. Recalling that two angles are said to be complementary if their sum is 90°, we shall now prove the theorem that *a trigonometric function of an angle is equal in value to the cofunction of its complementary angle.*

We shall prove the theorem for complementary angles that are positive and less than 90°.   The proof for other angles is very similar and need not be taken up here.

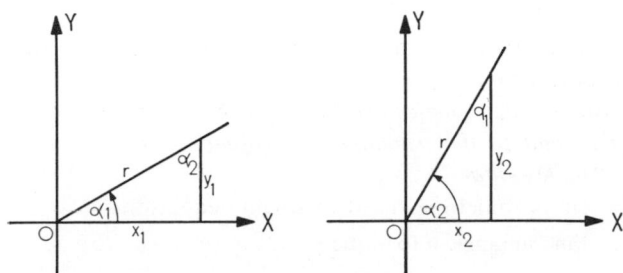

FIG. 4–18

Let α_1 and α_2 be any two complementary angles that are positive and less than 90°.   These angles may be placed in standard position, and equal distances r may be measured on the terminal side of each (Fig. 4–18). The two right triangles of reference formed by using r as hypotenuse are congruent, for the hypotenuse and an acute angle of one are equal, respectively, to the hypotenuse and an acute angle of the other.   Since the corresponding sides of congruent triangles are equal, it follows that $r = r$, $x_1 = y_2$, and $y_1 = x_2$.   Hence,

$$\sin \alpha_1 = \frac{y_1}{r} = \frac{x_2}{r} = \cos \alpha_2 ;$$

$$\cos \alpha_1 = \frac{x_1}{r} = \frac{y_2}{r} = \sin \alpha_2 ;$$

$$\tan \alpha_1 = \frac{y_1}{x_1} = \frac{x_2}{y_2} = \cot \alpha_2 ;$$

$$\cot \alpha_1 = \frac{x_1}{y_1} = \frac{y_2}{x_2} = \tan \alpha_2 ;$$

$$\sec \alpha_1 = \frac{r}{x_1} = \frac{r}{y_2} = \csc \alpha_2;$$

$$\csc \alpha_1 = \frac{r}{y_1} = \frac{r}{x_2} = \sec \alpha_2;$$

and the theorem is proved.

This theorem permits a considerable saving of space in compiling tables of trigonometric functions, for the functions of angles from 45° to 90° may be found by finding the functions of angles from 0° to 45°. For example,

$$\sin 72.3° = \cos (90° - 72.3°) = \cos 17.7°;$$

$$\cos 72.3° = \sin (90° - 72.3°) = \sin 17.7°;$$

$$\tan 72.3° = \cot (90° - 72.3°) = \cot 17.7°.$$

A converse of this theorem is also true. We shall state this converse theorem without proof.

If two positive, acute angles are known to have a trigonometric function of one angle equal to the trigonometric cofunction of the other, then the angles are complementary.

This theorem is sometimes used in solving equations.

Example. Find an angle θ for which

$$\csc (100° - 4\theta) = \sec (2\theta).$$

By the above theorem, if $100° - 4\theta$ and 2θ are acute angles,

$$100° - 4\theta + 2\theta = 90°,$$

$$-2\theta = -10°,$$

$$\theta = 5°.$$

We must check to see if $100° - 4\theta$ and 2θ are acute angles. Substituting $\theta = 5°$, we obtain

$$100° - 4\theta = 80°, \qquad 2\theta = 10°,$$

which are acute, positive, and complementary. Therefore $\theta = 5°$ is the desired angle.

Exercises

Find a trigonometric function of another positive angle less than or equal to 90° which has the same value as the given function.

1. $\sin 17°$.	**2.** $\cos 54°$.	**3.** $\csc 25°$.
4. $\tan 4°$.	**5.** $\sec 69°$.	**6.** $\cot 36.2°$.
7. $\csc 27.53°$.	**8.** $\sin 90°$.	**9.** $\cos 12.68°$.
10. $\sec 72.7°$.	**11.** $\csc 0.047°$.	**12.** $\tan 0°$.
13. $\sin 67°$.	**14.** $\cos 16.7°$.	**15.** $\sec 1.43°$.

Determine from each of the following equations, if possible, a value of θ which satisfies the equation. If θ makes the angles under consideration larger than 90°, verify the given equation by substitution.

16. $\sin \theta = \cos (\theta + 30°)$.

17. $\cos \theta = \sin (\theta + 30°)$.

18. $\sec 5\theta = \csc 4\theta$.

19. $\sin (3\theta - 10°) = \cos (\theta - 20°)$.

20. $\cot 2\theta = \tan (-\tfrac{3}{5}\theta)$.

21. $\csc (7\theta - 15°) = \sec (\theta - 25°)$.

22. $\cos \theta = \sin (\theta + 85°)$.

23. $\sec (4\theta + 42°) = \csc 4\theta$.

24. $\tan 6\theta = \cot 0°$.

25. $\cos 0 = \sin 90°$.

26. $\sin \dfrac{10°}{\theta} = \cos 80°$.

27. $\cot \dfrac{60°}{\theta} = \tan 78°$.

28. $\sec \left(\dfrac{3°}{\theta} + 20°\right) = \csc 30°$.

29. $\tan 89° = \cot \dfrac{1°}{\theta}$.

30. $\sin (\theta + 130°) = \cos (-9\theta + 24°)$.

31. $\cos (\theta - 10°) = \sin (5\theta + 10°)$.

32. $\tan (3\theta - 5°) = \cot (3\theta - 13°)$.

33. $\sec (\theta + 19°) = \csc (19° + \theta)$.

34. $\sin (\theta + 12°) = \cos (12° + \theta)$.

35. $\tan (3\theta - 10°) = \cot (20° - 3\theta)$.

4–10. Tables of Values of the Trigonometric Functions

We have already discussed several methods for obtaining the trigono-metric functions of a given angle. However, all the methods taken up were either inaccurate or restricted to special angles. Consequently, the usual practice is to compute the trigonometric functions for many angles between 0° and 90° by more accurate methods, which are beyond the scope of this book, and to compile these results in the form of a table to which the student may conveniently refer. Table 3 in the appendix is of this type.

This table consists of angles from 0° to 45° listed by tenths of a degree in the left-hand column, reading downward, and of angles from 45° to 90° listed by tenths of a degree in the right-hand column, reading upward. The angles horizontally opposite each other are complementary. For example, the angle 3° on the left has opposite it the angle 87°. In the four columns between are the sines, the cosines, the tangents, and the cotangents of these two angles. The same numbers have been used for the functions of each of the two complementary angles, for, as was shown in a previous section, the functions of an angle are equal to the cofunctions of its comple-ment. For the angles on the left, the trigonometric headings at the top of the table indicate the column in which each function may be found; the labels at the bottom identify the functions for the angles at the right.

In general, the tables of trigonometric functions are used for one of two purposes:

1. *To find the functional value when the angle is known.*
2. *To find the angle when the functional value is known.*

We shall illustrate these procedures separately, by the use of examples.
1. *Finding the functional values of a given angle.*

Example 1.   From the table, obtain cos 21.7°.
Locate 21.7° at the left on page 752, and note that the cosine is found in the fourth column.   We obtain

$$\cos 21.7° = 0.9291.$$

Example 2.   Find tan 63.6°.
Since this angle is greater than 45°, it is found in the right-hand column on page 752, and its tangent, according to the labels at the foot of the page, will be found in the third column, giving

$$\tan 63.6° = 2.014.$$

If the angle is given to the nearest hundredth of a degree, as, for example, in sin 20.27°, the nearest angle to it in the tables, 20.3°, may be used as an approximation.   However, if a more accurate approximation is desired, we may use the process of **interpolation.**

Interpolation, in this case, is the process of finding a function of an acute angle not listed in the table.   We assume that, for small differences in angles, the difference in their trigonometric functions are proportional. This assumption is not absolutely correct, but the subsequent error is negligible except for angles very close to 0° and 90°.

Example 3.   By interpolation, evaluate sin 41.74°.
From the table, the angles between which 41.74° lies have the following sines:

$$0.10° \left[0.04° \left[\begin{array}{l} \sin 41.70° = 0.6652 \\ \sin 41.74° = \\ \sin 41.80° = 0.6665 \end{array} \right] d \right] 0.0013$$

From the differences of the angles and the functions as indicated by the brackets on the left and right, respectively, we note that as the angle increases 0.10° the sine changes by 0.0013.   Therefore, as the angle increases 0.04°, the sine changes by $d = \dfrac{0.04}{0.10} (0.0013) = 0.00052$ units.

Interpolated values are not accurate to more decimal places than values given in the table.   Therefore we shorten d to four places, that is, to $d = 0.0005$, and hence

$$\sin 41.74° = 0.6652 + 0.0005 = 0.6657.$$

To save time, the computation of d can be made by omitting the decimal points, as in $d = \frac{4}{10} \times 13 = 5.2$ or 5 units.   Then the whole number 5 so obtained is added to the fourth decimal place of 0.6652.

Example 4. Evaluate cos 17.23°.
From the table

$$10 \left[3 \left[\begin{array}{l} \cos 17.20° = 0.9553 \\ \cos 17.23° = \\ \cos 17.30° = 0.9548 \end{array} \right] d \right] 5$$

Therefore $d = 0.3 \times 5 = 1.5$.   The question here arises as to whether to make $d = 1$ or 2.   In all such cases, *we shall adopt the convention of rounding off d to the nearest even number.*   In this case then, $d = 2$.   Noting that the cosine decreases from 0.9553 to 0.9548, we must in this case subtract $d = 2$ from 0.9553, obtaining

$$\cos 17.23° = 0.9553 - 0.0002 = 0.9551.$$

When angles are specified with limited accuracy, it is important to know how accurately the trigonometric functions of these angles may be written. Since a precise discussion of this question is beyond the scope of this chapter, we shall state the following convention, which will be followed in this book:

Accuracy in the angle to the nearest		*accuracy in the trigonometric functions to*	
1°		2	*significant figures*
0.1°	*is equivalent to*	3	*significant figures*
0.01°		4	*significant figures*

The reasonableness of this convention may be checked by a study of Table 3 in the appendix.   It should be noticed that this rule makes no statement about the number of significant figures in the specification of the angle.   For example, sin 9° and cos 81° (the angles in each case being known to the nearest degree) should both be written to two significant figures as 0.16.

Exercises

Find the values of the following functions from Table 3 in the appendix; use interpolation when necessary.   Regard each of the given angles as exact, and give the value as accurately as it can be read from Table 3.

1. sin 6.9°.	**2.** tan 0.8°.	**3.** cos 11.3°.
4. cot 28.4°.	**5.** cos 30.0°.	**6.** cot 45.0°.
7. sin 52.3°.	**8.** cos 74.6°.	**9.** tan 68.2°.
10. cot 73.5°.	**11.** sin 62.0°.	**12.** cos 75.8°.
13. sec 55.8°.	**14.** csc 23.7°.	**15.** sin 41.3°.
16. sin 12.17°.	**17.** cos 17.28°.	**18.** tan 45.00°.
19. sin 62.68°.	**20.** cot 74.12°.	**21.** tan 74.12°.
22. cos 32.32°.	**23.** sin 32.32°.	**24.** cos 30.02°.
25. tan 28.19°.	**26.** cos 46.81°.	**27.** sin 63.99°.
28. csc 63.99°.	**29.** sec 16.87°.	**30.** sin 88.63°.

Evaluate the following expressions, giving your results as precisely as the accuracy of the given data will allow. Assume that the numbers are all given in accordance with the conventions about significant figures.

31. $\sin 18.2° + \cos 78.6°$.

32. $2(3.1416)(\sin 17.38°)$.

33. $17.35 \cos 43.92°$.

34. $(\sin 21.83°)(\cos 87.50°) - 0.01468$.

35. $\dfrac{\sin 17.2°}{\sin 34.4°}$.

36. $\dfrac{1.842 \tan 38.68°}{\tan 27.24°}$.

37. $(1.02)^2 - 3.02(\sin 17.18°)$.

38. $24.42 \cos 81.3°$.

39. $81.4 \sin 31.12° + 81.4 \cos 58.88°$.

40. $\dfrac{(\sin 16.85°)(\tan 52.15°)}{16.75(\cos 34.18°)}$.

41. $1835.6 \cos 38.3°$.

42. $3.27 + 4.58 \cos 63°$.

43. $\dfrac{\sin 34.4°}{\sin 17.2°}$.

44. $\dfrac{2.54 \cos 69.82°}{\cos 22.76°}$.

45. $6.0000(3.12)(60.2) \sin 17.32°$.

46. $(37.482 \sin 12.18°)(\cos 38.55°)$.

2. Finding the angle having a given functional value.

Example 5. If $\tan \theta = 3.398$ and θ is in the first quadrant, find θ.

The tangents for angles between 0° and 90° degrees are listed in the second and third columns of the table. Locating 3.398 in the third column in page 751, we know that for this column the label *tan* is at the bottom. Hence the angle is found in the right-hand column and is 73.6°.

If the angle cannot be read directly from the table, interpolation is frequently used. With four-place tables interpolation for more than hundredths of a degree is not warranted.

Example 6. Find θ in the first quadrant if $\cos \theta = 0.4302$.

From the table, the cosines between which 0.4302 lies are

$$10\left[\; x\left[\begin{array}{l} \cos 64.50° = 0.4305 \\ \cos \theta \quad\;\; = 0.4302 \\ \cos 64.60° = 0.4289 \end{array}\right]\!3\;\right]16$$

Noting the differences denoted by the brackets, we find that, as the cosine changes from 0.4305 to 0.4302, the change in the angle will be $x = (\frac{3}{16})(10) = 1.9$. Actually this change is 0.019° or, to the nearest hundredth, 0.02°. Thus

$$\theta = 64.50° + 0.02° = 64.52°.$$

Where the proportional part x in a computation comes out to be half-way between two possible hundredths of a degree, the nearest even hundredth is used for x. To illustrate, suppose in a computation

$$x = \tfrac{6}{8} \cdot 10 = 7.5 \text{ or } 0.075°.$$

Then the nearest even hundredth, namely, $x = 8$ or 0.08°, is used.

Exercises

From Table 3 in the appendix find the positive acute angle θ having the given functional value.   Use interpolation when the given function is not found exactly in the table, and give θ to the nearest hundredth of a degree.

1. $\sin \theta = 0.09932$.
2. $\cos \theta = 0.8396$.
3. $\tan \theta = 0.5117$.
4. $\cot \theta = 1.6709$.
5. $\tan \theta = 1.2708$.
6. $\cos \theta = 0.8829$.
7. $\sin \theta = 0.9361$.
8. $\cot \theta = 0.3076$.
9. $\tan \theta = 9.677$.
10. $\tan \theta = 1.0000$.
11. $\cos \theta = 0.6909$.
12. $\sin \theta = 0.8111$.
13. $\cot \theta = 0.3561$.
14. $\cos \theta = 0.7906$.
15. $\cos \theta = 0.7908$.
16. $\cos \theta = 0.7904$.
17. $\sin \theta = 0.7904$.
18. $\cos \theta = 0.8315$.
19. $\sin \theta = 0.5235$.
20. $\tan \theta = 0.4071$.
21. $\tan \theta = 0.4067$.
22. $\tan \theta = 0.4068$.
23. $\cos \theta = 0.4652$.
24. $\cot \theta = 0.3273$.
25. $\sin \theta = 0.9481$.
26. $\cos \theta = 0.5591$.
27. $\tan \theta = 1.6710$.
28. $\sec \theta = 2.0684$.
29. $\sec \theta = 1.515$.
30. $\csc \theta = 1.515$.
31. $\sin \theta = 1.515$.
32. $\tan \theta = 106.12$.
33. $\cos \theta = 0.00430$.
34. $\sec \theta = 0.8518$.
35. $\sin \theta = 0.9995$.
36. $\csc \theta = 1.3675$.

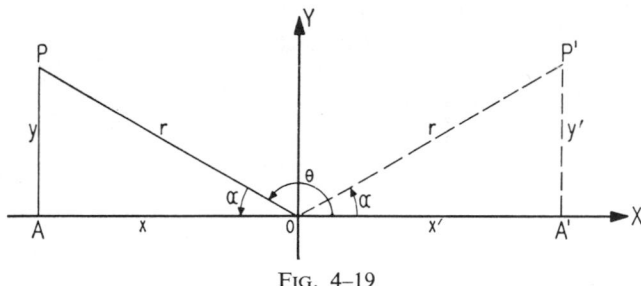

FIG. 4–19

4–11. The Functions of Angles not in the First Quadrant

As we have seen, tables of trigonometric functions are made up only for angles between $0°$ and $90°$.   The next problem that arises is the determination of the trigonometric ratios of angles that do not lie in the first quadrant.   It is convenient to develop a method for obtaining the functions of any angle in terms of the functions of a positive, acute angle which can be found in the table.

Consider any angle θ in standard position (Fig. 4–19).   In the reference triangle ($\triangle OPA$) of θ, the acute angle α with vertex at the origin is called the acute angle associated with θ.   This angle is always taken as positive. Thus, for example, the acute angles associated with $125°$, $217°$, $331°$, and $-47°$ are $55°$, $37°$, $29°$, and $47°$, respectively. It is possible to demonstrate that the trigonometric functions of any angle θ may be found in terms of the functions of this positive, acute angle α.

Consider a second quadrant angle θ, with its reference triangle OPA, and the positive, acute angle α associated with θ (Fig. 4–19). The trigonometric functions of θ may be found by placing α in standard position and constructing its triangle of reference $OP'A'$ with hypotenuse of length r. The two triangles of reference are therefore congruent, the hypotenuse

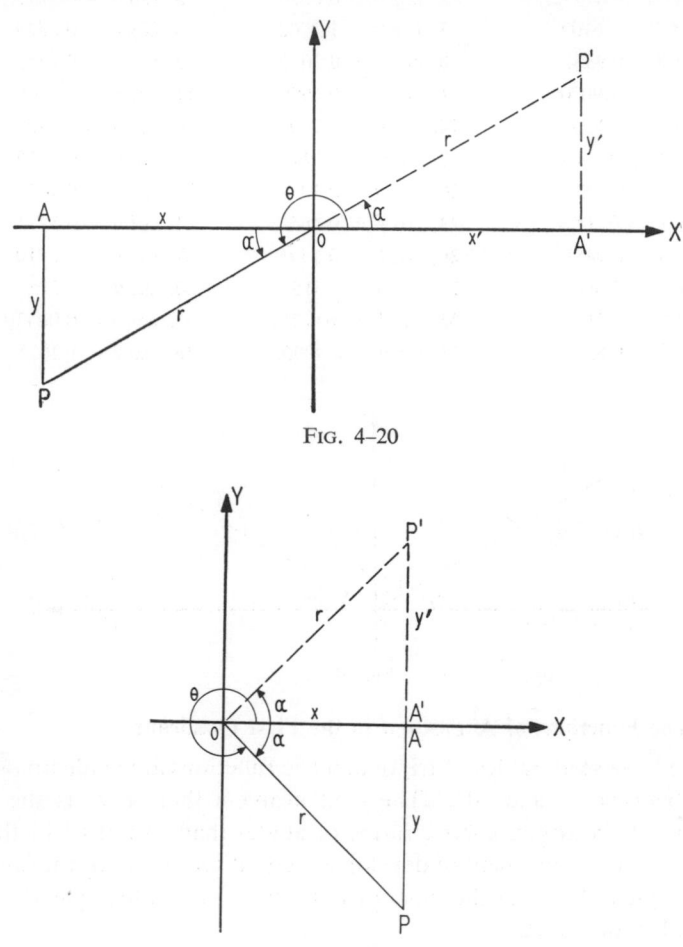

FIG. 4–20

FIG. 4–21

and an acute angle of one being equal to the hypotenuse and an acute angle of the other. Hence the absolute value of each function of θ is equal to the corresponding function of α. For example, since $\sin \theta$ is positive and $\sin \alpha$ is positive, $\sin \theta = \sin \alpha$. Again, since $\cos \theta$ is negative and $\cos \alpha$ is positive, $\cos \theta = -\cos \alpha$. In short, if θ is a second quadrant angle and α is the acute angle associated with θ, then:

θ in Quadrant II

sin θ = sin α	csc θ = csc α
cos θ = −cos α	sec θ = −sec α
tan θ = −tan α	cot θ = −cot α

When θ is a third or fourth quadrant angle (Figs. 4–20, 4–21), the same reasoning is followed, resulting in the relations:

θ in Quadrant III	*θ in Quadrant IV*
sin θ = −sin α	sin θ = −sin α
cos θ = −cos α	cos θ = cos α
tan θ = tan α	tan θ = −tan α
cot θ = cot α	cot θ = −cot α
sec θ = −sec α	sec θ = sec α
csc θ = −csc α	csc θ = −csc α

In every case the function of the given angle θ is equal to the same function of the acute angle α associated with θ, with the proper sign prefixed.

Thus we may state the following rule.

To evaluate the trigonometric function of an angle θ which is not a positive acute angle:

1. *Determine the acute angle α associated with θ, and in the table find the value of the same function of α.*

2. *From the function and quadrant of the given angle θ, determine the sign of the final result.*

3. *The desired trigonometric function of θ is then given by the number found in (1) prefixed by the sign obtained in (2).*

Example 1.   Evaluate sin 173.3°.

Since the sine is positive in the second quadrant, and since the acute angle associated with 173.3° is $\alpha = 180° - 173.3° = 6.7°$, then by the rule,

$$\sin 173.3° = +\sin 6.7°.$$

From the tables, sin 6.7° = 0.11667.   Therefore,

$$\sin 173.3° = 0.11667.$$

Example 2.   Evaluate cos 233.1°.

Since the cosine is negative in the third quadrant, and since $\alpha = 53.1°$ is the acute angle associated with 233.1°, then by the rule,

$$\cos 233.1° = -\cos 53.1°.$$

Since cos 53.1° = 0.6004, cos 233.1° = −0.6004.

Example 3.   Evaluate cot $\dfrac{7\pi}{4}$.

Since $7\pi/4$ radians is a fourth quadrant angle, the cotangent is negative.   Also,

$\alpha = 2\pi - 7\pi/4 = \pi/4$ is the acute angle associated with $7\pi/4$. Therefore, from the rule,

$$\cot \frac{7\pi}{4} = -\cot \frac{\pi}{4} = -1.$$

Example 4. Evaluate sin 590°.
Since 590° is in the third quadrant, and since $\alpha = 50°$,

$$\sin 590° = -\sin 50° = -0.7660.$$

Example 5. Evaluate tan $(-38.2°)$.
The angle $\theta = -38.2°$ is in the fourth quadrant, and its associated angle $\alpha = 38.2°$. Hence tan $(-38.2°) = -\tan 38.2° = -0.7869.$

Exercises

Using Table 3 in the appendix, find the values of the following functions, interpolating when necessary. Regard each of the given angles as exact, and give the value as accurately as it can be read from Table 3.

1. sin 143°.
2. cos 167°.
3. tan 142°.
4. sec 171°.
5. cot $(-119°)$.
6. csc 138°.
7. cos 154.2°.
8. tan $(-247.1°)$.
9. cos 202.4°.
10. sin 200°.
11. csc $(-200°)$.
12. tan 320.6°.
13. cos 275.3°.
14. sin 359.3°.
15. cos 333.3°.
16. sin 420.7°.
17. cos 538.2°.
18. sin 722.6°.
19. cos 52.4°.
20. cos $(-52.4°)$.
21. sin $(-138.5°)$.
22. tan 4331.4°.
23. sin 162.34°.
24. cos 166.66°.
25. tan 166.66°.
26. cot 302.09°.
27. sin $(-198.38°)$.
28. sin $(-536.18°)$.
29. sin 536.18°.
30. csc 536.18°.

Evaluate the following expressions, giving your results as accurately as the accuracy of the given data will allow. Assume that the numbers are all given in accordance with the conventions about significant figures.

31. cos 163.42° · cos 202.8° − sin 163.42° · cos 202.8°.
32. 0.78 · sin 261.92° · cos 261.92°.
33. 3.00 · sin $(-42°)$ · cos $(-42°)$.
34. $\dfrac{6.2 \cos 142.19°}{\cos 22.67°}$.
35. 314.2 tan 458.31°.
36. 1620 sec 410.4°.
37. $E_1 \cos 118.2° - E_2 \sin (-118.2°)$.
38. sin x cos 273.72° + sin 273.72°.
39. 29.2 cos $\frac{85}{33}$.
40. $\dfrac{\sin \frac{9}{25} - 1.82 \cos \frac{9}{25}}{\cot \frac{25}{9}}$.
41. $6\pi \sin \dfrac{12\pi}{7}$.

4–12. Angles for a given Trigonometric Function

In the preceding section it was found that the trigonometric functions of angles not in the first quadrant have the same absolute value as the same functions of their associated acute angles. We now investigate the inverse procedure, that is, the determination of all angles, particularly those between 0° and 360°, having a given trigonometric function.

If α is a given positive acute angle, then in each quadrant there is one and only one angle between 0° and 360° for which α is the associated acute angle. For example, $\alpha = 35°$ is the acute angle associated with 35° (itself), 145°, 215°, and 325° (Fig. 4–22). Since by the previous section

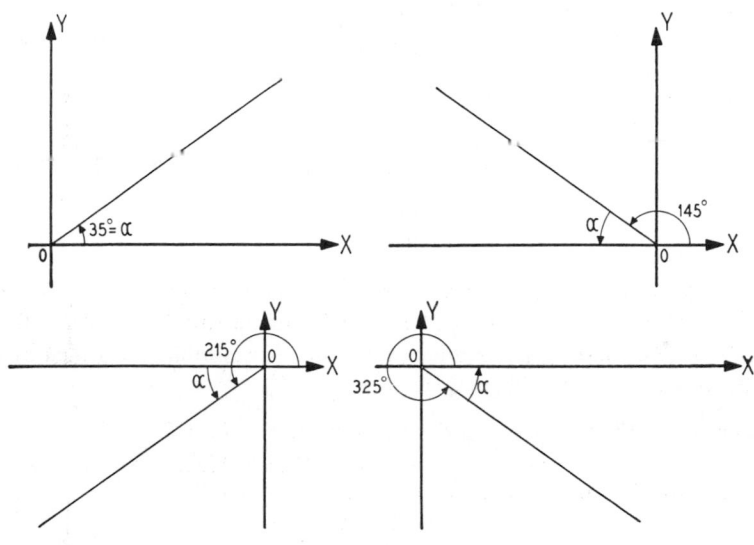

FIG. 4–22

the trigonometric functions of these angles are equal in absolute value to the same functions of α, it is obvious that, if α is a positive acute angle, then in each quadrant there is one and only one angle between 0° and 360° whose trigonometric functions are equal in absolute value to the corresponding functions of α.

Hence, to find all positive angles between 0° and 360° having a given, signed value for a particular trigonometric function:

1. Find the first quadrant angle α having the given functional value without regard to sign.

2. From the sign of the given value and the given function determine the quadrants in which the angles may lie.

3. In each possible quadrant, find the angle for which α is the associated acute angle.

Example 1. Find all angles θ between 0° and 360° such that $\cos \theta = -0.4540$.

The angle in the first quadrant having 0.4540 as its cosine value is 63°. Since the cosine is negative only in the second and third quadrants, we must find the one angle in each of these quadrants having 63° as the associated acute angle. The desired angles are

$$180° - 63° = 117°$$

and

$$180° + 63° = 243°.$$

If it is desired to find other angles, negative or greater than 360°, having cosines of −0.4540, it is merely necessary to find other angles having the same terminal sides as 117° and 243°.

Example 2. Find four other angles satisfying the condition that $\cos \theta = -0.4540$ of example 1.

Four other angles having the same terminal sides as 117° and 243° are:

(a) $117° + 360° = 477°.$

(b) $243° + 360° = 603°.$

(c) $117° - 360° = -243°.$

(d) $243° - 360° = -117°.$

As an abbreviation for the expression "θ is an angle between 0° and 360°" we often write $0° < \theta < 360°$. If θ can be equal to 0° as well, we write $0° \leq \theta < 360°$. The student may infer the similar meanings of the expressions $0° < \theta \leq 360°$ and $0° \leq \theta \leq 360°$.

Example 3. Find $0° \leq \theta < 360°$ such that $\sin \theta = 0.4325$.

The first quadrant angle having a sine of 0.4325 is 25.62°. Since the sine is positive in the first and second quadrants, the values of θ are

$$\theta = 25.62, \quad \theta = 180 - 25.62° = 154.38°.$$

Exercises

In each case find all angles θ, where $0° \leq \theta < 360°$, having the given functional value. Interpolate where necessary, giving angles to the nearest hundredth of a degree.

1. $\sin \theta = 0.5563.$ **2.** $\cos \theta = 0.7108.$ **3.** $\tan \theta = 2.793.$

4. $\sin \theta = -0.5563.$ **5.** $\cos \theta = -0.7108.$ **6.** $\tan \theta = -2.793.$

7. $\cot \theta = -0.3096.$ **8.** $\sin \theta = -\dfrac{\sqrt{3}}{2}.$ **9.** $\cos \theta = \dfrac{\sqrt{3}}{2}.$

10. $\cos \theta = 0.0000.$ **11.** $\sin \theta = \pm 1.0000.$ **12.** $\tan \theta = \pm 1.746.$

13. $\sin \theta = \pm 0.5344.$ **14.** $\cos \theta = 0.8784.$ **15.** $\sin \theta = 0.8784.$

16. $\tan \theta = -2.650.$ **17.** $\cot \theta = -2.546.$ **18.** $\cos \theta = \pm 0.7412.$

19. $\sin \theta = \pm 0.1832.$ **20.** $\sec \theta = -2.6857.$ **21.** $\csc \theta = \pm 0.3620.$

22. $\cos \theta = -0.1270.$ **23.** $\sin \theta = 142.64.$ **24.** $\tan \theta = 142.64.$

In each case find six angles having the given functional value.

25. $\sin = 0.5195.$ **26.** $\cos \theta = -0.9444.$ **27.** $\tan \theta = 0.9131.$

28. $\sin \theta = -0.9393.$ **29.** $\cot \theta = -0.3769.$ **30.** $\cos \theta = -0.9502.$

Find all angles θ that satisfy the given condition and for which $0° \leq \theta < 360°$.   Give your result as accurately as the given data will allow.

31. $\sin \theta = \dfrac{4.62 \sin 48.2°}{10.72}$.

Since

$$\sin 48.2° = 0.7455,$$

$$\sin \theta = \frac{4.62 \times 0.7455}{10.72} = 0.321.$$

Thus

$$\theta = 18.7° \text{ or } 161.3°.$$

32. $\sin \theta = \dfrac{5.24 \sin 37.48°}{31.32}$.

33. $\tan \theta = \dfrac{2.731}{7.618} \tan 71.42°$.

34. $\cos \theta = \dfrac{(2.54)^2 + (1.02)^2 - (0.76)}{(2.000)(8.1)^2(3.14)}$.

35. $\sin \theta = \dfrac{6.31 \sin 76.53°}{1.872}$.

36. $\cot \theta = \dfrac{1.000}{\tan 41.85°}$.

37. $\cot \theta = \dfrac{12.46}{86.34 \tan 14.35°}$.

38. $\tan \theta = \dfrac{86.34}{12.46} \tan 14.35°$.

39. $13.47 \cos \theta = \dfrac{8.16}{13.9} \sin 42.3°$.

40. $\dfrac{\tan \theta}{6.37} = 3.92 \dfrac{\sin 41.3°}{\cos 78.88°}$.

41. $4.12 \sec \theta = \dfrac{31.28}{8.02} \csc 40.12°$.

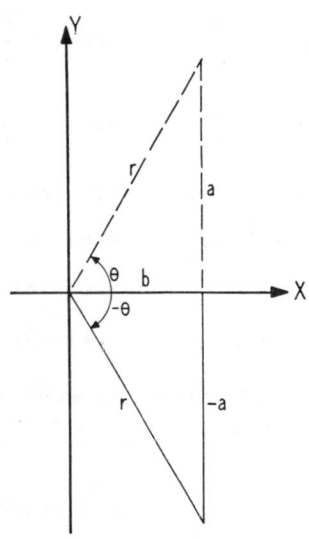

Fig. 4–23

4–13. Functions of Negative Angles

In Sec. 4–11 we discussed a general method for evaluating the trigonometric functions of all angles, including those that are negative.   In this section we shall develop a set of formulas which permit a slightly different approach to the same problem when the angles involved are negative.

Let $(-\theta)$ be a given negative angle whose terminal side lies in the fourth quadrant. Drawing a triangle of reference (Fig. 4–23) with sides, r, $-a$, b, where r, a, b are positive, we find

$$\sin(-\theta) = -\frac{a}{r}, \qquad \csc(-\theta) = -\frac{r}{a};$$

$$\cos(-\theta) = \frac{b}{r}, \qquad \sec(-\theta) = \frac{r}{b}; \qquad (1)$$

$$\tan(-\theta) = -\frac{a}{b}, \qquad \cot(-\theta) = -\frac{b}{a}.$$

Now, since $(-\theta)$ is negative, θ is positive. If θ is placed in standard position and a length r is measured on its terminal side, then the triangle of reference for θ is congruent to the triangle for $(-\theta)$, and thus the sides of the new triangle have the lengths given in the figure. Hence

$$\sin\theta = \frac{a}{r}, \qquad \csc\theta = \frac{r}{a};$$

$$\cos\theta = \frac{b}{r}, \qquad \sec\theta = \frac{r}{b}; \qquad (2)$$

$$\tan\theta = \frac{a}{b}, \qquad \cot\theta = \frac{b}{a}.$$

Equating the results from equations 1 and 2, we obtain

$$\mathbf{\sin(-\theta) = -\sin\theta,} \qquad \mathbf{\csc(-\theta) = -\csc\theta,}$$

$$\mathbf{\cos(-\theta) = \cos\theta,} \qquad \mathbf{\sec(-\theta) = \sec\theta,} \qquad (3)$$

$$\mathbf{\tan(-\theta) = -\tan\theta,} \qquad \mathbf{\cot(-\theta) = -\cot\theta.}$$

Equations 3, which we have proved to hold when $(-\theta)$ is a fourth quadrant angle, can be shown in the same manner to hold for $(-\theta)$ in any other quadrant.

Example 1. Evaluate the six trigonometric functions of $-38.2°$.

Using the proper formulas from (3,) we get

$$\sin(-38.2°) = -\sin 38.2° = -0.6184,$$

$$\cos(-38.2°) = \cos 38.2° = 0.7859,$$

$$\tan(-38.2°) = -\tan 38.2° = -0.7869,$$

$\cot(-38.2°) - -\cot 38.2° = -1.2708,$

$\sec(-38.2°) = \sec 38.2° = \dfrac{1}{\cos 38.2°} = \dfrac{1}{0.7859} = 1.272,$

$\csc(-38.2°) = -\csc 38.2° = -\dfrac{1}{\sin 38.2°} = -\dfrac{1}{0.6184} = -1.617.$

Example 2.   Evaluate $\cos(-163.7°)$.

From the proper formula in (3), $\cos(-163.7°) = \cos 163.7°$, and by Sec. 4–11, since the acute angle associated with 163.7° is 16.3°, $\cos 163.7° = -\cos 16.3° = -0.9598$.   Therefore, $\cos(-163.7°) = -0.9598$.

Exercises

Evaluate the following functions by the methods of this section.   Regard each of the given angles as exact, and give the value of the function as accurately as it can be read from Table 3.

1. $\sin(-26.15°)$.

2. $\cos(-26.15°)$.

3. $\tan(-26.15°)$.

4. $\cot(-153.85)°$.

5. $\sec(-12.3°)$.

6. $\csc(-90.0°)$.

7. $\sin(-538.6°)$.

8. $\cos(294.5°)$.

9. $\tan(-225°)$.

10. $\cot(-67.67°)$.

11. $\sin(-125.62°)$.

12. $\tan\left(-\dfrac{\pi}{4}\right)$.

13. $\sin\left(-\dfrac{\pi}{2}\right)$.

14. $\cos(-\pi)$.

15. $\cot\left(-\dfrac{\pi}{2}\right)$.

Prove the following:

16. $I_1 \sin\theta + I_1 \sin(-\theta) + \frac{3}{2}I_1 = \frac{3}{2}I_1$.

For example, since $\sin(-\theta) = -\sin\theta$, this expression becomes

$$I_1 \sin\theta - I_1 \sin\theta + \tfrac{3}{2}I_1 = \tfrac{3}{2}I_1.$$

17. $5.62[\sin(-\theta) + j\cos(-\theta)] = 5.62(-\sin\theta + j\cos\theta)$.

18. $5.62(\sin\theta + j\cos\theta) = 5.62[\sin\theta + j\cos(-\theta)]$.

19. $\dfrac{1}{\sqrt{3}}\left[\cos\left(-\dfrac{\theta}{6}\right) + j\sin\left(-\dfrac{\theta}{6}\right)\right] = \dfrac{\sqrt{3}}{3}\left(\cos\dfrac{\theta}{6} - j\sin\dfrac{\theta}{6}\right)$.

20. $3\tan^2\phi = -3\tan(-\phi)\tan\phi$.

21. If $\cos(-\phi) = 0.8028$ and $0° \le \phi < 360°$, then $\phi = 36.6°$ or $323.4°$.

22. If $\sin(-\phi) = 0.8028$ and $0° \le \phi < 360°$, then $\phi = 233.4°$ or $306.6°$.

23. If $\tan(-\theta) = -0.7080$ and $0° \le \theta < 360°$, then $\theta = 35.3°$ or $215.3°$.

24. If $\tan(-\theta) = 0.7080$ and $0° \le \theta < 360°$, then $\theta - 144.7°$ or $324.7°$.

4–14. The Trigonometric Functions for an Acute Angle of a Right Triangle

In discussing the trigonometric functions of one of the acute angles of a right triangle, it is often advantageous to use a modification of the original definitions.

If α is an acute angle of a right triangle, then

$$\sin \alpha = \frac{\text{side opposite } \alpha}{\text{hypotenuse}}, \qquad \csc \alpha = \frac{\text{hypotenuse}}{\text{side opposite } \alpha};$$

$$\cos \alpha = \frac{\text{side adjacent } \alpha}{\text{hypotenuse}}, \qquad \sec \alpha = \frac{\text{hypotenuse}}{\text{side adjacent } \alpha};$$

$$\tan \alpha = \frac{\text{side opposite } \alpha}{\text{side adjacent } \alpha}, \qquad \cot \alpha = \frac{\text{side adjacent } \alpha}{\text{side opposite } \alpha}.$$

These statements are immediately obvious if the right triangle is placed on the coordinate axes with α in standard position and the original definitions (Fig. 4–24) are applied.

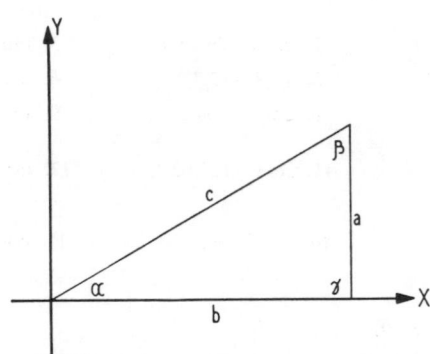

FIG. 4–24

In the following discussions, angles of a right triangle are denoted, as a rule, by α, β, γ, and the sides opposite these angles by a, b, c, respectively. γ is always the right angle.

Exercises

Give the six trigonometric functions for each of the acute angles, α and β, of a right triangle having sides:

1. $a = 4, b = 3, c = 5.$

2. $a = 4, b = 5, c = \sqrt{41}.$

3. $a = 12, b = 16, c = 20.$

4. $a = 15, b = 8, c = 17.$

5. $a = 9, b = 40, c = 41.$

6. $a = 2, b = 2.$

7. $a = 4, b = 7.$

8. $a = 13, b = 17.$

9. $a = 5, c = 11.$

10. $b = 5, c = 11.$

11. $a = 3, c = 3\sqrt{3}.$

12. $b = \sqrt{5}, c = \sqrt{27}.$

13. $b = 9, c = 9.$

14. $a = 28, c = 29.$

Obtain α and β correct to the nearest hundredth of a degree for the right triangle in:

15. Exercise 1.   **16.** Exercise 5.   **17.** Exercise 6.

18. Exercise 8.   **19.** Exercise 10.   **20.** Exercise 12.

Obtain α and β correct to the nearest hundredth of a degree for the right triangle having sides:

21. $a = 5, b = 3$.   **22.** $a = 7, c = 11$.   **23.** $b = 1, c = 1.2$.

24. $a = 4, b = 5$.   **25.** $b = 40, c = 70$.

26. $a = 3, b = 4$ (without finding c).

27. $b = 8, c = 12$ (without finding a).

28. $a = 13, b = 17$ (without finding c).

29. $a = 3.2, c = 81$ (without finding b).

30. $b = 1.8, c = 36$ (without finding a).

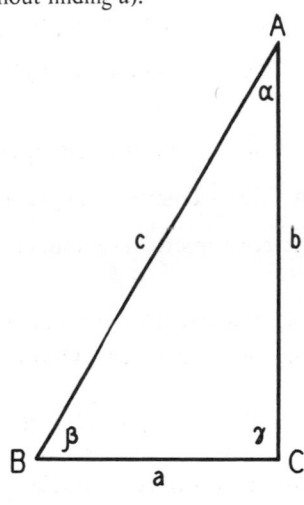

Fɪɢ. 4–25

4–15. The Solution of Right Triangles

In plane geometry it is shown that if two sides, or one side and one acute angle, of a right triangle are given, the triangle can be constructed, and the unknown sides and angles found by measurement. The same results can be obtained much more accurately by means of the modified definitions of the trigonometric functions given in equations 1 of Sec. 4–14. Each of these expressions involves three parts of the triangle. By selecting an expression involving the two known parts and an unknown part which is to be found, an equation is obtained that can be solved for the unknown part.

Briefly, the following rules can be used in solving for an unknown part of a right triangle.

1. *To find the acute angle* α, *knowing the acute angle* β, use the formula

$$\alpha = 90° - \beta.$$

2. *To find an unknown acute angle, knowing two sides but not the other acute angle,* select from equations 1 of Sec. 4–14 the proper relation involving the unknown angle and the two known sides.

3. *To find an unknown side, knowing one side and one acute angle,* from the relations below select the one most easily used.

Unknown side = hypotenuse · sine of angle opposite unknown side,

    = hypotenuse · cosine of angle adjacent to unknown side.

Unknown side = known side · tangent of angle opposite unknown side,

    = known side · cotangent of angle adjacent to unknown side.

Hypotenuse = known side ÷ sine of angle opposite known side,

    = known side ÷ cosine of angle adjacent to known side.

These relations are merely convenient restatements of those in equations 1 of Sec. 4–14.

Example 1.   Given a right triangle *ABC* in which $a = 10$, $\alpha = 30°$, find *c*.

Since the side *a* and α, the angle opposite, are known, the fifth relation in rule 3 is used, giving

$$c = \frac{a}{\sin \alpha} \quad \text{or} \quad c = \frac{10}{\frac{1}{2}} = 20.$$

When all three of the unknown parts of a triangle have been computed from the two known parts, it is desirable to check the accuracy of these results.   This check is easily accomplished as follows:

1. When one side and one angle are given, then the computed angle and the computed two sides are related by a trigonometric formula which may be used as a check.   For example, if *a* and α are given, then we may substitute the computed values of β, *b*, and *c* in the formula $\sin \beta = b/c$ as a check.

2. When two sides are given, the first unknown angle can be computed from a trigonometric relation and the second by subtracting the first from 90°.   The check may then be accomplished by substituting the second angle, the computed side, and a known side into a convenient trigonometric relation.   The example below shows how this may be done.

It is, of course, conceivable that compensating errors could be made in the computed values which would not be detected by the check procedure.

However, this possibility is so slight that if this check procedure yields agreement it may fairly be assumed that the computed values are correct.

Obviously, in solving a right-triangle problem it is advisable to make a sketch and carry out the computations in an orderly fashion. Example 2 shows one way that this may be done.

Example 2. In the right triangle ABC, $a = 7.320$, $c = 10.67$. Find the other parts of the triangle.

(1) *Given*   *To find*   *Sketch*

   $a = 7.320$ $b =$

   $c = 10.67$ $\alpha =$

                           $\beta =$

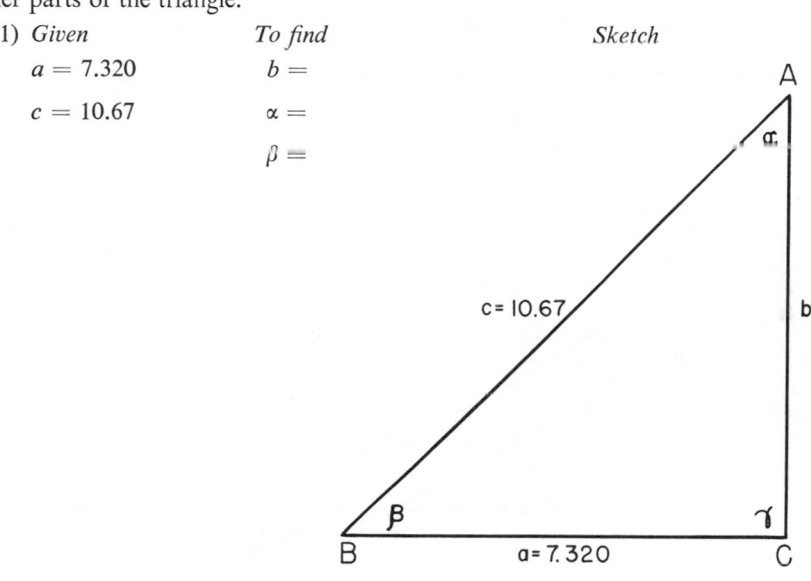

FIG. 4–26

(2) *Formulas:*

   (a) To find α, $\sin \alpha = a/c$.
   (b) To find β, $\beta = 90° - \alpha$.
   (c) To find b, $b = c \cos \alpha$.

(3) *Substitution and Computations:*

   (a) $\sin \alpha = 7.320/10.67 = 0.6860$, $\alpha = 43.32°$.
   (b) $\beta = 90° - 43.32° = 46.68°$.
   (c) $b = 10.67 \times \cos 43.32° = 10.67 \times 0.7276 = 7.763$.

(4) *Check:* Using $b = 7.763$ and $\beta = 46.68°$ (the second angle computed) in the relation $\tan \beta = b/a$, we obtain

$$\tan 46.68° = \frac{7.763}{7.320},$$

$$1.0605 = 1.0605,$$

which shows that the computed results are correct.

When the data given in a right-triangle problem are specified to a limited degree of accuracy, it is important to carry out the computations and present the results in such a manner that the accuracy of these results is consistent with the accuracy of the given data. Thus the conventions about such computations discussed in Chapter 1, and Secs. 4–2, 4–3, and 4–10 of this chapter should be used. Example 2 above derives results to a degree of accuracy consistent with the given data: Four significant figures are stated for the lengths of the given sides; the computed length is given to four significant figures, and the computed angles are given to the nearest hundredth of a degree. Further, the check procedure yields agreement to four significant figures.

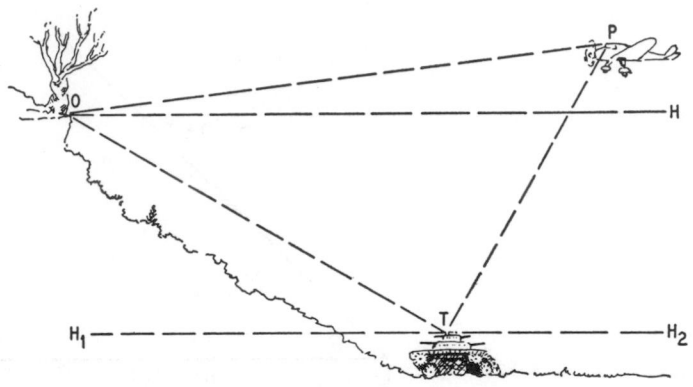

Fig. 4–27

The solution of right triangles has many practical applications, particularly in obtaining lengths of inaccessible distances by means of angles and other lengths that are more easily measured. In preparation for these applications, it will be valuable to discuss some of the terminology that is frequently used.

The acute angle between the line of sight from the observer O to the object P and the horizontal line which is in the same vertical plane as the line of sight is called the angle of **elevation** or the angle of **depression** of P from O, according as P is at a higher or lower elevation (height above sea level) than O. For example, if a man O at the top of a hill (Fig. 4–27) observes an enemy tank T in the valley below an enemy plane P in the sky above, $\angle HOT$ is the angle of depression of T from O, $\angle HOP$ is the angle of elevation of P from O, $\angle H_1TO$ is the angle of elevation of O from T, $\angle H_2TP$ is the angle of elevation of P from T, etc.

In navigation, if an observer at O sights an object W, and ON is the line from O directly north, then angle NOW, always measured in a clockwise direction, is called the **bearing** of W from (at) O. Thus (Fig. 4–28), if from a lighthouse O the keeper spots a submarine S directly southwest, then at O the bearing of S is 225°. Also, at O, the bearings of a whale W and a transport T as shown would be 76.2° and 124.7°, respectively.

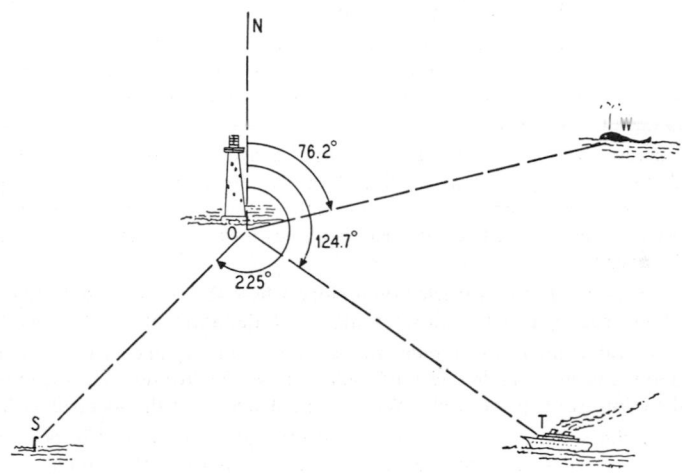

FIG. 4–28

Exercises

Solve each of the following right triangles for the three unknown parts, giving your results as precisely as the degree of accuracy of the given data will allow, and check your results.

1. $a = 9.22$, $c = 17.6$.

2. $b = 16.3$, $\beta = 28.8°$.

3. $a = 317.2$, $b = 214.4$.

4. $b = 4.186$, $\alpha = 78.71°$.

5. $a = 22.24$, $\alpha = 8.60°$.

6. $a = 83$, $b = 27$.

7. $a = 46.11$, $b = 212.4$.

8. $a = 124.2$, $\beta = 65°$.

9. $a = 38.1$, $c = 46.6$.

10. $b = 0.8915$, $\alpha = 43.68°$.

11. $a = 0.41$, $b = 1.34$.

12. $a = 32.81$, $c = 38.212$.

13. $b = 8,622$, $c = 12,424$.

14. $a = 12.012$, $\alpha = 72.86°$.

15. $b = 38.16$, $\beta = 14.32°$.

16. $a = 120.1$, $\beta = 16.32°$.

17. $\alpha = 57.6°$, $\beta = 32.4°$.

18. $\alpha = 63.6°$, $\beta = 26.4°$.

19. $b = 38.6$, $c = 29.4$.

20. $a = 18.2$, $c = 11.4$.

Give your results for the following problems as precisely as the degree of accuracy of the given data will allow.

21. An insurance company estimates that the maximum safe angle that a ladder can make with the ground is 65°. What is the shortest ladder a painter should use to paint at a height of 37 ft?

22. According to the criterion stated in exercise 21, will a fireman be safe in using a 54-ft ladder to reach a window that is 48.5 ft above the ground?

23. The shadow of a telegraph pole measures 76.4 ft when the sun has an angle of elevation of 35.8°. Find the height of the pole.

24. A meteorological balloon is released at a point A and ascends vertically. Shortly after its release an observer at a point B, which (to the nearest foot) is 830 ft from A, records the angle of elevation of the balloon as 28.6°. How high was the balloon at the instant of this observation?

25. A surveyor wishes to measure the width of a river, but the current prevents him from making a crossing. From a point A on the north bank, he sights a point B directly across the river, and then measures off on the north bank a distance AC of 380.0 ft perpendicular to the line AB. He then finds the angle ACB to be 27.46°. How wide is the river?

26. A railroad track is constructed on a slope where the vertical rise is 1.94 ft for a horizontal distance of 100.0 ft. At what angle with the horizontal is the track built?

27. The movable antenna of a radar transmitter in an airplane flying at an altitude, measured to the nearest hundred feet, of 6000 ft above the ground scans the ground 28° in each direction from the vertical. What is the diameter of the scanned surface?

28. At a radio station a vertical radiator 364 ft high is to be held in position on a level plain by guy wires extending from the top of the radiator to the ground. If proper precautions against failure due to wind pressure require that the angle formed by each guy wire and the axis of the structure be at least 42.5°, find the minimum length possible for each guy wire. How far from the base of the radiator must guy wires of the minimum length be secured to the ground?

29. For the situation described in exercise 28, determine whether or not guy wires 328 ft long will safely hold a 240-ft vertical radiator in place.

30. What is the length of the horizontal shadow of the Bunker Hill Monument when the sun has an angle of elevation of 36.4°, if the monument is 221 ft high? By how many feet does the shadow decrease in length if the angle of elevation increases by 2.2°?

31. A draftsman sets the legs of a pair of dividers so that the angle between them is 57.4°. What is the distance between the points if the legs are 5.84 in. long?

32. The Great Pyramid of Cheops at Gizeh is a regular pyramid with a square base 9068.8 in. on a side and a height of 5776 in. What angle does a face of the pyramid make with the base?

33. A chord of a circular piece of boiler plate is 2.62 ft long, and the angle subtended by the chord at the center is 58.5°. What is the radius of the plate?

34. A pendulum 39.4 in. long swings through an angle of 3.42° on each side of its vertical position. Find the horizontal distance between the two extreme positions of the end of the pendulum.

35. If a railroad embankment is 92 ft wide at the base, 56 ft wide at the top, and 14 ft high, and if its two sides have the same slope, find the angle each side makes with the horizontal.

36. The pilot of a helicopter hovering 1220 ft above a landing-field control tower observes that the angle of depression of the top of a water tank 90.0 ft high is 6.0°. If the foot of the water tank is on a level with the landing field, what is the distance from the control tower to the water tank?

37. If the rafters of a roof form the equal legs of an isosceles triangle with a horizontal base, what angle does a rafter 28.60 ft long make with the horizontal if the top end is fastened at a height of 8.50 ft above the bottom end? Find the width of the roof.

38. If a boat travels west from a dock for a distance of 5.64 miles, then turns and travels due north for 2.84 miles, what is the bearing of the dock from the boat?

39. A point source of light is focused on a mirror so that its reflection shows on a screen. The light, mirror, and image are at the vertices of an isosceles triangle, with the mirror at the vertex where the two equal sides of the triangle meet. The distance from the mirror to the light is 5.5 ft, and the distance from the light to the image is 2.8 ft. The angle of incidence of a light beam on a mirror is equal to the angle of reflection. Using this law, find the angle of incidence of the light beam on the mirror.

40. When the moon is 238,000 miles from the earth, the diameter of the moon subtends an angle of 0.53° as seen from the earth. Find the diameter of the moon in miles.

41. An airliner climbs at an angle of 18° while cruising at a speed of 240 mph. How many minutes will it take for the plane to climb 8500 ft?

42. In making a blind landing at the Indianapolis Airport an airplane approached the landing field at an altitude of 1208 ft above the airport. When 4.500 miles from the runway the plane began to descend. If the plane flew in a straight line, what angle did its line of flight make with the horizontal if the plane reached the end of the runway at a point exactly 20 ft above the runway?

43. What is the width, measured from one side to the opposite one, of a machine nut which is a regular hexagon if each side is 0.375 in. long?

44. A sailboat travels due southeast at 12.0 mph. At 2:00 P.M. the bearing of a lighthouse is 116.2°, and at 3:35 P.M. it is 26.2°. How far was the boat from the lighthouse at 2:00 P.M.?

45. From the top of the Tillamook Rock light station off the Oregon coast, an attendant 133 ft above the sea the determines the angle of depression of a tug to be 27.3°. If the tug steams directly away from the light station at a constant speed of 12.0 mph, what will the angle of depression be after 50.0 min? What will the angle of elevation of the light station be to an observer on the tug after a run of 40.0 min?

46. An airport operator sights an airliner flying toward his field. He finds that the angle of elevation of the plane is 5.8° and that the plane is at an altitude, measured to the nearest 100 ft, of 10,000 ft above the level of the field. Exactly 5 min later the angle of elevation is 12.4° and the altitude of the plane remains the same. How far did the plane travel between the two observations? What is its speed in miles per hour? In how many minutes will it be directly overhead?

47. Two airway beacon lights are 24.0 miles apart. While directly above one of the beacons, the pilot observes that the other light has an angle of depression of 5.2°. Exactly 3 min later, after flying at a constant speed and altitude in the direction of the other beacon, the pilot observes that its angle of depression has changed to 10.8°. If the two beacons are on the same level, find the altitude of the airplane in feet and its speed in miles per hour.

48. An observer at *A* is 9260 ft north of an observer at *B*. Simultaneously they observe an airplane directly south. At *A* and *B* the angles of elevation of the airplane are 21.4° and 28.8°, respectively. What is the altitude of the airplane, and how far is it from each observer?

49. Two speedboats set out from a dock at the same time. One boat steers a course bearing 26.3° at 42.0 mph. The other boat steers a course bearing 96.2° at 36.0 mph. How far apart will the boats be after 24.0 min? What will the bearing of the first boat be from the second? (*Hint*: Drop a perpendicular from the final position of one boat to the path of the other boat.)

50. A Coast Guard cutter leaves a buoy, bearing 254.3° at 52.0 mph. Eight minutes later a second cutter leaves the same buoy but bears 278.6° at 48.0 mph. What will the distance between the cutters be half an hour after the second cutter leaves the buoy. (*Hint*: Drop a perpendicular from the final position of one cutter to the path of the other cutter.)

Progress Report

In this chapter we discussed:

1. The degree and radian measurement of angles and the conversion from one unit of measurement to the other.

2. The definitions of the trigonometric functions.

3. The evaluation of the trigonometric functions of angles when one function is known and when the triangle of reference is either a 30°-60° or 45° right triangle.

4. The reduction of the problem of evaluating a trigonometric function of any angle to the problem of evaluating the same function of a corresponding first quadrant angle.

5. The use of tables of trigonometric functions and the application of interpolation to these tables.

6. Finding angles when their trigonometric functions are known.

7. The relations between the functions of complementary angles.

8. The trigonometric functions of negative angles.

9. The solution of right triangles.

10. Conventions about the accuracy of the computations encountered in this chapter, and the importance of stating results to a degree of precision commensurate with the accuracy of the given data.

5 ———————————————————

The Graphs of the
Trigonometric Functions

The graphs of various types of simple functions were studied in Chapter 3. It was shown there how useful graphs can be in revealing properties of functions. In this chapter we shall plot the graphs of the trigonometric functions and study some of the properties of these functions. We shall also study some combinations of trigonometric functions and draw their graphs.

A knowledge of the trigonometric functions and their properties is important in many fields of science and engineering. Besides being indispensable in many fields of pure and applied mathematics, trigonometric functions also arise in the study of many natural phenomena. For example, the electric power supplied by public utilities is produced by rotating machines which produce voltages and currents which can easily be described in terms of sine and cosine functions. The motion of a particle floating on the surface of water waves can be described in terms of trigonometric functions. Mechanical vibrations, such as the motion of a pendulum or the sometimes destructive vibrations of improperly balanced machinery, can also be described in terms of trigonometric functions. The motions of vibrating membranes, strings, reeds, and diaphragms can be resolved into combinations of sine waves superimposed on each other, and these motions produce correspondingly varying sound waves in the surrounding air. The grooves in a phonograph record and the sound track at the edge of a motion-picture film are representations of such superimposed sine waves. Radar, radio, and television stations transmit signal energy in the form of electric waves composed of varying combinations of sine waves.

5–1. Introduction

The six trigonometric ratios were defined in the preceding chapter for every angle. Thus the sine ratio defines a correspondence between angles and numbers as follows. To every angle there corresponds a certain

181

number, termed the sine of the angle. A similar correspondence between angles and numbers is defined by each of the other five trigonometric ratios.

In order to measure an angle, a unit of measurement must be chosen. In Sec. 4–3, two different units were discussed, the degree and the radian, and the relation between them was given by the formula

$$1 \text{ degree} = \frac{\pi}{180} \text{ radians.} \tag{1}$$

Thus, if θ is any number,

$$\theta \text{ degrees} = \frac{\pi}{180} \theta \text{ radians.} \tag{2}$$

When it is important that a given expression indicate the units of angular measure, we make use of the following agreement, which is adopted for the sake of convenience.

Consider an angle that contains x units.

(a) If the unit is the degree, the measure of the angle is written with the degree sign (°), as x°.

(b) If the unit is the radian, the measure of the angle is written without any sign, as x.

Thus relation 1 can be written

$$1° = \frac{\pi}{180} \tag{3}$$

and relation 2 as

$$\theta° = \frac{\pi}{180} \theta. \tag{4}$$

From equation 4, for example, 90° can be given in radians by $90° = \frac{\pi}{180} \cdot 90$ $= \frac{\pi}{2} = 1.5708$.

In trigonometric relations in which the unit of angular measurement is of no immediate interest, as in Chapter 10, the convention given above is ignored.

5–2. The Graph of $y = \sin x$

To each angle, and therefore to the number giving its measure, there corresponds a number which is the sine of the angle. Therefore, the functional notation and the graphical representation given in Chapter 3 can be used. Thus, if $x°$ denotes a variable angle measured in degrees, $y = \sin x°$ is a function of the angle $x°$, and we may write

$$y = f(x°) = \sin x°.$$

The graph of the function $y = \sin x°$ can be drawn (see Fig. 5–1), using the values given in the following table.

$x°$	−180°	−150°	−120°	−90°	−60°	−30°	0°	30°	60°	90°	120°	150°	180°
$\sin x°$	0	−0.50	−0.87	−1.00	−0.87	−0.50	0	0.50	0.87	1.00	0.87	0.50	0

In finding the above table of values for $\sin x°$, the angle $x°$ was measured in degrees. In this chapter, however, we shall use almost exclusively the radian measure of an angle, since this is more convenient for theoretical demonstrations and very useful in practical work.

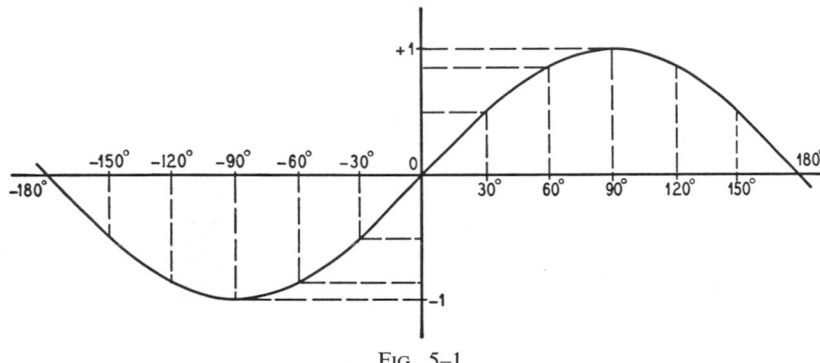

FIG. 5–1

If x is an angle measured in radians, $y = \sin x$ is a function of the angle x and we have, as above,

$$y = f(x) = \sin x.$$

The graph of the function $y = \sin x$ will now be plotted from a table of values in which the angle x is measured in radians. In order to construct such a table of pairs of numbers from which to plot the graph of $y = \sin x$, it is convenient to pick the values of x corresponding to the principal angles. Without using the tables in the appendix, the values of y can be found by the method explained in Sec. 4–8. The table of values from which the curve of Fig. 5–2 was plotted is shown on p. 184.

In Fig. 5–2 the unit on the x axis is the same as the unit on the y axis, whereas in Fig. 5–1 the unit on the y axis is much larger than the unit on the x axis.

For values of x from 2π to 4π the function repeats the cycle of values given in the table, and hence the curve repeats between 2π and 4π the pattern shown in Fig. 5–2 between 0 and 2π. The reader should verify this statement by extending the table of values given on p. 184. The graph

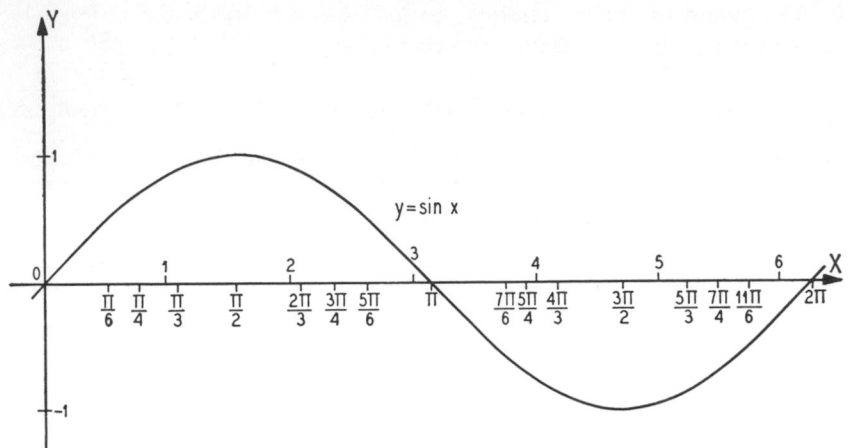

Fig. 5–2

x	y	x	y
0	$0 = 0.000$	$\dfrac{7\pi}{6}$	$-\dfrac{1}{2} = -0.500$
$\dfrac{\pi}{6}$	$\dfrac{1}{2} = 0.500$	$\dfrac{5\pi}{4}$	$-\dfrac{\sqrt{2}}{2} = -0.707$
$\dfrac{\pi}{4}$	$\dfrac{\sqrt{2}}{2} = 0.707$	$\dfrac{4\pi}{3}$	$-\dfrac{\sqrt{3}}{2} = -0.866$
$\dfrac{\pi}{3}$	$\dfrac{\sqrt{3}}{2} = 0.866$	$\dfrac{3\pi}{2}$	$-1 = -1.000$
$\dfrac{\pi}{2}$	$1 = 1.000$	$\dfrac{5\pi}{3}$	$-\dfrac{\sqrt{3}}{2} = -0.866$
$\dfrac{2\pi}{3}$	$\dfrac{\sqrt{3}}{2} = 0.866$	$\dfrac{7\pi}{4}$	$-\dfrac{\sqrt{2}}{2} = -0.707$
$\dfrac{3\pi}{4}$	$\dfrac{\sqrt{2}}{2} = 0.707$	$\dfrac{11\pi}{6}$	$-\dfrac{1}{2} = -0.500$
$\dfrac{5\pi}{6}$	$\dfrac{1}{2} = 0.500$	2π	$0 = 0.000$
π	$0 = 0.000$		

also repeats this pattern between -2π and 0, and continues to repeat it in both directions from the origin as far as we choose to plot the graph. A more extended graph is shown in Fig. 5–3. From these graphs it is seen that the maximum value of the function is $+1$, and the minimum is -1.

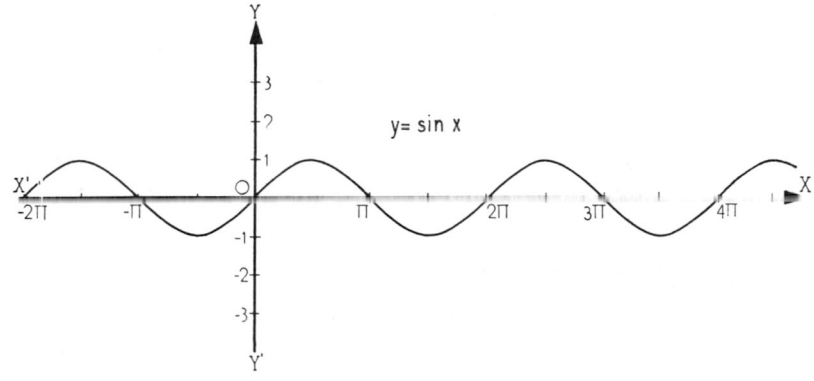

FIG. 5–3

5–3. The Graph of $y = \cos x$

The curve is shown in Fig. 5–4. The reader should construct a table of values and verify, by plotting the curve himself, that the curve shown

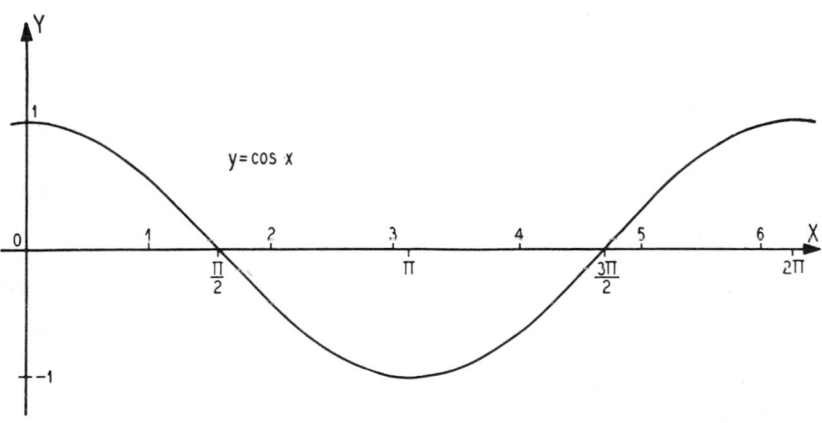

FIG. 5–4

is the correct one. That the pattern of the curve repeats is shown in Fig. 5–5. From these graphs it is seen that the maximum value of the function is $+1$ and the minimum value is -1.

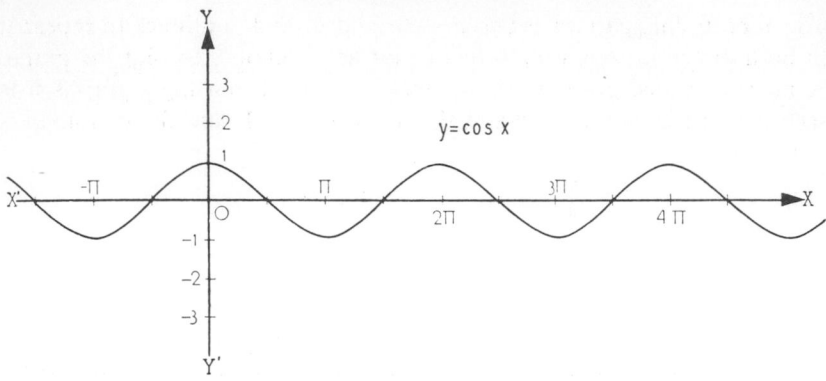

FIG. 5–5

5–4. The Graph of $y = \tan x$

A table of values of y for values of x corresponding to the principal angles is given below. The reader should verify the values given in this table by the method of Sec. 4–8.

TABLE 1

x	$-\dfrac{\pi}{2}$	$-\dfrac{\pi}{3}$	$-\dfrac{\pi}{4}$	$-\dfrac{\pi}{6}$	0	$\dfrac{\pi}{6}$	$\dfrac{\pi}{4}$
y	Not defined	-1.732	-1.000	-0.577	0.000	0.577	1.000

x	$\dfrac{\pi}{3}$	$\dfrac{\pi}{2}$	$\dfrac{2\pi}{3}$	$\dfrac{3\pi}{4}$	$\dfrac{5\pi}{6}$	π	$\dfrac{7\pi}{6}$
y	1.732	Not defined	-1.732	-1.000	-0.577	0.000	0.577

x	$\dfrac{5\pi}{4}$	$\dfrac{4\pi}{3}$	$\dfrac{3\pi}{2}$	$\dfrac{5\pi}{3}$	$\dfrac{7\pi}{4}$	$\dfrac{11\pi}{6}$	2π
y	1.000	1.732	Not defined	-1.732	-1.000	-0.577	0.000

An examination of this table reveals that, if the graph is to be accurately drawn for values of x between $\pi/3$ and $2\pi/3$, the behavior of the function in this interval must be studied in greater detail. In order to investigate the behavior of this function here, Table 2 is constructed from tables of trigonometric functions. By the method of Sec. 4–11, it is possible to obtain Table 3 from Table 2.

From Table 2, as x approaches $\pi/2 = 1.570796$ from the left, through values less than $\pi/2$, $\tan x$ increases very rapidly, getting extremely large when x is close to $\pi/2$. Similarly, from Table 3, as x approaches $\pi/2$ from the right, through values greater than $\pi/2$, $\tan x$ decreases very rapidly, assuming extremely large negative values when x is close to $\pi/2$. Since

| | TABLE 2 | | | TABLE 3 | |
| | Corresponding Angle in | | | Corresponding Angle in | |
x	Degrees	y	x	Degrees	y
1.047	60°	1.732	2.094	120°	−1.732
1.222	70°	2.747	1.920	110°	−2.747
1.309	75°	3.732	1.833	105°	−3.732
1.396	80°	5.671	1.745	100°	−5.671
1.449	83°	8.144	1.693	97°	−8.144
1.484	85°	11.43	1.658	95°	−11.43
1.518	87°	19.08	1.623	93°	−19.08
1.536	88°	28.64	1.606	92°	−28.64
1.553	89°	57.29	1.588	91°	−57.29
1.562	89.5°	114.6	1.580	90.5°	−114.6
1.567	89.8°	286.5	1.574	90.2°	−286.5
1.5706	89.99°	5730.	1.5710	90.01°	−5730.

$\tan \pi/2$ is not defined, we conclude that the curve is not connected over
this point.   Discussions paralleling this one can be given for the behavior
of $\tan x$ near $x = -\pi/2$, and $x = 3\pi/2$.   The curve is plotted in Fig. 5–6.
A dotted line is drawn perpendicular to the x axis at each point where
$\tan x$ is not defined.

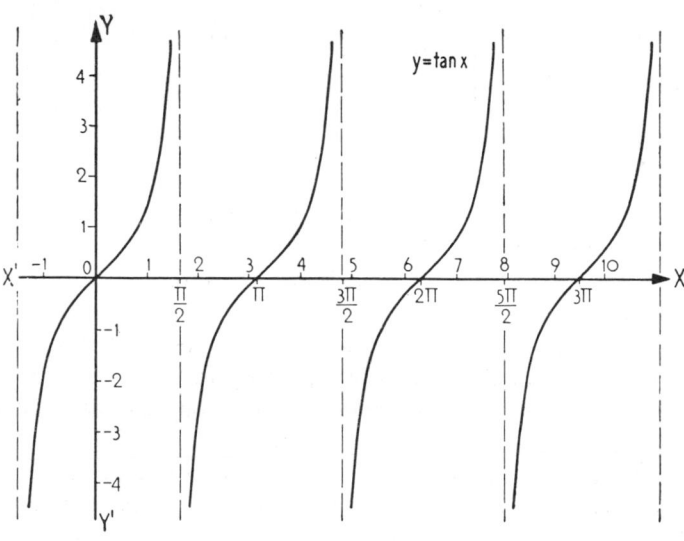

FIG. 5–6

To describe the behavior of the function $y = \tan x$ for values of x
close to $\pi/2$ we say that *the function tan x becomes positively infinite as x
approaches $\pi/2$ from the left, and that it becomes negatively infinite as x
approaches $\pi/2$ from the right.*   It is also sometimes stated by saying that

tan x approaches positive infinity as x approaches $\pi/2$ *from the left.* The phrase **positive infinity** can be replaced by the symbol $+\infty$, and then the second statement given above, although read as before, can be written as *tan x approaches* $+\infty$ *as x approaches* $\pi/2$ *from the left.* Similarly we can write that *tan x approaches* $-\infty$ *as x approaches* $\pi/2$ *from the right.*

An elementary function $f(x)$ may show, in the neighborhood of a point x_0, a behavior similar to that of tan x at $x = \pi/2$, in four ways.

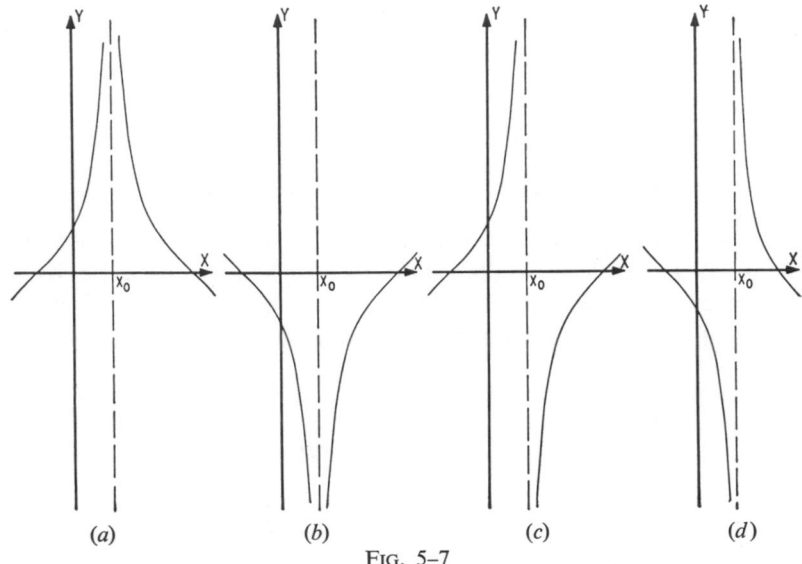

(a)   (b)   (c)   (d)

Fig. 5–7

(a) $f(x)$ may become positively infinite as x approaches x_0 from both the right and the left (Fig. 5–7a).

(b) $f(x)$ may become negatively infinite as x approaches x_0 from both the right and the left (Fig. 5–7b).

(c) $f(x)$ may become positively infinite as x approaches x_0 from the left and negatively infinite as x approaches x_0 from the right (Fig. 5–7c)

(d) $f(x)$ may become positively infinite as x approaches x_0 from the right and negatively infinite as x approaches x_0 from the left (Fig. 5–7d).

In all these cases the absolute value of $f(x)$ becomes positively infinite as x approaches x_0 from both right and left. *In all these cases we say that* $f(x)$ *becomes infinite or approaches infinity at the point* x_0. *We abbreviate this statement by the notation* $f(x_0) = \infty$. Thus we may write

$\tan \dfrac{\pi}{2} = \infty$, and the graphs in the next section will show that $\cot 0 = \infty$,

$\sec \dfrac{\pi}{2} = \infty$, and $\csc 0 = \infty$.

The symbol ∞ is not a number.   It furnishes simply a convenient abbreviation which can be used in describing the behavior of a function.

The four types of approach to infinity described above and illustrated in Fig. 5–7 are frequently shown by elementary functions.   It must not be inferred, however, that these are the only ways that a function may approach infinity: There are a number of other more complicated ways. However, functions exhibiting types of approach to infinity other than those described above are beyond the scope of this book.

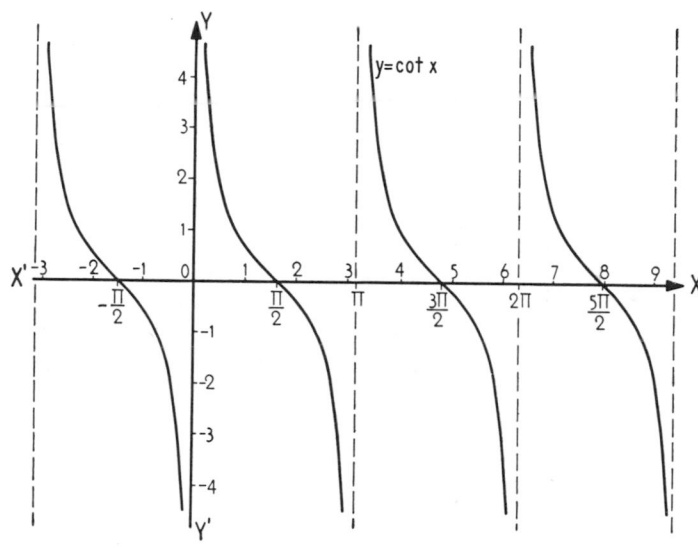

FIG. 5–8

5–5. The Graphs of the Cotangent, Secant, and Cosecant Curves

The graph of $y = \cot x$ is shown in Fig. 5–8.   A discussion similar to that for the tangent curve can be made to establish that the function has the graph shown.   Similar discussions apply for the functions $y = \sec x$ and $y = \csc x$, whose graphs are shown in Fig. 5–9 and Fig. 5–10, respectively.

Exercises

1. Verify that Fig. 5–3 gives the correct graph of $y = \sin x$ by extending the table of values in Sec. 5–2 to include values of x between -2π and 0, and between 2π and 6π. Use the values of x corresponding to the principal angles in setting up the table.   Explain why the pattern of the curve given for x between 0 and 2π is repeated to the right and left.

2. Construct a table of values for $y = \cos x$ for the same values of x given in the table in Sec. 5–2.   Measure off these values on Fig. 5–4 to verify that this curve is correctly plotted.

3. Extend the table of values for $y = \cos x$ obtained in exercise 2 to include values of x between -2π and 0, and between 2π and 6π.   Use the values of x corresponding to

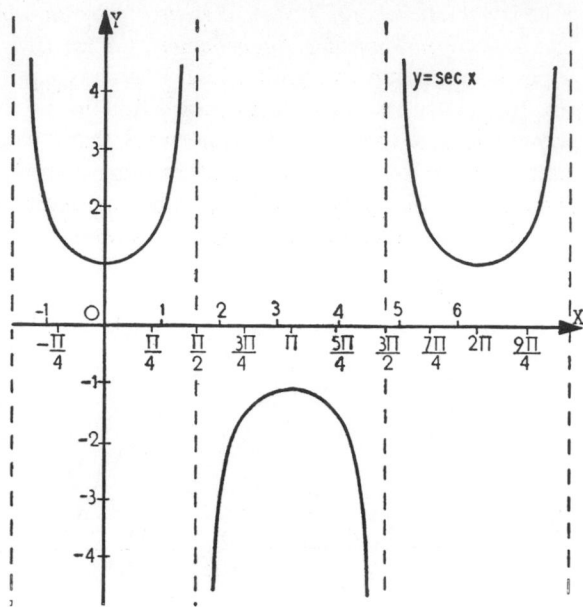

FIG. 5–9

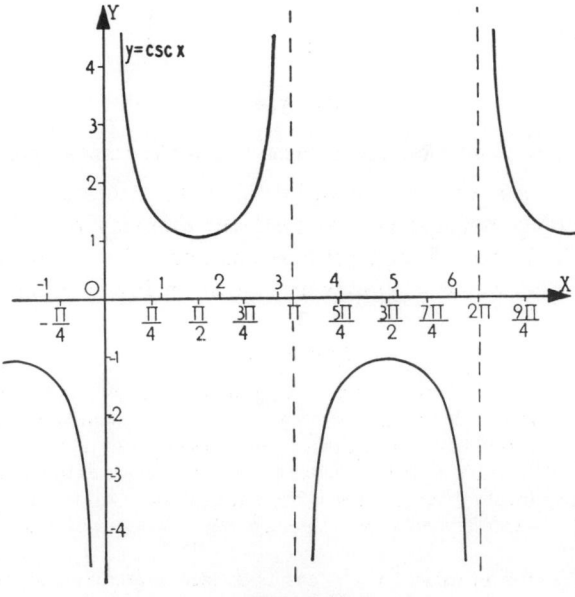

FIG. 5–10

the principal angles in setting up the table.   Use this table to verify that Fig. 5–5 is correctly drawn.   Explain why the pattern of the curve given for x between 0 and 2π is repeated to the right and left.

4. Verify Table 1 of Sec. 5–4 by constructing the angles corresponding to the given values of x and by finding from these figures the values of tan x.

5. Study the behavior of tan x between $4\pi/3$ and $5\pi/3$ by constructing tables like Table 2 and Table 3 of Sec. 5–4.   What conclusions do you draw from these data? Verify that Fig. 5–6 agrees with your conclusions.

6. Prepare a table of values for $y = \cot x$ like Table 1 of Sec. 5–4.   Measure off the values you obtain on Fig. 5–8 to verify that this curve is correctly plotted.

7. Study the behavior of cot x between $-\pi/6$ and $\pi/6$ by constructing tables like Table 2 and Table 3 of Sec. 5–4.   Does Fig. 5–8 bear out your conclusions from these tables?

8. Prepare a table of values for y sec x like Table 1 of Sec. 5–4   Measure off the values you obtain on Fig. 5–9 to verify that this curve is correctly plotted.

9. Study the behavior of sec x between $\pi/3$ and $2\pi/3$ by constructing tables like Table 2 and Table 3 of Sec. 5–4.   Does Fig. 5–9 bear out your conclusions from these tables?

10. Prepare a table of values for $y = \csc x$ like Table 1 of Sec. 5–4.   Measure off the values you obtain on Fig. 5–10 to verify that this curve is correctly plotted.

11. Study the behavior of csc x between $-\pi/6$ and $\pi/6$ by constructing tables like Table 2 and Table 3 of Sec. 5–4.   Does Fig. 5–10 bear out your conclusions from these tables?

Plot the curve of each of the following functions on a full page, using exactly the same units on both axes.   To plot the curves, make a table using the values of x corresponding to the principal angles between $-\pi/2$ and $5\pi/2$.

12. $y = \sin x$.   **13.** $y = \cos x$.   **14.** $y = \tan x$.
15. $y = \cot x$.   **16.** $y = \sec x$.   **17.** $y = \csc x$.

5–6. The Trigonometric Functions as Coordinates of Points

Consider the angle θ in standard position shown in Fig. 5–11, θ being measured in radians.   The terminal side of θ which falls in the first quadrant intersects the circle with unit radius and center at the origin at $P(x, y)$.   This circle is called the **unit circle.**   By dropping the perpendicular PL from P to the x axis we obtain a triangle of reference OPL for the angle θ.   Now, since the circle has unit radius, $r = \sqrt{x^2 + y^2} = 1$, and the arc PM has length θ, then by definition

$$\sin\theta = \frac{y}{r} = \frac{y}{1} = y, \qquad \cos\theta = \frac{x}{r} - \frac{x}{1} = x.$$

Thus the ordinate of P is sin θ, the abscissa of P is cos θ, and we can write P(cos θ, sin θ).   The argument and its results are precisely the same, regardless of the quadrant in which the terminal side of θ falls.   The reader may verify this by repeating it exactly as given above in connection with the triangles OPL in Figs. 5–12, 5–13, and 5–14.

At M, the intersection of the unit circle and the positive x axis, erect a perpendicular.   The intersection of this perpendicular and the terminal side of θ is $Q(x_1, y_1)$.   By a theorem of plane geometry QM is then tangent

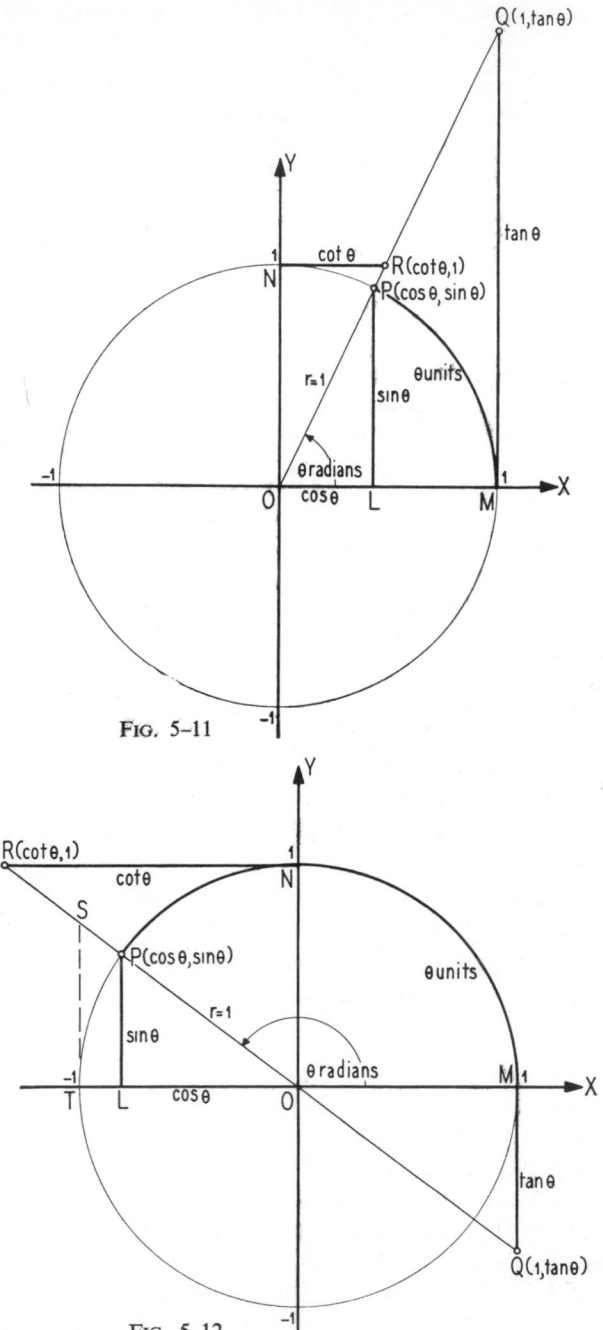

Fig. 5–11

Fig. 5–12

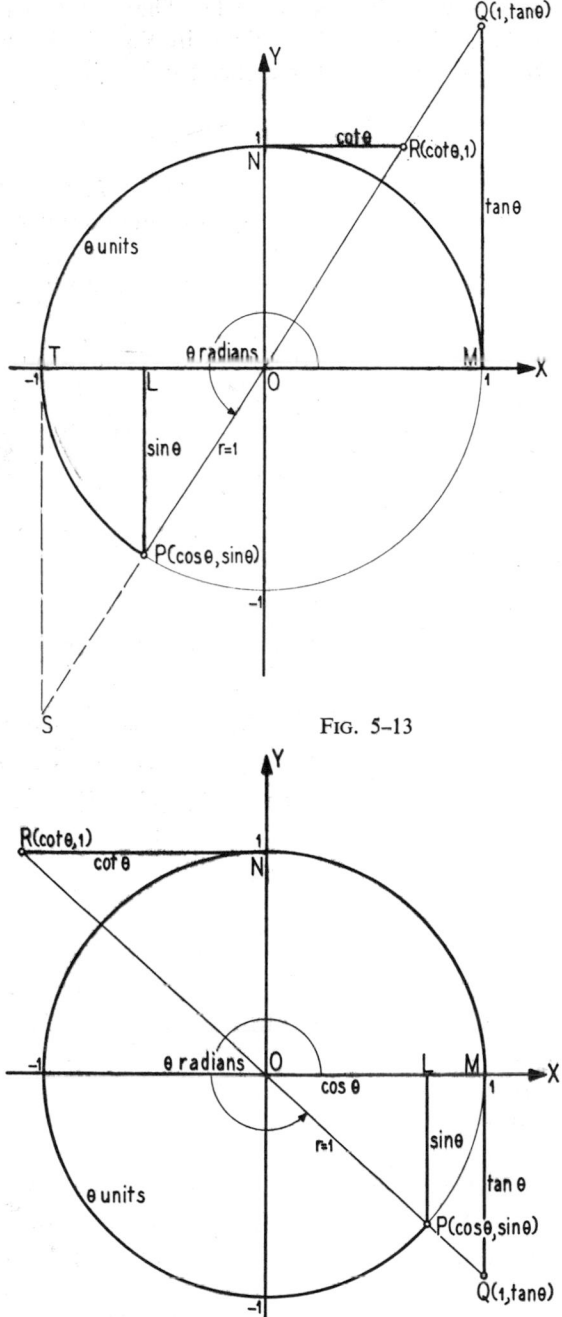

FIG. 5–13

FIG. 5–14

to the circle. We see also that $x_1 = 1$. These statements apply to Figs. 5–11, 5–12, 5–13, and 5–14. Now, in Fig. 5–11 and Fig. 5–14, the triangle OQM is a triangle of reference for θ. Then

$$\tan \theta = \frac{y_1}{x_1} = \frac{y_1}{1} = y_1.$$

In Figs. 5–12 and 5–13, erect a perpendicular at T, the intersection of the unit circle and the negative x axis. Let the intersection of the terminal side of θ and this perpendicular be $S(x_2, y_2)$. We see that $x_2 = -1$, and that the triangle OST forms a triangle of reference for θ. Then

$$\tan \theta = \frac{y_2}{x_2} = \frac{y_2}{-1} = -y_2.$$

Obviously triangle OST is congruent to triangle OQM; consequently the side ST has the same length as the side MQ. Then the ordinate of Q and the ordinate of S have the same absolute value but differ in sign, or $y_1 = -y_2$. Since this is so,

$$\tan \theta = -y_2 = y_1.$$

Hence, in all four cases,

$$\tan \theta = y_1,$$

or *the ordinate of Q is tan θ, and we may write $Q(1, \tan \theta)$.*

At N, the intersection of the positive y axis and the unit circle, erect a perpendicular. The intersection of this perpendicular and the terminal side of θ is $R(x_3, y_3)$. We see that RN is tangent to the circle, and that, for any θ, $y_3 = 1$. It can be shown easily that triangle OPL is similar to triangle ORN. Using triangle OPL as a reference triangle for θ, we get

$$\left| \cot \theta \right| = \left| \frac{x}{y} \right| = \left| \frac{NR}{ON} \right| = \left| \frac{x_3}{1} \right|,$$

or

$$\left| \cot \theta \right| = \left| x_3 \right|.$$

In the first and third quadrants cot θ is positive, and we see that x_3 is positive from Fig. 5–11 and Fig. 5–13; in the second and fourth quadrants cot θ is negative, and we see that x_3 is negative from Fig. 5–12 and Fig. 5–14. Since $\left| \cot \theta \right| = \left| x_3 \right|$, and x_3 and cot θ have always the same sign, we may write

$$\cot \theta = x_3.$$

Thus the abscissa of R is cot θ, and we can write $R(\cot \theta, 1)$.

The reader can show easily that sec θ equals the length of OQ, if we measure OQ as positive when Q and P are on the same side of the origin and as negative when the origin lies between P and Q. It can also be shown that csc θ equals the length OR, if we measure OR as positive when *R* and *P* are on the same side of the origin and as negative when the origin lies between *R* and *P*.

In summary, we have:

sin θ equals the ordinate of the point where the terminal side of θ intersects the unit circle.

cos θ equals the abscissa of the point where the terminal side of θ intersects the unit circle.

tan θ equals the ordinate of the point where the terminal side of θ intersects the line that is tangent to the unit circle at the point where this circle intersects the positive x axis.

cot θ equals the abscissa of the point where the terminal side of θ intersects the line that is tangent to the unit circle at the point where this circle intersects the positive y axis.

sec θ equals the length of the segment from the origin to the point where the terminal side of θ intersects the line that is tangent to the unit circle at the point where this circle intersects the positive x axis, this length being positive if θ terminates on this segment, negative otherwise.

csc θ equals the length of the segment from the origin to the point where the terminal side of θ intersects the line that is tangent to the unit circle at the point where this circle intersects the positive y axis, this length being positive if θ terminates on this segment, negative otherwise.

In this way all the trigonometric functions can be interpreted as line lengths, indicated by the dark lines on the figures. This interpretation is called **the line representation of the trigonometric functions.**

Historically, definitions of the trigonometric functions were given much as we have established them in this section long before the ratio definitions were established. The origin of the names tangent and secant becomes apparent from this interpretation of the functions. It also indicates why the trigonometric functions are called the **circular functions.** This graphical representation of the functions will be of much use in studying their properties.

5–7. Geometrical Method of Plotting the Graphs of the Trigonometric Functions

The information developed in the preceding section can be used very effectively in constructing the graphs of the trigonometric functions. To illustrate, we shall plot the graph of $z = \sin \theta$. Place the x–y coordinate system and the z–θ coordinate system (as shown in Fig. 5–15) so that the x axis and the θ axis are on the same line. Then use the same unit on the axes of both coordinate systems. Draw a unit circle with center at the origin of the x–y system. Now if an angle θ is drawn in standard position on the x y system, $\sin \theta$ is the ordinate of the point where the terminal side

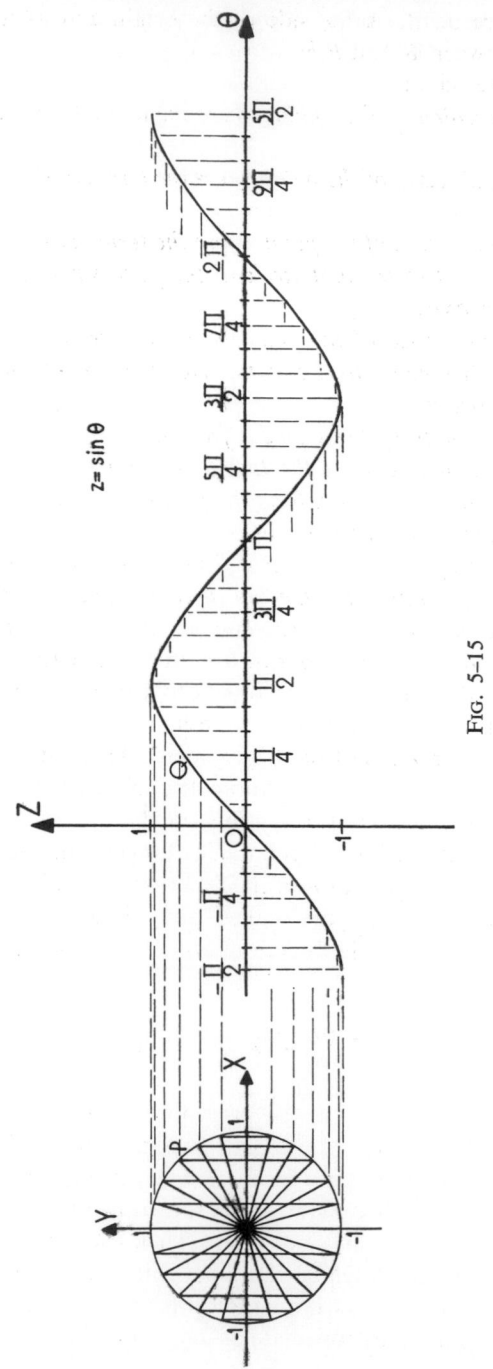

$z = \sin \theta$

Fig. 5–15

of θ intersects the unit circle.   Thus this ordinate value is the distance above the θ axis at which the point of the graph of $z = \sin \theta$ should be found that corresponds to the number θ on the θ axis.   The graph is constructed as shown in Fig. 5–15.

Consider, for example, $\theta = \pi/4$.   The terminal side of this angle intersects the unit circle at P.   Draw a line through P parallel to the horizontal axes.   Draw another line perpendicular to the θ axis at $\theta = \pi/4$.   The intersection Q of these two lines is a point on the graph of $z = \sin \theta$.

This method, with suitable modifications, may be used to plot each of the trigonometric functions.

Exercises

Use the method described in this section to plot the graphs given in the exercises below.   Plot each graph on a full page, using ruler and compasses.

1. Plot the graph of $z = \sin \theta$, as shown in Fig. 5–15.

2. Plot the graph of $z = \cos \theta$.   In order to use the method described, it will be convenient to rotate the x and y axes of Fig. 5–15 through $+90°$.

3. Plot the graph of $z = \tan \theta$.

4. Plot the graph of $z = \cot \theta$.   It will be convenient to use the same arrangement of axes as in exercise 2.

5. Plot the graph of $z = \sec \theta$.   In this case a direct construction cannot be made as in the previous exercises.   However, a compass or divider can be used to transfer the distances from diagram to graph.

6. Plot the graph of $z = \csc \theta$.   The remarks in exercise 5 apply here.

5–8. Variation of the Trigonometric Functions

From the graph of the sine function in Fig. 5–2, it is apparent that, as θ increases from 0 to $\pi/2$, $\sin \theta$ increases from 0 to 1.   As θ increases from $\pi/2$ to π, $\sin \theta$ decreases from 1 to 0.   As θ increases from π to $3\pi/2$, $\sin \theta$ decreases from 0 to -1.   As θ increases from $3\pi/2$ to 2π, $\sin \theta$ increases from -1 to 0.   These facts should be carefully verified also from figures like those of Figs. 5–11, 5–12, 5–13, and 5–14.   Figure 5–15 can also be used in this connection.   Similar studies will verify the information in the following table.

	From	To	From	To	From	To	From	To
θ	0	$\dfrac{\pi}{2}$	$\dfrac{\pi}{2}$	π	π	$\dfrac{3\pi}{2}$	$\dfrac{3\pi}{2}$	2π
$\sin \theta$	0	1	1	0	0	-1	-1	0
$\cos \theta$	1	0	0	-1	-1	0	0	1
$\tan \theta$	0	$+\infty$	$-\infty$	0	0	$+\infty$	$-\infty$	0
$\cot \theta$	$+\infty$	0	0	$-\infty$	$+\infty$	0	0	$-\infty$
$\sec \theta$	1	$+\infty$	$-\infty$	-1	-1	$-\infty$	$+\infty$	1
$\csc \theta$	$+\infty$	1	1	$+\infty$	$-\infty$	-1	-1	$-\infty$

Exercises

1. Make a table of the variations of the trigonometric functions like the table of Sec. 5–8 as θ ranges from 2π to 4π. Compare this table with the table in Sec. 5–8. Explain the similarities you notice.

2. Make a table as directed in exercise 1 as θ ranges from -2π to 0, and discuss as above. Explain the similarity of the tables arrived at in exercises 1 and 2.

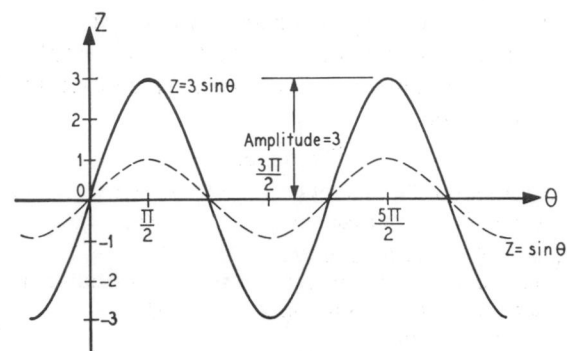

Fig. 5–16

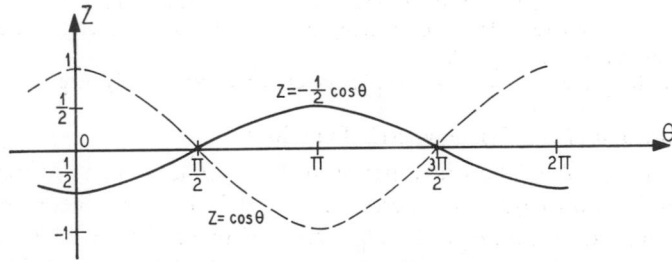

Fig. 5–17

5–9. Amplitude of the Sine and Cosine

This chapter has indicated that the largest value assumed by $\sin \theta$ and $\cos \theta$ is $+1$, and that the smallest value assumed by these functions is -1. Thus the maximum absolute value of these functions is $+1$.

Consider the function

$$z = A \sin \theta, \tag{1}$$

where A is a constant. By this relation the values of $\sin \theta$ are multiplied by A, and hence the graph of equation 1 is obtained by plotting the usual sine curve with the ordinates multiplied by A. Since the maximum absolute value of $\sin \theta$ is 1, the maximum absolute value of $A \sin \theta$, called its **amplitude** or **peak value,** is $|A|$. In like manner the amplitude of $A \cos \theta$ is also $|A|$.

The maximum absolute value of $z = A \sin \theta$ *or* $z = A \cos \theta$, *called the* amplitude, *is* $|A|$.

Since the other trigonometric functions become both positively and negatively infinite, it serves no useful purpose to define an *amplitude* for these functions.

The graph of $z = A \sin \theta$ or $z = A \cos \theta$ can be drawn by simply sketching the usual sine or cosine curve with the ordinates multiplied by A. Since the general shapes of these curves are well known, the sketch can be made very easily without plotting any points. Care should be taken, however, that the curves cross the horizontal axis in the proper places. Similar remarks apply to the other functions.

Example 1. Sketch the graph of $z = 3 \sin \theta$.

The amplitude of this function is 3, and the ordinates of its graph are 3 times those of the usual sine curve. Hence we can sketch the curve by drawing a sine curve, making it 3 times as tall as usual. The graph is shown in Fig. 5–16.

Example 2. Sketch the graph of $z = -\frac{1}{2} \cos \theta$.

The amplitude of this function is $\frac{1}{2}$. Since the minus sign is present, the cosine curve is turned upside down, with its ordinates halved. The graph is shown in Fig. 5–17.

Exercises

State the amplitude of each function, if it has one. Sketch the graph of each function, labeling the axes carefully, especially those points at which the curve crosses either axis.

1. $z = 4 \sin \theta$.	**2.** $z = -4 \sin \theta$.	**3.** $z = 4 \cos \theta$.
4. $z = -3 \cos \theta$.	**5.** $z = 6 \sin \theta$.	**6.** $z = -2 \cos \theta$.
7. $z = \frac{1}{2} \cos \theta$.	**8.** $z = -\frac{2}{3} \sin \theta$.	**9.** $z = 12 \cos \theta$.
10. $z = 3 \sin \theta$.	**11.** $z = -5 \cos \theta$.	**12.** $z = 7 \sin \theta$.
13. $z = 2 \tan \theta$.	**14.** $z = -2 \tan \theta$.	**15.** $z = 3 \cot \theta$.
16. $z = 2 \sec \theta$.	**17.** $z = 3 \sec \theta$.	**18.** $z = 2 \csc \theta$.
19. $z = \frac{4}{5} \sec \theta$.	**20.** $z = -2 \cot \theta$.	**21.** $z = 4 \tan \theta$.
22. $z = -6 \csc \theta$.	**23.** $z = 9 \tan \theta$.	**24.** $z = -7 \cot \theta$.

5–10. The Periods of the Trigonometric Functions

We have already noticed that the pattern of certain parts of the graphs of the trigonometric functions is repeated in other sections of the graph. This section will study this phenomenon more closely. The discussion will be based on the following definition.

If $f(x)$ *is such that* $f(x + P) = f(x)$ *for all values of* x, *then* $f(x)$ *is said to be periodic and have a period* P. *The smallest positive period is called the* **fundamental period.**

Figure 5–18 shows that θ and $\theta + 2\pi$ have the same terminal side. This terminal side cuts the unit circle at Q, and the ordinate of Q is then sin θ and also sin $(\theta + 2\pi)$. Hence sin θ = sin $(\theta + 2\pi)$, and, since Fig. 5–18 gives us representatives of all possible positions of the terminal side, this equality is true for all values of θ. *Hence sin θ is periodic with period 2π.*

It can also be shown that sin θ has period -2π, 4π, 6π, and in general any integral multiple of 2π. However 2π is the smallest positive number that is a period of sin θ, and hence 2π *is the fundamental period of sin θ.*

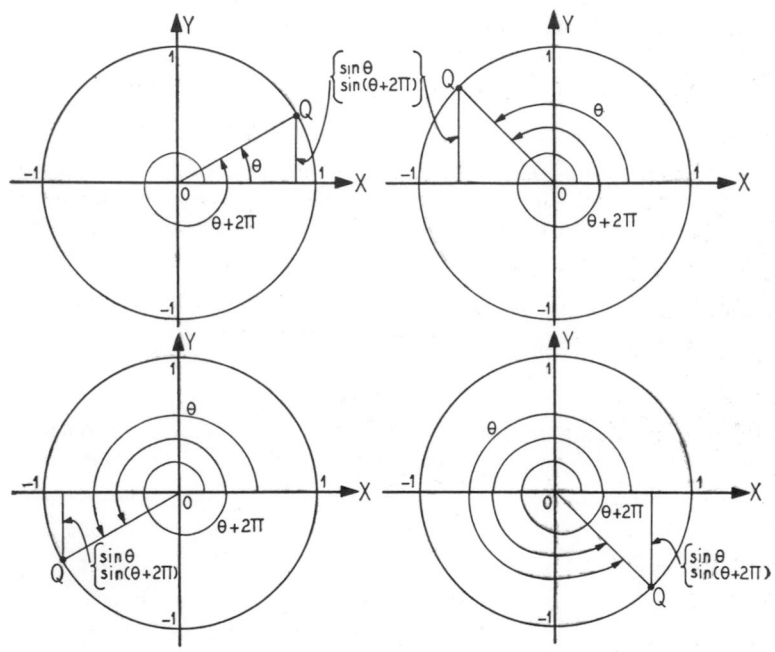

FIG. 5–18

Similar arguments will establish the information given in the table below.

Function	sin θ	cos θ	tan θ	cot θ	sec θ	csc θ
Fundamental period	2π	2π	π	π	2π	2π

Since we are most often concerned with the fundamental period of a trigonometric function, the adjective *fundamental* is often omitted, the context indicating the proper meaning to be attached to the term "period."

Consider the curve $z = $ sin θ in Fig. 5–19. Since the equation sin $(\theta + 2\pi) = $ sin θ holds for all values of θ, the ordinate at θ on the graph is the same as the ordinate at $(\theta + 2\pi)$. In this way we see that the curve between any angle α and $\alpha + 2\pi$ is repeated between $(\alpha + 2\pi)$ and $(\alpha + 4\pi)$.

Thus any section of horizontal length 2π can be used as a pattern for the rest of the curve. The period of the other trigonometric functions may also be interpreted in this way. The reader should construct figures for the other functions like Fig. 5–19, using the fundamental periods.

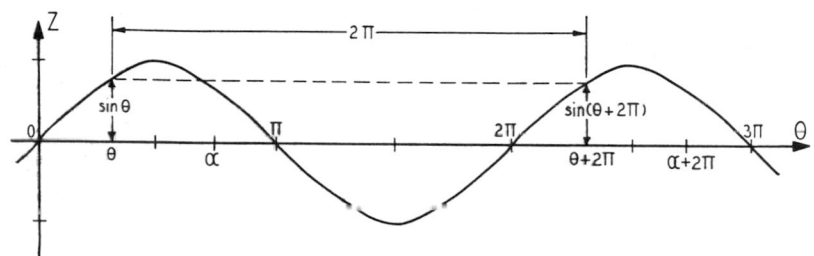

Fig. 5–19

Exercises

By using the line representation of the trigonometric functions as in Fig. 5–18, show that the following functions have the periods given.

1. $\sin \theta$, 2π.	2. $\cos \theta$, 2π.	3. $\tan \theta$, π.	4. $\cot \theta$, π.
5. $\sec \theta$, 2π.	6. $\csc \theta$, 2π.	7. $\sin \theta$, 4π.	8. $\cos \theta$, 4π.
9. $\tan \theta$, 2π.	10. $\cot \theta$, 2π.	11. $\sec \theta$, 4π.	12. $\csc \theta$, 4π.
13. $\sin \theta$, 6π.	14. $\cos \theta$, -2π.	15. $\sin \theta$, -2π.	16. $\tan \theta$, $-\pi$.
17. $\cot \theta$, $-\pi$.	18. $\sec \theta$, -2π.	19. $\csc \theta$, -2π.	20. $\sin \theta$, 8π.
21. $\cos \theta$, 12π.			

In the following exercises, n is any integer.

22. $\sin \theta$, $2n\pi$.	23. $\cos \theta$, $2n\pi$.	24. $\tan \theta$, $n\pi$.	25. $\cot \theta$, $n\pi$.
26. $\sec \theta$, $2n\pi$.	27. $\csc \theta$, $2n\pi$.		

5–11. The Period of sin ωx

Consider the function $\sin \omega x$ where ω is a constant greater than zero. Since the sine function has period 2π,

$$\sin \omega x = \sin (\omega x + 2\pi), \tag{1}$$

or

$$\sin \omega x = \sin \omega \left(x + \frac{2\pi}{\omega} \right). \tag{2}$$

Equation 2 may be interpreted, if we let $f(x) = \sin \omega x$, as meaning that

$$f(x) = f \left(x + \frac{2\pi}{\omega} \right).$$

Thus sin ωx has a period of 2π/ω.   Furthermore, it can be shown that this is the fundamental period of the function.

In like manner the information in the following table can be established:

Function	sin ωx	cos ωx	tan ωx	cot ωx	sec ωx	csc ωx
Fundamental period	$\dfrac{2\pi}{\omega}$	$\dfrac{2\pi}{\omega}$	$\dfrac{\pi}{\omega}$	$\dfrac{\pi}{\omega}$	$\dfrac{2\pi}{\omega}$	$\dfrac{2\pi}{\omega}$

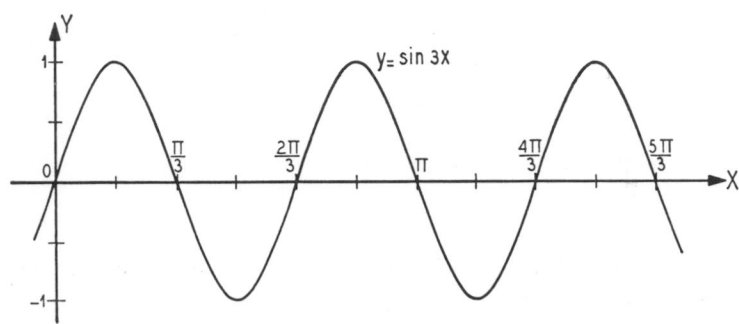

Fig. 5–20

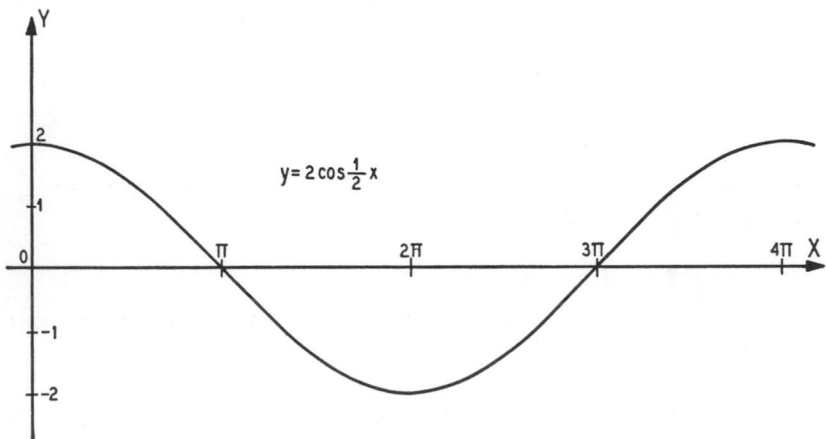

Fig. 5–21

Now sin x and sin ωx run through the same cycle of values, but sin x completes a cycle in a horizontal distance of 2π, and sin ωx in a horizontal distance of $2\pi/\omega$.   Since both cross the y axis at the origin, the graph of the second function may be sketched very easily from what we already know about the first.   The following examples will illustrate.

Example 1. Sketch the curve $y = \sin 3x$.

Here $\omega = 3$, and the function has period $2\pi/3$. This means that one cycle of the sine curve appears in a horizontal length of $2\pi/3$. The y axis is crossed at the origin. Thus to plot the graph we can lay off on the x axis multiples of the length $2\pi/3$. Each of these segments can be divided in four equal parts by points corresponding to where the curve crosses the axis or reaches a maximum or minimum. The curve then can be sketched easily, as shown in Fig. 5–20.

Example 2. Sketch the curve $y = 2 \cos \frac{1}{2}x$.

Since $\omega = \frac{1}{2}$, the function has period $\dfrac{2\pi}{\frac{1}{2}} = 4\pi$. Its amplitude is 2. The curve crosses the y axis at the point $y = 2$. The graph is plotted in Fig. 5–21.

Exercises

State the fundamental period of each function and its amplitude, if it has one. Sketch the graph of each function, labeling the axes carefully, especially those points at which the curve crosses either axis.

1. $y = \sin 2x$.
2. $y = \cos 2x$.
3. $y = \tan 2x$.

4. $y = \cot 2x$.
5. $y = \sec 2x$.
6. $y = \csc 2x$.

7. $y = 2 \sin 3x$.
8. $y = -2 \sin 3x$.
9. $y = 3 \sin 4x$.

10. $y = \frac{1}{2} \cos 2x$.
11. $y = \frac{1}{2} \sin \frac{1}{2}x$.
12. $y = -2 \cos \frac{1}{2}x$.

13. $y = 3 \sin 6\theta$.
14. $y = 2 \tan 3\theta$.
15. $z = \sin \pi\theta$.

16. $z = \cos \pi\theta$.
17. $z = \tan \pi\theta$.
18. $z = \cot \pi\theta$.

19. $z = \sec \pi\theta$.
20. $z = \csc \pi\theta$.
21. $y = \frac{1}{3} \sin \pi\theta$.

22. $y = 2 \sin 3\pi x$.
23. $y = 2 \cos 3\pi x$.
24. $y = -2 \cos 3\pi x$.

25. $z = \frac{1}{3} \tan 2x$.
26. $z = 3 \cot 2x$.
27. $z = 5 \cos 4x$.

28. $z = 2 \sec 5x$.
29. $z = 5 \sec 2x$.
30. $z = 7 \csc 3x$.

5–12. Phase and Displacement

Consider the function

$$y = \sin (\omega x + \alpha) \tag{1}$$

where ω is a positive constant and α is a constant that may be either positive or negative.

*The expression $\omega x + \alpha$ is called the **phase** of the function. The quantity α, the value of the phase for $x = 0$, is called the **initial phase** of the function.*

Now

$$\sin (\omega x + \alpha) = \sin \omega \left(x + \frac{\alpha}{\omega} \right),$$

which can be used to advantage in setting up the table below. As x increases from $-\dfrac{\alpha}{\omega}$ to $\dfrac{\pi}{2\omega} - \dfrac{\alpha}{\omega}$, $\sin(\omega x + \alpha)$ increases from 0 to 1. Examining the rest of the table in this way, we conclude that, as x increases from $-\dfrac{\alpha}{\omega}$ to $\dfrac{2\pi}{\omega} - \dfrac{\alpha}{\omega}$, the function $\sin(\omega x + \alpha)$ completes one cycle of values. Hence it has a period $2\pi/\omega$, as we already know, and intersects the x axis during this cycle at $-\dfrac{\alpha}{\omega}, \dfrac{1}{\omega}(\pi - \alpha),$ and $\dfrac{1}{\omega}(2\pi - \alpha)$.

x	$x + \dfrac{\alpha}{\omega}$	$\omega\left(x + \dfrac{\alpha}{\omega}\right)$	$\sin \omega \left(x + \dfrac{\alpha}{\omega}\right)$
$-\dfrac{\alpha}{\omega}$	0	0	0
$\dfrac{\pi}{2\omega} - \dfrac{\alpha}{\omega}$	$\dfrac{\pi}{2\omega}$	$\dfrac{\pi}{2}$	1
$\dfrac{\pi}{\omega} - \dfrac{\alpha}{\omega}$	$\dfrac{\pi}{\omega}$	π	0
$\dfrac{3\pi}{2\omega} - \dfrac{\alpha}{\omega}$	$\dfrac{3\pi}{2\omega}$	$\dfrac{3\pi}{2}$	-1
$\dfrac{2\pi}{\omega} - \dfrac{\alpha}{\omega}$	$\dfrac{2\pi}{\omega}$	2π	0

Hence the graph of equation 1 is the graph of $y = \sin \omega x$ displaced along the x axis a distance α/ω. The displacement is to the left if $\alpha > 0$, to the right if $\alpha < 0$. The quantity α/ω is called the **displacement.**

Similar arguments can be made for the other functions with the phase $\omega x + \alpha$ to show that the displacement in each case is α/ω, to the left if $\alpha > 0$, to the right if $\alpha < 0$.

5–13. Sketching the Graphs of Trigonometric Functions

The information we have gained about the graphs of the trigonometric functions is summarized in the following table.

Function	Amplitude	Period	Displacement
$A \sin(\omega x + \alpha)$			
$A \cos(\omega x + \alpha)$	$\|A\|$	$\dfrac{2\pi}{\omega}$	$\dfrac{\alpha}{\omega}$
$A \sec(\omega x + \alpha)$			
$A \csc(\omega x + \alpha)$			To the left if $\alpha > 0$.
$A \tan(\omega x + \alpha)$		$\dfrac{\pi}{\omega}$	To the right if $\alpha < 0$.
$A \cot(\omega x + \alpha)$			

$A \neq 0, \omega > 0, \alpha$ any value.

The information for the sine and cosine is also summarized in Fig. 5-22.

Using this information, it is possible to sketch easily and quickly the graphs of trigonometric functions. The procedure given below is suggested for the curves for $A \sin (\omega x + \alpha)$ and $A \cos (\omega x + \alpha)$. The reader will infer immediately the slight modifications necessary in sketching the other functions.

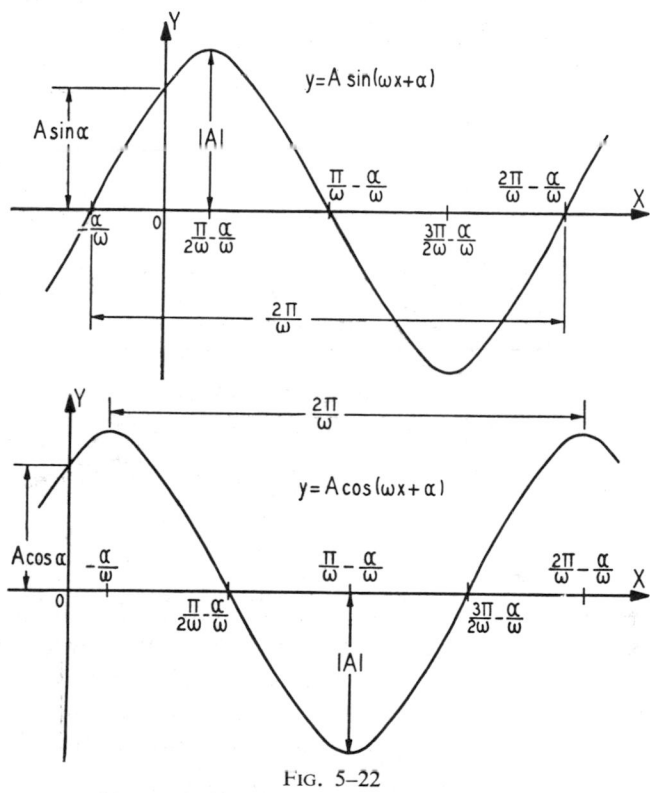

Fig. 5–22

To plot $y = A \sin (\omega x + \alpha)$ *or* $y = A \cos (\omega x + \alpha)$:
1. *List the following data:*
   (a) Amplitude, $|A|$.
   (b) Period, $2\pi/\omega$.
   (c) Initial phase, α.
   (d) Displacement, α/ω.
2. *Locate* $-\alpha/\omega$ *on the x axis and the point one period to its right, and then divide this segment into four equal parts. In this way we locate the*

five points with coordinates $-\dfrac{\alpha}{\omega}, \dfrac{1}{\omega}\left(\dfrac{\pi}{2} - \alpha\right), \dfrac{1}{\omega}(\pi - \alpha), \dfrac{1}{\omega}\left(\dfrac{3\pi}{2} - \alpha\right),$

$\dfrac{1}{\omega}(2\pi - \alpha)$. Between the end points of this segment, the function completes the cycle that $A \sin \omega x$ (or $A \cos \omega x$) completes between 0 and $2\pi/\omega$.

3. *Locate $\pm A$ on the y axis.*

4. *Locate the maximum and minimum points on the curve, using the ordinates obtained in step 3 and the appropriate abscissas obtained in step 2.* The abscissas obtained in step 2 and not used here are the points at which the curve crosses the x axis.

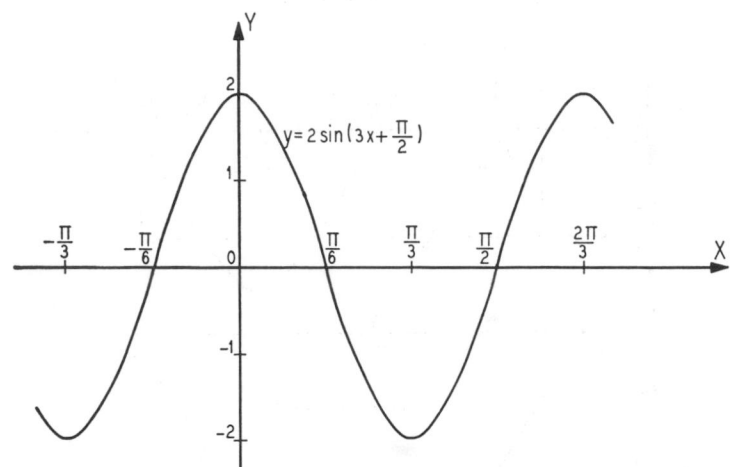

$$y = 2\sin\left(3x + \frac{\pi}{2}\right)$$

Fig. 5–23

5. *Sketch the graph of the function by drawing it freehand through the maximum and minimum points and the x intercepts.* By drawing the curve in lightly at first with short strokes of a pencil and then darkening as the shape becomes satisfactory, very good results can be achieved.

6. *If more than one period of the function is desired, repeat the pattern obtained between* $-\dfrac{\alpha}{\omega}$ *and* $\dfrac{1}{\omega}(2\pi - \alpha)$ *to the right and left.*

Example 1.   Sketch the curve of $y = 2 \sin (3x + \pi/2)$.

For this example:

$$\text{Amplitude} \quad = 2.$$

$$\text{Period} \quad = \frac{2\pi}{3}.$$

$$\text{Initial phase} \quad = \frac{\pi}{2}.$$

$$\text{Displacement} = \frac{\pi}{6}.$$

First we locate $-\pi/6$ on the x axis, and then $2\pi/3 - \pi/6 = \pi/2$. This segment is divided into four parts by the points 0, $\pi/6$, $\pi/3$. The maximum point is $(0, 2)$, and the minimum point is $(\pi/3, -2)$, and the curve crosses the x axis at $-\pi/6$, $\pi/6$, and $\pi/2$. The graph is shown in Fig. 5–23, with the unit on the x axis twice that on the y axis. For $x = 0$, $y = 2 \sin \dfrac{\pi}{2} = 2$, which is verified by the graph.

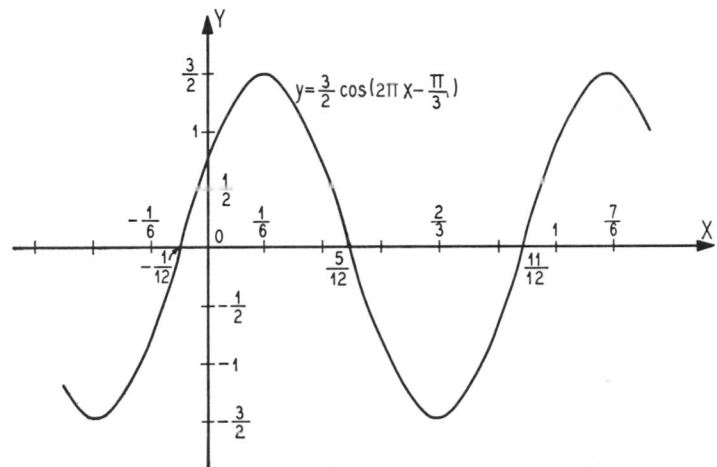

$$y = \tfrac{3}{2} \cos (2\pi x - \tfrac{\pi}{3})$$

FIG. 5–24

Example 2. Sketch the curve of $y = \tfrac{3}{2} \cos (2\pi x - \pi/3)$.

For this example:

$$\text{Amplitude} \quad = \tfrac{3}{2}.$$

$$\text{Period} \quad = 1.$$

$$\text{Initial phase} \quad = -\frac{\pi}{3}.$$

$$\text{Displacement} = -\tfrac{1}{6}.$$

First we locate $+\tfrac{1}{6}$ on the x axis, and then $\tfrac{7}{6}$. This segment is divided into four parts by the points $\tfrac{5}{12}$, $\tfrac{2}{3}$, and $\tfrac{11}{12}$. The maximum points are $(\tfrac{1}{6}, \tfrac{3}{2})$ and $(\tfrac{7}{6}, \tfrac{3}{2})$, and the minimum point is $(\tfrac{2}{3}, -\tfrac{3}{2})$. The curve crosses the x axis at $\tfrac{5}{12}$ and $\tfrac{11}{12}$. The graph is shown in Fig. 5–24 with the unit on the x axis three times that on the y axis. For $x = 0$, $y = \tfrac{3}{2} \cos (-\pi/3) = \tfrac{3}{2} \cdot \tfrac{1}{2} = \tfrac{3}{4}$, which is verified by the graph.

Exercises

For each function given below, list (*a*) the amplitude (if it has an amplitude), (*b*) the period, (*c*) the initial phase, and (*d*) the displacement. Sketch the curve according to the directions given above, labeling both axes carefully.

1. $y = 3 \sin (2x + \pi)$.

2. $y = 3 \sin (2x - \pi)$.

3. $y = 2 \cos (3x + \pi)$.

4. $y = \sin (3x + \pi)$.

5. $y = 2 \cos \left(2x + \dfrac{\pi}{2} \right)$.

6. $y = 2 \cos \left(2x - \dfrac{\pi}{2} \right)$.

7. $y = 3 \sin \left(3x - \dfrac{\pi}{4} \right)$.

8. $y = -\cos \left(2x + \dfrac{\pi}{3} \right)$.

9. $y = -4 \cos (4x - \pi)$.

10. $y = \tan (3x + \pi)$.

11. $y = \tan (x - \pi)$.

12. $y = \cot \left(2x + \dfrac{\pi}{2} \right)$.

13. $y = \cot \left(3x - \dfrac{\pi}{4} \right)$.

14. $y = 3 \tan \left(2x - \dfrac{\pi}{3} \right)$.

15. $y = 2 \cot \left(4x - \dfrac{\pi}{2} \right)$.

16. $y = \sec \left(x + \dfrac{\pi}{4} \right)$.

17. $z = 4 \sec (3\theta + \pi)$.

18. $z = \csc \left(\theta - \dfrac{\pi}{3} \right)$.

19. $z = 2 \csc \left(2\theta + \dfrac{\pi}{6} \right)$.

20. $z = 2 \sec \left(2\theta + \dfrac{\pi}{12} \right)$.

21. $z = 3 \csc \left(3\theta - \dfrac{\pi}{3} \right)$.

22. $z = \sin \left(\dfrac{3}{2} x + \dfrac{5}{8} \pi \right)$.

23. $y = 2 \cos \left(\dfrac{1}{4} x + \dfrac{5}{12} \pi \right)$.

24. $y = \dfrac{9}{5} \sin \left(\dfrac{3}{4} x - \dfrac{\pi}{12} \right)$.

25. $y = 5 \sin \left(\pi x - \dfrac{\pi}{3} \right)$.

26. $y = 3 \cos \left(3\pi x + \dfrac{\pi}{2} \right)$.

27. $y = 4 \sin \left(2\pi x + \dfrac{\pi}{6} \right)$.

28. $y = 2 \cos \left(4\pi x - \dfrac{\pi}{3} \right)$.

29. $z = \dfrac{3}{2} \sin (6\pi x + 2\pi)$.

30. $z = \tan \left(2\pi x + \dfrac{3\pi}{2} \right)$.

31. $z = 2 \cot \left(3\pi x + \dfrac{3\pi}{2} \right)$.

32. $z = 4 \sec \left(\pi \theta + \dfrac{\pi}{4} \right)$.

33. $z = 2 \csc \left(2\pi \theta - \dfrac{\pi}{6} \right)$.

34. $z = -\sin \left(\pi \theta + \dfrac{\pi}{6} \right)$.

35. $z = -4 \cos \left(\pi \theta - \dfrac{\pi}{9} \right)$.

36. $z = -\sin \left(\dfrac{3\theta}{4} - \dfrac{\pi}{4} \right)$.

5–14. Plotting Graphs by Composition of Ordinates

When a function is given that is the sum or difference of two or more trigonometric functions, its graph may be plotted easily by using the information we have obtained so far. The graph of each trigonometric function in the sum or difference is plotted, and then the corresponding ordinates are added (or subtracted) graphically by a compass or dividers to obtain the ordinates of the desired graph. This process is called the **composition of ordinates.** The following examples will illustrate the method.

Example 1.   Plot the graph of $y = 2 \sin x + \cos 2x$.

In order to plot the given function we shall plot the graph of $2 \sin x$ and the graph of $\cos 2x$ separately. They are shown in Fig. 5–25. Then, by using a compass or dividers, the ordinates of $2 \sin x + \cos 2x$ are found by adding and subtracting the ordinates of the graphs already plotted.

Example 2.   Plot the graph of $y = x + \sin x$.

We plot $y = x$ and $y = \sin x$ and then add the ordinates to obtain the graph of the given function, shown in Fig. 5–26.

Exercises

Plot the following graphs by composition of ordinates.

1. $y = \sin x - \cos x$.

2. $y = \sin x + 2 \cos x$.

3. $y = 2 \sin x + \cos x$.

4. $y = 2 \sin x + \cos 2x$.

5. $y = x + \sin 2x$.

6. $y = x + \cos x$.

7. $y = x - \sin x$.

8. $y = x - \cos x$.

9. $y = \sin x - 2x$.

10. $y = \cos x - 2x$.

11. $y = \frac{1}{4} \sin x + 2 \cos 3x$.

12. $y = 2 \cos 3x + \sin \frac{1}{3}x$.

13. $y = \sin 2\pi x + 2 \cos 2\pi x$.

14. $y = 3 \sin 2\pi x + \sin \frac{1}{4}x$.

15. $y = 6 \sin 4\pi x - 3 \cos 2\pi x$.

16. $y = 4 \sin 4x + 2 \cos 2x$.

17. $y = \frac{1}{3}x^2 - \sin 3\pi x$.

18. $y = \frac{1}{4}x^2 + \cos 3\pi x$.

19. $y = \sin x + \cos \left(x + \dfrac{\pi}{6} \right)$.

20. $y = \sin \left(2x - \dfrac{\pi}{4} \right) + \cos x$.

21. $y = \sin \left(x + \dfrac{\pi}{9} \right) + \cos \left(x - \dfrac{\pi}{9} \right)$.

22. $y = \sin \left(x + \dfrac{\pi}{9} \right) - \cos \left(x - \dfrac{\pi}{9} \right)$.

23. $y = 2 \sin 2x + \sin \left(2x + \dfrac{3\pi}{4} \right)$.

24. $y = \sin \left(\dfrac{3}{2} \pi x - \pi \right) + \cos \left(\dfrac{\pi}{4} x \right)$.

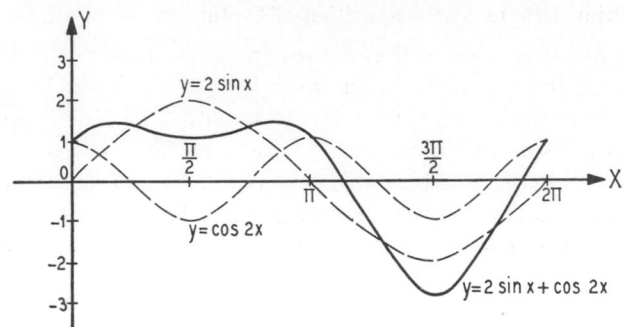

FIG. 5–25

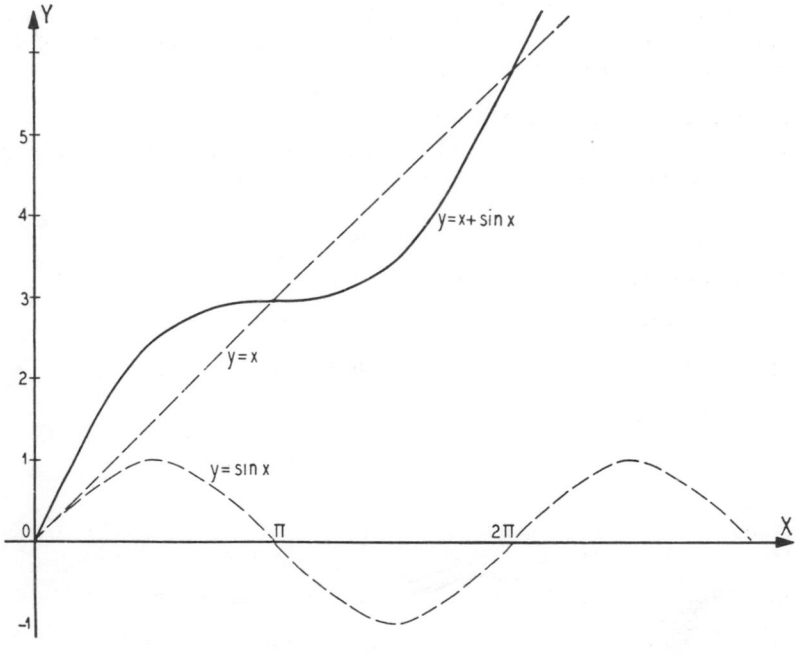

FIG. 5–26

5–15. Applications of Graphs of Trigonometric Functions: The Simple Pendulum

One of the applications of the trigonometric functions in science and engineering is to the study of periodic motions. Periodic motions, as the name suggests, are motions that are repeated after regular intervals of time. Some examples of such motion are the motion of the pendulum of a clock, the vibration of a taut string, the vibration of a reed, and the motion of a planet around the sun.

Each of the above motions can be expressed mathematically as some periodic function of time, but only a few of them can be expressed as simple trigonometric functions of the form $y = A \sin \omega t$. Others can be expressed as simple trigonometric functions only if simplifying assumptions are made, whereas still others may be very complicated periodic functions of time.

Consider the simple pendulum of length L ft shown in Fig. 5–27, which is suspended from the point P. The line PQ is the position of rest of the pendulum. The pendulum is started swinging by drawing it to the right

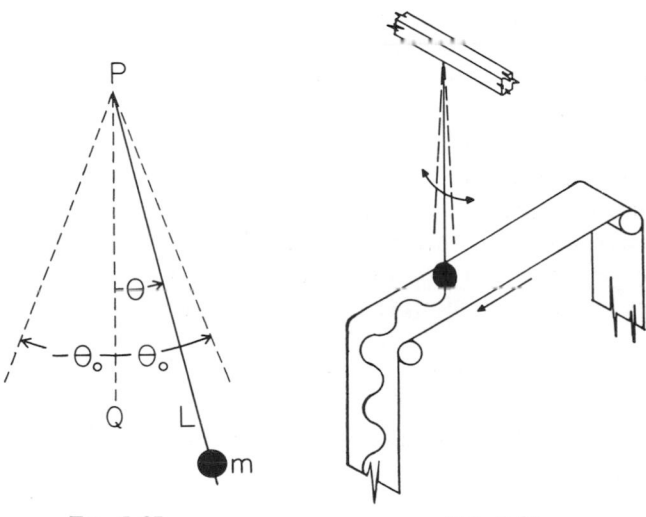

FIG. 5–27   FIG. 5–28

an angle of θ_0 radians and then releasing it. Then, (a) if θ_0 is reasonably small, say less than 0.3 radians, (b) if we neglect air resistance and frictional effects, and (c) if we agree to measure the variable position angle θ of the pendulum as positive to the right of PQ and negative to the left, the position of the pendulum t sec after it has been released can be shown to be very closely approximated by

$$\theta = \theta_0 \cos \sqrt{\frac{g}{L}}\, t, \tag{1}$$

where $g = 32.2$ is the acceleration of gravity in feet per second per second. If L is measured in centimeters, then g is measured in centimeters per second per second, and has the value 980.2.

The period T, the amount of time necessary to complete one swing from θ_0 to the left and back again, is the period of the function 1 and is given by

$$T = 2\pi \sqrt{\frac{L}{g}} \text{ sec.} \qquad (2)$$

We see that the motion of the pendulum given by function 1 is independent of the mass m at the end of the pendulum, and that the period given by function 2 depends only on the length of the pendulum.

By attaching a marking device to the lower end of the pendulum, a graph very similar to a sine curve will be drawn on a strip of paper which moves at a constant velocity perpendicular to the plane in which the pendulum is swinging (Fig. 5–28).

A complete swing of the pendulum from the position $\theta = \theta_0$ to the opposite position $\theta = -\theta_0$ and again to the original position $\theta = \theta_0$ is called a **cycle**, and the number of cycles in a unit of time is called **frequency.** These terms are used not only for motion of the pendulum but also for every periodic motion.   A cycle is the smallest part of a periodic motion by whose repetition the motion is produced.   If T is the period, the time needed for one cycle, and f the frequency, the number of cycles per unit of time, then we have the following fundamental relations:

$$fT = 1, \qquad f = \frac{1}{T}, \qquad T = \frac{1}{f}.$$

The period in seconds of a pendulum is given by formula 2.   The frequency therefore is

$$f = \frac{1}{2\pi} \sqrt{\frac{g}{L}} \text{ cycles per second.}$$

5–16. Projection of a Rotating Spoke

Consider the wheel of radius 10 ft shown in Fig. 5–29, whose axle is taken as the center of a Cartesian coordinate system.   The projection of the spoke OS on the line LL' which is perpendicular to the x axis is QR.

Suppose the spoke starts from a position OT making an angle $\pi/6$ with the positive x axis and rotates in a counterclockwise direction at a uniform speed of 3 revolutions per second (Fig. 5–29).   Since in each revolution it rotates through 2π radians, its angular velocity is 6π radians per second, and in t sec after starting the spoke will have rotated through an angle of $6\pi t$ radians.   Hence, the angle in standard position, of which OS is the terminal side, is $(6\pi t + \pi/6)$.   Thus the ordinate of S is $10 \sin (6\pi t + \pi/6)$.

If Q is the projection of O on LL', and R is the projection of S, then QR is the projection of OS, and, further, has a length $|10 \sin (6\pi t + \pi/6)|$.

If we agree to measure distances up from Q as positive and those down from Q as negative, then the length p of the projection of OS on $L'L$ is given by

$$p = 10 \sin\left(6\pi t + \frac{\pi}{6}\right). \tag{1}$$

This function has amplitude 10, period $\frac{1}{3}$ sec, initial phase $\pi/6$ sec, and displacement $\frac{1}{36}$ sec. Hence by plotting the function 1, we have a curve showing the behavior of the projection as time goes on.

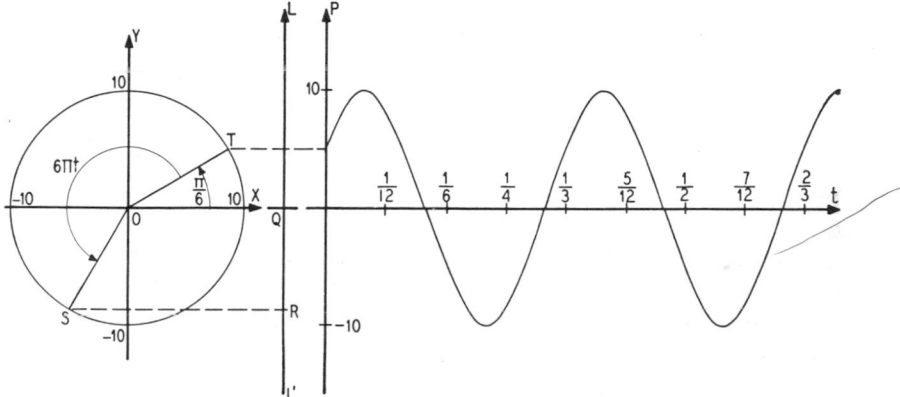

FIG. 5–29

In order to derive a general formula, suppose that the angular velocity is ω radians per second, the amplitude is A units of length, and the initial phase is α. Then,

$$p = A \sin(\omega t + \alpha).$$

The period $T = 2\pi/\omega$ sec; the frequency $f = \omega/2\pi$ cycles per second. The formula for the frequency f is often written $\omega = 2\pi f$.

5–17. Alternating Current

The results of Sec. 5–16 can now be applied in order to examine the voltage produced by a rotating a-c generator. This voltage varies from instant to instant in the same way as the projection of a spoke of the rotating part on a straight line, which was discussed in the previous section. If e is the value of the voltage, measured in volts, at a certain moment (the so-called instantaneous voltage), E_m the **maximum** or **peak voltage**, ω the angular velocity in radians of the conductor in which the voltage is induced, and t is the time measured in seconds, then we have the formula

$$e = E_m \sin(\omega t + \alpha). \tag{1}$$

The initial phase α is often called **phase angle.**   The curve corresponding to equation 1 is called a **sine wave.**   Such a wave is plotted in Fig. 5–30 for the case $\alpha = 0$.   The frequency is, according to Sec. 5–16, $f = \omega/2\pi$.

In reality the voltage produced by an a-c generator does not change exactly like a sine curve, but the curves obtained by plotting the voltage

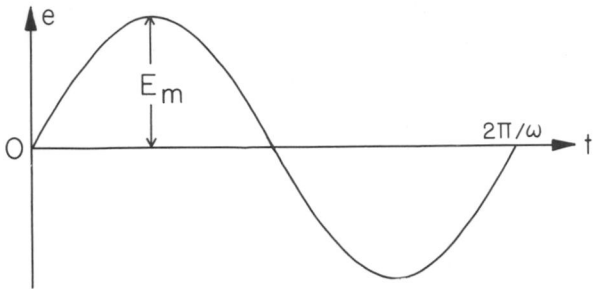

FIG. 5–30

against the time are often very similar to sine waves, and are then called **sinusoidal.**   The wave form can be observed on an oscilloscope, an extremely sensitive electronic device which projects the wave, using an electron stream, onto a fluorescent screen.

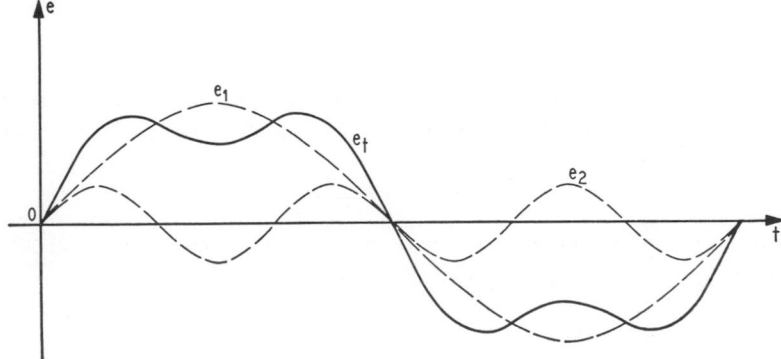

FIG. 5–31

An alternating voltage $e = E_m \sin \omega t$ produces in a circuit an electric current whose instantaneous value is

$$i = I_m \sin (\omega t + \alpha). \tag{2}$$

This current is represented by a sine wave of amplitude I_m, the same frequency $\omega/2\pi$ as the voltage, and a **phase difference** α compared with the voltage e.   The current is said to **lead** the voltage if α is positive and to **lag** behind the voltage if α is negative.

If two or more voltages e_1, e_2, $\cdots$ are produced in the same circuit, the resultant voltage is the sum of them, and the resultant wave can be found by plotting the waves corresponding to $e_1, e_2, \cdots$ and adding their ordinates according to Sec. 5–14.

Example. Two voltages present in a circuit are given by

$$e_1 = E_{m1} \sin \omega t \quad \text{and} \quad e_2 = E_{m2} \sin 3\omega t.$$

The resultant wave is the sum

$$e_t = e_1 + e_2 \quad \text{or} \quad e_t = E_{m1} \sin \omega t + E_{m2} \sin 3\omega t.$$

This wave is plotted in Fig. 5–31 by first plotting the two components and then adding their ordinates. In the laboratory the resultant wave can be shown on an oscilloscope. The student should supply labels at the points where the curves cross the t axis and locate the points E_{m1}, E_{m2}, and $E_{m1} + E_{m2}$ (as well as the corresponding negative points) on the e axis.

Exercises

1. A pendulum weighing 80 grams is suspended by a thin rod 95 cm long. The pendulum is pulled aside 7.0° from the vertical and allowed to swing. If the acceleration of gravity is 980 cm per sec per sec, find the period of the pendulum and its frequency. The angle θ which the rod makes with the vertical line is a function of the time t since the pendulum was released. Plot the graph of this function.

2. A pendulum 3.6 ft long is made to oscillate by pulling it aside exactly 10° from the vertical. If the acceleration of gravity at the point of the experiment is 32 ft per sec per sec, find the period of the pendulum, and plot the graph of the angle θ, which the pendulum makes with a vertical line, against time.

3. What is the length of a pendulum whose frequency is 120 cycles per minute at a place where the acceleration of gravity is 980 cm per sec per sec? Plot, against the time, the graph of the angle between the pendulum and a vertical line.

4. The driving wheel of the valve-action mechanism of Fig. 5–32 makes exactly 60 rpm. Describe the motion of the valve-actuating arm by expressing the distance y between the crankpin and the horizontal axis as a function of the time. The distance between the centers of the wheel and the crankpin is 2 in. (Use the discussion of Sec. 5–16.)

What is the angular velocity ω if the frequency f is:

5. 60 cycles per second.

6. 400 cycles per second.

7. 10 kilocycles per second.

8. 32 megacycles per second.

What is the frequency f if the angular velocity ω is:

9. 377 radians per second.

10. 5026 radians per second.

11. 6.28×10^4 radians per second.

12. 2.03×10^7 radians per second.

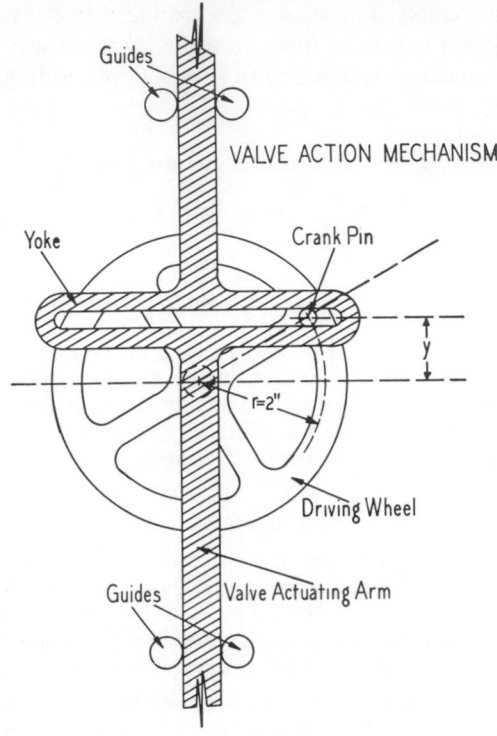

Guides

VALVE ACTION MECHANISM

Yoke   Crank Pin

y

r=2"

Driving Wheel

Guides   Valve Actuating Arm

Fig. 5–32

Compute the period T of an oscillatory motion of frequency f and corresponding angular velocity ω if:

13. $f = 60$ cycles per second.
14. $f = 1000$ cycles per second.
15. $f = 720$ kilocycles per second.
16. $f = 4.3$ megacycles per second.
17. $f = 5$ cycles per minute.
18. $f = 12$ cycles per hour.
19. $\omega = 377$ radians per second.
20. $\omega = 9425$ radians per second.
21. $\omega = 0.28$ radian per second.
22. $\omega = 6 \times 10^5$ radians per second.

Plot the following waves. Note that, as is frequent in engineering practice, the angle ωt is in radians while the phase angle is in degrees. Thus, for example, $\omega t + 60°$ must be changed to either $\omega t + \dfrac{\pi}{3}$ radians or $\dfrac{180}{\pi} \omega t + 60$ degrees for consistent units.

23. $e = E_m \sin \omega t$, where $E_m = 141$ volts and $\omega = 377$ radians per second.

24. $i = I_m \sin \omega t$, where $I_m = 8$ amperes and $\omega = 377$ radians per second.

25. $e = E_m \sin(\omega t + 45°)$, where $E_m = 310$ volts and $\omega = 157$ radians per second.

26. $i = I_m \sin(\omega t - 30°)$, where $I_m = 6.3$ amperes and $\omega = 3.14 \times 10^4$ radians per second.

27. Plot the wave of a current that lags 45° behind the voltage of exercise 23 if the maximum current I_m is 5 amperes.

28. Plot the wave of a current that leads the voltage of exercise 23 by 30° if the maximum current is 14 amperes.

29. Plot the wave of a voltage that leads the current of exercise 24 by 15° if the peak voltage is 85 volts.

Plot the following waves:

30. $e_t = e_1 + e_2 = E_{m1} \sin \omega t + E_{m2} \sin (\omega t + 90°)$, where $E_{m1} = E_{m2} = 60$ volts and $\omega = 20$ radians per second.

31. $i_t = i_1 + i_2 = I_{m1} \sin (\omega t + 60°) + I_{m2} \sin (\omega t + 30°)$, where $I_{m1} = I_{m2} = 10$ amperes and $\omega = 377$ radians per second.

32. $e_t = e_1 + e_2 + e_3 = E_{m1} \sin \omega t + E_{m2} \sin 2\omega t + E_{m3} \sin 3\omega t$, where $E_{m1} = E_{m2} = E_{m3} = 100$ volts and $\omega = 377$ radians per second.

5–18. Graphs of Parametric Equations Involving Trigonometric Functions

In Sec. 3–10 we considered graphs in the xy plane specified by a pair of parametric equations. These equations gave x and y each in terms of an elementary algebraic expression involving the parameter. In this section we shall consider some graphs that arise when the parametric expressions specifying x and y involve trigonometric functions. In plotting these graphs it is important to apply the knowledge of the properties of the trigonometric functions that was gained earlier in this chapter, as the following examples will illustrate.

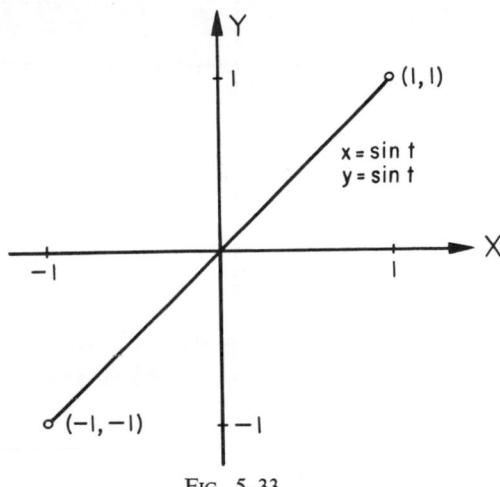

FIG. 5–33

Example 1. Plot the graph specified by $x = \sin t$, $y = \sin t$.

Since the expressions for x and y are identical, it is clear that this graph lies on the line $y = x$. However, since $\sin t$ lies between -1 and $+1$, both x and y lie between -1 and $+1$. Thus the graph that we are seeking is the segment of the line $y = x$ that connects the point $(-1, -1)$ and the point $(1, 1)$, as shown in Fig. 5–33. The student should verify this conclusion by plotting the values of x and y that correspond to various values of t.

Example 2.   Plot the graph specified by $x = 6 \sin \theta$, $y = 6 \csc \theta$.

Since $\csc \theta$ is the reciprocal of $\sin \theta$, it can be seen easily that this graph lies on the curve given by $y = 36/x$.   However, since $\sin \theta$ lies between -1 and $+1$, values of x occur only between -6 and $+6$; since $\csc \theta$ lies outside the interval between -1 and $+1$, values of y occur only outside the interval between -6 and $+6$.   Also, it will be useful to notice that $\sin \theta$ and $\csc \theta$ always have the same sign and therefore that corresponding values of x and y always have the same sign.   It follows that the graph lies entirely in the first and third quadrants. As a consequence of these arguments we infer that the graph must lie in the cross-hatched areas of Fig. 5–34.   The student should verify that the graph shown in

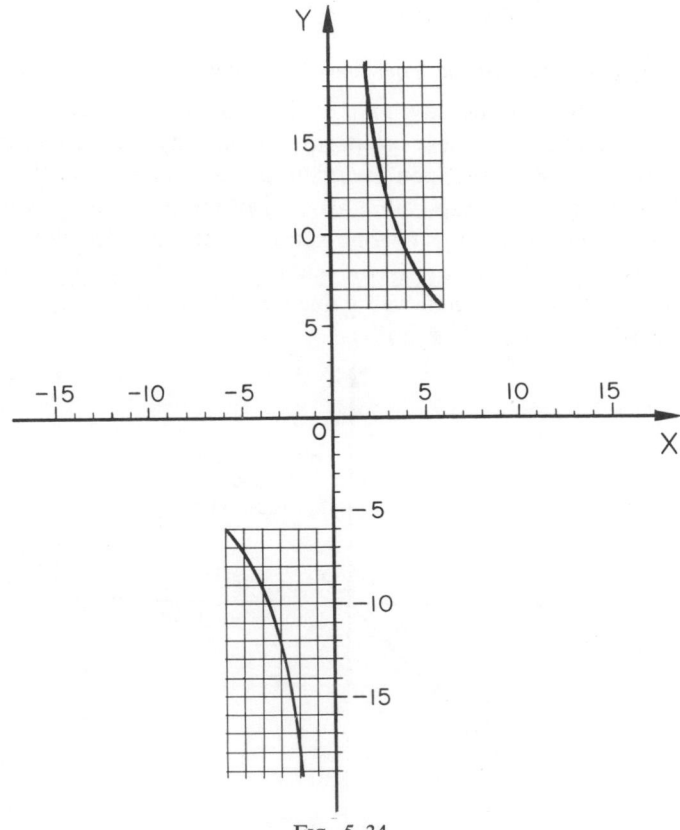

FIG. 5–34

Fig. 5–34 is correct by plotting the values of x and y corresponding to a comprehensive selection of values of θ.

A type of graph frequently seen on a laboratory oscilloscope can be represented by parametric equations of the form

$$x = A \sin \omega t,$$

$$y = B \sin (n\omega t + \alpha),$$

where A and B are amplitudes, ω is an angular velocity, t is time, n is an integer, and α is the phase angle.   Graphs of this type are called **Lissajous figures.**   These figures assume various shapes, depending on the values of A, B, n, and α that occur.   In fact, a laboratory technician familiar with the properties of Lissajous figures can frequently estimate the values of these constants from an observation of the graphs on his oscilloscope.

Example 3.   Plot the graph specified by $x = \sin \omega t$, $y = \sin 2\omega t$.

We first note that both x and y lie between -1 and $+1$, so that the graph lies entirely within the square shown in Fig. 5–35.   In order to plot the graph we

Point Number	ωt	$2\omega t$	x	y
1	0°	0°	0.00	0.00
2	10°	20°	0.17	0.34
3	20°	40°	0.34	0.64
4	30°	60°	0.50	0.87
5	40°	80°	0.64	0.98
6	50°	100°	0.77	0.98
7	60°	120°	0.87	0.87
8	70°	140°	0.94	0.64
9	80°	160°	0.98	0.34
10	90°	180°	1.00	0.00

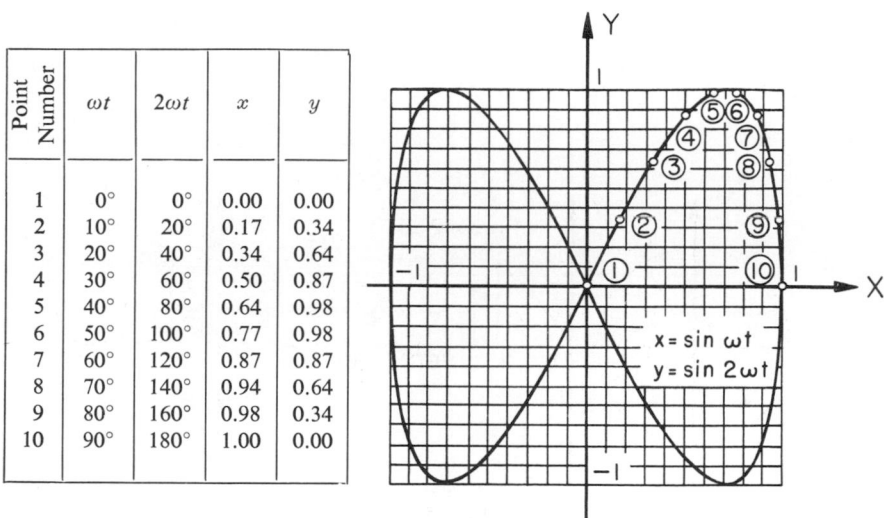

FIG. 5–35

choose values for ωt and $2\omega t$ which make it easy to read the corresponding values of x and y, respectively, from Table 3 in the appendix.   The table of values accompanying the graph gives only the pairs of values of x and y for the first quadrant portion of the graph; the student should complete this table for the other quadrants in order to check the correctness of the graph shown in Fig. 5–35.

Using the interpretation of the sine function as the projection of a rotating spoke explained in Sec. 5–16, it is possible to visualize the construction of the graph of Fig. 5–35 as shown in Fig. 5–36.   This latter figure shows a graphical construction of the points numbered 1 to 5 of Fig. 5–35.   While the two curves for $x - \sin \omega t$ and $y - \sin 2\omega t$ are not necessary to the construction, they are shown in Fig. 5–36 as an aid to understanding the process; they may be omitted, of course, in carrying out the construction.

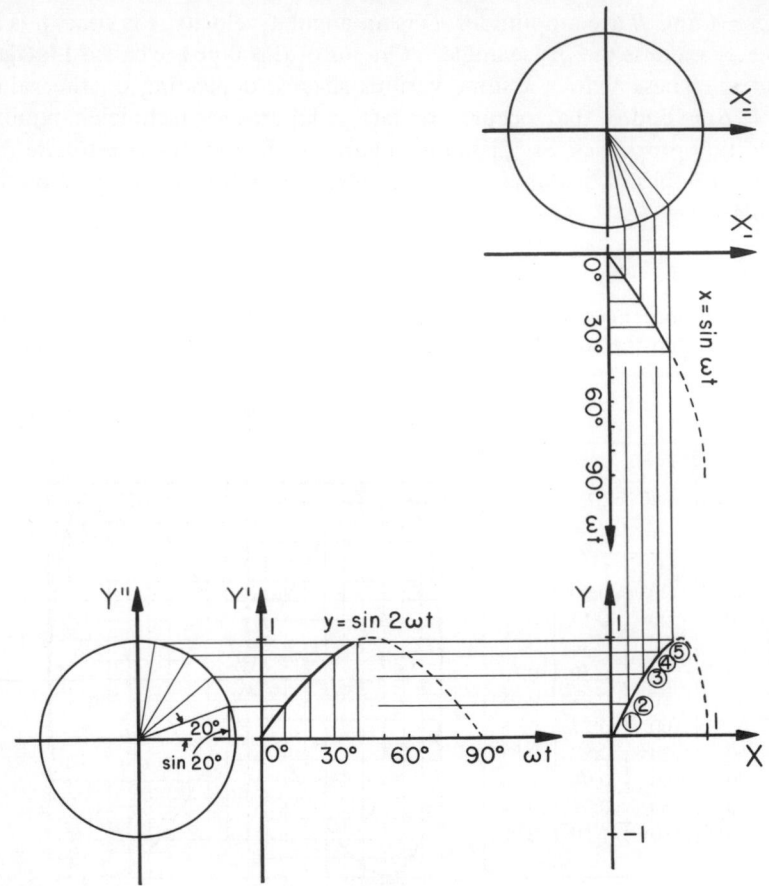

Fig. 5–36

Exercises

Plot the graph corresponding to each pair of parametric equations.

1. $x = \sin \theta$,
$y = -\sin \theta$.

2. $x = -\cos \theta$,
$y = 2 \cos \theta$.

3. $x = 4 \sin t$,
$y = 4 \cos t$.

4. $x = \sin^2 t$,
$y = \cos^2 t$.

5. $x = 1 + \sin \theta$,
$y = 2 - \sin \theta$.

6. $x = 3 \cos \theta$,
$y = -3 \sin \theta$.

7. $x = 3 \sin \theta$,
$y = 2 \csc \theta$.

8. $x = 2 \sec \theta$,
$y = 4 \cos \theta$.

9. $x = \tan \theta$,
$y = \tan \theta$.

10. $x = \sec \theta$,
$y = -\sec \theta$.

Draw each of the following Lissajous figures:

11. $x = \sin \omega t$,
$y = \sin (\omega t + 45°)$.

12. $x = \sin 6t$,
$y = \sin (6t + 90°)$.

13. $x = 2 \sin \omega t$,
$y = 3 \sin (\omega t + 90°)$.

14. $x = 2 \sin \omega t$,
$y = 3 \sin 2\omega t$.

15. $x = 10 \sin \omega t,$
$y = 10 \sin 3\omega t.$

16. $x = \sin \omega t,$
$y = \sin (3\omega t + 90°).$

17. $x = 4 \sin \omega t,$
$y = 4 \sin 4\omega t.$

18. $x = 10 \sin \omega t,$
$y = 10 \sin (4\omega t + 90°).$

19. $x = \sin \omega t,$
$y = \sin (\omega t + 30°).$

20. $x = \sin 36t,$
$y = \sin (72t + 60°).$

21. $x = 2 \sin 6t,$
$y = \sin \left(6t + \frac{\pi}{3} \right).$

22. $e_1 = 10 \sin 377t,$
$e_2 = 10 \sin (377t + 60°).$

23. $e_1 = 10 \sin 377t,$
$e_2 = 10 \sin (377t - 45°).$

24. $e_1 = 10 \sin 377t,$
$e_2 = 10 \sin 1508t.$

25. $e_1 = 10 \sin 377t,$
$e_2 = 10 \sin (754t + 30°).$

26. $e_1 = 10 \sin 377t,$
$e_2 = 10 \sin (1131t + 90°).$

Progress Report

In this chapter we examined the properties of the six trigonometric functions. In this connection we considered:

1. The graphs of the six trigonometric functions.

2. The line representation of the trigonometric functions.

3. The amplitude of the sine and cosine functions.

4. The periods of the trigonometric functions.

5. The phase, initial phase, and displacement of the trigonometric functions.

6. Sketching trigonometric functions with various amplitudes, periods, and displacements.

7. Sketching graphs of sums and differences by composition of ordinates.

After a discussion of applications of trigonometric functions to the motion of a pendulum, to the projection of a rotating spoke, and to a-c problems, the chapter concluded with a brief consideration of the graphs of some parametric equations involving simple trigonometric functions.

6

Simple Properties of Vectors

We have been interested so far in only one property of physical quantities, namely, magnitude. However, many physical quantities have a second property, direction, which it is important to consider. For instance, not only is the speed of the wind important to an aviator but also its direction: a head wind will decrease his speed relative to the ground, a tail wind will increase his speed relative to the ground, and a cross wind will not only affect his speed relative to the ground but also will cause him to travel in a direction different from the one in which his plane is headed. A shell fired from a gun is urged in the direction in which the gun is aimed by the force of the exploding powder; it is also urged earthward by the force of gravity, retarded by the resistance of the air, and perhaps even affected by the wind. All these forces and their directions influence the path of the projectile and consequently help to determine where it will strike. Displacement, motion, velocity, momentum, acceleration, and force are all physical quantities of which direction is an important property.

6–1. Scalars

Physical quantities, such as mass and temperature, that are characterized adequately for our purposes by the property of magnitude alone, are called **scalar quantities.** The numbers that represent them are called **scalars.** In order to solve problems concerning magnitudes alone we have used numbers as a tool, together with the science of operations with numbers which is called algebra.

In order to solve problems involving physical quantities with both magnitude and direction, we shall develop the **vector** as a tool. Physical quantities with both magnitude and direction are called **vector quantities.** The science of operations with vectors is called **vector algebra.**

Let us suppose an aviator flies a certain distance at a uniform speed of 180 mph. Since his speed is fully specified by its magnitude alone, it is a scalar quantity. However, suppose that we desire to specify not only his speed, but also the direction in which he is flying, by saying that he is

flying eastward at a speed of 180 mph.   Thus, we get a quantity called
velocity, which has both magnitude and direction.   Consequently, velocity
is a vector quantity and will be represented by a vector.

6–2. Vectors

A straight line segment to which a direction has been given is called a
directed line segment.   Usually the direction is indicated by an arrowhead,
as shown on the segment PQ in Fig. 6–1.   Obviously, there are only two
possible directions that can be assigned to a given segment.

*A vector is a directed line segment.   The magnitude or absolute value of
a vector is its (positive) length; its direction is given by its position in the
plane and the direction assigned to the segment.*   In the plane of Fig. 6 1,
the segment PQ with direction as shown by the arrowhead is a vector.
It is sometimes helpful to think of a vector as being drawn in the given
direction from an **initial point** to a **terminal point.**   In Fig. 6–1, P is the
initial point and Q the terminal point.   *Thus a vector can be specified by
giving its initial point and its terminal point.*   The distance between these
points is the magnitude, and the direction of the segment is from the initial
to the terminal point.

It is useful to have a notation for vectors.   We shall use boldfaced
capitals (in figures underlined capitals) for this purpose, as **A** in Fig. 6–1.

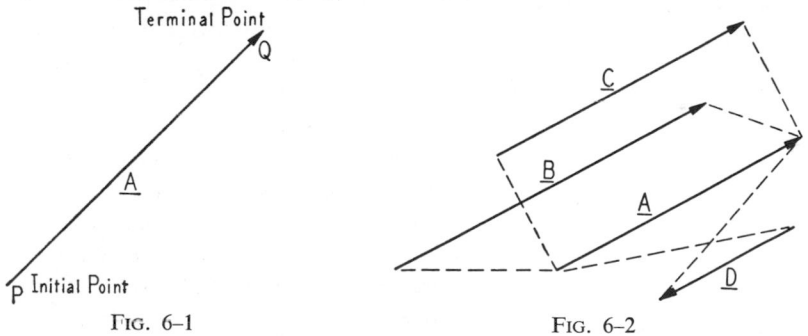

FIG. 6–1   FIG. 6–2

Scalars will be denoted by lower-case letters such as a, b, and capitals
which are not boldfaced, as A, B.   *The magnitude of a vector is a positive
scalar.*   The magnitude of the vector **A** is denoted either by the corre-
sponding capital A not boldfaced or by parallel bars $|A|$.   In this
chapter it is supposed that all vectors are located in one and the same
plane.

*Two vectors are said to have the same direction if they are parallel and
such that the segment joining their initial points does not intersect the segment joining their terminal points.*   In Fig. 6–2, all the vectors **A, B, C, D**

are parallel. The segment joining the initial points of **A** and **B** does not intersect the segment joining the terminal points, so that **A** and **B** have the same direction. Likewise **A** and **C** have the same direction, and thus **A, B,** and **C** have the same direction. However, **A** and **D** do not have the same direction, since the segment joining their initial points intersects the segment joining their terminal points. Two vectors, such as **A** and **D,** which are parallel but do not have the same direction, are said to be **opposite in sense.**

Two vectors are equal if they have the same magnitude and the same direction. In Fig. 6–2 the vectors **A** and **C** are equal, written as **A** = **C.** Since through any point in the plane a line can be drawn parallel to a given line, *any point in the plane can be taken as the initial point of a vector equal to any given vector.*

In any operation with vectors, a given vector may be replaced by any other equal vector. Usually convenience dictates the choice of such a substitute vector. If **A** is a given vector, we shall follow the practice of denoting also by **A** any vector equal to **A.** Any point in the plane is thus the initial point of a vector **A** equal to a given vector **A.**

6–3. Addition and Subtraction of Vectors

The sum of two vectors is defined as follows:

The sum of **A** *and* **B** *is the vector* **C** *whose initial point is the initial point*

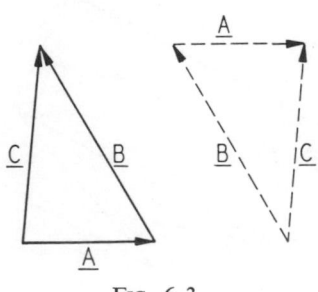

FIG. 6–3

of **A** *and whose terminal point is the terminal point of a vector equal to* **B** *drawn so that its initial point is the terminal point of* **A.** Such an addition is shown in Fig. 6–3 by the full lines. The dotted lines show how a vector equal to **C** can be found by performing the process in the reverse order. *The order in which vectors are added is immaterial.* To denote that **C** is the sum of **A** and **B** we write

$$C = A + B.$$

Several vectors can be added as shown in Fig. 6–4. The figure also illustrates the fact that the order of vector addition has no effect on the result. The sum of several vectors is often called the **resultant** of the vectors.

A second definition of vector addition can be formulated which is equivalent to the one already given:

To add **A** *and* **B,** *draw a vector* **B** *whose initial point is the initial point of* **A,** *and complete the parallelogram of which these vectors form two sides.*

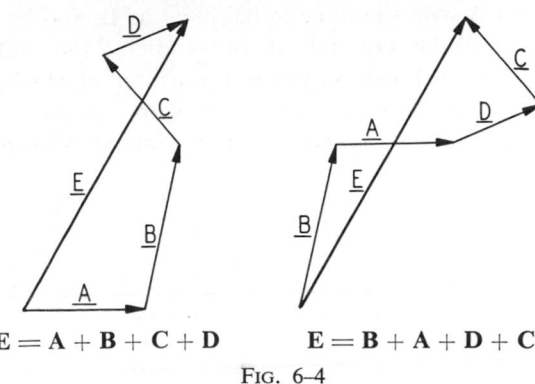

$$\mathbf{E} = \mathbf{A} + \mathbf{B} + \mathbf{C} + \mathbf{D} \qquad \mathbf{E} = \mathbf{B} + \mathbf{A} + \mathbf{D} + \mathbf{C}$$

FIG. 6–4

The diagonal vector **C** *with initial point the initial point of* **A** *and* **B** *and terminal point at the opposite vertex of the parallelogram is the sum of* **A** *and* **B**.   Such an addition is shown in Fig. 6–5.

The vector —**A** *is a vector of the same magnitude as* **A** *and parallel to* **A**, *but opposite in sense.*

The result of subtracting **B** *from* **A** *is obtained by adding* **A** *and* —**B**. Fig. 6–6 shows a vector **B** subtracted from **A**, which we denote by **A** — **B**.

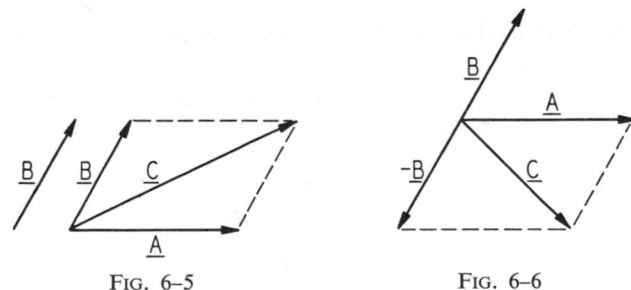

FIG. 6–5   FIG. 6–6

The reader may also devise a convenient direct parallelogram rule for the subtraction of vectors.   He may also verify that if several vectors are to be subtracted from a given vector the order in which they are subtracted is immaterial.

If a is a positive scalar, the vector $a \cdot \mathbf{A}$ *is a vector with the same direction as* **A** *and with magnitude aA.   If a is a negative scalar, the vector* $a \cdot \mathbf{A}$ *is equal to the vector* $|a| \cdot (-\mathbf{A})$.

The vector of zero magnitude is considered to have no direction.   It is called the **null vector,** and denoted by 0.

The direction of a vector **B** can be specified conveniently by an angle that the vector **B** makes with a given vector **A**.   This angle is formed by choosing a vector **A** with the same initial point as **B,** and by letting **A** be the initial side and **B** the terminal side of the angle; the angle is positive if measured counterclockwise, negative if measured clockwise. Usually, as here, the vector **A** is chosen horizontal with direction to the right. Several vectors with directions specified in this way are shown in Fig. 6–7.

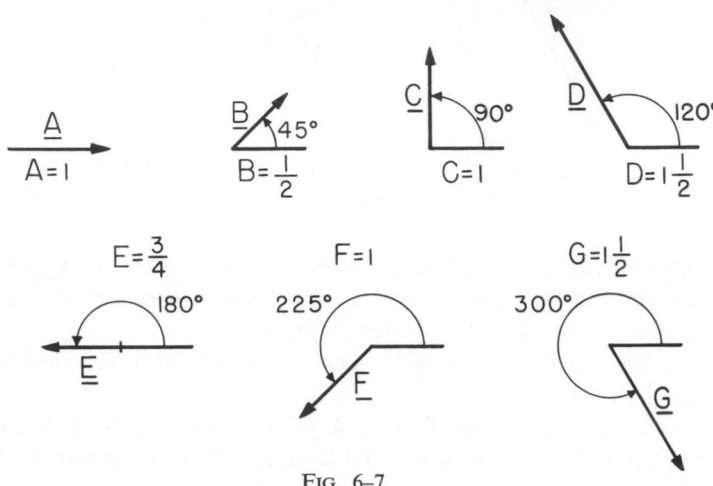

FIG. 6–7

We see then that a vector can be specified by giving
   (a) Its magnitude.
   (b) The angle (as described above) that it makes with some given vector.
   (c) Its initial or terminal point.

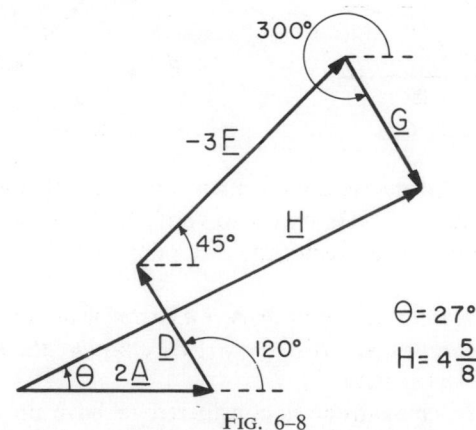

FIG. 6–8

Two vectors, having *a* and *b* the same, are equal, and thus, for most operations, *c* is of no interest, since any vector can be replaced by an equal in operations.   The vectors of Fig. 6–7 are given by their magnitudes, the angles they make with the horizontal, and their initial points.   However, in operations with these vectors other initial points may be chosen at pleasure.

Example.   Find graphically the vector **H** = 2**A** + **D** − 3**F** + **G** where **A**, **D**, **F**, and **G** are given in Fig. 6–7.

The graph is drawn in Fig. 6–8 by the use of a ruler and a protractor.   After the construction has been made, a measurement shows that the magnitude of **H** is $4\frac{5}{8}$, and that **H** makes an angle of 27° with the horizontal vector **A**.

Exercises

Given the vectors **A**, **B**, **C**, **D**, **E**, **F**, **G** of Fig. 6–7, find graphically each of the following vectors **H**.   Make your constructions carefully, using a ruler and a protractor.   Find the magnitude of **H** and the positive angle between 0° and 360° that it makes with the horizontal vector **A** by measurement from your construction.

1. **H** = **C** + **F**.   2. **H** = **B** + **D**.
3. **H** = 2**E** + 3**B**.   4. **H** = 2**E** + **D**.
5. **H** = **C** − **F**.   6. **H** = **D** − 2**B**.
7. **H** = 3**A** − 4**B**.   8. **H** = 3**F** − 2**G**.
9. **H** = 3**F** − 4**C**.   10. **H** = 2**D** + 3**F**.
11. **H** = **A** + **C** + **D**.   12. **H** = 3**A** + **D** + 2**F**.
13. **H** = 3**E** − 2**G** + 2**F**.   14. **H** = 2**G** − 3**F** + 2**E**.
15. **H** = 2**C** − **D** + 2**E**.   16. **H** = **A** + **B** + **C** + **D**.
17. **H** = **F** + **G** − **E** + **D**. 18. **H** = 3**E** − 2**C** + **D** + **F**.
19. **H** = 2**D** + 3**G** − 4**F** + 5**E**.   20. **H** = **C** + **F** − 2**G** − 3**E**.

In the following exercises find the result by drawing a vector diagram.

21. If a man leaves home and drives 36 miles north and then 25 miles east, what are his distance and bearing from home?

22. In hunting a submarine, a destroyer steams 45 miles north, 22 miles northeast, and then 32 miles south.   How far is the destroyer from its original position?

23. In rowing across a stream 2.5 miles wide a fisherman is carried 1.5 miles downstream by the current.   How far did he travel in making the crossing?

24. A train, in order to go from city *A* to city *B*, must travel around the lake near which both cities are located.   The train must travel 25 miles north, 35 miles northeast, 25 miles east, 15 miles south, and finally 5 miles southwest.   An airplane may go directly from *A* to *B*.   What would the cost of the air trip be if the fare is based on rate of 11 cents a mile?

25. Two trains travel between towns *A* and *B*.   The first train goes directly in a straight line from *A* to *B*.   The second train also goes to towns *C*, *D*, and *E*, which are located off the direct route.   Thus, the second train travels 42 miles northeast to *C*, then 65 miles north to *D*, 15 miles west to *E*, and finally 45 miles northeast to *B*.   What distance must the first train travel in going directly from *A* to *B*?   What is the bearing of *B* from *A*?

26. A small boat is propelled across a stream in a direction perpendicular to the banks of the stream at a rate of 4.5 mph.   If the stream is flowing at a rate of 2.5 mph, find the resulting velocity of the boat with respect to the earth. Since velocity is a vector quantity, both its magnitude and direction must be found.

27. A pilot points his aircraft toward the north. If it is flying at a speed of 175 mph and there is a 35-mph cross wind from the west, find the magnitude and direction of the velocity of the aircraft with respect to the earth.

28. If, a few seconds after being dropped from a B-29 bomber, a bomb has a horizontal velocity of 275 mph and a vertical velocity toward the earth of 75 mph, find the magnitude and direction of the resulting velocity of the bomb.

29. If one man is pulling eastward on a rope tied to a post with a force of 75 lb and another man is pulling northeastward on a different rope tied to the same pole with a force of 85 lb, find the magnitude and direction of the resulting force on the pole.

30. Four guy wires are attached to a vertical pole; they make angles of 35° with the pole and are evenly spaced around it. If each wire exerts a force of 125 lb at the point where it joins the pole, how much force in addition to that due to the weight of the pole must be borne by the foundation on which it rests?

6–4. Resolution of a Vector into Components

If a vector **A** is the sum of several vectors **B, C, D,** and so on, then these latter vectors are said to be **components** of **A.** Thus, in the example of Sec. 6–3, **H** has the components 2**A, D,** −3**F,** and **G.**

Given a vector **A,** it is possible to find a set of vectors whose sum is **A** in many ways. In other words, **A** has many sets of components. Figure 6–9 shows three sets of components for a vector **A.** In fact, we may draw

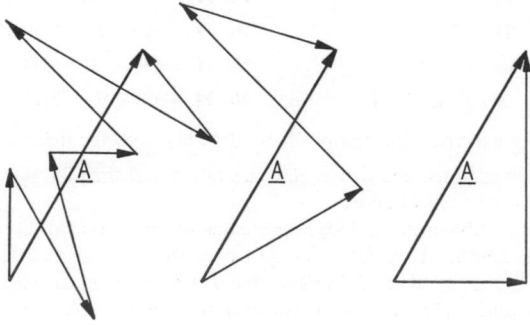

Fig. 6–9

components at random under the simple restrictions that the initial point of the first component is the initial point of **A,** that the initial point of each succeeding vector is the terminal point of its predecessor, and that the last component must have as its terminal point the terminal point of **A.** Thus every component but the last one may be chosen at our pleasure.

The process of finding some vectors of which a given vector is the sum is called **resolution into components.**

A vector may be resolved into components parallel to two given non-parallel vectors as shown in Fig. 6–10. **A** is the given vector, and we

desire components parallel to **B** and **C**. The components can be constructed easily by drawing lines through the initial and terminal points
parallel to **B** and **C**.   Three ways of finding these components are shown.

We shall use this process most often when two components of a given
vector are desired that are perpendicular to each other.   Usually one
component will be horizontal and the other vertical.   This will be discussed in the next section.

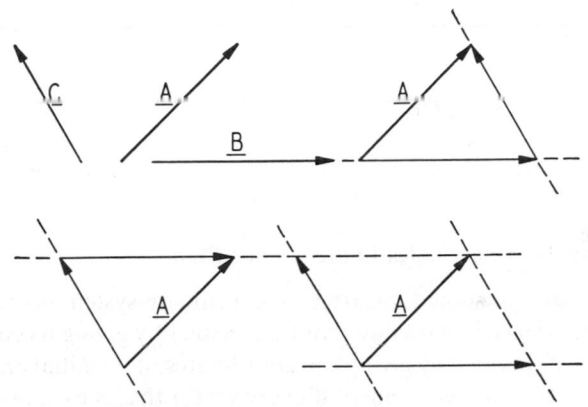

FIG. 6–10

Exercises

The vectors **A, B, C, D, E, F,** and **G** given below are those of Fig. 6–7.   In each of the
following problems make your constructions carefully with a ruler and a protractor.
Resolve the vector given first into components parallel to the last vectors.

1. E; D, F.

2. 3A; B, G.

3. 2F; A, C.

4. 3C; E, F.

5. C; A, D.

6. 3C; F, G.

7. 5E; B, D.

8. 4E; A, C.

9. 2C; B, D.

10. −3F; E, G.

11. −3B; A, D.

12. −3G; E, F.

13. −2A; D, E.

14. −4C; A, C.

15. 2D; D, E.

Resolve each of the vectors given below into horizontal and vertical components.

16. 3B.

17. −2B.

18. 3D.

19. 5F.

20. −3F.

21. 4G.

22. −4G.

23. −5D.

24. 2C.

25. −3A.

26. 2E.

27. −3E.

In the following exercises find the result by drawing a vector diagram.

28. A ship steams 56 miles on a bearing of 66°.   Find the distances east and north
sailed by the ship.

29. A man is pulling with a force of 65 lb on a rope that is fastened to the front of a sled and makes an angle of 35° with the horizontal. What force is being exerted to draw the sled along the ground? What force is lifting the front of the sled?

30. An airplane is flying on a bearing of 125° at 275 mph. Find the components of this velocity in the eastern and southern directions.

31. A platform is supported by two pipes of equal length that form the legs of an isosceles triangle with a horizontal base. The pipes come together at the vertex of the triangle to make an angle of 34°; the force **F** exerted by the platform at this point has a magnitude of 335 lb and a direction vertically downward. The force **F** creates two equal forces acting along the pipes; the sum of these forces is **F**. Determine the magnitude of the forces acting along the pipes and of their horizontal and vertical components. Show that the sum of the vertical components is **F** and that the sum of the horizontal components is 0.

32. A slingshot can be made from a rubber band fastened at the ends to the forks of a forked stick. If the rubber band is pulled back until the angle between its two sides is 28°, and the tension directed along the band on each side is 25 lb, what is the magnitude of the force available for hurling a stone forward?

6–5. Rectangular Coordinates in the Vector Plane

Suppose that a rectangular Cartesian coordinate system is placed in the vector plane. Since a point now can be specified by giving its coordinates, a vector can be specified by giving the coordinates of its initial and terminal points. However, it is convenient if every vector that is used in operations has its initial point at the origin of the coordinate system. *Every vector mentioned henceforth, unless specifically indicated otherwise, will have the origin as its initial point. With this agreement, a vector can be specified by the coordinates of its terminal point.*

Let **A** and **B** be two vectors whose initial points are at the origin and which lie along the x axis. Since the magnitude A of the vector **A** is positive, if **A** is to the right of the origin its terminal point is $(A, 0)$. If **A** extends to the left of the origin its terminal point is $(-A, 0)$. Thus, if **A** and **B** both extend to the right of the origin, the terminal point of **A** + **B** is $(A + B, 0)$. Likewise, if **A** and **B** both extend to the left of the origin, the terminal point of **A** + **B** is $(-A - B, 0)$. If **A** is to the right of the origin, and **B** is to the left, the terminal point of **A** + **B** is $(A - B, 0)$.

These remarks establish the following very useful result:

Given a set of vectors each of which:

(a) Has initial point at the origin.

(b) Extends along the x axis.

Then the vector which is the sum of the vectors of the set:

(a) Extends along the x axis.

(b) Has its initial point at the origin.

(c) Has a terminal point whose x coordinate is the sum of the x coordinates of the terminal points of the vectors of the set.

A similar statement holds, of course, for a sum of vectors that have initial points at the origin and extend along the y axis. What happens in subtraction of vectors can be easily inferred.

Let $P(x, y)$ be the terminal point of a vector **A** whose initial point is at

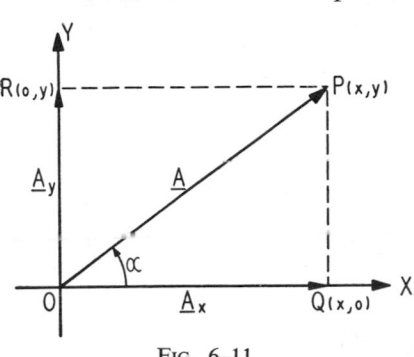

the origin, as shown in Fig. 6–11. By dropping perpendiculars from P to the axes it can be seen that **A** is the sum of two components $\mathbf{A}_x$ and $\mathbf{A}_y$ along the x and y axes, respectively. $\mathbf{A}_x$ is called the **x component** of **A**, and $\mathbf{A}_y$ is called the **y component** of **A**. We see that the terminal point of $\mathbf{A}_x$ is $Q(x, 0)$ and that the terminal point of $\mathbf{A}_y$ is $R(0, y)$.

FIG. 6–11

Let **A** and **B** be two vectors with initial points at the origin and terminal points $P_A(x_A, y_A)$ and $P_B(x_B, y_B)$, respectively. We may write these vectors as the sum of their x and y components:

Then,
$$\mathbf{A} = \mathbf{A}_x + \mathbf{A}_y, \quad \mathbf{B} = \mathbf{B}_x + \mathbf{B}_y.$$

$$\mathbf{C} = \mathbf{A} + \mathbf{B} = \mathbf{A}_x + \mathbf{A}_y + \mathbf{B}_x + \mathbf{B}_y = (\mathbf{A}_x + \mathbf{B}_x) + (\mathbf{A}_y + \mathbf{B}_y).$$

If we set $\mathbf{C}_x = \mathbf{A}_x + \mathbf{B}_x$ and $\mathbf{C}_y = \mathbf{A}_y + \mathbf{B}_y$, then $\mathbf{C} = \mathbf{C}_x + \mathbf{C}_y$, and, by the previous results of this section, the terminal point of $\mathbf{C}_x$ is $(x_A + x_B, 0)$, and the terminal point of $\mathbf{C}_y$ is $(0, y_A + y_B)$. Then, obviously, the terminal point of $\mathbf{C} = \mathbf{A} + \mathbf{B}$ is $P(x_A + x_B, y_A + y_B)$. A similar result can be established for subtraction.

These facts may be summarized as follows.

Given the vectors with initial points at the origin and terminal points as follows:

$$\mathbf{A}, P_A(x_A, y_A),$$
$$\mathbf{B}, P_B(x_B, y_B),$$
$$\mathbf{C}, P_C(x_C, y_C).$$

Then $\mathbf{A} + \mathbf{B} - \mathbf{C}$ is a vector with initial point at the origin and terminal point $P(x_A + x_B - x_C, y_A + y_B - y_C)$.

Example 1. Given **A**, **B**, **C**, and **D** with initial points at the origin and terminal points, respectively, as follows: $P_A(2, 5)$, $P_B(5, 3)$, $P_C(-3, 2)$, $P_D(-4, -2)$. Find the vector $\mathbf{E} = \mathbf{A} + \mathbf{B} + \mathbf{C} - \mathbf{D}$.

The terminal point of **E** has x coordinate $2 + 5 - 3 - (-4) = 8$ and y coordinate $5 + 3 + 2 - (-2) = 12$. Thus the terminal point of **E** is $P_E(8, 12)$. This result is verified in Fig. 6–12 by constructing the geometrical vector addition by dotted lines.

It was seen in Sec. 6–3 that a vector could be specified by giving its magnitude, its initial point, and the angle it makes with a given vector. Let **A** be a vector with initial point at the origin, magnitude A, and let **A**

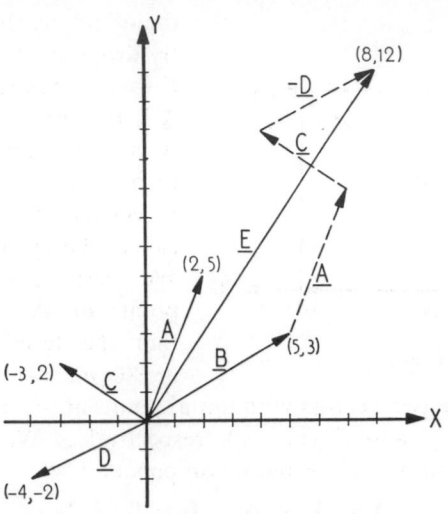

FIG. 6–12

form the terminal side of an angle α in standard position. The angle α is called the **direction angle** of **A**. If the terminal point of **A** is denoted by $P(x, y)$, the values of x and y can be found as follows:

$$\sin \alpha = \frac{y}{A}, \qquad \cos \alpha = \frac{x}{A}$$

from which

$$x = A \cos \alpha, \qquad y = A \sin \alpha. \tag{1}$$

Suppose that $P(x, y)$ is given and the values of A and α are desired (Fig. 6–11). From the Pythagorean theorem

$$A = \sqrt{x^2 + y^2}, \tag{2}$$

and from the definitions of the trigonometric functions

$$\tan \alpha = \frac{y}{x}. \tag{3}$$

Of course, equation 3 does not determine α uniquely, but the quadrant in which P lies determines the quadrant in which the terminal side of α should fall.

If the x and y components of $\mathbf{A}$ are A_x and A_y, respectively, we see that $A_x = |x|$ and $A_y = |y|$, whence $A_x{}^2 = x^2$ and $A_y{}^2 = y^2$. Then from equation 2,

$$A = \sqrt{A_x{}^2 + A_y{}^2}. \tag{4}$$

Obviously, the terminal point of the vector $-\mathbf{A}$ is $(-x, -y)$.

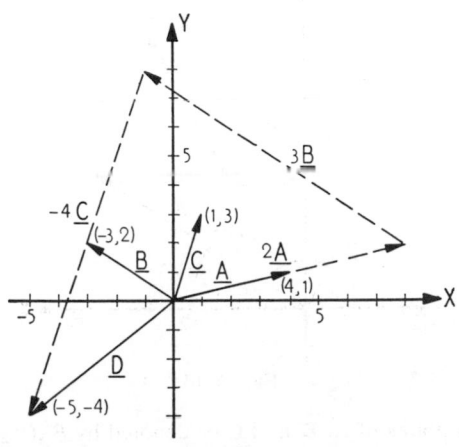

FIG. 6–13

Let a be a positive scalar. Then the vector $a\mathbf{A}$ has magnitude aA and the same direction angle α as $\mathbf{A}$. $P(x, y)$ is the terminal point of $\mathbf{A}$, and, if we let $P_a(x_a, y_a)$ be the terminal point of $a\mathbf{A}$, we have by equation 1

$$x = A\cos \alpha, \qquad y = A \sin \alpha;$$

$$x_a = aA \cos \alpha, \qquad y_a = aA \sin \alpha.$$

Solving for x_a and y_a, we obtain

$$x_a = ax, \qquad y_a = ay.$$

Thus the terminal point of $a\mathbf{A}$ is (ax, ay). Since, if a is a negative scalar, $a\mathbf{A} = |a|(-\mathbf{A})$, the terminal point of $a\mathbf{A}$ is $(|a|[-x], |a|[-y])$, which can be written (ax, ay). Hence, *if a is any scalar and $P(x, y)$ is the terminal point of a vector $\mathbf{A}$ with initial point at the origin, the terminal point of $a\mathbf{A}$ is (ax, ay).*

Example 2. Given the vectors $\mathbf{A}$, $\mathbf{B}$, and $\mathbf{C}$ with the terminal points $P_A(4, 1)$, $P_B(-3, 2)$, and $P_C(1, 3)$, respectively. Find the terminal point of $\mathbf{D} = 2\mathbf{A} + 3\mathbf{B} - 4\mathbf{C}$, and verify this result by adding the vectors graphically.

The x coordinate of the terminal point of $\mathbf{D}$ is $2 \cdot 4 + 3(-3) - 4 \cdot 1 = -5$, and the y coordinate is $2 \cdot 1 + 3 \cdot 2 - 4 \cdot 3 = -4$. Thus the terminal point of $\mathbf{D}$ is $(-5, -4)$. The graphical addition is shown in Fig. 6–13.

Example 3.   The vectors **A** and **B** have initial points at the origin and direction angles $\alpha = 33.2°$ and $\beta = 119.3°$, respectively.   If $A = 25.6$ and $B = 14.2$, find the magnitude of $\mathbf{C} = \mathbf{A} + \mathbf{B}$, and its direction angle, using the slide rule and obtaining results accurate to three significant figures.   Verify the result by a graphical construction.

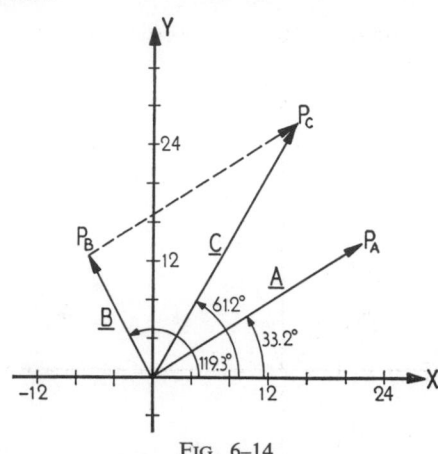

FIG. 6–14

Let the terminal points of **A**, **B**, and **C** be denoted by $P_A(x_A, y_A)$, $P_B(x_B, y_B)$, and $P_C(x_C, y_C)$, respectively.   Then

$$x_A = 25.6 \cos 33.2° = 25.6\,(0.837) = 21.4,$$
$$y_A = 25.6 \sin 33.2° = 25.6\,(0.548) = 14.0,$$
$$x_B = 14.2 \cos 119.3° = 14.2\,(-0.489) = -6.94,$$
$$y_B = 14.2 \sin 119.3° = 14.2\,(0.872) = 12.4.$$

It follows that

$$x_C = 21.4 - 6.94 = 14.46 = 14.5,$$
$$y_C = 14.0 + 12.4 = 26.4,$$

and that

$$C = \sqrt{(14.5)^2 + (26.4)^2} = \sqrt{210 + 697} = \sqrt{907} = 30.1.$$

If γ is the direction angle of **C**,

$$\tan \gamma = \frac{26.4}{14.5} = 1.82, \quad \text{and} \quad \gamma = 61.2°.$$

It is frequently more convenient (especially when the slide rule is being used for computing) to avoid the square root of a sum of squares that arises in computing C from the Pythagorean theorem, as was done above.   In this instance, after the angle γ has been computed, the side C may be determined by means of one of the formulas

$$C = \frac{y_C}{\sin \gamma}, \qquad C = \frac{x_C}{\cos \gamma},$$

which are better adapted to slide-rule computation than the square-root-of-the-sum-of-squares formula.   The student should verify that both of these formulas give the result $C = 30.1$ in this example.

The graphical construction is shown in Fig. 6–14.

Exercises

Each of the given vectors has initial point at the origin and is given by its terminal point. Given **A**: (3, 2); **B**: (1, 4); **C**: (0, 3); **D**: (−2, −2); **E**: (−5, 0); **F**: (−3, −2); **G**: (−1, −4); **H**: (2, −3); and **I**: (4, −1). Find the coordinates of the terminal point of each of the following vectors **J**, and verify your result by performing the operations graphically.

1. **J** = **B** + **C** + **D** + **E**.
2. **J** = **E** + **F** + **G** + **H**.
3. **J** = **A** + **C** + **E** + **G**.
4. **J** = **B** + **D** + **F** + **H**.
5. **J** = 3**A** + 4**B** + **C**.
6. **J** = 2**A** + 2**C** + 3**D**.
7. **J** = **B** − 2**C** + **D**.
8. **J** = 2**B** − 5**E** + 4**C**.
9. **J** = 2**F** + **G** + 3**I**.
10. **J** = 3**D** + **E** − 4**G**.
11. **J** = **H** − 3**I** + 4**F**.
12. **J** = 3**E** − 2**F** + 4**G**.
13. **J** = 3**C** − 5**D** + 4**E** + **G**.
14. **J** = 2**B** − 5**C** + 3**D** − **E**.
15. **J** = 3**F** − 4**A** + 2**C** + **I**.
16. **J** = 2**A** + **B** − 3**C** + 4**D**.
17. **J** = 2**A** − 5**I** + 3**B** − **D**.
18. **J** = 2**E** − 5**D** + 3**C** + **H**.
19. **J** = 2**H** + **I** − 5**C** + 4**D**.
20. **J** = **D** − 3**H** − 4**A** − 2**B**.

Each of the points given is the terminal point of a vector with initial point at the origin. Find the magnitude of the vector correct to three significant figures, and the direction angle which is between 0° and 360° correct to the nearest tenth of a degree. Use the slide rule in all computations.

21. (4, 5).
22. (3, −4).
23. (5, −12).
24. (8, −9).
25. (−7, −3).
26. (0, 8).
27. (7, 0).
28. (−7, −8).
29. (−3, −5).
30. (12, 27).
31. (13, −42).
32. (−27, 42).
33. (36, 27).
34. (−27, −39).
35. (−18, −14).
36. (25, 0).
37. (0, −32).
38. (−14, −14).

In each exercise below the magnitude and direction angle of a vector with initial point at the origin are given. Find the coordinates of the terminal point of each vector correct to three significant figures, using the slide rule in computations.

39. 20.0, 46.7°.
40. 32.0, 72.6°.
41. 14.6, 128.4°.
42. 3.75, 67.3°.
43. 4.37, 83.4°.
44. 9.69, 208.4°.
45. 127, 167.1°.
46. 369, 242.3°.
47. 529, 298.6°.
48. 2130, 302.5°.
49. 3240, 335.0°.
50. 5690, −31.2°.
51. 0.679, −67.8°.
52. 0.325, −136.8°.
53. 42.6, −198.2°.
54. 2.34, −256.3°.
55. 5.69, −405.2°.
56. 7.34, −436.2°.

Each vector given below has its initial point at the origin. The direction angle is denoted by the Greek letter corresponding to its name, as **A**, α. Find the magnitude and direction angle of the sum of the vectors given in each exercise. Use the slide rule, and obtain results accurate to three significant figures and the nearest tenth of a degree.

57. **A**: $A = 28.5$, $\alpha = 26.2°$,
    B: $B = 19.7$, $\beta = 67.5°$.
58. **A**: $A = 0.979$, $\alpha = 48.2°$,
    B: $B = 0.763$, $\beta = 126.7°$.

59. A: $A = 236$, $\alpha = 158.5°$,
   B: $B = 396$, $\beta = 227.8°$.

60. A: $A = 4.32$, $\alpha = 25.8°$,
   B: $B = 5.97$, $\beta = 149.7°$.

61. A: $A = 0.0763$, $\alpha = 227.4°$,
   B: $B = 0.0267$, $\beta = 332.4°$.

62. A: $A = 596$, $\alpha = -67.4°$,
   B: $B = 732$, $\beta = 87.5°$.

63. A: $A = 0.00692$, $\alpha = -25.4°$,
   B: $B = 0.00327$, $\beta = 78.6°$.

64. A: $A = 47{,}200$, $\alpha = -39.6°$,
   B: $B = 21{,}300$, $\beta = -208.4°$.

65. A: $A = 1230$, $\alpha = -325.4°$,
   B: $B = 673$, $\beta = 193.6°$.

66. A: $A = 12.4$, $\alpha = -436.7°$,
   B: $B = 9.67$, $\beta = 195.9°$.

67. A: $A = 12.4$, $\alpha = 28.3°$,
   B: $B = 19.6$, $\beta = 99.9°$,
   C: $C = 27.5$, $\gamma = 213.5°$.

68. A: $A = 379$, $\alpha = 316.4°$,
   B: $B = 226$, $\beta = -108.5°$,
   C: $C = 767$, $\gamma = 93.6°$.

69. A: $A = 0.629$, $\alpha = 143.4°$,
   B: $B = 0.219$, $\beta = 236.2°$,
   C: $C = 0.967$, $\gamma = 11.5°$.

70. A: $A = 0.000362$, $\alpha = -306.9°$,
   B: $B = 0.000231$, $\beta = 323.4°$,
   C: $C = 0.000467$, $\gamma = 15.6°$.

71. A: $A = 4.32$, $\alpha = 409.6°$,
   B: $B = 2.19$, $\beta = 206.3°$,
   C: $C = 0.923$, $\gamma = 103.7°$.

72. A: $A = 52.3$, $\alpha = 56.3°$,
   B: $B = 36.2$, $\beta = -310.4°$,
   C: $C = 119.0$, $\gamma = 427.3°$.

6–6. Applications of Vectors

Since a vector is a mathematical entity with magnitude and direction as fundamental properties, many physical quantities with these two properties are customarily represented by vectors. So far in this chapter we have seen that four physical entities can be represented by vectors: displacement, velocity, force, and the location of a point (or a body). More advanced texts in mathematics and the physical sciences give many other applications of vectors which are beyond the scope of this book. However, in order to give the student an idea of how vectors can be used in connection with somewhat more advanced applications, the forces in connection with the inclined plane and the engine speed governor will be discussed briefly in this section. Later, in Chapter 10, we will find that certain quantities associated with alternating current and voltage can be represented by vectors.

In order to solve problems involving the application of forces, it is necessary to understand several fundamental physical principles. One of the first laws of mechanics states that, *when a body is at rest or moving with a constant velocity, the resultant force on that body is zero.* Thus, when a book is lying on a table, the action of gravity exerts a force downward, and the table must exert an equal and opposite force on the book in order that the resultant force on the book will be equal to zero. It is an experimental fact that *the force due to the action of gravity acts vertically downward.*

With these two fundamental principles of mechanics and the technique of vector manipulation, the student is equipped to consider the problem of the inclined plane. Wedges, factory chutes, and V-type automobile engines involve these basic principles.

Example 1. A loading platform is 4 ft high, and a board is placed from the platform to the ground, thus becoming an inclined plane. The board rests on the ground at a distance of 4 ft measured horizontally from the edge of the platform as shown in Fig. 6–15. If a box weighing 100 lb is being pushed up

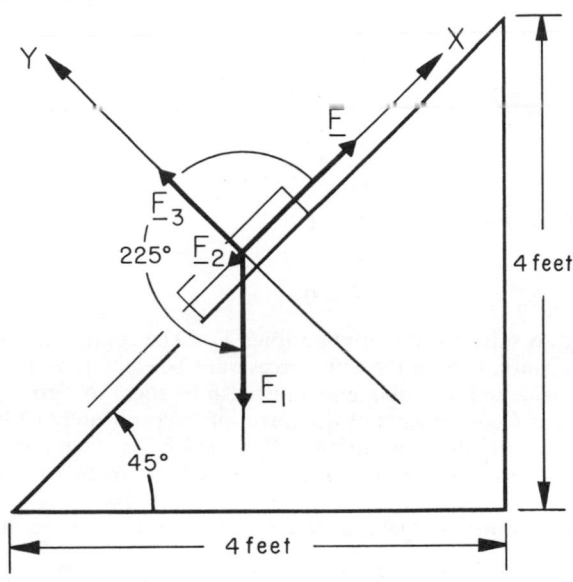

FIG. 6–15

the board with a constant velocity, and a frictional force of 10 lb acts in a direction opposite to that of the motion of the box, what is the magnitude and direction of the force being applied to the box?

It is convenient to superimpose a vector diagram on the drawing of the box on the inclined plane. Thus in Fig. 6–15 the downward force of 100 lb due to the action of gravity is represented by the vector F_1. The force that is being applied to push the box up the plane acts parallel to the plane and is represented by F. The given problem is to determine the magnitude of this vector. The frictional force of 10 lb acts in the opposite direction of F and is represented by F_2. Since a portion of the weight of the box is supported by the inclined plane, there is an additional force F_3, perpendicular to the plane, whose magnitude will be determined.

Since the box is assumed to be moving up the plane at a constant velocity, we may apply the law of mechanics which states that the resultant force acting on the box is zero; in other words, $F_1 + F_2 + F_3 + F = 0$.

It is convenient to introduce a rectangular coordinate system with origin at the common initial point of the four vectors and with positive x axis in the direction of **F**. The x axis is therefore parallel to the inclined plane, as shown in Fig. 6–15. Since the force **F**$_3$ is perpendicular to the inclined plane, the vector lies along the y axis. The force **F**$_1$ makes an angle of 45° with the inclined plane and hence makes an angle of 225° with the positive x axis. Thus the vector **F**$_1$ will have terminal point

$$x_1 = 100 \sin 225° = -70.7 \text{ lb},$$
$$y_1 = 100 \cos 225° = -70.7 \text{ lb}.$$

The terminal point of the frictional force vector **F**$_2$ will be given by

$$x_2 = -10 \text{ lb},$$
$$y_2 = 0.$$

The terminal point of **F**$_3$, the force due to the action of the inclined plane, will be

$$x_3 = 0,$$
$$y_3 = +F_3 \text{ lb}.$$

The applied force **F** will have terminal point

$$x = +F \text{ lb},$$
$$y = 0.$$

We are asked to solve for the applied force **F**. The condition that must be satisfied is that the resultant of the four forces must be equal to zero. If a force is equal to zero its x and y components must also be equal to zero. Therefore, the resultant x and y components of the forces of the problem must be equal to zero, to satisfy the required conditions. If we solve for these components in terms of the two unknowns F and F_3, two equations in two unknowns are obtained.

Adding the x coordinates algebraically and equating to zero, we obtain

$$x_1 + x_2 + x_3 + x = -70.7 - 10 + 0 + F = 0, \tag{1}$$
$$y_1 + y_2 + y_3 + y = -70.7 + 0 + F_3 + 0 = 0. \tag{2}$$

Solving equation 1 for F gives
$$F = 80.7 \text{ lb}. \tag{3}$$

Solving equation 2 for F_3 yields
$$F_3 = 70.7 \text{ lb}.$$

The problem asked for the applied force **F**, and hence, the answer is that $F = 80.7$ lb, *directed upward along the inclined plane.*

There is an important force associated with rotating machinery called **centrifugal force**. This force occurs whenever there is rotation of any type and is always directed outward from the axis of rotation. If a stone is fastened to the end of a string and whirled about, the stone tends to fly outward. This tendency to fly outward is due to the action of the centrifugal force. The inward constraining force exerted by the string on the stone keeps the stone in a circular path.

Another example of this force is clearly illustrated when a pail of water is whirled around in a vertical plane. The water will be forced to the bottom of the pail, and will not spill, even when the pail is upside down.

Centrifugal force depends on velocity and increases with it. An engine flywheel may fracture if rotated at too high a speed, because of the difference in the action of centrifugal force on various parts of the wheel, and this fact consequently must be considered in the design.

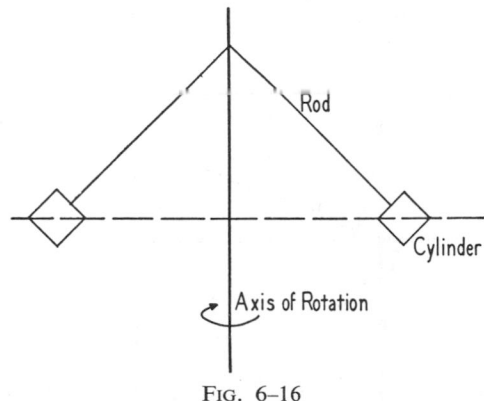

FIG. 6–16

The engine speed governor is one of the many engineering applications which utilize this force. Since the magnitude of the centrifugal force is a function of the speed, this force may be used to open or close a valve. Whenever the speed of the engine becomes higher than a predetermined value, the governor closes the valve in the steam supply, causing the engine to slow down. Exactly the opposite action occurs when an engine begins to run too slowly. A similar governing device coupled with frictional means is used to maintain constant speed of phonograph turntables.

Consider a simple type where the governor mechanism consists of two metallic cylinders, located directly opposite one another, which are rotated about an axis which cuts through the centre and is perpendicular to a line drawn between the two cylinders. The two cylinders are connected to the top of the shaft about which they rotate by means of rods from the top of the shaft to the cylinders, as indicated in Fig. 6–16.

The connection at the top of the shaft is flexible, in order that the cylinders may respond freely to the actions of gravity and of centrifugal force when they are rotating. A lever arrangement not shown in the figure transmits the movement of the cylinders to the valve of the steam engine.

Example 2.   The centrifugal force exerted by each brass cylinder of a steam-engine governor is 105 lb, and the force due to the action of gravity is 45 lb on each cylinder.   The governor is perfectly symmetrical, and its axis of rotation is vertical.   What will the angle be between the rod holding one of the cylinders and the shaft?

We first construct a vector diagram showing all the forces acting on the cylinder.   It is convenient to superimpose this diagram on an outline of the governor as shown in Fig. 6–17.   The centrifugal force $\mathbf{F_1}$ of magnitude 105 lb acts outward from and perpendicular to the axis of rotation, and the force $\mathbf{F_2}$ of 45 lb due to gravity acts vertically downward, as shown in the figure.   In addition there is a force $\mathbf{F}$ acting along the rod joining the cylinder to the shaft as shown.   The angle θ between the rod and the shaft is the angle that we wish to determine.

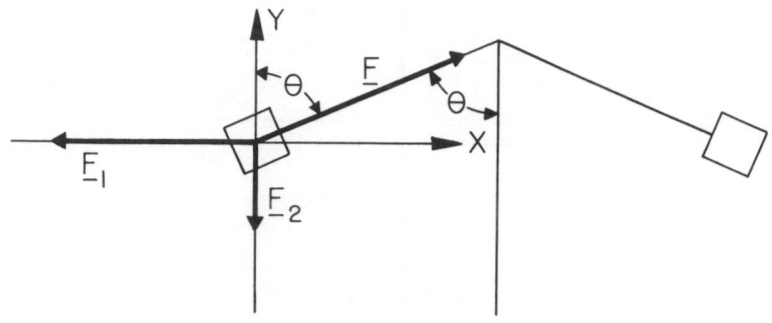

FIG. 6–17

It is also convenient to introduce a rectangular coordinate system with origin at the common initial point of the three vectors and x axis horizontal, as shown in Fig. 6–17.   Then the angle between $\mathbf{F}$ and the y axis can be seen to be θ. If the cylinder is not moving outward or inward (as we assumed) the resultant force acting on the cylinder is zero; in other words, $\mathbf{F_1} + \mathbf{F_2} + \mathbf{F} = 0$.

The terminal point of $\mathbf{F_1}$ is given by

$$x_1 = -105 \text{ lb}, \qquad y_1 = 0.$$

The terminal point of $\mathbf{F_2}$ is given by

$$x_2 = 0, \qquad y_2 = -45 \text{ lb}.$$

It is easy to establish that the terminal point of $\mathbf{F}$ is given by

$$x = F \sin \theta, \qquad y = F \cos \theta.$$

Since the resultant of these three forces must be equal to zero, the resultant of the vertical and horizontal components of these three forces must also be zero.   Thus

$$x_1 + x_2 + x = -105 + 0 + F \sin \theta = 0, \tag{1}$$

$$y_1 + y_2 + y = 0 - 45 + F \cos \theta = 0. \tag{2}$$

Rewriting equations 1 and 2, we obtain

$$F \sin \theta = 105, \tag{3}$$

$$F \cos \theta = 45. \tag{4}$$

Dividing equation 3 by equation 4, we obtain

$$\frac{\sin \theta}{\cos \theta} = \tan \theta = 2.3,$$

whence

$$\theta = 67°.$$

6–7. Variable Vectors

If a vector has its initial point at the origin of a system of rectangular coordinates, the vector can be specified by giving the coordinates (x, y) of its terminal point, as we saw in Sec. 6–5. Since the point can be thought of as specifying the vector or, conversely, the vector can be thought of as specifying the point, these two elements can be regarded in many situations as equivalent.

In Sec. 3–10 we saw how points in the plane can be located by giving their x and y coordinates as functions of a parameter. If this parameter is t, then we may write $x = x(t)$, $y = y(t)$, and a given value of t specifies a number pair (x, y) which in turn specifies a point in the plane. However, this point also specifies a vector $\mathbf{A}$. Thus the vector $\mathbf{A}$ depends on the parameter t, and, since it may change as t takes on different values, it is a variable vector. In order to display the parameter t on which $\mathbf{A}$ depends, the vector is often written as $\mathbf{A}(t)$. If a more complex description is desired, giving the functions $x(t)$ and $y(t)$ explicitly, the notation $\mathbf{A}(t)$, $P_A(x(t), y(t))$ may be used, in accordance with the notational conventions established in Sec. 6–5. This idea of a variable vector is fundamental to many higher applications of vectors, but further development of it is beyond the scope of this book.

Exercises

In each of the following exercises, give your results as precisely as the accuracy of the given data will allow. Construct a suitable graphical representation for each exercise.

1. If a man drives 76 miles west, then 37 miles northwest, and then 56 miles on a bearing of 342°, how far is he from his starting point?

2. A point B is 346 miles from a point A on a bearing of 136°. If a man starts from A and drives 214 miles east and then 85 miles on a bearing of 152°, how far and in what direction is he from B?

3. A destroyer leaves a cruiser and steams 52 miles on a bearing of 232° and then 38 miles on a bearing of 167°. In the meantime the cruiser steams 75 miles on a bearing of 185°. How far and in what direction is the cruiser from the destroyer?

4. A fighter plane leaves an aircraft carrier and flies 76 miles on a bearing of 37°. In the meantime the carrier cruises 22 miles on a bearing of 342°. What are the bearing and distance of the cruiser from the plane?

5. At a certain time two destroyers are 85 nautical miles apart, the second bearing 76° from the first. If the first destroyer steams on a bearing of 12° at 24 knots, and the second destroyer is instructed to meet the first destroyer in exactly 3 hr, at what speed and on what bearing should the second destroyer steam to make the meeting? (A knot is one nautical mile per hour.)

6. An airplane is flying with a speed through the air of 345 mph on a heading of 136° and there is a wind of 45 mph from 85°.   Find the resultant speed and direction of the plane with respect to the ground.

7. If an airplane is flying at an airspeed of 138 mph on a heading of 257° but is actually making good a speed of 125 mph and a direction of 242° with respect to the ground, what is the speed of the wind and from what direction is it blowing?

8. One destroyer is steaming at 25 knots on a bearing of 314° and another nearby destroyer is steaming at 21 knots on a bearing of 267°.   At what speed is the distance between the two destroyers increasing?   (A knot equals one nautical mile per hour.)

9. An aviator wishes to fly in a direction relative to the ground of 67°, and there is a wind of 55 mph from 136°.   If the airplane will cruise at an airspeed of 275 mph, in what direction must the airplane be headed, and what speed will the airplane make with respect to the ground?

10. A marine navigator finds that, although the speed of his tanker through the water has been 18 knots and his heading has been 129°, he is making good 21 knots on a bearing of 139° with respect to the ground.   What are the speed and direction of the current?

11. A box, standing on a board, weighs 218 lb and slides down with a constant speed if the board makes an angle of 38° with the horizontal.   Find the magnitude of the frictional force between the box and the board.

12. An inclined plane leading up to a loading platform makes an angle of 37° with the horizontal.   A box weighing 86 lb is being pushed up the board with a constant velocity. If a frictional force of 12 lb acts in a direction opposite to that of motion of the box, what are the magnitude and direction of the force applied to the box?

13. A man is pulling a sled at a constant speed along a level surface by means of a rope which makes an angle of 37° with the horizontal.   If a frictional force of 67 lb acts in a direction opposite to that of the motion of the sled, what force is the man exerting on the rope?   If an additional load added to the sled would not change the frictional force, would the man have to exert more pull on the rope to move the sled?   Would your answer be the same if the sled were being pulled uphill?

14. An object on an inclined plane weighs 76 lb and slides down the plane with a constant speed when the plane makes an angle of 23° with the horizontal.   Find the frictional force between the object and the plane.

15. An inclined plane makes an angle of 48° with the horizontal.   An object weighing 215 lb is being pushed up the plane with a constant velocity.   If a frictional force of 56 lb acts in a direction opposite to the direction of motion of the object, what are the magnitude and direction of the force applied to the object?

16. A man is pulling a box up an inclined plane by means of a rope fastened to the box which makes an angle of 27° with the inclined plane.   If the inclined plane makes an angle of 12° with the horizontal, if the force due to friction is 29 lb, and if the weight of the box is 235 lb, what force must be exerted by the man in order to move the box up the plane at a constant velocity?

17. For an engine speed governor of the type discussed in Sec. 6–6 the centrifugal force exerted by each cylinder is 76 lb, and the force due to gravity is 18 lb.   What is the angle between the rod holding one of the cylinders and the shaft which forms the axis of rotation?

18. For an engine speed governor of the type discussed in Sec. 6–6 the angle between the rod holding a cylinder and the shaft is 69° when the centrifugal force is 132 lb. What is the weight of the cylinder?

19. For an engine speed governor of the type discussed in Sec. 6–6 the weight of a cylinder is 15 lb. What centrifugal force is being exerted if the angle between the rod and the shaft is 32°?

20. For an engine speed governor of the type discussed in Sec. 6–6 the centrifugal force exerted by each cylinder is 129 lb and the force due to gravity is 32 lb. What is the angle between the rod holding one of the cylinders and the shaft?

21. For an engine speed governor of the type discussed in Sec. 6–6 the angle between a rod holding a cylinder and the shaft is 58° when the centrifugal force is 65 lb. What is the weight of the cylinder?

22. For an engine speed governor of the type discussed in Sec. 6–6 the weight of a cylinder is 23 lb. What centrifugal force is being exerted if the angle between the rod and the shaft is 42°?

23. An automobile in going around a curve at a constant speed has three forces acting on it: the force of gravity vertically downward, the force due to the friction between the tires and the road directed toward the center of the curve and parallel to the road, and the centrifugal force directed outward from the center of the curve in a horizontal direction. If the pavement around the curve is level, only the force of friction counteracts the centrifugal force. However, if the road is banked properly, the force due to the action of gravity assists the friction; the effects are similar to those of an inclined plane. If a road is banked at an angle of 18° and an automobile weighing 3540 lb goes around a curve so as to create a centrifugal force of 850 lb, what force must be overcome by friction if the automobile is not to skid?

24. For the situation described in exercise 23 a road is banked at an angle of 21° and an automobile weighing 4280 lb goes around a curve so as to create a centrifugal force of 1640 lb. What force must be overcome by friction if the automobile is not to skid?

25. A clothesline is 48 ft long and has its ends tied to posts at equal heights above the horizontal. A 23-lb weight at the center of the line pulls it sufficiently taut that we may consider the line as consisting of two straight segments. If the center of the line is 1.2 ft below a horizontal segment joining the ends of the line, how much force does the line exert on each pole?

26. For the situation described in exercise 25, plot a graph of the pull exerted on each pole as a function of the distance between the center of the line and the horizontal segment joining the ends of the line.

27. When there is a cross wind an airplane does not travel with respect to the ground in the direction in which it is headed. The angle between the heading of the airplane and the direction in which it is actually traveling with respect to the ground (called the track) is called the *drift angle*; this angle is called *right* if the track lies to the right of the heading, *left* if the track lies to the left of the heading. The navigator of an airplane has instruments which enable him to determine the drift angle, the heading, and the air speed (that is, the speed of the airplane with respect to the air). This information by itself is not sufficient to enable him to determine the speed and direction of the wind, since he does not know the ground speed (that is, the speed with respect to the ground). However, by observing the drift angle on two different headings (usually chosen about perpendicular to each other), the navigator can obtain enough information to determine the direction and speed of the wind; this method is called the *double-drift method*. For an airplane with an air speed of 175 mph the navigator observes a drift angle of 12° to the right from a heading of 47° and a drift angle of 17° to the left from a heading of 317°. What is the speed of the wind, and from what direction is it blowing? (Since it is standard practice in aerial navigation to solve such problems as this by graphical means, find your result by means of a carefully constructed graph.)

28. For an airplane with an air speed of 225 mph the navigator observes a drift angle of 8° to the left from a heading of 32° and a drift angle of 12° to the right from a heading of 302°. Determine the speed of the wind and the direction from which it is blowing by graphical means (see exercise 27.)

29. For an airplane with an air speed of 315 mph the navigator observes a drift angle of 5° to the right from a heading of 180° and a drift angle of 5° to the left from a heading of 225°. Determine the speed of the wind and the direction from which it is blowing by graphical means (see exercise 27).

30. For an airplane with an air speed of 135 mph the navigator observes a drift angle of 4° to the right from a heading of 52° and a drift angle of 14° to the right from a heading of 327°. Determine the speed of the wind and the direction from which it is blowing by graphical means (see exercise 27).

Progress Report

In this chapter we considered quantities that have direction (as well as magnitude) as a fundamental property, and developed a mathematical representation for such quantities. This representation was called a vector. The operations with vectors, including addition, subtraction, and multiplication by a scalar factor, were defined, and thus a vector algebra was constructed. The resolution of a vector into two components with given directions was considered. In particular the resolution of vectors into components along the x and y axis was considered both graphically and numerically and used as a convenient means for computing the resultant of a combination of vectors. Finally, the use of vectors in solving a number of simple applied problems was considered.

7

Algebraic Operations

In the previous chapters we considered certain simple algebraic expressions and attained a skill in manipulating them. However, in both theoretical and applied mathematics much more complex algebraic relations are encountered. Therefore, if we are to be able to solve a wide variety of practical and theoretical problems, it is necessary that we possess skill in manipulating these more complex algebraic relations. The purpose of this chapter is to develop this skill.

7–1. Special Products

Certain special products were given in Sec. 2–7. These are repeated here together with other new forms which are very useful in mathematics and its applications.

$$(a + b)(a - b) = a^2 - b^2. \tag{1}$$
$$(a + b)^2 = a^2 + 2ab + b^2. \tag{2}$$
$$(a - b)^2 = a^2 - 2ab + b^2. \tag{3}$$
$$a(x + y - z) = ax + ay - az. \tag{4}$$
$$(a + b)(x - y) = ax + bx - ay - by. \tag{5}$$
$$(x + a)(x + b) = x^2 + (a + b)x + ab. \tag{6}$$
$$(ax + b)(cx + d) = acx^2 + (ad + bc)x + bd. \tag{7}$$
$$(a + b + c)^2 = a^2 + b^2 + c^2 + 2ab + 2ac + 2bc. \tag{8}$$
$$(a + b)(a^2 - ab + b^2) = a^3 + b^3. \tag{9}$$
$$(a - b)(a^2 + ab + b^2) = a^3 - b^3. \tag{10}$$
$$(a + b)^3 = a^3 + 3a^2b + 3ab^2 + b^3. \tag{11}$$
$$(a - b)^3 = a^3 - 3a^2b + 3ab^2 - b^3. \tag{12}$$
$$(a + b)^4 = a^4 + 4a^3b + 6a^2b^2 + 4ab^3 + b^4. \tag{13}$$
$$(a - b)^4 = a^4 - 4a^3b + 6a^2b^2 - 4ab^3 + b^4. \tag{14}$$

The reader should satisfy himself concerning the correctness of the above forms by actual multiplication.

The above products can be described in words. For example, formula 8 states that *the square of a trinomial equals the sum of the squares of all the terms of the trinomial plus twice the product of each term by every term that follows it.* Similar statements may be supplied by the reader for the other formulas.

Algebraic computations may be shortened by the use of the above forms, as illustrated in the following examples.

Example 1. Find the product $(2L - 3)(4L + 5)$.

Using equation 7, where we substitute L for x, 2 for a, -3 for b, 4 for c, and 5 for d, we obtain
$$(2L - 3)(4L + 5) = 8L^2 - 2L - 15.$$

Example 2. Expand the expression $(Rx + Py + Qz)^2$.

Substituting Rx for a, Py for b, and Qz for c in equation 8, we obtain
$$(Rx + Py + Qz)^2 = R^2x^2 + P^2y^2 + Q^2z^2 + 2PRxy + 2QRxz + 2PQyz.$$

Example 3. Find the product $(x - 2y)(x^2 + 2xy + 4y^2)$.

Substituting x for a and $2y$ for b in equation 10, we obtain
$$(x - 2y)(x^2 + 2xy + 4y^2) = x^3 - 8y^3.$$

Example 4. Expand $(z + 2)^3$.

Substituting z for a and 2 for b in equation 11, we obtain
$$(z + 2)^3 = z^3 + 6z^3 + 12z + 8.$$

Example 5. Using formula 14, find $(0.99)^4$.

Substituting 1 for a and 0.01 for b in equation 14, we obtain
$$(0.99)^4 = (1 - 0.01)^4$$
$$= 1^4 - 4 \cdot 1^3 \cdot 0.01 + 6 \cdot 1^2 \cdot (0.01)^2 - 4 \cdot 1 \cdot (0.01)^3 + (0.01)^4 = 0.96.$$

Formulas 2, 3, 11, 12, 13, and 14 are particular cases of a general form called the **binomial expansion,** given by the following:

$$(a + b)^n = a^n + na^{n-1}b + \frac{n(n - 1)}{1 \cdot 2} a^{n-2}b^2$$

$$(15)$$

$$+ \frac{n(n - 1)(n - 2)}{1 \cdot 2 \cdot 3} a^{n-3}b^3 + \cdots + nab^{n-1} + b^n.$$

Formula 15 enables us to write a binomial raised to any power as a sequence of terms. We note that the binomial expansion has the following properties:

1. *The number of terms is $n + 1$.*
2. *The first term is a^n, and the last term is b^n.*

3. *The exponents of a decrease by* 1, *those of b increase by* 1 *from term to term, and the sum of these two exponents in each term is n.*

4. *The coefficients are symmetric in the sense that the coefficients of terms equidistant from the ends of the binomial expansion are the same.*

The reader should satisfy himself that these properties are displayed by formulas 2, 3, 11, 12, 13, and 14, which are particular cases of the binomial expansion.

Example 6.   Expand $(x^2 + 2y)^5$.

Substituting x^2 for a, $2y$ for b, and 5 for n in equation 15, we obtain

$$(x^2 + 2y)^5 = [(x^2) + (2y)]^5$$

$$= (x^2)^5 + 5(x^2)^4(2y) + \frac{5 \cdot 4}{1 \cdot 2}(x^2)^3(2y)^2 +$$

$$\frac{5 \cdot 4 \cdot 3}{1 \cdot 2 \cdot 3}(x^2)^2(2y)^3 + \frac{5 \cdot 4 \cdot 3 \cdot 2}{1 \cdot 2 \cdot 3 \cdot 4}(x^2)(2y)^4 + \frac{5 \cdot 4 \cdot 3 \cdot 2 \cdot 1}{1 \cdot 2 \cdot 3 \cdot 4 \cdot 5}(2y)^5$$

$$= x^{10} + 10x^8y + 40x^6y^2 + 80x^4y^3 + 80x^2y^4 + 32y^5.$$

Example 7.   Use the binomial expansion to find the value of $(1.01)^{10}$, correct to five significant figures.

Substituting 1 for a, 0.01 for b, and 10 for n in equation 15, we obtain

$$(1.01)^{10} = (1 + 0.01)^{10}$$

$$= 1 + 10(0.01) + \frac{10 \cdot 9}{1 \cdot 2}(0.01)^2 + \frac{10 \cdot 9 \cdot 8}{1 \cdot 2 \cdot 3}(0.01)^3$$

$$+ \frac{10 \cdot 9 \cdot 8 \cdot 7}{1 \cdot 2 \cdot 3 \cdot 4}(0.01)^4 + \cdots$$

$$= 1 + 0.10 + 0.0045 + 0.000120 + 0.00000210 + \cdots$$

$$= 1.1046.$$

It is clear that in order to obtain a result to five significant figures the terms beyond the fourth one may be neglected.

Exercises

Perform the indicated operations.

1. $(7P - 5)(7P + 5)$.

2. $(3a + 5b)^2$.

3. $(5k - 1)^2$.

4. $(8pq + 5)(-8pq + 5)$.

5. $(3a + 7x)(a - 5x)$.

6. $(3M + 1)(2N - p)$.

7. $(x^2 - 3a)^2$.

8. $(2r_1^2 + 3r_2^2)^2$.

9. $(x + y)(e_1 - e_2)$.

10. $(2x + 5)(x - 3)$.

11. $(a + 5)(a - 7)$.

12. $(3q + 1)(5q - 7)$.

13. $(5x^2 + 2)(3x^2 - 1)$.

14. $(3 - 2\alpha)(-5 - \alpha)$.

15. $(7R^2 + 2)(1 - 2R^2)$.

16. $(2a^2 + 3b^2)(-4a^2 + 5b^2)$.

17. $(S_1 + S_2 + S_3)^2$.

18. $(i_1 + i_2 - i_3)^2$.

19. $(a - b - c)^2$.

20. $(x + 2y + z)^2$.

21. $(p + q - 2)^2$.

22. $(r_1 + r_2 + 1)^2$.

23. $(3Z_1 - Z_2 + 2)^2$.

24. $(2E_1 + E_2 + 3)^2$.

25. $(w + 2v + 3z)^2$.

26. $(x^2 - xy + y^2)^2$.

27. $(L^2 + 2L + 3)^2$.

28. $(p^2 - 3pq + 2q^2)^2$.

29. $(2a + 3)(4a^2 - 6a + 9)$.

30. $(3x - 4)(9x^2 + 12x + 16)$.

31. $(a^2 - b^2)(a^4 + a^2b^2 + b^4)$.

32. $(1 + y + y^2)(1 - y)$.

33. $(r_1 + r_2)(r_1{}^2 - r_1r_2 + r_2{}^2)$.

34. $(e_1{}^2 + e_2{}^2)(e_1{}^4 - e_1{}^2e_2{}^2 + e_2{}^4)$.

35. $\left(\dfrac{1}{2} - xy\right)\left(\dfrac{1}{4} + \dfrac{xy}{2} + x^2y^2\right)$.

36. $\left(\dfrac{v_1{}^2}{9} - \dfrac{v_1v_2}{15} + \dfrac{v_2{}^2}{25}\right)\left(\dfrac{v_1}{3} + \dfrac{v_2}{5}\right)$.

37. $(4 + 3t)^3$.

38. $(1 - f)^4$.

39. $(m^2 - d)^4$.

40. $(2h + 3k)^3$.

41. $(E_1 - E_2)^5$.

42. $(I_1 + I_2)^6$.

43. $(x + y)^7$.

44. $(p - q)^7$.

45. $\left(\dfrac{x^2}{2} + \dfrac{y}{3}\right)^4$.

46. $\left(\dfrac{a}{4} - \dfrac{b}{3}\right)^4$.

47. $\left(x^2 - \dfrac{1}{x}\right)^5$.

48. $\left(m^3 + \dfrac{1}{m^2}\right)^3$.

49. $\left(\dfrac{x}{y} + \dfrac{y}{x}\right)^4$.

50. $\left(\dfrac{a}{b} - \dfrac{b}{a}\right)^3$.

51. $(2x^m + 3y^n)^2$.

52. $(2a^{m+n} - 3a^{m-n})^2$.

53. $(3x^n + 5y^m)(12x^n - 15y^m)$.

54. $(x^{2n-1} - 2y^{n+1})(3x^{2n-1} + 4y^{n+1})$.

55. $(a^{2m} + 1)(a^{4m} - a^{2m} + 1)$.

56. $(x^{3m} - y^{3n})(x^{6m} + x^{3m}y^{3n} + y^{6n})$.

Verify the following identities:

57. $(x^2 + y^2)(r^2 + s^2) - (xr + ys)^2 = (xs - yr)^2$.

58. $(x + y + c)^2 + (x - y)^2 + (x - c)^2 + (y - c)^2 = 3(x^2 + y^2 + c^2)$.

59. $(I_1 - I_2)^3 + 3(I_1 - I_2)^2(I_1 + I_2) + (I_1 + I_2)^3 + 3(I_1 - I_2)(I_1 + I_2)^2 = 8I_1{}^3$.

60. $(a+b+c)^3 = a^3+b^3+c^3+3a^2b+3ab^2+3a^2c+3ac^2+3b^2c+3bc^2+6abc$.

Use the binomial theorem to evaluate each of the following quantities correct to five significant figures.

61. $(1.02)^5$.

62. $(0.99)^8$.

63. $(0.98)^{10}$.

64. $(1.003)^7$.

65. $(2.02)^5$.

66. $(1.999)^8$.

67. $(0.96)^{10}$.

68. $(1.997)^6$.

69. $(1.01)^{20}$.

70. $(0.99)^{30}$.

71. $(2.001)^{10}$.

72. $(1.0005)^{50}$.

7–2. Factoring

In Sec. 7–1 we learned from formulas 1 through 15 how to write by inspection the products of certain special types of forms.   As in Sec. 2–8, we may now interpret formulas 1 through 14 of the preceding section as working from right to left and thus obtain the process of factoring.   In the problems of the present section we shall be given the product, and we shall have to discover its factors.   The products that we shall learn to factor are special products of the type given in Sec. 7–1.   To follow the work of the present article, the reader must be well acquainted with the formulas of the preceding section.   The ability to factor an expression depends on one's ability to identify it with one or more forms of these special products.   The reader's efficiency in factoring will increase with careful reading and working of the illustrative examples and exercises that follow.

Formula 7 of the preceding section enables us to factor trinomials like $mx^2 + px + q$.   Such expressions can sometimes be factored by inspection into two binomials $(ax + b)(cx + d)$, where $ac = m$, $bd = q$, and $ad + bc = p$.   As possibilities for a and c we take the various pairs of factors of m (the number that multiplies x^2); as possibilities for b and d we take the pairs of factors of q (the number free of x).   We then try these various possibilities to give $ad + bc = p$.   To illustrate this method consider:

Example 1.   Factor $5x^2 + 16x + 3$.

The factors of 5 are 1 and 5.   The factors of 3 are 1 and 3.   By trial and error we find that $(1 \cdot 1) + (3 \cdot 5) = 16$, and hence

$$5x^2 + 16x + 3 = (x + 3)(5x + 1).$$

Example 2.   Factor $21x^3 - 14x^2y - 56xy^2$.

Using formula 4 of Sec. 7–1, we have

$$21x^3 - 14x^2y - 56xy^2 = 7x(3x^2 - 2xy - 8y^2).$$

To factor the trinomial $3x^2 - 2xy - 8y^2$, we find that the factors of 3 are 1 and 3, and that the factors of 8 are 1 and 8 or 2 and 4.   By trial and error we find that $(1 \cdot 4) - (2 \cdot 3) = -2$, and hence

$$3x^2 - 2xy - 8y^2 = (3x + 4y)(x - 2y).$$

Finally we have the result

$$21x^3 - 14x^2y - 56xy^2 = 7x(3x + 4y)(x - 2y).$$

Obviously not every trinomial of this form can be factored in this manner.   Thus, for example, in $6R^2 + 5R + 4$ no such pair of binomials can be found.

Example 3. Factor $x^3 - 27y^3$.

Using formula 10 of Sec. 7–1, we have
$$x^3 - 27y^3 = x^3 - (3y)^3 = (x - 3y)(x^2 + 3xy + 9y^2).$$

Example 4. Factor $9R^2 + 16R^2I^2 + 25 - 24R^2I + 30R - 40RI$.

Using formula 8 of Sec. 7–1, we have
$$9R^2 + 16R^2I^2 + 25 - 24R^2I + 30R - 40RI$$
$$= (3R)^2 + (-4RI)^2 + 5^2 + 2(3R)(-4RI) + 2(3R)(5) + 2(-4RI)(5)$$
$$= (3R - 4RI + 5)^2.$$

Example 5. Factor $12a^2 - 8a^3 - 6a + 1$.

Rearranging the order of the terms and applying formula 12 of Sec. 7–1, we obtain
$$1 - 6a + 12a^2 - 8a^3 = 1^3 - 3(1)^2(2a) + 3 \cdot 1(2a)^2 - (2a)^3 = (1 - 2a)^3.$$

Exercises

Factor the following expressions:

1. $x^2 - 11x - 60$.
2. $3y^2 - 5y - 2$.
3. $10a^2 + 17a - 6$.
4. $28 - 15b + 2b^2$.
5. $6e^2 - 7e - 20$.
6. $3z^2 + 7z + 2$.
7. $(a + 2)^2 - (3a - 1)^2$.
8. $49p^2 - (4q + 3r)^2$.
9. $1 - y^3$.
10. $E_1^3 - E_2^3$.
11. $r^3 + s^3$.
12. $a^3 - 8b^3$.
13. $6y^2 - 3y - 3$.
14. $9e_1^2 + 6e_1e_2 + e_2^2$.
15. $7x^2 - x - 6$.
16. $3a^2 - 5ab - 2b^2$.
17. $a^4 + 11a^2 - 42$.
18. $x^4 + 13x^2 + 36$.
19. $8x^3 - 125y^3$.
20. $i_1^3 - 8i_2^3$.
21. $8m^3 + 27n^3$.
22. $729y^3 + 1$.
23. $a^3 + (b + c)^3$.
24. $(x - y)^3 + z^3$.
25. $x^6 - y^6$.
26. $a^6 + b^6$.
27. $18m^2 - 24mnp + 8n^2p^2$.
28. $(y + 1)^3 - (y - 1)^3$.
29. $(x + y + z)^2 - 36$.
30. $(e_1 + e_2)^2 - 4(e_3 - e_4)^2$.
31. $x^2+y^2+z^2+2xy+2xz+2yz$.
32. $a^2+b^2+4c^2-2ab+4ac-4bc$.
33. $12X^2 + 20X - 8$.
34. $21r^2 + 28r - 28$.
35. $2a^4 + 4a^2 + 2$.
36. $5x^4 + 4x^2y - 12y^2$.
37. $81P^4 - 16Q^4$.
38. $a^3x^6 + b^3x^9$.
39. $20x^2 - 5xy - 15y^2$.
40. $-10a^2 - 29ab + 21b^2$.
41. $10Z_1^2 - 11Z_1Z_2 - 6Z_2^2$.
42. $10v^2 - v - 3$.
43. $27 - (x + y + z)^3$.
44. $(u + 1)^3 + (u - 1)^3$.
45. $3\alpha^4 - 17\alpha^2 + 10$.
46. $5y^2 - 14yx - 3x^2$.
47. $12p^2 + 8p - 15$.
48. $4q^2 + 10q - 6$.
49. $3x^4 - 11x^2y + 6y^2$.
50. $6a^2 - 21ab^2 + 9b^4$.
51. $x^8 + 6x^4y - 91y^2$.
52. $36r_1^2 - 6r_1r_2 - 90r_2^2$.
53. $(r + s)^3 - (r - s)^3$.
54. $(x - y)^3 - 64$.

55. $(x_1 + x_2)^3 - (x_3 - x_4)^3$.

56. $(1 + S)^3 - (1 - S)^3$.

57. $64\alpha^3 + 27\beta^3$.

58. $a^9 + b^3$.

59. $(x + y)^3 + (z + w)^3$.

60. $512(x + 1)^3 + 1$.

61. $3(x + y)^2 + 7(x + y) + 2$.

62. $2(a - b)^2 - 13(a - b) + 20$.

63. $2(s_1 + s_2)^2 + 7(s_1 + s_2) + 5$.

64. $3(u + v)^2 + 16(u + v) - 12$.

65. $5x^{2n} - 7x^n - 6$.

66. $9y^{2m} + 6y^m - 8$.

67. $2x^{2n}y^{2m} + x^n y^m - 3$.

68. $2u^{2a}v^{2b} - 5u^a v^b - 12$.

69. $\dfrac{8}{L^3} - \dfrac{T^3}{27}$.

70. $\dfrac{1}{M^3} - \dfrac{27}{64}N^3$.

71. $r^2 + s^2 - 2rs - t^2$.

72. $u^3 - u + v - v^3$.

73. $z + z^3 - z^2 - 1$.

74. $r^2 + y + r - y^2$.

75. $1 - r_1{}^2 - 2r_1 r_2 - r_2{}^2$.

76. $p^2 - 1 + q^2 + 2pq$.

77. $x^2 - y^2 - z^2 + 2yz$.

78. $x^2 + y^2 - z^2 + 2xy$.

79. $R^2 - S^2 + R + S$.

80. $2r_1 r_2 - r_1{}^2 - r_2{}^2 + r_3{}^2$.

81. $m^3 + 8n^3 + 6m^2n + 12mn^2$.

82. $6x^3 + 12x^2y - 2x^2 - 4xy$.

83. $p^2 + 4r^2 + t^2 + 4pr + 4rt + 2pt$.

84. $x^2 + z^2 - 4yz + 2xz + 4y^2 - 4xy$.

85. $9x^2 + 4z^2 - 12xz + 16yz - 24xy + 16y^2$.

86. $K^2 - 2CL - 2CK + C^2 + 2KL + L^2$.

87. $9a^2 + 6ab + b^2 - 12ac - 4bc + 4c^2$.

88. $4x^2 + 9y^2 + z^2 - 12xy - 4xz + 6yz$.

89. $9x_1{}^2 + 16x_1{}^2 x_2{}^2 + 25 - 24x_1{}^2 x_2 + 30x_1 - 40x_1 x_2$.

90. $x^2 + m^2y^2 + n^2z^2 - 2mxy - 2nxz + 2mnyz$.

7–3. Operations with fractions

We can now add, subtract, multiply, and divide fractions whose denominators require factoring of the type explained in this chapter. The work of this present section is an extension to more complicated fractions of what was accomplished in Sec. 2–10 and Sec. 2–11. The reader should therefore review these sections before proceeding with the following illustrative examples and exercises.

Example 1. Simplifying,

$$\frac{1}{r + R} + \frac{r - R}{r^2 - rR + R^2} - \frac{r^2 - rR}{r^3 + R^3}$$

$$= \frac{1}{r + R} + \frac{r - R}{r^2 - rR + R^2} - \frac{r^2 - rR}{(r + R)(r^2 - rR + R^2)}$$

$$= \frac{r^2 - rR + R^2}{(r + R)(r^2 - rR + R^2)} + \frac{(r - R)(r + R)}{(r + R)(r^2 - rR + R^2)} - \frac{r^2 - rR}{(r + R)(r^2 - rR + R^2)}$$

$$= \frac{r^2 - rR + R^2 + r^2 - R^2 - r^2 + rR}{(r + R)(r^2 - rR + R^2)}$$

$$= \frac{r^2}{r^3 + R^3}.$$

Example 2. Simplifying,

$$\frac{a^3 - b^3}{a(a + b)} \cdot \frac{b^2 - a^2}{a^3b - a^4} \cdot \frac{a^4}{a^2 - b^2}$$

$$= \frac{(a - b)(a^2 + ab + b^2)}{a(a + b)} \cdot \frac{(b - a)(b + a)}{a^3(b - a)} \cdot \frac{a^4}{(a - b)(a + b)}$$

$$= \frac{a^2 + ab + b^2}{a + b} \cdot$$

Example 3. Simplifying,

$$\frac{6x^2 + 13xy - 5y^2}{2x^2 + 9xy + 9y^2} \div \frac{2x^2 - xy - 15y^2}{6x^2 + 13xy + 6y^2}$$

$$= \frac{6x^2 + 13xy - 5y^2}{2x^2 + 9xy + 9y^2} \cdot \frac{6x^2 + 13xy + 6y^2}{2x^2 - xy - 15y^2}$$

$$= \frac{(2x - 5y)(3x - y)}{(x + 3y)(2x + 3y)} \cdot \frac{(2x + 3y)(3x + 2y)}{(x - 3y)(2x + 5y)}$$

$$= \frac{(3x - y)(3x + 2y)}{(x + 3y)(x - 3y)} \cdot$$

Exercises

Perform the indicated operations and simplify:

1. $\dfrac{3x - 2y}{20} + \dfrac{x - 4y}{15}.$

2. $\dfrac{9 - 3p}{16q} + \dfrac{3 + 5p}{20q^2}.$

3. $\dfrac{11xy + 2}{x^2y^2} - \dfrac{5y^2 - 3}{xy^3} - \dfrac{6x^2 - 5}{x^3y}.$

4. $\dfrac{1}{a - 3c} - \dfrac{(a - 3c)^2}{a^3 - 27c^3}.$

5. $\dfrac{\mu + 1}{\mu^2 - 3\mu + 9} - \dfrac{\mu^2 + \mu + 1}{\mu^3 + 27}.$

6. $\dfrac{3}{i - r} + \dfrac{4r}{(i - r)^2} - \dfrac{5r^2}{(i - r)^3}.$

7. $\dfrac{35x^4}{4y^2z} \cdot \dfrac{8yz^3}{21x^7}.$

8. $\dfrac{a^4}{3b^2} \div 3b^2.$

9. $\dfrac{2a + ax}{x^2 + 4x + 4} \div \dfrac{2b - bx}{x^2 - 4}.$

10. $\dfrac{p^2q - 9q^3}{p^2 + 2pq - 3q^2} \cdot \dfrac{q - p}{3q - p}.$

11. $\dfrac{I_1^3 - I_2^3}{I_1^3 + I_2^3} \cdot \dfrac{I_1^2 - I_1I_2 + I_2^2}{I_1 - I_2}.$

12. $\dfrac{Y - 1}{XY + X} \cdot \dfrac{Y^3 + 1}{Y^3 - 1}.$

13. $\dfrac{\mu^4 - \beta^4}{(\mu - \beta)^2} \div \dfrac{\mu^2\beta + \beta^3}{\mu^3 - \beta^3}.$

14. $\dfrac{x^3 + a^3}{x^2 - 9a^2} \cdot \dfrac{x + 3a}{x + a}.$

15. $(E_g^2 - E_p^2) \div \dfrac{E_g^3 - E_p^3}{E_g + E_p}.$

16. $\left(a + b + \dfrac{b^2}{a}\right) \div (a^3 - b^3).$

17. $\dfrac{a^2 + ab + b^2}{a^3 - b^3} - \dfrac{1}{a - b}$.

18. $\dfrac{e_1{}^2 - e_1 e_2 + e_2{}^2}{e_1{}^3 + e_2{}^3} - \dfrac{1}{e_1 + e_2}$.

19. $\dfrac{uv^2}{u^3 - v^3} - \dfrac{v}{u^2 + uv + v^2}$.

20. $\dfrac{R^2 + 5}{27 - 8R^3} + \dfrac{3R - 2}{3 - 2R}$.

21. $\dfrac{P^2 - P - 1}{P^3 - 1} + \dfrac{1}{P - 1}$.

22. $\dfrac{xy}{x^3 - y^3} + \dfrac{2x}{x^2 + xy + y^2}$.

23. $\dfrac{a + r}{a^2 + ar + r^2} + \dfrac{a - r}{a^2 - ar + r^2}$.

24. $\dfrac{pq}{p^3 + q^3} - \dfrac{p}{p^2 - q^2} + \dfrac{1}{p + q}$.

25. $\left(1 - \dfrac{s^2}{r^2}\right) \div \left(1 - \dfrac{r^3}{s^3}\right)$.

26. $\dfrac{x^3 - y^3}{x + 2y} \cdot \dfrac{1}{3x^2 + 3xy + 3y^2}$.

27. $\dfrac{z - 3}{z^2 - 2z + 4} \div \dfrac{z^2 - 9}{z^3 + 8}$.

28. $\dfrac{12P^2 + 5P - 2}{6P^2 + 13P + 6} \cdot \dfrac{8P^2 + 10P - 3}{4P - 1}$.

29. $\left(\dfrac{x}{y} - \dfrac{y}{x}\right) \div \left(\dfrac{x^2}{y} - \dfrac{y^2}{x}\right)$.

30. $\dfrac{a^2 - z^2 + z - a}{2z + a} \cdot \dfrac{a - 2z}{a - z}$.

31. $x^2 + x + 1 + \dfrac{1}{x - 1}$.

32. $\dfrac{by}{y^3 + b^3} - \dfrac{y}{y^2 - b^2} + \dfrac{1}{y + b}$.

33. $\dfrac{I_1{}^6 - I_2{}^6}{I_1{}^4 - I_2{}^4} \div \dfrac{I_1{}^3 - I_2{}^3}{I_1 - I_2}$.

34. $\dfrac{u^3 - v^3}{(u^2 - v^2)^2} + \dfrac{uv(u - v)}{(v^2 - u^2)^2}$.

35. $\dfrac{T}{a^2 - aT + T^2} + \dfrac{1}{a + T} + \dfrac{a^2}{a^3 + T^3}$.

36. $\dfrac{E}{E^3 - 1} + \dfrac{E}{E^3 + 1} - \dfrac{E - 1}{E + 1}$.

37. $\dfrac{e_1}{e_1{}^3 + e_2{}^3} - \dfrac{e_2}{e_1{}^3 - e_2{}^3} + \dfrac{e_1{}^3 e_2 + e_1 e_2{}^3}{e_1{}^6 - e_2{}^6}$.

38. $\dfrac{x + y}{x^2 + xy + y^2} + \dfrac{x - y}{x^2 - xy + y^2} + \dfrac{2x^2 y^2}{x^6 - y^6}$.

39. $-\dfrac{2}{n^3 + n^2 + n + 1} + \dfrac{3}{n^3 - n^2 + n - 1}$.

40. $\dfrac{8\pi^3 + 1}{2\pi - 1} \cdot \dfrac{8\pi^3 - 1}{4\pi^2 + 2\pi + 1} \cdot \dfrac{\pi}{2\pi + 1}$.

41. $\dfrac{a^6 + b^6}{a^6 - b^6} \cdot \dfrac{a - b}{a + b} \div \dfrac{a^4 - a^2 b^2 + b^4}{a^4 + a^2 b^2 + b^4}$.

42. $\dfrac{t(s^3 - t^3)}{s(s + t)} \cdot \dfrac{(s^2 - t^2)^2}{s^2 + st + t^2} \cdot \dfrac{(s + t)^2}{(s - t)^2}$.

43. $\left(\dfrac{8X^3}{Y^3} - 1\right)\left(\dfrac{4X^2 + 2XY}{4X^2 + 2XY + Y^2} - 1\right)$.

44. $\dfrac{p^2 - q^2}{p^3 + q^3} \cdot \left(\dfrac{1}{p^2} - \dfrac{1}{pq} + \dfrac{1}{q^2}\right) \cdot \dfrac{p^3 q^3}{(p - q)^2}$.

45. $\dfrac{E^3 + 6E^2 + 12E + 8}{E^2 - 4E + 4} \div \dfrac{E^2 + 4E + 4}{E^3 - 6E^2 + 12E - 8}$.

46. $\dfrac{x^3 + 3x^2y + 3xy^2 + y^3}{x^2 - 2xy + y^2} \div \dfrac{x^2 + 2xy + y^2}{x^3 - 3x^2y + 3xy^2 - y^3}$.

47. $\dfrac{I_1^4 - I_1^2 I_2^2 + I_2^4}{I_1^6 - I_2^6} + \dfrac{I_1 + I_2}{I_1^3 - I_2^3} - \dfrac{I_1 - I_2}{I_1^3 + I_2^3} + \dfrac{1}{I_2^2 - I_1^2}$.

48. $\dfrac{E_1^4 - E_2^4}{E_1^3 + E_2^3} \cdot \left[1 + \dfrac{E_1 E_2}{(E_1 - E_2)^2} \right] \div \left[\dfrac{1}{E_1^2} + \dfrac{1}{E_2^2} \right]$.

49. $\dfrac{K^4 - 1}{K^3 - 1} \cdot \left[1 - \dfrac{K}{(K + 1)^2} \right] \div \left[\dfrac{1}{K} + K \right]$.

50. $\dfrac{1}{1 + R} + \dfrac{1 + R^2}{(1 + R)^3} - \dfrac{6R^2(1 - R)}{(1 + R)^4} - \dfrac{2(1 - R)}{(1 + R)^2}$.

7–4. Complex Fractions

A fraction whose numerator or denominator or both are fractions, is called a **complex fraction.** A **simple fraction** is one without a fraction in its numerator or denominator. A complex fraction, like a simple one, is of course a quotient, but it usually involves some other operations. Performing these operations is spoken of as **simplifying the fraction.**

In simplifying a complex fraction, first perform the indicated additions and subtractions; second, perform the multiplications and divisions.

Example 1. Simplify

$$\frac{x - \dfrac{1}{x^2}}{1 - \dfrac{1}{x}} .$$

First perform the subtraction indicated in the numerator and the subtraction indicated in the denominator. Thus we obtain

$$\frac{x - \dfrac{1}{x^2}}{1 - \dfrac{1}{x}} = \frac{\dfrac{x^3 - 1}{x^2}}{\dfrac{x - 1}{x}} .$$

Second perform the indicated division, and simplify by factoring. Thus

$$\frac{\dfrac{x^3 - 1}{x^2}}{\dfrac{x - 1}{x}} = \frac{x^3 - 1}{x^2} \cdot \frac{x}{x - 1}$$

$$= \frac{(x - 1)(x^2 + x + 1)}{x^2} \cdot \frac{x}{x - 1}$$

$$= \frac{x^2 + x + 1}{x} .$$

Simplifying complex fractions may involve several separate steps.   The above procedure can be repeated until a simple fraction is obtained.

Example 2.   Simplify

$$\frac{E+1}{E+1+\dfrac{1}{E-1+\dfrac{1}{E+1}}}.$$

Adding the last fraction makes the given expression equal to

$$\frac{E+1}{E+1+\dfrac{1}{\dfrac{E^2-1+1}{E+1}}} = \frac{E+1}{E+1+\dfrac{1}{\dfrac{E^2}{E+1}}}.$$

Dividing the last fraction reduces the last expression to

$$\frac{E+1}{E+1+\dfrac{E+1}{E^2}}.$$

Proceeding as in example 1, we reduce the last complex fraction to

$$\frac{E+1}{\dfrac{E^3+E^2+E+1}{E^2}} = \frac{E^2(E+1)}{E^3+E^2+E+1}$$

$$= \frac{E^2(E+1)}{E^2(E+1)+(E+1)}$$

$$= \frac{E^2(E+1)}{(E+1)(E^2+1)} = \frac{E^2}{E^2+1},$$

which is a simple fraction.

The **reciprocal of a number** is 1 divided by that number.   Thus the reciprocal of any number N is $1/N$.   Every number, with the exception of zero, has a reciprocal.

By taking the reciprocal of a fraction we obtain a complex fraction. Since $\dfrac{1}{\dfrac{a}{b}} = \dfrac{b}{a}$, *the reciprocal of a fraction is the fraction inverted* (see Sec. 1–10).

Example 3.   Simplify

$$\frac{1}{E+\dfrac{1}{I+\dfrac{1}{R}}}.$$

Proceeding as in example 2, we obtain

$$\cfrac{1}{E+\cfrac{1}{I+\cfrac{1}{R}}} = \cfrac{1}{E+\cfrac{1}{\cfrac{IR+1}{R}}}$$

$$= \cfrac{1}{E+\cfrac{R}{IR+1}} = \cfrac{1}{\cfrac{EIR+E+R}{IR+1}}$$

$$= \cfrac{IR+1}{EIR+E+R}.$$

Exercises

Simplify the following:

1. $\dfrac{\frac{1}{3}+\frac{1}{4}}{\frac{2}{3}}.$

2. $\dfrac{\frac{1}{5}-\frac{1}{3}}{\frac{1}{5}+\frac{1}{3}}.$

3. $\dfrac{R-\dfrac{1}{R}}{1-\dfrac{1}{R}}.$

4. $\dfrac{\dfrac{E}{R}+1}{1+\dfrac{R}{E}}.$

5. $\dfrac{\dfrac{x}{y}-\dfrac{y}{x}}{\dfrac{1}{y}-\dfrac{1}{x}}.$

6. $\dfrac{1}{1+\dfrac{1}{1+x}}.$

7. $\dfrac{\dfrac{ab}{3}-\dfrac{c}{2}}{\dfrac{2}{c}-\dfrac{3}{ab}}.$

8. $\dfrac{\dfrac{u^2}{v}+\dfrac{v^2}{u}}{\dfrac{u}{v}+\dfrac{v}{u}-1}.$

9. $\dfrac{\dfrac{1}{e}+\dfrac{1}{i}+\dfrac{1}{r}}{\dfrac{e}{i}+\dfrac{i}{r}+\dfrac{r}{e}}.$

10. $\dfrac{\dfrac{Z_1}{Z_1-Z_2}-\dfrac{Z_2}{Z_1+Z_2}}{\dfrac{Z_2}{Z_1-Z_2}-\dfrac{Z_1}{Z_1+Z_2}}.$

11. $\dfrac{\dfrac{1}{p-q}-\dfrac{p}{p^2-q^2}}{\dfrac{p}{pq+q^2}-\dfrac{q}{p^2+pq}}.$

12. $\dfrac{\dfrac{P+1}{P-1}+\dfrac{P-1}{P+1}}{\dfrac{P-1}{P+1}-\dfrac{P+1}{P-1}}.$

13. $\dfrac{\dfrac{s-r}{s+r}-\dfrac{s+r}{s-r}}{\dfrac{s-r}{s+r}+\dfrac{s+r}{s-r}}.$

14. $\dfrac{\dfrac{m^2+r^2}{m^2-r^2}-\dfrac{m^2-r^2}{m^2+r^2}}{\dfrac{m+r}{m-r}-\dfrac{m-r}{m+r}}.$

15. $\dfrac{\dfrac{C_1}{C_1 + C_2} + \dfrac{C_2}{C_1 - C_2}}{\dfrac{C_1{}^2 - C_2{}^2}{C_1{}^2 + C_2{}^2} - \dfrac{C_1{}^2 + C_2{}^2}{C_1{}^2 - C_2{}^2}}$.

16. $\dfrac{e - \dfrac{1}{e + \dfrac{1}{e}}}{e + \dfrac{1}{e - \dfrac{1}{e}}}$.

17. $\dfrac{\dfrac{1}{1 + E} + \dfrac{1 - E}{E}}{\dfrac{E}{1 + E} - \dfrac{1 - E}{E}}$.

18. $\dfrac{x - \dfrac{1}{2 - x} + 2}{x + \dfrac{1}{x + 2} - 2}$.

19. $\dfrac{1}{C - \dfrac{1}{C + \dfrac{1}{C}}}$.

20. $\dfrac{1}{L + \dfrac{1}{1 + \dfrac{L + 1}{3 - L}}}$.

21. $\dfrac{\dfrac{R_1}{R_2} - 1 + \dfrac{R_2}{R_1}}{R_1{}^3 + R_2{}^3}{R_1{}^2 R_2 + R_1 R_2{}^2}$.

21. $\dfrac{\dfrac{R_1}{R_2} - 1 + \dfrac{R_2}{R_1}}{\dfrac{R_1{}^3 + R_2{}^3}{R_1{}^2 R_2 + R_1 R_2{}^2}}$.

22. $\dfrac{1}{K + \dfrac{1}{1 + \dfrac{K + 1}{3 - K}}}$.

23. $1 + \dfrac{1}{1 + \dfrac{1}{1 + \dfrac{1}{x - 1}}}$.

24. $1 - \dfrac{1}{3 - \dfrac{1}{2 - \dfrac{x}{1 - x}}}$.

25. $\dfrac{E_1{}^2 - E_1 E_2 + E_2{}^2 - \dfrac{E_1{}^3 - E_2{}^3}{E_1 + E_2}}{E_1{}^2 + E_1 E_2 + E_2{}^2 + \dfrac{E_1{}^3 + E_2{}^3}{E_1 - E_2}}$.

26. $p + q + \dfrac{1}{p + q + \dfrac{1}{p + q - \dfrac{1}{p + q}}}$.

27. $\dfrac{n}{1 - \dfrac{n}{1 + n + \dfrac{n}{1 - n + n^2}}}$.

28. $z^3 + \dfrac{z^2}{z^2 + \dfrac{1}{z^3 - \dfrac{z^8 + z^3 - 1}{z^5}}}$.

29. $\dfrac{1}{C_1 - \dfrac{C_1{}^2 - 1}{C_1 + \dfrac{1}{C_1 - 1}}}$.

30. $1 + \dfrac{T}{1 - \dfrac{T}{T + 2}} - \dfrac{1}{1 + \dfrac{T}{1 - T + \dfrac{T}{T + 2}}}$.

7–5. Equations in One Unknown

Equations were introduced in Chapter 2. Before proceeding, the reader should review Sec. 2–12.

An equation is a statement of the equality of two expressions. If the two members of the equation are equal for every value of the symbols or variables involved for which the members have a meaning, the equation is called an **identical equation** or an **identity.** For example, all formulas 1 to 14 in Sec. 1 of this chapter are identities since they hold true for every value of the letters involved. At times, to emphasize that some equation is an identity, the symbol $\equiv$ is used instead of $=$. We may write

$$a^3 - b^3 \equiv (a - b)(a^2 + ab + b^2).$$

An equation whose members are not equal for all values of the letters is called a **conditional equation** or, briefly, an **equation.** Examples of equations are:

$3x - 2 = 10,$ true only if $x = 4$.

$x^2 - 4x + 3 = 0,$ true only if $x = 1$ or $x = 3$.

$\cos x = 1,$ true only if $x = 0°, 360°, 720°, \cdots$.

$x + 2 = x,$ not true for any value of x.

These examples illustrate the fact that a conditional equation may be true for one value, two values, an unlimited number of values, or for no value of the unknown.

When an equation is given, it is frequently of interest to solve it, i.e., to find values of the unknown that make the two members equal. Such values of the unknown are called **roots** or **solutions** of the equation. In the process of solving an equation we have to make certain changes and transformations. From this process, it is important that we obtain for roots only numbers that satisfy the original equation. These changes and transformations, when applied to an equation, result in a new equation. This new **derived equation** *is equivalent to the original equation if it contains all the roots of that equation and no others.* In Sec. 2–12 we indicated the operations on an equation which yield an equivalent equation. These operations are:

1. *Addition of the same expression to both members of the equation or subtraction of the same expression from both members of the equation.*

2. *Multiplication or division of both members of the equation by the same expression, provided this expression is not zero and does not involve the unknown.*

Applications of these operations were given in Sec. 2–12.

If an equation is multiplied or divided by an expression involving the unknown, the derived equation may not be equivalent to the original, as indicated in the following two rules.

3. *If both members of an equation are multiplied by an expression involving the unknown, roots not in the original equation may be introduced. These are called* **extraneous roots.**

4. *If both members of an equation are divided by an expression involving the unknown, the resulting equation may have fewer roots than the original equation.*

The next three examples will illustrate rule 3.

Example 1. Solve

$$\frac{5E^2 - 7E - 6}{E - 2} = 4. \tag{1}$$

Multiplying both sides by $E - 2$, an expression involving the unknown E, we obtain

$$5E^2 - 7E - 6 = 4E - 8,$$

and hence

$$5E^2 - 11E + 2 = 0. \tag{2}$$

This when factored gives

$$(5E - 1)(E - 2) = 0.$$

Proceeding as in Sec. 2–13, we obtain

$$5E - 1 = 0 \quad \text{or} \quad E - 2 = 0,$$

$$E = \tfrac{1}{5} \quad \text{and} \quad E = 2.$$

The solution shows $\frac{1}{5}$ and 2 as roots. But $E = 2$ cannot be a root of the given equation 1, since substitution of this value reduces the left member to $\frac{0}{0}$ which has no meaning (see Sec. 1–7). Substituting $E = \frac{1}{5}$ in the given equation reduces the left member to the right member, since

$$\frac{5 \cdot (\tfrac{1}{5})^2 - 7(\tfrac{1}{5}) - 6}{\tfrac{1}{5} - 2} = 4.$$

Hence we conclude that $E = \frac{1}{5}$ is the only solution of equation 1. The other answer, $E = 2$, is a root of the derived equation 2 but not of 1, and hence it is an extraneous root.

Example 2. Solve

$$\frac{1}{R - 2} + \frac{1}{R - 3} = \frac{2}{R - 4}.$$

To clear fractions, we multiply by $(R - 2)(R - 3)(R - 4)$, an expression involving the unknown, and obtain

$$(R - 3)(R - 4) + (R - 2)(R - 4) = 2(R - 2)(R - 3),$$

$$R^2 - 7R + 12 + R^2 - 6R + 8 = 2R^2 - 10R + 12,$$

$$3R = 8,$$

$$R = \tfrac{8}{3}.$$

Substituting $R = \frac{8}{3}$ into the given equation, we obtain

$$\frac{1}{\frac{8}{3} - 2} + \frac{1}{\frac{8}{3} - 3} = \frac{2}{\frac{8}{3} - 4},$$

$$\tfrac{3}{2} - 3 = -\tfrac{3}{2}.$$

The multiplier $(R - 2)(R - 3)(R - 4)$, used in clearing fractions of the given equation, is zero for $R = 2, 3, 4$. Since none of these turns up as a root in the solution, no extraneous roots were introduced, even though both members of the given equation were multiplied by an expression involving the unknown R.

Example 3.　Solve

$$\frac{2}{x - 1} + \frac{5}{x + 1} = \frac{4}{x^2 - 1}.$$

To clear fractions we multiply by $(x - 1)(x + 1)$, an expression involving the unknown x, and obtain

$$2(x + 1) + 5(x - 1) = 4,$$
$$2x + 2 + 5x - 5 = 4,$$
$$x = 1.$$

But $x = 1$ is not a solution of the given equation, since, by substituting $x = 1$, we obtain

$$\tfrac{2}{0} + \tfrac{5}{2} = \tfrac{4}{0},$$

which has no meaning, division by zero being excluded (see Sec. 1–7). Hence $x = 1$ is an extraneous root, and the given equation has no solution.

The next example will illustrate rule 4

Example 4.　Solve $(x + 3)(x - 1) = 5(x - 1)$.

Dividing both members by $x - 1$, an expression involving the unknown x, we have the equation

$$x + 3 = 5,$$

which is satisfied by

$$x = 2.$$

Now $x = 2$ satisfies the given equation, but so does $x = 1$. Division of both members by $x - 1$ loses the root $x = 1$.

The last example shows that if both members of an equation are divided by an expression involving the unknown, we may lose some roots. A root once lost may not be easily discovered; for that reason, in solving equations, a student should not divide by an expression involving the unknown unless he is certain that by doing so no root is lost. The equation in example 4 should be worked as follows:

$$(x + 3)(x - 1) = 5(x - 1),$$
$$(x + 3)(x - 1) - 5(x - 1) = 0,$$
$$(x - 1)(x + 3 - 5) = 0,$$
$$(x - 1)(x - 2) = 0,$$
$$x = 1 \quad \text{and} \quad x = 2.$$

Exercises

Solve the following equations:

1. $\dfrac{a}{a-3} = \dfrac{a-2}{a-4}$.

2. $\dfrac{7}{x+1} - \dfrac{2x}{x-5} + 2 = 0$.

3. $(x+2)(x-4) - 5(x-4) = 0$.

4. $(y-1)(y+5) = 3(y+5)$.

5. $2(u+1)(u-3) = (u+1)(u-5)$.

6. $(2b-1)(b+2) = (2b-1)(b-7)$.

7. $(3x-2)(x-6) = (3x-2)(x+12)$.

8. $(x-7)(2x+5) + (3x-1)(x-7) = 0$.

9. $\dfrac{e-2}{e+2} - \dfrac{e-3}{e-1} = \dfrac{5}{e-1}$.

10. $\dfrac{1}{l_6} - \dfrac{5}{l_9} = -\dfrac{1}{A-1}$.

11. $\dfrac{2}{t^2-4} + \dfrac{1}{t+2} - \dfrac{3}{t-2} = 0$.

12. $\dfrac{1}{x+2} - \dfrac{2}{x-2} = \dfrac{2}{x^2-4}$.

13. $\dfrac{1}{y} + \dfrac{1}{y-1} = \dfrac{1}{y(y-1)}$.

14. $\dfrac{20}{Z^2-25} - \dfrac{1}{Z+5} = \dfrac{2}{Z-5}$.

15. $\dfrac{1}{R+1} + \dfrac{1}{R-1} = \dfrac{2}{R^2-1}$.

16. $3 - \dfrac{6x}{2x-1} = \dfrac{2}{x+4}$.

17. $\dfrac{4}{x-4} - \dfrac{1}{x-2} = \dfrac{x}{x^2-6x+8}$.

18. $\dfrac{2}{v} + \dfrac{1}{v-1} = \dfrac{3v-4}{v(v-1)}$.

19. $\dfrac{3}{3m+4} - \dfrac{5}{3m-4} = \dfrac{4}{9m^2-16}$.

20. $\dfrac{2}{4-9p^2} - \dfrac{3}{2+3p} = \dfrac{3p}{4-9p^2}$.

21. $\dfrac{r^2-4}{r^2-9} - \dfrac{r+2}{r-3} = 0$.

22. $\dfrac{z^2-25}{z^2-1} - \dfrac{z+5}{z-1} = 0$.

23. $\dfrac{x^2+x-2}{x^2-9x+20} = \dfrac{x-1}{x-3}$.

24. $\dfrac{y^2+y-6}{y^2+5y+4} = \dfrac{y-2}{y+1}$.

25. $\dfrac{3}{(2x-3)(x+2)} = \dfrac{4}{(4x-5)(x+2)}$.

26. $\dfrac{L+2}{L+3} + \dfrac{L-2}{3-L} = \dfrac{5}{L^2-9}$.

27. $\dfrac{M-4}{2M-5} - \dfrac{M+2}{2M+5} = \dfrac{3M^2-10}{4M^2-25}$.

28. $\dfrac{1+3e}{5+7e} - \dfrac{9-11e}{5-7e} = \dfrac{14(2e-3)^2}{25-49e^2}$.

29. $\dfrac{2s+13}{2s^2+5s-3} + \dfrac{3}{s+3} - \dfrac{4}{2s-1} = 0$.

30. $\dfrac{1}{T-2} - \dfrac{1}{T+1} = \dfrac{1}{T-1} - \dfrac{1}{T+2}$.

31. $\dfrac{18}{E^2-3E-10} + \dfrac{8}{E+2} - \dfrac{27}{E-5} = 0$.

32. $\dfrac{1}{K-1} - \dfrac{2}{K-2} = \dfrac{3}{K-3} - \dfrac{4}{K-4}$.

33. $\dfrac{5}{5L+8} - \dfrac{4}{2L+3} = \dfrac{1}{5}\left(\dfrac{3}{L+3} - \dfrac{8}{L+2}\right)$.

34. $\dfrac{2}{2t-1} - \dfrac{1}{t-3} - \dfrac{1}{t} + \dfrac{2}{2t-5} = 0$.

35. $\dfrac{x+3}{x+1} + \dfrac{x-6}{x-4} = \dfrac{x+4}{x+2} + \dfrac{x-5}{x-3}$.

36. $\dfrac{2}{3Z+1} = \dfrac{6Z+1}{9Z^2-3Z+1} + \dfrac{15+6Z}{27Z^3+1}$.

37. $\dfrac{1-2y}{3(y^2+2y+4)} = \dfrac{1}{6(2-y)} - \dfrac{y^2+3}{2y^3-16}$.

7–6. Application to Solution of Problems

In setting up the equations for many problems, we often obtain equations that are expressed in the form of fractions of the type discussed in the preceding section. The problems that follow will lead to fractional equations whose solution will depend on the work just completed. Before proceeding, the reader should review Sec. 2–17 in which a discussion of the solution of word problems was given.

Example 1. The denominator of a fraction is two more than the numerator. If both the numerator and the denominator of the fraction are increased by one, the resulting fraction equals $\frac{2}{3}$. Find the fraction.

Let x be the numerator of the fraction. Then $x+2$ is the denominator. Increasing both of these expressions by 1, we obtain the equation

$$\frac{x+1}{x+2+1} = \frac{2}{3},$$

$$3x + 3 = 2x + 6,$$

$$x = 3.$$

The fraction is $\frac{3}{5}$.

To check the answer we have

$$\frac{3+1}{5+1} = \frac{2}{3}.$$

Example 2. After traveling 60 miles at a certain speed, a motorist increases his speed by 5 mph and travels an additional 50 miles. If the total time required for the 110 miles was 5 hr, find the speed for the first 60 miles.

Let x mph be the motorist's speed for the first 60 miles. Then $x+5$ mph is the speed for the last 50 miles. Remembering from Sec. 2–18 that time is equal to distance divided by speed, we have, as the equation to determine x,

$$\frac{60}{x} + \frac{50}{x+5} = 5.$$

This equation can be reduced to

$$x^2 - 17x - 60 = 0,$$

$$(x - 20)(x + 3) = 0,$$

$$x = 20 \quad \text{or} \quad x = -3.$$

The two solutions of the equation are $x = 20$ and $x = -3$, but only one of these solutions, $x = 20$, has a meaning in this problem. Since a speed cannot be a negative number, the negative root -3 is meaningless as far as our problem is concerned. Hence the speed during the first 60 miles was 20 mph. This result is correct since

$$\frac{60}{20} + \frac{50}{25} = 5.$$

Exercises

1. The sum of the numerator and denominator of a proper fraction is 42.   If each is increased by 9, the value of the fraction becomes $\frac{1}{2}$.   Find the fraction.

2. The difference between the numerator and denominator of a proper fraction is 6. If each is decreased by 5, the value of the fraction becomes $\frac{2}{3}$.   Find the fraction.

3. The denominator of a fraction exceeds the numerator by 37.   If 11 is added to the numerator and subtracted from the denominator, the fraction becomes $\frac{3}{4}$.   Find the fraction.

4. Find the number which, when subtracted from the numerator and added to the denominator, changes the fraction $\frac{11}{15}$ to $\frac{4}{9}$.

5. What number should be subtracted from both the numerator and denominator of $\frac{23}{33}$ to make the fraction equal to $\frac{2}{3}$?

6. The sum of the digits of a given two-place natural number is 10.   If the given number plus 1 is divided by the number obtained by reversing the digits of the given number, the quotient is 2.   Find the given number.

7. The unit's digit of a two-place natural number exceeds the ten's digit by 4.   If the number increased by 6 is divided by the sum of the digits the quotient is 4.   Find the number.

8. The unit's digit of a two-place natural number is 2 greater than the ten's digit. If the number is divided by the sum of its digits, the quotient is 4 and the remainder is 12. Find the number.

9. The unit's digit of a two-place natural number is 3 more than the ten's digit. The number divided by the sum of its digits gives 4 as the quotient and 6 as the remainder. What is the number?

10. In a given two-place natural number the ten's digit is 5 more than the unit's digit. The number divided by the sum of its digits gives 7 as the quotient and 6 as the remainder. What is the number?

11. The denominator of a proper fraction is 5 more than the numerator.   If the denominator is decreased by 20, the resulting fraction, increased by 1, is equal to twice the original fraction.   Find the original fraction.

12. Divide the number 209 into two parts which will be in the ratio of 6 to 13.

13. Divide the number N into two parts which will be in the ratio of a to b, thus obtaining a general formula for solving problems like the one in the preceding exercise.

14. A tank can be filled by one pipe in 6 hr and emptied by another in 9 hr. In what time can it be filled if both pipes are opened?

15. A reservoir can be filled by one pipe in 4 hr and by another in 5 hr and can be emptied by a third pipe in 10 hr. In what time will the reservoir be filled if all three pipes are open?

16. The cost of a certain number of castings was $60. If there had been one more for the same cost, the cost for each would have been $2 less. Find the number of castings.

17. A pontoon bridge across a river 300 ft wide requires a certain number of sections of standard length. If the sections were 5 ft shorter, 5 more would be required. What is the length of the sections?

18. An automobile traveled 160 miles at a uniform speed. If the speed had been 8 mph more, the journey could have been completed 1 hr sooner. Find the speed of the automobile.

19. The speed of the current of a river is 3 mph. A man can row downstream 18 miles and return in 8 hr. At what speed can he row in still water?

20. The speed of the current of a river is 3 mph. Some men in a motorboat found that it took them as long to travel 8 miles downstream as to travel 4 miles upstream. At what speed does the motorboat travel in still water?

21. Two rivers flow at speeds of 2 mph and 4 mph, respectively. A man finds that he can row 20 miles downstream on the second river in the same time that it takes him to row 15 miles downstream on the first river. At what speed can he row in still water?

22. A motor torpedo boat and a destroyer each ran 210 nautical miles. The torpedo boat ran 5 knots faster than the destroyer and required 1 hr less time for the trip. Find the speed of the motor torpedo boat. (A knot equals one nautical mile per hour.)

23. A motorist traveled 400 miles from his starting point. Returning by the same route, he increased his speed by 10 mph and made his return trip in 2 hr less time than the outbound trip. What was his speed each way?

24. The rear wheel of a wagon has a circumference 3 ft greater than that of the front wheel. The front wheel makes 20 revolutions more than the rear wheel while the wagon travels 1200 ft. Find the circumferences of the wheels.

7–7. Applications to Engineering Formulas

Some of the work of the present chapter will now be applied to formulas that are encountered in science and engineering. Before proceeding, the reader will do well to review Sec. 2–14.

Many formulas are in the form of complex fractions. To work with such formulas the information and experience obtained in Sec. 7–4 are very useful. For example, if the capacitances of three capacitors, connected in series, are, respectively, C_1, C_2, and C_3, then the resultant capacitance C is given by the formula

$$C = \frac{1}{\dfrac{1}{C_1} + \dfrac{1}{C_2} + \dfrac{1}{C_3}}. \tag{1}$$

When C_1, C_2, and C_3 are given, we can compute C from formula 1. When C, C_1, and C_2 are given, we may regard C_3 as unknown and solve for it in terms of the remaining letters. One thus obtains from formula 1

$$C = \frac{1}{\dfrac{C_2 C_3 + C_1 C_3 + C_1 C_2}{C_1 C_2 C_3}},$$

$$C = \frac{C_1 C_2 C_3}{C_2 C_3 + C_1 C_3 + C_1 C_2},$$

$$C(C_2 C_3 + C_1 C_3 + C_1 C_2) = C_1 C_2 C_3,$$

$$CC_2 C_3 + CC_1 C_3 + CC_1 C_2 = C_1 C_2 C_3,$$

$$CC_2 C_3 + CC_1 C_3 - C_1 C_2 C_3 = -CC_1 C_2,$$

$$C_3(CC_2 + CC_1 - C_1 C_2) = -CC_1 C_2,$$

$$C_3 = \frac{CC_1 C_2}{C_1 C_2 - CC_1 - CC_2}. \tag{2}$$

Formula 2 gives an expression for C_3 in terms of C, C_1, and C_2. The above process is called *solving formula 1 for* C_3. One can easily obtain similar expressions for C_1 and for C_2 in terms of the remaining letters; i.e., one can *solve for* C_1 and C_2.

In working with formulas it is important that we do not divide by zero. All the operations with the variables must be carefully examined and division by zero excluded. To exclude division by zero in formula 1, all the quantities C_1, C_2, and C_3 must be different from zero, which is written as $C_1 \neq 0$, $C_2 \neq 0$, and $C_3 \neq 0$, or $C_1 \cdot C_2 \cdot C_3 \neq 0$. These restrictions on the values of C_1, C_2, and C_3 have a physical significance, for, if one of them, say C_1, is zero, we then have an open circuit.

By starting with formula 1 we discover by simple algebraic steps another important fact about electricity represented by formula 2. It is quite possible that a radio mechanic may know the fact represented by formula 1, but if he is ignorant of elementary algebra he will not be able to discover formula 2, even though it is of practical importance to him.

The above example and the exercises that follow are an indication of one of the roles algebra plays in the sciences and engineering. A knowledge of algebra enables the engineer or scientist to derive many facts from a known formula, facts that otherwise would have to be discovered by experiment, often quite laborious.

Exercises

In the formulas below, consider all the quantities as given except the one to be solved for. In each case indicate the value or values, if any, of the quantities for which the expressions fail to have a meaning. A hint concerning the meaning of the formulas is given. If the reader is interested in a more detailed description of the formula, an appropriate textbook should be consulted. Simplify your answer.

Given	Solve for	Description
1. $PV = RT$	V, T	Characteristic equation for gases
2. $E = \dfrac{N \cdot \phi}{10^8 \cdot t}$	t	Voltage generated by a conductor
3. $\lambda = \dfrac{300 \cdot 10^6}{f}$	f	Length of a radio wave
4. $T_r = \dfrac{j_s J}{c}$	J	Torsion formula for round shafts
5. $r_s = \dfrac{\pi d^2}{4} f_s$	f_s, d^2	Shearing strength of a rivet
6. $X_c = \dfrac{1}{2\pi f C}$	f, C	Capacitive reactance
7. $C = \dfrac{8.84 KA}{10^8 d}$	d	Calculation of capacitance
8. $I = \dfrac{p^2}{2\rho_0 c}$	p_0	Intensity of sound
9. $y = \dfrac{Wl^3}{8EI}$	W, E	Maximum deflection of a cantilever beam
10. $L = \frac{1}{2}C_L \rho S V^2$	C_L, S	Lift of an airfoil
11. $T = \dfrac{2rfW}{3 \sin \alpha}$	f, r	Torque of a conical pivot
12. $\dfrac{R_1}{R_3} = \dfrac{R_2}{R_x}$	R_x	Wheatstone bridge equation
13. $\varepsilon_c = \dfrac{Q_1 - Q_2}{Q_1}$	Q_1, Q_2	Carnot efficiency
14. $I_a = \dfrac{h^3(B + 3b)}{12}$	B, b	Moment of inertia of a trapezoid
15. $f_c = \dfrac{f_s}{n} \cdot \dfrac{k}{1 - k}$	k	Fiber stress
16. $V = \dfrac{Q}{r_1} - \dfrac{Q}{r_2}$	r_1, r_2, Q	Potential difference between two point charges

	Given	Solve for	Description
17.	$W + JQ = J(E_2 - E_1)$	J, E_1	First law of thermodynamics
18.	$\lambda' = \dfrac{\lambda(v - j)}{v}$	λ, v	A formula from light spectroscopy
19.	$T = \dfrac{J_m(\omega_1 - \omega_2)}{t}$	ω_1	Torque about the axis of rotation
20.	$\delta = \dfrac{2\pi d}{\lambda}(n_1 - n_2)$	λ, n_2	Phase difference in the polarization of light
21.	$A = \dfrac{R_L}{r_p + R_L}$	R_L, r_p	Net gain of an amplifier tube
22.	$I - X = \dfrac{G_m}{G_s + G_m}$	G_s, G_m	Separating calorimeters
23.	$M = R_L\left(a + \dfrac{R_L}{2w}\right)$	w	A bending moment in mechanics
24.	$V = \dfrac{\pi r^2}{2}(h_1 + h_2)$	h_1, h_2	Volume of a truncated right circular cylinder
25.	$L = \dfrac{0.8r^2 N^2}{6r + 9l + 10t}$	l, t	Inductance of a multilayer coil
26.	$\beta = -\dfrac{v_1(p_2 - p_1)}{v_2 - v_1}$	p_1, v_2	Compressibility of water
27.	$\dfrac{1}{f} = (\mu - 1)\left(\dfrac{1}{r} - \dfrac{1}{r'}\right)$	f, r	Focal length of a lens
28.	$R = \left(\dfrac{E_b}{E} - 1\right)R_m$	E_b, E, R_m	Resistance measured with a voltmeter and a battery
29.	$\dfrac{U_2^2 - U_1^2}{64.34} = J(H_1 - H_2)$	J, H_1	Energy equation of thermodynamics
30.	$M = m + md_a\left(\dfrac{1}{d_m} - \dfrac{1}{d_w}\right)$	d_m	True mass of a body in a vacuum
31.	$\dfrac{1}{C_T} = \dfrac{1}{C_1} + \dfrac{1}{C_2} + \dfrac{1}{C_3}$	C_1, C_2, C_3	Total capacitance in series
32.	$\psi = \dfrac{ky}{\pi(x^2 + y^2)}$	x^2	Steam function
33.	$R = \dfrac{L_1}{K_1 A_1} + \dfrac{L_2}{K_2 A_2}$	L_1, A_2	Resistance to heat flow
34.	$\left(P + \dfrac{a}{V^2}\right)(V - b) = nRT$	b, P	Van der Waals' equation in the chemistry of gases

Given	Solve for	Description
35. $e_{\text{avg}} = L\dfrac{i_2 - i_1}{t_2 - t_1}$	t_1, t_2	Self-induced voltage of a coil
36. $I = \dfrac{E}{R + \dfrac{r}{n_p}}$	n_p, r, R	Ohm's law modified for n_p cells in parallel
37. $h_0 = h\left\{1 - \dfrac{(m - l)t}{1 + mt}\right\}$	t, l	Height of a column of mercury
38. $I = \dfrac{n_s E}{R + \dfrac{n_s r}{n_p}}$	n_s, n_p, r, R	Ohm's law modified for several cells, n_s in series and n_p in parallel
39. $A = \dfrac{\dfrac{W_A}{M_A}}{\dfrac{W_A}{M_A} + \dfrac{W_B}{M_B} + \dfrac{W_C}{M_C}}$	W_A, M_A	Molecular fraction in a mixture

Progress Report

In this chapter we extended our domain of knowledge and skill in the operations of:

1. Expanding products.
2. Factoring.
3. Simplifying simple and complex fractions.
4. Solving equations involving fractions.

Finally, these skills were applied to mathematical exercises and problems in applications.

8

Exponents and Radicals

In Sec. 1–11 we used positive and negative integral exponents. In practice, however, fractions are also used as exponents. Because of the importance of such exponents in mathematics and their applications in science and engineering, it is necessary to study them in detail. Before proceeding, the reader should carefully review Sec. 1–11.

8–1. Integral Exponents

We shall start by giving a brief summary of the work done in Sec. 1–11.

If a and b are any numbers, and m, n, and q are integers, we have the following definitions and laws.

DEFINITIONS

1. $\underbrace{a \cdot a \cdots a}_{n \text{ factors}} = a^n.$

2. $a^{-n} = \dfrac{1}{a^n}.$

3. $a^0 = 1.$

LAWS OF EXPONENTS

1. $a^n \cdot a^m = a^{m+n}.$

2. $\dfrac{a^n}{a^m} = a^{n-m}.$

3. $\dfrac{a^n}{a^m} = \dfrac{1}{a^{m-n}}.$

4. $(a^n)^m = a^{mn}.$

5. $(a \cdot b)^n = a^n \cdot b^n.$

6. $\left(\dfrac{a}{b}\right)^n = \dfrac{a^n}{b^n}.$

7. $(a^n b^m)^q = a^{nq} \cdot b^{mq}.$

Examples.

$$L^3 \cdot L^5 = L^{3+5} \qquad \text{Law 1}$$
$$= L^8.$$
$$(-\tfrac{5}{3}R^3)^2 = (-\tfrac{5}{3})^2(R^3)^2 \qquad \text{Law 7}$$
$$= \tfrac{25}{9}R^6. \qquad \text{Laws 4 and 6}$$
$$(x^2)^3(-2x)^4(-x^2)^2 = x^6(-2)^4x^4x^4 \qquad \text{Laws 7 and 4}$$
$$= 16x^{14}. \qquad \text{Law 1}$$
$$\frac{(6a^4b^3)^3}{(3a^2b^2)^2(2ab)^5} = \frac{216a^{12}b^9}{9a^4b^4 \cdot 32a^5b^5} \qquad \text{Laws 7 and 4}$$
$$= \frac{216a^{12}b^9}{288a^9b^9} \qquad \text{Law 1}$$
$$= \tfrac{3}{4}a^3. \qquad \text{Law 2}$$
$$\left(-\frac{R^2}{L^4}\right)^{-3}\left(\frac{R^3}{L^6}\right)^2 = \frac{1}{\left(-\dfrac{R^2}{L^4}\right)^3} \cdot \left(\frac{R^3}{L^6}\right)^2 \qquad \text{Definition 2}$$
$$= \frac{1}{-\dfrac{R^6}{L^{12}}} \cdot \frac{R^6}{L^{12}} \qquad \text{Laws 6 and 4}$$
$$= -\frac{L^{12}}{R^6} \cdot \frac{R^6}{L^{12}} \qquad \text{Inverting the fraction}$$
$$= -1. \qquad \text{Law 2, definition 3}$$

Exercises

Perform the indicated operations by the laws of exponents. Simplify the results when possible.

1. $L^6 \cdot L^2$.   2. $q^5 \cdot q^3$.   3. $xy^2 \cdot (xy)^2$.

4. $(\tfrac{1}{2}P)^4$.   5. $(3sr)^4$.   6. $(-\tfrac{1}{3}R^2)^3$.

7. $(Q_1{}^2Q_2{}^3)^4$.   8. $(-2A_m{}^2)^4$.   9. $(-3L_1L_2{}^2)^3$.

10. $\dfrac{N^7}{N^3}$.   11. $\dfrac{Z^{13}}{Z^5}$.   12. $\dfrac{R^3 \cdot r^7}{R^5 \cdot r^4}$.

13. $\dfrac{a^3b^4}{a^5b^7}$.   14. $\dfrac{S_1{}^3 \cdot S_2{}^5}{S_1 \cdot S_2{}^2}$.   15. $\dfrac{(F_1F_2)^3}{F_1{}^2F_2{}^3}$.

16. $(4x)^2(3x^2)^3$.   17. $(3r_1{}^2r_2)^3 \cdot r_2{}^2$.   18. $(x_1{}^2 \cdot x_2{}^3)^2 \cdot x_1x_2$.

19. $a^4 \cdot b^2 \cdot a^3 \cdot b^5$.   20. $(sr)^2 \cdot (s^2r)^3$.   21. $(2z_1z_2{}^2z_3{}^3)^3$.

22. $r^2 \cdot r^{-2}$.   23. $x^2 \cdot x^{-4} \cdot x^7$.   24. $Z^5 \cdot Z^{-3} \cdot Z^{-1}$.

25. $\dfrac{a^6}{a^{-2}}$.

26. $\left(\dfrac{x^{-2}\cdot y^3}{z}\right)^4$.

27. $\left(\dfrac{a^{-1}\cdot b^{-2}}{c}\right)^5$.

28. $(r_x r_y{}^2)^3\cdot(r_x{}^2 r_y)^{-2}$.

29. $(-2f_1 f_2)^{-2}\cdot f_1{}^3$.

30. $(1.2p^3)^{-2}\cdot p^5$.

31. $x^{m-2}\cdot x^2$.

32. $a^{n+1}a^{n-1}$.

33. $2(2b^m)^4$.

34. $\dfrac{u^{n+2}}{u^{n-2}}$.

35. $\dfrac{v^{1-m}}{v^{1+m}}$.

36. $\dfrac{z^{n^2}}{z^{n^2(n+1)}}$.

37. $y^{q+1}\cdot y^{q+3}\cdot y^{4(q-1)}$.

38. $(h^{n-1}\cdot h^{n+1})^2$.

39. $(p^{2n}\cdot p^{-n})^4$.

40. $\left(\dfrac{3x}{4y}\right)^2\left(\dfrac{8yz}{9x}\right)^3$.

41. $\left(\dfrac{-a}{bc}\right)^3\left(\dfrac{b}{a^2 c}\right)^5$.

42. $\left(\dfrac{-3u}{-vw}\right)^3\left(\dfrac{w^2}{u}\right)^2$.

43. $(x^y y^2)^{\,8}\cdot y^5$.

44. $x_1{}^m(3x_2{}^{2m})^{-3}\cdot x_2{}^m$.

45. $(r_x{}^{-m}\cdot r_y{}^m)^{-1}r_y{}^{2m}$.

46. $(5F-7)^0$.

47. $(R_1+R_2)^0(R_1 R_2)^{-2}$.

48. $5a^0(a+b)^{-1}$.

49. $\left(\dfrac{\sin\theta}{\cos\theta}\right)^{-1}\sin\theta$.

50. $\left(\dfrac{\cos A}{\sin A}\right)^{-2}(\cos A)^2$.

51. $\left(\dfrac{\tan B}{\cot B}\right)^3(\tan B)^{-3}$.

52. $25^{n+2}\cdot 5^{n-1}$.

53. $4^2\cdot 2^{3q}\cdot 8^{q+2}$.

54. $9^{2m}\cdot(3^{m+1})^m$.

55. $\dfrac{9^2\cdot 3^{3p}}{27^{p+2}}$.

56. $\dfrac{36^{p+2}}{6^{p-1}\cdot 6^{-1}}$.

57. $\dfrac{(4^{n-1})^n\cdot 8^{2n-1}}{16^n}$.

58. $(r_1-r_2)^{-2}(r_1{}^2-r_2{}^2)$.

59. $(s^2+s^{-2})(s^2-s^{-2})$.

60. $\left(\dfrac{\sin^2 A}{\cos A}\right)^3\left(\dfrac{\sin^3 A}{\cos^2 A}\right)^{-2}\dfrac{\sin A}{\cos A}$.

61. $\left(\dfrac{\sin\alpha}{\cos\beta}\right)^{-1}\left(\dfrac{\sin^2\alpha}{\cos^3\beta}\right)^{-3}\left(\dfrac{\cos\beta}{\sin\alpha}\right)^2$.

62. $\left(\dfrac{x^{2n-3}}{x^{2n+3}}\right)^{n+1}$.

63. $\left(\dfrac{z^m}{z^{2n-m}}\right)^{n+m}\left(\dfrac{z^{2n-m}}{z^m}\right)^{n-m}$.

64. $\left(\dfrac{y^m}{y^n}\right)^{m+n}\left(\dfrac{y^n}{y^p}\right)^{n-p}\left(\dfrac{y^p}{y^n}\right)^{m-p}$.

65. $\left(\dfrac{r^{2n-1}}{s^{n+3}}\cdot\dfrac{s^2}{r^n}\right)^{1+n}\left(\dfrac{s}{r}\right)^{n^2+1}$.

8–2. Roots and Radicals

When in the preceding section we wrote $y=x^n$, n any positive integer, we said that the number y is equal to the nth power of the number x. On the other hand, if x is a number such that $x^n=y$, then x is said to be an nth *root* of y. That is, *an nth root of a number y is a number whose nth power equals the number y.* Briefly

$$x \text{ is an } n\text{th root of } y \text{ if } x^n=y. \qquad (1)$$

Thus, since $6^2=36$, we say that 6 is a second or *square* root of 36. Similarly since $(2E^2)^3=8E^6$, the quantity $2E^2$ is a third or *cube* root of $8E^6$. The symbol $\sqrt[n]{}$ is used to denote the nth root. Thus, if $6^2=36$, then $6=\sqrt{36}$, and, if $(2E^2)^3=8E^6$, then $2E^2=\sqrt[3]{8E^6}$. The symbol $\sqrt{}$

is called the **radical** sign; the number n placed above the radical sign is called the **index** of the root. In the square root the index is omitted. The term **radical** applies also to expressions of the form $\sqrt[n]{y}$, where the number y is called the **radicand.**

Example 1. Since $3^2 = 9$, 3 is a square root of 9.

Since $(-2)^3 = -8$, -2 is a cube root of -8.
Since $(-3)^5 = -243$, -3 is a fifth root of -243.
Since $(-6)^2 = 36$ and $6^2 = 36$, -6 and 6 are square roots of 36.
Since $(-5)^4 = 625$ and $5^4 = 625$, -5 and 5 are fourth roots of 625.

From the preceding example it is obvious that sometimes there exists more than one nth root of a number. When, in a later section, imaginary numbers are introduced, it will be shown that every number has exactly n nth roots. Not all these roots are real numbers. For the present we shall be interested only in those roots that are real numbers. (Real numbers were introduced in Sec. 1–3.)

Since the square of a positive or negative real number is positive, it follows that negative numbers have no real square roots. Thus numbers like -9, -16, -25 have no real square roots.

From the last example it follows that a positive number has two square roots, numerically equal but opposite in sign. Thus the square roots of 9 are 3 and -3. *The positive square root is called the principal square root.* Thus 3 is the principal square root of 9 and is denoted by $\sqrt{9}$.

The term **principal root** is also used for roots of a higher index. Thus, the principal cube root of 8 is 2, denoted by $\sqrt[3]{8}$; similarly, the principal cube root of -8, denoted by $\sqrt[3]{-8}$, is -2.

To unify the discussion on principal roots in all cases, we say that *the principal nth root of the number y, denoted $\sqrt[n]{y}$, is positive when y is positive and is negative if y is negative and n odd.* If y is negative and n even, the nth root of y does not exist for the present; this will lead in a later chapter to the consideration of a new kind of numbers, the so-called imaginary numbers.

Example 2. $\sqrt{25} = 5$.

It should be emphasized that $\sqrt{25}$ is *not* equal to -5, even though -5 is one of the square roots of 25. When the negative square root of 25 is desired, the minus sign is explicitly attached, so that $-\sqrt{25} = -5$, whereas $\sqrt{25} = 5$.

Example 3. By inspection we find

$$\sqrt{49} = 7 \qquad \text{since} \qquad 7^2 = 49.$$
$$\sqrt[4]{81} = 3 \qquad \text{since} \qquad 3^4 = 81.$$
$$\sqrt[3]{-27} = -3 \quad \text{since} \quad (-3)^3 = -27.$$
$$\sqrt[5]{-1} = -1 \quad \text{since} \quad (-1)^5 = -1.$$

These examples illustrate the following:

> **For y positive and n even or odd, $\sqrt[n]{y}$ is positive.**
>
> **For y negative and n odd, $\sqrt[n]{y}$ is negative.**
>
> **For y negative and n even, $\sqrt[n]{y}$ does not exist.**

Example 4. By inspection we find

$$\sqrt{\tfrac{4}{9}} = \tfrac{2}{3} \qquad \text{since} \qquad (\tfrac{2}{3})^2 = \tfrac{4}{9}.$$
$$\sqrt[3]{-\tfrac{1}{64}} = -\tfrac{1}{4} \qquad \text{since} \qquad (-\tfrac{1}{4})^3 = -\tfrac{1}{64}.$$
$$\sqrt{16a^4b^6} = 4a^2b^3 \qquad \text{since} \qquad (4a^2b^3)^2 = 16a^4b^6.$$
$$\sqrt{9\sin^4\theta} = 3\sin^2\theta \qquad \text{since} \qquad (3\sin^2\theta)^2 = 9\sin^4\theta.$$

Exercises

Find the value of each of the following radicals:

1. $\sqrt{9}$. 2. $\sqrt{25}$. 3. $\sqrt{49}$.

4. $\sqrt{81}$. 5. $\sqrt{144}$. 6. $\sqrt{256}$.

7. $\sqrt{900}$. 8. $\sqrt{169}$. 9. $\sqrt{625}$.

10. $\sqrt[3]{1}$. 11. $\sqrt[3]{8}$. 12. $\sqrt[3]{64}$.

13. $\sqrt[3]{-1}$. 14. $\sqrt[3]{-27}$. 15. $\sqrt[3]{125}$.

16. $\sqrt[3]{-125}$. 17. $\sqrt[3]{1000}$. 18. $\sqrt[3]{-1000}$.

19. $\sqrt{\dfrac{4}{9}}$. 20. $\sqrt{\dfrac{9}{49}}$. 21. $\sqrt{\dfrac{144}{169}}$.

22. $\sqrt{441E^2}$. 23. $\sqrt{196r^2s^2}$. 24. $\sqrt{0.01L_x^2}$.

25. $\sqrt[3]{-\dfrac{8}{125}}$. 26. $\sqrt[3]{-\dfrac{1}{216}}$. 27. $\sqrt[3]{\dfrac{a^3}{27}}$.

28. $\sqrt[4]{16x^4}$. 29. $\sqrt[4]{81u^4}$. 30. $\sqrt[5]{32p^5q^5}$.

31. $\sqrt[5]{-243M_x^5}$. 32. $\sqrt[5]{-32R^{10}}$. 33. $\sqrt[6]{64a^6}$.

34. $\sqrt[4]{\dfrac{16}{81}x^4y^4}$. 35. $\sqrt[3]{-\dfrac{1}{8}L^6}$. 36. $\sqrt[5]{-\dfrac{1}{32}Q^5}$.

37. $\sqrt[3]{-8\sin^3\theta}$. 38. $\sqrt{36\cos^4\alpha}$. 39. $\sqrt{625\tan^2 A}$.

40. $\sqrt[4]{10{,}000\sin^8 B}$. 41. $\sqrt[3]{-343(\sin\beta)^6}$. 42. $\sqrt[5]{-32\cos^5\theta}$.

43. $\sqrt[5]{-L_x^{10}L_y^5}$. 44. $\sqrt[4]{E_1^8E_2^4}$. 45. $\sqrt[3]{-E_1^6E_2^3E_3^3}$.

Find the principal square root of:

46. 25.	**47.** 9.	**48.** 36.	**49.** 64.
50. 100.	**51.** 144.	**52.** 196.	**53.** 400.
54. 225.	**55.** 49.	**56.** 81.	**57.** 196.
58. 0.01.	**59.** 0.0001.	**60.** 0.0256.	**61.** $\dfrac{1}{4}$.
62. $\dfrac{1}{9}$.	**63.** $\dfrac{1}{16}$.	**64.** 2.89.	**65.** 0.0169.
66. 0.49.	**67.** $\dfrac{9}{25}$.	**68.** $\dfrac{36}{49}$.	**69.** $\dfrac{16}{121}$.

Find the principal cube root of:

70. 8.	**71.** −8.	**72.** 27.	**73.** −64.
74. 125.	**75.** −216.	**76.** 0.001.	**77.** −0.001.
78. 0.125.	**79.** $-\dfrac{1}{27}$.	**80.** $\dfrac{8}{125}$.	**81.** $-\dfrac{27}{64}$.

Find the real square roots of:

82. 25.	**83.** 64.	**84.** 100.	**85.** 81.
86. 144.	**87.** 256.	**88.** 0.09.	**89.** 0.16.
90. 0.49.	**91.** $\dfrac{16}{49}$.	**92.** $\dfrac{36}{121}$.	**93.** $\dfrac{9}{400}$.

8–3. Rational and Irrational Numbers

It was shown in Sec. 1–3 that, by starting with positive integers and the operations of addition, subtraction, multiplication, and division, we arrived at the system of rational numbers. It will be recalled that a rational number was defined to be one that can be expressed as a fraction m/n, where the numerator m and the denominator n are integers. Now the operation of extracting the nth root of positive rational numbers and the nth (n odd) root of negative rational numbers leads to new numbers like $\sqrt{2}$, $\sqrt[3]{5}$, etc., called **irrational numbers;** other processes beyond the scope of this book also lead to irrational numbers. The rational and irrational numbers make up the system of **real numbers.**

It will be seen in Chapter 12 that the nth (n-even) root of negative numbers will lead to a new type of numbers, the so-called imaginaries.

8–4. Fractional Exponents

In the present section we shall introduce fractional exponents. This will unite the subject of radicals and that of exponents under one set of laws.

At present a^2 means a taken twice as a factor, but to say that $a^{\frac{1}{2}}$ means a taken one-half time as a factor is unintelligible. Since $a^{\frac{1}{2}}$ is at present a meaningless symbol, we are free to define it in any manner we please. In mathematics a definition is said to be a good one if it is in harmony with the material previously developed. With this in mind we shall give meanings to fractional exponents that will conform to the laws of integral exponents summarized in Sec. 8–1.

To give a meaning to $a^{\frac{1}{2}}$ by assuming that law 1 of Sec. 8–1 holds, we write

$$a^{\frac{1}{2}} \cdot a^{\frac{1}{2}} = a^{\frac{1}{2}+\frac{1}{2}} = a^1 = a,$$

$$a^{\frac{1}{2}} \cdot a^{\frac{1}{2}} = a.$$

We see, then, that $a^{\frac{1}{2}}$ is one of the two equal factors of a, and hence

$$a^{\frac{1}{2}} = \sqrt{a}.$$

The right-hand side of the last equation is our definition of $a^{\frac{1}{2}}$.

Similarly, to give a meaning to $a^{\frac{1}{3}}$, we write

$$a^{\frac{1}{3}} \cdot a^{\frac{1}{3}} \cdot a^{\frac{1}{3}} = a^{\frac{1}{3}+\frac{1}{3}+\frac{1}{3}} = a^1 = a,$$

$$a^{\frac{1}{3}} \cdot a^{\frac{1}{3}} \cdot a^{\frac{1}{3}} = a.$$

The expression $a^{\frac{1}{3}}$ is now one of three equal factors of a, and hence

$$a^{\frac{1}{3}} = \sqrt[3]{a}.$$

In general, if n is a positive integer

$$a^{1/n} \cdot a^{1/n} \cdots a^{1/n} = a \quad (n \text{ factors}),$$

and hence

$$a^{1/n} = \sqrt[n]{a}. \tag{1}$$

In an analogous way we can define an expression like $a^{\frac{2}{3}}$ by writing

$$a^{\frac{2}{3}} \cdot a^{\frac{2}{3}} \cdot a^{\frac{2}{3}} = a^{\frac{2}{3}+\frac{2}{3}+\frac{2}{3}} = a^2,$$

$$a^{\frac{2}{3}} = \sqrt[3]{a^2}.$$

In general, if m and n are positive integers,

$$a^{m/n} \cdot a^{m/n} \cdots a^{m/n} = a^m \quad (n \text{ factors})$$

$$a^{m/n} = \sqrt[n]{a^m}. \tag{2}$$

The last equation defines a fractional exponent to be a radical. Thus we *define $a^{m/n}$ to be the principal nth root of a^m*. For $m = 1$ this equation reduces to $a^{1/n} = \sqrt[n]{a}$.

Section 1–11 contained a discussion of negative and zero exponents. This, with the material of the present section, enables us to work with expressions like a^x where a is any **real number** and x any **rational number**.

Summarizing all the definitions concerning exponents so far developed, we have:

DEFINITIONS

a any real number; m and n integers.

1. $a^n = a \cdot a \cdots a$ (**n factors**).

2. $a^{-n} = \dfrac{1}{a^n}.$

3. $a^0 = 1.$

4. $a^{1/n} = \sqrt[n]{a}.$

5. $a^{m/n} = \sqrt[n]{a^m}.$

Using these definitions we can now write expressions involving radicals and powers in various forms.

Examples.

$x \cdot x \cdot x = x^3.$	Definition 1	$(-1)^{\frac{1}{3}} = \sqrt[3]{(-1)}$	Definition 4
$a^{-4} = \dfrac{1}{a^4}.$	Definition 2	$= -1.$	
$(\frac{1}{5})^0 = 1.$	Definition 3	$5E^{-2}L^{-3} = \dfrac{5}{E^2L^3}.$	Definition 2
$8^{\frac{2}{3}} = \sqrt[3]{8^2}$	Definition 5	$4^{-\frac{3}{2}} = \dfrac{1}{4^{\frac{3}{2}}}$	Definition 2
$= \sqrt[3]{64}$		$= \dfrac{1}{\sqrt[2]{4^3}}$	Definition 5
$= 4.$			
$(5x + 9)^0 = 1.$	Definition 3	$= \dfrac{1}{\sqrt{64}}$	
$\dfrac{a^0}{b^{-5}} = a^0 \cdot b^5$	Definition 2	$= \frac{1}{8}.$	
$= b^5.$	Definition 3	$\dfrac{3x^2y^{-3}}{z^{-1}} = \dfrac{3x^2z}{y^3}.$	Definition 2

Exercises

Find the value of each expression by changing from a fractional exponent to a radical or from a negative to a positive exponent.

1. $36^{\frac{1}{2}}.$ 2. $81^{\frac{1}{2}}.$ 3. $8^{\frac{1}{3}}.$ 4. $27^{\frac{1}{3}}.$

5. $(-64)^{\frac{1}{3}}.$ 6. $16^{\frac{1}{4}}.$ 7. $32^{\frac{1}{5}}.$ 8. $8x^0.$

9. $(2a + 3b)^0.$ 10. $(0.0001)^{\frac{1}{2}}.$ 11. $(0.0049)^{\frac{1}{2}}.$ 12. $(0.008)^{\frac{1}{3}}.$

13. $5^{-1}.$ 14. $3^{-2}.$ 15. $(3.1)^{-2}.$ 16. $125^{-\frac{1}{3}}.$

17. $16^{-\frac{1}{4}}$.   **18.** $(0.216)^{-\frac{1}{3}}$.   **19.** $\left(\dfrac{1}{4}\right)^{\frac{1}{2}}$.   **20.** $\dfrac{3}{2^{-2}}$.

21. $\left(-\dfrac{1}{11}\right)^{-1}$.   **22.** $(-1)^{-\frac{3}{5}}$.   **23.** $1^{-1} + 3^{-2}$.   **24.** $(0.0036)^{-\frac{1}{2}}$.

25. $\left(\dfrac{1}{36}\right)^{-\frac{1}{2}}$.   **26.** $\left(-\dfrac{1}{7}\right)^{-1}$.   **27.** $\left(\dfrac{1}{125}\right)^{-\frac{1}{3}}$.

Change the following into identical expressions without zero or negative exponents, and simplify.

28. M^{-4}.   **29.** $5^{-2}y^{-3}$.   **30.** p^2R^{-5}.

31. $4(\sin\theta)^{-3}$.   **32.** $(3\cos\alpha)^{-2}$.   **33.** $6^{-1}r^0y^{-4}$.

34. $\dfrac{3^{-2}}{3^{-1}}$.   **35.** $\dfrac{R^0}{s^{-4}}$.   **36.** $\dfrac{5a^{-3}}{bc^{-5}}$.

37. $\dfrac{\tan\theta}{3x^{-2}y^{-3}}$.   **38.** $\dfrac{5e^{-2}}{(\sin\alpha)^{-3}}$.   **39.** $\dfrac{e_1^{-2}e_2^{-3}}{e_3^{-1}}$.

40. $\dfrac{5^{-1}x^{-2}y^{-3}}{x^{-4}y^5}$.   **41.** $\dfrac{3^0E_3{}^5}{2^{-3}E_1{}^{-1}E_2{}^{-3}}$.   **42.** $\dfrac{e^m + e^{-m}}{2}$.

43. $\dfrac{1}{x^{-1} - y^{-1}}$.   **44.** $\dfrac{1}{r^{-2} - s^{-2}}$.   **45.** $\dfrac{ab^{-2} + a^{-2}b}{a^{-1} + b^{-1}}$.

Write the following in forms without denominators, using negative exponents if necessary.

46. $\dfrac{a}{b}$.   **47.** $\dfrac{5}{x^3}$.   **48.** $\dfrac{1}{M^3}$.

49. $\dfrac{n}{N^2}$.   **50.** $\dfrac{a^2}{y^5}$.   **51.** $\dfrac{3}{L^4}$.

52. $\dfrac{x^3}{abc}$.   **53.** $\dfrac{1}{E_1E_2E_3}$.   **54.** $\dfrac{1}{r_1r_2r_3}$.

55. $\dfrac{a}{x^3y^3z^3}$.   **56.** $\dfrac{ax^{-3}}{y^{-2}}$.   **57.** $\dfrac{a}{7^{-1}p^{-2}q^3}$.

58. $\dfrac{a}{x+y}$.   **59.** $\dfrac{m+n}{p-q}$.   **60.** $\dfrac{e}{a^2+b^2}$.

Write the following expressions with radicals instead of fractional exponents.

61. $x^{\frac{1}{3}}$.   **62.** $a^{\frac{1}{4}}$.   **63.** $L^{\frac{3}{4}}$.

64. $3b^{\frac{1}{2}}$.   **65.** $az^{\frac{1}{3}}$.   **66.** $7S^{\frac{2}{3}}$.

67. $(2V)^{\frac{1}{3}}$.   **68.** $5(\sin\theta)^{\frac{1}{2}}$.   **69.** $(3M_1{}^3)^{\frac{1}{3}}$.

70. $(6\cos x)^{\frac{2}{3}}$.   **71.** $(9e_1e_2e_3)^{\frac{1}{4}}$.   **72.** $7a(\sin\alpha)^{\frac{2}{3}}$.

73. $(e_1 + e_2)^{\frac{1}{2}}$. **74.** $(r_1 + r_2 + r_3)^{\frac{1}{3}}$. **75.** $(L_x + L_y)^{\frac{3}{2}}$.

76. $(\sin \theta_1 + \sin \theta_2)^{\frac{2}{3}}$. **77.** $(a^2 + b^2)^{\frac{1}{2}}$. **78.** $(x^3 + y^3)^{\frac{2}{3}}$.

Using fractional exponents, write the following expressions without radicals.

79. $\sqrt{a}$. **80.** $\sqrt{x^3}$. **81.** $\sqrt[3]{L_1}$.

82. $\sqrt[4]{M_x}$. **83.** $\sqrt[3]{d^5}$. **84.** $\sqrt[4]{e^7}$.

85. $\sqrt[8]{b^7}$. **86.** $\sqrt{\sin \alpha}$. **87.** $\sqrt[3]{\cos \beta}$.

88. $\sqrt[3]{\tan^2 x}$. **89.** $\sqrt[8]{r^{12}}$. **90.** $\sqrt{a + b}$.

91. $\sqrt{R_1 + R_2}$. **92.** $\sqrt[3]{E_1 - E_2}$. **93.** $\sqrt[3]{L_1 + L_2 + L_3}$.

94. $\sqrt[3]{(x + y)^2}$. **95.** $\sqrt{\sin A + \cos A}$. **96.** $\sqrt[5]{(s_1 - 3s_2 + s_3)^3}$.

8–5. The Laws of Exponents

In Sec. 8–1 of this chapter a summary of the laws of exponents was given. These laws were true only for positive and negative integral values of the exponents. Later, in Sec. 8–4, fractional exponents were defined in such a way that they were in harmony with the first law of Sec. 8–1. Although we leave the proof to more advanced work in mathematics, it can be shown that our definition of fractional exponents of Sec. 8–4 is consistent with all the laws of exponents. To state this fact explicitly, we write:

Laws 1 to 7 of Sec. 8–1 are true when the exponents m, n, and q are any rational numbers.

The way the negative and fractional exponents have been introduced exemplifies an important characteristic of mathematics. This is known as the **principle of permanence of formal laws of algebra.** This same principle helped us when we enlarged our number system from positive integers to that of the real number system. Its object is to make a few laws suffice where otherwise we would need many. With appropriate definitions for the negative and fractional exponents, this principle achieves concision by applying a formula already in existence to a much greater variety of cases, thus making the introduction of new formulas unnecessary.

With this generalization in mind, the definitions given in Sec. 8–4 and the laws given in Sec. 8–1 form the basis for all the work in the remainder of this chapter and for any work on exponents in this book. Subsequent allusions to a definition or law of exponents will always refer to Secs. 8–4 and 8–1, and to Sec. 8–6, which follows.

Examples. Perform the indicated operations by the laws of exponents.

$$S^{\frac{1}{2}}S^{\frac{1}{3}} = S^{\frac{1}{2}+\frac{1}{3}} \qquad\qquad\qquad \text{Law 1}$$

$$= S^{\frac{5}{6}}.$$

$$\frac{R^{\frac{1}{2}}}{R^{\frac{1}{3}}} = R^{\frac{1}{2}-\frac{1}{3}} \qquad\qquad\qquad \text{Law 2}$$

$$= R^{\frac{1}{6}}.$$

$$(4^{\frac{3}{2}})^{-2} = 4^{\frac{3}{2}(-2)} \qquad\qquad\qquad \text{Law 4}$$

$$= 4^{-3}$$

$$= \tfrac{1}{64}. \qquad\qquad\qquad \text{Definition 2}$$

$$(9xy)^{\frac{1}{2}} = 9^{\frac{1}{2}}x^{\frac{1}{2}}y^{\frac{1}{2}} \qquad\qquad\qquad \text{Law 5}$$

$$= 3x^{\frac{1}{2}}y^{\frac{1}{2}}.$$

$$\left(\frac{M^2}{N^{-3}}\right)^{\frac{1}{3}} = \frac{(M^2)^{\frac{1}{3}}}{(N^{-3})^{\frac{1}{3}}} \qquad\qquad\qquad \text{Law 6}$$

$$= \frac{M^{\frac{2}{3}}}{N^{-1}} \qquad\qquad\qquad \text{Law 4}$$

$$= M^{\frac{2}{3}}N. \qquad\qquad\qquad \text{Definition 2}$$

$$(5^{\frac{1}{2}}x^{\frac{3}{4}}y)^4 = (5^{\frac{1}{2}})^4 \cdot (x^{\frac{3}{4}})^4 \cdot (y)^4 \qquad\qquad \text{Law 7}$$

$$= 25x^3y^4. \qquad\qquad\qquad \text{Law 4}$$

$$(x^{\frac{2}{3}} + x^{\frac{1}{3}}y^{\frac{1}{3}} + y^{\frac{2}{3}})(x^{\frac{1}{3}} - y^{\frac{1}{3}})$$

$$= x + x^{\frac{2}{3}}y^{\frac{1}{3}} + x^{\frac{1}{3}}y^{\frac{2}{3}} - x^{\frac{2}{3}}y^{\frac{1}{3}} - x^{\frac{1}{3}}y^{\frac{2}{3}} - y \qquad \text{Law 1}$$

$$= x - y.$$

$$\left(\frac{a^{-1} \cdot b}{a^3 \cdot b^{-3}}\right)^{\frac{1}{2}} = (a^{-4} \cdot b^4)^{\frac{1}{2}} \qquad\qquad\qquad \text{Law 2}$$

$$= a^{-2}b^2 \qquad\qquad\qquad \text{Law 7}$$

$$= \frac{b^2}{a^2}. \qquad\qquad\qquad \text{Definition 2}$$

Using negative or fractional exponents we can factor certain expressions. This will be illustrated by the following examples.

Example 1.

$$R_1 - R_2 = (R_1^{\frac{1}{2}})^2 - (R_2^{\frac{1}{2}})^2$$

$$= (R_1^{\frac{1}{2}} - R_2^{\frac{1}{2}})(R_1^{\frac{1}{2}} + R_2^{\frac{1}{2}}).$$

Example 2.

$$25x^{-4} - 9y^2 = (5x^{-2})^2 - (3y)^2$$

$$= (5x^{-2} - 3y)(5x^{-2} + 3y).$$

Example 3.

$$4L^2 + 4LR^{-\frac{1}{2}} + R^{-1} = (2L)^2 + 2 \cdot (2L)\,(R^{-\frac{1}{2}}) + (R^{-\frac{1}{2}})^2$$
$$= (2L + R^{-\frac{1}{2}})^2.$$

The laws of exponents are useful in evaluating certain numerical expressions.

Example 4.

$$(64)^{\frac{1}{3}} = (2^6)^{\frac{1}{3}} = 2^{6 \cdot \frac{1}{3}} = 2^2 = 4.$$

Example 5.

$$\left(\frac{1}{125}\right)^{\frac{2}{3}} = \left(\frac{1}{5^3}\right)^{\frac{2}{3}} = \frac{1}{5^{3 \cdot \frac{2}{3}}} = \frac{1}{5^2} = \frac{1}{25}.$$

Example 6.

$$(32)^{\frac{2}{5}} \cdot (16)^{-\frac{1}{4}} = (2^5)^{\frac{2}{5}} \cdot (2^4)^{-\frac{1}{4}} = 2^2 \cdot 2^{-1} = 2.$$

Exercises

Perform the indicated operations by the laws of exponents and express the results without negative or zero exponents.

1. $(xy^{\frac{1}{2}})^0.$

2. $a^3 a^{-\frac{1}{2}}.$

3. $b^0(c^{\frac{1}{3}})^6.$

4. $(2v^{-\frac{1}{2}})^4.$

5. $3M^{\frac{1}{2}}M^{-\frac{3}{2}}.$

6. $(81p^4q^2)^{\frac{1}{2}}.$

7. $\left(\dfrac{r^0 s^{-2}}{r^{-2} s^0}\right)^{\frac{1}{2}}.$

8. $\left(\dfrac{m^3}{n^6}\right)^{-\frac{1}{3}}.$

9. $\dfrac{L^{\frac{3}{4}}}{L^{\frac{1}{2}}}.$

10. $(5^{\frac{1}{2}}x^{\frac{3}{4}}y)^4.$

11. $(2e_1^{\frac{2}{3}}e_2^{-\frac{3}{5}})^5.$

12. $(25Z^{-2})^{-\frac{3}{2}}.$

13. $(8K^3K^{-6})^{\frac{1}{3}}.$

14. $(x^{\frac{2}{3}}x^{-\frac{1}{6}})^6.$

15. $(27u^6v^3)^{\frac{2}{3}}.$

16. $\left(\dfrac{32h^5}{p^{10}}\right)^{-\frac{1}{5}}.$

17. $\left(\dfrac{n^x}{n^{-x}}\right)^{-1}.$

18. $\left(\dfrac{3a^{-4}}{y^{5m}}\right)^{-m}.$

19. $(x^3y^{2n}z^{n-1})^n.$

20. $(a^{n+1}a^{n-1})^{\frac{1}{2}}.$

21. $x^{m(m-1)}(x^{-m})^{-1}.$

22. $(-125e^{12})^{\frac{1}{3}}(-64e^{15})^{-\frac{1}{3}}.$

23. $(x^{\frac{1}{4}})^{\frac{2}{3}}(x^{-\frac{4}{3}})^{\frac{1}{4}}(x^3)^{-\frac{1}{9}}.$

24. $(a^{\frac{1}{2}} + b^{\frac{1}{2}})(a^{\frac{1}{2}} - b^{\frac{1}{2}}).$

25. $(x^{\frac{1}{2}} + x^{-\frac{1}{2}})^2.$

26. $(E_1^{-\frac{1}{2}} + E_2^{-\frac{1}{2}})^2.$

27. $(u^{\frac{1}{4}} + v^{\frac{1}{4}})(u^{\frac{1}{4}} - v^{\frac{1}{4}}).$

28. $(R_1^{-1} + R_2^{-1})(R_1 + R_2).$

29. $(x^n + x^{-n})(x^n - x^{-n}).$

30. $(x^{\frac{1}{2}} + y^{\frac{1}{2}})(x^{-\frac{1}{2}} - y^{-\frac{1}{2}}).$

31. $(r^{\frac{1}{2}} - 3r^{-\frac{1}{2}})(r^{\frac{1}{2}} - 2r^{-\frac{1}{2}}).$

32. $(x^{\frac{1}{3}} + y^{\frac{1}{3}})(x^{\frac{2}{3}} - x^{\frac{1}{3}}y^{\frac{1}{3}} + y^{\frac{2}{3}}).$

33. $(a^{\frac{1}{3}} - b^{\frac{1}{3}})(a^{\frac{2}{3}} + a^{\frac{1}{3}}b^{\frac{1}{3}} + b^{\frac{2}{3}}).$

34. $(L^{\frac{1}{2}} + L^{\frac{1}{4}}M^{\frac{1}{4}} + M^{\frac{1}{2}})(L^{\frac{1}{4}} - M^{\frac{1}{4}}).$

35. $(x + y^{\frac{1}{2}})(x^2 - xy^{\frac{1}{2}} + y).$

Find one value of:

36. $8^{\frac{1}{3}}$.

37. $27^{\frac{1}{3}}$.

38. $4^{-\frac{1}{2}}$.

39. $(-2)^{-3}$.

40. $(-27)^{\frac{1}{3}}$.

41. $32^{\frac{3}{5}}$.

42. $(0.01)^{-\frac{1}{2}}$.

43. $(0.008)^{-\frac{1}{3}}$.

44. $216^{\frac{2}{3}}$.

45. $(-27)^{\frac{2}{3}}$.

46. $125^{-\frac{2}{3}}$.

47. $8^{-\frac{5}{3}}$.

48. $9^{\frac{3}{2}} \cdot 3^{-3}$.

49. $8^{\frac{2}{3}}(16)^{-\frac{1}{2}}$.

50. $16^{-\frac{3}{4}}(\frac{1}{8})^{-\frac{5}{3}}$.

Factor each of the following expressions into two factors, using negative or fractional exponents.

51. $a^{\frac{2}{3}} - 4b^{\frac{2}{3}}$.

52. $9x^{\frac{4}{5}} - 4y^{\frac{2}{5}}$.

53. $R^2 - L^{-2}$.

54. $x^{\frac{3}{4}} - 27$.

55. $c_1^2 - 2c_1 c_2^{-1} + c_2^{-2}$.

56. $25P^c - 9R^a$.

57. $p^8 - 2p^4 q^{\frac{1}{3}} + q^{\frac{2}{3}}$.

58. $x^{\frac{3}{5}} + 125$.

59. $z^{\frac{2}{3}} - 4z^{\frac{1}{3}} + 3$.

60. $9x^{-2} - 6x^{-1}y^{-2} + y^{-4}$.

61. $4s_1^{\frac{2}{3}} + 20s_1^{\frac{1}{3}}s_2^{\frac{1}{3}} + 25s_2^{\frac{2}{3}}$.

62. $z_1 - 3z_1^{\frac{1}{2}}z_2^{\frac{1}{2}} - 4z_2$.

63. $3h^{-2} + h^{-1}k - 2k^2$.

8–6. Changes in Radical Form

By means of definition 5, fractional exponents were identified with radicals. Although it is possible with the aid of this definition to express a radical as a power, it is convenient to retain the radical form in many operations. Definition 5, together with the laws of exponents, makes it possible to write any radical in a variety of equivalent forms. These can be condensed in the following laws of radicals which follow directly from the laws of exponents.

LAWS OF RADICALS

8. $(\sqrt[n]{a})^n = \sqrt[n]{a^n} = a$.

9. $\sqrt[n]{a}\,\sqrt[n]{b} = \sqrt[n]{ab}$.

10. $\sqrt[m]{\sqrt[n]{a}} = \sqrt[mn]{a}$.

11. $\dfrac{\sqrt[n]{a}}{\sqrt[n]{b}} = \sqrt[n]{\dfrac{a}{b}}$.

12. $\sqrt[km]{a^{kn}} = \sqrt[m]{a^n}$.

These laws enable us to write radicals in a number of equivalent forms. The four ways in which radicals are ordinarily changed are given below with accompanying illustrative examples.

(a) *Removing factors from the radicand.* This is usually accomplished by making use of law 9, for any factor that is a perfect nth power can be thus removed from the radicand.

Examples.

$$\sqrt{28} = \sqrt{2^2 \cdot 7}$$

$$= \sqrt{2^2} \cdot \sqrt{7} \qquad \text{Law 9}$$

$$= 2\sqrt{7}. \qquad \text{Law 8}$$

$$\sqrt[3]{40x^5y^3} = \sqrt[3]{2^3 \cdot x^3 \cdot y^3 \cdot 5 \cdot x^2} \qquad \text{Law 1}$$

$$= \sqrt[3]{2^3 \cdot x^3 \cdot y^3} \cdot \sqrt[3]{5x^2} \qquad \text{Law 9}$$

$$= 2xy\sqrt[3]{5x^2}. \qquad \text{Laws 8 and 9}$$

$$\sqrt[4]{162E^5R^6L^7} = \sqrt[4]{3^4 \cdot 2E^4R^4L^4ER^2L^3} \qquad \text{Law 1}$$

$$= \sqrt[4]{3^4E^4R^4L^4} \cdot \sqrt[4]{2ER^2L^3} \qquad \text{Law 9}$$

$$= 3ERL\sqrt[4]{2ER^2L^3}. \qquad \text{Laws 8 and 9}$$

$$\sqrt{a^2b^4 + 9b^4} = \sqrt{b^4(a^2 + 9)}$$

$$= \sqrt{b^4}\ \sqrt{a^2 + 9} \qquad \text{Law 9}$$

$$= b^2\sqrt{a^2 + 9}. \qquad \text{Law 8}$$

$$\sqrt{\sin^3 \theta - \sin^2 \theta \cos \theta} = \sqrt{\sin^2 \theta\ (\sin \theta - \cos \theta)}$$

$$= \sqrt{\sin^2 \theta}\ \sqrt{\sin \theta - \cos \theta} \qquad \text{Law 9}$$

$$= \sin \theta\sqrt{\sin \theta - \cos \theta}. \qquad \text{Law 8}$$

(b) *Introducing quantities under the radical.* Any quantity multiplying a radical may be introduced under the radical sign. To accomplish this, the quantity has to be raised to a power corresponding to the index of the radical.

Examples.

$$7\sqrt{2} = \sqrt{7^2} \cdot \sqrt{2} \qquad \text{Law 8}$$

$$= \sqrt{49 \cdot 2} \qquad \text{Law 9}$$

$$= \sqrt{98}.$$

$$4ab\sqrt[3]{5ab} = \sqrt[3]{(4ab)^3} \cdot \sqrt[3]{5ab} \qquad \text{Law 8}$$

$$= \sqrt[3]{(4ab)^3 \cdot 5ab} \qquad \text{Law 9}$$

$$= \sqrt[3]{64a^3b^3 \cdot 5ab} \qquad \text{Laws 7 and 4}$$

$$= \sqrt[3]{320a^4b^4}. \qquad \text{Law 1}$$

(c) *Eliminating fractions from the radicand.* In order to simplify computations, a radical should usually be written without fractions under the radical sign. A fraction may be eliminated from the radicand by multiplying numerator and denominator of this fraction by a quantity that will make the denominator a perfect nth power, where n is the index of the radical.

Examples.

$$\sqrt{\tfrac{2}{5}} = \sqrt{\tfrac{2}{5} \cdot \tfrac{5}{5}} \qquad\qquad \text{Multiplication by } \tfrac{5}{5} = 1$$

$$= \sqrt{\frac{10}{5^2}} \qquad\qquad \text{Definition 1}$$

$$= \frac{\sqrt{10}}{\sqrt{5^2}} \qquad\qquad \text{Law 11}$$

$$= \tfrac{1}{5}\sqrt{10}. \qquad\qquad \text{Law 8}$$

$$\sqrt{\frac{E}{R}} = \sqrt{\frac{E}{R}\frac{R}{R}} \qquad\qquad \text{Multiplication by } \frac{R}{R} = 1$$

$$= \sqrt{\frac{ER}{R^2}} \qquad\qquad \text{Definition 1}$$

$$= \frac{\sqrt{ER}}{\sqrt{R^2}} \qquad\qquad \text{Law 11}$$

$$= \frac{1}{R}\sqrt{ER}. \qquad\qquad \text{Law 8}$$

$$\sqrt[3]{\frac{4ab}{9xy^7}} = \sqrt[3]{\frac{4ab}{9xy^7} \cdot \frac{3x^2y^2}{3x^2y^2}} \qquad\qquad \text{Multiplication by } \frac{3x^2y^2}{3x^2y^2} = 1$$

$$= \sqrt[3]{\frac{12abx^2y^2}{3^3x^3y^9}} \qquad\qquad \text{Law 1}$$

$$= \frac{\sqrt[3]{12abx^2y^2}}{\sqrt[3]{3^3x^3y^9}} \qquad\qquad \text{Law 11}$$

$$= \frac{1}{3xy^3}\sqrt[3]{12abx^2y^2}. \qquad\qquad \text{Law 8}$$

$$\sqrt{\frac{\sin\theta}{\cos\theta}} = \sqrt{\frac{\sin\theta}{\cos\theta} \cdot \frac{\cos\theta}{\cos\theta}} \qquad\qquad \text{Multiplication by } \frac{\cos\theta}{\cos\theta} = 1$$

$$= \sqrt{\frac{\sin\theta \cdot \cos\theta}{\cos^2\theta}} \qquad\qquad \text{Law 1}$$

$$= \frac{\sqrt{\sin\theta \cdot \cos\theta}}{\sqrt{\cos^2\theta}} \qquad\qquad \text{Law 11}$$

$$= \frac{1}{\cos\theta}\sqrt{\sin\theta \cdot \cos\theta}. \qquad\qquad \text{Law 8}$$

This procedure is known as **rationalizing the denominator,** for in the final form of the expression no radical appears in the denominator.

(*d*) *Reducing the radical to one with a lower index.* By means of law 12 the index of many radicals may be reduced.

Examples.

$$\sqrt[6]{27} = \sqrt[6]{3^3} \qquad\qquad \text{Definition 1}$$

$$= \sqrt[2]{3^1} \qquad\qquad \text{Law 12}$$

$$= \sqrt{3}.$$

$$\sqrt[4]{49x^2y^2} = \sqrt[4]{(7xy)^2} \qquad\qquad \text{Law 7}$$

$$= \sqrt[2]{(7xy)^1} \qquad\qquad \text{Law 12}$$

$$= \sqrt{7xy}.$$

$$\sqrt[4]{\sin^2 A \cdot \cos^2 A} = \sqrt[4]{(\sin A \cdot \cos A)^2} \qquad\qquad \text{Law 7}$$

$$= \sqrt{\sin A \cos A}. \qquad\qquad \text{Law 12}$$

It is not possible to make a general definition of the **simplest** form of a radical that will be applicable in all problems. In most problems, however, *the changes in cases a, c, and d above are said to simplify radicals.*

Exercises

Simplify by removing factors from the radicands.

1. $\sqrt{90}$.

2. $\sqrt{162}$.

3. $\sqrt[3]{24}$.

4. $\sqrt[3]{250}$.

5. $\sqrt[4]{48}$.

6. $\sqrt[5]{64}$.

7. $\sqrt[3]{-40}$.

8. $\sqrt[5]{-96}$.

9. $\sqrt[3]{a^8}$.

10. $\sqrt{12L^3}$.

11. $\sqrt{20E^7}$.

12. $\sqrt[3]{-2x^7}$.

13. $\sqrt[3]{54 \sin \alpha}$.

14. $\sqrt{3 \cos^3 \theta}$.

15. $\sqrt{28(\tan A)^3}$.

16. $\sqrt[4]{81r_1^{\,5}r_2^{\,11}}$.

17. $\sqrt[3]{0.008x^5y^6}$.

18. $\sqrt[3]{-128p^6q^7}$.

19. $\sqrt{18a^6b^4 \sin^3 \theta}$.

20. $\sqrt{x^2 + x^2y}$.

21. $\sqrt{4 - 4M}$.

22. $\sqrt[3]{M^6 - M^3N}$.

23. $\sqrt{(x^2 + y^2)^3}$.

24. $\sqrt[3]{54m^3 + 27m^3n^3}$

Change each expression to a radical whose coefficient is 1.

25. $3\sqrt{2}$.

26. $2\sqrt[3]{5}$.

27. $2\sqrt[4]{3}$.

28. $x\sqrt{yz}$.

29. $5\sqrt{2R}$.

30. $2E\sqrt[3]{3E}$.

31. $3R_1^{\,2}\sqrt[4]{2R_1^{\,3}}$.

32. $\tfrac{2}{3}u^2\sqrt{12u}$.

33. $(a + b)\sqrt{a + b}$.

34. $5x^2 \sqrt{\dfrac{3y}{5x}}$.

35. $6r \sqrt{\dfrac{5s}{3r}}$.

36. $\dfrac{m^3}{2n} \sqrt{\dfrac{2n^3}{m}}$.

37. $\cos \theta \sqrt[3]{\dfrac{\sin \theta}{\cos^2 \theta}}$.

38. $\dfrac{b^2}{3a} \sqrt[3]{\dfrac{3a}{b^3}}$.

39. $\dfrac{p}{p-q} \sqrt[4]{\dfrac{p-q}{p^3}}$.

Rationalize the denominators, i.e., eliminate fractions from the radicand, and remove factors from the radicands.

40. $\sqrt{\tfrac{1}{2}}$.

41. $\sqrt{\tfrac{1}{3}}$.

42. $\sqrt{\tfrac{2}{5}}$.

43. $\sqrt[3]{\tfrac{1}{4}}$.

44. $\sqrt[3]{\tfrac{5}{9}}$.

45. $\sqrt[4]{\tfrac{7}{8}}$.

46. $\dfrac{4a}{\sqrt{3}}$.

47. $\sqrt{\dfrac{L_1}{L_2}}$.

48. $\sqrt{\dfrac{1}{6p^3}}$.

49. $\sqrt{\dfrac{3\pi}{5y}}$.

50. $\dfrac{\sqrt{2}}{3\sqrt{3}}$.

51. $\sqrt[3]{\dfrac{m^3 n^2}{3}}$.

52. $\sqrt[3]{\dfrac{27}{16R^4}}$.

53. $\sqrt{\dfrac{a}{\sin A}}$.

54. $\sqrt{\dfrac{\cos A}{\sin A}}$.

55. $\sqrt[4]{\dfrac{\tan \theta}{2}}$.

56. $\sqrt[5]{\dfrac{z}{16}}$.

57. $\sqrt[3]{\dfrac{-16E^5}{9R^2}}$.

58. $\sqrt{\dfrac{\cot \alpha}{\tan \alpha}}$.

59. $\sqrt{\dfrac{3a+1}{8b^3}}$.

60. $\sqrt{\dfrac{R_1 R_2}{R_1 + R_2}}$.

Reduce each radical to one with the lowest possible index, and remove all of the factors from the radicands that you can.

61. $\sqrt[4]{x^6}$.

62. $\sqrt[5]{a^{10}}$.

63. $\sqrt[4]{64}$.

64. $\sqrt[6]{9}$.

65. $\sqrt[9]{27}$.

66. $\sqrt[6]{625}$.

67. $\sqrt[4]{4b^6}$.

68. $\sqrt[6]{27E^3}$.

69. $\sqrt[4]{36R^6L^4}$.

70. $\sqrt[6]{\sin^3 \theta}$.

71. $\sqrt[4]{a^2 \tan^2 A}$.

72. $\sqrt[6]{8x^3 y^3 z^9}$.

73. $\sqrt[4]{\dfrac{36u^2v^2}{49z^2}}$.

74. $\sqrt[6]{\dfrac{27a^3b^3}{8x^3y^3}}$.

75. $\sqrt[6]{\dfrac{a^6 \sin^3 A}{b^6 \sin^3 B}}$.

76. $\sqrt[4]{144 \sin^2 \theta \cos^2 \theta}$.

77. $\sqrt[6]{125(E_1 - E_2)^3}$.

78. $\sqrt[4]{(r_1 - r_2)^2 (s_1 + s_2)^2}$.

Using the changes discussed in cases (a), (c), and (d), simplify the following radicals.

79. $\dfrac{1}{\sqrt{3}}$.

80. $\sqrt{\dfrac{2x^2}{5}}$.

81. $\sqrt{\dfrac{a^3b^3}{9}}$.

82. $\sqrt{\dfrac{2m}{7n^2}}$.

83. $\dfrac{3\sqrt{15}}{2\sqrt{3}}$.

84. $\dfrac{21}{2\sqrt{35}}$.

85. $\sqrt{\dfrac{4}{uv^2}}$.

86. $\sqrt[3]{\dfrac{1}{2c^5}}$.

87. $\sqrt[3]{\dfrac{8E}{5}}$.

88. $\sqrt[4]{\dfrac{16p^4}{9q^5}}$.

89. $\sqrt{\dfrac{4x^3y}{9a^2b}}$.

90. $\sqrt[5]{\dfrac{r^6s}{81}}$.

91. $\sqrt[4]{\frac{16}{25}L^8}$.

92. $\sqrt{\frac{12}{49}R_1{}^3R_2{}^3}$.

93. $\sqrt[5]{128Z^{-6}}$.

94. $\sqrt[5]{\dfrac{32a^3b^6}{81a^2c^9}}$.

95. $\sqrt[3]{\dfrac{(r-s)^3t^3}{16}}$.

96. $\sqrt{\dfrac{e_1{}^2e_2}{e_1+e_2}}$.

97. $\sqrt[3]{1+\dfrac{1}{R}}$.

98. $\sqrt{1+\dfrac{L^2}{1+2L}}$.

99. $\sqrt[3]{\dfrac{x}{8}+\dfrac{3}{y^3}}$.

100. $\sqrt{(E^3+R^3)(E+R)^3}$.

101. $\sqrt{(a+b)(a^2-ab-2b^2)}$.

102. $\sqrt{\dfrac{a}{x^2}+\dfrac{2a}{xy}+\dfrac{a}{y^2}}$.

103. $\sqrt{5-\dfrac{10u}{v^2}+\dfrac{5u^2}{v^4}}$.

8–7. Addition and Subtraction of Radicals

Terms containing the same radicals can be added or subtracted. Radicals are the **same** if they can be simplified so as to have the same radicand and the same index. In adding or subtracting expressions containing radicals, *simplify each radical by the methods of Sec. 8–6, and collect all multiples of the same radical.*

Examples.

$2\sqrt{6}+9\sqrt{\tfrac{2}{3}}-\sqrt[4]{36}$

$\qquad = 2\sqrt{6}+9\sqrt{\tfrac{2}{3}\cdot\tfrac{3}{3}}-\sqrt[4]{6^2}$ $\left.\begin{array}{l}\\ \\ \\ \end{array}\right\}$ Case c of Sec. 8–6 for the second term and case d for the third term

$\qquad = 2\sqrt{6}+3\sqrt{6}-\sqrt{6}$

$\qquad = 4\sqrt{6}.$ Collecting multiples of the same radical

$3\sqrt{125}+\sqrt{75}-3\sqrt{20}-\sqrt{675}$

$\qquad = 3\sqrt{5\cdot5^2}+\sqrt{3\cdot5^2}-3\sqrt{5\cdot2^2}-\sqrt{3\cdot15^2}$ $\left.\begin{array}{l}\\ \\ \end{array}\right\}$ Case a of Sec. 8–6 for all terms

$\qquad = 15\sqrt{5}+5\sqrt{3}-6\sqrt{5}-15\sqrt{3}$

$\qquad = 9\sqrt{5}-10\sqrt{3}.$ Collecting multiples of the same radical

$\sqrt[6]{16E^2}-E\sqrt[3]{4E}+\sqrt[9]{64E^3}$

$\qquad = \sqrt[6]{(4E)^2}-E\sqrt[3]{4E}+\sqrt[9]{(4E)^3}$ $\left.\begin{array}{l}\\ \\ \end{array}\right\}$ Case d for the first and last terms

$\qquad = \sqrt[3]{4E}-E\sqrt[3]{4E}+\sqrt[3]{4E}$

$\qquad = (2-E)\sqrt[3]{4E}.$ Collecting multiples of the same radical and factoring

Exercises

Perform the indicated additions and subtractions where possible.

1. $2\sqrt{50}-7\sqrt{2}+\sqrt{18}$.

2. $5\sqrt{54}+\sqrt{150}-7\sqrt{96}$.

3. $4\sqrt{108}+6\sqrt{75}-\sqrt{192}$.

4. $\sqrt[3]{250}+2\sqrt[3]{16}+12\sqrt[3]{2}$.

5. $8\sqrt[3]{81}-\sqrt[3]{24}-\sqrt[3]{375}$.

6. $\sqrt[3]{5}+6\sqrt[6]{25}-2\sqrt[3]{40}$.

7. $\sqrt{25L} + \sqrt{9L} - 2\sqrt{4L}.$

8. $3\sqrt{49a^2 \sin \theta} - \sqrt{16b^2 \sin \theta}.$

9. $\sqrt[3]{40x} + 3\sqrt[3]{135a^3x}.$

10. $R\sqrt{R} + \sqrt{121R^3} - \sqrt[4]{16R^6}.$

11. $4\sqrt{\tfrac{3}{2}} + 7\sqrt{24} - \sqrt{150}.$

12. $25\sqrt{\tfrac{3}{5}} + \sqrt{60} - 27\sqrt{\tfrac{5}{3}}.$

13. $2\sqrt{\tfrac{1}{7}} + \sqrt{63} - 3\sqrt{112}.$

14. $\sqrt[4]{48} - \sqrt[8]{9} + 6\sqrt[4]{\tfrac{1}{27}}.$

15. $\sqrt{4 \sin^3 A} - \sqrt{9a^2 \sin A \cdot \cos^2 A}.$

16. $\sqrt[4]{81 z^5} + \sqrt[4]{16 z} - \sqrt[4]{625 z^5}.$

17. $\sqrt{\dfrac{E_1 - E_2}{E_1 + E_2}} - \sqrt{\dfrac{E_1 + E_2}{E_1 - E_2}}.$

18. $\sqrt{\dfrac{E_1 - E_2}{E_1 + E_2}} + \sqrt{\dfrac{E_1 + E_2}{E_1 - E_2}}.$

19. $\sqrt[3]{\tfrac{9}{2}} + 6\sqrt[3]{\tfrac{1}{3}} - \sqrt[3]{\tfrac{3}{6}}.$

20. $\sqrt{\tfrac{1.8}{7}} + \sqrt{\tfrac{25}{14}} - 6\sqrt[4]{\tfrac{4.9}{4}}.$

21. $\sqrt{\dfrac{u^3}{2v}} + v\sqrt{\dfrac{u^3}{2v^3}} + \dfrac{v^2}{2}\sqrt{\dfrac{2u^3}{v^5}} - \dfrac{3}{2}\sqrt[4]{\dfrac{u^2}{4v^2}}.$

22. $\sqrt{r^3s^2 + r^2s^3} - \sqrt{(r+s)^3} + \sqrt{4r + 4s}.$

23. $2\sqrt[3]{81a^4b^2} + 3a\sqrt[6]{576a^8b^4} - 5\sqrt[3]{24a^4b^2} + a^2\sqrt[3]{3ab^2}.$

24. $\sqrt[3]{64pq^3 + 64q^4} - \sqrt[3]{p^4 + p^3q} + \sqrt[9]{(p+q)^{12}}.$

25. $\sqrt{50b^4 - 75b^2y} - \sqrt{32b^2y^4 - 48y^5} + \sqrt{2b^4y^2 - 3b^2y^3}.$

8-8. Multiplication of Radicals

When the indices of the radicals are alike, they are multiplied according to law 9 (Sec. 8–6). Thus, to find the product of two or more radicals of the same index, multiply the coefficients to obtain the coefficient of the product, and multiply the radicals by means of law 9 to obtain the radical of the product. Simplify the result when necessary.

Examples.

$$5\sqrt{2} \cdot 3\sqrt{5} = 15\sqrt{2} \cdot \sqrt{5} \qquad \text{Multiplying the coefficients}$$

$$= 15\sqrt{10}. \qquad \text{Law 9}$$

$$2\sqrt[3]{3} \cdot 5\sqrt[3]{4} \cdot 3\sqrt[3]{18} = 2 \cdot 5 \cdot 3\sqrt[3]{3}\,\sqrt[3]{4}\,\sqrt[3]{18} \qquad \text{Multiplying the coefficients}$$

$$= 30\sqrt[3]{216} \qquad \text{Law 9}$$

$$= 30 \cdot 6 \qquad \text{Law 8}$$

$$= 180.$$

$$\sqrt[4]{8R^3}\,\sqrt[4]{4R^2E} = \sqrt[4]{32R^5E} \qquad \text{Law 9}$$

$$= 2R\sqrt[4]{2RE}. \qquad \text{Simplifying}$$

To multiply radicals of different indices, we must first express them (by law 12) as radicals of a common index and proceed as above. This common index is the lowest common multiple (LCM) of the indices of the original radicals. In performing operations with radicals it is sometimes easier to replace the radicals by fractional exponents and to use the laws of exponents.

Examples.

$$\sqrt{2} \cdot \sqrt[3]{4} = \sqrt[2]{2} \cdot \sqrt[3]{4} \qquad \text{The LCM of the indices 2 and 3 is 6}$$

$$= \sqrt[6]{2^3} \cdot \sqrt[6]{4^2} \qquad \text{Law 12}$$

$$= \sqrt[6]{2^3 \cdot 4^2} \qquad \text{Law 9}$$

$$= 2\sqrt[6]{2}. \qquad \text{Simplifying}$$

$$\sqrt[3]{4n} \sqrt[4]{8n^3} = \sqrt[12]{(4n)^4} \cdot \sqrt[12]{(8n^3)^3} \qquad \text{The LCM of the indices 3 and 4 is 12}$$

$$= \sqrt[12]{(2^2 n)^4 \cdot (2^3 n^3)^3} \qquad \text{Law 9}$$

$$= 2n \sqrt[12]{32n}. \qquad \text{Simplifying}$$

$$\sqrt{\sin \theta} \sqrt[4]{\sin^3 \theta} = \sqrt[4]{\sin^2 \theta} \cdot \sqrt[4]{\sin^3 \theta} \qquad \text{Law 12}$$

$$= \sin \theta \sqrt[4]{\sin \theta}. \qquad \text{Law 9 and Simplifying}$$

$$\sqrt[8]{x^2} \sqrt[3]{x^2} = \sqrt[8]{\sqrt[3]{x^6 \cdot x^2}} \qquad \text{Case } b \text{ of Sec. 8–6}$$

$$= \sqrt[24]{x^8} \qquad \text{Laws 1 and 10}$$

$$= \sqrt[3]{x}. \qquad \text{Law 12}$$

Exercises

Find the following products and simplify.

1. $2\sqrt{3} \cdot 7\sqrt{6}.$ **2.** $3\sqrt{2} \cdot 5\sqrt{2}.$ **3.** $6\sqrt{3} \cdot 8\sqrt{2}.$

4. $5\sqrt{11} \cdot 2\sqrt{11}.$ **5.** $4\sqrt{6a} \cdot 5\sqrt{8a}.$ **6.** $7\sqrt[3]{3} \cdot 2\sqrt[3]{9}.$

7. $6\sqrt[3]{4} \cdot 3\sqrt[3]{8}.$ **8.** $\sqrt[5]{16} \cdot 3\sqrt[5]{4}.$ **9.** $x\sqrt[3]{x^2} \cdot \sqrt[3]{x^2}.$

10. $3\sqrt[4]{8x^2 y^3} \cdot \sqrt[4]{2x^3 y^2}.$ **11.** $(3\sqrt{2L})^2.$ **12.** $(3\sqrt[3]{5})^3.$

13. $3\sqrt{\tfrac{27}{5}} \cdot 4\sqrt{\tfrac{3}{5}}.$ **14.** $\sqrt{\tfrac{12}{7}} \sqrt{\tfrac{21}{2}}.$ **15.** $\sqrt[3]{\tfrac{3}{4}} \cdot \tfrac{1}{3}\sqrt[3]{\tfrac{28}{2}}.$

16. $\sqrt{3L} \cdot \sqrt[3]{9L^2}.$ **17.** $\sqrt[4]{4} \cdot \sqrt{6}.$ **18.** $\sqrt{6p} \sqrt[4]{24p^3 q}.$

19. $\sqrt[4]{s^3} \sqrt[6]{3s^5}.$ **20.** $\sqrt[5]{u^3 v^4} \sqrt[6]{u^4 v^5}.$ **21.** $\sqrt[5]{a^3 b^4 c^2} \cdot \sqrt[4]{a^2 b^3 c}.$

22. $\sqrt[3]{\tfrac{3}{2}} \sqrt{\tfrac{2}{3}}.$ **23.** $3\sqrt[4]{\tfrac{1}{7}} \cdot \sqrt[6]{\tfrac{3}{8}}.$ **24.** $\sqrt[3]{\tfrac{4}{9}} \cdot \sqrt[4]{\tfrac{7}{12}}.$

25. $\sqrt[3]{\sqrt[5]{32x^{30}}}.$ **26.** $\sqrt[4]{\sqrt{(R_1 - R_2)^8}}.$ **27.** $\sqrt[3]{27\sqrt{\sin^9 \theta}}.$

28. $(\sqrt{2} + \sqrt{3})^2.$ **29.** $(\sqrt{x} - \sqrt{y})^2.$ **30.** $(\sqrt{a} + \sqrt{b})^2.$

31. $(\sqrt{5} + \sqrt{3})(\sqrt{5} - \sqrt{3})$.

32. $(\sqrt{E_1} + \sqrt{E_2})(\sqrt{E_1} - \sqrt{E_2})$.

33. $(\sqrt{6} + 2\sqrt{3})(\sqrt{3} - 4\sqrt{6})$.

34. $(4\sqrt{5} - \sqrt{2})(3\sqrt{5} + 2\sqrt{2})$.

35. $\sqrt{e_1 - e_2}\,\sqrt{e_1 + e_2}$.

36. $\sqrt{7} + \sqrt{3} \cdot \sqrt{7} - \sqrt{3}$.

37. $\sqrt{8 - \sqrt{19}} \cdot \sqrt{8 + \sqrt{19}}$.

38. $\sqrt[4]{8 \sin^3 A} \cdot \sqrt{2 \sin A}$.

39. $\sqrt[3]{\sin^2 \theta}\,\sqrt[4]{\sin^3 \theta \cos^2 \theta}$.

40. $\sqrt[3]{9 \tan^2 \theta} \cdot \sqrt[4]{3 \tan^3 \theta}$.

41. $\sqrt{2}(2\sqrt{18} - \sqrt{72} + 5\sqrt{8})$.

42. $\sqrt{LR} \cdot \sqrt[3]{L^2 R} \cdot \sqrt[4]{L^3 R^3}$.

43. $\sqrt[3]{a^2 b}\,\sqrt[4]{a^3 b^2}\,\sqrt[6]{a^5 b^5}$.

44. $(\sqrt{5} + \sqrt{x + 6})^2$.

45. $(\sqrt{M - 1} - \sqrt{M + 1})^2$.

46. $(\sqrt[3]{p} + \sqrt[3]{q})^2$.

47. $(\sqrt{x_1} + \sqrt{x_2} + \sqrt{x_3})^2$.

48. $(\sqrt[3]{y^2} - 5)(\sqrt[3]{y} + 3)$.

49. $\sqrt[3]{(E_1 + E_2)^2 (E_1 - E_2)} \cdot \sqrt[4]{(E_1 + E_2)^3 (E_1 - E_2)^2}$.

50. $(\sqrt{x + y} + 3\sqrt{y})(\sqrt{x + y} - 3\sqrt{y})$.

51. $(\sqrt{2a + 3b} - 7\sqrt{c})(\sqrt{2a + 3b} + 7\sqrt{c})$.

52. $(\sqrt{6} + \sqrt{3} - \sqrt{2})(\sqrt{6} + \sqrt{3} + \sqrt{2})$.

53. $(2\sqrt{5} - \sqrt{7} + 3\sqrt{3})(2\sqrt{5} + \sqrt{7} + 3\sqrt{3})$.

8–9. Division of Radicals

When the dividend and the divisor are monomials containing radicals of the same index, the division may be performed by the use of law 11. When the radicals are of different indices, express them as radicals of the same index by the method of the preceding section, or change to fractional exponents.

Examples.

$$\frac{\sqrt{28}}{\sqrt{12}} = \sqrt{\frac{28}{12}} \qquad\qquad \text{Law 11}$$

$$= \sqrt{\tfrac{7}{3}}$$

$$= \tfrac{1}{3}\sqrt{21}. \qquad\qquad \text{Rationalizing}$$

$$\frac{\sqrt{4ER}\,\sqrt[3]{2ER}}{\sqrt[6]{4E^5 R^3}} \qquad\qquad \text{The LCM of the indices 2, 3, and 6 is 6}$$

$$= \frac{\sqrt[6]{(4ER)^3}\,\sqrt[6]{(2ER)^2}}{\sqrt[6]{4E^5 R^3}} \qquad\qquad \text{Law 12}$$

$$= \sqrt[6]{\frac{64E^3 R^3 \cdot 4E^2 R^2}{4E^5 R^3}} \qquad\qquad \text{Laws 9 and 11}$$

$$= \sqrt[6]{64R^2} \qquad\qquad\qquad \text{Laws 1 and 3}$$

$$= 2\sqrt[3]{R}. \qquad\qquad\qquad \text{Simplifying}$$

$$\frac{\sqrt{\frac{3}{4}}}{\sqrt{\frac{8}{5}}} = \sqrt{\frac{\frac{3}{4}}{\frac{8}{5}}} \qquad\qquad\qquad \text{Law 11}$$

$$= \sqrt{\frac{3}{4} \cdot \frac{5}{8}} \qquad\qquad\qquad \text{Inverting the divisor}$$

$$= \tfrac{1}{4}\sqrt{\tfrac{15}{2}} \qquad\qquad\qquad \text{Simplifying}$$

$$= \tfrac{1}{8}\sqrt{30}. \qquad\qquad\qquad \text{Rationalizing}$$

One important case in division of radicals occurs when the divisor is a binomial in which one or both of the terms contain a square root.　In such a case the division is performed by rationalizing the divisor.　This is done as follows.　*Multiply numerator and denominator of the resulting fraction by the denominator with the sign between its terms changed.　Simplify the numerator and the denominator.*　Similar methods can be applied to other special forms.

Examples.

$$\frac{2}{3 + \sqrt{5}}$$

$$= \frac{2}{3 + \sqrt{5}} \cdot \frac{3 - \sqrt{5}}{3 - \sqrt{5}} \qquad\qquad \text{Multiplying divisor and dividend by } 3 - \sqrt{5}$$

$$= \frac{2(3 - \sqrt{5})}{9 - 5} \qquad\qquad\qquad \text{Multiplying}$$

$$= \tfrac{1}{2}(3 - \sqrt{5}). \qquad\qquad\qquad \text{Reducing}$$

$$\frac{\sqrt{5} + 6\sqrt{2}}{2\sqrt{5} - 3\sqrt{2}}$$

$$= \frac{\sqrt{5} + 6\sqrt{2}}{2\sqrt{5} - 3\sqrt{2}} \cdot \frac{2\sqrt{5} + 3\sqrt{2}}{2\sqrt{5} + 3\sqrt{2}} \qquad \text{Multiplying divisor and dividend by } 2\sqrt{5} + 3\sqrt{2}$$

$$= \frac{2 \cdot 5 + 3\sqrt{10} + 12\sqrt{10} + 3 \cdot 6 \cdot 2}{20 - 18} \qquad \text{Multiplying}$$

$$= \tfrac{1}{2}(46 + 15\sqrt{10}). \qquad\qquad\qquad \text{Reducing}$$

$$\frac{\sqrt{1 + a} - \sqrt{1 - a}}{\sqrt{1 + a} + \sqrt{1 - a}}$$

$$= \frac{\sqrt{1 + a} - \sqrt{1 - a}}{\sqrt{1 + a} + \sqrt{1 - a}} \cdot \frac{\sqrt{1 + a} - \sqrt{1 - a}}{\sqrt{1 + a} - \sqrt{1 - a}} \qquad \text{Multiplying divisor and dividend by } \sqrt{1 + a} - \sqrt{1 - a}$$

$$= \frac{(1+a) - 2\sqrt{1-a^2} + (1-a)}{(1+a) - (1-a)} \qquad \text{Multiplying}$$

$$= \frac{2 - 2\sqrt{1-a^2}}{2a} \qquad \text{Reducing}$$

$$= \frac{1 - \sqrt{1-a^2}}{a}.$$

Exercises

Divide and reduce to simplest form,

1. $\dfrac{\sqrt{24}}{\sqrt{6}}.$

2. $\dfrac{\sqrt{50}}{\sqrt{2}}.$

3. $\dfrac{\sqrt{27ab}}{\sqrt{3ab}}.$

4. $\dfrac{\sqrt[3]{8L^2}}{\sqrt[3]{2L}}.$

5. $\dfrac{\sqrt[3]{135 \sin^2 \theta}}{\sqrt[3]{5 \sin \theta}}.$

6. $\dfrac{\sqrt[4]{a^5 b^7 c^8}}{\sqrt[4]{a^3 b^3 c^2}}.$

7. $\dfrac{2 \cos A}{\sqrt{\cos A}}.$

8. $\dfrac{6xy^2}{\sqrt{12xy^3}}.$

9. $\dfrac{2\sqrt{2}}{3\sqrt{3}}.$

10. $\dfrac{8\sqrt{45}}{5\sqrt{32}}.$

11. $\dfrac{\sqrt[3]{2}}{\sqrt{3}}.$

12. $\dfrac{4}{\sqrt[3]{2}}.$

13. $\dfrac{\sqrt{10}}{\sqrt[3]{2}}.$

14. $\dfrac{\sqrt{M}}{\sqrt[5]{M^2}}.$

15. $\dfrac{\sqrt[3]{r^2 s^2}}{\sqrt{rs}}.$

16. $\dfrac{\sqrt[6]{4pq^3 y^2}}{\sqrt[4]{2p^3 q}}.$

17. $\dfrac{\sqrt{6} - \sqrt{15}}{\sqrt{3}}.$

18. $\dfrac{2\sqrt{21} + 4\sqrt{35}}{2\sqrt{7}}.$

19. $\dfrac{1}{2 - \sqrt{6}}.$

20. $\dfrac{2}{5 + \sqrt{3}}.$

21. $\dfrac{4 + \sqrt{3}}{2 - \sqrt{3}}.$

22. $\dfrac{1}{\sqrt{E_1} + \sqrt{F_2}}.$

23. $\dfrac{1}{\sqrt{R_1} - \sqrt{R_2}}.$

24. $\dfrac{a}{\sqrt{x} - \sqrt{y}}.$

25. $\dfrac{\sqrt{e_1} + \sqrt{e_2}}{\sqrt{e_1} - \sqrt{e_2}}.$

26. $\dfrac{\sqrt{8} + \sqrt{3}}{\sqrt{8} - \sqrt{3}}.$

27. $\dfrac{\sqrt{i_1} - \sqrt{i_2}}{\sqrt{i_1} + \sqrt{i_2}}.$

28. $\dfrac{\sqrt{30z}\,\sqrt[4]{24z^2}\,\sqrt[3]{75z}}{\sqrt[6]{5z}}.$

29. $\dfrac{\sqrt[3]{mn}\,\sqrt[4]{mn^2}}{\sqrt[3]{m^2 n^2}}.$

30. $\dfrac{5\sqrt{3} - \sqrt{2}}{\sqrt{3} + 3\sqrt{2}}.$

31. $\dfrac{3\sqrt{7} + 2\sqrt{3}}{5\sqrt{7} - \sqrt{3}}.$

32. $\dfrac{H}{\sqrt{H} - \sqrt{HL}}.$

33. $\dfrac{\sqrt{I_p}}{\sqrt{I_p + 1} + 1}.$

34. $\dfrac{\sqrt{2} + \sqrt{L}}{\sqrt{L} + 2\sqrt{2}}.$

35. $\dfrac{4\sqrt{3} - 3\sqrt{2}}{\sqrt{6}} \div \dfrac{\sqrt{10}}{4\sqrt{3} + 3\sqrt{2}}.$

36. $\dfrac{\sqrt{u + v} + \sqrt{u - v}}{\sqrt{u + v} - \sqrt{u - v}}.$

37. $\dfrac{\sqrt{\sin A} - \sqrt{\cos A}}{\sqrt{\sin A} + \sqrt{\cos A}}.$

38. $\sqrt{\dfrac{a + \sqrt{a^2 - b^2}}{a - \sqrt{a^2 - b^2}}}.$

39. $\dfrac{\sqrt{x^2 + y^2} + \sqrt{x^2 - y^2}}{2\sqrt{x^2 + y^2} + 3\sqrt{x^2 - y^2}}.$

8–10. Equations Involving Radicals

An equation in which the unknown quantity appears under a radical sign is called a **radical equation.** The radical equations needed in elementary mathematics and its applications contain no radicals other than square roots. We shall therefore limit our discussion to such equations only.

If the equation contains only *one* square radical, *transpose the terms of the equation so that the radical term stands alone in one member; then square both members, and solve the resulting equation.*

The process of squaring is equivalent to a multiplication. In Sec. 7–5 we saw that, if both members of an equation are multiplied by an expression involving the unknown, roots not in the original equation may be introduced. These are called extraneous roots and must be rejected. Therefore, it is essential to test each root obtained by substituting in the *original* equation.

The following examples should make the above procedure clear.

Example 1. Solve the equation $\sqrt{3x - 5} - 2 = 0.$

Arrange the equation so that the radical $\sqrt{3x - 5}$ will stand alone on one side of the equation:

$$\sqrt{3x - 5} = 2.$$

Squaring:

$$3x - 5 = 4$$

and

$$x = 3.$$

Substituting $x = 3$ in the original equation, we get

$$\sqrt{9 - 5} - 2 = 0.$$

Therefore $x = 3$ is a root of the given equation.

Example 2. Solve the equation $\sqrt{R - 4} + 1 = 0.$

Transposing, we get

$$\sqrt{R - 4} = -1.$$

Squaring:

$$R - 4 = 1.$$

Thus:

$$R = 5.$$

Substituting $R = 5$ in the original equation, we get $\sqrt{5-4} + 1 \neq 0$ or $1 + 1 \neq 0$, and therefore $R = 5$ is an extraneous root and must be rejected. We conclude that the original equation has no solution. That this must be so is obvious from a careful inspection of the equation $\sqrt{R-4} + 1 = 0$. In order that this equation be satisfied, $\sqrt{R-4}$ must equal -1, which is not possible according to Sec. 8–2, since $\sqrt{R-4}$ is the positive root of $R-4$.

Example 3.   Solve the equation $\sqrt{4x+1} + 5 = x$.

Transposing:
$$\sqrt{4x+1} = x - 5.$$

Squaring:
$$4x + 1 = x^2 - 10x + 25.$$

Collecting terms:
$$x^2 - 14x + 24 = 0.$$

Factoring:
$$(x-2)(x-12) = 0.$$

Therefore:
$$x = 2 \quad \text{or} \quad x = 12.$$

Substituting $x = 2$ in the given equation, we find $\sqrt{8+1} + 5 \neq 2$, and therefore $x = 2$ is extraneous and must be rejected. Substituting $x = 12$ in the original equation gives
$$\sqrt{48+1} + 5 = 12.$$

Therefore $x = 12$ is a root of the given equation.

To solve an equation containing two or more square radicals, the general procedure is to isolate one radical at a time on one side of the equation and continue as with an equation containing only one radical. This process is repeated until all radicals are eliminated.

Example 4.   Solve the equation $\sqrt{E-12} - \sqrt{E} = -2$.

Arrange the equation so that the more complicated radical term $\sqrt{E-12}$ will stand alone on one side of the equation.
$$\sqrt{E-12} = \sqrt{E} - 2.$$

Squaring:
$$E - 12 = E - 4\sqrt{E} + 4.$$

Transposing and dividing by 4:
$$\sqrt{E} = 4.$$

Squaring:
$$E = 16.$$

Substituting $E = 16$ in the original equation gives
$$\sqrt{16-12} - \sqrt{16} = -2.$$

Therefore $E = 16$ is a root of the given equation.

Example 5. Solve the equation $\sqrt{x-1} + \sqrt{3x+3} = 4$.

Transposing:
$$\sqrt{3x+3} = 4 - \sqrt{x-1}.$$

Squaring:
$$3x + 3 = 16 - 8\sqrt{x-1} + x - 1.$$

Collecting terms:
$$4\sqrt{x-1} = 6 - x.$$

Squaring:
$$16(x-1) = 36 - 12x + x^2.$$

Collecting terms:
$$x^2 - 28x + 52 = 0.$$

Factoring:
$$(x-26)(x-2) = 0.$$

Therefore:
$$x = 26 \quad \text{or} \quad x = 2.$$

Substituting $x = 26$ in the given equation, we find
$$\sqrt{25} + \sqrt{81} \neq 4,$$

and therefore $x = 26$ is extraneous and must be rejected as a root.
Substituting $x = 2$ gives
$$\sqrt{1} + \sqrt{9} = 4.$$

Therefore $x = 2$ is a root of the given equation.

Exercises

Solve the following equations:

1. $\sqrt{x+3} = 5$.

2. $\sqrt{y} - 4 = 3$.

3. $5 + \sqrt{3a} = 4$.

4. $\sqrt{2L-5} = 9$.

5. $5\sqrt{u-3} = \sqrt{u+9}$.

6. $\sqrt{R^2-5} - R + 1 = 0$.

7. $\sqrt{3E+4} - E = 0$.

8. $\sqrt{z^2-7} + z - 7 = 0$.

9. $2u = 3 + \sqrt{u^2+6u-6}$.

10. $2\sqrt{x^2-x+6} = 7 - x$.

11. $\sqrt{K+20} - \sqrt{K-1} = 3$.

12. $\sqrt{10N-1} - \sqrt{N} = 2$.

13. $\sqrt{x} + \sqrt{32+x} = 16$.

14. $\sqrt{v+4} + \sqrt{v} = 4$.

15. $\sqrt{4y-11} + 1 = 2\sqrt{y}$.

16. $\sqrt{5p+10} - \sqrt{5p} = 2$.

17. $i + 3\sqrt{i} - 10 = 0$.

18. $\sqrt{2z-3} - \sqrt{2z} = 1$.

19. $\sqrt{L-3} + \sqrt{L+4} = 7$.

20. $\sqrt{x+8} + \sqrt{x+1} = 5$.

21. $\sqrt{7+3m} = 2 - \sqrt{m+1}$.

22. $\sqrt{2x+3} + \sqrt{x-2} + 2 = 0$.

23. $\sqrt{K-2} + \sqrt{2K-3} = 1.$

24. $\sqrt{z+5} = 1 + \sqrt{z-2}.$

25. $\dfrac{3}{\sqrt{4Q+1}} + \dfrac{\sqrt{4Q+1}}{3} = 2.$

26. $\sqrt{T} + \sqrt{T+4} = \dfrac{2}{\sqrt{T}}.$

27. $\dfrac{\sqrt{t}-3}{\sqrt{t}+3} = \dfrac{\sqrt{t}-1}{\sqrt{t}-2}.$

28. $\dfrac{\sqrt{1+R_1} - \sqrt{1-R_1}}{\sqrt{1+R_1} + \sqrt{1-R_1}} = \dfrac{1}{2}.$

29. $\sqrt{2r+1} = \sqrt{r+1} + \sqrt{r}.$

30. $\sqrt{16M-7} - \sqrt{4M+1} = 2\sqrt{M-1}.$

31. $\sqrt{x - \sqrt{x+2}} = 2.$

32. $\sqrt{\sqrt{y+3} + 2y + 5} - 3 = 0.$

33. $\sqrt{p-1} + \sqrt{p-4} = \sqrt{2p-1}.$

34. $\sqrt{E_1 + 5} = \sqrt{E_1 - 3} + \sqrt{E_1}.$

8–11. Numerical Computations

It was indicated in Sec. 8–3 that the operations of extracting roots lead to numbers like $\sqrt{2}$, $\sqrt[3]{5}$, called irrational numbers. It is shown in books on more advanced mathematics that such numbers cannot be represented in the form of fractions like m/n, where m and n are integers. However, these irrational numbers can be approximated as closely as we please by rational numbers, i.e., fractions. Several methods are available for such computations. For our purpose it will be sufficient to use the slide rule for finding square and cube roots as explained in Chapter 1. If in computing square roots greater accuracy is required, the student should use Table 1 in the appendix.

In evaluating square roots, it is often necessary to modify or simplify the radicand. The following examples will indicate some of the devices employed. The student should compare the answers of the following examples with those obtained by using the slide rule.

Example 1. Evaluate $\sqrt{120}$, using tables.

Although the number 120 is in the table, we can also find $\sqrt{120}$ by writing

$$\sqrt{120} = \sqrt{4 \cdot 30} = 2\sqrt{30} = 2 \cdot 5.477 = 10.954.$$

Example 2. Evaluate $\sqrt{3\frac{1}{2}}$, using tables.

We have $\sqrt{3\frac{1}{2}} = \sqrt{\dfrac{7}{2}} = \dfrac{\sqrt{7}}{\sqrt{2}}$, and, although both $\sqrt{7}$ and $\sqrt{2}$ can now be found, it is advisable to rationalize our expression so as to avoid division by $\sqrt{2} = 1.414.$

Thus:

$$\sqrt{3\frac{1}{2}} = \dfrac{\sqrt{7}}{\sqrt{2}} = \dfrac{\sqrt{7}}{\sqrt{2}} \cdot \dfrac{\sqrt{2}}{\sqrt{2}} = \dfrac{\sqrt{14}}{2} = \dfrac{3.742}{2} = 1.871.$$

Example 3. Evaluate $\dfrac{3}{5 - \sqrt{7}}$, using tables.

To avoid division by $5 - \sqrt{7} = 5 - 2.646 = 2.354$, rationalize the given expression. Thus:

$$\frac{3}{5 - \sqrt{7}} = \frac{3(5 + \sqrt{7})}{(5 - \sqrt{7})(5 + \sqrt{7})} = \frac{15 + \sqrt{63}}{18}$$

$$= \frac{15 + 7.937}{18} = \frac{22.937}{18} = 1.274.$$

Exercises

Using the table or the slide rule, evaluate the following expressions:

1. $\sqrt{160} - \sqrt{120}$.

2. $\dfrac{1}{\sqrt{3}}$.

3. $\dfrac{3}{\sqrt{2}}$.

4. $\dfrac{1}{\sqrt{8}}$.

5. $\dfrac{8}{\sqrt{5}}$.

6. $\dfrac{5\sqrt{3}}{\sqrt{2}}$.

7. $\dfrac{3\sqrt{11}}{\sqrt{7}}$.

8. $\dfrac{\sqrt{5}\sqrt{7}}{\sqrt{10}}$.

9. $\dfrac{\sqrt{12}}{\sqrt{5} \cdot \sqrt{6}}$.

10. $\dfrac{5}{1 + \sqrt{2}}$.

11. $\dfrac{1}{1 - \sqrt{10}}$.

12. $\dfrac{1}{3 + 2\sqrt{2}}$.

13. $\dfrac{1}{\sqrt{11} - \sqrt{5}}$.

14. $\dfrac{\sqrt{13} + \sqrt{3}}{\sqrt{2}}$.

15. $\dfrac{\sqrt{7}}{\sqrt{3} - \sqrt{2}}$.

16. $\dfrac{\sqrt{35} - \sqrt{17}}{\sqrt{3}}$.

17. $\dfrac{4 + \sqrt{5}}{4 - \sqrt{5}}$.

18. $\dfrac{3\sqrt{5} - \sqrt{2}}{3\sqrt{5} + \sqrt{2}}$.

19. $\dfrac{\sqrt{7} + \sqrt{13}}{\sqrt{5} - \sqrt{2}}$.

20. $\dfrac{1 + \sqrt{\frac{1}{2}}}{1 - \sqrt{\frac{1}{3}}}$.

21. $\dfrac{1}{\sqrt{5} - \sqrt{3} + \sqrt{2}}$.

8–12. Applications to Formulas and Problems

Some of the results obtained in this chapter will now be applied to the transformation of formulas that are encountered in engineering and in physics. It is of great importance to be able to transpose a formula from one form to another.

For example, in radio engineering the inductance of a single-layer-wound air-core coil is given by the equation

$$L = \frac{(rN)^2}{9r + 10l}$$

where N is the number of turns, r the radius of coil in inches, l the length of winding in inches, and L the inductance in microhenrys. When r, N, and

l are given, we can compute L from the above formula.   When L, r, and l are given, we may regard N as the unknown and solve for it in terms of the known quantities.   We thus obtain

$$r^2N^2 = L(9r + 10l),$$

$$N^2 = \frac{L(9r + 10l)}{r^2},$$

$$N = \frac{1}{r}\sqrt{L(9r + 10l)}.$$

In the last step only the positive square root was taken, since N is the number of turns and hence should be positive.   Thus by transformation, a desired quantity in a formula can be isolated.

Exercises

In the formulas below, express the required quantity in terms of the others:

Given	Solve for	Description
1. $A = \pi r^2$	r	Area of a circle
2. $S = 4\pi r^2$	r	Surface of a sphere
3. $P = \dfrac{E^2}{R}$	E	Electric power
4. $E = \frac{1}{2}mv^2$	v	Kinetic energy
5. $V = \pi r^2 h$	r	Volume of a circular cylinder
6. $F = \dfrac{mv^2}{r}$	v	Centripetal force
7. $D = 1.063\sqrt{h}$	h	A formula used in navigation
8. $c^2 = a^2 + b^2$	a, b	Pythagorean theorem
9. $s = \frac{1}{2}gt^2$	t	Freely falling body
10. $R = \dfrac{K \cdot l}{d^2}$	d	Resistance of a wire
11. $V = a^3$	a	Volume of a cube
12. $H = \dfrac{d^2 n}{2.5}$	d	Horsepower for an automobile engine
13. $V = \frac{4}{3}\pi r^3$	r	Volume of a sphere
14. $F = \dfrac{M_1 M_2}{D^2}$	D	Force between two magnets

Given	Solve for	Description
15. $I = \sqrt{\dfrac{W}{R}}$	W, R	Power law for d-c circuits
16. $t = 2\pi \sqrt{\dfrac{L}{g}}$	L, g	Time for one complete swing of a pendulum
17. $\lambda = 1884\sqrt{LC}$	L, C	Wavelength in terms of inductance and capacitance.
18. $S = \dfrac{n \cdot W \cdot 33 \cdot 10^8}{r^4}$	r	Steel springs
19. $K = \dfrac{M}{\sqrt{L_1 \cdot L_2}}$	L_1, L_2	Coefficient of coupling between two coils
20. $P_{\max} = \dfrac{(\mu E_s)^2}{4r_p}$	E_s	Vacuum tube formula for maximum power output
21. $V = \sqrt{\dfrac{g\lambda}{2\pi}}$	λ	Velocity of a wave in deep water
22. $L = \dfrac{(rN)^2}{9r + 10l}$	r	Inductance of a single-layer-wound air-core coil
23. $M = \dfrac{V}{\sqrt{Rg\gamma t}}$	R, t	Flight Mach number
24. $L_{\text{avg}} = \dfrac{1.26N^2A\mu}{10^8 \cdot l}$	N	Inductance of long coils
25. $P = \dfrac{3.095Lwz}{g \cdot D^2}$	D	Determination of viscosity
26. $Z = \dfrac{RX}{\sqrt{R^2 + X^2}}$	R, X	Impedance of a resistance and reactance in parallel
27. $S_p = \dfrac{ae^2\beta}{4\pi r^2 c^2} \sin^2 \theta$	$r, \sin\theta$	Theory of X rays
28. $Z = \sqrt{R^2 + (X_L - X_C)^2}$	R	Impedance of an inductance, resistance, and capacitance in series
29. $\dfrac{Q_1}{Q_2} = \left(\dfrac{V_1}{V_2}\right)^\alpha$	V_1, V_2	Theory of illumination
30. $r = \sqrt{F}\left(\dfrac{1}{d_1} + \dfrac{1}{d_2}\right) \cdot 10^{-3}$	F	Radio-frequency resistance of a copper concentric transmission line
31. $L = \dfrac{0.315a^2N^2}{6a + 9b + 10c}\mu h$	N	Inductance of a coil

Given	Solve for	Description
32. $S_p = \dfrac{c}{j} \sqrt{M^2 + T^2}$	M, T	Intensity of the shearing stress of a belt-driven shaft
33. $I_g = \dfrac{s_1{}^4 - s_2{}^4}{12}$	s_1, s_2	Moment of inertia
34. $I_L = \dfrac{E_0}{\sqrt{2(R_0{}^2 + w^2 L_0{}^2)}}$	R_0, L_0	Alternating current of a class-C amplifier
35. $V_T = V_0 \sqrt{1 + \dfrac{T}{273}}$	T	Variation of velocity of sound with temperature
36. $P = R_l \left(\dfrac{\mu E_s}{r_p + R_l} \right)^?$	E_s	Vacuum-tube formula for power output
37. $k_g = \sqrt{\dfrac{s_1{}^2 + s_2{}^2}{12}}$	s_1, s_2	Radius of gyration
38. $k_g = \sqrt{\dfrac{6R^2 - a^2}{24}}$	R	Radius of gyration
39. $V_x = W \left(\dfrac{1}{3} - \dfrac{x^2}{l^2} \right)$	x, l	Vertical shear in a simple beam
40. $I = \dfrac{E}{\sqrt{R^2 + \left(2\pi FL - \dfrac{1}{2\pi fC} \right)^2}}$	R	Ohm's law for a-c circuits
41. $m = \dfrac{m_0}{\sqrt{1 - \left(\dfrac{v}{c} \right)^2}}$	$\dfrac{v}{c}$	Mass of the electron
42. $I_p + I_g = K \left(E_g + \dfrac{E_p}{\mu} \right)^{\frac{3}{2}}$	E_g, E_p	Triode formula
43. $r_p = \dfrac{\sqrt{a + b}[a + b(\mu + 1)]^{\frac{3}{2}} \cdot 10^6}{A_1 \sqrt{E_p + E_g}}$	$E_p + E_g$	Plate resistance of a tube

8–13. Related Graphs: The Power Function

Graphs of various algebraic functions were discussed in Chapter 3.   In the present section we shall draw the graphs of the function

$$y = x^n$$

for different values of n.   This function is sometimes called a **power function**.

x	$y = x^2$
-3	9
-2	4
-1	1
0	0
1	1
2	4
3	9

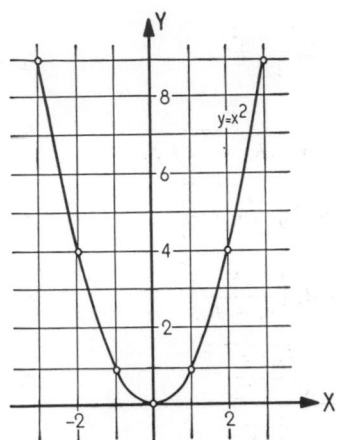

Fig. 8–1.   Graph of $y = x^2$.

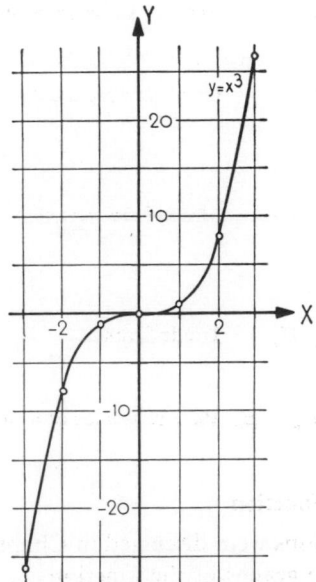

x	$y = x^3$
-3	-27
-2	-8
-1	-1
0	0
1	1
2	8
3	27

Fig. 8–2.   Graph of $y = x^3$.

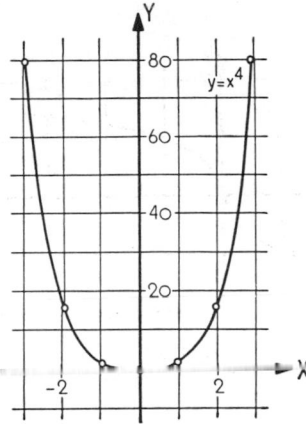

x	$y = x^4$
-3	81
-2	16
-1	1
0	0
1	1
2	16
3	81

FIG. 8–3.   Graph of $y = x^4$.

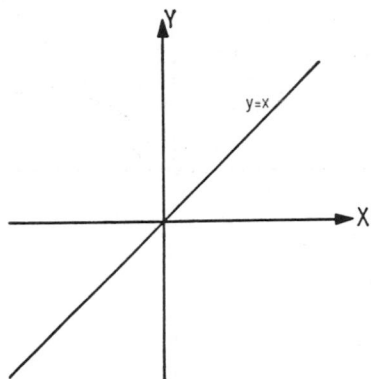

FIG. 8–4.   Graph of $y = x$.   This is a straight line passing through the origin.

x	$y = \sqrt{x}$
5	2.23
4	2
3	1.73
2	1.41
1	1
0	0

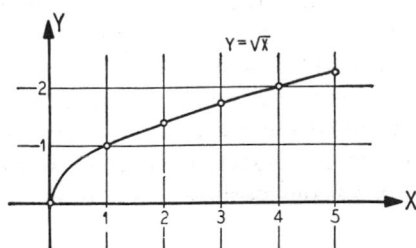

FIG. 8–5.   Graph of $y = \sqrt{x}$.

In this case the function $y = \sqrt{x}$ is defined for positive values of x only.

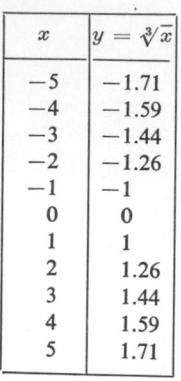

x	$y = \sqrt[3]{x}$
-5	-1.71
-4	-1.59
-3	-1.44
-2	-1.26
-1	-1
0	0
1	1
2	1.26
3	1.44
4	1.59
5	1.71

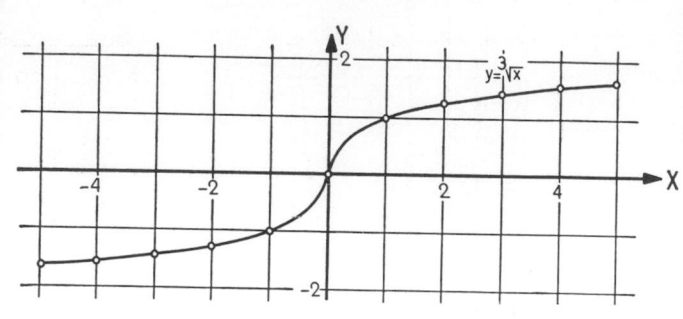

FIG. 8–6.   Graph of $y = \sqrt[3]{x}$.

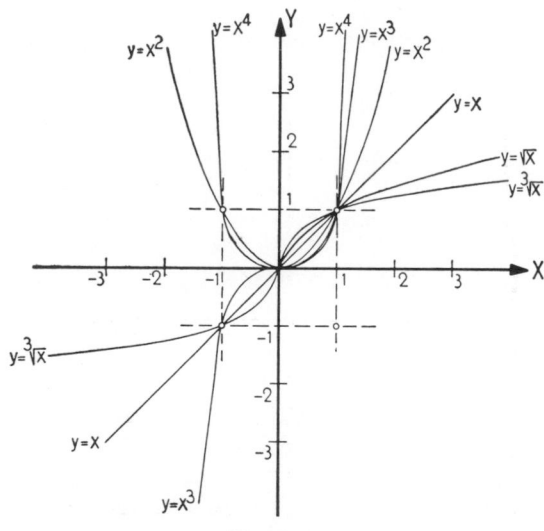

FIG. 8–7

We now combine the six graphs given in this section in one figure. Thus in Fig. 8–7 are shown the graphs of $y = x^n$ for $n = \frac{1}{3}$, $\frac{1}{2}$, 1, 2, 3, 4. The student will be able to deduce for himself the relative positions and forms of the graphs of $y = x^n$ as n takes on various possible values. These graphs occur frequently, and it is therefore worth while to remember the graphs given in this section and to be able to sketch quickly the graphs of $y = x^n$ for various positive values of n.

Exercises

Draw graphs of each of the following functions.

1. $y = \sqrt[4]{x}$.   **2.** $y = x^{\frac{2}{3}}$.   **3.** $y = x^5$.   **4.** $y = x^6$.

8–14. Related Graphs: The Exponential Function

Up to the present we have considered the exponent n in the equation

$$y = x^n$$

as a constant and the base x as a variable.   These roles can now be interchanged; i.e., consider the base as a constant and the exponent as a variable.   When this is done, we obtain the equation

$$y = a^x,$$

called an **exponential function**. We shall consider only the exponential functions whose bases are positive numbers different from unity.

In this connection it should be pointed out that the exponent x in $y = a^x$, can assume also irrational values.   The quantity $y = a^x$ can, in this case, be computed with any desired accuracy by replacing the irrational number x by a rational approximation: For example, $3^{\sqrt{2}}$ is approximately $3^{1.41}$ because 1.41 is an approximation of $\sqrt{2}$.

Exponential functions occur frequently in engineering and it is important to be able to sketch them quickly.   Consider, for example, the two functions

$$y = 2^x \text{ and } y = 3^x$$

for which we obtain the following table of values.

x	$y = 2^x$		x	$y = 3^x$
-3	$\frac{1}{8}$		-3	$\frac{1}{27}$
-2	$\frac{1}{4}$		-2	$\frac{1}{9}$
-1	$\frac{1}{2}$		-1	$\frac{1}{3}$
0	1		0	1
1	2		1	3
2	4		2	9
3	8		3	27

The corresponding graphs are drawn in Fig. 8–8.

The functions of the form $y = 2^x$ and $y = 3^x$ are exponential functions $y = a^x$. All such functions have the property that they intersect the y axis in one point P whose coordinates are $(0, 1)$.   At the point P (in Fig. 8–8) draw tangent lines to the two curves $y = 2^x$ and $y = 3^x$.   These tangent lines form the angles α and β with the x axis, called **angles of inclination**.

Measuring, the student will find that $\alpha < 45°$ and $\beta > 45°$. For different values of a, graphs of a^x may be plotted. The graphs will resemble those given in Fig. 8–8 and will intersect at point P.   The tangent lines to these curves at the point will have various angles of inclination.

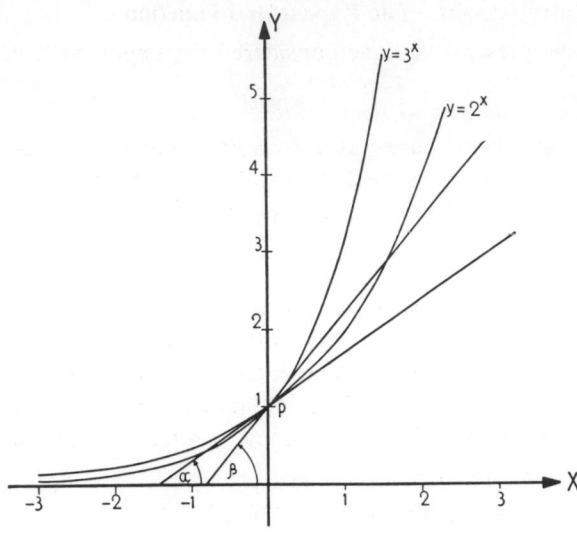

FIG. 8–8

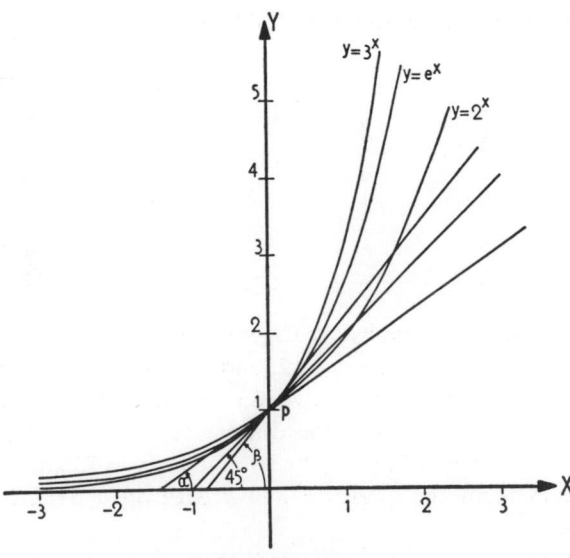

FIG. 8–9

We shall call the particular curve for which this angle of inclination is 45° **the standard curve.**   From Fig. 8–8 it is obvious that this standard curve is of the form $y = a^x$ where $2 < a < 3$.   In books on more advanced mathematics it is shown that for this standard curve $a = 2.718$, approximately.   Because of the importance of this number, a special notation e is introduced.   *Thus $e = 2.718$ and $y = e^x$ is the standard curve described above.*   In Fig. 8–9 $y = 2^x$, $y = 3^x$, and $y = e^x$ are all plotted on the same set of axes.

The student should also plot the functions $y = 2^{-x}$, $y = e^{-x}$, $y = 3^{-x}$ and compare them with Fig. 8–9.

Exercises

Construct the graphs of the following functions.

1. $y = 2^{-x}$.   2. $y = 4^x$.   3. $y = 5^x$.

4. $y = 3^{-x}$.   5. $y = 4^{-x}$.   6. $y = e^{-x}$.

7. $y = (\frac{1}{2})^x$.   8. $y = (\frac{1}{3})^x$.   9. $y = 2^{-x^2}$.

Progress Report

In Chapters 1 and 2 integral exponents were introduced, and rules governing computations with them were derived.   In this chapter the concept of exponents was extended to include rational numbers.   This extension was performed in such a manner that the laws for operating with exponents that were developed earlier hold in the new situations.

The following topics were also considered:

1. Roots and radicals.

2. Operations with radicals.

3. Equations involving radicals.

4. Applications to numerical computation and to the transformation of formulas.

5. Graphs of the power function and the exponential function.

9

Logarithms

Both in mathematics and in its applications to the other sciences and engineering, many problems involve long and tedious computations. Logarithms, which will be studied in this chapter, can be utilized to simplify many of these computations. Too, when adequate tables of logarithms are available, results of a high degree of accuracy can be obtained by their use.

On the other hand, logarithms have other applications than the practical one of simplifying computations: They are employed in theoretical discussions in science and engineering, they occur in formulas from applications of science and engineering, and they serve in describing natural phenomena quantitatively (for example, the response of a human sense to a stimulus is proportional to the logarithm of the stimulus). Thus a knowledge of the properties and uses of logarithms is essential for the scientist and engineer.

9-1. Introduction

The discussions in this chapter will be based on those of the preceding chapter on exponents.

Consider the two numbers $a = B^m$ and $c = B^n$. In writing $a = B^m$ we are associating an exponent m with the number a, and in writing $c = B^n$ we are associating an exponent n with the number c. By the laws of exponents in Chapter 8,

$$ac = B^m \cdot B^n = B^{m+n}, \tag{1}$$

$$\frac{a}{c} = \frac{B^m}{B^n} = B^{m-n}, \tag{2}$$

$$a^k = (B^m)^k = B^{mk}. \tag{3}$$

From equation 1 we see that the multiplication of a and c corresponds to the addition of the associated exponents m and n. From equation 2 we see that the division of a by c corresponds to the subtraction of n from m. From equation 3 we see that the operation of raising a to the kth power corresponds to multiplying the exponent m (which is associated with a) by k.

It is easier to add than multiply, easier to subtract than divide, and easier to multiply than raise to a power. Therefore, provided that we had some way of associating with any number a an exponent m such that $a = B^m$, formula 1 would supply a convenient method for replacing multiplication by addition, formula 2 would supply a method for replacing division by subtraction, and formula 3 would supply a method for replacing raising to a power by multiplication. For example, let us choose $B = 2$. Then, for $a = 2^m$, Table 1 can be easily compiled.

TABLE 1

$a = 2^m$	$\frac{1}{32}$	$\frac{1}{16}$	$\frac{1}{8}$	$\frac{1}{4}$	$\frac{1}{2}$	1	2	4	8	16	32	64	128	256	512
m	-5	-4	-3	-2	-1	0	1	2	3	4	5	6	7	8	9

The use of this table is suggested in the following examples.

Example 1.

$$\tfrac{1}{32} \times 128 = 2^{-5} \times 2^7 \qquad \text{(By Table 1)}$$
$$= 2^{7-5} \qquad \text{(By the laws of exponents)}$$
$$= 2^2 = 4. \qquad \text{(By Table 1)}$$

Example 2.

$$8^3 = (2^3)^3 \qquad \text{(By Table 1)}$$
$$= 2^9 \qquad \text{(By the laws of exponents)}$$
$$= 512. \qquad \text{(By Table 1)}$$

Of course, if the process suggested in this way is to have any practical value, the table must give exponents m for any a that occurs in computations. It will now be shown how a simple table of this kind for positive two-digit numbers can be constructed on the basis of what the reader already knows.

In Chapter 8 it was shown that, when a positive number B is given, $a = B^m$ has a definite meaning for any number m. If m is regarded as the independent variable and a as the dependent variable, this function can be plotted as shown in Sec. 8–14. Figure 9–1 shows the graph of the function $a = 2^m$.

From Fig. 9–1, we see that (*a*) for every value of m there is a corresponding value of a, (*b*) the values of a are always positive, (*c*) as m increases, *a* increases, and (*d*) as *m* decreases, *a* decreases, approaching zero as *m* becomes very small, i.e., negative with a very large absolute value.

Suppose we now attempt to find an exponent *m* such that $5 = 2^m$. In Fig. 9–1 the straight line perpendicular to the *a* axis at $a = 5$ intersects the curve at *P*, whose abscissa is about $m = 2.3$. It can be concluded, therefore, that $5 = 2^{2.3}$ approximately. Actually the exact value of *m* is irrational, and in calculus methods are developed to compute *m* to as many digits as desired, but this rough approximation is sufficient here.

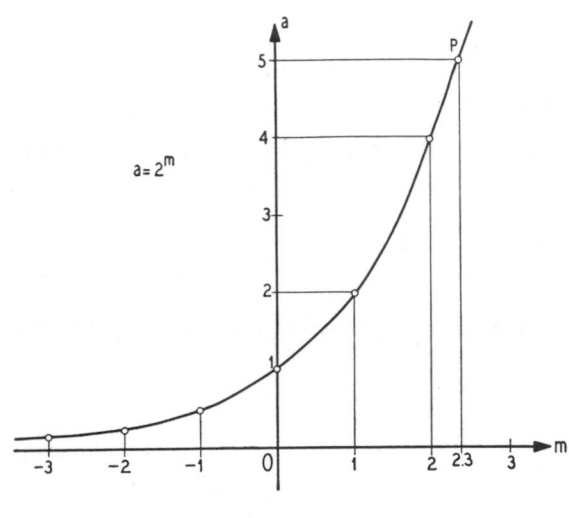

FIG. 9–1

By drawing a careful graph like Fig. 9–1 to a large scale, the exponents in Table 2 can be easily determined.

Computations may be performed by the use of Table 2 as illustrated in the following examples.

Example 3.

$$67 \times 84 = 6.7 \times 10 \times 8.4 \times 10$$

$$= 6.7 \times 8.4 \times 10^2$$

$$= 2^{2.74} \times 2^{3.07} \times 10^2 \qquad \text{(By Table 2)}$$

$$= 2^{2.74+3.07} \times 10^2 \qquad \text{(By the laws of exponents)}$$

$$= 2^{5.81} \times 10^2.$$

TABLE 2
$$a = 2^m$$

a	m	a	m	a	m	a	m	a	m	a	m
1.0	0.00	2.5	1.32	4.0	2.00	5.5	2.46	7.0	2.81	8.5	3.09
1.1	0.14	2.6	1.38	4.1	2.04	5.6	2.49	7.1	2.83	8.6	3.10
1.2	0.26	2.7	1.43	4.2	2.07	5.7	2.51	7.2	2.85	8.7	3.12
1.3	0.38	2.8	1.49	4.3	2.10	5.8	2.54	7.3	2.87	8.8	3.14
1.4	0.49	2.9	1.54	4.4	2.14	5.9	2.56	7.4	2.89	8.9	3.15
1.5	0.58	3.0	1.59	4.5	2.17	6.0	2.59	7.5	2.91	9.0	3.17
1.6	0.68	3.1	1.63	4.6	2.20	6.1	2.61	7.6	2.93	9.1	3.18
1.7	0.77	3.2	1.68	4.7	2.23	6.2	2.63	7.7	2.94	9.2	3.20
1.8	0.85	3.3	1.72	4.8	2.26	6.3	2.66	7.8	2.96	9.3	3.22
1.9	0.93	3.4	1.77	4.9	2.29	6.4	2.68	7.9	2.98	9.4	3.23
2.0	1.00	3.5	1.81	5.0	2.32	6.5	2.70	8.0	3.00	9.5	3.25
2.1	1.07	3.6	1.85	5.1	2.35	6.6	2.72	8.1	3.02	9.6	3.26
2.2	1.14	3.7	1.89	5.2	2.38	6.7	2.74	8.2	3.04	9.7	3.28
2.3	1.20	3.8	1.93	5.3	2.40	6.8	2.77	8.3	3.05	9.8	3.29
2.4	1.26	3.9	1.96	5.4	2.43	6.9	2.79	8.4	3.07	9.9	3.31
2.5	1.32	4.0	2.00	5.5	2.46	7.0	2.81	8.5	3.09	10.0	3.32

Since the exponent 5.81 is not in the tables and since $10 = 2^{3.32}$, we can write $5.81 = 2.49 + 3.32$ and we have

$$67 \times 84 = 2^{2.49+3.32} \times 10^2$$

$$= 2^{2.49} \times 2^{3.32} \times 10^2 \quad \text{(By the laws of exponents)}$$

$$= 2^{2.49} \times 10 \times 10^2 \quad\quad\quad\quad \text{(By Table 2)}$$

$$= 5.6 \times 10^3. \quad\quad\quad\quad\quad\quad \text{(By Table 2)}$$

A slide-rule computation easily verifies this to be the correct result.

Example 4.

$$67^3 = (6.7 \times 10)^3$$

$$= 6.7^3 \times 10^3 \quad\quad\quad\quad\quad \text{(By the laws of exponents)}$$

$$= (2^{2.74})^3 \times 10^3 \quad\quad\quad\quad\quad \text{(By Table 2)}$$

$$= 2^{8.22} \times 10^3 \quad\quad\quad\quad \text{(By the laws of exponents)}$$

$$= 2^{1.58+2(3.32)} \times 10^3$$

$$= 2^{1.58} \times (2^{3.32})^2 \times 10^3 \text{ (By the laws of exponents)}$$

$$= 2^{1.58} \times 10^2 \times 10^3 \quad\quad\quad\quad \text{(By Table 2)}$$

$$= 3.0 \times 10^5. \quad\quad\quad\quad\quad\quad \text{(By Table 2)}$$

Example 5.

$$(5.6)^{3.2} = (2^{2.49})^{3.2} \qquad\qquad\qquad \text{(By Table 2)}$$
$$= 2^{2.49\times 3.2} \qquad\qquad \text{(By the laws of exponents)}$$
$$= 2^{7.97}$$
$$= 2^{1.33+2(3.32)}$$
$$= 2^{1.33} \times (2^{3.32})^2 \qquad \text{(By the laws of exponents)}$$
$$= 2.5 \times 10^2. \qquad\qquad\qquad \text{(By Table 2)}$$

Example 6.

$$\frac{510}{67} = \frac{5.1 \times 10^2}{6.7 \times 10}$$
$$= \frac{5.1}{6.7} \times 10$$
$$= \frac{2^{2.35}}{2^{2.74}} \times 10 \qquad\qquad\qquad \text{(By Table 2)}$$
$$= 2^{2.35-2.74} \times 10 \qquad \text{(By the laws of exponents)}$$
$$= 2^{-0.39} \times 2^{3.32} \qquad\qquad\qquad \text{(By Table 2)}$$
$$= 2^{3.32-0.39} \qquad \text{(By the laws of exponents)}$$
$$= 2^{2.93}$$
$$= 7.6. \qquad\qquad\qquad\qquad\qquad \text{(By Table 2)}$$

Examples 4 and 6 can be readily checked on a Mannheim slide rule. However, example 5 cannot be checked on a Mannheim rule.

From these considerations we see clearly how the use of exponents simplifies long computations by reducing the complexity of the operations involved. The purpose of this chapter is to develop a system which has the same properties as the one above but which can be used for practical computations when more accuracy is desired than can be obtained by slide-rule computations. Thus we shall reduce, for example, the multiplication of 895.2 by 95.34 to the addition of two numbers. In this way tedious computations can be greatly simplified, with a corresponding saving in in time and energy. Although we shall concentrate on computations in which four-digit accuracy is required, the methods developed will apply no matter what degree of accuracy is required.

Exercises

Perform the following computations by the use of Table 2. Express your results in scientific notation correct to two significant digits.

1. 8.7×3.2.

2. 6.6×41.

3. 2.7×390.

4. $53 \times 1.4 \times 29$.

5. $(7.2)^3$.

6. $(2.2)^{5.2}$.

7. $(48)^2 \times (2.7)^{3.1}$.

8. $(6.3)^{1.8} \times (1.2)^4$.

9. $86 \div 27$.

10. $530 \div 32$.

11. $32 \div 530$.

12. $92 \div 840$.

13. $(4.4)^3 \div 780$.

14. $(27)^{2.2} \div (4.8)^{3.1}$.

15. $(1.7)^2 (8.3)^3 \div (5.3)^2$.

9–2. The Definition of a Logarithm

It is natural to surmise that the graph of the function $a = B^m$, for any B greater than 1, resembles the graph of $a = 2^m$ shown in Fig. 9–1 (see also Fig. 8–9), and that therefore the function $a = B^m$ has properties similar to those noted in the preceding section for the function $a = 2^m$:

(a) For every value of m there is a corresponding value of a.

(b) The values of a are positive.

(c) As m increases, a increases.

(d) As m decreases, a decreases, approaching zero as m is negative and its absolute value becomes very large.

It can be shown that this surmise is correct, and that the function $a = B^m$ does have the four properties listed above. Therefore it is clear that a system of computation like that set forth in Sec. 9–1 can be devised for any positive number B greater than 1. The number B is called the **base** of the system. For any given B, the fundamental question in constructing the system is the same: Given any a, can an m be found such that $a = B^m$?

From b above we see that a must be positive. As in Sec. 9–1, we see that, if $a > O$ and $B > 1$ are given, *then an exponent m can be found such that* $a = B^m$.

In Fig. 9–1, $a = B^m$ was considered as a function with m as the independent variable and a as the dependent variable. If a is given and the corresponding value of m is desired, we are considering a as the independent variable and m as the dependent variable. We thus have the **inverse function,** $m = f(a)$. For simplicity in theoretical discussions and in later computational work we shall introduce a notation for this inverse function.

If $a = B^m$, we say that a is equal to B raised to the exponent m. Inversely, m is the exponent to which B must be raised to obtain a. If the term **exponent** is replaced by **logarithm,** we may say that *m is the logarithm to which B must be raised to obtain a.* This latter statement is conveniently abbreviated by

$$m = \log_B a. \tag{1}$$

For the sake of brevity, equation 1 is often read "*m equals the logarithm to the base B of a*". To summarize:

In Symbols	In Words
$B^m = a$	B raised to the exponent m equals a.
$m = \log_B a$	$\begin{cases} m \text{ equals the exponent to which } B \text{ must be raised to get } a. \\ m \text{ equals the logarithm to the base } B \text{ of } a. \end{cases}$

By the definition above, only positive numbers have logarithms. In advanced treatises logarithms are defined for negative numbers, but such an extension of the definition is beyond the scope of this book. The logarithms of positive numbers are sufficient for all computational work.

Example.   The exponential expressions given below are restated to the right in logarithmic notation.

$$2^3 = 8 \quad \text{is equivalent to} \quad \log_2 8 = 3.$$

$$2^2 = 4 \quad \text{is equivalent to} \quad \log_2 4 = 2.$$

$$2^1 = 2 \quad \text{is equivalent to} \quad \log_2 2 = 1.$$

$$2^0 = 1 \quad \text{is equivalent to} \quad \log_2 1 = 0.$$

$$2^{-1} = \tfrac{1}{2} \quad \text{is equivalent to} \quad \log_2 \tfrac{1}{2} = -1.$$

$$2^{-2} = \tfrac{1}{4} \quad \text{is equivalent to} \quad \log_2 \tfrac{1}{4} = -2.$$

Table 1 of Sec. 9–1 gives values of a and m for the relation $a = 2^m$. Since $m = \log_2 a$, the columns headed with an m could also be labeled $\log_2 a$.   The same can be said of Table 2.   The information stated above can be read from Table 1.

Information gained from Table 2 can be stated in two ways as follows:

$$2.5 = 2^{1.32} \quad \text{is equivalent to} \quad \log_2 2.5 \quad = 1.32.$$

$$5.7 = 2^{2.51} \quad \text{is equivalent to} \quad \log_2 5.7 \quad = 2.51.$$

Likewise we see that:

$$1000 = 10^3 \quad \text{is equivalent to} \quad \log_{10} 1000 = 3.$$

$$100 = 10^2 \quad \text{is equivalent to} \quad \log_{10} 100 \quad = 2.$$

$$0.001 = 10^{-3} \quad \text{is equivalent to} \quad \log_{10} 0.001 = -3.$$

$$9 = 3^2 \quad \text{is equivalent to} \quad \log_3 9 \quad = 2.$$

$$125 = 5^3 \quad \text{is equivalent to} \quad \log_5 125 \quad = 3.$$

Exercises

Write each of the following exponential expressions in logarithmic notation.

1. $2^3 = 8.$ **2.** $7^2 = 49.$ **3.** $4^3 = 64.$

4. $12^2 = 144.$ **5.** $12^{-2} = \tfrac{1}{144}.$ **6.** $7^3 = 343.$

7. $7^{-3} = \tfrac{1}{343}.$ **8.** $10^2 = 100.$ **9.** $10^{-2} = 0.01.$

10. $10^{-3} = 0.001.$ **11.** $10^4 = 10,000.$ **12.** $10^{-1} = \tfrac{1}{10}.$

13. $6^{-4} = \tfrac{1}{1296}.$ **14.** $2^{10} = 1024.$ **15.** $13^3 = 2197.$

16. $3^5 = 243.$ **17.** $5^3 = 125.$ **18.** $3^{-5} = \tfrac{1}{243}.$

19. $2^{1.20} = 2.3.$ **20.** $2^{-1.20} = \dfrac{1}{2.3}.$ **21.** $2^{3.32} = 10.$

Write each of the following logarithmic expressions in exponential form.

22. $\log_2 64 = 6.$ **23.** $\log_4 16 = 2.$ **24.** $\log_3 27 = 3.$

25. $\log_4 256 = 4.$ **26.** $\log_3 243 = 5.$ **27.** $\log_{10} 100 = 2.$

28. $\log_5 125 = 3.$ **29.** $\log_7 2401 = 4.$ **30.** $\log_{11} 121 = 2.$

31. $\log_{10} 1000 = 3.$ **32.** $\log_2 \frac{1}{8} = -3.$ **33.** $\log_3 \frac{1}{81} = -4.$

34. $\log_7 \frac{1}{343} = -3.$ **35.** $\log_6 216 = 3.$ **36.** $\log_6 \frac{1}{216} = -3.$

37. $\log_6 \frac{1}{6} = -1.$ **38.** $\log_{10} 0.01 = -2.$ **39.** $\log_{1.3} 1.69 = 2.$

40. $\log_{1.4} \dfrac{1}{1.96} = -2.$ **41.** $\log_{10} \sqrt{10} = \frac{1}{2}.$ **42.** $\log_{10} \sqrt[4]{10} = \frac{1}{4}.$

Find the values of the following logarithms.

43. $\log_2 512.$ **44.** $\log_3 729.$ **45.** $\log_8 512.$

46. $\log_2 \frac{1}{32}.$ **47.** $\log_4 \frac{1}{256}.$ **48.** $\log_7 343.$

49. $\log_6 \sqrt{6}.$ **50.** $\log_5 \sqrt[3]{5}.$ **51.** $\log_3 \frac{1}{343}.$

52. $\log_9 3.$ **53.** $\log_9 81.$ **54.** $\log_3 1.$

55. $\log_4 2.$ **56.** $\log_8 \frac{1}{512}.$ **57.** $\log_8 4.$

58. $\log_5 \frac{1}{625}.$ **59.** $\log_7 1.$ **60.** $\log_2 2.$

Find the number x such that:

61. $\log_2 x = 4.$ **62.** $\log_4 x = 2.$ **63.** $\log_7 x = 2.$

64. $\log_6 x = 3.$ **65.** $\log_6 x = -1.$ **66.** $\log_9 x = -2.$

67. $\log_5 x = 4.$ **68.** $\log_2 x = -6.$ **69.** $\log_{1.2} x = 2.$

70. $\log_{2.20} x = 3.$ **71.** $\log_7 x = 1.$ **72.** $\log_7 x = 0.$

73. $\log_9 x = 0.$ **74.** $\log_3 x = 5.$ **75.** $\log_3 x = -5.$

76. $\log_3 x = \frac{1}{5}.$ **77.** $\log_{12} x = 2.$ **78.** $\log_{12} x = 0.$

9–3. Simple Properties of Logarithms

In all the discussions that follow in this chapter, the reader should constantly remind himself that *a logarithm is an exponent.*

If $a = B^m$, then $m = \log_B a$.   Obviously, then

$$B^{\log_B a} = a. \qquad (1)$$

Similarly, replacing a by B^m in $\log_B a$, we have

$$\log_B B^m = m. \qquad (2)$$

Examples.

$$10^{\log_{10} 8} = 8.$$

$$3^{\log_3 9^2} = 9^2 = 81.$$

$$\log_2 2^3 = 3.$$

$$\log_3 9^{-3} = \log_3 (3^2)^{-3} = \log_3 3^{-6} = -6.$$

$$\log_5 \sqrt[3]{5} = \log_5 5^{\frac{1}{3}} = \tfrac{1}{3}.$$

$$\log_6 \frac{1}{\sqrt{6}} = \log_6 6^{-\frac{1}{2}} = -\tfrac{1}{2}.$$

Exercises

Find the values of the following expressions.

1. $\log_2 8^{-3}$.

2. $\log_4 4^{\frac{1}{3}}$.

3. $\log_5 125^{-1}$.

4. $\log_2 4^{\frac{1}{4}}$.

5. $\log_6 216^{\frac{2}{3}}$.

6. $\log_3 1$.

7. $\log_5 5^{-3}$.

8. $\log_6 (6^2 \times 6^3)$.

9. $\log_B B^4$.

10. $\log_e e^2$.

11. $\log_e e^{2x+1}$

12. $\log_e e^{\cos \theta}$.

13. $\log_{10} (10^6 \times 10^{-8})$.

14. $\log_B (B^3)^m$.

15. $\log_e e$.

16. $\log_e e^{x+1}$.

17. $\log_e (e^{x+1})^2$.

18. $\log_e (e^{2x} \cdot e^{-3x})$.

19. $\log_5 \dfrac{1}{\sqrt[3]{5}}$.

20. $\log_{10} (\sqrt[5]{10} \times 10^2)$.

21. $\log_{10} \left(\dfrac{10^3 \times 10^{-6}}{10^{-5}} \right)$.

22. $3^{\log_3 5}$.

23. $10^{\log_{10} 15}$.

24. $10^{\log_{10} 36}$.

25. $12^{\log_{12} 4}$.

26. $10^{\log_{10} x}$.

27. $e^{\log_e 5}$.

28. $e^{\log_e 3x}$.

29. $6^{\log_6 (x^3 - 1)}$.

30. $B^{\log_B (x^2 + 2x)}$.

31. $B^{\log_B (x^2 - y^2)}$.

32. $10^{\log_{10} 10^2}$.

33. $e^{\log_e e^2}$.

34. $10^{\log_{10} \sqrt[3]{x+1}}$.

35. $e^{\log_e 1}$.

36. $B^{\log_B (x^2 + y^0)}$.

9–4. The Graph of the Function $m = \log_B a$.

In Fig. 9–1 the function $a = 2^m$ was plotted, where m is the independent variable and a the dependent variable. The inverse function is $m = \log_2 a$. In this case a is the independent variable and m the dependent variable. Table 1 of Sec. 9–1 gives the information necessary to plot the graph of $m = \log_2 a$, shown in Fig. 9–2.

In like manner, the graph of $m = \log_{10} a$ can be plotted from the following table:

a	0.001	0.01	0.1	1	10	100	1000
$m = \log_{10} a$	-3	-2	-1	0	1	2	3

The graph is shown in Fig. 9–2.

Similar tables and graphs can be obtained for any function of the form $m = \log_B a$, where $B > 1$.

A study of the graphs in Fig. 9–2 yields the following information about the functions $m = \log_B a$, where $B > 1$:

(a) For every positive value of the independent variable a there is a corresponding value of the dependent variable m.

(b) When $a > 1$, $m > 0$; when $a = 1$, $m = 0$; and, when $a < 1$, $m < 0$.

(c) When a increases, m increases. When a approaches zero, m becomes negatively infinite.

(d) Since the graph extends only to the right of the y axis, there are no logarithms of negative numbers by this definition.

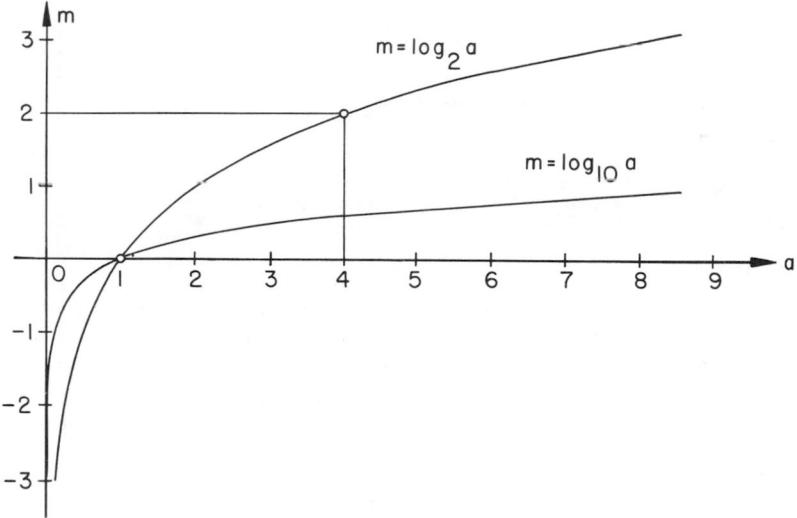

Fig. 9–2

Exercises

Plot the functions given in each exercise on a common coordinate system with x as the independent variable and y as the dependent variable.

1. $y = 4^x$ and $y = \log_4 x$.

2. $y = 5^x$ and $y = \log_5 x$.

3. $y = 10^x$ and $y = \log_{10} x$.

4. $y = 1.2^x$ and $y = \log_{1.2} x$.

5. $y = 1.5^x$ and $y = \log_{1.5} x$.

6. $y = 3^x$ and $y = \log_3 x$.

7. Use the graph of $y = \log_4 x$ to find approximate values of $\log_4 3$, $\log_4 5$, $\log_4 6$, $\log_4 2.5$, and $\log_4 1.5$.

8. Use the graph of $y = \log_5 x$ to find approximate values of $\log_5 3$, $\log_5 4$, $\log_5 6$, $\log_5 2.5$, and $\log_5 1.5$.

9. What can be inferred from Fig. 9–2 about the relation between the values of $\log_2 x$ and $\log_{10} x$?

10. In this chapter we have considered only bases for logarithms that are greater than 1. To show that this is a convenience instead of a necessity, plot $y = 0.5^x$, and then define and plot $y = \log_{0.5} x$. What differences do you observe between logarithms to the base 0.5 and logarithms to bases greater than 1?

9–5. Properties of Logarithms

The equations

$$\log_B a = m, \qquad \log_B c = n \tag{1}$$

are equïvalent to the equations

$$a = B^m, \qquad c = B^n. \tag{2}$$

Since

$$ac = B^m B^n = B^{m+n},$$

it follows that

$$\log_B ac = \log_B B^{m+n} = m + n = \log_B a + \log_B c. \tag{3}$$

This proves:

Property 1. *The logarithm of a product is equal to the sum of the logarithms of the factors, all logarithms being taken to the same base. Symbolically,*

$$\log_B (ac) = \log_B a + \log_B c.$$

Since

$$\frac{a}{c} = \frac{B^m}{B^n} = B^{m-n},$$

it follows that

$$\log \left(\frac{a}{c} \right) = \log_B B^{m-n} = m - n = \log_B a - \log_B c,$$

giving:

Property 2. *The logarithm of a quotient is equal to the logarithm of the numerator minus the logarithm of the denominator, all logarithms being taken to the same base. Symbolically,*

$$\log_B \left(\frac{a}{c} \right) = \log_B a - \log_B c.$$

Since
$$a^k = (B^m)^k = B^{mk},$$
it follows that
$$\log_B a^k = \log_B B^{mk} = mk = k \log_B a,$$
giving:

Property 3. *The logarithm of a power of a number is equal to the exponent times the logarithm of the number, all logarithms being taken to the same base.* *Symbolically,*
$$\log_B a^k = k \log_B a.$$

An important special case of the last property is the following:

$$\log_B \sqrt[q]{a} = \frac{1}{q} \log_B a.$$

There is no simple formula for the logarithm of a sum in terms of the logarithms of the members of the sum.   The value of $\log_B (a + b)$ cannot be computed if only $\log_B a$ and $\log_B b$ are known.

Example 1.   Express $\log_B 2\pi \sqrt{\dfrac{L}{g}}$ as the sum of logarithms of first powers of single letters or numbers.

$$\log_B 2\pi \sqrt{\frac{L}{g}} = \log_B 2 + \log_B \pi \sqrt{\frac{L}{g}} \qquad \text{(By property 1)}$$

$$= \log_B 2 + \log_B \pi + \log_B \sqrt{\frac{L}{g}} \qquad \text{(By property 1)}$$

$$= \log_B 2 + \log_B \pi + \log_B \sqrt{L} - \log_B \sqrt{g} \quad \text{(By property 2)}$$

$$= \log_B 2 + \log_B \pi + \tfrac{1}{2} \log_B L - \tfrac{1}{2} \log_B g. \quad \text{(By property 3)}$$

Example 2.   Express $\frac{1}{3} \log_{10} 2 - \log_{10} 3 + 3 \log_{10} 5$ as a single logarithm.

$$\tfrac{1}{3} \log_{10} 2 - \log_{10} 3 + 3 \log_{10} 5$$

$$= \log_{10} 2^{\frac{1}{3}} - \log_{10} 3 + \log_{10} 5^3 \qquad \text{(By property 3)}$$

$$= \log_{10} \frac{2^{\frac{1}{3}}}{3} + \log_{10} 5^3 \qquad \text{(By property 2)}$$

$$= \log_{10} \frac{2^{\frac{1}{3}} \cdot 5^3}{3} . \qquad \text{(By property 1)}$$

Example 3. Given $\log_{10} 2 = 0.3010$, $\log_{10} 3 = 0.4771$, and $\log_{10} 7 = 0.8451$, find the value of $\log_{10} \dfrac{12}{7}$.

$$\log_{10} \frac{12}{7} = \log_{10} \frac{3 \times 2^2}{7} = \log_{10} 3 + 2 \log_{10} 2 - \log_{10} 7$$

$$= 0.4771 + 2(0.3010) - 0.8451 = 0.4771 + 0.6020 - 0.8451 = 0.2340.$$

Example 4. Find the value of $\log_2 8^3$.

$$\log_2 8^3 = 3 \log_2 8 = 3 \log_2 2^3 = 3 \times 3 = 9.$$

Also

$$\log_2 8^3 = \log_2 (2^3)^3 = \log_2 2^9 = 9.$$

Example 5. Find the value of $\log_2 (16 \times 8)$.

$$\log_2 (16 \times 8) = \log_2 16 + \log_2 8 = \log_2 2^4 + \log_2 2^3 = 4 + 3 = 7.$$

Exercises

Express each of the following logarithms as the sum of logarithms of first powers of single letters or numbers.

1. $\log_B LMN$.

2. $\log_B \dfrac{L}{MN}$.

3. $\log_B \dfrac{LM}{N}$.

4. $\log_B M^6$.

5. $\log_B (M^2 N^4)$.

6. $\log_B \dfrac{1}{T^n}$.

7. $\log_B T^{-n}$.

8. $\log_2 \sqrt{M^3}$.

9. $\log_3 \sqrt[3]{C^5 D^7}$.

10. $\log_e \sqrt[q]{C^n}$.

11. $\log_{10} \dfrac{L^n \cdot \sqrt{S}}{T^m}$.

12. $\log_5 \sqrt{\dfrac{CD^2}{E^3}}$.

13. $\log_5 \sqrt[5]{\dfrac{E^3}{CD^2}}$.

14. $\log_2 (\sqrt[3]{x} \cdot \sqrt[4]{y})$.

15. $\log_e (\sqrt[m]{x} \cdot y^{-n})$.

Express each of the following as a single logarithm.

16. $\log_2 3 + \log_2 5 + \log_2 7$.

17. $\log_{10} L + \log_{10} M + \log_{10} N$.

18. $\log_B C + \log_B D$.

19. $2 \log_{10} x + 3 \log_{10} y$.

20. $9 \log_3 Q - 7 \log_3 R + 5 \log_3 S$.

21. $\log_e Q + \log_e R + \log_e S$.

22. $\frac{1}{3} \log_e Q - \frac{1}{5} \log_e R + \frac{1}{7} \log_e S$.

23. $\frac{1}{2} \log_e 2x + \log_e 3y - \frac{1}{3} \log_e 2z$.

24. $\frac{1}{10} \log_{10} 2 + \frac{1}{5} \log_{10} 3 + \frac{1}{10} \log_{10} 5$.

25. $\log_a 13 - \log_a 12 + \frac{1}{2} \log_a 6 - 3 \log_a 4$.

26. $\frac{1}{5} \log_e (x - a) + \log_e (x - b) - \log_e (x + c) - \log_e (x - 1)$.

27. $2 \log_B x + \frac{1}{2} \log_B y + 3 \log_B z$.

28. $\log_4 Q - \log_4 4R - \frac{1}{4} \log_4 S$.

29. $m \log_e x + \log_e y - \dfrac{n}{4} \log_e z$.

30. $\log_B \sin^2 x - \log_B \cos^2 x$.

Given $\log_{10} 2 = 0.3010$, $\log_{10} 3 = 0.4771$, $\log_{10} 5 = 0.6990$, and $\log_{10} 7 = 0.8451$. Compute the values of the following expressions:

31. $\log_{10} 6$.

32. $\log_{10} 12$.

33. $\log_{10} \sqrt{5}$.

34. $\log_{10} \frac{1}{3}$.

35. $\log_{10} 64$.

36. $\log_{10} \frac{3 \cdot 5}{6}$.

37. $\log_{10} \sqrt{30}$.

38. $\log_{10} \sqrt[7]{21}$.

39. $\log_{10} (18 \sqrt[5]{30})$.

40. $\log_{10} 294^3$.

41. $\log_{10} \dfrac{81}{\sqrt{7}}$.

42. $\log_{10} \dfrac{5^{1.5} \sqrt{3}}{7^3 \cdot \sqrt[3]{2}}$.

43. $\log_{10} \dfrac{28^3}{\sqrt[6]{21}}$.

44. $\log_{10} \dfrac{\sqrt[5]{7}}{\sqrt[3]{2}}$.

45. $\log_{10} \dfrac{3 \sqrt[3]{5}}{210}$.

Evaluate each of the following expressions:

46. $\log_2 \sqrt[4]{4}$.

47. $\log_4 \sqrt[3]{32}$.

48. $\log_5 125^{2.8}$.

49. $\log_8 \sqrt[6]{\frac{1}{4}}$.

50. $\log_{64} 16$.

51. $\log_{32} 16^4$.

52. $\log_7 \sqrt{49^3}$.

53. $\log_{2 \sqrt{2}} 8$.

54. $\log_{3 \sqrt{3}} 27^{\frac{1}{4}}$.

55. $\log_{\frac{4}{3}} (\frac{9}{16})^{-2}$.

56. $\log_3 (81 \times 9^2)^3$.

57. $\log_5 (125)^{\frac{1}{4}} + \log_{3 \sqrt{2}} (324)^{\frac{1}{4}}$.

58. $\log_4 (\frac{1}{2})^{-3} - \log_9 (27)^{-4}$.

59. $\log_{10} (100)^{\frac{5}{3}} - \log_9 \sqrt{81^3}$.

60. $\log_{25} 5 + \log_{36} 6 + \log_{49} \frac{1}{7}$.

61. Given $T = T_0 e^{-kt}$, show that $\log_e T = \log_e T_0 - kt$.

62. Given $V = V_0 (1 - e^{-\frac{Rt}{L}})$, show that $\log_e(V_0 - V) = \log_e V_0 - \frac{Rt}{L}$.

63. Given $q = q_0 e^{-\frac{t}{RC}}$, show that $\log_e q = \log_e q_0 - \frac{t}{CR}$.

64. Given $\log_e (q_0 - q) = \log_e q_0 - \frac{t}{CR}$, show that $q = q_0(1 - e^{-\frac{t}{CR}})$.

65. Given $\log_e i = \log_e q_0 - \log_e C - \log_e R - \frac{t}{CR}$, show that $i = \frac{q_0}{CR} e^{-\frac{t}{CR}}$.

9–6. Simple Logarithmic Computations

Since logarithms are exponents, computations by their aid can be performed as suggested by the exponential computations of Sec. 9–1. In that section we used the relation $a = 2^m$, whence $m = \log_2 a$. Thus, if we replace m by $\log_2 a$ in Tables 1 and 2 of that section, these tables become tables of logarithms to the base 2. The following examples indicate the parallel between logarithmic and exponential computations, using these tables.

Example 1.   Multiply 64×8, using Table 1 of Sec. 9–1.

Exponential Computation

$$64 \times 8 = 2^6 \times 2^3 \qquad\qquad\qquad \text{(Table 1)}$$
$$= 2^{6+3} \qquad\qquad\qquad \text{(Laws of exponents)}$$
$$= 2^9$$
$$= 512. \qquad\qquad\qquad\qquad \text{(Table 1)}$$

Logarithmic Computation

$$\log_2 (64 \times 8) = \log_2 64 + \log_2 8 \qquad\qquad \text{(Property 1)}$$
$$\log_2 64 = 6 \qquad\qquad\qquad\qquad \text{(Table 1)}$$
$$\log_2 8 = 3 \qquad\qquad\qquad\qquad \text{(Table 1)}$$
$$\overline{}$$
$$\log_2 (64 \times 8) = 9$$
$$64 \times 8 = 512. \qquad\qquad\qquad\qquad \text{(Table 1)}$$

Example 2.   Multiply $2.5 \times 1.9 \times 5.2$ using Table 2 of Sec. 9–1.

Exponential Computation

$$2.5 \times 1.9 \times 5.2$$
$$= 2^{1.32} \times 2^{0.93} \times 2^{2.38} \qquad\qquad \text{(Table 2)}$$
$$= 2^{1.32+0.93+2.38} \qquad\qquad \text{(Laws of exponents)}$$
$$= 2^{4.63}$$
$$= 2^{1.31+3.32}$$
$$= 2^{1.31} \times 2^{3.32}$$
$$= 2.5 \times 10. \qquad\qquad\qquad \text{(Table 2)}$$

Logarithmic Computation

$$\log_2 (2.5 \times 1.9 \times 5.2)$$
$$= \log_2 2.5 + \log_2 1.9 + \log_2 5.2. \qquad \text{(Property 1)}$$
$$\log_2 2.5 = 1.32 \qquad\qquad\qquad \text{(Table 2)}$$
$$\log_2 1.9 = 0.93 \qquad\qquad\qquad \text{(Table 2)}$$
$$\log_2 5.2 = 2.38 \qquad\qquad\qquad \text{(Table 2)}$$
$$\overline{}$$
$$4.63 = 1.31 + 3.32.$$
$$\log 2.5 = 1.31 \qquad\qquad\qquad \text{(Table 2)}$$
$$\log 10 = 3.32 \qquad\qquad\qquad \text{(Table 2)}$$
$$\log (2.5 \times 10) = 4.63. \qquad\qquad \text{(Property 1)}$$
$$2.5 \times 1.9 \times 5.2 = 2.5 \times 10.$$

We will find in the following sections that logarithmic notation in extended computations simplifies the mechanical details of the work very considerably. The examples given above are too simple, however, to illustrate this fact.

Exercises

Perform the following computations by means of logarithms to the base 2 as given in Table 2 of Sec. 9–1.   Compare your computations with the exponential computations of these same exercises which you performed previously.

Exercises 1–15 of Sec. 9–1.

9–7. Systems of Logarithms

The logarithms to a given base of all positive numbers constitute a **system of logarithms.**   For illustrative purposes up to this point we have used the base 2.   In practical applications, however, two other bases are most frequently used,

The **common system of logarithms** employs the base 10.   Since 10 is also the base of the ordinary number system, many simplifications in the use of logarithms occur when this base is used.   Therefore this base is the most convenient for computational purposes.   It is sometimes called the **Briggsian system** after its inventor Henry Briggs (1560–1631).

The **natural system of logarithms** employs as its base the number e discussed in Chapter 8.   This number is irrational, and its value to 10 decimal places is 2.71828 18285.   This system is the most convenient for theoretical purposes and will be met in the study of the calculus.   It is sometimes called the **Napierian system,** after John Napier (1550–1617).

When the base is not expressed, the base 10 is understood.   Thus $\log 2 = 0.3010$ means $\log_{10} 2 = 0.3010$.   The word logarithm will also mean common logarithm unless otherwise stated.   For the natural system we write $\log_e a$ as $\ln a$.   Since these conventions are not entirely general, it is well to check the convention in any given text.

As we have seen, a table of logarithms must be at hand if they are to be used for numerical work.   Very extended tables have been computed for common logarithms and for natural logarithms.   The methods used in computing these tables depend on calculus and will not be discussed here.

9–8. Common Logarithms: Characteristic and Mantissa

We saw in Chapter 1 that a given positive number N can be written in scientific notation as

$$N = M \times 10^n, \tag{1}$$

where M is between 1 and 10, and n is an integer which may be positive, negative, or zero.   Then

$$\log N = \log (M \times 10^n) = \log M + \log 10^n,$$

and thus

$$\log N = n + \log M. \tag{2}$$

The graph of the function $m = \log a$ in Fig. 9–2 shows that $\log 1 = 0$, $\log 10 = 1$, and that the logarithm of a number between 1 and 10 is between 0 and 1.   Then, since M is between 1 and 10, $\log M$ is between 0 and 1. Actually, $\log M$ may be an irrational number, but practically we shall consider only the first few places of its decimal equivalent.   Thus equation 2 states that *the logarithm of a positive number N can be written as the sum of an integer n (positive, negative, or zero) and a decimal fraction which is positive and between 0 and 1.*

The integral part n of the logarithm is called the **characteristic**.   The portion of the logarithm that is the decimal fraction between 0 and 1 is called the **mantissa.**

Example.   Given that $\log 2.56 = 0.4082$, we find that:

$\log 25.6 \quad = \log (10 \times 2.56) \quad = \log 10 \quad + \log 2.56 = \quad 1 + 0.4082.$

$\log 256 \quad = \log (100 \times 2.56) \quad = \log 100 \quad + \log 2.56 = \quad 2 + 0.4082.$

$\log 2560 \quad = \log (1000 \times 2.56) = \log 1000 + \log 2.56 = \quad 3 + 0.4082.$

$\log 0.256 \quad = \log (0.1 \times 2.56) \quad = \log 0.1 \quad + \log 2.56 = -1 + 0.4082.$

$\log 0.0256 \quad = \log (0.01 \times 2.56) \quad = \log 0.01 \quad + \log 2.56 = -2 + 0.4082.$

$\log 0.00256 = \log (0.001 \times 2.56) = \log 0.001 + \log 2.56 = -3 + 0.4082.$

Summarizing this information in a table, we have:

Number	Characteristic	Mantissa
2560	3	0.4082
256	2	0.4082
25.6	1	0.4082
2.56	0	0.4082
0.256	-1	0.4082
0.0256	-2	0.4082
0.00256	-3	0.4082

9–9. Rules for Finding the Characteristic

The definition of scientific notation supplies the rule for finding the characteristic of such numbers.

If N is given in scientific notation as

$$N = M \times 10^n, \tag{1}$$

then the characteristic of log N is the integer n.

Since the characteristic of $\log N$ is the integer n of equation 1 the rule of Sec. 1–12 can be used to find its value for numbers given in positional notation.   For the sake of convenience it will be restated here.

Let N be a positive number given in positional notation.

(a) If $N > 1$, the characteristic of log N is 1 less than the number of digits to the left of the decimal point of N.

(b) *If $N < 1$, the characteristic of log N is a negative number whose absolute value is one more than the number of zeros between the decimal point and the leftmost significant digit.*

We see, then, that the characteristic of a logarithm depends only on the position of the decimal point, and not on the digits in a given number.

When the characteristic is negative, it will often be more convenient to use an equivalent form, writing the characteristic as the difference of two positive numbers, the number to be subtracted being a multiple of 10. Thus we replace -1 by $9 - 10$, -2 by $8 - 10$, -4 by $6 - 10$, etc.

Example.

N	Characteristic of log N
2.56	0
256	2
0.256	-1 or $9 - 10$
0.00256	-3 or $7 - 10$
35670	4
0.00008	-5 or $5 - 10$
2.3×10^4	4
2.3×10^{-3}	-3 or $7 - 10$
2.3×10^{-11}	-11 or $9 - 20$
2.5×10^{-31}	-31 or $9 - 40$

Exercises

Determine the characteristics of the logarithms of the following numbers.

1. 3.9.

2. 46.7.

3. 182.

4. 0.571.

5. 0.0382.

6. 0.00045.

7. 38.512.

8. 66,985.

9. 0.1728.

10. 0.000043.

11. 1.752.

12. 0.0613.

13. π.

14. 8π.

15. e^2.

16. 6.4×10.

17. 9.13×10^3.

18. 4.78×10^9.

19. 3×10^{-4}.

20. 8.52×10^{-6}.

21. 3.1416×10^{-1}.

22. 9.1×10^{-7}.

23. 9.10×10^{-7}.

24. 7.272×10^3.

25. 1.327×10^6.

26. 6.281×10^{-9}.

27. 3.54×10^{-12}.

28. 2.126×10^{16}.

29. 2.126×10^{-16}.

30. 8.63×10^{-21}.

Place the decimal point in the sequence of digits 3741 as determined by the following characteristics:

31. 4.

32. 2.

33. 5.

34. 1.

35. 0.

36. 7.

37. $9 - 10$.

38. $7 - 10$.

39. $2 - 10$.

40. 10.

41. $4 - 10$.

42. 6.

43. 12.

44. $3 - 20$.

45. $8 - 40$.

9–10. Properties of the Mantissa

We have seen that the characteristic depends only on the position of the decimal point. Now we shall show that the mantissa is independent of the position of the decimal point.

If two positive numbers N_1 and N_2 have the same significant digits arranged in the same order, then, regardless of the position of the decimal point in each number, we say that N_1 and N_2 have the *same sequence of digits*. For example, the numbers 85.6, 856, 0.00856, 8560, and 85,600,000 have the same sequence of digits.

The two numbers N_1 and N_2 can be written in scientific notation as

$$N_1 = M_1 \times 10^{n_1}, \tag{1}$$

$$N_2 = M_2 \times 10^{n_2}, \tag{2}$$

where M_1 and M_2 are between 1 and 10, and n_1 and n_2 are integers. Now, if N_1 and N_2 have the same sequence of integers, then $M_1 = M_2$. The following example illustrates this fact.

Example 1.

N (in positional notation)	$N = M \times 10^n$ (in scientific notation)	M	n
85.6	8.56×10	8.56	1
856	8.56×10^2	8.56	2
0.00856	8.56×10^{-3}	8.56	-3
8560	8.56×10^3	8.56	3
85,600,000	8.56×10^7	8.56	7

All of the numbers N above have the same sequence of digits, and hence all the numbers M are the same.

From equations 1 and 2,

$$\log N_1 = n_1 + \log M_1,$$

$$\log N_2 = n_2 + \log M_2,$$

and thus the characteristics of $\log N_1$ and $\log N_2$ are n_1 and n_2 respectively. The mantissa of $\log N_1$ is $\log M_1$, and the mantissa of $\log N_2$ is $\log M_2$. If N_1 and N_2 have the same sequence of digits, $M_1 = M_2$, and therefore $\log M_1 = \log M_2$.

Since the mantissa is positive by its definition, we now have these two important properties of the mantissa:

Property 1. The mantissa is always a positive number.

Property 2. The common logarithms of numbers that have the same sequence of digits have the same mantissas.

The example in Sec. 9–8 illustrates property 2.   This property of the mantissa, which holds only for common logarithms, makes common logarithms superior for computational purposes to natural logarithms or to logarithms with any other base.

Example 2.   Given that $\log 8.56 = 0.9325$, we may obtain the information set forth below:

N	Characteristic of log N	Mantissa of log N	Log N
85.6	1	0.9325	1.9325
856	2	0.9325	2.9325
0.00856	7 − 10	0.9325	7.9325 − 10
8560	3	0.9325	3.9325
85,600,000	7	0.9325	7.9325
8.56×10^{-18}	2 − 20	0.9325	2.9325 − 20

Log $0.00856 = 7.9325 - 10$ has actually the value -2.0675, but written in this way the logarithm does not exhibit its characteristic and mantissa.   Hence we always write $\log 0.00856$ as $7.9325 - 10$, so that the characteristic $7 - 10$ and the mantissa 0.9325 are clearly shown.

Example 3.   If the logarithm of a number is -2.3765, what are the characteristic and mantissa?

The entire quantity -2.3765 is negative, and the mantissa is always positive. If we rewrite -2.3765 as $7.6235 - 10$, then it can be seen that the characteristic is $7 - 10$, and the mantissa is 0.6235.

Exercises

State the characteristic and mantissa of each of the following logarithms.

1. 6.4134.   **2.** 4.51724.   **3.** 2.4816.

4. 5.1894.   **5.** 0.37216.   **6.** 0.00749.

7. 9.1736 − 10.   **8.** 1.3284 − 10.   **9.** 7.3619 − 10.

10. −2.1462.   **11.** −9.8974.   **12.** 0.3165.

13. −3.4437 + 10.   **14.** −6.1982 + 10.   **15.** −4.2016 + 10.

16. −9.2367 + 10.   **17.** 9.2367 − 10.   **18.** −9.2367.

19. 2.1735 − 10.   **20.** 2.1735 − 20.   **21.** −8.5713 + 20.

Given $\log 316 = 2.4997$, $\log 4.82 = 0.6830$, $\log 79.5 = 1.9004$, and $\log 53.7 = 1.7300$, find the logarithms of the following numbers:

22. 3.16.   **23.** 482.   **24.** 5370.

25. 7.95.   **26.** 7950.   **27.** 0.316.

28. 0.482.   **29.** 0.00795.   **30.** 0.0537.

31. 0.795.   **32.** 31,600.   **33.** 4820.

34. 0.00316.   **35.** 53,700.   **36.** 48.2.

37. 7,950,000.   **38.** 7.95×10^2.   **39.** 5.37×10^7.

40. 4.82×10^{-6}.   **41.** 3.16×10^{10}.   **42.** 7.95×10^{-4}.

9–11. Using a Table of Logarithms

The mantissas of the logarithms of most numbers are unending decimal fractions. Methods are developed in advanced treatises for computing the values of these mantissas to as many decimal places as desired. Such computed values are usually given in a table of mantissas, called also a **table of logarithms.** These are called four-place, five-place, etc., according to the number of decimal places in the tabulated mantissas. While the mantissas are decimal fractions, the decimal point is usually omitted in the table for convenience in printing.

A four-place table is sufficient for most elementary purposes, and such a table is given in the appendix of this book (Table 2). We shall illustrate the use of this table, and the student can adapt these methods to the use of other tables. In the four-place table, the mantissas are given for the logarithms of all numbers that are written with a sequence of three significant digits. The table is arranged so that the first two significant digits of the given number are located in the leftmost column of the table, and the third digit is located at the top. For example, if the given number has the sequence of digits 738, then 73 is located in the column to the left and 8 is located at the top of the table. Going down from this 8 to the line corresponding to 73, the mantissa 8681 is found.

To find the logarithm of a number:

(*a*) *Determine the characteristic by the rules of Sec. 9–9.*

(*b*) *Find the mantissa corresponding to the given sequence of digits by using a table of logarithms.*

Example 1. To find log 73.8:

(*a*) By the rule of Sec. 9–9 the characteristic is 1.

(*b*) As we saw above the mantissa is 8681.

Hence

$$\log 73.8 = 1.8681.$$

In the same way it is found that

$$\log 7.38 = 0.8681,$$
$$\log 73800 = 4.8681,$$
$$\log 0.738 = 9.8681 - 10,$$
$$\log 0.000738 = 6.8681 - 10.$$

The process of finding the number corresponding to a given logarithm is the inverse of the process given above. The number obtained is sometimes called the **antilogarithm** (abbreviated **antilog**) of the given logarithm.

To find a number corresponding to a given logarithm:

(*a*) *Determine the sequence of digits corresponding to the given mantissa by using the table.*

(*b*) *Place the decimal point in the position given by the characteristic and the rules of Sec. 9–9.*

Example 2.   Find x such that $\log x = 1.5453$.

From the table it is found that 5453 is the common mantissa of all numbers with the sequence of digits 351.   Because a positive characteristic of a logarithm is 1 less than the number of places to the left of the decimal point of the corresponding number, x has two places to the left of the decimal, for the characteristic of $\log x$ is 1.   Hence

$$x = 35.1.$$

In the same way it is found that, if

$$\log x = 2.5453, \qquad x - 351;$$
$$\log x = 4.5453, \qquad x = 35100;$$
$$\log x = 8.5453 - 10, \quad x = 0.0351;$$
$$\log x = 9.5453 - 10, \quad x = 0.351;$$
$$\log x = 7.5453 - 10, \quad x = 0.00351.$$

Exercises

Using a four-place table, find the logarithm of each of the following numbers:

1. 4.58.	**2.** 7.36.	**3.** 27.5.
4. 89.3.	**5.** 913.	**6.** 4820.
7. 0.174.	**8.** 0.00639.	**9.** 89,700.
10. 0.000219.	**11.** 219,000.	**12.** 0.0781.
13. 3.48×10^3.	**14.** 7.62×10^{-4}.	**15.** 5.85×10^6.

Using a four-place table, find the number of three significant digits corresponding to each of the following logarithms:

16. 1.4378.	**17.** 2.9175.	**18.** 4.7160.	**19.** 0.2304.
20. $9.8633 - 10$.	**21.** 8.5514	**22.** $8.7466 - 10$.	**23.** 3.5999.
24. 0.3962.	**25.** 5.3222.	**26.** $7.7110 - 10$.	**27.** 1.8007.
28. 4.0414.	**29.** 0.8000.	**30.** $9.9903 - 10$.	

9–12. Interpolation

If a number has four significant figures, the mantissa of its logarithm cannot be found directly from a four-place table.   The mantissa of the logarithm of a four-digit number lies between the mantissas of two consecutive three digit numbers which are given in the table.   Thus, to find the mantissa of the logarithm of a four-digit number, the process of **interpolation** can be employed.   This process was used in Chapter 4 in connection with the trigonometric tables.

The following examples illustrate the process of interpolation:

Example 1. Find log 35.64.

The characteristic of this logarithm is 1. The logarithm of the number with the digits 356 has the mantissa 5514; the logarithm of the number with the digits 357 has the mantissa 5527. We can then set up a scheme similar to that used in Chapter 4:

$$0.10 \left[0.04 \begin{bmatrix} \log 35.60 = 1.5514 \\ \log 35.64 = \ \ \cdots \\ \end{bmatrix} x \atop \log 35.70 = 1.5527 \right] 0.0013$$

Since the logarithm increases by 0.0013 as the number increases by 0.10, then, as the number increases by 0.04, the logarithm will increase by

$$x = \frac{0.04}{0.10} \times 0.0013 = 0.00052.$$

Since this last statement is only approximately true (see below), and since the four-place mantissas given in the table are rounded off from longer decimals, we round off the result of this interpolation to four places and call $x = 0.0005$. Adding this amount to log 35.60, we obtain

$$\log 35.64 = 1.5519.$$

A shortened scheme can be set up as follows.

$$\begin{array}{cc} \text{Number} & \text{Mantissa} \\ 10 \left[4 \begin{bmatrix} 3560 \\ 3564 \\ \end{bmatrix} \atop 3570 \right. & \left. \begin{matrix} 5514 \\ \cdots \\ 5527 \end{matrix} \right] x \ 13 \end{array}$$

$$x = \tfrac{4}{10} \times 13 = 5.2 \quad \text{or} \quad 5.$$

Then

$$\log 35.64 = 1.5519.$$

Example 2. Find log 0.01947.

Setting up the simplified scheme, we have

$$\begin{array}{cc} \text{Number} & \text{Mantissa} \\ 10 \left[7 \begin{bmatrix} 1940 \\ 1947 \\ \end{bmatrix} \atop 1950 \right. & \left. \begin{matrix} 2878 \\ \cdots \\ 2900 \end{matrix} \right] x \ 22 \end{array}$$

$$x = \tfrac{7}{10} \times 22 = 15.4 \quad \text{or} \quad 15.$$

Adding 15 to 2878, we obtain 2893. Since the characteristic of log 0.01947 is −2 or 8 − 10, then

$$\log 0.01947 = 8.2893 - 10.$$

Example 3.   Find x such that log $x = 4.8079$.

The mantissa of this logarithm is 8079.   From the table we find that 8079 lies between 8075 and 8082, which correspond to the numbers 642 and 643, respectively.   Setting up the scheme, we have

$$10\begin{bmatrix} x\begin{bmatrix} 6420 \\ \ldots \\ 6430 \end{bmatrix} & \begin{bmatrix} 8075 \\ 8079 \\ 8082 \end{bmatrix}\!\!\!\begin{array}{c} 4 \end{array} \end{bmatrix}7$$

If we proceed as before,

$$x = \tfrac{4}{7} \times 10 = 5\tfrac{5}{7} \quad \text{or} \quad 6,$$

approximately.   Adding 6420 and 6, we obtain 6426, the sequence of digits in x. Since the characteristic of log x is 4, x has five digits in front of the decimal point.   Consequently

$$x = 64{,}260.$$

Example 4.   Find x such that log $x = 8.4864 - 10$.

Setting up the scheme, we have

$$10\begin{bmatrix} x\begin{bmatrix} 3060 \\ \ldots \\ 3070 \end{bmatrix} & \begin{bmatrix} 4857 \\ 4864 \\ 4871 \end{bmatrix}\!\!\!\begin{array}{c} 7 \end{array} \end{bmatrix}14$$

$$x = \tfrac{7}{14} \times 10 = 5.$$

The sequence of digits is 3065, and since the characteristic is $8 - 10$,

$$x = 0.03065.$$

Example 5.   Find log 557.5.

$$10\begin{bmatrix} 5\begin{bmatrix} 5570 \\ 5575 \\ 5580 \end{bmatrix} & \begin{bmatrix} 7459 \\ \ldots \\ 7466 \end{bmatrix}\!\!\!\begin{array}{c} x \end{array} \end{bmatrix}7$$

$$x = \tfrac{5}{10} \times 7 = 3.5.$$

Since the result of the interpolation is 3.5, we round off to 4, in accordance with our convention of rounding off to the even digit.   *This is done before the correction is added to the smaller mantissa.*   Hence the result is

$$\log 557.5 = 2.7463.$$

By the process of interpolation, a four-place mantissa can be found for a number of four significant figures when a four-place table is issued. Similarly, five-place mantissas can be found by interpolation from five-place tables when numbers of five significant figures are given, etc.   That accurate results by interpolation are limited to one digit more than the number of digits for which the table is tabulated is a consequence of two facts:

(*a*) The mantissas given in the tables are approximations, rounded off from longer decimals.

(*b*) The process of interpolation is based on the assumption that, for small differences in numbers, the changes in the mantissas are proportional to the changes in the numbers. This assumption is not strictly true, but the results based on it are sufficiently accurate for practical purposes if we consider numbers of four significant digits with four-place tables, numbers of five significant figures with five-place tables, etc. For this assumption to be precisely true, the curve of $y = \log x$ would have to be a straight line. Figure 9–2, of course, shows that this curve is not a straight line. However, small sections of it are very nearly straight, and this assumption merely states that they are close enough to being straight that we may consider them so for practical purposes.

After some practice, the student will be able to perform the process of interpolation mentally. Some tables contain aids to interpolation in the form of tables of *proportional parts*. When the process of interpolation is well understood, the use of such aids is immediately clear. Therefore they will not be discussed here.

Exercises

Find the value of the following.

1. log 3.686.	**2.** log 75.82.	**3.** log 103.7.
4. log 18.89.	**5.** log 236,800.	**6.** log 8,251,000.
7. log 0.4728.	**8.** log 0.6312.	**9.** log 0.08465.
10. log 0.00002913.	**11.** log 0.003072.	**12.** log 0.05999.
13. log 17.48.	**14.** log 1.748.	**15.** log 749.6.
16. log 3006.	**17.** log 0.4982.	**18.** log 800,100.
19. log 0.007823.	**20.** log 1.297.	**21.** log 63.55.
22. $\log (4.128 \times 10^2)$.	**23.** $\log (4.817 \times 10^3)$.	**24.** $\log (1.395 \times 10^8)$.
25. $\log (7.218 \times 10^{-4})$.	**26.** $\log (5.555 \times 10^{-8})$.	**27.** $\log (6.241 \times 10^5)$.
28. $\log (1.138 \times 10^{12})$.	**29.** $\log (9.274 \times 10^{-2})$.	**30.** $\log (3.447 \times 10^2)$.
31. $\log (2.885 \times 10^{-8})$.	**32.** $\log (7.111 \times 10)$.	**33.** $\log (8.427 \times 10^{-3})$.

Find the numbers whose logarithms are given by:

34. 0.9518.	**35.** 2.6708.	**36.** 3.3191.	**37.** 6.6216.
38. 1.8707.	**39.** 0.2460.	**40.** 9.3460 − 10.	**41.** 6.3516 − 10.
42. 8.9022 − 10.	**43.** 7.5400 − 10.	**44.** 9.0842 − 10.	**45.** 4.5748 − 10.
46. 2.9610.	**47.** 1.7296.	**48.** 6.9995.	**49.** 0.0092.
50. 3.0007.	**51.** 0.6836.	**52.** 9.4241 − 10.	**53.** 7.9705 − 10.
54. 6.9706 − 10.	**55.** 8.5976 − 10.	**56.** 3.1400 − 20.	**57.** 9.8780 − 30.

9–13. Computations by Means of Logarithms

The examples of this section are designed to illustrate a number of important details of procedure in logarithmic computations. It is very important that work of this kind be carried out neatly and systematically. Several convenient ways of arranging the work are indicated in the examples. Before using the tables, the student should first analyze the problem to see what operations are involved. Then he should prepare a skeleton outline of the work to be performed, in which each operation is indicated and in which a place is provided for each number and logarithm entering into the computation. Finally, this outline can be filled in from the tables and the work completed. This procedure is the most conducive to speed and accuracy.

Example 1. Compute $89.46 \times 0.04137 \times 0.3214$ by logarithms.

If N denotes the result of this computation, then, by property 1 of Sec. 9–5,

$$\log N = \log 89.46 + \log 0.04137 + \log 0.3124.$$

Before using the table, we prepare the following outline:

$$\log 89.46 =$$
$$\log 0.04137 =$$
$$\log 0.3124 =$$
$$\overline{}$$
$$\log N =$$
$$N =$$

The characteristics can be filled in easily, and finally the table is used to complete the work. The completed computation is given below.

$$\log 89.46 = 1.9516$$
$$\log 0.04137 = 8.6167 - 10$$
$$\log 0.3124 = 9.4947 - 10$$
$$\overline{}$$
$$\log N = 20.0630 - 20 = 0.0630$$
$$N = 1.156.$$

An abbreviated form is very useful. In it the numbers are always to the left, their corresponding logarithms on the same horizontal line to the right. This form for this example is given below:

Numbers	Logarithms
89.46	1.9516
0.04137	$8.6167 - 10$
0.3124	$9.4947 - 10$
N	$20.0630 - 20 = 0.0630$
	$N = 1.156.$

Example 2.   Compute $\dfrac{0.08942 \times 3.592}{105.2 \times 0.5127}$ by logarithms.

If N denotes the result of this computation, then, by property 2 of Sec. 9–5,

$$\log N = \log \text{numerator} - \log \text{denominator},$$

where $\log \text{numerator} = \log 0.08942 + \log 3.592,$

$$\log \text{denominator} = \log 105.2 + \log 0.5127.$$

The computation is carried out below:

$$\log 0.08942 = \quad 8.9514 - 10$$
$$\log 3.592 = \quad 0.5553$$

$$\log \text{numerator} = \quad 9.5067 - 10 = 9.5067 - 10$$

$$\log 105.2 = \quad 2.0220$$
$$\log 0.5127 = \quad 9.7099 - 10$$

$$\log \text{denominator} = 11.7319 - 10 = 1.7319$$

$$\log N \qquad\qquad\qquad = 7.7748 - 10$$
$$N = 0.005954.$$

The abbreviated form for this computation can be set up as follows:

Numbers	Logarithms	
0.08942	$8.9514 - 10$	
3.592	0.5553	
	$9.5067 - 10$	$9.5067 - 10$
105.2	2.0220	
0.5127	$9.7099 - 10$	
	$11.7319 - 10$	1.7319
N		$7.7748 - 10$

$$N = 0.005954.$$

Example 3.   Compute $47.82/815.1$ by logarithms.

If N denotes the result of this computation,

$$\log N = \log 47.82 - \log 815.1.$$

Then from

Numbers	Logarithms
47.82	1.6796
815.1	2.9112

we see that the logarithm 2.9112 which is to be subtracted is larger than 1.6796.

If we replace 1.6796 by $11.6796 - 10$, the computation can be carried out as follows:

Numbers	Logarithms
47.82	$11.6796 - 10$
815.1	2.9112
N	$8.7684 - 10$

$$N = 0.05867.$$

Example 4.   Compute $\sqrt[3]{0.1084} \times (0.4231)^5$ by logarithms.

If N denotes the result of this computation,

$$\log N = \log \sqrt[3]{0.1084} + \log (0.4231)^5,$$

and, by property 3 of Sec. 9–5,

$$\log N = \tfrac{1}{3} \log 0.1084 + 5 \log 0.4231.$$

Now $\log 0.1084 = 9.0350 - 10$.   To find $\tfrac{1}{3} \log 0.1084$, it is more convenient to write $\log 0.1084 = 29.0350 - 30$, and then $\tfrac{1}{3} \log 0.1084 = \tfrac{1}{3} (29.0350 - 30)$ $= 9.6783 - 10$.

The complete computation is carried out below.

Numbers		Logarithms	
0.1084	29.0350 − 30		
$\sqrt[3]{0.1084}$		9.6783 − 10	
0.4231	9.6264 − 10		
$(0.4231)^5$	48.1320 − 50	8.1320 − 10	
N		17.8103 − 20	
N		7.8103 − 10	
	$N = 0.006461.$		

Example 5.   Compute $\dfrac{(0.04163)^3 \times \sqrt[5]{0.001574}}{(0.3157)^2 \times 5.123}$ by logarithms.

If N denotes the result of this computation, then

$$\log N = \log \text{ numerator} - \log \text{ denominator},$$

$$\log N = [3 \log 0.04163 + \tfrac{1}{5} \log 0.001574] - [2 \log 0.3157 + \log 5.123].$$

The computation is tabulated below.

Numbers		Logarithms	
0.04163	8.6194 − 10		
$(0.04163)^3$	25.8582 − 30	5.8582 − 10	
0.001574	7.1970 − 10		
$\sqrt[5]{0.001574}$	1.4394 − 2	9.4394 − 10	
Numerator		15.2976 − 20	15.2976 − 20

Numbers		Logarithms	
0.3157	9.4993 − 10		
$(0.3157)^2$	18.9986 − 20	8.9986 − 10	
5.123		0.7095	
Denominator		9.7081 − 10	9.7081 − 10
N			5.5895 − 10
	$N = 0.00003886.$		

Exercises

Compute each of the following expressions by a four-place table of logarithms.   Each number is assumed to be accurate to four significant figures, and hence each result should have four significant figures.

1. $6.735 \times 2.414 \times 890.0$.

2. 39.48×0.005465.

3. $0.06818 \times 46.73 \times 0.5291$.

4. $8.762 \times 0.0003153 \times 0.2985$.

5. $7.634 \times 0.002541 \times 310.4$.

6. $18.35 \times 6.263 \times 437.5$.

7. $0.009820 \times 0.002735 \times 17.63$.

8. $2.208 \times 5.358 \times 937.9$.

9. $0.7003 \times 21.83 \times 0.004276$.

10. $36.28 \times 0.0008173 \times 653.0$.

11. $\dfrac{7.324}{2.561}$.

12. $\dfrac{92.10}{8.740}$.

13. $\dfrac{1}{3.982}$.

14. $\dfrac{1}{668.5}$.

15. $\dfrac{0.004920}{0.03675}$.

16. $\dfrac{0.02164}{8.247}$.

17. $\dfrac{7368}{0.002716}$.

18. $\dfrac{0.0004657}{39.22}$.

19. $\dfrac{7.549 \times 0.001835}{5.584}$.

20. $\dfrac{2.193 \times 36.43}{4937}$.

21. $\dfrac{9.462 \times 6427}{18.95}$.

22. $\dfrac{0.07663 \times 21.88}{4919}$.

23. $\dfrac{3.160}{47.38 \times 19.00}$.

24. $\dfrac{8472}{3110 \times 0.0001810}$.

25. $\dfrac{1283}{67.94 \times 46.66}$.

26. $\dfrac{89.98}{0.7416 \times 392.7}$.

27. $\dfrac{2.967 \times 4.500}{1.874 \times 9.833}$.

28. $\dfrac{37.66 \times 428.3}{7.129 \times 54.75}$.

29. $\dfrac{5.556 \times 0.04138}{8.742 \times 0.0002788}$.

30. $\dfrac{2.394 \times 6.888}{37.16 \times 5.473}$.

31. $\dfrac{92.74 \times 3845}{0.01728 \times 6.458}$.

32. $\dfrac{89.76 \times 317.5}{5821 \times 9839}$.

33. $(2.168)^2$.

34. $(72.38)^4$.

35. $(46.85)^2 \times (0.07618)^3$.

36. $(912.7)^5 \times (834.6)^3$.

37. $(0.03651)^2 \times (8.724)^2 \times 92.05$.

38. $(13.89)^2 \times (0.6313)^3 \times (0.7815)^5$.

39. $(1750)^{\frac{1}{2}}$.

40. $(87.16)^{\frac{1}{3}}$.

41. $\sqrt{6.372}$.

42. $\sqrt[3]{9802}$.

43. $\sqrt[3]{0.2002}$.

44. $\sqrt{0.03194}$.

45. $\sqrt[3]{127.6}$.

46. $\sqrt[3]{3.850}$.

47. $(8.144)^{\frac{3}{4}}$.

48. $(0.4444)^{\frac{1}{11}}$

49. $\sqrt{64.64} \times \sqrt[3]{27.27}$.

50. $\sqrt[3]{0.5913} \times \sqrt[5]{7862}$.

51. $(54.88)^{\frac{3}{5}}$.

52. $(17.62)^{\frac{3}{2}}$.

53. $(4,924)^{\frac{2}{5}}$.

54. $(0.005218)^{\frac{1}{6}}$.

55. $(66.75)^{0.2}$.

56. $(390.3)^{0.3}$.

57. $(8156)^{0.7}$.

58. $(4.878)^{0.8}$.

59. $(5.535)^{1.4}$.

60. $(73.16)^{2.3}$.

61. $(2.540)^{\frac{1}{2}} \times (69.17)^{\frac{1}{3}}$.

62. $(129.1)^{\frac{1}{4}} \times (83.07)^{0.3}$.

63. $\left(\dfrac{57.68}{89.15}\right)^2$.

64. $\left(\dfrac{32.72}{40.06}\right)^3$.

65. $\dfrac{\sqrt[3]{403.8}}{\sqrt{27.82}}$.

66. $\sqrt[4]{\dfrac{0.5734}{6.420}}$.

67. $\sqrt{\dfrac{\sqrt[3]{67.45}}{39.01}}$.

68. $\sqrt[3]{\dfrac{111.8}{\sqrt{92.84}}}$.

69. $\sqrt[5]{\dfrac{735.2}{(43.13)^3}}$.

70. $\sqrt[4]{\dfrac{2.631 \times 0.09245}{(60.40)^2}}$.

71. $\dfrac{37.08 \times \sqrt{6027}}{\sqrt{7.603 \times 55.55}}$.

72. $\dfrac{82.54 \times \sqrt{27.06} \times \sqrt[3]{1730}}{912.5 \times 30.03}$.

73. $\dfrac{61.13 \times \sqrt[3]{54.04}}{39.08 \times \sqrt{85.26} \times \sqrt[3]{289.5}}$.

74. $\dfrac{1.008 \times \sqrt[3]{2.009}}{\sqrt{8.374 \times 911.5}}$.

75. $\sqrt{\dfrac{6.643 \times 0.05802}{700.8}}$.

76. $\sqrt[3]{\dfrac{37.13 \times 20.94}{0.05480 \times 0.03706}}$.

77. $\sqrt[4]{\dfrac{5.120 \times 3.807}{9.613 \times 4.440}}$.

78. $\sqrt[3]{\dfrac{0.006218 \times 0.07409}{3.165 \times 5.140}}$.

79. $\sqrt{\dfrac{3.943 \times 8.080}{0.04362 \times \sqrt[3]{283.5}}}$.

80. $\sqrt{\dfrac{3.182}{9.009}} \sqrt[3]{\dfrac{8643}{52.17}}$.

9–14. Accuracy in Logarithmic Computations

All the numbers in the computations of the previous section were accurate to four significant figures. Therefore, by the rules of Sec. 1–16, the results of these multiplications and divisions will be accurate to four digits. We have seen that the ordinary logarithmic computation with four-place tables gives precisely this accuracy. The following examples illustrate the use of logarithms in computations in which the question of the accuracy of the result arises.

Example 1. Compute $\dfrac{3.464 \times 9.8765}{46.574}$.

Assuming that the numbers given are accurate in accordance with the agreements of Sec. 1–13, we see that there is one four-digit number and that there are two five-digit numbers. We can expect a result accurate to four digits. In order to use four-place logarithms, it will be convenient, without affecting the accuracy materially, to round off all the numbers to four digits and then compute as usual.

$$N = \frac{3.464 \times 9.876}{46.57} .$$

Numbers	Mantissas
3.464	0.5396
9.876	0.9946
Numerator	11.5342 − 10
46.57	1.6681
N	9.8661 − 10

$$N = 0.7347.$$

We can assure ourselves that no major error has occurred during the computation by computing $\dfrac{3.46 \times 9.88}{46.6}$ by the slide rule. The result is 0.734.

Example 2. Compute $\dfrac{0.462 \times 9.567}{8.4163}$.

Since there is a three-digit number in this computation, we can expect a result accurate to at most three digits. It is convenient, then, if logarithms are to be used, to round off all the numbers to three digits before computing:

$$N = \frac{0.462 \times 9.57}{8.42} .$$

Exercises

Compute each of the following quantities by logarithms, assuming that each number given has the number of significant digits specified by the usual agreements.

1. $36.17 \times 58.8238 \times 74.059$.

2. $2.9306 \times 7.572 \times 0.64055$.

3. $9.12 \times 1.06317 \times 5.56$.

4. $0.04382 \times 8.87 \times 6.09$.

5. $1.99 \times 0.0001346 \times 572$.

6. $823.8 \times 3.555 \times 0.084725$.

7. $\dfrac{48.37 \times 0.031643}{0.64815}$.

8. $\dfrac{6.325 \times 0.19083}{76.054}$.

9. $\sqrt{1.18 \times 2.643}$.

10. $\sqrt[3]{5.748 \times 0.00825}$.

11. $\sqrt[4]{0.00168 \times 7.450}$.

12. $\sqrt{3.002 \times 0.002404}$.

13. $\sqrt{\dfrac{2.7681 \times 0.07936}{53.148}}$.

14. $\sqrt[3]{\dfrac{40048 \times 0.066}{0.02370 \times 8.9474}}$.

15. $\sqrt[5]{\dfrac{30.4 \times 92.18}{4070}}$.

16. $\sqrt{\dfrac{3.003}{16.92 \times 4.83}}$.

17. $\sqrt[3]{\dfrac{19.73 \times 652}{0.3124 \times 50.83}}$.

18. $\sqrt[4]{\dfrac{319.6}{(54.3)^3}}$.

19. $\sqrt{\dfrac{84.48 \times 0.193}{(60.30)^2}}$.

20. $\sqrt[3]{\dfrac{(9120)^2}{(63.7)^3}}$.

21. $\sqrt[5]{78.3 \times (8.456)^2 \times 23.10}$.

22. $\sqrt[7]{(49.85)^2 \times (34.7)^2 \times (0.04092)^3}$.

23. The horsepower of the prony brake is given by the expression $H = 2\pi PLN/33{,}000$ where P is the pressure in pounds at the end of the lever arm, L is the length in feet of the lever arm, and N is the speed in revolutions per minute. In a given engine test, the pressure was 73.8 lb, the lever arm had a length of 3.85 ft, and the speed was 224 rpm. What horsepower was developed by the engine?

24. The wavelength of an electric circuit at resonance is given by the formula $\lambda = 1884 \sqrt{LC}$ meters, where L is the inductance in microhenrys and C is the capacitance in microfarads. In a certain circuit the inductance was measured and found to be 8.0 microhenrys. The capacitance was 0.38 microfarad. Find the wavelength of the circuit at resonance.

25. The time required for a bomb released from an airplane to reach the ground is given by the formula $T = \sqrt{2S/g}$ where T is the time in seconds, S is the height in feet of the plane above the ground, and g is the acceleration due to gravity which is 32.16 ft per sec per sec. If a plane is at a height of 3.860 miles, what is the elapsed time in seconds from the release of the bomb until it strikes?

26. The load P in pounds supported by a helical compression spring is given by the formula $P = 0.196\,\dfrac{d^3}{r}\,f$, where d is the diameter in inches of the wire, r is the mean radius in inches of the coil, and f is the fiber stress in pounds. What is the load on a spring that has a wire diameter of 0.325 in., a mean coil radius of 1.10 in., and a fiber stress measured to the nearest 10 lb of 45,000 lb?

9–15. Hints about Logarithmic Computations

The negative exponent is defined by the relation

$$a^{-b} = \frac{1}{a^b}.$$

We can use this definition to compute numbers with negative exponents by logarithms as shown in the following example.

Example 1. Compute $(0.6192)^{-2}$.

$$N = (0.6192)^{-2} = \frac{1}{(0.6192)^2}$$

$$\log N = \log 1 - 2 \log 0.6192.$$

Numbers	Logarithms	
1	0	10.0000 − 10
0.6192	9.7918 − 10	
$(0.6192)^2$	19.5836 − 20	9.5836 − 10
N		0.4164

$$N = 2.609.$$

We have defined logarithms only for positive numbers. However, they can be used for computations with negative numbers as well, for, in any operation with negative numbers involving multiplication, division, raising to powers, or extraction of roots, the sign of the result can be determined independent of any calculation, and then the absolute value of the result can be found by logarithms. The following example will illustrate.

Example 2. Compute $N = \sqrt[3]{\dfrac{(-0.8162)(-4.673)}{-5.160}}$.

Since each of the three factors in the radicand is negative, the radicand itself is negative; and, since the cube root of a negative number is negative, the result is negative. The result is, then, the negative of

$$M = \sqrt[3]{\frac{0.8162 \times 4.673}{5.160}},$$

which we shall compute as usual.

Numbers	Logarithms	
0.8162	9.9118 − 10	
4.673	0.6696	
Numerator	10.5814 − 10	
5.160	0.7126	
Radicand	9.8688 − 10	29.8688 − 30
M		9.9563 − 10

$$M = 0.9042,$$

and

$$N = -0.9042.$$

Since there is no formula for log $(M + N)$ in terms of log M and log N, computations involving sums or differences must be made arithmetically, as indicated in the following example.

Example 3.   Compute $(0.8921)^{-\frac{1}{2}} + (1.231)^3$.
Each term must be computed separately, and the results added.

$$(0.8921)^{-\frac{1}{2}} = \frac{1}{(0.8921)^{\frac{1}{2}}}$$

$$\log (0.8921)^{-\frac{1}{2}} = \log 1 - \tfrac{1}{2} \log (0.8921)$$
$$\log (1.231)^3 = 3 \log 1.231.$$

Numbers	Logarithms	
1		10.0000 − 10
0.8921	9.9504 − 10	
$(0.8921)^{\frac{1}{2}}$	4.9752 − 5	9.9752 − 10
		0.0248

$$(0.8921)^{-\frac{1}{2}} = 1.059.$$

Numbers	Logarithms
1.231	0.0903
$(1.231)^3$	0.2709

$$(1.231)^3 = 1.866.$$

The final result is
$$1.059 + 1.866 = 2.925.$$

The identity $a^2 - b^2 = (a + b)(a - b)$ can be used to advantage in certain logarithmic computations, as indicated in the following example.

Example 4.   Compute $N = \sqrt{(45.63)^2 - (31.57)^2}$ by logarithms.
Using the identity

$$\sqrt{(45.63)^2 - (31.57)^2} = \sqrt{(45.63 + 31.57)(45.63 - 31.57)},$$
we obtain
$$N = \sqrt{77.20 \times 14.06}.$$

This expression is easily computed by logarithms, for
$$\log N = \tfrac{1}{2} [\log 77.20 + \log 14.06].$$

Numbers	Logarithms
77.20	1.8876
14.06	1.1480
	3.0356
N	1.5178

$$N = 32.95.$$

Exercises

Compute each of the following by logarithms. Obtain results accurate to as many significant digits as possible, assuming that the quantities are accurate only to the given number of significant digits.

1. $(5.471)^{-2}$.

2. $(5.471)^{-\frac{1}{2}}$.

3. $(8.835)^{-3}$.

4. $(12.69)^{-5}$.

5. $(0.04386)^{-\frac{1}{3}}$.

6. $(71.34)^{-\frac{3}{4}}$.

7. $(1.064)^{-4}$.

8. $(0.8367)^{-2.4}$.

9. $(312.6)^{-\frac{1}{6}}$.

10. $(0.005328)^{-\frac{1}{3}}$.

11. $(0.9414)^{-0.4}$.

12. $(64.3)^{-1.8}$.

13. $(1.13)^{-1}$.

14. $(937.0)^{-0.1}$.

15. $(3.17)^{-2} \times (5.326)^{-\frac{1}{2}}$.

16. $(0.3748)^{\frac{1}{3}} \times (0.0246)^{-\frac{1}{3}}$.

17. $(682)^{-\frac{1}{4}} \times (1.73)^{-\frac{3}{4}}$.

18. $(12.5)^{-2} \times (682.8)^{\frac{3}{2}}$.

19. $\dfrac{(20.6)^{-3} \times (18.0)^{-0.3}}{(75.4)^{-1}}$.

20. $\dfrac{27.4}{(38.5)^{-2} \times (63.7)^{\frac{1}{3}}}$.

21. $(-3.469)(-8.008)(-7.236)$.

22. $43.7 \times (-36.2)$.

23. $(-1.064)^2$.

24. $(-74.38)^3$.

25. $(-0.006314)^3$.

26. $(-35.75)^5$.

27. $\sqrt[3]{-461.1}$.

28. $\sqrt{(-88.2)(-63.6)}$.

29. $\sqrt[5]{-3.752}$.

30. $(-3.752)^{-\frac{1}{5}}$.

31. $(8842.8)^{\frac{1}{8}}$.

32. $-\sqrt{29.36}$.

33. $\sqrt[3]{(-5.436)^2}$.

34. $\sqrt{(-71.81)^4}$.

35. $\left(-\dfrac{62.13}{0.384}\right)^3$.

36. $\left[\dfrac{25.70}{(-9.502)^2}\right]^{\frac{1}{3}}$.

37. $\sqrt{\dfrac{(-4.062)(97.18)^3}{(-604.7)}}$.

38. $\sqrt[3]{112(-3.84)^5}$.

39. $(2.947)^{-2} + (625.5)^3$.

40. $(918.7)^2 - (746.1)^{-2}$.

41. $9.16[38.0 + (45.4)^{-2}]$.

42. $(89.5)^{\frac{3}{8}} + 129$.

43. $(2.040)^4 - (7.365)^5$.

44. $(62.07)^{-\frac{1}{2}} + (47.36)^{-\frac{1}{3}}$.

45. $\dfrac{18.5}{(46.3)^2 - (62.7)^3}$.

46. $\dfrac{7.48}{(9.52)^{\frac{1}{4}} + (4.46)^{\frac{3}{4}}}$.

47. $\dfrac{(72.41)^{\frac{1}{2}} + (80.25)^{\frac{2}{3}}}{716.6}$.

48. $\dfrac{(34.65)^{\frac{1}{5}} + (32.46)^{\frac{1}{8}}}{81.16}$.

49. $\dfrac{(24.17)^{-\frac{1}{3}}}{76.03}$.

50. $\dfrac{48.82}{(17.73)^{-\frac{1}{5}}}$.

51. $(46.37)^2 - (62.53)^2$.

52. $(76.84)^2 - (0.3216)^2$.

53. $\sqrt{(9.17)^2 - (4.42)^2}$.

54. $\sqrt{(0.571)^2 - (0.384)^2}$.

55. $\sqrt{(189.6)^2 - (184.4)^2}$.

56. $\sqrt{(8.68)^2 - (7.17)^2}$.

57. $\sqrt[3]{(515)^2 - (842)^2}$.

58. $\sqrt[3]{(431)^2 - (672)^2}$.

59. $\sqrt[3]{(73.7)^2 - (48.5)^2}$.

60. $\sqrt[3]{(4.18)^2 - (3.94)^2}$.

61. $14.8\sqrt{(76.5)^2 - (63.7)^2}$.

62. $9.02\sqrt{(6.36)^2 - (5.95)^2}$.

63. $7.10 \cdot 8.14\sqrt{(126)^2 - (85.3)^2}$.

64. $6.81\sqrt[3]{(70.7)^2 - (69.8)^2}$.

65. $\dfrac{11.8\sqrt{(48.0)^2 - (37.8)^2}}{36.2}$.

66. $\dfrac{9.140\sqrt[5]{(8376)^2 - (5973)^2}}{63.80}$.

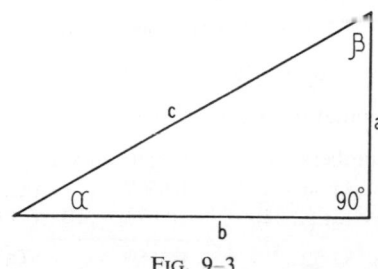

FIG. 9–3

9–16. Solution of Right Triangles with the Use of Logarithms

Logarithms can be used to simplify computations involving trigonometric functions. In this section we will use logarithms to perform
computations like those considered in Sec. 4–15.

From Fig. 9–3 we see that

$$a = c \sin \alpha. \tag{1}$$

From this relation it follows that

$$\log a = \log c + \log \sin \alpha. \tag{2}$$

In order to compute log sin α, we could first find sin α in the table of
trigonometric functions and then find the logarithm of the number sin α.
This procedure would necessitate using the table of trigonometric functions
and the table of logarithms.   To avoid the inconvenience of using two
tables, we have at our disposal a table in which the logarithms of the
trigonometric functions are given.   In the appendix, Table 4 is a table of
this kind.   For example, from Table 3, the table of natural functions, we
find that

$$\sin 40° = 0.6428$$

and from the table of logarithms

$$\log \sin 40° = \log 0.6428 = 9.8081 - 10.$$

However, this last value can be found directly from Table 4. It should be noted that the -10 of the negative characteristics of logarithms of trigonometric functions is omitted from Table 4 to save space. It will have to be supplied, of course, in computations.

Example 1. In a right triangle $a = 44.74$ and $\alpha = 52.35°$. Find the remaining parts of the triangle.

As in Sec. 4–15, we have the following formulas:

$$b = a \cot \alpha, \quad c = \frac{a}{\sin \alpha}, \quad \beta = 90° - \alpha.$$

Taking the logarithms of the first two expressions we have,

$$\log b = \log a + \log \cot \alpha,$$

$$\log c = \log a - \log \sin \alpha.$$

The logarithmic computation is shown below.

Numbers	Logarithms	
$a = 44.74$	1.6507	(Table 2)
$\cot \alpha = \cot 52.35°$	$9.8873 - 10$	(Table 4)
$b = 34.52$	1.5380	(Table 2)
$a = 44.74$	$11.6507 - 10$	(Table 2)
$\sin \alpha = \sin 52.35°$	$9.8986 - 10$	(Table 4)
$c = 56.51$	1.7521	(Table 2)

Hence

$$b = 34.52, \quad c = 56.51, \quad \beta = 37.65°.$$

These computations may be checked by substituting them in the formula $b = c \sin \beta$, as discussed in Sec. 4–15. However, the value for the angle β is easily verified since it was obtained by subtracting α from $90°$, so that we are primarily concerned with checking the values computed for b and c. These two values can be checked by means of the Pythagorean theorem in the form

$$a = \sqrt{c^2 - b^2} = \sqrt{(c + b)(c - b)}.$$

Since

$$c = 56.51 \quad \text{and} \quad b = 34.52,$$

we have

Numbers	Logarithms
$c + b = 91.03$	1.9592
$c - b = 21.99$	1.3422
$(c + b)(c - b)$	3.3014
$\sqrt{(c + b)(c - b)}$	1.6507

The logarithm 1.6507 found for $\sqrt{c^2 - b^2}$ is equal to the value of log a found in the preceding computations, and hence we have a check on the work of this example.

Example 2. Compute the coordinates of the terminal point of a vector **A** whose magnitude $A = 57.25$, whose direction angle $\alpha = 123.28°$, and whose initial point is at the origin.

From Sec. 6–5, we have

$$x_A = A \cos \alpha, \; y_A = A \sin \alpha;$$
$$\log |x_A| = \log A + \log |\cos \alpha|\;;$$
$$\log |y_A| = \log A + \log |\sin \alpha|\,.$$

Since $\sin 123.28° = \sin 56.72°$ and $\cos 123.28° = -\cos 56.72°$, we have
$$|\sin 123.28°| = \sin 56.72° \quad \text{and} \quad |\cos 123.28°| = \cos 56.72°.$$

The computation follows.

Numbers	Logarithms	Numbers	Logarithms
$A = 57.25$	1.7578	$A = 57.25$	1.7578
$\lvert \cos \alpha \rvert = \cos 56.72°$	$9.7394 - 10$	$\lvert \sin \alpha \rvert = \sin 56.72°$	$9.9222 - 10$
$\lvert x_A \rvert = 31.42$	1.4972	$\lvert y_A \rvert = 47.87$	1.6800

Since x_A is negative and y_A positive,

$$x_A = -31.42, \; y_A = 47.87.$$

Since
$$A^2 = x_A{}^2 + y_A{}^2,$$
$$y_A{}^2 = A^2 - x_A{}^2 = (A + x_A)(A - x_A)$$
and
$$|y_A| = \sqrt{(A + x_A)(A - x_A)}.$$

The computation follows:

$$A = 57.25, \; x_A = -31.42.$$

	Numbers	Logarithms
$A - x_A = 88.67$		1.9478
$A + x_A = 25.83$		1.4121
$(A + x_A)(A - x_A)$		3.3599
$\lvert y_A \rvert = \sqrt{(A + x_A)(A - x_A)}$		1.6800

Since $\log |y_A| = 1.6800$ here as well as in the first computation, we have a check on our work.

In computations of this kind, where lengths are measured to four significant digits and angles to the nearest hundredth of a degree, four-place logarithms give results to the same degree of accuracy.

Exercises

Solve the following right triangles and check. Give your results to the degree of accuracy of the given data.

1. $a = 24, \; \alpha = 56°.$ **2.** $a = 49.3, \; \beta = 20.4°.$

3. $c = 676, \; \alpha = 31.8°.$ **4.** $c = 0.862, \; \beta = 66.8°.$

5. $c = 68.37, \beta = 54.26°$.   **6.** $c = 218.3, \alpha = 41.50°$.

7. $a = 0.3475, \alpha = 24.16°$.   **8.** $a = 1945, \beta = 13.62°$.

9. $b = 2.743, \beta = 33.64°$.   **10.** $b = 0.06472, \alpha = 85.07°$.

11. $a = 6194, b = 8008$.   **12.** $a = 37,030, c = 48,220$.

13. $b = 3.16 \times 10^8, c = 5.30 \times 10^8$. **14.** $a = 0.0032, b = 0.019$.

Each vector **A** given below has its initial point at the origin. If A is the magnitude of **A** and α its direction angle, find by using logarithms the coordinates of the terminal point of **A**. Give your results to the degree of accuracy of the given data.

15. $A = 4.8, \alpha = 26°$.   **16.** $A = 37.5, \alpha = 64.4°$.

17. $A = 86.7, \alpha = 142°$.   **18.** $A = 814, \alpha = 245°$.

19. $A = 7891, \alpha = 204.6°$.   **20.** $A = 0.01372, \alpha = 314.2°$.

21. $A = 0.0429, \alpha = 344.8°$.   **22.** $A = 2461, \alpha = 192.12°$.

23. $A = 563.8, \alpha = 137.62°$.   **24.** $A = 18.37, \alpha = 263.14°$.

25. $A = 96.75, \alpha = 42.78°$.   **26.** $A = 623.3, \alpha = 175.6°$.

Find the magnitude A and direction angle α of the vector **A** whose initial point is at the origin and whose terminal point is (x_A, y_A). Give your results to the degree of accuracy of the given data.

27. $x_A = 18.2, y_A = 3.65$.   **28.** $x_A = 246, y_A = 853$.

29. $x_A = -4.72, y_A = 8.04$.   **30.** $x_A = 0.324, y_A = -0.0906$.

31. $x_A = -8.601, y_A = -6.043$.   **32.** $x_A = -12.88, y_A = 7.364$.

33. $x_A = 726.1, y_A = -309.8$.   **34.** $x_A = -9305, y_A = -4142$.

Using the method described in Sec. 6–5, find the sum of the vectors in each exercise. Give your results to the degree of accuracy of the given data.

35. **A**: $A = 206, \alpha = 38.2°$.   **36.** **A**: $A = 54.2, \alpha = 116°$.
       B: $B = 188, \beta = 73.6°$.   **B**: $B = 70.2, \beta = 187°$.

37. **A**: $A = 893, \alpha = 216.2°$.   **38.** **A**: $A = 1.17, \alpha = 37.6°$.
       B: $B = 364, \beta = 240.5°$.   **B**: $B = 4.03, \beta = 168.7°$.

39. **A**: $A = 420.6, \alpha = 63.36°$.   **40.** **A**: $A = 0.06317, \alpha = 69.2°$.
       B: $B = 604.5, \beta = 92.03°$.   **B**: $B = 0.08604, \beta = 192.5°$.

41. **A**: $A = 167, \alpha = 29.5°$.   **42.** **A**: $A = 27.7, \alpha = 187.0°$.
       B: $B = 246, \beta = 43.6°$.   **B**: $B = 53.0, \beta = 210.4°$.
       C: $C = 623, \gamma = 192.8°$.   **C**: $C = 12.4, \gamma = 326.2°$.

43. **A**: $A = 4.173, \alpha = 0.00°$.   **44.** **A**: $A = 0.01672, \alpha = 110.00°$.
       B: $B = 6.242, \beta = 30.05°$.   **B**: $B = 0.03403, \beta = 162.18°$.
       C: $C = 2.650, \gamma = 62.82°$.   **C**: $C = 0.02809, \gamma = 283.52°$.
       D: $D = 3.006, \delta = 87.03°$.   **D**: $D = 0.06163, \delta = 340.04°$.

45. Find the height of a tower if the length of its shadow is measured as 742 ft at a moment when the elevation of the sun is $40.3°$.

46. A sailor in a crow's nest 87.2 ft above the surface of the water observes that the angle of depression of a piece of wreckage floating on the water is $28°$. What is the distance from the ship to the wreckage?

47. A certain jet-assisted airplane can climb at an angle of $36.2°$ while making 406 mph. How long will it take for the plane to reach an altitude of 4600 ft?

48. A television transmitter antenna is located on the top of a building. From a point 6650 ft from the building, the angles of elevation of the top of the building and the top of the antenna are 2.88° and 3.24°, respectively. How high are the building and the antenna?

Additional exercises can be formed by selecting problems from Secs. 4–15 and 6–5 for logarithmic computation.

9–17. Logarithms to Other Bases than 10

Suppose it is required to find $\log_a N$ when the available table of logarithms has the base b. By the definition of logarithms

$$N = a^{\log_a N}.$$

Then

$$\log_b N = \log_b(a^{\log_a N}) = (\log_a N)(\log_b a).$$

From this last relation, dividing by $\log_b a$, we have

$$\log_a N = \frac{\log_b N}{\log_b a}$$

or

$$\log_a N = \frac{1}{\log_b a}\, \log_b N. \tag{1}$$

The special case of this relation in which we are primarily interested is the one in which $a = e$ and $b = 10$, where $e = 2.71828 \cdots$. We have then

$$\log_e N = \frac{1}{\log_{10} e}\, \log_{10} N. \tag{2}$$

Since $\log_{10} e = 0.4343$, $1/\log_{10} e = 2.3026$, and equation 2 can be written

$$\log_e N = 2.3026 \log_{10} N. \tag{3}$$

Using equation 3, we can thus compute logarithms to the base e from a table of logarithms to the base 10. From equation 2, we have

$$\log_{10} N = 0.4343 \log_e N. \tag{4}$$

Using the conventions of writing $\log_{10} N$ as $\log N$, and $\log_e N$ as $\ln N$, equations 3 and 4 can be written

$$\ln N = 2.3026 \log N, \tag{5}$$

$$\log N = 0.4343 \ln N. \tag{6}$$

The values of $\log N$ are found on most slide rules and, therefore, formula 5 can be used to find $\ln N$ with the slide rule alone without tables.

The so-called log log scales which are found on certain types of slide rules permit us to find directly the natural logarithm of a number without any computations.

Exercises

Using Table 2 in the appendix, evaluate each of the following to four significant figures. Verify the first three figures by the slide rule.

1. ln 83.60.   **2.** ln 3.570.   **3.** ln 710.4.

4. ln 0.09133.   **5.** ln 2761.   **6.** ln 0.0008066.

7. ln 7.069.   **8.** ln 31.05.   **9.** ln 0.6298.

Find N in each of the following to three significant figures:

10. ln $N = 7.604$.   **11.** ln $N \doteq 6.316$.   **12.** ln $N = 0.928$.

13. ln $N = 0.562$.   **14.** ln $N = 2.301$.   **15.** ln $N = -2.301$.

16. ln $N = -0.734$.   **17.** ln $N = -6.057$.   **18.** ln $N = -3.898$.

9–18. Exponential Equations

An equation in which the unknown occurs in an exponent is called an **exponential equation.** Thus, $2^x = 7$, $5^{x^2-1} = 31$ are examples of exponential equations.

There is no general method for solving exponential equations. Many exponential equations, for example,

$$2^x + 3^x = 10,$$

cannot be solved by methods studied in this book. There is one case, however, that may be solved easily by taking the logarithm of each member. This is the one in which the equations have the form

$$a^x = b,$$

where a and b are given and x is to be found. Taking the logarithm of each member, we obtain

$$x \log a = \log b,$$

and hence

$$x = \frac{\log b}{\log a}.$$

Example 1. Solve $2^x = 30$ for x.

Take the logarithm of each member and obtain

$$\log 2^x = \log 30,$$

$$x \log 2 = \log 30,$$

and

$$x = \frac{\log 30}{\log 2} = \frac{1.4771}{0.3010}.$$

If two- or three-place accuracy is desired, the last division can be performed on the slide rule.   When greater accuracy is wanted, we use logarithms and obtain:

Numbers	Logarithms
1.4771	$10.1694 - 10$
0.3010	$9.4786 - 10$
4.907	0.6908

The answer, to four significant figures, is 4.907.

Example 2.   Solve $3^{x-2} = 12$ for x.

Taking the logarithm of each member and proceeding as in the previous example, we obtain

$$\log 3^{x-2} = \log 12,$$

$$(x - 2) \log 3 = \log 12,$$

$$x - 2 = \frac{\log 12}{\log 3} = \frac{1.0792}{0.4771} = 2.262,$$

$$x = 4.262.$$

Exercises

Solve the following equations.

1. $10^x = 40.$ **2.** $100^x = 30.$ **3.** $4^x = 18.$

4. $7^x = 26.$ **5.** $0.6^x = 0.4.$ **6.** $0.35^x = 6.1.$

7. $2.0^x = 5.8.$ **8.** $3.08^x = 42.5.$ **9.** $3.08^{-x} = 42.5.$

10. $12^{3x} = (3^x)60.$ **11.** $\dfrac{1.15^{x-1}}{0.0403} = 7.628.$ **12.** $4.92^{x+8} = 3.07^x.$

13. $1.6^{x-4} = 1.$ **14.** $4^{2x} \cdot 2^{3x-1} = 3^x \cdot 5^{x+1}.$

15. $e^x = 62 \ (e = 2.718).$ **16.** $e^{-4x} = 0.4.$

9–19. The Slide Rule and Logarithmic Paper

The construction and use of the Mannheim slide rule are based on the results that have been established in this chapter.

Suppose that the distance between the two index marks on the C or D scales is L.   Then L is the length of the scale.   The location of the marks $1, 2, 3, \cdots, 10$ is found as follows.   1 is located at the left index mark. The point marked 2 is found so that the length from 1 to 2 is $L \log 2$. In the same way, the length from 1 to 3 is $L \log 3$, the length from 1 to 4 is $L \log 4$ and so on.

Thus, if the left index of the C scale is set at a point, marked with p on the D scale, and a point, marked with q, is located on the C scale, then the distance from the left index of the D scale to the point q on the C scale is

$$L \log p + L \log q = L(\log p + \log q) = L \log pq.$$

Therefore, the mark on the D scale opposite q on the C scale is pq. Thus, the slide rule is simply a mechanical device for adding and subtracting logarithms.

The scales A and B are constructed in like manner, with a basic length that is one-half the length of the scales C or D. The distance from the index to a certain number, say 5, on scale A is half as great as the distance from the index to the mark 5 on scale D, and, therefore, corresponds to the number $\sqrt{5}$ on scale D.

Thus *logarithmic scales*, on which the points, 1, 2, 3, $\cdots$ are not marked at equal distances from each other, but at distances from the origin of the scale that are proportional to log 1, log 2, log 3, etc., are the basis of the slide rule. A little experimentation will soon reveal how the other scales on any given rule are constructed.

Scales of the same kind can be used in order to simplify the construction of graphs of certain functions. Graphs are usually plotted on cross-section paper on which horizontal and vertical lines are printed at equidistant intervals, so that the whole area is covered with small squares. On one type of *semilogarithmic paper*, the scale on the horizontal axis is replaced by a logarithmic scale, and the vertical lines are printed in distances corresponding to this scale, with the horizontal lines at equidistant intervals as on ordinary cross-section paper. On the other type of semilogarithmic paper there is a logarithmic scale on the vertical axis and an equidistant scale on the horizontal axis.

On *logarithmic paper*, both systems of lines are printed corresponding to a logarithmic scale. A few examples will show how to use these papers.

Example 1. Construct a graph of the function

$$y = 2 \times 1.6^x. \tag{1}$$

This equation can be written in the equivalent form

$$\log y = \log 2 + x \log 1.6;$$

and, if $\log y = u$,

$$u = 0.3010 + 0.2041x. \tag{2}$$

Since u is a linear function of x, the graph of equation 2 is therefore a straight line, which can be constructed easily. In order to locate the values of y on the graph without computations, a logarithmic scale is used on the vertical axis, so that the numbers 2, 3, 4, $\cdots$ are located at the distances $u = \log 2$, $u = \log 3$, $u = \log 4$, $\cdots$.

The most convenient way to plot the graph is to find two of its points. If $x = 0$, $y = 2$; and, if $x = 4$, $y = 2 \times 1.6^4 = 2 \times 6.55 = 13.10$. The graph is plotted on semilogarithmic paper in Fig. 9–4.

This graph can now be used to solve certain problems. If $y = 2 \times 1.6^{2.5}$ has to be computed, point A of the graph gives approximately the answer $y = 6.5$. If we desire the value of x such that $2 \times 1.6^x = 10$, point B of the graph shows that $x = 3.4$ is the answer.

Example 2. Plot a graph corresponding to the equation $x^2 y^3 = 1$.

Computing the logarithms on both sides of this equation, we have $2 \log x + 3 \log y = \log 1 = 0$. This is a linear equation between $\log x$ and $\log y$. Therefore if we plot $\log y$ against $\log x$ we obtain a straight line. We plot this graph by plotting x and y on logarithmic scales.

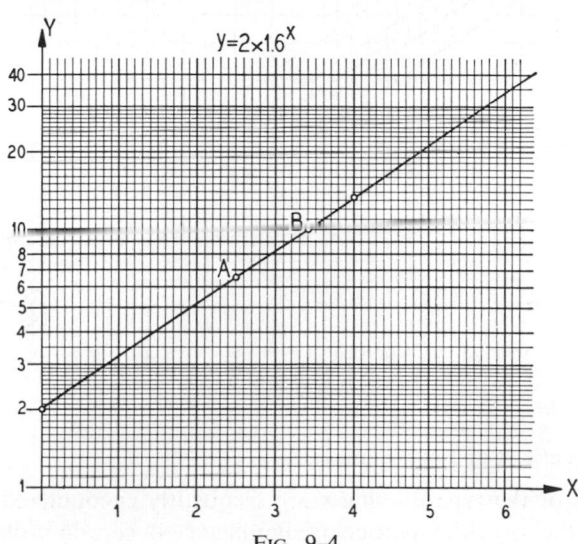

Fig. 9–4

In order to plot the graph, two sets of corresponding values of x and y are used. For example:

$$x_1 = 1, \quad y_1 = 1;$$
$$x_2 = 4, \quad y_2 = \sqrt[3]{\tfrac{1}{16}} = 0.397.$$

The graph is plotted in Fig. 9–5.

It is very easy to construct a system of graphs corresponding to the equations $x^2 y^3 = c$ for different values of c. From the corresponding equation $2 \log x + 3 \log y = \log c$ it can be inferred that, on logarithmic paper, all these graphs are parallel straight lines. They are easily constructed, observing that for $y = 1$, $x = \sqrt{c}$.

In Fig. 9–5 the graphs corresponding to the values $c = 1, 2, 3, 4, 9$ are plotted.

Exercises

Plot the graphs of the following equations on logarithmic or semilogarithmic paper.

1. $u = 2.8 \times 10^v$.
2. $y = 4.3 \times 9.6^x$.
3. $u = 7 \times 1.68^{2.4v}$.

4. $m \times 4.84^{0.62n} = 10$.
5. $x = 0.55 \cdot y^2$.
6. $pv^{1.41} = 1$.

7. $y = \dfrac{206}{3^x}$.
8. $300\sqrt{LC} = 2$.
9. $M = \dfrac{40}{3.66^{3n}}$.

10. Construct a graphical table for the equation $xy = C$ for the integral values of C from -10 to $+10$.

11. Make the same construction as in exercise 10 for the equation $x = C/y^2$.

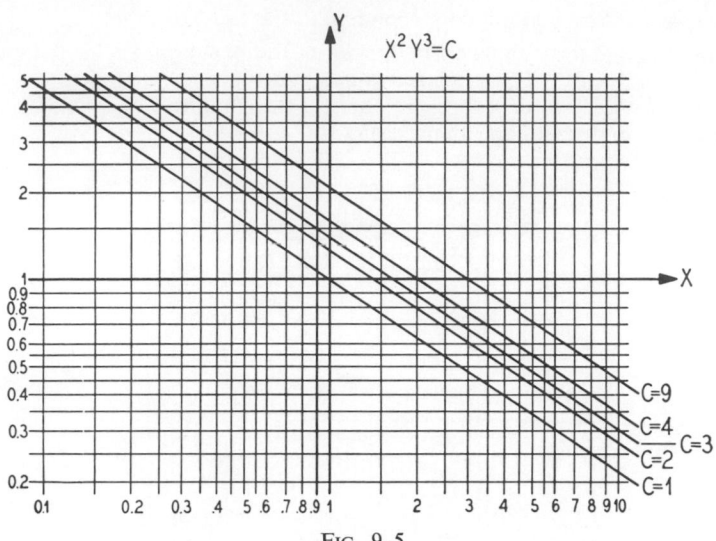

FIG. 9–5

9–20. Applications of Logarithms

Relations of the type $y = \log x$ are frequently encountered in science, thus giving the logarithm particular significance in certain problems. For example, the response of the human ear to sound is proportional to the logarithm of the power producing the sound.

The power used to produce a sound A is called the power level P_A of the sound and is measured in convenient power units, usually in watts. It is difficult to define a unit for the loudness of a sound; it is easier to compare the loudness of two sounds and to introduce a unit for the difference in loudness between the sounds. This unit is the **decibel** and is defined as follows.

If P_A and P_B are the power levels of two sounds A and B, then the difference in loudness between the sounds A and B, measured in decibels (denoted by db) is $10 \log \dfrac{P_A}{P_B}$. This value is often called **decibel gain or loss.**

Thus the number of decibels corresponding to a power ratio $P_A/P_B = 10 : 1$ is $10 \log \frac{10}{1} = 10$. In this way the following table can be constructed.

Power Ratio	Db	Power Ratio	Db
0.1 : 1	−10	10 : 1	10
1 : 1	0	100 : 1	20
2 : 1	3	1000 : 1	30

Experiment shows that a difference of 1 decibel is the smallest change in sound intensity that the ear can detect. The usefulness of the decibel in engineering problems is shown by the illustrative problem below.

Example 1. If the power output of a public-address system were increased from 20 to 22 watts, would the change in loudness be discernible to the ear? Finding the decibel gain, we obtain

decibel gain $= 10 \log \frac{22}{20} = 10 \log 1.1 = 10 \cdot 0.041 = 0.41$ db.

The change is less than 1 decibel and is therefore not apparent to the ear.

The number of decibels corresponding to a given power ratio can be used to characterize this power ratio. In electric circuits, it may also be used to measure current or voltage ratios. For this purpose the formula $P = I^2 R = E^2/R$ is used where I is the current, E the voltage, and R the resistance. The expression for the number of decibels corresponding to the power ratio P_A/P_B may then be rewritten as follows:

$$\text{decibel gain (or loss)} = 10 \log \frac{P_A}{P_B} = 10 \log \frac{I_A{}^2 R_A}{I_B{}^2 R_B} = 10 \log \frac{\dfrac{E_A{}^2}{R_A}}{\dfrac{E_B{}^2}{R_B}}. \quad (1)$$

If the resistances R_A and R_B are equal, these expressions become

$$\text{decibel gain (or loss)} = 10 \log \frac{I_A{}^2}{I_B{}^2} = 20 \log \frac{I_A}{I_B}, \quad (2)$$

$$\text{decibel gain (or loss)} = 10 \log \frac{E_A{}^2}{E_B{}^2} = 20 \log \frac{E_A}{E_B}. \quad (3)$$

Since the decibel gain or loss depends on a power, current, or voltage *ratio* in the above cases, the units in which the quantities are given are immaterial as long as they are the same in both cases.

Example 2. A motion-picture sound amplifier delivers a voltage of 80 volts at a frequency of 800 cycles and 20 volts at a frequency of 90 cycles across a constant resistance. What is the decibel loss in the amplifier between these two frequencies?

The decibel loss of an amplifier delivering various voltages across a constant resistance is given by the expression 3. The decibel loss which occurs in the amplifier between the two frequencies is then given by

decibel loss $= 20 \log \dfrac{E_A}{E_B} = 20 \log \dfrac{80}{20} = 20 \log 4 = 20 \cdot 0.602 = 1.204$ db.

Example 3. The resistance of a tungsten lamp filament varies with the temperature and is given by the relation $R_1/R_2 = (T_1/T_2)^{1.2}$ where R_1 is the resistance in ohms of the filament at room temperature T_1, measured in degrees Kelvin, and

R_2 and T_2 are the operating resistance and temperature. A given lamp was measured at a room temperature of 293° Kelvin and found to have a resistance of 24.0 ohms. What is the operating temperature at a measured operating resistance of 348 ohms?

The operating temperature in terms of the other factors may be obtained from the given equation $(T_1/T_2)^{1.2} = R_1/R_2$. By inversion

$$\left(\frac{T_2}{T_1}\right)^{1.2} = \frac{R_2}{R_1} \quad \text{or} \quad \frac{T_2}{T_1} = \left(\frac{R_2}{R_1}\right)^{1/1.2},$$

whence

$$T_2 = T_1 \left(\frac{R_2}{R_1}\right)^{1/1.2}.$$

Substituting the specific values given in the example, we obtain

$$T_2 = 293 \left(\frac{348}{24.0}\right)^{1/1.2} = 293(14.5)^{1/1.2}.$$

Solving by logarithms gives

Numbers	Logarithms
14.5	1.1614
$14.5^{1/1.2}$	0.9678
293	2.4669
$T_2 = 2720$	3.4347

The operating temperature T_2 of the filament is therefore 2720° Kelvin.

Exercises

1. A loudspeaker supplied with 1.4 watts produces a proper volume of sound output in a room. If the power is increased to 2.3 watts, how many decibels have been added to the original output?

2. If the output varies 4.0 db above and 2.0 db below the original output of 1.4 watts in problem 1, between what power levels does the amplifier operate?

3. The range in decibels of speech may be as much as 24. What is the ratio of power in speech?

4. The current through a resistance is 28 milliamperes. What is the change in decibels if the current is increased to 62 milliamperes?

5. A television amplifier tube can deliver a power output of 2.8 watts with a plate voltage of 185 volts. When the plate voltage is increased to 250 volts, the power output is increased to 4.3 watts. How many decibels increase in power does this represent?

6. How many additional watts must be supplied to a 20-watt motion-picture sound amplifier if the output is to be raised 8.2 decibels?

7. The energy loss in iron due to repeated magnetizations, such as occurs in transformer cores, is given by the Steinmetz hysteresis equation $W = \eta B^{1.6}$ where W is the loss in energy in ergs in 1 cu cm of iron for each cycle of magnetization, and η is the hysteresis coefficient which for a particular kind of iron has a value of 2.3×10^{-3}. The maximum magnetic-induction density in maxwells per square centimeter is designated by B. In testing a sample of iron the energy loss per cubic centimeter was measured to be 1840 ergs. If the hysteresis coefficient is as given above, determine the maximum magnetic-induction density.

8. A tungsten lamp filament was measured at a room temperature of 293° Kelvin and found to have a resistance of 26.5 ohms. What is the operating temperature at a measured operating resistance of 382 ohms?

9. The cold resistance of a tungsten lamp filament at a room temperature of 295° Kelvin is 14.0 ohms. The temperature of the filament under normal operating conditions was determined and found to be 2730° Kelvin. What is the resistance of the filament under operating conditions?

Progress Report

In this chapter we considered logarithms. The logarithm to the base *B* of *a* was defined as the exponent to which *B* must be raised to obtain *a*. After the establishment of this definition, the following topics were discussed:

1. The properties of logarithms.

2. The use of common logarithms in computations; the characteristic and the mantissa and the use of a table of logarithms.

3. The solution of right triangles by means of logarithms.

4. Logarithms to bases other than 10.

5. The application of logarithms to the solution of exponential equations and a few problems in applications.

6. The use of logarithmic scales on the slide rule and on logarithmic graph paper.

10

The Fundamental Relations of Trigonometry

In Chapter 7 identical and conditional equations were discussed. Both types of equations occur frequently in trigonometry where the variables involved are angles and functions of angles. Most of this chapter will be devoted to the discussion of trigonometric identities, that is, trigonometric relations that are true for all values of the variables (angles) involved. Toward the end of this chapter, a few conditional trigonometric equations will be discussed. Finally, inverse trigonometric functions will be introduced.

Trigonometric identities have many uses in mathematics. Most important is their use in the simplification of trigonometric expressions and formulas which occur in science and engineering.

10–1. The Fundamental Trigonometric Identities

There are several trigonometric identities which are so frequently used that they are commonly known as the **fundamental identities.** These are

$$\sin \theta = \frac{1}{\csc \theta}, \qquad \csc \theta = \frac{1}{\sin \theta}.$$

$$\cos \theta = \frac{1}{\sec \theta}, \qquad \sec \theta = \frac{1}{\cos \theta}. \tag{1}$$

$$\tan \theta = \frac{1}{\cot \theta}, \qquad \cot \theta = \frac{1}{\tan \theta}.$$

$$\tan \theta = \frac{\sin \theta}{\cos \theta}.$$

$$\cot \theta = \frac{\cos \theta}{\sin \theta}. \tag{2}$$

$$\sin^2 \theta + \cos^2 \theta = 1.$$

$$1 + \tan^2 \theta = \sec^2 \theta. \tag{3}$$

$$1 + \cot^2 \theta = \csc^2 \theta.$$

All these relations are proved by using the original definitions of the trigonometric functions as given in Chapter 4. For example, three of the identities of group 1 are proved immediately from the definitions in the following manner (see Fig. 4–6 of Chapter 4):

$$\frac{1}{\csc\theta} = \frac{1}{\dfrac{r}{y}} = \frac{y}{r} = \sin\theta.$$

$$\frac{1}{\sec\theta} = \frac{1}{\dfrac{r}{x}} = \frac{x}{r} = \cos\theta.$$

$$\frac{1}{\cot\theta} = \frac{1}{\dfrac{x}{y}} = \frac{x}{y} = \tan\theta.$$

The other three identities of group 1 are proved in the same manner. Similar proofs apply to the identities of group 2. For example,

$$\frac{\sin\theta}{\cos\theta} = \frac{\dfrac{y}{r}}{\dfrac{x}{r}} = \frac{y}{r}\cdot\frac{r}{x} = \frac{y}{x} = \tan\theta.$$

In order to prove the third set of identities, we again use the fundamental definitions, this time in connection with the Pythagorean theorem. For every angle θ the sides of the reference triangle satisfy the relation

$$y^2 + x^2 = r^2. \tag{4}$$

Dividing both sides of the equation by r^2, we obtain

$$\frac{y^2}{r^2} + \frac{x^2}{r^2} = 1 \text{ or } \left(\frac{y}{r}\right)^2 + \left(\frac{x}{r}\right)^2 = 1.$$

Now, substituting the definitions $\sin\theta = y/r$, $\cos\theta = x/r$ in this equation, we obtain the first identity of group 3, namely,

$$\sin^2\theta + \cos^2\theta = 1.$$

This relation is true for every angle θ.

The second and third identities are proved in a similar manner, dividing equation 4 by x^2 and y^2, respectively.

Since the fundamental identities hold for all values of θ, they are frequently useful in the simplification of trigonometric expressions. Consequently, it is of great value for the student of applied mathematics to be able to manipulate and use these identities as the occasion demands.

Example 1. In a certain problem, to find a length L when given an angle θ and another length a, an engineer developed a formula

$$L = \frac{a^2 \sin^2 \theta + a^2 \cos^2 \theta}{\csc \theta \tan \theta}.$$

Can we, by use of identities, simplify this formula?

Factoring the numerator, and substituting in the denominator the identities

$$\csc \theta = \frac{1}{\sin \theta}, \qquad \tan \theta = \frac{\sin \theta}{\cos \theta},$$

we obtain

$$L = \frac{a^2(\sin^2 \theta + \cos^2 \theta)}{\dfrac{1}{\sin \theta} \cdot \dfrac{\sin \theta}{\cos \theta}}.$$

But from equation 3 of this section $\sin^2 \theta + \cos^2 \theta = 1$. Therefore

$$L = \frac{a^2 \cdot 1}{1 \cdot \dfrac{1}{\cos \theta}} = a^2 \cos \theta.$$

Example 2. Simplify

$$\frac{1 + \cos \beta}{\sin \beta} + \frac{\sin \beta}{1 + \cos \beta}.$$

Adding the two fractions by finding the common denominator, we obtain

$$\frac{1 + 2 \cos \beta + \cos^2 \beta + \sin^2 \beta}{\sin \beta (1 + \cos \beta)}.$$

When the identity $\sin^2 \beta + \cos^2 \beta = 1$ is substituted, this becomes

$$\frac{2 + 2 \cos \beta}{\sin \beta (1 + \cos \beta)}.$$

Factoring the numerator and substituting $\dfrac{1}{\sin \beta} = \csc \beta$, we get

$$\frac{2(1 + \cos \beta)}{\sin \beta (1 + \cos \beta)} = \frac{2}{\sin \beta} = 2 \csc \beta.$$

Example 3. Express in terms of $\cos \alpha$:

$$K = \frac{1 - 2 \sin^2 \alpha}{1 + \tan^2 \alpha}.$$

Substituting $\sin^2 \alpha = 1 - \cos^2 \alpha$ (derived from $\sin^2 \alpha + \cos^2 \alpha = 1$) and $\sec^2 \alpha = 1 + \tan^2 \alpha$, we obtain

$$K = \frac{1 - 2(1 - \cos^2 \alpha)}{\sec^2 \alpha} = \frac{2 \cos^2 \alpha - 1}{\sec^2 \alpha}.$$

But $\sec^2 \alpha = \dfrac{1}{\cos^2 \alpha}$. Therefore

$$K = \frac{2 \cos^2 \alpha - 1}{\dfrac{1}{\cos^2 \alpha}} = 2 \cos^4 \alpha - \cos^2 \alpha.$$

Example 4.   Assuming that θ is an angle in the third quadrant, use the fundamental identities to express $\sin \theta$ as a function of $\tan \theta$.

First we observe that

$$\sin \theta = \sin \theta \cdot \frac{\cos \theta}{\cos \theta} = \tan \theta \cos \theta, \tag{5}$$

so that we can express $\sin \theta$ as a function of $\tan \theta$ if we can express $\cos \theta$ as a function of $\tan \theta$.   Since $1 + \tan^2 \theta = \sec^2 \theta$, we have $\cos^2 \theta = \dfrac{1}{1 + \tan^2 \theta}$, whence

$$\cos \theta = \frac{1}{\pm\sqrt{1 + \tan^2 \theta}}. \tag{6}$$

Since θ is in the third quadrant (see Sec. 4 5), we choose the negative sign in equation 6.   Combining this result with equation 5, we obtain

$$\sin \theta = -\frac{\tan \theta}{\sqrt{1 + \tan^2 \theta}}.$$

Since the procedure depends on the type of result desired, there are no general instructions that can be given for the manipulation of trigonometric expressions.   At first the student cannot do much more than proceed by trial and error.   Skill will develop with experience and practice.   There is no necessarily correct method of procedure.   Often several methods will yield the same result.

Exercises

Verify that each of the fundamental identities holds for the given angle.

1. $\theta = 45°$.   2. $\theta = 60°$.   3. $\theta = 150°$.   4. $\theta = 210°$.

5. $\theta = 110°$.   6. $\theta = 58°$.   7. $\theta = 147°$.   8. $\theta = -36°$.

9. $\theta = 0°$.   10. $\theta = 180°$.   11. $\theta = 90°$.   12. $\theta = 28.65°$.

Simplify each of the following trigonometric expressions as much as possible.

13. $\sin \theta \cot \theta$.   14. $\cos \theta \tan \theta$.

15. $\sin \theta \cos^2 \theta \sec^3 \theta$.   16. $\sin^3 \theta \csc^2 \theta \sec \theta$.

17. $\sin^2 \theta \,(1 + \csc^2 \theta)$.   18. $\csc \theta \sec^2 \theta \cos^3 \theta$.

19. $\cos \theta \tan \theta \csc \theta$.   20. $\cos^2 \alpha(1 + \cot^2 \alpha)$.

21. $\dfrac{\sin \theta}{\cos \theta} + \dfrac{\cos \theta}{\sin \theta}$.   22. $\dfrac{F \csc A}{1 + \cot^2 A}$.

23. $\dfrac{\sin \alpha}{\csc \alpha} - \cos \alpha \sec \alpha$.   24. $\dfrac{I \cos \theta}{2} + \dfrac{1}{2 \sec \theta}$.

25. $\dfrac{E_m(1 + \tan^2 A)}{\sec A}$.   26. $\dfrac{\cot 45° + \cot^2 \theta}{\tan 45° + \tan^2 \theta}$.

27. $\dfrac{\sin \theta \cot \theta}{\cos \theta \tan \theta}$.   28. $\dfrac{\sin \alpha \csc \alpha}{\cos \alpha \sec \alpha}$.

29. $\dfrac{\sec A}{\cot A + \tan A}$.

30. $\dfrac{\tan A + \cot A}{\csc A}$.

31. $\dfrac{\csc^2 \theta (1 - \sin^2 \theta)}{\cot \theta}$.

32. $\dfrac{\tan^2 A - \sin^2 A}{\tan^2 A \sin^2 A}$.

33. $\dfrac{\sec^2 \alpha + 2 \tan \alpha}{1 + \tan \alpha}$.

34. $\dfrac{\sec^2 \alpha (1 - \cos^2 \alpha)}{\tan \alpha}$.

35. $\dfrac{\csc A}{\csc A - 1} + \dfrac{\csc A}{\csc A + 1}$.

36. $\dfrac{1 - \tan A}{1 + \tan A} - \dfrac{\cot A - 1}{\cot A + 1}$.

37. $\dfrac{\sin \theta - \cos \theta}{\tan \theta \csc \theta - \sec \theta \cot \theta}$.

38. $\dfrac{\tan \theta \cot \theta - \sin^2 \theta}{\cos \theta}$.

39. $\dfrac{\cos^4 \theta - \sin^4 \theta + 2 \sin^2 \theta}{\sec \theta}$.

40. $\dfrac{1 - \cot^4 \theta - 2 \csc^2 \theta}{\csc^2 \theta}$.

Prove the following fundamental identities.

41. $\csc \theta = \dfrac{1}{\sin \theta}$.

42. $\sec \theta = \dfrac{1}{\cos \theta}$.

43. $\cot \theta = \dfrac{1}{\tan \theta}$.

44. $\cot \theta = \dfrac{\cos \theta}{\sin \theta}$.

45. $1 + \tan^2 \theta = \sec^2 \theta$.

46. $1 + \cot^2 \theta = \csc^2 \theta$.

By use of the fundamental identities express each of the six trigonometric functions in terms of the given function, assuming that θ is in the given quadrant.

47. $\sin \theta$, θ in I.   **48.** $\cos \theta$, θ in II.   **49.** $\tan \theta$, θ in II.

50. $\cot \theta$, θ in III.   **51.** $\sec \theta$, θ in IV.   **52.** $\csc \theta$, θ in IV.

53. $\sin \theta$, θ in III.   **54.** $\cos \theta$, θ in IV.   **55.** $\tan \theta$, θ in III.

56. $\cot \theta$, θ in IV.   **57.** $\sec \theta$, θ in II.   **58.** $\csc \theta$, θ in II.

10–2. Some Trigonometric Formulas

Following the fundamental identities are three groups of important identities known as the **addition formulas, the double-angle formulas,** and **the half-angle formulas.**

The Addition Formulas.   We shall accept without proof the following formulas:

$$\sin (A + B) = \sin A \cos B + \cos A \sin B.$$
$$\sin (A - B) = \sin A \cos B - \cos A \sin B. \tag{1}$$

$$\cos (A + B) = \cos A \cos B - \sin A \sin B.$$
$$\cos (A - B) = \cos A \cos B + \sin A \sin B. \tag{2}$$

$$\tan (A + B) = \frac{\tan A + \tan B}{1 - \tan A \tan B}.$$
$$\tan (A - B) = \frac{\tan A - \tan B}{1 + \tan A \tan B}. \tag{3}$$

The Double-Angle Formulas. If the two angles considered in formulas 1 and 2 are equal (i.e., if $A = B$), then the addition formulas 1 and 2 become, respectively,

$$\sin (A + A) = \sin A \cos A + \cos A \sin A,$$

$$\cos (A + A) = \cos A \cos A - \sin A \sin A,$$

or

$$\sin 2A = 2 \sin A \cos A,$$

$$\cos 2A = \cos^2 A - \sin^2 A. \tag{4}$$

The Half-Angle Formulas. Two more expressions can be derived from the second of formulas 4 by the substitutions $\cos^2 A - 1$　　$\sin^2 A$ and $\sin^2 A = 1 - \cos^2 A$. The results of these substitutions are

$$2 \sin^2 A = 1 - \cos 2A,$$

$$2 \cos^2 A = 1 + \cos 2A. \tag{5}$$

Formulas 5 are true for all values of A. They remain true, therefore, if A is replaced by $A/2$ and $2A$ by A.

$$2 \sin^2 \frac{A}{2} = 1 - \cos A.$$

$$2 \cos^2 \frac{A}{2} = 1 + \cos A. \tag{6}$$

Using formulas 6, it is easy to compute $\sin \frac{A}{2}$ and $\cos \frac{A}{2}$ in terms of A:

$$\sin \frac{A}{2} = \pm \sqrt{\frac{1 - \cos A}{2}},$$

$$\cos \frac{A}{2} = \pm \sqrt{\frac{1 + \cos A}{2}}. \tag{7}$$

The sign $+$ is chosen in these formulas if $\sin \frac{A}{2} \left(\text{or} \cos \frac{A}{2} \right)$ is positive; the sign $-$ is chosen if $\sin \frac{A}{2} \left(\text{or} \cos \frac{A}{2} \right)$ is negative.

The above formulas are often useful in the manipulation of trigonometric expressions. Corresponding formulas for the other trigonometric functions are given in exercises of this section; the proofs are left to the student.

Various examples using formulas 1 through 7 of this section will now be shown.

Example 1. Evaluate sin 90°, using 90° = 60° + 30°.

Substituting $A + B = 60° + 30°$ in the addition formula of 1, we obtain

$$\sin 90° = \sin (60° + 30°) = \sin 60° \cos 30° + \cos 60° \sin 30°;$$

evaluating, we get

$$\sin 90° = \frac{\sqrt{3}}{2} \cdot \frac{\sqrt{3}}{2} + \frac{1}{2} \cdot \frac{1}{2} = \frac{4}{3} + \frac{1}{4} = 1,$$

which is the correct value.

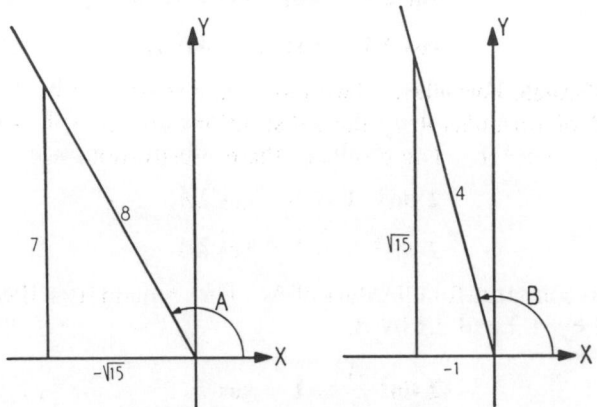

Fig. 10–1

Example 2. Given $\sin A = \frac{7}{8}$ and $\cos B = -\frac{1}{4}$, where both A and B are in the second quadrant, evaluate $\cos (A - B)$ and $\tan (A + B)$.

The formulas to be used are

$$\cos (A - B) = \cos A \cos B + \sin A \sin B,$$

$$\tan (A + B) = \frac{\tan A + \tan B}{1 - \tan A \tan B}.$$

However, these require the evaluation of other functions of A and B, which is easily accomplished by forming the triangles of reference in the proper quadrants (Fig. 10–1).

Now, substituting the proper values, we get

$$\cos (A - B) = \left(\frac{-\sqrt{15}}{8}\right)\left(-\frac{1}{4}\right) + \left(\frac{7}{8}\right)\left(\frac{\sqrt{15}}{4}\right) = \frac{\sqrt{15}}{4 \cdot 8}(1 + 7) = \frac{\sqrt{15}}{4},$$

$$\tan (A + B) = \frac{\left(-\frac{7}{\sqrt{15}}\right) + \left(-\frac{\sqrt{15}}{1}\right)}{1 - \left(-\frac{7}{\sqrt{15}}\right)\left(-\frac{\sqrt{15}}{1}\right)} = \frac{\frac{-7 - 15}{\sqrt{15}}}{1 - 7} = \frac{22}{\sqrt{15}} \cdot \frac{1}{6} = \frac{11\sqrt{15}}{45}.$$

Example 3.   Express sin 6x in terms of functions of (*a*) 3x, (*b*) 12x.

Substituting in the first of formulas 4 and 7 yields

$$(a)\ \sin 6x = 2 \sin 3x \cos 3x,$$

$$(b)\ \sin 6x = \pm\sqrt{\frac{1 - \cos 12x}{2}}.$$

Example 4.   Given that $\cos 30° = \dfrac{\sqrt{3}}{2}$, find (*a*) cos 60°, (*b*) cos 15°.

By the formulas of this section

$$(a)\ \cos 60° = 2\cos^2 30° - 1 = 2\left(\frac{\sqrt{3}}{2}\right)^2 - 1 = \frac{3}{2} - 1 = \frac{1}{2},$$

$$(b)\ \cos 15° = +\sqrt{\frac{1 + \cos 30°}{2}} = +\sqrt{\frac{1 + \dfrac{\sqrt{3}}{2}}{2}} = \frac{1}{2}\sqrt{2 + \sqrt{3}}.$$

Why was the plus sign chosen before the radical?

Example 5.   Show that

$$E_1E_2 \sin \omega t \cos \omega t - E_1E_2 \cos(\omega t + 90°) = \frac{E_1E_2}{2}[\sin 2\omega t + 2 \sin \omega t].$$

Multiplying the first term of the left-hand side by $\frac{2}{2}$ and expanding the second term by the proper formula, we obtain

$$E_1E_2\left[\frac{2 \sin \omega t \cos \omega t}{2} - (\cos \omega t \cos 90° - \sin \omega t \sin 90°)\right].$$

Substituting sin 2ωt for 2 sin ωt cos ωt and evaluating, we get

$$E_1E_2\left(\frac{\sin 2\omega t}{2} - 0 + \sin \omega t\right) = \frac{E_1E_2}{2}(\sin 2\omega t + 2 \sin \omega t).$$

Exercises

Using the fundamental identities and formulas 1 through 7 of this section, prove the following formulas:

1. $\tan(A - B) = \dfrac{\tan A - \tan B}{1 + \tan A \tan B}.$

2. $\cot(A + B) = \dfrac{\cot A \cot B - 1}{\cot B + \cot A}.$

3. $\cot(A - B) = \dfrac{\cot A \cot B + 1}{\cot B - \cot A}.$

4. $\dfrac{1}{\sec(A \pm B)} = \dfrac{\csc A \csc B \mp \sec A \sec B}{\csc A \csc B \sec A \sec B}.$

5. $\dfrac{1}{\csc(A \pm B)} = \dfrac{\sec A \csc B \pm \csc A \sec B}{\csc A \csc B \sec A \sec B}.$

6. $\tan 2A = \dfrac{2 \tan A}{1 - \tan^2 A}.$

7. $\cot 2A = \dfrac{\cot^2 A - 1}{2 \cot A}$.

8. $\cos 2A = \dfrac{\csc^2 A - \sec^2 A}{\csc^2 A \sec^2 A}$.

9. $\sin 2A = \dfrac{2}{\sec A \csc A}$.

10. $\tan A = \pm \sqrt{\dfrac{1 - \cos 2A}{1 + \cos 2A}}$.

11. $\tan A = \dfrac{1 - \cos 2A}{\sin 2A}$.

12. $\tan A = \dfrac{\sin 2A}{1 + \cos 2A}$.

13. Develop a formula for $\cot A$ in terms of functions of the angle $2A$.

14. Develop a formula for $\sec A$ in terms of functions of the angle $2A$.

15. Develop a formula for $\csc A$ in terms of functions of the angle $2A$.

16–30. For each of the identities in exercises 1 through 15, determine the values of the unknown angles for which the equality fails.

By the addition formulas 1 and 2 find, without tables, the sine, cosine, and tangent of:

31. $60°$, using $60° = 30° + 30°$.

32. $60°$, using $60° = 90° - 30°$.

33. $75°$.

34. $15°$.

35. $90°$.

36. $210°$.

37. $-330°$.

38. $\left(\dfrac{\pi}{2} - \dfrac{\pi}{3}\right)$.

39. $-\dfrac{\pi}{6}$.

40. $\dfrac{5\pi}{6}$.

Expand and simplify.

41. $\cos\left(\dfrac{\pi}{2} + \alpha\right)$.

42. $\cos\left(\dfrac{3\pi}{2} - \alpha\right)$.

43. $\sin(\theta - 45°)$.

44. $\sin(45° - \theta)$.

45. $\cos\left(\phi - \dfrac{\pi}{6}\right)$.

46. $\tan\left(\dfrac{2\pi}{3} - \phi\right)$.

47. $\sin(-A) = \sin(0° - A)$.

48. $\cos(-A)$.

49. $\tan(-A)$.

50. $\cos[\pi + (\alpha + \beta)]$.

Evaluate $\sin(A + B)$, $\cos(A + B)$, $\sin(A - B)$, and $\cos(A - B)$ in each of the following cases:

51. Given $\sin A = \frac{1}{2}$, $\sin B = \frac{1}{3}$, A and B in the first quadrant.

52. Given $\sin A = \frac{2}{3}$, $\cos B = \frac{5}{8}$, A in first, B in fourth quadrant.

53. Given $\tan A = \frac{3}{4}$, $\cot B = 6$, A and B both acute angles.

54. Given $\csc A = 6$, $\cos B = -\frac{2}{3}$, $\cos A$ negative, $\sin B$ negative.

55. Given $\sin A = \frac{2}{5}$, $\cos B = \frac{4}{5}$, A and B in same quadrant.

56. Given $\sin A = \frac{1}{6}$, $\cos B = -\frac{5}{13}$, A and B in same quadrant.

57. Given $\cot A = \frac{5}{12}$, $\sec B = \frac{5}{2}$, $0° < A < 180°$, $90° < B < 360°$.

58. Given $\tan A = -\frac{3}{2}$, $\cos B = -\frac{12}{13}$, $90° < A < 270°$, $180° < B < 360°$.

59. Given $\sec A = -2$, $\tan B = 3$, $\sin A > 0$, $\cos B < 0$.

60. Given $\tan A = -\frac{8}{5}$, $\cos B = -\frac{4}{13}$, A and B in same quadrant.

In exercises 61–70 express the sine and cosine of the given angle in terms of functions of an angle (*a*) half as large, (*b*) twice as large.

61. 40°.

Example:

(*a*) $\sin 40° = 2 \sin 20° \cos 20°$,

   $\cos 40° = \cos^2 20° - \sin^2 20°$.

(*b*) $\sin 40° = \sqrt{\dfrac{1 - \cos 80°}{2}}$,

   $\cos 40° = \sqrt{\dfrac{1 + \cos 80°}{2}}$.

62. 116°.	**63.** 220°.	**64.** $5x$.	**65.** 3θ.	**66.** π.
67. 100°.	**68.** $\dfrac{3\pi}{4}$.	**69.** $\dfrac{\theta}{5}$.	**70.** $\frac{1}{4}A$.	**71.** $\frac{2}{3}\omega t$.

Using the double-angle formulas, find the sine and cosine of

72. 90°.   **73.** 60°.   **74.** 120°.   **75.** 180°.   **76.** 360°.   **77.** 0°.

Using the half-angle formulas, find the sine and cosine of

78. 30°.   **79.** 15°.   **80.** 45°.   **81.** 90°.   **82.** 180°.   **83.** 0°.

Find $\sin 2A$, $\cos 2A$, $\sin \dfrac{A}{2}$, $\cos \dfrac{A}{2}$.

84. Given $\sin A = \frac{1}{3}$, $90° < A < 270°$.

85. Given $\cos A = \frac{3}{4}$, $360° < A < 540°$.

86. Given $\tan A = -5$, $180° < A < 360°$.

87. Given $\cot A = -\frac{12}{5}$, $-90° < A < 90°$.

88. Given $\sec A = \frac{3}{2}$, $-360° < A < -180°$.

89. Given $\csc A = \frac{13}{12}$, $90° < A < 270°$.

90. Given $\sin A = -\frac{3}{8}$, $270° < A < 450°$.

91. Given $\cos A = -1$, $0° < A < 270°$.

In each of the following exercises transform the expression on the left-hand side of the equal sign into that on the right-hand side. These are all transformations taken from engineering applications.

92. $K \sin^2 2x \left(\dfrac{\sec^2 x}{4 \sin^2 x} \right) = K.$

93. $\dfrac{R(I_m \sin \alpha)^2}{\pi} = \dfrac{RI_m{}^2}{2\pi} (1 - \cos 2\alpha).$

94. $\dfrac{f_x \sin \alpha \cos \theta}{\beta - \alpha} - \dfrac{f_x \cos \alpha \sin \theta}{\alpha - \beta} = \dfrac{f_x}{\beta - \alpha} \sin (\alpha + \theta).$

95. $d_m(1 - \sin 2\theta) \left(\dfrac{\cot \theta + 1}{\cot \theta - 1} \right) = d_m \cos 2\theta.$

96. Given $p = ei$, where $e = E_m \sin \omega t$ and $i = I_m \sin \omega t$. Show that

$$p = \frac{E_m I_m}{2} (1 - \cos 2\omega t).$$

97. Show that

$$E_m[\sin \omega t - \sin (\omega t + 90°)] = \sqrt{2}\, E_m \sin (\omega t - 45°).$$

Hint: From Chapter 4, $\sin (\omega t + 90°) = \cos \omega t.$

98. Show that $I_m \sin (\omega t + 120°) - I_m \sin \omega t = \sqrt{3}\, I_m \sin (\omega t + 150°).$

99. Show that $\dfrac{EI \cos \theta}{2\pi} - \dfrac{EI \cos (2\alpha + \theta)}{2\pi} = \dfrac{EI}{2\pi} (\cos \theta - \cos 2\alpha \cos \theta + \sin 2\alpha \sin \theta).$

100. Show that $E_m \sin (\omega t - 240°) = E_m \sin (\omega t + 120°).$

101. If $P_r = EI \cos \theta$, $P_x = EI \sin \theta$, show that

$$\sqrt{P_r{}^2 + P_x{}^2} = EI.$$

102. Show that $RI_m \sin \omega t - RI_m \sin (\omega t + 240°) = RI_m \left(\dfrac{3}{2} \sin \omega t + \dfrac{\sqrt{3}}{2} \cos \omega t \right)$

$$= \sqrt{3}\, RI_m \sin (\omega t + 30°).$$

103. Show that

$$[A \sin (\alpha + \theta_1)][B \sin (\alpha + \theta_2)] = AB \left(\frac{1 - \cos 2\alpha}{2} \cos \theta_1 \cos \theta_2 + \right.$$

$$\left. \frac{\sin 2\alpha}{2} \cos \theta_1 \sin \theta_2 + \frac{\sin 2\alpha}{2} \cos \theta_2 \sin \theta_1 + \frac{1 + \cos 2\alpha}{2} \sin \theta_1 \sin \theta_2 \right).$$

10–3. Other Trigonometric Formulas

The following formulas are also used in mathematics and engineering for the manipulation of trigonometric expressions.

The product formulas,

$$\sin A \sin B = \tfrac{1}{2} [\cos (A - B) - \cos (A + B)],$$

$$\sin A \cos B = \tfrac{1}{2} [\sin (A - B) + \sin (A + B)], \qquad (1)$$

$$\cos A \cos B = \tfrac{1}{2} [\cos (A - B) + \cos (A + B)],$$

are useful in expressing products of sines and cosines in terms of sums of those functions.

The proofs follow readily from the addition formulas. For example, to prove the first of the above identities, the addition formula

$$\cos (A + B) = \cos A \cos B - \sin A \sin B$$

is subtracted from

$$\cos (A - B) = \cos A \cos B + \sin A \sin B,$$

giving

$$\cos (A - B) - \cos (A + B) = 2 \sin A \sin B,$$

which may be solved for the desired result:

$$\sin A \sin B = \tfrac{1}{2}[\cos (A - B) - \cos (A + B)].$$

In similar fashion the other product identities can be readily derived by adding or subtracting the proper pair of formulas in (1) and (2) of Sec. 10–2.

The sum formulas,

$$\sin A + \sin B = 2 \sin \left(\frac{A + B}{2}\right) \cos \left(\frac{A - B}{2}\right),$$

$$\sin A - \sin B = 2 \cos \left(\frac{A + B}{2}\right) \sin \left(\frac{A - B}{2}\right),$$

$$\cos A + \cos B = 2 \cos \left(\frac{A + B}{2}\right) \cos \left(\frac{A - B}{2}\right),$$

$$\cos A - \cos B = -2 \sin \left(\frac{A + B}{2}\right) \sin \left(\frac{A - B}{2}\right),$$

$$\sin A + \cos B = 2 \sin \left(\frac{A - B}{2} + 45°\right) \cos \left(\frac{A + B}{2} - 45°\right),$$

$$\sin A - \cos B = 2 \cos \left(\frac{A - B}{2} + 45°\right) \sin \left(\frac{A + B}{2} - 45°\right),$$

(2)

are utilized when it is necessary to change a sum of sines and cosines into a product of similar functions. These formulas can be derived from the product formulas discussed in the first part of this section.

Example 1. Express $\sin 7y \cdot \cos \frac{y}{2}$ as a sum of trigonometric functions.

Using the second formula of (1), and taking $A = 7y$, $B = y/2$, we get

$$\sin 7y \cdot \cos \frac{y}{2} = \frac{1}{2}\left[\sin\left(7y - \frac{y}{2}\right) + \sin\left(7y + \frac{y}{2}\right)\right] = \frac{1}{2}\left(\sin \frac{13y}{2} + \sin \frac{15y}{2}\right).$$

Example 2.   Express [cos $(\omega t - \pi/2)$ − cos ωt] as a product of trigonometric functions.

For the difference of cosines, the fourth identity of group 2 is used.   Taking $A = (\omega t - \pi/2)$, $B = \omega t$, we get

$$\cos\left(\omega t - \frac{\pi}{2}\right) - \cos \omega t = -2 \sin\left(\frac{2\omega t - \frac{\pi}{2}}{2}\right) \sin\left(\frac{-\frac{\pi}{2}}{2}\right)$$

$$= -2 \sin\left(\omega t - \frac{\pi}{4}\right) \sin\left(-\frac{\pi}{4}\right),$$

and, since $\sin\left(-\dfrac{\pi}{4}\right) = -\sin\dfrac{\pi}{4} = -\dfrac{\sqrt{2}}{2}$,

$$\cos\left(\omega t - \frac{\pi}{2}\right) - \cos \omega t = \sqrt{2} \sin\left(\omega t - \frac{\pi}{4}\right).$$

An alternative method for achieving the final result would have been the substitution sin $\omega t = \cos (\pi/2 - \omega t) = \cos (\omega t - \pi/2)$ in the given expression, resulting in sin ωt − cos ωt.   This could then have been expressed as a product by the last identity of group 2.

Exercises

Derive the formulas:

1. $\sin A \cos B = \frac{1}{2}[\sin (A - B) + \sin (A + B)]$.

2. $\cos A \cos B = \frac{1}{2}[\cos (A - B) + \cos (A + B)]$.

3. $\sin A \sin B = \frac{1}{2}[\cos (A - B) - \cos (A + B)]$.

4. $\cos A + \cos B = 2 \cos \left(\dfrac{A + B}{2}\right) \cos \left(\dfrac{A - B}{2}\right)$.

5. $\cos A - \cos B = -2 \sin \left(\dfrac{A + B}{2}\right) \sin \left(\dfrac{A - B}{2}\right)$.

6. $\sin A + \cos B = 2 \sin \left(\dfrac{A - B}{2} + 45°\right) \cos \left(\dfrac{A + B}{2} - 45°\right)$.

7. $\sin A - \cos B = 2 \cos \left(\dfrac{A - B}{2} + 45°\right) \sin \left(\dfrac{A + B}{2} - 45°\right)$.

Write the following as sums, using the product formulas 1 of Sec. 10–3:

8. $\sin 65° \cos 20°$.   **9.** $\cos 48° \cos 36°$.   **10.** $\sin 4\alpha \sin 6\alpha$.

11. $\sin 5\omega t \cos 2\omega t$.   **12.** $\cos \dfrac{x}{7} \cos x$.   **13.** $\sin \dfrac{\pi}{3} \sin \dfrac{\pi}{9}$.

14. $\sin \alpha \sin (\alpha + \beta)$.   **15.** $\cos x \cos (-x)$.   **16.** $\cos (\omega t + \alpha) \cos (\omega t - \alpha)$.

17. $\sin (\alpha + \beta + \theta) \cos (\alpha - \beta + \theta)$.   **18.** $\cos (3\alpha - \beta) \sin (2\alpha + 2\beta)$.

19. $\cos \left(\dfrac{\pi}{3} - \alpha\right) \sin \left(\alpha - \dfrac{\pi}{3}\right)$.

Write the following as products, using the sum formulas 2 of Sec. 10–3:

20. $\sin 60° + \sin 30°$.

21. $\cos \omega t - \cos 3\omega t$.

22. $\cos \dfrac{\alpha}{3} + \cos \dfrac{\alpha}{4}$.

23. $\sin \dfrac{\pi}{4} - \sin \dfrac{3\pi}{2}$.

24. $\sin (2\alpha - \theta) - \sin (\theta - 3\alpha)$.

25. $\cos (\theta - 30°) + \cos (\theta + 30°)$.

26. $\cos \left(\dfrac{\pi}{6} - \alpha\right) - \cos \left(\dfrac{\pi}{6} + \alpha\right)$.

27. $\sin \left(\omega t + \dfrac{\theta}{3}\right) + \sin \left(\dfrac{\theta}{4} - \omega t\right)$.

28. $\sin (3\alpha + 2\beta) + \sin 2\alpha$.

29. $\cos 5\theta - \cos (3\theta + \alpha)$.

30. $\sin \left(\theta + \dfrac{5\pi}{2}\right) + \cos \left(3\theta - \dfrac{\pi}{4}\right)$.

31. $\cos \left(\dfrac{\omega t}{2} + \dfrac{\pi}{6}\right) - \sin \left(\omega t + \dfrac{2\pi}{3}\right)$.

In each of the following exercises convert the left-hand side of each of the equalities into the right-hand side by means of the identities in this chapter, particularly those of Sec. 10–3.

32. $\sin 60° + \sin 30° = \sqrt{2} \cos 15°$.

33. $\sin 150° + \cos 60° = 2 \cos 60°$.

34. $\dfrac{3 \cos 125° \cos 35°}{\cos 160°} = \dfrac{3}{2}$.

35. $\cos (x - 2\pi) + \cos x = 2 \cos x$.

36. $\dfrac{\sin 3\theta + \sin 7\theta}{\cos 7\theta + \cos 3\theta} = \tan 5\theta$.

37. $\dfrac{\sin 3\alpha - \sin 5\alpha}{\cos 5\alpha - \cos 3\alpha} = \cot 4\alpha$.

38. $\dfrac{\sin 6\omega t \cos 3\omega t}{\sin 3\omega t + \sin 9\omega t} = \dfrac{1}{2}$.

39. $\dfrac{\sin 36° + \sin (-144°)}{\cos 36° - \cos (-144°)} = 0$.

40. $\cos 208° \cos 28° = -\sin^2 118°$.

41. $\sin^2 13° + \sin 103° \sin 77° = 1$.

42. $\dfrac{\sin 10\theta - \sin 6\theta + \sin 4\theta}{4 \sin 2\theta} = \cos 3\theta \cos 5\theta$.

43. $\dfrac{\tan \beta}{\cot \alpha \cot \gamma} = \tan \alpha + \tan \beta + \tan \gamma$, where $\alpha + \beta + \gamma = 180°$.

44. $\dfrac{\sec A + \tan A - 1}{\sec A + \tan A + 1} = \tan \dfrac{A}{2}$.

Follow the above procedure in exercises 45 through 51. These are manipulations used by physicists and engineers.

45. If $p = ei$, where $e = E_m \sin \alpha$ and $i = I_m \sin (\alpha + \theta)$, show that

$$p = \dfrac{E_m I_m}{2} \cos \theta - \dfrac{E_m I_m}{2} \cos (2\alpha + \theta).$$

46. Express as a product of trigonometric functions:

$$f = 8 \cos \omega t + 8 \cos (2\omega t + \alpha).$$

47. If $\cos \theta_n = \dfrac{A_n}{\sqrt{A_n{}^2 + B_n{}^2}}$, show that, if $0° \leq \theta_n \leq 90°$ and $B_n > 0$,

$$A_n \sin n\alpha + B_n \cos n\alpha = \sqrt{A_n{}^2 + B_n{}^2} \sin (n\alpha + \theta_n).$$

48. Show that

$$\omega^2 R \sin \lambda \cos \lambda = \frac{\omega^2 R}{2 \cot \lambda} (1 + \cos 2\lambda).$$

49. Given $s_1 = 65\sqrt{3} \sin \omega t$ and $s_2 = 65 \sqrt{3} \sin \left(\omega t + \dfrac{\pi}{3} \right)$, show that

$$s_1 + s_2 = 195 \sin \left(\omega t + \frac{\pi}{6} \right).$$

50. If $w = fd$, where $f = F_m \sin \alpha$ and $d = D_m \sin (\alpha + 90°)$, show that

$$w = \tfrac{1}{2} F_m D_m \sin 2\alpha.$$

51. In an electric circuit, the transferred power P is given by

$$P = \frac{E^2 Z_L \cos \theta}{(Z \cos \phi + Z_L \cos \theta)^2 + (Z \sin \phi + Z_L \sin \theta)^2}.$$

This has a maximum value when $Z_L = Z$. Show that

$$P_{\max} = \frac{E^2 \cos \theta}{2Z[1 + \cos (\phi - \theta)]}$$

by substituting $Z_L = Z$ in the formula for P.

10–4. A Useful Application of the Trigonometric Identities

In Chapter 5 it was shown that a function which is the sum or difference of two or more functions may be graphed by composition of ordinates. Various questions arise in connection with this process, particularly when the elementary functions used are periodic. One question of particular interest to the scientist and engineer is: "Under what conditions will a curve that is a sum or difference of sine and cosine curves again be a sine or cosine curve?"

The question may immediately be simplified by noting from the addition formulas that

$$\cos x = \sin \left(x + \frac{\pi}{2} \right) = -\sin \left(x - \frac{\pi}{2} \right), \qquad \sin x = -\sin (x + \pi),$$

and that therefore *by a change in the initial phases any sum or difference of sine and cosine functions may be rewritten as a sum of sine functions with positive coefficients.*

Example 1.   The   function $f(t) = \sin 2t - 4 \cos 2t + \cos (2t - \pi/8) - 3 \sin (2t - \pi/3)$ may be rewritten with sine functions with positive coefficients as follows:

$$f(t) = \sin 2t + 4 \sin\left(2t - \frac{\pi}{2}\right) + \sin\left(2t - \frac{\pi}{8} + \frac{\pi}{2}\right) + 3 \sin\left(2t - \frac{\pi}{3} + \pi\right)$$

$$= \sin 2t + 4 \sin\left(2t - \frac{\pi}{2}\right) + \sin\left(2t + \frac{3\pi}{8}\right) + 3 \sin\left(2t + \frac{2\pi}{3}\right).$$

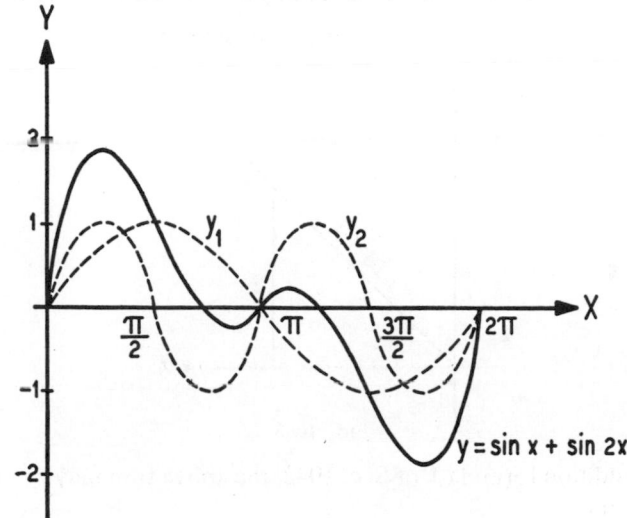

FIG. 10–2

Hence, the question now is: "Under what conditions will a curve that is the sum of sine curves again be a sine curve?" We shall distinguish two cases.   First, the sum of two sine functions with unequal periods, and, second, the sum of two sine functions with equal periods.

For the first case, the following example shows that the sum of two sine functions with unequal periods is not necessarily a sine curve.

Example 2.   By composition of ordinates show that
$$y = \sin 2x + \sin x$$
is not a sine function.

Graphing $y_1 = \sin x$ and $y_2 = \sin 2x$ (Fig. 10–2) whose periods are 2π and π, respectively, we see that the composite function $y = \sin x + \sin 2x$ obviously is not a sine function.

In the second case a very useful theorem can be stated:

The sum of two sine functions which have positive coefficients and the same period,
$$A_1 \sin (\omega t + \alpha_1) + A_2 \sin (\omega t + \alpha_2),$$
is again a sine function
$$A \sin (\omega t + \alpha)$$

with amplitude $A = \sqrt{M^2 + N^2}$ and initial phase α, α being an angle in standard position whose terminal side passes through (M, N), where M and N are given by the relations

$$M = A_1 \cos \alpha_1 + A_2 \cos \alpha_2,$$

$$N = A_1 \sin \alpha_1 + A_2 \sin \alpha_2.$$

The proof of this theorem is not difficult.   Consider the sum function

$$f(t) = A_1 \sin (\omega t + \alpha_1) + A_2 \sin (\omega t + \alpha_2).$$

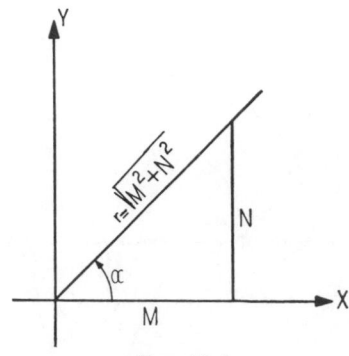

FIG. 10–3

By using addition formula 1 of Sec. 10–2, the above sum may be expanded into the form

$$f(t) = A_1 \sin \omega t \cos \alpha_1 + A_1 \cos \omega t \sin \alpha_1 + A_2 \sin \omega t \cos \alpha_2 +$$

$$A_2 \cos \omega t \sin \alpha_2$$

$$= (A_1 \cos \alpha_1 + A_2 \cos \alpha_2) \sin \omega t + (A_1 \sin \alpha_1 + A_2 \sin \alpha_2) \cos \omega t,$$

and, if we set

$$M = A_1 \cos \alpha_1 + A_2 \cos \alpha_2,$$

$$N = A_1 \sin \alpha_1 + A_2 \sin \alpha_2,$$

$$(1)$$

this may be written

$$f(t) = M \sin \omega t + N \cos \omega t, \qquad (2)$$

where M and N are determined by A_1, A_2, α_1, α_2.   We can now find an angle α such that (M, N) is on its terminal side.   From the triangle of reference of α (Fig. 10–3)

$$\sin \alpha = \frac{N}{\sqrt{M^2 + N^2}},$$

$$(3)$$

$$\cos \alpha = \frac{M}{\sqrt{M^2 + N^2}}.$$

Thus, if we multiply and divide the right-hand side of equation 2 by $\sqrt{M^2 + N^2}$,

$$f(t) = \sqrt{M^2 + N^2} \left(\frac{M}{\sqrt{M^2 + N^2}} \sin \omega t + \frac{N}{\sqrt{M^2 + N^2}} \cos \omega t \right),$$

and, by equation 3,

$$f(t) = \sqrt{M^2 + N^2} \, (\sin \omega t \cos \alpha + \cos \omega t \sin \alpha).$$

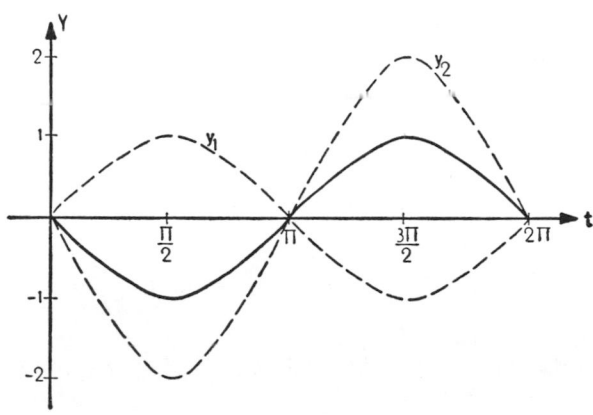

FIG. 10–4

Using addition formula 1 of Sec. 10–2 yields

$$f(t) = \sqrt{M^2 + N^2} \sin (\omega t + \alpha);$$

we thus obtain

$$A_1 \sin (\omega t + \alpha_1) + A_2 \sin (\omega t + \alpha_2) = \sqrt{M^2 + N^2} \sin (\omega t + \alpha),$$

where M and N are given by equations 1 and α is given by equations 3. In short, *the sum of the two given sine functions is again a sine function with amplitude $\sqrt{M^2 + N^2}$ and initial phase α.*

It is obvious that the sum of any number of sine functions with the same period is again a sine function with the same period, for by taking two functions at a time and applying the theorem just discussed the sum can finally be reduced to a single sine function.

Example 3.   Express $\sin t + 2 \sin (t + \pi)$ as a sine function.

For purposes of verification, we will first perform the addition graphically by composition of ordinates.   Adding the ordinates of graphs $y_1 = \sin t$ and $y_2 = 2 \sin (t + \pi)$, we obtain the graph $y = \sin t + 2 \sin (t + \pi)$, which appears to be also the graph of $y = \sin (t + \pi)$, (Fig. 10–4).

Let us perform this operation by the methods of this section.   For this example $\omega = 1$, $A_1 = 1$, $A_2 = 2$, $\alpha_1 = 0$, $\alpha_2 = \pi$.   Hence by relations 1,

$$M = 1 \cos 0 + 2 \cos \pi = 1 - 2 = -1,$$

$$N = 1 \sin 0 + 2 \sin \pi = 0 + 0 = 0.$$

Thus $\alpha = +\pi$; also

$$A = \sqrt{M^2 + N^2} = \sqrt{1 + 0} = 1.$$

Hence, by the theorem the sum is

$$A \sin(\omega t + \alpha) = \sin(t + \pi),$$

as was predicted from the graph.

Example 4.   Given that

$$e = E_{m_1} \sin \omega t + E_{m_2} \sin(\omega t - 90°),$$

show that

$$e = E_m \sin(\omega t - B),$$

where

$$E_m = \sqrt{E_{m_1}^2 + E_{m_2}^2} \quad \text{and} \quad \tan B = \frac{E_{m_2}}{E_{m_1}}.$$

(Assume that $E_{m_1} \geq 0$ and $E_{m_2} \geq 0$.)

To use the theorem of this section, take $\omega = \omega$, $A_1 = E_{m_1}$, $A_2 = E_{m_2}$, $\alpha_1 = 0$, $\alpha_2 = -90°$.   Then, by relations 1,

$$M = E_{m_1} \cos 0° + E_{m_2} \cos(-90°) = E_{m_1},$$

$$N = E_{m_1} \sin 0° + E_{m_2} \sin(-90°) = -E_{m_2}.$$

Therefore, in the wave $A \sin(\omega t + \alpha)$ obtained by addition,

$$A = \sqrt{M^2 + N^2} = \sqrt{E_{m_1}^2 + E_{m_2}^2},$$

and

$$\tan \alpha = \frac{N}{M} = -\frac{E_{m_2}}{E_{m_1}},$$

giving

$$e = \sqrt{E_{m_1}^2 + E_{m_1}^2} \sin(\omega t + \alpha).$$

If we take $\alpha = -B$, then

$$e = \sqrt{E_{m_1}^2 + E_{m_2}^2} \sin(\omega t - B)$$

where $\tan(-B) = -\dfrac{E_{m_2}}{E_{m_1}}$ or $\tan B = \dfrac{E_{m_2}}{E_{m_1}}.$

In this section we have shown that, regardless of the amplitudes and initial phases involved, a sum of sine functions all with the same period is a sine function with this period and that a sum of sine functions with unequal periods is not necessarily a sine function.   Whether or not a sum of sine functions with unequal periods always leads to a function other than a sine function has not been considered.   However, a consideration of this question is beyond the scope of this book.

The relations between the amplitudes and initial phases stated by the theorem of this section can be given an interesting vectorial interpretation. Consider the sum

$$A_1 \sin (\omega t + \alpha_1) + A_2 \sin (\omega t + \alpha_2), \quad A_1 > 0, \quad A_2 > 0,$$

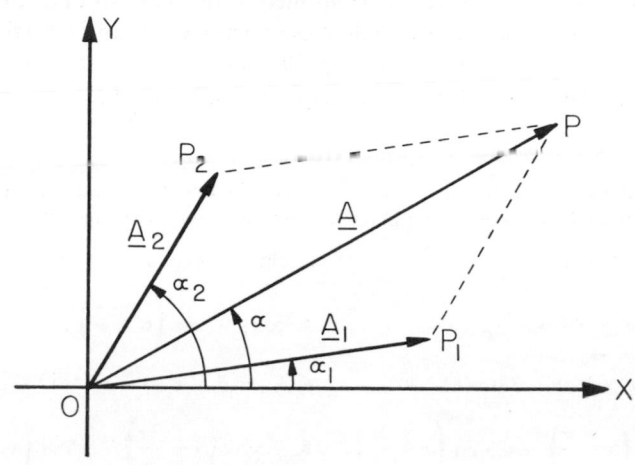

FIG. 10–5

and draw the vectors $\mathbf{A}_1$ and $\mathbf{A}_2$ with magnitudes A_1, A_2 and direction angles α_1, α_2, respectively, and with initial points at the origin (see Fig. 10–5). If $\mathbf{A} = \mathbf{A}_1 + \mathbf{A}_2$, then $P_1 (A_1 \cos \alpha_1, A_1 \sin \alpha_1)$ and $P_2(A_2 \cos \alpha_2, A_2 \sin \alpha_2)$ are the terminal points of $\mathbf{A}_1$ and $\mathbf{A}_2$, respectively, and hence the terminal point $P(M, N)$ of $\mathbf{A}$ is given by

$$M = A_1 \cos \alpha_1 + A_2 \cos \alpha_2,$$
$$N = A_1 \sin \alpha_1 + A_2 \sin \alpha_2.$$

It follows that the magnitude of $\mathbf{A}$ is $A = \sqrt{M^2 + N^2}$ and that the terminal side of the direction angle α passes through (M, N). We may now restate the theorem as follows:

The sum of two sine functions with positive coefficients and the same period,

$$A_1 \sin (\omega t + \alpha_1) + A_2 \sin (\omega t + \alpha_2) = A \sin (\omega t + \alpha), \qquad (4)$$

where A and α are the magnitude and direction angle of a vector $\mathbf{A} = \mathbf{A}_1 + \mathbf{A}_2$, $\mathbf{A}_1$ and $\mathbf{A}_2$ having magnitudes A_1, A_2 and direction angles α_1, α_2, respectively.

This statement of the theorem has an advantage over the earlier one in that it makes the content of the theorem easier to remember.   Figure 10–5 is called the *amplitude and initial-phase vector diagram for the sum* 4. The reader should construct the amplitude and initial-phase vector diagrams for examples 3 and 4.

The theorem of this section is basic in the analysis of a-c circuits, since the voltages and currents in such circuits are sinusoidal. In addition to the obvious application of this theorem to the addition and subtraction of these currents and voltages, it can be applied to the analysis of a variety of circuits. The amplitude and initial-phase vector diagram is particularly valuable in this connection. The application of this theorem to the analysis of a simple circuit will be discussed in Chapter 12.

Exercises

Express each sum as a single sine function, finding the amplitude and initial phase of this function with the help of an amplitude and initial-phase vector diagram.

1. $\sin x + 3 \sin x$.

2. $\sin x + 4 \cos x$.

3. $3 \sin \omega t - \cos \omega t$.

4. $\cos \theta + \cos \left(\theta + \dfrac{\pi}{4} \right)$.

5. $2 \sin (x + 30°) + 3 \sin (x - 30°)$.

6. $\tfrac{2}{3} \sin (\theta - 60°) - \tfrac{3}{2} \cos (\theta - 30°)$.

7. $4.2 \sin \left(x - \dfrac{\pi}{6} \right) - 6.3 \sin \left(x + \dfrac{2\pi}{3} \right)$.

8. $7.5 \cos \left(x + \dfrac{3\pi}{4} \right) + 1.5 \sin \left(x - \dfrac{\pi}{4} \right)$.

9. $80 \sin (\theta - 54°) - 63 \cos (\theta - 38°)$.

10. $9.8 \cos (\theta - 32°) + 7.4 \cos (\theta + 32°)$.

11. $4.3 \sin (\omega t - 28.4°) - 2.6 \sin (\omega t + 14.2°)$.

12. $14 \sin (x + 16.8°) + 18 \cos (x - 22.6°)$.

13. $9.6 \sin \left(\omega t + \dfrac{\pi}{4} \right) + 9.6 \sin \left(\omega t - \dfrac{3\pi}{4} \right)$.

14. $2R \sin (\omega t + 42.4°) + 3S \sin (\omega t - 48.6°)$.

15. $6 \cos (x - 17.4°) - 7 \cos (x + 19.4°)$.

16. $3 \sin \left(\omega t + \dfrac{5\pi}{18} \right) - 2 \cos \left(\omega t + \dfrac{7\pi}{18} \right)$.

17. $8.4 \cos (\omega t - 36°) + 5.5 \sin (\omega t + 54°)$.

18. $12.2 \sin (x + 17.4°) - 16.8 \cos (x - 24.2°)$.

19. $\sin \omega t + 2 \cos (\omega t - 30°) + 3 \sin (\omega t - 60°)$.

20. $2 \sin \omega t - 3 \sin (\omega t + 20°) + 2 \cos (\omega t - 20°)$.

21. Given that

$$e_0 = I_m R \sin \omega t + I_m X_L \cos \omega t.$$

Show that
$$e_0 = I_m Z \sin(\omega t + \theta)$$
where
$$Z = \sqrt{R^2 + X_L^2}$$
and
$$\tan \theta = \frac{X_L}{R}.$$

10–5. Trigonometric Equations

In trigonometry as in algebra there exist identical equalities and conditional equalities, the former being true for all values of the angles
involved, the latter being true for only certain specific angles. For
example, suppose that power in a particular circuit is given experimentally
by
$$P = 110 \cos \theta,$$

and it is desired to find for what values of θ the power will be 55. The
conditional equation
$$55 = 110 \cos \theta \quad \text{or} \quad \cos \theta = \tfrac{55}{110} = 0.5$$

is obtained. This is certainly true only for certain values of θ, namely,
$$\theta = \frac{\pi}{3}, \ \frac{5\pi}{3}, \ \frac{7\pi}{3}, \ \frac{11\pi}{3}, \ \text{etc.}$$

There are many types of trigonometric equations. Frequently the
equations are of such a nature that the methods of algebra may be applied
directly.

Example 1. Find all values of θ, $0° \le \theta < 360°$, that satisfy the equation
$$3 \sin \theta + 2 \sin \theta = 4.$$

This is solved in much the same way as the algebraic equation $3x + 2x = 4$.
Combining the terms on the left-hand side, we get
$$5 \sin \theta = 4, \quad \text{and} \quad \sin \theta = \tfrac{4}{5} = 0.8000.$$
Then from the tables
$$\theta = 53.13°, \ 126.87°.$$

Example 2. Find all values of x, $0 \le x < 360°$, for which $\cos 3x = 0.8910$.

Since $\cos 3x$ is not equal to $3 \cos x$, we cannot here divide both sides of the
equation by 3. Instead, with the aid of the table we find the possible values for
the angle $3x$, noting that to obtain values of x between 0 and 360° we must find
values of $3x$ between 0 and $3 \cdot 360° = 1080°$. Thus
$$3x = 27°, \ 333°, \ 387°, \ 693°, \ 747°, \ 1053°,$$
and
$$x = 9°, \ 111°, \ 129°, \ 231°, \ 249°, \ 351°.$$

Example 3.   Find all values of θ, $0° \leq \theta < 360°$, for which $2 \sin^2 \theta + 3 \sin \theta = 2$.

This equation is solved in the same way that the equation $2x^2 + 3x = 2$ would be solved in algebra.   Thus, subtracting 2 from both sides, we get

$$2 \sin^2 \theta + 3 \sin \theta - 2 = 0.$$

Factoring, we obtain

$$(2 \sin \theta - 1)(\sin \theta + 2) = 0.$$

Now, since the product of the two factors may be zero only when one of the factors is zero, this equation will be true if either

$$2 \sin \theta - 1 = 0 \quad \text{or} \quad \sin \theta + 2 = 0.$$

From these conditions

$$\sin \theta = \tfrac{1}{2}, \quad \text{and therefore} \quad \theta = 30°, 150°,$$

or

$$\sin \theta = -2, \quad \text{which is impossible.}$$

Hence the permissible values of θ are $30°$ and $150°$.

Frequently, before algebraic methods can be used, the form of an equation must be changed by use of the identities already discussed in this chapter.   No general procedure can be given that will be successful in all cases.   In many equations which involve several different trigonometric functions, it is often expedient to write the equation so that only one function is involved.

Example 4.   Find all values of θ, $0° \leq \theta < 360°$, that satisfy the equation $\sin \theta = \csc \theta$.

To write the equation in terms of only one trigonometric function, substitute the identity $\csc \theta = 1/\sin \theta$.   Thus

$$\sin \theta = \frac{1}{\sin \theta}, \quad \text{and} \quad \sin^2 \theta = 1.$$

Transposing yields

$$\sin^2 \theta - 1 = 0,$$

and factoring

$$(\sin \theta - 1)(\sin \theta + 1) = 0.$$

If each factor is set equal to zero,

$$\sin \theta = 1, \quad \text{and therefore} \quad \theta = 90°,$$

$$\sin \theta = -1, \quad \text{and therefore} \quad \theta = 270°.$$

Thus $\theta = 90°, 270°$ are the only angles between $0°$ and $360°$ for which $\sin \theta = \csc \theta$.

In many cases, the method of solving a trigonometric equation must be suggested by the solver's intuition.   Each problem may require a new method of attack.

Example 5.   Find $0° \le \theta < 360°$ such that $\sin 2\theta = \cos \theta$.

Transposing gives

$$\sin 2\theta - \cos \theta = 0,$$

and substituting $2 \sin \theta \cos \theta = \sin 2\theta$ yields

$$2 \sin \theta \cos \theta - \cos \theta = 0.$$

Factoring the left-hand side, we get

$$\cos \theta (2 \sin \theta - 1) = 0.$$

Thus

$$\cos \theta = 0,$$

$$2 \sin \theta - 1 = 0, \quad \text{and therefore} \quad \sin \theta = \tfrac{1}{2},$$

from which

$$\theta = 90°, 270° \quad \text{and} \quad \theta = 30°, 150°.$$

Thus the solution to our problem is $\theta = 30°, 90°, 150°, 270°$.

Example 6.   Find $0° \le x < 360°$ such that $2 \sin x - 3 \cos x = -2$.

There are two methods by which this equation may be solved.

Method 1.   Substituting the identity $\sin x = \pm \sqrt{1 - \cos^2 x}$, we have

$$\pm 2\sqrt{1 - \cos^2 x} = 3 \cos x - 2.$$

Squaring both sides gives

$$4 - 4 \cos^2 x = 9 \cos^2 x - 12 \cos x + 4,$$

from which

$$13 \cos^2 x - 12 \cos x = 0.$$

Factoring, we get

$$\cos x(13 \cos x - 12) = 0,$$

whence

$$\cos x = 0, \qquad \cos x = \tfrac{12}{13} = 0.9231,$$

giving

$$x = 90°, 270° \quad \text{and} \quad x = 22.61°, 337.39°.$$

However, since both sides of the equation were squared, extraneous roots may have been introduced.   On checking, we find that

$$x = 22.61° \quad \text{and} \quad 270°$$

are the only roots that satisfy the original equation.

Method 2.   Since the left-hand side of the equation written as

$$2 \sin x + 3 \sin (x - 90°)$$

is the sum of two sine functions with the same period, the method of the previous section may be used to write this as a single sine function.   Hence $\omega = 1$, $A_1 = 2$, $A_2 = 3$, $\alpha_1 = 0°$, $\alpha_2 = -90°$, and hence

$$M = 2 \cos 0° + 3 \cos (-90°) = 2,$$

$$N = 2 \sin 0° + 3 \sin (-90°) = -3.$$

Thus for the sum function

$$A = \sqrt{M^2 + N^2} = \sqrt{4 + 9} = \sqrt{13},$$

$$\tan \alpha = \frac{-3}{2}, \quad \text{and therefore} \quad \alpha = 303.69°.$$

Substituting the sum function $\sqrt{13} \sin (x + 303.69°)$ in the left-hand side of the given equation for $2 \sin x - 3 \cos x$, we have

$$\sqrt{13} \sin (x + 303.69°) = -2.$$

Now, dividing both sides by $\sqrt{13}$ gives

$$\sin (x + 303.69°) = -\frac{2}{\sqrt{13}} \quad \text{or} \quad -\frac{2\sqrt{13}}{13} = -0.5547.$$

Thus the angle $x + 303.69° = 213.69°, 326.31°$, and hence

$$x = \begin{cases} 213.69° - 303.69° = -90° \quad (\text{or } 213.69° - 303.69° + 360° = 270°), \\ 326.31° - 303.69° = 22.62°. \end{cases}$$

Exercises

In the following exercises, find all the angles greater than or equal to 0° and less than 360° for which the given equation is true:

1. $\sin x = 0.6018.$

2. $\cos x = 0.9171.$

3. $3 \sin \alpha = -2.4606.$

4. $0.9 \tan \alpha = 1.6443.$

5. $3 \cos x + 4 \cos x = -5.6320.$

6. $0.2 \cos \alpha + 0.8 \sin (90° - \alpha) = 0.9406.$

7. $4.5 \cos \theta - 0.7458 = 3.2 \cos \theta.$

8. $\csc \theta = -3.8213.$

9. $\dfrac{0.6872}{\sec \beta} = 0.8130.$

10. $\dfrac{4 \cos \beta}{6.880} = -0.1640.$

11. $\sin 2x = 0.4206.$

12. $\cos 3x = 1.3204.$

13. $\cos \dfrac{x}{4} = 0.5304.$

14. $\tan \dfrac{2x}{3} = -1.7500.$

15. $\dfrac{\sin \dfrac{\alpha}{2}}{4} = -0.1003.$

16. $\dfrac{5 \cos 2.1\, t}{6} = 0.6837.$

17. $2 \sin 2x + 3 \sin 2x = 3.6024.$

18. $12 \cos 2x = 3.1862.$

19. $3 \csc \dfrac{\theta}{4} = 3.41.$

20. $2 \sin \dfrac{\theta}{2} = \cos \dfrac{\theta}{2}.$

21. $(\cos x + \tfrac{1}{2})(\cos x - \tfrac{1}{2}) = 0.$

22. $\sin^2 x - \tfrac{1}{9} = 0.$

23. $\sin^2 \alpha - 1 = 0.$

24. $\cos^2 \alpha - \tfrac{5}{16} = 0.$

25. $(3 \cos x - \sqrt{5})(3 \cos x + \sqrt{3}) = 0.$

26. $(\csc \theta + \sqrt{2})(\csc - 1) = 0.$

27. $\sin^2 \alpha + 0.8211 = 0.$

28. $\sin^2 \alpha - 0.8211 = 0.$

29. $2 \sin^2 \theta + \sqrt{2} \sin \theta = 0.$

30. $\cos \theta - \dfrac{1}{\cos \theta} = 0.$

31. $3 \csc^2 \alpha + 8 \csc \alpha = 3.$

32. $3 \sec^2 \alpha - 3 \sec \alpha = 6.$

33. $2 \sin^2 x + 5 \sin x = 3.$

34. $5 \tan^2 x + 14 = 17 \tan x.$

35. $2.8 - 4.0 \cos^2 \theta = 1.4 \sin \theta.$

36. $\cot 3\theta\,(2 \sin 2\theta - 1) = 0.$

37. $\sin \theta\,(\cot \theta + \tan \theta) - 2 = 0.$

38. $2 \cos \theta + \dfrac{5 \sin \theta}{\sec \theta} = 0.$

39. $3 \sin 4x \tan 2x - \tan 2x = 0.$

40. $\cos 4x - \cos 2x = 0.$

41. $3 \sin 2\alpha + 4 \cos 2\alpha = 6.$

42. $4 \cos 3\alpha - 2 \sin 3\alpha - 2.464 = 0.$

43. $2 \cos^2 \theta + \dfrac{\cos \theta}{\sec \theta} = 3 - \dfrac{3 \sin \theta}{\csc \theta}.$

44. $4 \sin \theta + 3 \sin \left(\theta + \dfrac{\pi}{2}\right) = 2.469.$

45. $\cos 2t + 5 \cos t + 3 = 0.$

46. $2.5 \cos 4x - 6 \cos 2x + 3.5 = 0.$

47. $\sin^2 \alpha - \cos^2 \alpha = -\sin \alpha.$

48. $3 \sin (\theta + 45°) + 4 \cos (\theta + 45°) = 3.$

49. $\cos \theta \cos \dfrac{\theta}{2} - \sin \theta \sin \dfrac{\theta}{2} = 0.$

50. $2 \sin^2 \dfrac{\theta}{2} + \cos \theta = 1.$

51. $4 \sin^2 x + 6 \cos x = 0.$

52. $\sqrt{6} \sin x - \sqrt{3} \cos x = 1.$

53. $\cos 3\theta - \cos \theta = 0.$

54. $\sin \theta - \sin \dfrac{\theta}{2} = 0.$

55. $\cos 2\alpha = \cos 4\alpha + 2 \sin 3\alpha.$

56. $2 \cos 2\alpha = \cos 5\alpha + \cos \alpha.$

57. $2 \sin^2 2x - 3 \sin 2x + 1 = 0.$

58. $5 \sin (x + 30°) + 8 \cos (x - 180°) - 2 \cos x = 7.$

10–6. The Inverse Trigonometric Functions

In algebra when a relation $y = f(x)$ is solved for x in terms of y, obtaining say, $x = g(y)$, then $g(y)$ is called the **inverse function** of $f(x)$. Conversely, $f(x)$ may be called the inverse of $g(y)$. For example, consider $y = 2x + 3$. This may be solved for x to give $x = \dfrac{y - 3}{2}$. Then $x = \dfrac{y - 3}{2}$ is called the inverse of $y = 2x + 3$, and $y = 2x + 3$ may be called the inverse of $\dfrac{y - 3}{2}$ (see Sec. 3–5.)

The equation

$$x = \sin y \qquad\qquad (1)$$

determines a value of x for every value of y. In other words, equation 1 gives x as a function of y. It can also be used to determine y as a function of x. For instance, if $x = \frac{1}{2}$ is given, the corresponding values of y are $\pi/6,\ 5\pi/6,\ 13\pi/6$, etc.

In this way y is a function of x when x and y are related by equation 1. In order to express this function, we say that

$$y \text{ is an angle whose sine is } x. \tag{2}$$

This statement is abbreviated by

$$y = \text{arcsin } x. \tag{3}$$

Thus the inverse function of equation 1 is given by equation 3. The relation 3 is preferably read by the defining phrase 2, but the phrases *y is equal to the arc sine of x* and *y is equal to the inverse sine of x* are also in common use.

Some textbooks use the notation $\sin^{-1} x$ in place of arcsin x. This notation, however, is not recommended because $\sin^{-1} x$ may be mistaken for $(\sin x)^{-1} = \dfrac{1}{\sin x}$.

Example 1.   Evaluate $\arcsin \dfrac{1}{2}$.

If we set $y = \arcsin \dfrac{1}{2}$, by definition this means that

$$\sin y = \frac{1}{2},$$

and hence

$$y = \frac{\pi}{6},\ \frac{5\pi}{6},\ \text{etc.,}$$

or

$$\arcsin \frac{1}{2} = \frac{\pi}{6},\ \frac{5\pi}{6},\ \text{etc.}$$

Example 2.   Evaluate arcsin 0.6157.

As above, set $\theta = $ arcsin 0.6157.   Then by the definition,

$$\sin \theta = 0.6157.$$

Therefore

$$\theta = 38°,\ \ 142°,\ \ \text{etc.,}$$

or

$$\text{arcsin } 0.6157 = 38°,\ \ 142°,\ \ \text{etc.}$$

The same notations apply to the other five trigonometric functions when written in the inverse form.   Thus the accompanying table lists, in the two right-hand columns, equivalent notations for expressing the relation given in the left-hand column.

Relation Expressed	Direct Notation	Inverse Notation
y is the angle whose sine is x	$x = \sin y$	$y = \arcsin x$
y is the angle whose cosine is x	$x = \cos y$	$y = \arccos x$
y is the angle whose tangent is x	$x = \tan y$	$y = \arctan x$
y is the angle whose cotangent is x	$x = \cot y$	$y = \operatorname{arccot} x$
y is the angle whose secant is x	$x = \sec y$	$y = \operatorname{arcsec} x$
y is the angle whose cosecant is x	$x = \csc y$	$y = \operatorname{arccsc} x$

From examples 1 and 2 it is evident that the inverse function

$$y = \arcsin x \tag{3}$$

has for one value of x many values of y. Although to a given angle y there corresponds only one value x for $\sin y$, to a given sine value x there correspond infinitely many angles y. This is also true for the other functions. It is convenient to select one of the values from the multiple values of $\arcsin x$ as the **principal value.** We denote this principal value by capitalizing the initial letter of the function, e.g., *Arcsin x* or *Sin*$^{-1}$ *x*. Similarly we select a single principal value from the multiple values of each of the other inverse trigonometric functions.

The principal values of $\arcsin x$, $\arccos x$, $\arctan x$, and $\operatorname{arccot} x$ are defined by the following table.

Function	Principal Value Lies in the Interval
$y = \arcsin x$	$-\dfrac{\pi}{2} \leq y \leq \dfrac{\pi}{2}$
$y = \arccos x$	$0 \leq y \leq \pi$
$y = \arctan x$	$-\dfrac{\pi}{2} < y < \dfrac{\pi}{2}$
$y = \operatorname{arccot} x$	$0 < y < \pi$

That the restriction of y to the interval given in this table for each function defines a single value can be seen readily from the graphs of the trigonometric functions given in Chapter 5. Since the principal values of $\operatorname{arcsec} x$ and $\operatorname{arccsc} x$ are seldom used, they will not be defined here.

Example 3. Find the values of

$$y = \operatorname{Arcsin} \tfrac{1}{2}, \quad y - \operatorname{Arctan} 1, \quad y = \operatorname{Arccos}\left(-\frac{\sqrt{3}}{2}\right).$$

The notation indicates that principal values are desired. Rewriting each of the functions in direct notation and finding the angle in the proper interval, we have:

$$y = \text{Arcsin } \tfrac{1}{2}. \qquad \sin y = \tfrac{1}{2}, \qquad y = \frac{\pi}{6};$$

$$y = \text{Arctan } 1, \qquad \tan y = 1, \qquad y = \frac{\pi}{4};$$

$$y = \text{Arccos}\left(-\frac{\sqrt{3}}{2}\right), \qquad \cos y = -\frac{\sqrt{3}}{2}, \qquad y = \frac{5\pi}{6}.$$

Example 4. Find the values of arcsin 0.
The notation indicates that all the possible values are desired. Setting $y = $ arcsin 0, and rewriting it in the form

$$\sin y = 0,$$

we have

$$y = 0, \pi, 2\pi, 3\pi, \text{ etc.}$$

Exercises

Solve for y:

1. $x = \cos y.$ **2.** $z = 2 \sin y.$ **3.** $x = \tan y.$

4. $x = \cos 2y.$ **5.** $2z = \tan 2y.$ **6.** $\dfrac{x}{3} = \sec 3y.$

7. $z = \sin (y + \pi).$ **8.** $\dfrac{m}{n} = \cos \dfrac{y}{a}.$ **9.** $\tan y = \dfrac{x + 2}{3}.$

10. $\sec\left(y + \dfrac{\pi}{2}\right) = \dfrac{x - 2}{3}.$ **11.** $\dfrac{m - n}{x} = \sin 3y.$

12. $a = b \cos y.$ **13.** $\tan 4y = 4x.$

14. $z = A \sec 3y.$ **15.** $A \sin y = B.$

16. $x = 2 \cot \dfrac{3y}{2}.$ **17.** $2 \cos \dfrac{y}{3} = 3x - 2.$

18. $\dfrac{x}{2} = 2 \sin\left(\dfrac{y}{2} + 1\right).$ **19.** $M + 2N = \tfrac{1}{2} \cos (y - 1).$

20. $\tfrac{1}{2} \tan \dfrac{3(y + 1)}{2} = \dfrac{a}{b + c}.$ **21.** $\dfrac{2a - b}{c} = \sec \dfrac{7(y - 1)}{3}.$

Solve for x:

22. $\arcsin x = \alpha.$ **23.** $\arccos x = y.$ **24.** $\arctan x = \dfrac{\beta}{2}.$

25. $\theta = \arccos \dfrac{7ax}{3}.$ **26.** $\omega t + \theta = \arcsin \dfrac{2x}{\sqrt{2I_m}}.$ **27.** $\text{arcsec } 2mx = \dfrac{\theta}{3}.$

28. $\arcsin \dfrac{x}{y} = \dfrac{3\theta}{2}.$ **29.** $\dfrac{9M - 2}{N} = \arctan \dfrac{x}{4}.$ **30.** $\dfrac{\arccos x}{7} = 7\theta.$

31. $\alpha = \arcsin \dfrac{Ax}{B}$.

32. $\dfrac{a^3}{4} \arccos \dfrac{a}{x} = \dfrac{b}{y}$.

33. $\dfrac{\theta - 1}{2} = \arctan (x-y)$.

34. $3\alpha = 2 \operatorname{arccot} \dfrac{7x}{5}$.

35. $377t = 30° + \arcsin \dfrac{2x}{\sqrt{2}\,E_m}$.

36. $\theta = \arctan \dfrac{x}{A}$.

37. $\theta = \arctan \left(\dfrac{-x}{A} \right)$.

38. $\arcsin \dfrac{x_0 - x}{M} = \phi$.

39. $\phi = \arccos \dfrac{1 - x}{N} + y$.

40. $\theta = \dfrac{1}{\sqrt{3}(a^2 + b^2)} \arctan \dfrac{2x}{a - b}$.

41. $\theta = \arcsin \left(\dfrac{5x}{3} - \dfrac{\pi}{2} \right)$,

Evaluate the following by first rewriting as direct trigonometric functions.   Find six possible values of the unknown if principal values are not indicated.

42. $y = \operatorname{Arcsin} \tfrac{1}{2}$.

43. $y = \operatorname{Arctan} 1$.

44. $\theta = \operatorname{arccot} (-1)$.

45. $\theta = \arcsin (-\tfrac{1}{2})$.

46. $y = \operatorname{arccsc} 2$.

47. $y = \operatorname{Arccot} \sqrt{3}$.

48. $\beta = \operatorname{arcsec} \dfrac{2}{\sqrt{3}}$.

49. $\alpha = \operatorname{Arcsin} \dfrac{\sqrt{2}}{2}$.

50. $\theta = \operatorname{Arcsin} 1$.

51. $\theta = \arccos 0$.

52. $y = \arctan \left(-\dfrac{\sqrt{3}}{3} \right)$.

53. $y = \arccos \left(-\dfrac{\sqrt{3}}{2} \right)$.

54. $\alpha = \arcsin 0$.

55. $\alpha = \operatorname{Arctan} 0$.

56. $x = \arccos (-1)$.

57. $2x = \arccos (-1)$.

58. $\theta + \dfrac{\pi}{6} = \operatorname{Arcsin} 1$.

59. $\dfrac{\theta}{2} - \dfrac{2\pi}{3} = \operatorname{Arccos} \tfrac{1}{2}$.

60. $5 \left(y - \dfrac{\pi}{8} \right) = \operatorname{Arctan} 1$.

61. $y = \pi + 4 \operatorname{Arcsin} (-\tfrac{1}{2})$.

62. $\theta - \dfrac{\pi}{3} = \tfrac{1}{3} \operatorname{Arcsin} \left(-\dfrac{\sqrt{3}}{2} \right)$.

63. $y + 2\pi = 6 \operatorname{Arccos} \left(-\dfrac{\sqrt{2}}{2} \right)$.

64. $y = \operatorname{Arccos} 0.4741$.

65. $y = \operatorname{Arcsin} 0.3272$.

66. $\theta = \arctan (-2.888)$.

67. $\theta = \operatorname{arccsc} 1.626$.

68. $2y = \operatorname{arcsec} (-1.103)$.

69. $3y = \operatorname{Arccos} 0.9207$.

70. $\arctan 3.271 = \dfrac{\theta}{2} - 31°$.

71. $\arcsin 0.8306 = 3\theta$.

72. $\operatorname{Arccot} (-0.7692) = \dfrac{5\theta}{4}$.

73. $4 \operatorname{Arccos} \iota \quad `88 = 16\theta$.

74. $\operatorname{Arccos} 0.7315$.

75. $\operatorname{arccsc} 1.408$.

76. $\operatorname{arcsec} (-3.446)$.

77. $\operatorname{Arctan} 42.03$.

78. $33° - 2$ Arcsin (-0.9988).   **79.** 2.6 Arccos $0.9504 - 21°$.

Evaluate the following expressions.

80. Arcsin $1 -$ Arccos $\frac{1}{2}$.   **81.** 2 Arcsin $\left(-\dfrac{\sqrt{3}}{2} \right) +$ Arccos $\dfrac{\sqrt{2}}{2}$.

82. Arctan $0.6201 - 2$ Arccos (-0.7005).

83. sin (Arcsin 0.3371).

84. cos [Arccos (-0.1959)].   **85.** tan [Arccos (-0.1959)].

86. sec [Arccos (-0.1959)].   **87.** cot [Arccot (-1.5)] $+ 3$.

88. 27 Arccot $(-16.500) - 14$ Arcsin 0.02062.

89. sin (Arctan $\sqrt{3} -$ Arccot $\sqrt{3}$).   **90.** cos [Arcsin $\frac{1}{2} +$ arcsec (-2)].

91. tan $[(\frac{1}{2}$ Arcsin $\frac{2}{3}) - 112°]$.   **92.** sin $\left[\dfrac{\pi}{4} + \frac{1}{3}\ \text{arcsec}\ \frac{7}{5} \right]$.

93. 3 Arcsin $\frac{3}{7} + 4$ (Arccos $\frac{7}{9}$).   **94.** Arccot $\frac{5}{3} +$ Arctan $(-\frac{3}{2}) - \dfrac{\pi}{4}$.

Progress Report

Trigonometric functions and their graphs were considered in Chapters 4 and 5. In this chapter we studied two kinds of trigonometric equations:
1. Identities.
2. Conditional equations.

The identities were used to simplify or change the form of many trigonometric expressions, and their usefulness in applications was illustrated. Several common types of conditional trigonometric equations were solved.

Finally, inverse trigonometric functions and their principal values were discussed.

11

The Oblique Triangle

In Chapter 4 we discussed the use of trigonometric functions in finding unknown sides and angles of right triangles. In this chapter we will develop methods for finding unknown sides and angles of triangles that are not right triangles; such triangles are called **oblique** triangles. These methods will be based on the law of sines and the law of cosines, formulas that apply to all triangles. The chapter also considers the application of these methods to vector problems.

11–1. The Law of Sines

Throughout this chapter we shall label the sides of a triangle a, b, and c, the opposite angles will be labeled, respectively α, β, and γ, and the vertices of these angles will be labeled, respectively A, B, and C. This is done in Fig. 11–1.

The law of sines states that *in any triangle the sides are proportional to the sines of the opposite angles.* Expressed symbolically,

$$\frac{a}{\sin \alpha} = \frac{b}{\sin \beta} = \frac{c}{\sin \gamma}. \tag{1}$$

This law is readily verified. Consider any oblique triangle ABC (Fig. 11–1). In the figure we have the only two possible types of oblique triangles, the first with three acute angles, the second with one obtuse and two acute angles. The steps in the proof below are valid for both triangles.

From the vertex C draw the altitude h, denoting by D the point at which h meets the opposite side or its projection. Then

$$\sin \alpha = \frac{h}{b} \quad \text{or} \quad h = b \sin \alpha,$$

$$\sin \beta = \frac{h}{a} \quad \text{or} \quad h = a \sin \beta.$$

From these two equalities

$$a \sin \beta = b \sin \alpha.$$

Dividing both sides by the quantity $\sin \alpha \sin \beta$, we obtain

$$\frac{a}{\sin \alpha} = \frac{b}{\sin \beta}. \tag{2}$$

Similarly, the relation

$$\frac{b}{\sin \beta} = \frac{c}{\sin \gamma} \tag{3}$$

may be verified by using the altitude from vertex A to the opposite side. When equations 2 and 3 are combined, the law of sines as stated in equation 1 is obtained.

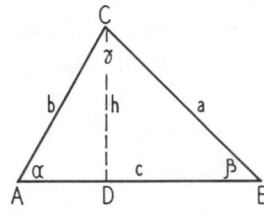

FIG. 11–1

11–2. The Law of Cosines

The law of cosines states that *in any triangle the square of any side is equal to the sum of the squares of the other two sides less twice the product of those sides and the cosine of their included angle.* Expressed symbolically, we have three forms:

$$a^2 = b^2 + c^2 - 2bc \cos \alpha, \tag{1}$$

$$b^2 = a^2 + c^2 - 2ac \cos \beta, \tag{2}$$

$$c^2 = a^2 + b^2 - 2ab \cos \gamma. \tag{3}$$

Consider any oblique triangle ABC (Fig. 11–2). From the vertex C in each triangle draw the altitude h, denoting by D the point at which h meets the opposite side or its projection.

Now for both cases

$$a^2 = h^2 + (DB)^2,$$

and

$$h^2 = b^2 - (DA)^2.$$

Substituting for h^2 from the second relation into the first, we get

$$a^2 = b^2 - (DA)^2 + (DB)^2. \tag{4}$$

Equation 4 applies to both triangles. The remainder of the proof must be carried out separately for each figure.

Fig. 11–2a	Fig. 11–2b
$DB = c - DA$	$DB = c + DA$

Substituting for DB in (4),

$$a^2 = b^2 - (DA)^2 + (c - DA)^2$$

or

$$a^2 = b^2 + c^2 - 2c(DA).$$

From the figure,

$$DA = b \cos \alpha,$$

whence

$$a^2 = b^2 + c^2 - 2bc \cos \alpha.$$

Substituting for DB in (4),

$$a^2 = b^2 - (DA)^2 + (c + DA)^2$$

or

$$a^2 = b^2 + c^2 + 2c(DA).$$

From the figure,

$$DA = b \cos (\angle DAC)$$
$$= b \cos (180° - \alpha)$$
$$= -b \cos \alpha,$$

whence

$$a^2 = b^2 + c^2 - 2bc \cos \alpha.$$

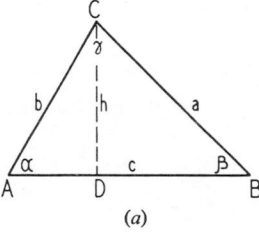 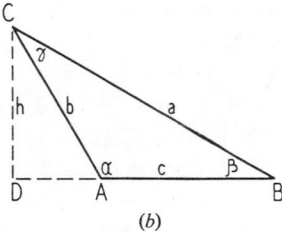

(a)   (b)

Fig. 11–2

Thus in both cases we have the result given in equation 1.   Relations 2 and 3 may be obtained in like manner by using other altitudes of the triangles.

11–3. The Classification of Oblique-Triangle Problems

A triangle has three sides and three angles, a total of six parts.   When any three are given, at least one of which is a side, the triangle can be constructed by means of a ruler and compass.   When three angles are given, the triangle is not uniquely determined.   For more details see the appendix.

In plane geometry, the ruler and compass serve as the tools for constructing a triangle if any three parts, at least one of which is a side, are given.   From the constructed triangle, we can find the missing parts by measurement.   In trigonometry the law of sines and the law of cosines are the tools for computing the missing parts of the triangle.   Thus:

1. In geometry we *construct*; in trigonometry we *compute*.

2. The tools of geometry are *ruler and compass*; the tools of trigonometry are *formulas*.

To use the law of sines, any two of the three ratios in equation 1 of Sec. 11–1 may be selected to give an equation; for example,

$$\frac{a}{\sin \alpha} = \frac{c}{\sin \gamma}.$$

Any such equation contains four parts of the triangle, in this case a, c, α, γ. Therefore, if any three of these parts are known, the fourth may be found by solving the equation. In like manner, each form of the cosine law involves four parts of the triangle. For example,

$$b^2 = a^2 + c^2 - 2ac \cos \beta$$

involves b, a, c, β. Here again, if any three of these parts are known, the fourth may be found by solving the equation. Thus, in general, if three parts of a triangle are known, at least one of which is a side, and it is desired to find a fourth part, the proper procedure is to select a form of the sine law or the cosine law which contains the four parts mentioned. After substitution of the known values in the pertinent equation, the unknown element can be found.

In order to systemize the procedures for solving triangles, we shall classify all oblique-triangle problems into the following four cases.

Case 1. Given two angles and any side.

Case 2. Given two sides and the included angle.

Case 3. Given two sides and an angle opposite one.

Case 4. Given three sides.

In some problems the given parts may be of such a nature that no triangle is possible. In other problems the given parts determine one or two triangles. In many instances these possibilities can be recognized after a sketch of the triangle has been carefully made with a ruler and a protractor. Hence, before attempting to solve a problem by means of trigonometry, it is advisable to construct the possible figure or figures from the given parts. Finer criteria, to be used when a sketch is unsatisfactory, will be given as the various cases are discussed.

The computations may be performed in any convenient manner. If only two- or three-figure accuracy is required, a slide rule having trigonometric scales is most satisfactory. For greater accuracy, logarithms should be used. The computations in the examples of this chapter will be performed with logarithms.

There are various methods of checking the accuracy of computations of this kind. Measurement of the unknown sides and angles in the carefully drawn sketch is usually enough of a check, although it will detect only major errors. Other more accurate methods of checking these computations are discussed in books on trigonometry, but they will not be considered here.

11–4. Case 1: Given Two Angles and Any Side

This case may be summarized in the following way.

(*a*) *Number of triangles*:

(1) *None if the sum of the two given angles is greater than or equal to* 180°.

(2) *One if the sum of the two given angles is less than* 180°.

(*b*) *Method of solution*: *Law of sines*.

The number of triangles. Since by one of the fundamental theorems about triangles (see the appendix), the sum of the three angles of a triangle is 180°, no triangle can have two angles with a sum greater than 180°. If a triangle can be thought of as having an angle of 0°, it can be seen that such an angle will require the two sides adjacent to it to coincide, and that therefore the geometric figure representing such a "triangle" is a segment. Thus in this book we will require that each angle of a triangle be greater than zero in order for the geometric figure to be given the name triangle. It follows that when the sum of two given angles is 180°, there can be no triangle with these angles, since the third angle of the triangle would have to be 0°.

When two angles and a side of a triangle are given, the third angle can be obtained by subtracting the sum of the given angles from 180°, thus making all three angles known immediately. In particular, the two angles adjacent to the given side are known and have a sum less than 180°. It follows that the lines along which the two unknown sides lie are not parallel (why?) and therefore intersect in a point that is the vertex of the third angle and that determines the lengths of the unknown sides. Thus a given side and any two given angles with sum less than 180° determine a triangle. It can be shown that this triangle is unique and that therefore these data determine exactly one triangle.

The method of solution. When the sum of the two given angles is less than 180°, the difference between 180° and this sum gives the third angle. With the three angles of the triangle known, each of the unknown sides can be found by means of the sine-law relation involving an unknown side and the known side.

Example. Given an oblique triangle in which $\alpha = 77.00°$, $\beta = 72.00°$, and $c = 10.20$. Find γ, a, b.

The triangle is sketched in Fig. 11–3. To find γ we have at once

$$\gamma = 180.00° - (77.00° + 72.00°) = 180.00° - 149.00°$$

whence

$$\gamma = 31.00°.$$

Now to find b and a we use

$$\frac{b}{\sin \beta} = \frac{c}{\sin \gamma} \quad \text{and} \quad \frac{a}{\sin \alpha} = \frac{c}{\sin \gamma}$$

whence

$$b = \frac{c \sin \beta}{\sin \gamma} = \frac{10.20 \sin 72.00°}{\sin 31.00°},$$

$$a = \frac{c \sin \alpha}{\sin \gamma} = \frac{10.20 \sin 77.00°}{\sin 31.00°}.$$

Performing the computations by logarithms we have the following:

Numbers	Logarithms	Numbers	Logarithms
10.20	1.0086	10.20	1.0086
sin 72.00°	9.9782 − 10	sin 77.00°	9.9887 − 10
10.20 sin 72.00°	10.9868 − 10	10.20 sin 77.00°	10.9973 − 10
sin 31.00°	9.7118 − 10	sin 31.00°	9.7118 − 10
b	1.2750	a	1.2855
$b = 18.83$		$a = 19.30$	

That these results are correct can be checked roughly by measuring the figure.

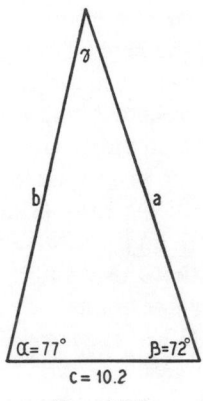

FIG. 11–3

When, as in this example, lengths are given accurate to four significant figures, and angles to the nearest hundredth of a degree, computation by four-place logarithms gives results of the same degree of accuracy. When lengths are given accurate to three figures and angles to the nearest tenth of a degree, the logarithmic computations can be carried out with a four-place table precisely as in the example above and the answers rounded off to the same degree of accuracy. Thus in the example, if the data given were $\alpha = 77.0°$, $\beta = 72.0°$, and $c = 10.2$, the work would be the same, but the result would be rounded off and given as $\gamma = 31.0° \; b = 18.8$, and $a = 19.3$. A similar remark can be made if the data are given to two significant figures and the nearest degree. The slide rule can, of course, be used if the desired degree of accuracy is three significant figures and the nearest tenth of a degree, or less (see Sec. 4–15).

Exercises

Solve the following triangles for the three unknown parts. Give results that are as accurate as the data will permit, assuming that they are accurate in accordance with the agreements about significant digits.

1. $\alpha = 32°, \beta = 76°, a = 24.$

2. $\alpha = 53°, \beta = 45°, b = 81.$

3. $\beta = 66.0°, \gamma = 45.5°, b = 40.0$

4. $\alpha = 82.4°, \gamma = 37.3°, c = 40.0.$

5. $\beta = 8.4°, \alpha = 6.2°, b = 12.6.$

6. $\gamma = 74.7°, \beta = 69.3°, a = 90.2.$

7. $\alpha = 50.0°, \gamma = 33.5°, a = 41.6.$

8. $\alpha = 8.6°, \beta = 7.3°, c = 39.6.$

9. $\gamma = 53.67°, \beta = 84.00°, a = 13.10.$

10. $\beta = 42.04°, \gamma = 19.70°, a = 6.056.$

In each of the following triangles find the unknown part listed. Give results that are as accurate as the data will permit.

11. $\alpha = 37°, \gamma = 74°, b = 46, \beta = \quad .$

12. $\beta = 124.3°, \gamma = 25.6°, c = 18.2, b = \quad .$

13. $\gamma = 77.2°, \beta = 89.4°, b = 9.58, c = \quad .$

14. $\beta = 18.2°, \gamma = 130.5°, b = 31.3, a = \quad .$

15. $\alpha = 128.3°, \gamma = 11.8°, a = 2.89, b = \quad .$

16. $\alpha = 128.3°, \gamma = 11.8°, b = 406, c = \quad .$

17. $\gamma = 165.4°, \alpha = 6.2°, c = 21.7, a = \quad .$

18. $\beta = 69.6°, \gamma = 80.3°, a = 341.5, b = \quad .$

19. $\beta = 126.00°, \alpha = 32.35°, c = 137.9, b = \quad .$

20. $\alpha = 84.42°, \gamma = 65.83°, b = 0.3603, a = \quad .$

21. State the law of sines, assuming that one of the angles of the triangle is a right angle.

22. State the law of cosines, assuming that one of the angles of the triangle is a right angle.

11–5. Case 2: Given Two Sides and the Included Angle

In this case we have the following summary:

(a) *Number of triangles*: *One, if the given angle is less than* 180°; *none otherwise.*

(b) *Method of solution*: *Law of cosines, followed by the law of sines.*

The number of triangles.   Since each angle of a triangle must be greater than zero and the sum of the three angles is 180°, every angle of a triangle is less than 180°. Thus if two sides and an angle are given there can be a triangle with these two sides including the given angle only if the given angle is less than 180°.   It is obvious that, if the angle is less than 180°, a triangle can be formed by drawing a line connecting the ends of the two sides that are not joined at the vertex of the given angle.   It can be shown that this triangle is unique.

   The method of solution.   The third side can be found by the law of cosines.   Then with the three sides known the law of sines can be used to determine the two unknown angles.   For example, if a, b, and γ are known, c can be found by the cosine law in the form

$$c^2 = a^2 + b^2 - 2ab \cos \gamma.$$

Knowing a, c, and γ, we can find α by the sine law, or, knowing b, c, and γ, we can find β by the sine law.   Of course, α and β can also be found by using other forms of the cosine law, but the sine law is more convenient for computations.   In general, whenever either the law of sines or the law of cosines can be used, we shall employ the former, for the law of sines is much better adapted for logarithmic or slide-rule computations.   An advantage of computing *both* α and β by the sine law is that the relation $\alpha + \beta + \gamma = 180°$ can then be used as a check on the accuracy of the computations.

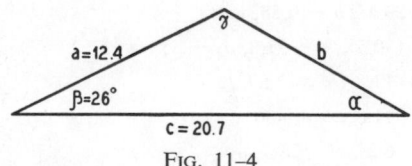

FIG. 11–4

   Example.   Given an oblique triangle in which $a = 12.4$, $c = 20.7$, $\beta = 26.0°$. Find α, γ, b.
   First we make an accurate sketch (Fig. 11–4).   Using the law of cosines to find b, we write,

$$b^2 = a^2 + c^2 - 2ac \cos \beta.$$

The computation carried out to three significant numbers gives

$$a^2 = 12.4^2 = 154, \qquad c^2 = 20.7^2 = 428.$$

The term $2ac \cos \beta$ is computed by logarithms:

Numbers	Logarithms
2	0.3010
$a = 12.4$	1.0934
$c = 20.7$	1.3160
$\cos \beta = \cos 26.0°$	$9.9537 - 10$
$2ac \cos \beta = 461$	2.6641

From the computed values

$$b^2 = 154 + 428 - 461 = 121, \qquad b = 11.0.$$

To find α and γ we use the law of sines.

$$\frac{a}{\sin \alpha} = \frac{b}{\sin \beta}, \qquad \frac{c}{\sin \gamma} = \frac{b}{\sin \beta},$$

whence

$$\sin \alpha = \frac{a \sin \beta}{b} = \frac{12.4 \sin 26.0°}{11.0},$$

$$\sin \gamma = \frac{c \sin \beta}{b} = \frac{20.7 \sin 26.0°}{11.0}.$$

The corresponding logarithmic computation gives:

Numbers	Logarithms		Numbers	Logarithms
12.4	1.0934		20.7	1.3160
sin 26.0°	9.6418 − 10		sin 26.0°	9.6418 − 10
12.4 sin 26.0°	10.7352 − 10		20.7 sin 26.0°	10.9578 − 10
11.0	1.0414		11.0	1.0414
sin α	9.6938 − 10		sin γ	9.9164 − 10
α = 29.6°			γ = 124.4°	

In finding γ from log sin γ, there is some question whether γ is acute or obtuse, that is, whether $\gamma = 55.6°$ or $\gamma = 180° - 55.6° = 124.4°$, since both have the same sine and log sine. However, Fig. 11–4 shows that γ is obtuse.
To check, $\alpha + \beta + \gamma = 29.6° + 26.0° + 124.4° = 180.0°$, whence the computations are accurate.

When the angle in question is very nearly a right angle, it might be difficult to determine from the figure whether to choose the acute or obtuse value of the angle. In such cases, *solve for the angle opposite the shortest side first. It will always be an acute angle.*

Exercises

Solve the following triangles for the three unknown parts, giving results that are as accurate as the data will permit:

1. $\alpha = 21°, b = 30, c = 30.$ **2.** $\beta = 36°, a = 14, c = 20.$

3. $\alpha = 75°, b = 5.0, c = 4.1.$ **4.** $\gamma = 54°, b = 78, a = 37.$

5. $\gamma = 23°, a = 76, b = 108.$ **6.** $\alpha = 66.2°, c = 14.5, b = 30.4.$

7. $\beta = 89.4°, c = 9.34, a = 2.22.$ **8.** $\gamma = 27.3°, a = 485, b = 702.$

9. $\alpha = 101.68°, b = 940.4, c = 1314.$ **10.** $\beta = 89.21°, c = 74.03, a = 50.84.$

In each of the following triangles determine the fourth part listed, giving results that are as accurate as the data will permit:

11. $\alpha = 32°, b = 4.8, c = 3.5, a = \quad .$

12. $\beta = 54°, a = 16, c = 48, b =$.
13. $\alpha = 81.4°, b = 138, c = 92, a =$.
14. $\gamma = 27.9°, a = 0.018, b = 0.104, \alpha =$.
15. $\gamma = 53.8°, a = 0.843, b = 1.121, c =$.
16. $\beta = 42.8°, c = 742.4, a = 605.7, \gamma =$.
17. $\alpha = 128.3°, b = 2.35, c = 0.749, \beta =$.
18. $\gamma = 83.5°, b = 1604, a = 1370, \alpha =$.
19. $\beta = 70.83°, a = 0.1852, c = 0.4556, b =$.
20. $\alpha = 36.62°, b = 36.74, c = 2026, \gamma =$.

11–6. Case 3: Given Two Sides and an Angle Opposite One

In this case the given parts may form two, one, or no triangles, depending on the relative size and position of the parts. When a figure will not show clearly the number of triangles possible, the table given below may be used. The case in which there are two triangles is often referred to as the **ambiguous case**.

Consider a triangle in which a, b, and α are given and the problem is to determine the other parts. Using the law of sines to find angle β, we have

$$\sin \beta = \frac{b}{a} \sin \alpha.$$

We shall use this expression for $\sin \beta$ to enumerate several cases arising from this problem.

(a) *Number of triangles.* Assuming that a, b, and α are given, we have the following possibilities:

When $\alpha \geq 90°$		When $\alpha < 90°$			
If $a > b$, there is *one* triangle and β is acute.	If $a \leq b$, there is *no* triangle.	If $a \geq b$, there is *one* triangle and β is acute.	If $a < b$, and		
			If $\sin \beta < 1$, there are two triangles, β being acute in one and obtuse in the other.	If $\sin \beta = 1$, there is one triangle and $\beta = 90°$.	If $\sin\beta > 1$, there is no triangle.

(*b*) *Method of solution*: *Law of sines.*   When parts other than a, b, α are given, the table must be interpreted with α replaced by the given angle, a by the side opposite, and b by the second given side.

Number of triangles.   We shall illustrate geometrically the various possibilities given in the above table for $\alpha < 90°$.   A similar discussion for the case $\alpha \geq 90°$ is left for the student.

The construction must be performed in the following way (Fig. 11–5).

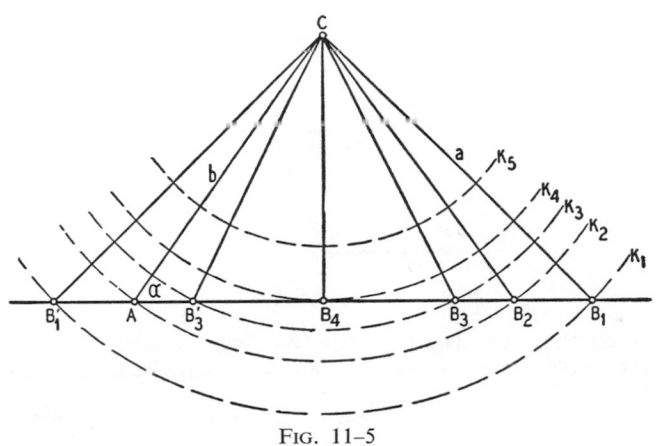

Fig. 11–5

1. Draw the given angle α.
2. Lay off length b on one side of α.
3. Draw an arc of radius a from the end point C of b.   The result depends on the relative lengths of a and b.   The different possibilities are plotted in Fig. 11–5 in order to indicate clearly how the number of solutions depends on the relative sizes of a and b.

The case $a > b$ is represented by the circle K_1.   This circle meets the horizontal side of the angle α in B_1 and B_1'.   The triangle AB_1C is the only triangle formed with the parts a, b, α.

If the length a is chosen so that $a = b$, the circle K_2 is obtained.   In this case, the one triangle AB_2C is formed by the given parts.

If $a < b$, three different cases are possible, corresponding to the circles K_3, K_4, and K_5.   The circle K_3 meets the horizontal side of α in the two points B_3 and B_3', and the two triangles AB_3C and $AB_3'C$ are formed from the given parts.   If a is sufficiently small, a circle like K_5 is found that does not meet the horizontal side of α, and there is no triangle.   There is also an intermediate case in which the circle has only a contact point with the straight line.   As shown by the circle K_4, only one triangle (triangle AB_4C) is obtained in this case.

Method of solution.   When two sides and an angle opposite one are given, the remaining parts of the triangle can be computed by the use of the law of sines, as illustrated in the following examples.

Example 1. Solve the triangle $\beta = 44°$, $b = 17$, $a = 10$, assuming that a has two significant digits.

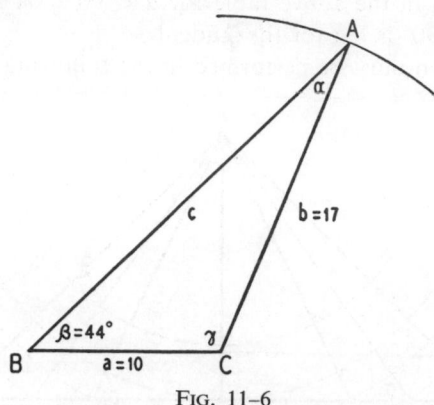

FIG. 11–6

From a sketch of the triangle (Fig. 11–6) we can conclude that only one triangle is possible and α is acute. To find α, we have by the law of sines

$$\sin \alpha = \frac{a}{b} \sin \beta = \frac{10}{17} \sin 44°.$$

Using logarithmic computations, we have

Numbers	Logarithms
10	1.0000
sin 44°	9.8418 − 10
10 sin 44°	10.8418 − 10
17	1.2304
sin α	9.6114 − 10

$$\alpha = 24°$$

To find the angle γ we have

$$\gamma = 180° - (44° + 24°) = 180° - 68°$$

whence

$$\gamma = 112°.$$

Finally, to obtain c we write

$$c = \frac{b \sin \gamma}{\sin \beta} = \frac{17 \sin 112°}{\sin 44°}.$$

Numbers	Logarithms
17	1.2304
sin 112°	9.9672 − 10
17 sin 112°	11.1976 − 10
sin 44°	9.8418 − 10
c	1.3558

$$c = 23.$$

Example 2.   Solve the triangle $\gamma = 52°$, $c = 5.1$, $a = 8.2$.

From the sketch (Fig. 11–7), we can conclude that there is no solution.   We shall now arrive at this conclusion computationally with the aid of the table of various cases given above.   In this example, $\gamma < 90°$ and $c < a$, and so it is necessary to evaluate

$$\sin \alpha = \frac{a}{c} \sin \gamma = \frac{8.2}{5.1} \sin 52° = 1.27 > 1,$$

which shows that there is no solution and no such triangle is possible.

Example 3.   Given $\gamma = 30.02°$, $c = 5.123$, $b = 10.04$.   Find the other parts of the triangle.

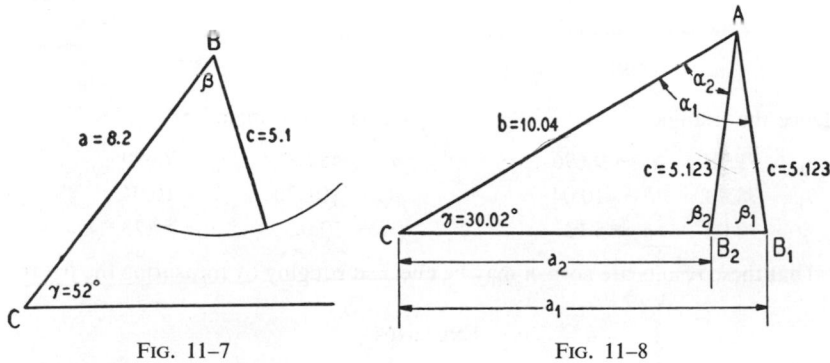

FIG. 11–7   FIG. 11–8

Making the sketch (Fig. 11–8), we find that there are two such triangles. Using the law of sines we have

$$\sin \beta = \frac{b}{c} \sin \gamma = \frac{10.04}{5.123} \sin 30.02°.$$

Computing:

Numbers	Logarithms
10.04	1.0017
sin 30.02°	9.6993 − 10
10.24 sin 30.02°	10.7010 − 10
5.123	0.7095
sin β	9.9915 − 10

Since there are two triangles, there are two angles β such that $\log \sin \beta = 9.9915 - 10$.

$\beta_1 = 78.70°$ $\beta_2 = 101.30°$

For the third angle α_1 we have   For the third angle α_2 we have

$\alpha_1 = 180.00° - (\gamma + \beta_1)$ $\alpha_2 = 180.00° - (\gamma + \beta_2)$

$\quad = 180.00° - 108.72°$ $\quad = 180.00° - 131.32°$

$\alpha_1 = 71.28°.$ $\alpha_2 = 48.68°.$

Finally, to find a_1 we write

$$a_1 = \frac{c \sin \alpha_1}{\sin \gamma} = \frac{5.123 \sin 71.28°}{\sin 30.02°}.$$

Finally, to find a_2 we write

$$a_2 = \frac{c \sin \alpha_2}{\sin \gamma} = \frac{5.123 \sin 48.68°}{\sin 30.02°}.$$

Computing:

Numbers	Logarithms
5.123	0.7095
$\sin 71.28°$	$9.9764 - 10$
5.123 sin 71.28°	$10.6859 - 10$
$\sin 30.02°$	$9.6993 - 10$
a_1	0.9866

$$a_1 = 9.696$$

Computing:

Numbers	Logarithms
5.123	0.7095
$\sin 48.68°$	$9.8757 - 10$
5.123 sin 48.68°	$10.5852 - 10$
$\sin 30.02°$	$9.6993 - 10$
a_2	0.8859

$$a_2 = 7.690$$

Hence the triangle is:

$\alpha_1 = 71.28°, \quad a_1 = 9.696$
$\beta_1 = 78.70°, \quad b = 10.04$
$\gamma = 30.02°, \quad c = 5.123$

Hence the triangle is:

$\alpha_2 = 48.68°, \quad a_2 = 7.690$
$\beta_2 = 101.30°, \quad b = 10.04$
$\gamma = 30.02°, \quad c = 5.123$

That these results are correct may be checked roughly by measuring the figure.

Exercises

In each of the following exercises, determine the number of triangles possible, and find the missing parts, giving results as accurate as the given data will permit:

1. $\alpha = 30°, a = 1.6, b = 7.0$.

2. $\gamma = 58°, a = 38, c = 42$.

3. $\beta = 112°, b = 36, a = 71$.

4. $b = 34.0, c = 15.5, \gamma = 20.6°$.

5. $b = 40.0, a = 31.2, \beta = 66.0°$.

6. $a = 282, \beta = 62.4°, b = 467$.

7. $c = 91.0, \gamma = 65.5°, b = 97.5$.

8. $a = 316, b = 359, \alpha = 121.2°$.

9. $\gamma = 53.8°, a = 5.13, c = 4.14$.

10. $a = 7.035, c = 4.946, \gamma = 37.56°$.

In each of the following exercises find the fourth part listed, giving results as accurate as the given data will permit. There may be two, one, or no possible values.

11. $\beta = 32.3°, c = 18.1, b = 10.7, \alpha = $.

12. $a = 134, b = 112, \alpha = 96.4°, \beta = $.

13. $b = 29.8, \gamma = 73.4°, c = 41.2, \alpha = $.

14. $c = 118, b = 94, \beta = 137.2°, \gamma = $.

15. $\alpha = 12.15°, a = 0.936, b = 1.758, \beta = $

16. $b = 12.6, c = 17.8, \beta = 41.7°, a = $.

17. $\alpha = 49.7°, a = 18.6, c = 24.4, \gamma = $.

18. $c = 4736, a = 4883, \alpha = 74.52°, \beta = $.

19. $\gamma = 155.00°, b = 1039, c = 2321, \alpha = $.

20. $a = 274, \alpha = 26.03°, b = 1612, c = $.

11–7. Case 4: Given Three Sides

The possibilities in this case may be summarized as follows.

(a) *Number of triangles*:

(1) *None, if the sum of the two smaller sides is less than or equal to the third side.*

(2) *One, if the sum of the two smaller sides is greater than the third side.*

(b) *Method of solution: Law of cosines, followed by the law of sines.*

Number of triangles. From plane geometry we know that in any triangle the sum of the lengths of any two sides is greater than the length of the other side. In testing three given sides to see if they have this property, we need only compare the sum of the lengths of the two smaller sides with the length of the largest side (why?). It follows that if the sum of the two smaller given sides of a triangle is not greater than the given third side, no triangle is determined by the given data. On the other hand, if the sum of the two smaller sides is greater than the third side, a triangle is determined which can be shown to be unique. As an aid in solving problems, the student should remind himself of the ruler and compass construction for a triangle when three sides are given.

Method of solution. Since none of the angles are known, it is first necessary to use the law of cosines to find one angle. Then a second angle may be found by the law of sines, and the third angle by subtracting the sum of the other two from 180°.

An angle of a triangle is always smaller than 180°. From this fact it follows that, when the cosine of an angle is given, the angle can be uniquely determined. When the cosine is positive, the angle is acute; when the cosine is negative, the angle is obtuse. In following the procedure given above, it is most convenient to find the obtuse angle first, if the triangle contains such an angle. The remaining two angles are acute, and to find them we can use the law of sines conveniently, without any fear of the ambiguity mentioned in Sec. 11–5. When using the law of cosines, therefore, *the largest angle (that is, the angle opposite the largest side) should be found first.*

Example. Given $a = 7.30$, $b = 10.1$, $c = 12.7$. Find α, β, γ.

The sketch of the triangle is given in Fig. 11–9. Since c is the longest side, γ is the largest angle, and we find it first by the law of cosines,

$$c^2 = a^2 + b^2 - 2ab \cos \gamma,$$

which gives us

$$\cos \gamma = \frac{a^2 + b^2 - c^2}{2ab} = \frac{(7.30)^2 + (10.1)^2 - (12.7)^2}{2(7.30)(10.1)}.$$

Performing the computations, we obtain

$$\cos \gamma = -0.0406,$$
$$\gamma = 180° - 87.7°,$$
$$\gamma = 92.3°.$$

To find β, we use the law of sines,

$$\sin \beta = \frac{b \sin \gamma}{c} = \frac{10.1 \sin 92.3°}{12.7}.$$

Numbers	Logarithms
10.1	1.0043
sin 92.3°	9.9996 − 10
10.1 sin 92.33°	11.0039 − 10
12.7	1.1038
sin β	9.9001 − 10

$$\beta = 52.6°$$

Finally,

$$\alpha = 180° - (\alpha + \beta) = 180° - (92.3° + 52.6°),$$
$$\alpha = 35.1°.$$

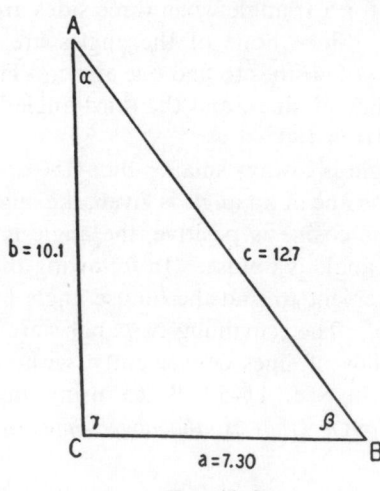

Fig. 11–9

Exercises

Find the angles of each of the following triangles, giving results as accurate as the given data will permit:

1. $a = 4.5, b = 18, c = 12.$ 2. $a = 72, b = 48, c = 164.$

3. $a = 76, b = 108, c = 42.$ 4. $a = 16.8, b = 24.3, c = 10.1.$

5. $a = 13.1, b = 19.3, c = 15.7$.   **6.** $a = 96.0, b = 128, c = 166$.

7. $a = 279, b = 557, c = 358$.   **8.** $a = 273, b = 131, c = 162$.

9. $a = 0.9824, b = 0.5759, c = 0.4261$.   **10.** $a = 7435, b = 1056, c = 7435$.

In each of the following exercises find the angle listed, giving results as accurate as the given data will permit:

11. $a = 15, b = 11, c = 14, \alpha =$.

12. $a = 56, b = 72, c = 31, \gamma =$.

13. $a = 298, b = 144, c = 125, \alpha =$.

14. $a = 3.02, b = 1.73, c = 4.95, \beta =$.

15. $a = 213, b = 641, c = 635, \alpha =$.

16. $a = 738, b = 315, c = 807, \gamma =$.

17. $a = 1076, b = 1250, c = 1280, \gamma =$.

18. $a = 38.16, b = 13.09, c = 38.16, \beta =$.

11–8. Application to Vector Analysis

The methods of solving oblique-triangle problems permit simple solutions of many problems that would require involved procedures if the methods of solving right triangles were used. Any problem that can be set up in terms of triangles may now be solved, provided, of course, that enough parts are known.

The solution of oblique triangles places at our disposal a second method applicable to some of the operations with vectors which were discussed in Chapter 6. The problems of finding the resultant of two given vectors (Sec. 6–3) and of finding the components of a given vector in two given directions (Sec. 6–4) may be carried out by solving triangles, as shown in the following illustrations.

Example 1. Find the resultant of the two vectors

$$\textbf{D}: D = 10.0, \quad \delta = 31.0°; \qquad \textbf{E}: E = 15.0, \quad \epsilon = 296.0°;$$

where the italicized capital letters denote the magnitudes of the vectors and the corresponding Greek letters denote the direction angles.

The graphical solution of this problem is given in Fig. 11–10. The vectors **D** and **E** with their resultant $\textbf{L} = \textbf{D} + \textbf{E}$ form an oblique triangle. Denote the angles of this triangle by α, β, and γ. Our object is to find the magnitude and direction angle of the vector **L**, that is, the length L and the angle $\lambda = 360° - (\alpha - \delta)$.

From the triangle, we have

$$\beta = \epsilon - (180.0° + \delta) = 296.0° - 180.0° - 31.0°,$$

$$\beta = 85.0°.$$

To find L we use the law of cosines and obtain

$$L^2 = D^2 + E^2 - 2DE \cos \beta = 100 + 225 - 300 \cos 85.0° = 299,$$

$$L = 17.3.$$

To find α we use the law of sines and obtain

$$\sin \alpha = \frac{E \sin \beta}{L} = \frac{15.0 \sin 85.0°}{17.3}.$$

Numbers	Logarithms
15.0	1.1761
$\sin 85.0°$	$9.9983 - 10$
$15.0 \sin 85.0°$	$11.1744 - 10$
17.3	1.2380
$\sin \alpha$	$9.9364 - 10$

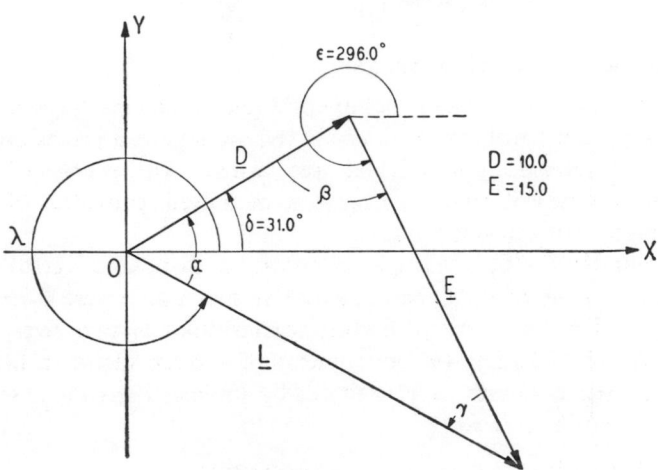

Fig. 11–10

Since from the figure α is obviously an acute angle, we have

$$\alpha = 59.7°.$$

Therefore

$$\alpha - \delta = 59.7° - 31.0° = 28.7°,$$

whence, for λ, the direction angle of **L**, we have

$$\lambda = 360.0° - 28.7° = 331.3°.$$

Thus $\mathbf{L} = \mathbf{D} + \mathbf{E}$ is the vector given by

$$\mathbf{L}: L = 17.3, \qquad \lambda = 331.3°.$$

Example 2.   Resolve the vector **L**: $L = 200$, $\lambda = 63°$ into two components **D** and **E** whose direction angles are $0°$ and $122°$, respectively.   Assume that L is given to two significant figures.

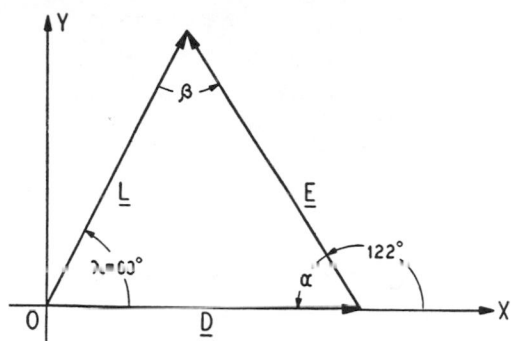

FIG. 11–11

We first construct the diagram shown in Fig. 11–11.   From the figure and the law of sines

$$E = \frac{L \sin \lambda}{\sin \alpha} = \frac{200 \sin 63°}{\sin 58°} = 210.$$

Since

$$\beta = 180° - (\alpha + \lambda) = 180° - 121° = 59°,$$

we have also

$$D = \frac{L \sin \beta}{\sin \alpha} = \frac{200 \sin 59°}{\sin 58°} \doteq 200.$$

Hence the components of **L** are

$$\textbf{D:} \ D = 200, \quad \delta = 0°; \qquad \textbf{E:} \ E = 210, \quad \epsilon = 122°.$$

Exercises

In each problem find the magnitude and direction angle of the resultant of the two given vectors.   State your results as accurately as the given data will permit.

1. **A**: $A = 14$, $\alpha = 32°$; **B**: $B = 18$, $\beta = 76°$.

2. **A**: $A = 5.1$, $\alpha = 66°$; **B**: $B = 8.3$, $\beta = 47°$.

3. **A**: $A = 124$, $\alpha = 55°$; **B**: $B = 146$, $\beta = 225°$.

4. **A**: $A = 2.37$, $\alpha = 168°$; **B**: $B = 2.69$, $\beta = 211°$.

5. **A**: $A = 603$, $\alpha = 292°$; **B**: $B = 446$, $\beta = 245°$.

6. **A**: $A = 157$, $\alpha = 103.6°$; **B**: $B = 308$, $\beta = 342.3°$.

7. **A**: $A = 1.42$, $\alpha = 70.3°$; **B**: $B = 5.60$, $\beta = 235.8°$.

8. **A**: $A = 80.7$, $\alpha = 149.2°$; **B**: $B = 36.3$, $\beta = 8.4°$.

9. **A**: $A = 381$, $\alpha = 12.5°$; **B**: $B = 436$, $\beta = 192.5°$.

10. **A**: $A = 286$, $\alpha = 322°$; **B**: $B = 303$, $\beta = 79°$.

11. **A**: $A = 0.843$, $\alpha = 142.5°$; **B**: $B = 0.671$, $\beta = 193.5°$.

12. A: $A = 446$, $\alpha = 227.3°$; **B:** $B = 589$, $\beta = 341.8°$.

13. A: $A = 1604$, $\alpha = 133.8°$; **B:** $B = 1860$, $\beta = 27.7°$.

14. A: $A = 398$, $\alpha = 5.4°$; **B:** $B = 572$, $\beta = 352.3°$.

15. A: $A = 63.5$, $\alpha = 212.7°$; **B:** $B = 130.4$, $\beta = 275.8°$.

16. A: $A = 92.36$, $\alpha = 87.2°$; **B:** $B = 88.03$, $\beta = 281.1°$.

In each problem find the components of the given vector in either the directions (a) 10°, 112°, (b) 60°, 330°, and (c) 90°, 240° or the opposite ones. State results that are as accurate as the given data will permit, assuming the given component directions to be exact.

17. A: $A = 19$, $\alpha = 42°$.   **18. A:** $A = 68$, $\alpha = 27°$.

19. A: $A = 210$, $\alpha = 78°$.   **20. A:** $A = 104$, $\alpha = 55°$.

21. A: $A = 38.3$, $\alpha = 216°$.   **22. A:** $A = 0.836$, $\alpha = 174°$.

23. A: $A = 562$, $\alpha = 96°$.   **24. A:** $A = 409$, $\alpha = 261°$.

25. A: $A = 7.18$, $\alpha = 245°$.   **26. A:** $A = 376$, $\alpha = 188°$.

27. A: $A = 1420$, $\alpha = 136°$.   **28. A:** $A = 82.04$, $\alpha = 117°$.

29. A: $A = 54.03$, $\alpha = 118.5°$.   **30. A:** $A = 0.7739$, $\alpha = 291.8°$.

31. A: $A = 626.4$, $\alpha = 32.3°$.   **32. A:** $A = 9861$, $\alpha = 320.5°$.

33. A surveyor and a chainman at points A and B, 1298 ft apart on the east bank of a river, sight a point C on the west bank of the river. If angle $CAB = 47.7°$ and angle $CBA = 96.8°$, how far are the surveyor and the chainman from C?

34. An airplane, flying above level ground, is directly above a straight course on which there are two beacons, $M1$, and $M2$, that are 28.3 miles apart. The pilot observes that the angle of depression of the beacon $M1$ is 44° and the angle of depression of $M2$ is 67°. Find the distance from the airplane to each beacon, and also the height of the airplane above the ground.

35. If town A is 11.2 miles from B and 4.2 miles from C, and if C is 13.7 miles from B, find the angles at which straight roads joining the three towns will intersect.

36. A navigator in a blimp flying due south above water observes that the angles of depression of two buoys 5280 ft apart on his course are 68.4° and 27.9°, respectively. Find the distance between the blimp and the nearest buoy, and also the distance of the blimp above the water.

37. A highway engineer is asked to build a bridge from point A to point B across a swamp. At a point C which is 1076 ft from A and 1250 ft from B the angle ACB is 66.2°. How long must the bridge be?

38. The lens of a motion-picture projector at a drive-in theater is located at a point 162 ft from the bottom of the screen and 178 ft from the top of the screen. If the image on the screen subtends an angle of 8.36°, what is the height of the image?

39. To find the distance of a buoy A from a point C on shore, a base line CB, 1314 ft long, is laid off on the shore, and it is found that the angle $ACB = 31.47°$ and the angle $ABC = 101.68°$. Find the distance CA.

40. A man in a small boat finds the angle of elevation of a marker on land to be 14.3°. After he rows 1880 ft toward the marker, the angle of elevation becomes 16.8°. How high is the marker?

41. An airplane based at A flies due west 107.6 miles to B, and then turns and flies 125.0 miles to C. The angle $ABC = 66.2°$. How far is the airplane from its base at the end of the journey?

42. An observer in a skyscraper 1240 ft above ground takes a telescopic measurement on a distant building. The angles of depression for the top and base of the building are 40.7° and 43.9°, respectively. How high is the building, assuming that the bases of the building and the skyscraper are on the same level?

43. Three airplane beacons, *A*, *B*, and *C*, are so located that *C* is 76 miles from *A* and 42 miles from *B*, and *A* is 108 miles due north from *B*. How far is *C* north of an east-west line through *B*?

44. The angle of intersection of two radio beams at beacon *A* is 72.8°. Along one beam the distance to town *B* is 28.5 miles, and along the other the distance to town *C* is 30.1 miles. How far is *C* from *B*?

45. A farmer has a triangular field with sides 468, 1203, and 1472 ft long. What is the area of the field?

46. The sides of a V-shaped irrigation canal are 6.42 ft and 7.36 ft long, and the angle between the sides is 74.7°. What is the cross-sectional area of the canal?

47. A telegraph pole, extending exactly 50 ft above the ground, is erected vertically at a point *P* on a hillside which slopes at a uniform angle of 53.1° with the horizontal. If the pole is to be supported by guy wires fastened exactly 10 ft below its top, how far down the hill from *P* must a 145-ft guy wire be secured?

48. In exercise 47, determine how long a guy wire which is secured up the hill must be if it makes the same angle with the pole as the downhill guy wire.

49. A painting exactly 10 ft long is seen by an observer whose eyes are 4.5 ft from one end of the painting and 12 ft from the other. Through what angle does the observer view the painting?

50. Two horizontal tunnels are constructed from a common point *A* at the bottom of a vertical shaft. Tunnel *AB* is 1750 ft long. Tunnel *AC* is 950 ft long and makes an angle of 84° with tunnel *AB*. Find the distance between the ends of the tunnels.

51. An airplane flies from *A* to *B*, a distance of 40.0 miles, and then from *B* to *C*, a distance of 108.2 miles. If the angle *ABC* is exactly 152°, how far is *C* from *A*?

52. If a tender leaves a ship on a course making an angle of 43° with the course of the ship, how far will the tender be from the ship in exactly 2 hr and 30 min if both travel on straight lines, the tender averaging 18 knots and the ship 22 knots?

53. An observer can see two guns, one at *A* and the other at *B*. He observes that the interval between the flash and the report is 8.3 sec for *A* and 7.3 sec for *B*. If the angle at his eye subtended by the guns is 50.7°, find the distance from *B* to *A*. Sound travels at approximately 1100 ft per sec.

54. An airplane travels 310 mph in a straight line bearing 146°. At the end of 2.20 hr how far south and how far east of the starting point will the plane be?

55. A speedboat travels 6.62 miles on a line bearing 206.3° and then 14.13 miles on a line bearing 251.8°. How far and in what direction must the boat travel to get back to its starting point by means of a straight course?

56. An airplane pilot flying on a straight course and at a constant altitude observes two light beacons *A* and *B* which are known to be 52 miles apart. The bearing of *A* is 85.5°; of *B*, 128.2°. Exactly 10 min later the bearing of *A* is 38.1°; of *B*, 83.6°. How fast and on what bearing is the airplane flying?

57. A submarine is at rest 7.0 miles outside a harbor, and the bearing of the harbor from the submarine is 121.5°. An enemy ship sails from the harbor at exactly 20 mph along a line bearing 255.0°. If the submarine starts as soon as the ship leaves the harbor, at what speed and on what bearing should the submarine travel to meet the ship after the ship has gone 11.0 miles?

58. A meteorological balloon is observed simultaneously from two stations, 12,800 ft apart. At station A the horizontal angle between the balloon and station B is 68.3°, and the angle of elevation of the balloon is 21.9°. At B the horizontal angle between A and the balloon is 42.7°. How high is the balloon?

59. A vertical pole that supports a television antenna is mounted on the ridge of a roof whose sides make angles of 38.7° with the horizontal. When the angle of elevation of the sun is 62.0°, the pole casts a shadow 18.5 ft long perpendicular to the ridge of the roof. What is the length of the pole?

60. A mountain top is observed by two surveyors A and B who are 13,640 ft apart and at the same height above sea level. To surveyor A the horizontal angle between the mountain top and surveyor B is 58.34°, and the angle of elevation of the mountain top is 18.62°. Surveyor B finds that the horizontal angle between surveyor A and the mountain top is 52.64°. Find the height of the mountain above the level of the surveyors.

61. For the situation described in the previous exercise, how high would the mountain have been if surveyor B had observed that the top had an angle of elevation of 16.38°?

Solve the following exercises by using vectors. Give all results as accurately as the given data will allow.

62. Two forces, one of 92 lb, the other of 85 lb, act simultaneously on an object. The angle between their directions is 43°. Find the resultant force and the angle it makes with each given force.

63. A 3.5-lb force and a 5.7-lb force are acting on a weight in such a way that the angle between their directions is 112°. Find the resultant force and the angle it makes with each given force.

64. Two forces are acting simultaneously on an object. The angle between the forces is 146°. If one of the forces is 160 lb and the resultant force is 310 lb, find the second force and the angle that the resultant force makes with each component.

65. If a force of 580 lb and an unknown force at an angle of 21° act on an object to produce a resultant force of 980 lb, find the magnitude of the unknown force and the angle that the resultant makes with each of the other forces.

66. A tractor pulls a harrow in such a way that it exerts a force of 525 lb at an angle of 18.5° with the horizontal. Find the horizontal and vertical components of this force.

67. A telegraph pole is erected by means of an auxiliary pole and lines. Find the horizontal and vertical components of a force of 406 lb acting at 37.5° with the horizontal at the point where the auxiliary pole is applied.

68. A casting weighing 2270 lb is hoisted vertically by a crane. If the casting is suspended by two cables making angles of 135° and 148° with the downward vertical, what is the tension in each of the cables?

69. It is known that a 260-lb force will be required at the end of a rope to move an engine casting straight ahead. If two men, one who can exert a force of 165 lb and the other a force of 155 lb, wish to exert a force of exactly 260 lb straight ahead, at what angle from this direction should each pull on the rope?

70. A man finds that the force required to move his lawn mower is 65 lb exerted along the handle and that the handle makes an angle of 52° with the horizontal. If he adjusts the handle to make an angle of 42° with the horizontal, how much force must be exerted along the handle to make the mower move?

71. A rope on which a weight of 250 lb is fastened is passed through a pulley. If the free end of the rope makes an angle of 37° with the vertical, what horizontal force must be exerted on the free end of the rope to lift the weight from the ground?

72. The load on a cable of a dredge consists of a clamshell bucket which with its load of marl weighs 3200 lb. What is the magnitude of the horizontal force that must be applied to the bucket to hold it in such a position that the cable will make an angle of 21.5° with the vertical?

73. An automobile weighing 2300 lb is standing on an 18° slope. Neglecting friction, how much force must be exerted to prevent the car from rolling downhill?

74. Find the largest weight a tractor can pull up a hill that slopes 18.3° from the horizontal if the tractor can pull with a force of 980 lb.

75. An automobile wrecker wishes to drag a 2200-lb wreck up an incline to the road. The wreck is in a ditch which is 12 ft deep. What is the length of the shortest inclined plane the wrecker truck can use if its maximum pulling force is 800 lb and if the friction of the wreck on the plane may be safely ignored?

76. A keg of nails weighing 52 lb is placed on a roof which is inclined 26° with the horizontal. What is the magnitude of the force due to friction that prevents the keg from sliding down the roof?

77. A certain derrick consists of a boom which makes an angle of 32° with the vertical and a steel cable which extends from the weight vertically to the upper end of the boom and then horizontally to a windlass. Find the compression in the boom if the weight being lifted is 2500 lb.

78. A weight of 280 lb fastened at the end of a 90-ft cable swings as a pendulum through a total arc of 30°. What is the change in the tension in the cable as the pendulum swings from 0° to 15°?

79. The speed of an airplane in still air is 310 mph. If the plane encounters a head wind of exactly 30 mph that makes an angle of 21.5° with the desired line of flight, at what angle with the intended line of flight must the pilot head the plane? Under these circumstances how long will it take the plane to fly 840 miles?

80. In still air a certain transport plane can fly 320 mph and reaches its objective O from its base B in 54.0 min. The bearing of O from B is 116°. In what direction must the pilot head the plane and how long must he fly in going from B to O if there is a wind from the east of 45 mph?

81. A man wishes to cross a river from A to a point B 3.5 miles upstream. The speed of his boat in still water is 12 mph, and the speed of the current of the river is 2.5 mph. If the river is 0.8 mile wide, at what angle must he travel with the banks of the river? How far will he travel through the water and how long will the trip take him?

82. Three boats X, Y, and Z are at a point A on the east bank of a river, in which current flows south at the rate of 2.2 mph. All three boats have a speed of 14.5 mph in still water. These boats are to meet at an island B, 6200 ft directly across the river from A. X is to proceed to a point 2400 ft directly upstream from B and then come down to the island. Y is to proceed directly across to B, and Z is to proceed to a point 1800 ft downstream from B and then approach the island from the south. At what angle with the river bank must each boat direct its course? If it is desired that all three boats reach the island at midnight, at what time must each boat leave A?

Progress Report

In plane geometry the ruler and compass serve as the tools for constructing a triangle if any three parts, at least one of which is a side, are given. From the constructed triangle, the missing parts are found by measurement. However, the method of construction is not very accurate, and therefore in this chapter we developed methods for computing the unknown parts of a triangle when three parts, at least one of which is a side, are given.   It was convenient to divide the possible problems of this type into four cases:

1. Given two angles and any side.
2. Given two sides and the included angle.
3. Given two sides and an angle opposite one.
4. Given three sides.

The formulas called the law of sines and the law of cosines were derived and used to compute the unknown parts of the triangles.

These methods were then applied to vector problems and to problems from other scientific and engineering applications.

12

The *J* Operator

In this chapter we shall develop the operations with complex numbers and the operations with their graphical representations. We shall find that for complex numbers the graphical operations of addition and subtraction are like the addition and subtraction of vectors. Consequently, complex numbers are frequently applied to vector problems, especially in the theory of sinusoidal alternating currents. Further, the operations of multiplication and division for complex numbers can be used advantageously in the theory of alternating currents. Thus complex numbers are a most important theoretical and practical tool for the electrical engineer. Indeed, applications of complex numbers in science and engineering are so widespread that a knowledge of these numbers and their properties is essential for all scientists and engineers.

12–1. Imaginary Numbers

In Chapter 1 it was shown how the various operations led to the consideration of different kinds of numbers. Thus in Sec. 1–2 subtraction brought forth negative numbers while division introduced fractions. Later on in Sec. 1–3 the extraction of roots forced the consideration of irrationals which, together with the integers and fractions, formed the real numbers.

From Sec. 1–3 the student will remember that the class of irrational numbers included nth roots of many positive rational numbers and nth roots (where n is odd) of many negative rational numbers. However, the operation of taking the nth root (where n is even) of negative numbers was carefully avoided. This last operation will now be introduced.

The extraction of a square root of a negative number like $\sqrt{-4}$ is not possible, since there is no real number which when squared will yield -4. The square of any real number, whether positive or negative, is a positive number. The same holds true for the fourth, sixth, or any even-indexed root of a negative number, for example, $\sqrt[4]{-81}$ and $\sqrt[6]{-64}$. Hence, if we wish to extract the roots of all numbers, positive and negative, we must make a new definition which will introduce a new type of number into our number system.

Consider the following square roots of negative numbers:

$$\sqrt{-4}, \quad \sqrt{-9}, \quad \sqrt{-3}, \quad -\sqrt{-16}, \quad \sqrt{-\tfrac{4}{25}}.$$

Let us suppose that these as yet undefined symbols obey, as far as possible, the laws of radicals given in Sec. 8–6. Hence it follows that:

$$\sqrt{-4} = \sqrt{4(-1)} = \sqrt{4} \cdot \sqrt{-1} = 2\sqrt{-1}.$$

$$\sqrt{-9} = \sqrt{9(-1)} = \sqrt{9} \cdot \sqrt{-1} = 3\sqrt{-1}.$$

$$\sqrt{-3} = \sqrt{3(-1)} = \sqrt{3} \cdot \sqrt{-1} = 1.73\sqrt{-1}.$$

$$-\sqrt{-16} = -\sqrt{16(-1)} = -\sqrt{16} \cdot \sqrt{-1} = -4\sqrt{-1}.$$

$$\sqrt{-\tfrac{4}{25}} = \sqrt{\tfrac{4}{25}(-1)} = \sqrt{\tfrac{4}{25}} \cdot \sqrt{-1} = \tfrac{2}{5}\sqrt{-1}.$$

The symbol $\sqrt{-1}$ appears in every one of these results.

This symbol $\sqrt{-1}$ *is called the* **imaginary unit,** *or* **j operator,** *and will be denoted by* **j.** *The fundamental property of this symbol is that*

$$j^2 = -1.$$

Using this symbol, we can now write the expressions above in the form $\sqrt{-4} = j2$, $\sqrt{-9} = j3$, $\sqrt{-3} = j1.73$, $-\sqrt{-16} = -j4$, $\sqrt{-\tfrac{4}{25}} = j\tfrac{2}{5}$. Most mathematics texts write the real coefficient before the j as, for example, $5j$ and $\tfrac{2}{5}j$, but, since it is quite common in scientific and engineering books to let the j precede, we shall follow the latter practice.

While j is called the imaginary unit, any multiple of j is called an **imaginary number.** Thus from the above examples $j2$, $j3$, $j1.73$, $-j4$, $j\tfrac{2}{5}$ are imaginary numbers. We now assume that imaginary numbers obey the laws of addition, subtraction, multiplication, division, and raising to a power stated in Chapter 1. Thus we have

$$j^3 = j^2 \cdot j = -1\,(j) = -j,$$
$$j^4 = j^2 \cdot j^2 = -1\,(-1) = 1.$$

We can now write down a table for the powers of j, where $j = \sqrt{-1}$:

$$j = j$$
$$j^2 = -1$$
$$j^3 = -j$$
$$j^4 = 1$$

Since $j^4 = 1$, we also have $j^5 = j^4 \cdot j = j$, $j^6 = j^4 \cdot j^2 = j^2$, $j^7 = j^4 \cdot j^3 = j^3$, $j^8 = j^4 \cdot j^4 = j^4$, etc., and hence that the successive integral powers of j run through the cycle j, -1, $-j$, 1 given in the above table.

It was stated above that it is customary to write j instead of $\sqrt{-1}$. It should be pointed out here that mathematicians use the letter i for $\sqrt{-1}$, but scientists and engineers usually use j to avoid confusion with various other notations for which the letter i is used. Since this book is primarily intended for the scientific and engineering student, the j notation will be employed.

The choice of the term **imaginary unit** for $\sqrt{-1}$ is unfortunate from the point of view of the scientific and engineering student, for this quantity is not imaginary in the sense that the adjective is used in everyday language. The reason for the choice of this term lies in the historical development of mathematics; it was introduced hundreds of years ago when to extract the square root of a negative number was thought to be inconceivable. From our standpoint it might have been better to call it the **j number.**

In dealing with square roots of negative numbers it is convenient to introduce j for $\sqrt{-1}$. This will be done in the example that follows.

Example 1.

$$\sqrt{-\tfrac{4}{9}} = \sqrt{\tfrac{4}{9}} \cdot \sqrt{-1} = j\tfrac{2}{3}.$$

$$\sqrt{-0.01} = \sqrt{0.01} \cdot \sqrt{-1} = j0.1.$$

$$\sqrt{-45} = \sqrt{45} \cdot \sqrt{-1} = j3\sqrt{5} = j6.71.$$

$$\sqrt{-7} = \sqrt{7} \cdot \sqrt{-1} = j\sqrt{7} = j2.65.$$

This can be condensed in the following rule:

$$\textbf{For } a > 0, \quad \sqrt{-a} = j\sqrt{a},$$

i.e., *the square root of a negative number $\sqrt{-a}$ can be written as j times the square root of the corresponding positive number $\sqrt{a}$.*

Multiples of j can be added like any other quantity. This is made plain by the next example.

Example 2.

$$j2 + j7 - j = j8.$$

$$j\tfrac{1}{3} + j\tfrac{1}{4} - j\tfrac{1}{5} = j\,\frac{20 + 15 - 12}{60} = j\,\frac{23}{60}.$$

$$\sqrt{-9} + \sqrt{-25} - \sqrt{-4} = j3 + j5 - j2 = j6.$$

$$j^3 5 + j8 = -j5 + j8 = j3.$$

Exercises

Simplify the following by expressing each number in one of the forms, j, -1, $-j$, 1:

1. j^3.   **2.** j^2.   **3.** j^6.   **4.** j^5.   **5.** j^4.

6. j^7.   **7.** j^8.   **8.** j^{10}.   **9.** j^{12}.   **10.** j^9.

11. j^{13}.   **12.** j^{14}.   **13.** j^{18}.   **14.** j^{15}.   **15.** j^{21}.

Express each number in terms of j, and simplify.

16. $\sqrt{-36}$.   **17.** $\sqrt{-81}$.   **18.** $\sqrt{-100}$.   **19.** $\sqrt{-64}$.

20. $-\sqrt{-25}$.   **21.** $-\sqrt{-49}$.   **22.** $\sqrt{-0.09}$.   **23.** $\sqrt{-0.01}$.

24. $-\sqrt{-1.44}$.   **25.** $\sqrt{-18}$.   **26.** $-\sqrt{-27}$.   **27.** $\sqrt{-40}$.

28. $\sqrt{-\frac{9}{49}}$.   **29.** $\sqrt{-\frac{25}{121}}$.   **30.** $-\sqrt{-\frac{75}{16}}$.   **31.** $\sqrt{-\frac{48}{49}}$.

32. $\sqrt{-64} + \sqrt{-9}$.   **33.** $\sqrt{-169} - \sqrt{-81}$.

34. $-\sqrt{-225} + \sqrt{-49}$.   **35.** $\sqrt{-\frac{36}{25}} - \sqrt{-\frac{4}{25}}$.

Using the table of square roots, write the following in terms of j:

36. $\sqrt{-3}$.   **37.** $\sqrt{-6}$.   **38.** $-\sqrt{-2}$.   **39.** $\sqrt{-5}$.

40. $-\sqrt{-7}$.   **41.** $\sqrt{-11}$.   **42.** $-\sqrt{-15}$.   **43.** $-\sqrt{-20}$.

44. $\sqrt{-3.74}$.   **45.** $-\sqrt{-4.31}$.   **46.** $\sqrt{-83.4}$.   **47.** $-\sqrt{-94.7}$.

Simplify.

48. $j8 - j3 + j2$.   **49.** $j^33 + j7$.   **50.** $j^5 - j - j^312$.

51. $j^64 + j^25 - 17$.   **52.** $j^86 - j^23$.   **53.** $j^4 + j^8 + 12$.

12–2. Complex Numbers

Real numbers were introduced early in this book. Imaginary numbers (i.e., multiples of j) were defined in the preceding section. To make the real and imaginary numbers part of one system of numbers we define new quantities in the following way.

If a and b are real numbers, $a + jb$ is called a **complex number.** *The number a is called the* **real part** *of the complex number $a + jb$, the number jb is called the* **imaginary part.** Thus $2 + j3$, $5 - j9$, $\frac{1}{3} + j\frac{2}{3}$, $2.16 - j0.09$, $3 - j2$ are examples of complex numbers.

Real and imaginary numbers are a part of the system of complex numbers. For, when $b = 0$ and a is any real number, then $a + jb$ gives us the real numbers; on the other hand, when $a = 0$ and b is any real number, then $a + jb$ gives us the imaginary numbers (multiples of j). We can summarize this by saying that the real numbers and the imaginary numbers are subdivisions of the complex numbers.

We have come now to the end of the road in defining new numbers. It will be well to stop and review the main features of this travel. *Starting with the natural numbers* 1, 2, 3, 4, $\cdots$ *which are suggested by counting*:

1. *Subtraction led to negative numbers.*
2. *Division brought forth fractions (rational numbers).*
3. *Square roots of positive numbers produced irrational numbers.*
4. *Square roots of negative numbers produced complex numbers.*

It is shown in more advanced mathematics textbooks that, in working with the elementary operations of mathematics, no other numbers than the complex numbers are required. It can also be shown that the operations with complex numbers can be defined so that the numbers obey the fundamental laws of operation stated for real numbers in Sec. 1–8. In the remaining sections of the present chapter the student will learn how to use the complex numbers.

Complex numbers are of great importance in mathematics and in the applications of mathematics to practical problems. This is particularly true of electrical engineering, where alternating currents and voltages can be handled by methods that will be developed in this chapter. In fact, many problems in a-c theory would be very difficult to solve without the use of complex numbers.

12–3. Equality of Complex Numbers

In this section and in the next few sections we shall define equality, sum, and product in such a way that the complex numbers obey the same laws as the real numbers. We start with a definition of equality of two complex numbers.

If two complex numbers $x + jy$ and $a + jb$ are equal, then $x = a$ and $y = b$.

Symbolically:

$$\text{If } x + jy = a + jb, \text{ then } x = a, \text{ and } y = b.$$

This can also be stated by saying that complex numbers are made equal by equating their real parts and also their imaginary parts. For example, if $x + jy = 5 - j3$, then $x = 5$, and $y = -3$.

In particular, when equating a complex number to zero, the above definition becomes:

$$\text{If } a + jb = 0, \text{ then } a = 0, \text{ and } b = 0.$$

Using this definition we can solve equations containing complex quantities.

Example. What values of x and y satisfy the equation $x + y + jx - jy = 2 + j3 + j$?

Simplifying and separating reals and imaginaries by the above definition, we get

$$(x + y) + j(x - y) = 2 + j4,$$

$$x + y = 2, \qquad x - y = 4.$$

Solving these equations simultaneously by the method of Sec. 2–16, we find

$$x = 3, \qquad y = -1.$$

The following definition is useful in working with complex quantities. *Two complex numbers such as*

$$a + jb \text{ and } a - jb,$$

whose real parts are the same, and for which the coefficients of j are equal in absolute value but opposite in sign, are called **conjugate complex numbers,** *and each is the conjugate of the other.*

$$a + jb \text{ and } a - jb \text{ are conjugate.}$$

Examples of conjugate complex numbers are given by:

$$3 + j4 \quad \text{and} \quad 3 - j4,$$

$$-1 - j \quad \text{and} \quad -1 + j,$$

$$j5 \quad \text{and} \quad -j5,$$

$$3 \quad \text{and} \quad 3.$$

The last example illustrates the fact that a *real number is its own conjugate.*

Throughout this chapter it is assumed that except for $j = \sqrt{-1}$ all literal numbers are real.

Exercises

Find the conjugates of the following complex numbers:

1. $3 + j2$.
2. $7 - j4$.
3. $-1 + j$.
4. $-5 - j9$.

5. $-7 - j$.
6. $j5$.
7. j.
8. $-j6$.

9. $-j$.
10. 5.
11. -15.
12. $m + jn$.

13. $x - jy$.
14. $-c + jd$.
15. $3 + 4\sqrt{-1}$.
16. $1 - \sqrt{-25}$.

Find the values of x and y that satisfy the following equations:

17. $x + jy = 4 + j3$.

18. $3x + jy = 9$.

19. $2x - j5y = 6 + j10$.

20. $x + y + j(x - y) = 1 + j7$.

21. $2x - 5y + j(x + 3y) = -1 + j5$.

22. $4x + 9y - j(2x - y) = 3 - j7$.

12–4. Operations with Complex Numbers

A complex number, when reduced to the form $a + jb$, may be regarded as a binomial. Thus the addition, subtraction, and multiplication of complex numbers are reduced to the corresponding operations with binomials in which one term is real and the other imaginary.

Example 1. Examples of the addition and subtraction of complex numbers are given by the following:

$$(2 + j3) + (4 - j) = 2 + j3 + 4 - j = 6 + j2.$$

$$(5 - j2) - (7 - j9) = 5 - j2 - 7 + j9 = -2 + j7.$$

$$(\tfrac{1}{2} + j\tfrac{1}{3}) + (1 - j\tfrac{1}{2}) - (\tfrac{1}{6} - j) = \tfrac{1}{2} + j\tfrac{1}{3} + 1 - j\tfrac{1}{2} - \tfrac{1}{6} + j = \tfrac{4}{3} + j\tfrac{5}{6}.$$

$$(7 - j3) + (j - 7) = 7 - j3 + j - 7 = -j2.$$

Example 2. In multiplying complex numbers, the student must make use of the fact that $j^2 = -1$.

$$(2 + j5)(1 + j3) = 2 + j6 + j5 + j^2 15 = 2 + j11 - 15 = -13 + j11.$$

$$(1 - j3)(-2 + j) = -2 + j7 - j^2 3 = -2 + j7 + 3 = 1 + j7.$$

We can now put the definitions of addition, subtraction, and multiplication of complex numbers in the following symbolic way:

Addition (or subtraction): $(a + jb) + (c + jd) = (a + c) + j(b + d)$.

Multiplication: $(a + jb)(c + jd) = (ac - bd) + j(ad + bc)$.

It is of interest to apply the above definitions of addition, subtraction, and multiplication to conjugate complex numbers $a + jb$ and $a - jb$. One thus obtains:

Sum of conjugates: $(a + jb) + (a - jb) = 2a$.
Difference of conjugates: $(a + jb) - (a - jb) = j2b$.
Product of conjugates: $(a + jb)(a - jb) = a^2 + b^2$.

Therefore the sum or the product of two conjugate complex numbers is a real number.

The fact that the product of two conjugate complex numbers is a real number leads to the following method for division of complex numbers. *Division by a complex number is performed by multiplying numerator and denominator by the conjugate of the denominator:*

$$\frac{a + jb}{c + jd} = \frac{(a + jb)(c - jd)}{(c + jd)(c - jd)} = \frac{(ac + bd) + j(bc - ad)}{c^2 + d^2}.$$

Example 3. Perform the following divisions.

$$\frac{3+j4}{1+j} = \frac{3+j4}{1+j} \cdot \frac{1-j}{1-j} = \frac{3+j-j^2 4}{1+1} = \frac{3+j+4}{2} = \frac{7}{2} + j\frac{1}{2}.$$

$$\frac{3-j2}{8-j6} = \frac{3-j2}{8-j6} \cdot \frac{8+j6}{8+j6} = \frac{24+j2+12}{64+36} = \frac{36+j2}{100} = 0.36 + j0.02.$$

The student will notice that this method for division by a complex number corresponds to the method of Sec. 8–9 for rationalizing a denominator.

The work of this section can be summarized by saying that *complex numbers may be used in algebraic operations in the same way that real numbers are used, provided that j^2 is replaced by* −1.

Remark and warning. It has been explained so far in this chapter that the square roots of negative numbers are expressed in terms of the symbol *j* where $j^2 = -1$. For example, the roots of the equation $x^2 = -4$ are $x = \pm\sqrt{-4}$ or $x = \pm j2$. It should be emphasized, however, that we use the form $\pm j2$ in preference to $\pm\sqrt{-4}$. Had we used the less desirable notation $\pm\sqrt{-4}$ for the roots of the equation $x^2 = -4$, we might be tempted to take the following erroneous step in checking the root $x = \sqrt{-4}$:

$$(\sqrt{-4})^2 = \sqrt{-4} \cdot \sqrt{-4} = \sqrt{(-4)(-4)} = \sqrt{16} = 4,$$

while the given equation is $x^2 = -4$ and not $x^2 = 4$. The reason for this error is that the rules for operating with square roots of positive numbers cannot be entirely extended to square roots of negative numbers. To prevent such errors, we write $\pm j2$ in place of $\pm\sqrt{-4}$ and obtain $(j2)^2 = j^2 4 = -4$, which yields the correct result.

In the light of this discussion, *the student should always introduce j in place of $\sqrt{-1}$ when working with square roots of negative real numbers.*

Example 4. Multiply $(3 - \sqrt{-16})(2 + \sqrt{-9})$.

$$(3 - \sqrt{-16})(2 + \sqrt{-9}) = (3 - j4)(2 + j3) = 6 + j - j^2 12$$
$$= 6 + j + 12 = 18 + j.$$

Exercises

Perform the indicated operations and simplify to the form $a + jb$.

1. $(3 + j4) + (1 - j5)$.   2. $(15 - j7) + (-6 + j)$.

3. $(2 - j9) - (1 + j4)$.   4. $(-1 + j12) - (-18 - j3)$.

5. $\left(5 - \sqrt{-36}\right) + \left(13 + \sqrt{-49}\right).$

6. $\left(4 + \sqrt{-9}\right) - \left(5 - \sqrt{-16}\right).$

7. $(0.7 + j0.9) + (-0.2 + j1.3).$

8. $(2.7 - j1.3) - (5.2 - j3.5).$

9. $(0.35 - j2.71) + (0.35 + j2.71).$

10. $(0.5 + j0.6) - (0.5 - j0.6).$

11. $3\sqrt{-4} + 5\sqrt{-25} - 7\sqrt{-9}.$

12. $\frac{1}{2}\sqrt{-36} - \frac{1}{3}\sqrt{-81} + \frac{2}{7}\sqrt{-49}.$

13. $(3 + j2)(3 - j2).$

14. $(1 + j)(1 - j).$

15. $\left(3\sqrt{-25}\right)^2.$

16. $\left(-2\sqrt{-7}\right)^2.$

17. $\sqrt{-2}\,\sqrt{-8}.$

18. $\sqrt{-3}\,\sqrt{-27}.$

19. $(\frac{1}{2} - j\frac{1}{3}) + (\frac{1}{3} + j\frac{1}{2}).$

20. $(\frac{2}{5} - j\frac{1}{4}) - (\frac{3}{5} - j\frac{3}{8}).$

21. $(5 + j3)(4 - j).$

22. $(6 - j7)(8 + j3).$

23. $(3 + j4)^2.$

24. $(2 - j5)^2.$

25. $\sqrt{-6}\,\sqrt{-3}\,\sqrt{-2}.$

26. $\sqrt{-10}\,\sqrt{-2}\,\sqrt{-5}.$

27. $\left(4 + 3\sqrt{-9}\right)\left(3 - 2\sqrt{-4}\right).$

28. $\left(1 - 2\sqrt{-16}\right)\left(1 + 2\sqrt{-16}\right).$

29. $\left(1 + j\sqrt{5}\right)^2.$

30. $\left(2 - j\sqrt{3}\right)^2.$

31. $\left(7 + \sqrt{-18}\right)^2.$

32. $\left(-3 + \sqrt{-20}\right)^2.$

33. $\left(3 - 4\sqrt{-5}\right)\left(-1 + \sqrt{-20}\right).$

34. $\left(\frac{1}{2} + \sqrt{-\frac{3}{4}}\right)^2.$

35. $(1 + j)^3.$

36. $(3 - j2)^3.$

37. $\left(\sqrt{2} - j\right)^3.$

38. $(1 + j)^4.$

39. $(j^4 - j^3 2 + j^2 5 + j3)^2.$

40. $(j - j^2 + j^3 4 + j^4)^2.$

41. $(j^2 + j^3 + j^4)^3.$

42. $(j^3 + j^5)^4.$

43. $\dfrac{1}{1 - j}.$

44. $\dfrac{1}{2 + \sqrt{-9}}.$

45. $\dfrac{2}{-3 + j}.$

46. $\dfrac{5}{4 - j^3}.$

47. $\dfrac{1 - j}{1 + j}.$

48. $\dfrac{5 + j^3}{j}.$

49. $\dfrac{5 + \sqrt{-4}}{7 - \sqrt{-36}}.$

50. $\dfrac{4 - \sqrt{-9}}{-2 + \sqrt{-1}}.$

51. $\dfrac{2+j}{3-j2} \cdot \dfrac{2-j}{-1+j}.$

52. $\dfrac{4+j3}{-1+j} \cdot \dfrac{-5+j2}{-1-j}.$

53. $\dfrac{1}{2 \div 3\sqrt{-18}}.$

54. $\dfrac{2+j3}{1+j} \cdot \dfrac{1-j5}{4-j3}.$

12–5. Graphical Representation of Complex Numbers

In Chapter 1, real numbers were represented graphically as points on a line. In order to represent complex numbers, one line will not suffice, and

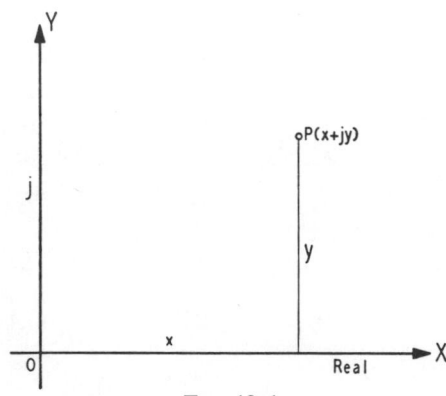

Fig. 12–1

it becomes necessary to use two dimensions. The complex number $x + jy$ may be represented on a rectangular coordinate system in the plane by a

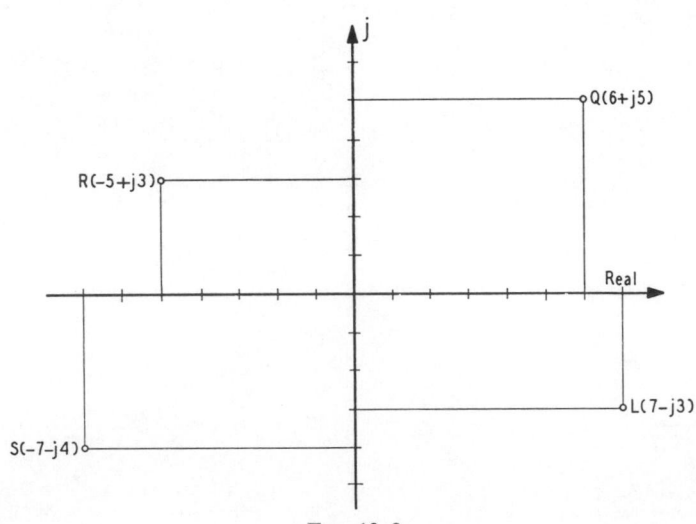

Fig. 12–2

point whose abscissa is x and whose ordinate is y (see Fig. 12–1).   One speaks of P as the point $x + jy$.   When complex numbers are so represented, the horizontal axis is the **axis of real numbers** or the **real axis,** and the vertical axis is called the **axis of imaginary numbers** or the **imaginary axis.**   The entire plane when used for the representation of complex numbers is called the **complex plane.**

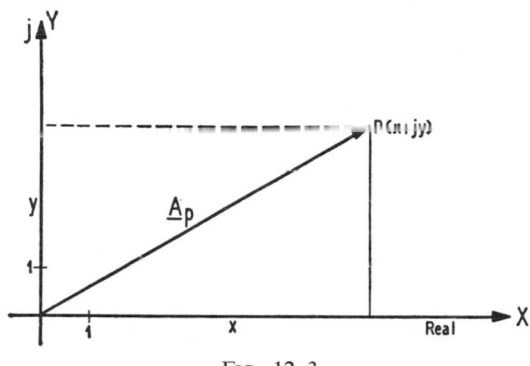

FIG. 12–3

Thus in Fig. 12–2, the complex numbers $6 + j5$, $-5 + j3$, $-7 - j4$, and $7 - j3$ are represented by the points Q, R, S, and L, respectively.

If in Fig. 12–1 a vector $\mathbf{A}_P$ is drawn from the origin to point $P(x + jy)$, one can then think of $x + jy$ as the directed distance or vector from the origin to that point (see Fig. 12–3).

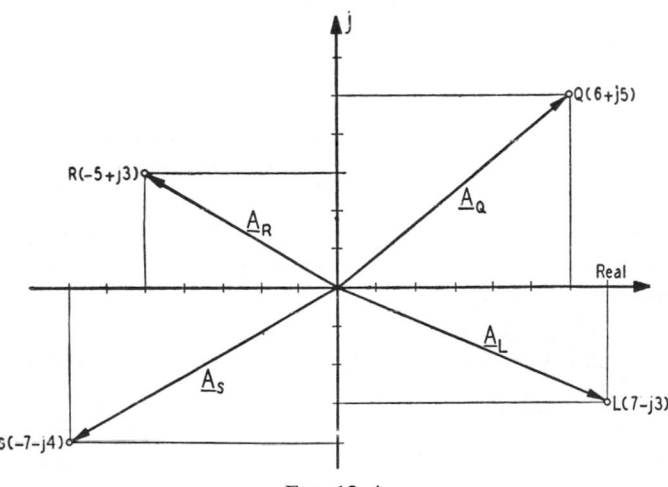

FIG. 12–4

A complex quantity may thus be used to represent a vector, and it is in this way that complex numbers are used in electrical engineering.

Thus in Fig. 12–2, the complex quantity $6 + j5$ can represent a vector A_Q, and the other complex numbers similarly, as may be seen in Fig. 12–4.

The addition and subtraction of complex numbers explained in the preceding section will now be performed graphically. Complex quantities are added, according to Sec. 12–4, by adding first the real parts and then the imaginary parts. This procedure corresponds to the law given in

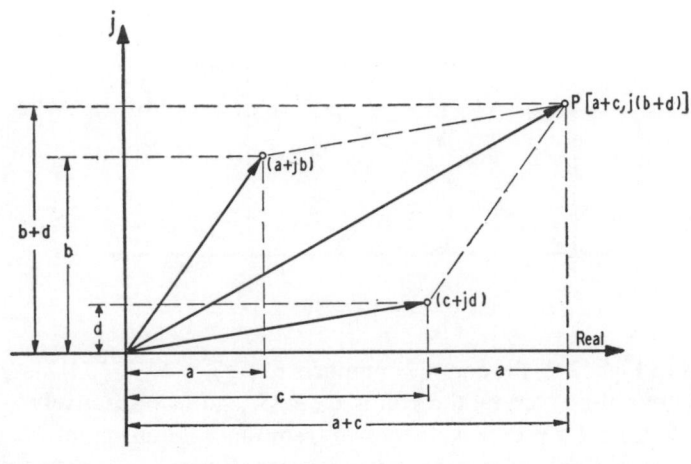

FIG. 12–5

Sec. 6–5 for the addition of vectors. This law stated that the sum of two vectors can be found by adding first the x coordinates and then the y coordinates. Thus the addition of two complex quantities $a + jb$ and $c + jd$ can be performed graphically by adding the corresponding vectors as shown in Fig. 12–5. Hence the point P represents the complex number $(a + c) + j(b + d)$ which is the sum of the given complex numbers $a + jb$ and $c + jd$.

Thus, *to add two complex numbers graphically, complete the parallelogram which has as adjacent sides the lines drawn from the origin to the points representing the two complex numbers. The fourth vertex of the parallelogram will be the point representing the sum of the two complex numbers.*

This method can be extended to find the sum of more than two complex numbers by first adding two of the numbers, then adding their sum to a third, and so on.

To subtract one complex number $c + jd$ from $a + jb$ graphically we merely add $a + jb$ and $-c - jd$ graphically.

Example 1.   Add graphically $(7 + j5) + (9 - j2)$.

$$(7 + j5) + (9 - j2) = 16 + j3.$$

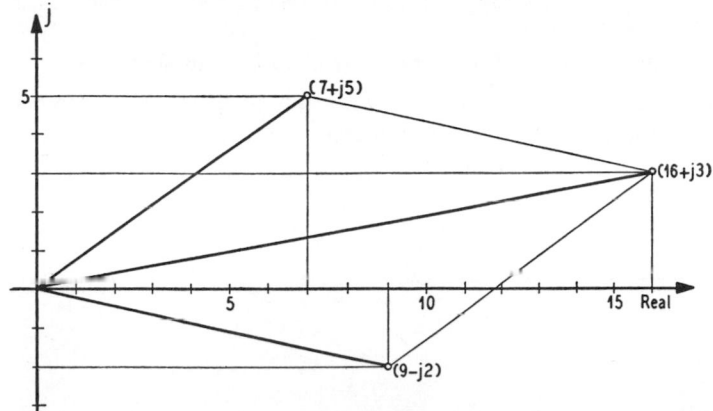

FIG. 12–6

Example 2.   Subtract graphically $(3 + j5) - (7 + j2)$.
This subtraction may be replaced by the addition $(3 + j5) + (-7 - j2)$.

$$(3 + j5) - (7 + j2) = -4 + j3.$$

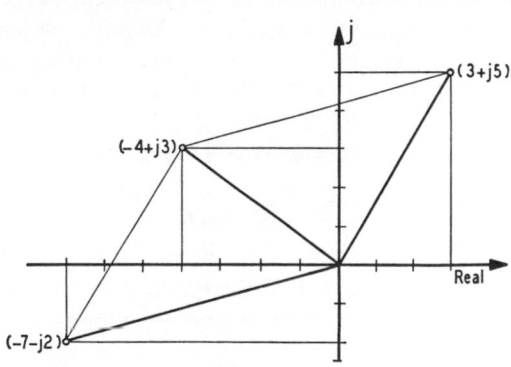

FIG. 12–7

Exercises

Perform the indicated operations graphically.   Use graph paper.

1. $(2 + j3) + (1 + j4)$.
2. $(4 - j) + (5 + j2)$.
3. $(3) + (-2 - j5)$.
4. $(-1 + j3) + (j2)$.
5. $(3 - j2) - (1 + j)$.
6. $(8 + j5) - (4 - j7)$.
7. $(-1) - (-3 + j5)$.
8. $(j2) - (-4 + j)$.
9. $(3 + j) + (-4 + j5) - (-1 - j3)$.
10. $(-2 + j3) - (2 - j5) + (1 - j6)$.

11. Show graphically that the sum of two conjugate complex numbers is a real number.

12. Show graphically that the difference of two conjugate complex numbers is an imaginary number.

12–6. Trigonometric and Polar Forms of Complex Numbers

In Sec. 12–5, the vectorial representation of a complex number was given. In the present section a method of writing the complex number itself in terms of the length and direction angle of its vector will be given.

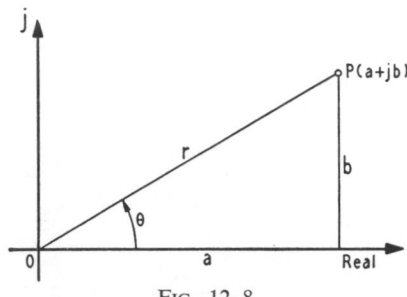

FIG. 12–8

Let the complex number $a + jb$ be represented by the point P in Fig. 12–8. Join the point P to the origin. Denote the length of OP by r and an angle that OP makes with the positive direction of the real axis by θ. From the right triangle in Fig. 12–8 we obtain the following equations:

$$a = r \cos \theta, \quad b = r \sin \theta. \tag{1}$$

$$r = \sqrt{a^2 + b^2}, \tan \theta = \frac{b}{a}. \tag{2}$$

These equations hold, irrespective of the quadrant in which θ terminates.

If we use the equations in (1), the complex number can now be written in the form

$$a + jb = r(\cos \theta + j \sin \theta). \tag{3}$$

The form $r(\cos \theta + j \sin \theta)$ is called the **trigonometric form** of the complex number, while the form $a + jb$ is spoken of as the **rectangular form** of the complex number.

In the trigonometric form, r is sometimes called the **modulus** or the **absolute value** (magnitude) of the complex number, and the angle θ is called the **amplitude** or **argument** of the complex number. In our work *we shall simply call r the* **magnitude** *of the complex number and* θ *the* **angle** *of the complex number.*

A very convenient way of writing a complex number is

$$r\underline{/\theta}. \qquad (4)$$

This form is commonly called the **polar form** and is particularly useful when the complex number represents a vector.   This polar form $r\underline{/\theta}$ must be interpreted as a complex number having a magnitude r and making an angle of θ degrees with the axis of reals.   In this notation the magnitude r and the angle θ do *not* form a product.   The form $r\underline{/\theta}$ is simply a shorthand notation for $r (\cos \theta + j \sin \theta)$.

We now summarize the various forms in which a complex number has so far been represented

Rectangular Form	Trigonometric Form	Polar Form
$a + jb$	$r(\cos \theta + j \sin \theta)$	$r\underline{/\theta}$

It is important that the student be able to change one form of the complex number into another.   This will be illustrated by the following examples.

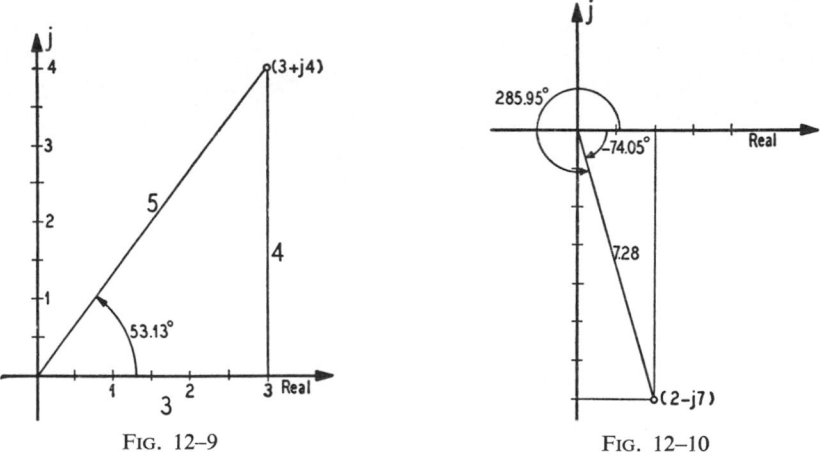

FIG. 12–9   FIG. 12–10

Example 1.   Express the complex number $3 + j4$ in trigonometric and polar forms.

Using equations 2 with $a = 3$ and $b = 4$ and the trigonometric tables, we obtain

$$r = \sqrt{3^2 + 4^2} = 5,$$

$$\tan \theta = \tfrac{4}{3} = 1.3333, \qquad \theta = 53.13°.$$

Hence

$$3 + j4 = 5(\cos 53.13° + j \sin 53.13°) = 5\underline{/53.13°},$$

which gives, respectively, the rectangular, trigonometric, and polar forms.   The graphical representation is given in Fig. 12–9.

Example 2. Express the complex number $2 - j7$ in trigonometric and polar forms.

Equations 2 with $a = 2$ and $b = -7$ yield

$$r = \sqrt{2^2 + (-7)^2} = \sqrt{53} = 7.28,$$

$$\tan \theta = \frac{-7}{2} = -3.50.$$

Since a is positive and b is negative, θ is in the fourth quadrant. From the trigonometric table we find

$$\tan 74.05° = 3.50.$$

Therefore

$$\theta = 360° - 74.05° = 285.95° \quad \text{or} \quad \theta = -74.05°.$$

Thus

$$2 - j7 = 7.28(\cos 285.95° + j \sin 285.95°) = 7.28\underline{/285.95°},$$

or

$$2 - j7 = 7.28[\cos (-74.05°) + j \sin (-74.05°)] = 7.28\underline{/-74.05°}.$$

These are, respectively, the rectangular, trigonometric, and polar forms of the complex number under consideration. The graphical representation is given in Fig. 12–10.

We summarize the method illustrated in the last two examples by stating that, *in order to transform the rectangular form of a complex number into the trigonometric or polar form, expressions 2 of this section should be used.*

By equation 3 any nonzero real number a can be regarded as a complex number of magnitude $|a|$ and of angle either $0°$ or $180°$, depending on whether a is positive or negative, respectively. Therefore

$$a = |a|(\cos 0° + j \sin 0°) = |a|\underline{/0°}, \quad \text{when} \quad a > 0,$$

$$a = |a|(\cos 180° + j \sin 180°) = |a|\underline{/180°}, \quad \text{when} \quad a < 0,$$

give, respectively, the rectangular, trigonometric, and polar forms of any nonzero real number a.

Similarly, any imaginary number jb, where $b \neq 0$, can be regarded as a complex number of magnitude $|b|$ and of angle either $90°$ or $270°$, depending on whether b is positive or negative, respectively. Therefore,

$$jb = |b|(\cos 90° + j \sin 90°) = |b|\underline{/90°}, \quad \text{when} \quad b > 0,$$

$$jb = |b|(\cos 270° + j \sin 270°) = |b|\underline{/270°}, \quad \text{when} \quad b < 0,$$

give, respectively, the rectangular, trigonometric, and polar forms of any imaginary number jb.

If zero is thought of as a complex number, it is clear that equations 2 do not determine an angle for the trigonometric and polar forms. However, this ambiguity can be avoided simply by always making use of this number in its rectangular form.

Example 3. Express the complex number $4\underline{/132.7°}$ in trigonometric and rectangular forms.

Changing into the trigonometric form we have at once

$$4\underline{/132.7°} = 4(\cos 132.7° + j\sin 132.7°).$$

From the trigonometric table, we find

$$\cos 132.7° = -0.6782, \sin 132.7° = 0.7349.$$

Therefore

$$4(\cos 132.7° + j\sin 132.7°) = 4(-0.6782 + j0.7349) = -2.7128 + j2.9396,$$

which is the rectangular form of the given complex number. Thus $4\underline{/132.7°} = 4(\cos 132.7° + j\sin 132.7°) = -2.71 + j2.94.$ The graphical representation is given in Fig. 12–11.

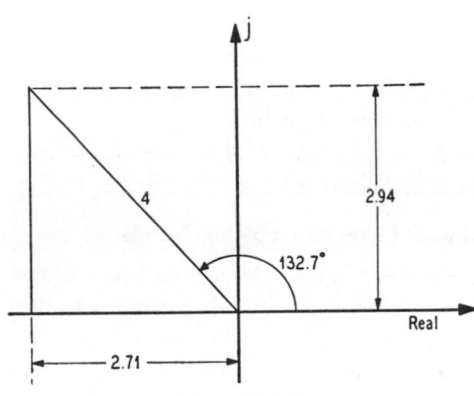

FIG. 12–11

Thus, as illustrated by the above example, *to transform the polar or trigonometric form of a complex number into the rectangular form, evaluate the sine and cosine of the angle by trigonometric tables and simplify the result.*

Exercises

In each example write the complex number in trigonometric and polar form. Plot the corresponding point.

1. $1 + j$. 2. $-1 + j$. 3. 1. 4. -1.

5. 3. 6. -8. 7. j. 8. $-j$.

9. $j3$. 10. $3 - j5$. 11. $-2 + j9$. 12. $-1 - j5$.

13. $1 + j\sqrt{3}$. 14. $\frac{1}{2}(\sqrt{3} + j)$. 15. $1.6 + j2.8$. 16. $-2.4 + j3.5$.

17. $\frac{1}{2} - j\frac{1}{3}$. 18. $5.2 - j7.8$.

In each example find the rectangular form of the complex number. Plot the corresponding point.

19. $6(\cos 90° + j \sin 90°)$.

20. $3\,\underline{/270°}$.

21. $2(\cos 180° - j \sin 180°)$.

22. $4\,\underline{/45°}$.

23. $\frac{1}{2}(\cos 0° + j \sin 0°)$.

24. $\frac{1}{3}\,\underline{/60°}$.

25. $5(\cos 30° - j \sin 30°)$.

26. $12\,\underline{/-90°}$.

27. $8(\cos 27° + j \sin 27°)$.

28. $9\,\underline{/42°}$.

29. $2.4(\cos 71.2° - j \sin 71.2°)$.

30. $5.2\,\underline{/-65.3°}$.

31. $6.12\,\underline{/123.54°}$.

32. $7.08\,\underline{/143.25°}$.

33. $1.71\,\underline{/-82.53°}$.

34. $12.07\,\underline{/215.71°}$.

35. Show that $r\,\underline{/\theta}$ and $r\,\underline{/-\theta}$ are conjugates.

36. Show graphically that the magnitude of the sum of two complex numbers is less than or equal to the sum of their magnitudes.

37. Show graphically that the magnitude of the difference of two complex numbers is greater than or equal to the difference of their magnitudes.

12–7. The Exponential Form of Complex Numbers

In Sec. 8–4 the idea of an exponent was enlarged to include expressions of the form a^x where x was any rational number. In Sec. 8–14 we gave a meaning to a^x where x and a have any real values. Having introduced complex numbers, it would be very desirable to use them for exponents. One is free to give any meaning to the expression a^{x+jy}. In view of the discussion given in Sec. 8–5, it is important to define a^{x+jy} in such a fashion that the laws of exponents remain true and so that it will be necessary to introduce as few additional new formulas as possible. If we require the ordinary laws of exponents to hold, then a^{x+jy} must equal $a^x \cdot a^{jy}$, and, since a^x has already been defined, we can give a meaning to a^{x+jy} by giving meaning to a^{jy}. This is done in more advanced treatises on complex numbers; here we will define this expression only for $a = e$, where e is the number discussed in Sec. 8–14.

To define the expression $e^{j\theta}$ we write

$$e^{j\theta} = \cos \theta + j \sin \theta. \tag{1}$$

If the number θ is thought of as an angle (as is frequently the case), it is understood to be given in radian measurement.

Using trigonometric identities, advanced mathematics books show that expressions of the form $e^{j\theta}$, as defined in equation 1, obey all the laws of exponents stated in Chapter 8.

The importance of equation 1 here lies in the fact that we can use it now to give another form of the complex number. Comparing the definition of $e^{j\theta}$ with equation 3 in Sec. 12–6, one obtains

$$a + jb = r \,(\cos\theta + j\sin\theta) = re^{j\theta} = r\,\underline{/\theta}. \qquad (2)$$

The expression $re^{j\theta}$ is the **exponential form** *of a complex number.* The polar form $r\underline{/\theta}$ is now a shorthand notation for the trigonometric form $r\,(\cos\theta + j\sin\theta)$ or for the exponential form $re^{j\theta}$.

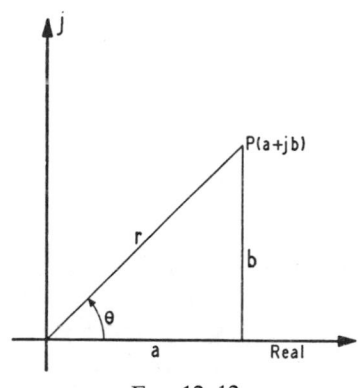

FIG. 12–12

The following table with the graphical representation of Fig. 12–12 gives a summary of the various forms in which a complex number can be represented.

Form	Complex Number
Rectangular	$a + jb$
Trigonometric	$r(\cos\theta + j\sin\theta)$
Exponential	$re^{j\theta}$
Polar	$r\underline{/\theta}$

In the trigonometric and polar forms θ may be expressed in terms of radians or degrees (most of the electrical engineering books use degrees). In the exponential form, θ must be given in radians, although books on electrical engineering often express θ in degrees. However, when θ is expressed in degrees, it is to be understood that the corresponding angle in radians is intended. For example, since $\theta = 54.3° = 0.948$ radian, engineering books write $e^{j54.3°}$ for $e^{j0.948}$. Thus $e^{j54.3°}$ and $\underline{/54.3°}$ are simply notations for the correct $e^{j0.948}$.

It is important that the student be able to change one form of the complex number into another. This will be illustrated by the following examples.

Example 1.   Express the complex number $3 + j4$ in the different forms.

Using the results obtained in example 1 of the preceding section, we obtain
$$3 + j4 = 5(\cos 53.13° + j \sin 53.13°) = 5e^{j0.927} = 5\underline{/53.13°},$$
since $53.13° = 0.927$ radians (see Table 6).

Example 2.   Similarly from example 2 of Sec. 12–6, one obtains
$$2 - j7 = 7.28[\cos(-74.05°) + j \sin(-74.05°)] = 7.28e^{j1.292} = 7.28 \underline{/-74.05°},$$
since $74.05° = 1.292$ radians.

Example 3.   Express the complex number $10e^{j0.37}$ in rectangular form. Using Table 5, we get
$$10e^{j0.37} = 10(\cos 0.37 + j \sin 0.37) = 10(0.932 + j0.362)$$
$$= 9.32 + j3.62.$$

Example 4.   Express the complex number $e^{j\pi}$ in rectangular form. We have at once $e^{j\pi} = \cos \pi + j \sin \pi = -1$.

Exercises

In each example write the complex number in exponential form.

1. 1.   2. -1.   3. j.   4. $-j$.   5. $1 + j$.

6. $1 - j$.   7. $-1 + j$.   8. $-j5$.   9. $\sqrt{3} + j$.   10. $1 - j\sqrt{3}$.

11. $3 - j8$.   12. $2.4 + j7.1$.   13. $-3.7 - j5.4$.

14. $3.5\underline{/152.3°}$.   15. $11.2\underline{/-75.1°}$.   16. $5\left(\cos\dfrac{\pi}{2} - j \sin \dfrac{\pi}{2}\right)$.

17. $3(\cos 41° + j \sin 41°)$.

In each example write the complex number in rectangular form.

18. $e^{j2\pi}$.   19. $e^{j(\pi/2)}$.   20. $e^{j(\pi/4)}$.   21. $e^{-j\pi}$.

22. $e^{-j2\pi}$.   23. $e^{-j(\pi/2)}$.   24. $e^{-j(\pi/4)}$.   25. $e^{j5\pi}$.

26. $e^{-j7\pi}$.   27. $e^{j1.5}$.   28. $e^{j3.8}$.   29. $e^{-j2.7}$.

30. $e^{-j0.7}$.   31. $e^{-j(3\pi/2)}$.   32. $e^{j5.9}$.

12–8. Multiplication and Division of Complex Numbers in Polar Form

In Sec. 12–4, the operations with complex numbers in the rectangular form $a + jb$ were discussed.   Multiplication and division of complex numbers are much simpler when the numbers are given in a polar form.

Let $r_1\underline{/\theta_1}$ and $r_2\underline{/\theta_2}$ be two complex numbers in their polar form.   By multiplication,

$$r_1\underline{/\theta_1} \cdot r_2\underline{/\theta_2} = r_1 e^{j\theta_1} \cdot r_2 e^{j\theta_2}$$
$$= r_1 r_2 e^{j\theta_1 + j\theta_2} = r_1 r_2 e^{j(\theta_1 + \theta_2)}$$
$$= r_1 r_2 \underline{/\theta_1 + \theta_2}.$$
$$r_1\underline{/\theta_1} \cdot r_2\underline{/\theta_2} = r_1 r_2 \underline{/\theta_1 + \theta_2}. \tag{1}$$

Therefore, the product of two complex numbers is a complex number whose magnitude is the product of the magnitudes of the numbers and whose angle is the sum of their angles.

It can readily be seen that this holds for the product of any number of complex quantities. Thus in the case of three complex numbers we have

$$r_1\underline{/\theta_1} \cdot r_2\underline{/\theta_2} \cdot r_3\underline{/\theta_3} = r_1 r_2 r_3\underline{/\theta_1 + \theta_2 + \theta_3}.$$

Similarly, we have for the quotient of two complex numbers

$$\frac{r_1\underline{/\theta_1}}{r_2\underline{/\theta_2}} = \frac{r_1 e^{j\theta_1}}{r_2 e^{j\theta_2}} = \frac{r_1}{r_2} e^{j\theta_1 - j\theta_2} = \frac{r_1}{r_2} e^{j(\theta_1 - \theta_2)} = \frac{r_1}{r_2}\underline{/\theta_1 - \theta_2}.$$

$$\frac{r_1\underline{/\theta_1}}{r_2\underline{/\theta_2}} = \frac{r_1}{r_2}\underline{/\theta_1 - \theta_2}. \tag{2}$$

Therefore, the quotient of two complex numbers $r_1\underline{/\theta_1}$ and $r_2\underline{/\theta_2}$ is a complex number whose magnitude is the quotient r_1/r_2 of the magnitudes of the numbers and whose angle is the difference $\theta_1 - \theta_2$ of their angles.

In order to use the methods of this section in multiplying and dividing complex numbers, they must first be expressed in polar form (see Sec. 12–6).

Example 1. Find the product $(3 + j4)(2 - j7)$.

Using the results obtained in examples 1 and 2 of Sec. 12–6, we have

$$(3 + j4)(2 - j7) = 5.00\underline{/53.13°} \cdot 7.28\underline{/285.95°}$$

$$= 5.00 \cdot 7.28\underline{/53.13° + 285.95°}$$

$$= 36.40\underline{/339.08°} = 36.40\underline{/-20.92°}$$

$$= 36.40(\cos 20.92° - j\sin 20.92°)$$

$$= 36.40(0.9341 - j0.3570)$$

$$= 34.00 - j12.99.$$

The answer can be given in the form

$$36.40\underline{/-20.92°} \quad \text{or} \quad 34.00 - j12.99.$$

Direct multiplication (method of Sec. 12–4) gives

$$(3 + j4)(2 - j7) = 34 - j13.$$

Example 2. Perform the multiplication and division in

$$\frac{(67.4 - j37.7)\,(35.0 + j75.0)}{(45.0 - j12.9)\,(14.6 + j8.86)}.$$

Converting from rectangular form to polar form, by methods of Sec. 12–6, we obtain

$$(67.4 - j37.7) = 77.2\underline{/-29.2°},$$

$$(35.0 + j75.0) = 82.8\underline{/65.0°},$$

$$(45.0 - j12.9) = 46.8\underline{/-16.0°},$$

$$(14.6 + j8.86) = 17.1\underline{/31.2°}.$$

The given expression is equal to

$$\frac{77.2\underline{/-29.2°} \cdot 82.8\underline{/65.0°}}{46.8\underline{/-16.0°} \cdot 17.1\underline{/31.2°}} = \frac{77.2 \cdot 82.8\underline{/-29.2° + 65.0°}}{46.8 \cdot 17.1\underline{/-16.0° + 31.2°}} = 7.99\frac{\underline{/35.8°}}{\underline{/15.2°}}$$

$$= 7.99\underline{/35.8° - 15.2°} = 7.99\underline{/20.6°}$$

$$= 7.99(\cos 20.6° + j \sin 20.6°)$$

$$= 7.99(0.9361 + j0.3518) = 7.48 + j2.81.$$

Hence the answer is $7.99\underline{/20.6°}$ in polar form and $7.48 + j2.81$ in rectangular form.

12–9. Powers and Roots of Complex Numbers

The method of the previous section may now be applied to powers and roots of complex numbers.

Let $r\underline{/\theta}$ be any complex number and n any positive integer; then

$$(r\underline{/\theta})^n = (re^{j\theta})^n = r^n e^{jn\theta} = r^n\underline{/n\theta}.$$

$$(r\underline{/\theta})^n = r^n \underline{/n\theta}. \tag{1}$$

Therefore *when a complex number is raised to the nth power, the result is a complex quantity whose magnitude is the nth power of the original magnitude and whose angle is n times the original angle.*

Relation 1 holds for both integral and fractional values of n. Using fractional values, we obtain a simple method for finding the roots of complex quantities. Thus one of the mth roots of a complex number is given by

$$\sqrt[m]{r\underline{/\theta}} = (r\underline{/\theta})^{\frac{1}{m}} = \sqrt[m]{r} \underline{/\frac{\theta}{m}}. \tag{2}$$

When the complex number is given in its trigonometric form, relations 1 and 2 can be written in the form

$$[r(\cos \theta + j \sin \theta)]^n = r^n(\cos n\theta + j \sin n\theta) \tag{3}$$

known as *DeMoivre's theorem,* and it holds true for all real values of n.

Example 1. Find $(4.60 + j2.82)^3$.

Converting into polar form, we have

$$(4.60 + j2.82)^3 = (5.40\underline{/31.5°})^3 = 5.40^3\underline{/3 \cdot 31.5°} = 157.5\underline{/94.5°}.$$

In polar form the answer is $157.5\underline{/94.5°}$.   Let the student express this answer also in rectangular form.

Example 2.   Find $\sqrt{4.60 + j2.82}$.

Using the result of example 1 and relation 2, we have

$$\sqrt{4.60 + j2.82} = (5.40\underline{/31.5°})^{\frac{1}{2}} = \sqrt{5.40}\underline{/15.8°} = 2.32\underline{/15.8°}.$$

Since

$$\underline{/31.5°} = \underline{/360° + 31.5°} = \underline{/391.5°},$$

we also have

$$\sqrt{4.60 + j2.82} = (5.40\underline{/391.5°})^{\frac{1}{2}} = 2.32\underline{/195.8°}.$$

Hence $2.32\underline{/15.8°}$ and $2.32\underline{/195.8°}$ are the two square roots of $4.60 + j2.82$.

Following the method of the last example, it can be shown that *any complex number $r\underline{/\theta}$ has n distinct nth roots given by the formula*

$$\sqrt[n]{r}\ \underline{\bigg/\dfrac{\theta + k \cdot 360°}{n}}$$

where k takes the values $0, 1, 2, \cdots, n - 1.$

12–10. Graphical Representation

In Sec. 12–5 the graphical representation of addition and subtraction of complex numbers (in rectangular form) was given.   We shall now give the graphical representation of the operations discussed in Secs. 12–8 and 12–9.

Let P_1 and P_2 (Fig. 12–13) be the points representing the complex numbers $r_1\underline{/\theta_1}$ and $r_2\underline{/\theta_2}$.   Join point A (1, 0) to P_1.   Construct $\angle P_2OP$ equal to θ_1.   Construct $\angle OP_2P$ equal to $\angle OAP_1$.   From the fact that $\triangle OAP_1$ is similar to $\triangle OP_2P$, it can be shown that $OP = r_1r_2$.   Hence point P represents the complex number $r_1r_2\underline{/\theta_1 + \theta_2}$, which is the product of $r_1\underline{/\theta_1}$ and $r_1\underline{/\theta_2}$.

Similar graphical representation can be given to division and also to raising to a power and extracting roots of complex numbers.

In Sec. 12–5 it was indicated that in engineering a complex quantity is used to represent a vector.   In particular, making use of the table for the values of the powers of j (Sec. 12–1), we can represent:

> 1 by a vector of length 1 and an angle   0°,
> j by a vector of length 1 and an angle   90°,
> j^2 by a vector of length 1 and an angle 180°,
> j^3 by a vector of length 1 and an angle 270°,
> j^4 by a vector of length 1 and an angle 360°.

From Fig. 12–13 it follows that when a complex quantity is multiplied by j its vector is rotated through an angle of 90° in a counterclockwise direction. Therefore *j can be considered as an operator that rotates a vector in the complex plane through a positive angle of* 90°.

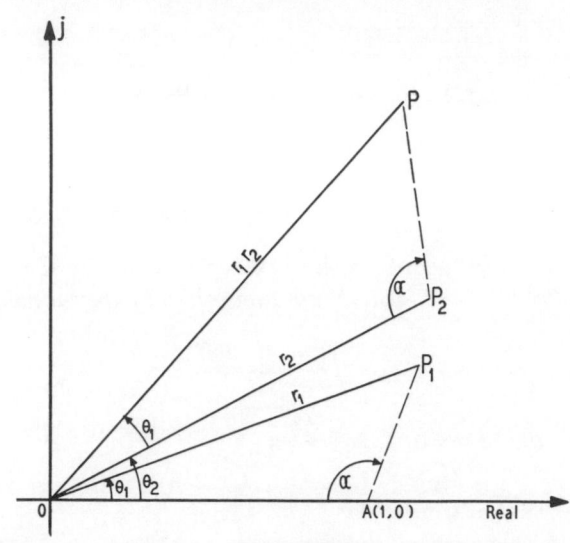

Fig. 12–13

Similar interpretations are given to the powers of j as follows:
j^2 corresponds to a 180° counterclockwise rotation.
j^3 corresponds to a 270° counterclockwise rotation or to a 90° clockwise rotation.

Exercises

Perform the indicated operations, first reducing the numbers to polar form (if necessary). In the odd-numbered exercises express the results also in rectangular form.

1. $4\underline{/35°} \cdot 5\underline{/18°}$.

2. $5.7\underline{/62.3°} \cdot 3.2\underline{/13.5°}$.

3. $3\underline{/25°} \cdot 2\underline{/13°} \cdot 7\underline{/37°}$.

4. $3.2\underline{/22.3°} \cdot 1.3\underline{/15.4°} \cdot 2.5\underline{/39.7°}$.

5. $(2 + j3)(4 + j)$.

6. $(4.3 - j6.8)(7.5 + j5.2)$.

7. $j(3 - j)(-1 + j4)$.

8. $(2.3 - j4.1)(3.7 + j5.2)(-0.5 + j1.2)$.

9. $(67.2\underline{/34.1°}) \div (18.5\underline{/12.7°})$.

10. $(2.85\underline{/43.7°}) \div (3.72\underline{/61.3°})$.

11. $(2 + j5) \div (1 - j)$.

12. $(2.3 + j6.5) \div (-3.6 + j4.7)$.

13. $(2 + j)^6$.

14. $(7.2 + j1.7)^3$.

15. $(11.2 - j17.5)^4$.

16. $(-1 + j)^{12}$.

17. $\dfrac{4.3\underline{/8.2^\circ} \cdot 3.1\underline{/-12.5^\circ}}{1.7\underline{/21.4^\circ} \cdot 2.3\underline{/31.2^\circ}}$.

18. $\dfrac{(12.7 + j7.31)(-5.21 + j18.8)}{(23.4 - j17.5)}$.

19. $\dfrac{8.2}{(9.1 + j3.9)(-5.7 + j7.5)}$.

20. $\dfrac{(2 - j)^3(-1 + j)^2}{(1 + j2)(3 - j)^2}$.

Perform the following multiplications graphically, and check the results algebraically.

21. $j(5 + j3)$.

22. $j(1 - j)$.

23. $j \cdot 4\underline{/30^\circ}$.

24. $j \cdot 2\underline{/-90^\circ}$.

25. $(1 + j)(2 + j3)$.

26. $(2 - j)(4 + j)$.

27. $j^2(1 - j2)$.

28. $j^3 \cdot 2\underline{/-270^\circ}$.

29. $j^3(2 - j)$.

Find the following roots. Express the results in rectangular form.

30. $\sqrt{j}$.

31. $\sqrt{-j}$.

32. $\sqrt{1 + j}$.

33. $\sqrt{2.3 - j3.8}$.

34. $\sqrt[3]{1 + j2}$.

35. $\sqrt[3]{j}$.

36. $\sqrt[3]{7.8 + j4.7}$.

37. $\sqrt[4]{j}$.

38. $\sqrt[4]{-1}$.

39. Prove formula 1 of Sec. 12–8 by writing the numbers in trigonometric form and by using some of the trigonometric identities from Chapter 10.

40. Prove formula 2 of Sec. 12–8 by writing the numbers in trigonometric form and by using some of the trigonometric identities from Chapter 10.

12–11. Summary of Operations with Complex Numbers

A complex quantity may be represented in a rectangular or polar form. One form may be used more advantageously than the other in various operations.

1. *Addition of complex numbers.* Write the numbers in the rectangular form. Add the real parts together, and then add the j parts together.

2. *Subtraction of complex numbers.* Write the numbers in the rectangular form. Subtract the real parts, and then subtract the j parts.

3. *Multiplication of complex numbers.* Write the numbers in the polar form. Multiply the magnitudes together, and add the angles.

4. *Division of complex numbers.* Write the numbers in the polar form. Take the quotient of the magnitudes, and subtract the angles.

5. *Powers of complex numbers.* Write the numbers in the polar form. Take the power of the magnitude, and multiply the angle by the exponent of the power.

6. *Roots of complex numbers.* Write the numbers in the polar form. Extract the root of the magnitude, and divide the angle by the index of the root.

The answer to a problem involving operations with complex numbers may be given in either the rectangular or the polar form. In engineering applications, the answer is usually given in the polar form.

Example 1.

$20\underline{/23°} + 11\underline{/-37°} - 15\underline{/47°} = (18.4 + j7.81) + (8.78 - j6.62) - (10.2 + j11.0)$
$$= (18.4 + 8.78 - 10.2) + j(7.81 - 6.62 - 11.0)$$
$$= 17.0 - j9.81.$$

The student should convert the answer into the polar form.

Example 2. Using some of the results obtained in example 2 of Sec. 12–8 in converting from the rectangular to polar form, we have

$$\frac{(67.4 - j37.7)^2(14.6 + j8.86)}{(45.0 - j12.9)^3} = \frac{(77.2\underline{/-29.2°})^2(17.1\underline{/31.2°})}{(46.8\underline{/-16.0°})^3}$$

$$= \frac{5960\underline{/-58.4°} \cdot 17.1\underline{/31.2°}}{103{,}000\underline{/-48.0°}}$$

$$= \frac{5960 \cdot 17.1\underline{/-27.2°}}{103{,}000\underline{/-48.0°}} = 0.989\underline{/20.8°}.$$

The student should convert this answer into the rectangular form.

Exercises

Perform the indicated operations. Express the results in rectangular and polar forms.

1. $(3.2 + \sqrt{-7.6}) + (4.5 - \sqrt{-9.3}).$

2. $2.9\underline{/37°} - 5.7\underline{/78°}.$

3. $2e^{j(\pi/4)} - e^{-j(\pi/6)} + 3e^{j(\pi/3)}.$

4. $12e^{j1.3} + 17e^{-j2.0}.$

5. $7.1 - j3.7 + 3.1(\cos 62° + j \sin 62°).$

6. $8.2\underline{/-16°} + 1.3\underline{/71°}.$

7. $15e^{j0.72} - (2.3 - j7.5).$

8. $2.8e^{-j1.2} + 7.2\underline{/83°}.$

9. $5.6(\cos 21° - j \sin 21°) + 9.1\underline{/-14°}.$

10. $1.9\underline{/-24°} - 3.4\underline{/-37°}.$

11. $(7 - j2)^3(-4 + j3).$

12. $(3 + j)(1 - j)^4.$

13. $(3.2 - j4.5)^2 - (7.2 - j1.3).$

14. $(2 + j)^3 + (4 - j3).$

15. $3.2e^{j0.78} \cdot 7.3e^{-j2.13}.$

16. $(2.4e^{-j0.63})^2 \cdot (1.8e^{j0.42})^3.$

17. $\dfrac{j2.8}{(6.3 - j2.5)^2}.$

18. $\dfrac{3}{(2 - j5)^3(7 + j)^2}.$

19. $\dfrac{3.7\underline{/13°} \cdot 5.2\underline{/-24°}}{7.3\underline{/42°} \cdot 1.3\underline{/34°}}.$

20. $\dfrac{[2.6(\cos 18° + j \sin 18°)]^4}{[1.7(\cos 23° - j \sin 23°)]^3}.$

21. $(1 + j)^{40}.$

22. $(1 - j)^{30}.$

23. $(5.2\underline{/15°})^3(3.7\underline{/-18°})^2.$

24. $(7.8\underline{/271°})^2(3.1\underline{/137°})^3.$

25. $\dfrac{(5.5\underline{/-12°})^2(8.2\underline{/153°})^3}{(2.6\underline{/71°})^4}.$

26. $\dfrac{(6.2e^{j1.2})^2(3.1e^{-j0.3})^3}{(2.2e^{j0.2})^4}.$

27. $\dfrac{(1.7 - j2.3)^3(-2.8 + j3.1)^4}{(4.2 + j5.3)^4}$.

28. $\dfrac{(-2.6 + j4.1)^2(3.7 - j5.4)^3}{(1.8 + j2.9)(8.3 - j9.1)^2}$.

29. $\sqrt[3]{-j}$.

30. $\sqrt{1 - j}$.

31. $\sqrt{7.4 + j5.2}$.

32. $\sqrt[3]{2 - j3}$.

33. $\sqrt{8.3 \underline{/128°}}$.

34. $\sqrt[3]{8 \underline{/-23°}}$.

35. $\sqrt[4]{-j}$.

36. $\sqrt{9.3e^{j1.3}}$.

37. $\sqrt[3]{27(\cos 30° - j \sin 30°)}$.

38. $\sqrt{11.2(\cos 162° + j \sin 162°)}$.

39. Using equation 1 of Sec. 12–7, show that

(a) $\cos \theta = \dfrac{e^{j\theta} + e^{-j\theta}}{2}$,

(b) $\sin \theta = \dfrac{e^{j\theta} - e^{-j\theta}}{j2}$,

(c) $e^{jn\theta} + e^{-jn\theta} = 2 \cos n\theta$,

(d) $e^{jn\theta} - e^{-jn\theta} = j2 \sin n\theta$.

12–12. The Application of Complex Numbers to the Analysis of A-C Circuits

One of the most important applications of complex numbers occurs in the solution of problems dealing with alternating currents and voltages. In this section we will give a brief discussion of such an application.

Electric current frequently is generated by a machine which produces rapid fluctuations in the magnitude of the current so that the current flowing at any time t in a circuit is of the form $I \sin \omega t$, where ω depends on the speed of rotation of the machine and I is the maximum value the current attains. A generator of this type is called a sinusoidal generator. Experiments show that a simple circuit in which such an alternating current flows can have three different fundamental properties, resistance, inductance, and capacitance, and, further, that a circuit can be designed so that each of these three properties predominates in a separate piece of equipment.

Consider a simple circuit in which there is a sinusoidal generator, a resistor, an inductor, and a capacitor (also frequently called a condenser), and suppose that these elements are connected so that the same current flows through each element. Such a circuit is called a series circuit and is wired as shown in the schematic diagram of Fig. 12–14, in which the lines connecting the four basic elements represent wires.

The fact that the student is probably not familiar with the electrical terms used so far in the discussion will not be a handicap, even though they will not be given their physical definitions here, since the mathematical properties we are interested in will be fully defined later in the discussion. The details of the basic physical meanings of these terms may be sought in any textbook in a-c circuits.

Suppose now that the generator in the circuit of Fig. 12–14 causes a current of the form $I \sin \omega t$ to flow through each of the elements of the circuit. Then across the resistor there will be a voltage drop which can be measured; similar voltage drops also exist across the inductor and the capacitor. About these voltage drops experiments have established the following facts:

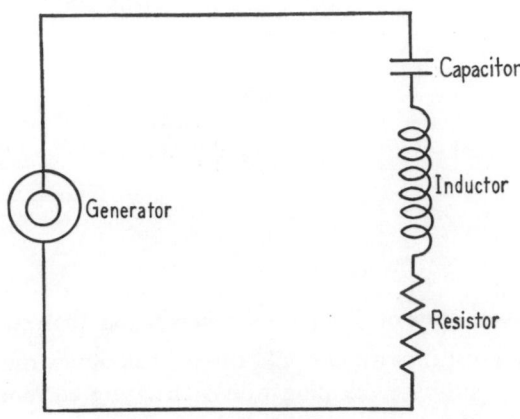

FIG. 12–14

1. The voltage drop across the resistor is $IR \sin \omega t$, where R is a quantity called the resistance, which depends on the physical design of the resistor.

2. The voltage drop across the inductor is $IX_L \sin (\omega t + 90°)$, where X_L is a quantity called the inductive reactance which depends on the physical design of the inductor.

3. The voltage drop across the capacitor is $IX_C \sin (\omega t - 90°)$, where X_C is a quantity called the capacitive reactance which depends on the design of the capacitor.

It follows that the total voltage drop across all three elements is the sum of these terms:

$$IR \sin \omega t + IX_L \sin (\omega t + 90°) + IX_C \sin (\omega t - 90°). \qquad (1)$$

A law due to Kirchhoff states that the voltage produced by the generator must equal this sum.

If the current I is measured in a unit called the ampere, and if R, X_L, and X_C are measured in a unit called the ohm, then the voltage drops are given in a unit called the volt.

We may now apply the results of Sec. 10–4 to the sum 1. These results state that, if we consider three vectors

$$\mathbf{E}_R: \quad IR, \qquad 0°,$$
$$\mathbf{E}_L: \quad IX_L, \qquad 90°,$$
$$\mathbf{E}_C: \quad IX_C, \qquad -90°,$$

and let $\mathbf{E} = \mathbf{E}_R + \mathbf{E}_L + \mathbf{E}_C$, then the sum 1 equals $E \sin(\omega t + \varepsilon)$, where E is the magnitude of $\mathbf{E}$ and ε is its direction angle. It is easy to see from Fig. 12–15 that

$$E = I\sqrt{R^2 + (X_L - X_C)^2}, \qquad \tan \varepsilon = \frac{X_L - X_C}{R}.$$

Let us now superimpose a complex plane on the vector plane of Fig. 12–15 so that the horizontal and vertical axes coincide and have the same directions. Then the terminal points of the three vectors $\mathbf{E}_R$, $\mathbf{E}_L$, $\mathbf{E}_C$

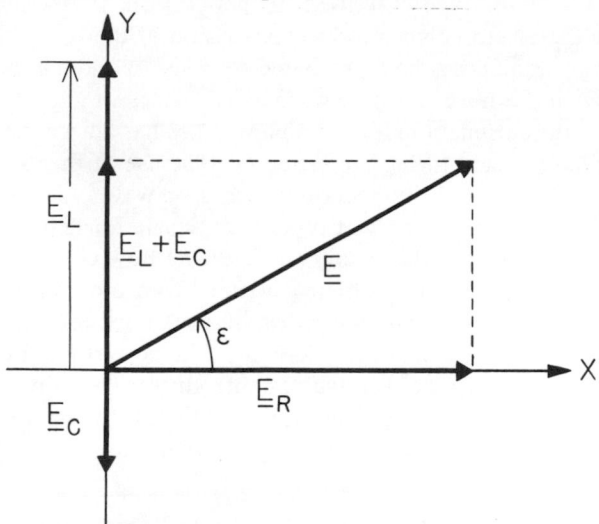

Fig. 12–15

correspond to the complex numbers $E_R = IR$, $jE_L = jIX_L$, $-jE_C = -jIX_C$, respectively, and the vector sum $\mathbf{E}$ has a terminal point corresponding to the sum of these three complex numbers. Thus it is common in practice to identify the complex numbers and the corresponding vectors by writing

$$\mathbf{E}_R = E_R, \quad \mathbf{E}_L = jE_L, \quad \mathbf{E}_C = -jE_c,$$

whence

$$\mathbf{E} = E_R + jE_L - jE_C = IR + jI(X_L - X_C). \tag{2}$$

If we set $\mathbf{Z} = R + j(X_L - X_C)$, then this complex number is called the impedance, and, if we let $\mathbf{I} = I$, then from equation 2,

$$E = IZ, \tag{3}$$

a direct analogue of Ohm's law, a fundamental law governing simple d-c circuits (see Sec. 2–18).

Of course, the term vector cannot be used in the sense of Chapter 6 in connection with equation 3, for no vector product was defined in that chapter. Furthermore the vector products defined in advanced works do not behave like the complex numbers in the product 3. Therefore, in order to have meaning, equation 3 must be thought of as expressing a relation between quantities given in the form of complex numbers. However, in texts on electrical engineering, the term vector is quite frequently used in this connection as synonymous with complex number, rather than in the sense of Chapter 6.

In problems of the type described in this section, it is customary to consider the current to be given by an expression of the form I sin ωt, so that the current can always be represented by a vector $\mathbf{I}$ along the positive x axis. Then, if a voltage vector has a positive direction angle the voltage is said to lead the current, and if the voltage vector has a negative direction angle the voltage is said to lag the current. The use of this terminology here arises from its use in connection with sine waves (see Sec. 5–17).

This brief account of the application of complex numbers to the analysis of a simple circuit is intended merely to indicate how complex numbers can be brought to bear on problems arising from a-c circuits. Many volumes have been written which exploit this technique for more complicated circuits, since it is compact, simple, and powerful. For further information the student should consult an introductory text on a-c circuits. The example that follows will show how the complex-number technique of analyzing an a-c circuit can be used in a specific case.

Example. A 5-ohm resistance is connected in series with an 18-ohm inductive reactance and a 6-ohm capacitive reactance across a 60-cycle, 110-volt a-c supply line. Calculate the magnitudes of the impedance, the resistance voltage drop, the reactance voltage drop, and the current.

The impedance of the circuit is given by

$$\mathbf{Z} = 5 + j(18 - 6) = 5 + j12 = 13\underline{/67.4°}.$$

Since $\mathbf{E} = \mathbf{IZ} = I(5 + j12) = 13I\underline{/67.4°}$, by Kirchhoff's law it follows that this is equal to the supply line voltage, which has a magnitude of 110 volts. Thus

$$\mathbf{E} = 110\underline{/67.4°} = 110(\cos 67.4° + j \sin 67.4°) = 110(0.3843 + j0.9232)$$
$$= 42.27 + j101.55.$$

It follows that the magnitude of the resistance voltage drop is 42.27 volts and the magnitude of the reactance voltage drop is 101.55 volts.   It is further clear

that the magnitude of the current is $\dfrac{110}{13}$ = 8.46 amperes.   This last value can

be checked from the relation $\mathbf{I} = \dfrac{\mathbf{E}}{\mathbf{Z}}$.   Substituting the values for $\mathbf{E}$ and $\mathbf{Z}$
obtained above gives

$$\mathbf{I} = \frac{42.27 + j101.55}{5 + j12} = \frac{(42.27 + j101.55)(5 - j12)}{(5 + j12)(5 - j12)} = \frac{1430}{169} = 8.46,$$

which can be written in the polar form

$$\mathbf{I} = \frac{110\underline{/67.4^\circ}}{13\underline{/67.4^\circ}} = 8.46\underline{/0^\circ}.$$

Exercises

1. An a-c magnetic chuck operating on 115 volts has a resistance of 12 ohms and an inductive reactance of 16 ohms.   Calculate the magnitudes of the impedance of the chuck and of the current flowing through it.

2. An alternating voltage of 440 volts is applied to a brake solenoid having a resistance of 26 ohms and an inductive reactance of 72 ohms.   What is the magnitude of the current flowing through the solenoid?

3. An inductive reactance of 14 ohms is connected in series with a resistance of 9 ohms across a 120-volt 60-cycle a-c circuit.   What is the voltage drop across the reactance?

4. A circuit consisting of an 8-ohm resistor and a capacitor having a capacitive reactance of 46 ohms connected in series is placed across the timing contacts of a seconds pendulum.   If the voltage applied across the contacts is 110 volts, what are the magnitudes of the impedance, resistance voltage drop, reactance voltage drop, and current in the circuit?

5. Two resistors having values of 32 and 14 ohms, respectively, are connected in series, and they are connected in series with an inductive reactance of 26 ohms across a 220-volt a-c circuit.   What is the magnitude of the current that flows in the circuit, and what is the voltage drop across the 32-ohm resistor?

6. A capacitive reactance of 2680 ohms is connected in series with a resistance of 1250 ohms across a 10,000-cycle 0.8 volt circuit.   What are the voltage drop in millivolts across the condenser and the current in milliamperes through the capacitor?

7. A simple series circuit connected to a 110-volt 60-cycle power source contains a resistance of 18 ohms, an inductive reactance of 22 ohms, and a capacitive reactance of 12 ohms.   What are the magnitudes of the impedance and current in the circuit?   What are the voltage drops across each element in the circuit?

8. A simple series circuit connected to a 440-volt 60-cycle power source contains a resistance of 56 ohms, an inductive reactance of 38 ohms, and a capacitive reactance of 96 ohms.   What are the magnitudes of the impedance and current in the circuit?   What are the voltage drops across each element in the circuit?

9. The impedance of a series circuit connected to a 440-volt 60-cycle power source is given by $132\underline{/38.4^\circ}$.   If the inductive reactance is 115 ohms, compute the capacitive reactance and the resistance of the circuit.

10. The impedance of a series circuit connected to a 220-volt 60-cycle power source is given by $88\underline{/66.3^\circ}$.   If the capacitative reactance is 36 ohms, compute the inductive reactance and the resistance of the circuit.

Progress Report

In this chapter an extension of the system of real numbers to the system of complex numbers was considered. After introducing the imaginary unit and defining complex numbers, we considered:

1. The graphical representation of complex numbers.

2. The trigonometric, polar, and exponential forms of complex numbers.

3. The addition, subtraction, multiplication, and division of complex numbers.

4. Powers and roots of complex numbers.

5. The application of complex numbers to simple a-c circuit problems.

13————————

Linear Equations and Determinants

Students of engineering, physics, and other sciences deal with equations at every stage of their work. In particular, systems of linear equations are used frequently in mechanics, electricity, statistics, and other applied sciences. This chapter gives some of the graphical and algebraic methods for solving such equations. It also contains an introductory discussion of the theory of determinants and their application to systems of linear equations.

13–1. Systems of Linear Equations with Two and Three Unknowns

Linear equations with two unknown quantities were discussed in Sec. 2–16. The system of two linear equations with two unknowns was solved by deriving from the given system of two equations a single equation containing only one of the two unknowns. This procedure is called elimination of one of the unknowns, and this unknown is said to be eliminated. Two elimination methods were discussed.

A. Elimination by substitution. This method consists in solving one equation for one of the unknowns and substituting the result in the second equation.

Example 1. Find x and y from the system of the two equations

$$x + y = 2, \tag{1}$$
$$x + 2y = 3. \tag{2}$$

Equation 1 solved for x gives

$$x = 2 - y, \tag{3}$$

and this value when substituted in equation 2 gives an equation with only one unknown quantity

$$2 - y + 2y = 3,$$

whence

$$y = 1.$$

The substitution of this value in equation 3 gives $x = 2 - 1 = 1$. Thus $x = 1$, $y = 1$ is the solution of the given system of two simultaneous equations.

B. Elimination by addition or subtraction.

Example 2. We shall study the same system as before.

$$x + y = 2, \tag{1}$$
$$x + 2y = 3. \tag{2}$$

If equation 1 is subtracted from equation 2, x is eliminated, and the result is $y = 1$. The substitution of this value in equation 1 gives $x + 1 = 2$, whence $x = 1$ as before.

The procedure in this case was extremely simple because the unknown x had the same coefficient in both equations. Otherwise, one or both equations must be multiplied by numerical factors which are chosen so that x or y has the same numerical coefficient in both equations.

Example 3. Solve the system

$$5x - 2y = 4, \tag{4}$$
$$2x + 3y = 10. \tag{5}$$

If equation 4 is multiplied by 3 and equation 5 by 2, the following equivalent system is obtained:

$$15x - 6y = 12,$$
$$4x + 6y = 20.$$

The unknown y can be eliminated by adding these two equations, thus obtaining

$$19x = 32 \quad \text{or} \quad x = \tfrac{32}{19}.$$

In order to find y, the variable x can be eliminated in a similar way. Multiplying the first equation by 2, the second by 5, and subtracting the first from the second, we obtain

$$10x - 4y = 8$$
$$10x + 15y = 50$$
$$\overline{}$$
$$19y = 42$$
$$y = \tfrac{42}{19}.$$

This method of elimination can easily be extended to systems of more than two unknowns. If, for example, a system of three equations with three unknowns x, y, and z is given, the system can be solved in the following way:

1. Eliminate one of the unknowns, say x, by combining two of the given equations by addition or subtraction after having multiplied them by factors such that x has the same numerical coefficient in both equations. The result is an equation containing y and z.

2. Eliminate the same unknown x, using another combination of two of the given equations. The result is again a linear equation in y and z.

3. The two linear equations obtained in the first two steps form a system of two equations with two unknowns which can be solved by the methods described before.

4. The values of y and z obtained in the third step are substituted in one of the given equations in order to find the unknown quantity x.

Example 4. Solve the following system of three simultaneous equations with three unknowns.

(A) $x + 2y + z = 3,$

(B) $2x + y + 3z = 7,$

(C) $3x + 3y - 4z = 2.$

The equations are denoted by letters A, B, C so that it is possible to indicate schematically the operations performed with these equations.

In order to eliminate x from the first two equations, multiply equation A by -2 and add the result to equation B:

(−2A) $-2x - 4y - 2z = -6$

(B) $2x + y + 3z = 7$

(−2A + B = B′) $-3y + z = 1.$

For brevity, the equation $-2A + B$ is denoted by B′. In order to have a second equation for y and z, the variable x is eliminated from A and C as follows:

(−3A) $-3x - 6y - 3z = -9$

(C) $3x + 3y - 4z = 2$

(−3A + C = C′) $-3y - 7z = -7.$

Having eliminated x we have now the following system of equations:

(B′) $-3y + z = 1,$

(C′) $-3y - 7z = -7.$

The unknown y is eliminated by subtraction, and we have

(B′ − C′) $8z = 8$

or $z = 1.$

The substitution of this value in B′ gives

$$-3y + 1 = 1,$$

and

$$-3y = 0,$$
$$y = 0.$$

The result of the substitution of the values $y = 0$, $z = 1$ in equation A is

$$x + 0 + 1 = 3$$

or

$$x = 2.$$

The given system of equations has, therefore, the solution

$$x = 2, \quad y = 0, \quad z = 1.$$

A check for the computation is obtained by substituting these values in B and C:

$$2 \cdot 2 + 0 + 3 \cdot 1 = 7,$$
$$3 \cdot 2 + 3 \cdot 0 - 4 \cdot 1 = 2.$$

Exercises

Solve the following systems of equations:

1. $x + y = 5,$
$x + z = 8,$
$y + z = 9.$

2. $3V_1 = 12,$
$3V_2 + 5V_1 = 11,$
$4V_3 + 7V_2 + 9V_1 = 23.$

3. $p - r = 2,$
$q + 3r = 1,$
$p - 2q = 7.$

4. $I_1 - I_2 - 2I_3 = -1,$
$10I_1 - 2I_2 + 5I_3 = 5,$
$5I_1 - 2I_2 = 0.$

5. $3u + v = 10,$
$5v + 3w = 2,$
$18u + 8w = 46.$

6. $10a + 3b + c = 20,$
$6a - 10b + 6c = -16,$
$5a - 3b - 4c = 19.$

7. $2R_1 + R_2 - R_3 = 1,$
$3R_1 - R_2 - R_3 = -2,$
$8R_1 - 4R_2 + 2R_3 = 6.$

8. $e_1 + e_2 + e_3 = 5,$
$3e_1 + 4e_2 - 3e_3 = 5,$
$7e_1 + 9e_2 - 11e_3 = 3.$

9. $-A + B + C = 30,$
$A - B + C = 24,$
$A + B - C = 20.$

10. $2A - B + 2C = 2,$
$10A + 2B - 5C = 1,$
$6A - B + 6C = 12.$

11. $3E_1 - 5E_2 - 5E_3 = 3,$
$E_1 + E_2 - 2E_3 = 7,$
$2E_1 - 3E_2 - 2E_3 = 0.$

12. $r_1 + 7r_2 + 10r_3 = 8,$
$-2r_1 + 6r_2 - r_3 = 2,$
$8r_1 - 12r_2 + 12r_3 + 4 = 0.$

13. $2x + 3y - 4z = -4,$
$x + y + 2z = \tfrac{2}{3},$
$12(3x - 2y + z) = 29.$

14. $10x - 2y + 2z = 14,$
$x + y + \tfrac{2}{3}z = 1,$
$3x - 4y - z = 7.$

15. $7R_1 + 5R_2 - R_3 = 12,$
$5R_1 + 2R_2 + 4R_3 - 18 = 0,$
$R_1 - \tfrac{1}{3}R_2 + R_3 - 3 = 0.$

16. $2x - 3y + z = 3,$
$x + 4y - 3z = -5,$
$3x + y + 4z = 4.$

17. $5x_1 - 15x_2 - 5x_3 = 8,$
$20x_1 - 3x_2 + x_3 = 7,$
$5x_1 - 4x_2 - \tfrac{1}{3}x_3 = 1\tfrac{1}{3}.$

18. $3I_1 + 6I_2 + 5I_3 - 7 = 0,$
$I_1 + 2I_2 + 3I_3 - 6 = 0,$
$-2I_1 + 4I_2 + 7I_3 + 5 = 0.$

19. $x + y + z = -1,$
$y + z + t = -3,$
$z + t + x = 0,$
$t + x + y = -2.$

20. $x_1 + 4x_2 + x_3 - x_4 = -7,$
$-2x_1 + 3x_2 + x_3 - 4x_4 = -4,$
$-x_1 - 6x_2 + 2x_3 + 2x_4 = 14,$
$2x_1 - 5x_2 + 2x_3 + 5x_4 = 11.$

13–2. Solution of Systems of Linear Equations by the Doolittle Method

A great amount of numerical work is involved in the solution of a system of linear equations if the coefficients are numbers with two or more significant digits.   The amount of work is immensely increased if systems of four or more equations have to be solved.   Since such systems arise in many practical applications, it is important to have a method that meets the following requirements:

1. The work should be arranged so that what has already been done is obvious at every step of the computation, and so that there is no doubt concerning the next step to be performed.

2. The method should be such that the number of necessary multiplications and divisions is as small as possible.

3. The method should be suited to the use of the slide rule or a computing machine.

4. A check for each step of the work should be available.

The mathematician Gauss showed how the method of elimination by adding and subtracting can be arranged so that all but the second of the conditions enumerated above are satisfied.   A modification of his method, which saves a certain amount of work, is due to the American mathematician Doolittle.   This method, known as the Doolittle method, will be discussed in this section.   It will be explained by means of examples.

An important feature of the Doolittle method is the way each step of the computation is checked.   In this check, the sum of all the coefficients and the right-hand member of each equation is computed.   This sum is called the **check sum** of the equation.

Example 1.   The check sum of the equation
$$3x - 4y + 5z = 10 \text{ is } 3 - 4 + 5 + 10 = 14.$$

The check sums are always written on the same line as the equation itself, separated by a vertical line.

Example 2.   A system of three equations with its check sums is written in this way.

$$
\begin{array}{r|r}
x + 2y + z = 3 & 7 \\
2x + y + z = 7 & 11 \\
3x + 3y - 4z = 2 & 4
\end{array}
$$

The check sums are treated exactly as the coefficients of the corresponding equations.   Thus, if the equation is multiplied by a factor, the check sum is to be multiplied by the same factor.   Or, if two equations are added or subtracted, the corresponding check sums are to be added or subtracted.   *The result of the operations carried out with the check sums of the original equations has to be equal to the check sum of the final equation. This fact is used as the check on the computations.*

Example 3. If the first equation of example 2 and its check sum are multiplied by 3, the result is

$$3x + 6y + 3z = 9 \quad | \quad 21$$

The check sum of the new equation is $3 + 6 + 3 + 9 = 21$ and is equal to the number that was obtained by multiplying the original check number 7 by 3.

Example 4. If the first two equations of example 2 and their check sums are added, the equation

$$3x + 3y + 2z = 10 \quad | \quad 18$$

is obtained. Its check sum is $3 + 3 + 2 + 10 = 18$ and is equal to the sum of the check sums 7 and 11.

How this check is used throughout the computation for each single step will be seen from the following example in which Doolittle's method for solving a system of linear equations is explained.

Solve the system of three simultaneous equations:

$$(A) \quad | \quad 3.42x - 5.93y + 1.44z = -2.20 \quad | \quad -3.27$$
$$(B) \quad | \quad 1.95x + 4.68y - 2.73z = 3.00 \quad | \quad +6.90$$
$$(C) \quad | \quad 2.14x + 3.64y + 2.92z = 13.58 \quad | \quad +22.28$$

The check sum of each equation is written in the rightmost column, to the right of the vertical line. In the leftmost column, each equation is denoted by a letter in order that we shall be able to indicate the operations performed with the equations.

All the computations in this example are made with the slide rule.

Step 1. The first equation is divided by the coefficient 3.42 of x in order to obtain an equivalent equation in which the coefficient of x is equal to 1. The divided equation is given by

$$\left(A' = \frac{A}{3.42}\right) \quad | \quad x - 1.734y + 0.421z = -0.643 \quad | \quad -0.956.$$

The check sum -0.956 is found by dividing the check sum -3.27 of the original equation by 3.42. The check consists in investigating whether the number -0.956 obtained in this way is equal to the check sum of the new equation. This check sum is

$$1 - 1.734 + 0.421 - 0.643 = -0.956$$

which is equal to the number -0.956 found before. It is, therefore, not likely that a numerical error has been made so far.

Step 2. In order to eliminate x, equation A' is multiplied by -1.95, and the result is added to equation B. This computation is performed below.

(B)	$1.95x + 4.68y - 2.73z = 3.00$	$+6.90$
$(-1.95A')$	$-1.95x + 3.38y - 0.82z = 1.25$	$+1.86$
$(B - 1.95A')$	$8.06y - 3.55z = 4.25$	8.76

After a line is written down, the check sum test has to be made immediately. Thus, for the last equation, the number 8.76 was obtained by adding the check sums $6.90 + 1.86$, and this has to be compared with the check sum of the new equation

$$8.06 - 3.55 + 4.25 = 8.76.$$

If this check sum had another value, then the computation of the last line would have to be investigated for errors. A small difference of one or two units in the rightmost place of the check sum may result from the fact that all the coefficients are incomplete numbers, computed only with a certain number of digits. It is obvious that the sum of several incomplete numbers may differ from the correct result by a few units in the rightmost place.

The last equation contains only y and z. Dividing by 8.06, the coefficient of y, we have the following equivalent equation in which the first coefficient is one:

$$\left(B' = \frac{B - 1.95A'}{8.06}\right) \quad \Big| \qquad y - 0.440z = 0.528 \quad \Big| \quad 1.087.$$

Step 3. The two equations A' and B' are now used in order to eliminate x and y from the equation C. This is done in the following way:

First write equation C,

$$(C) \quad \Big| \quad 2.14x + 3.64y + 2.92z = 13.58 \quad \Big| \quad +22.28$$

and write below the product of equation A' and the number -2.14, the negative coefficient of x in equation C. This product is

$$(-2.14A') \quad \Big| \quad -2.14x + 3.71y - 0.90z = 1.38 \quad \Big| \quad +2.05.$$

If the equations C and $(-2.14A')$ are added, the unknown x is eliminated, and the unknown y has the coefficient $3.64 + 3.71 = 7.35$. In order to eliminate y simultaneously, equation B' is multiplied by -7.35, giving

$$(-7.35B') \quad \Big| \qquad -7.35y + 3.24z = -3.88 \quad \Big| -7.99.$$

If now the last three equations are added, x and y are eliminated, and the following equation results:

$$(C - 2.14A' - 7.35B') \quad \Big| \qquad\qquad 5.26z = \quad 11.08 \ \Big| \ 16.34.$$

Dividing this equation by 5.26 gives

$$(C') \quad \Big| \qquad\qquad\qquad z = \quad 2.11 \quad \Big| \quad 3.11.$$

The check sum test indicates no error, because $1 + 2.11 = 3.11$. The value of z obtained in the last step is now substituted in equation B' in order to find y. The result is

$$y = 0.528 + 0.440z = 0.528 + 0.440 \times 2.11 = 1.456.$$

Finally, the values of z and y are substituted in equation A', from which

$$x = -0.643 - 0.421z + 1.734y$$
$$= -0.643 - 0.421 \times 2.11 + 1.734 \times 1.456$$
$$= 0.99.$$

Writing all the results with the same number of decimal places, the solution is

$$x = 0.99, \qquad y = 1.46, \qquad z = 2.11.$$

A check for the whole computation can be made by substituting these values in equation C. The substitution in equation A or B would not furnish a complete test because these equations were used for the computation for y and x.

The whole computation, as described, can be done in much less space than has been used here, omitting all explanations and writing only the coefficients instead of the whole equations. In this way it is possible to have a layout for the whole procedure which can be filled in systematically. The following table gives the computation carried out before, but omits everything that is not required for the computation itself or the check.

Name of Equation	Coefficient of			Right Member	Check Sum
	x	y	z		
A	$+3.42$	-5.93	$+1.44$	-2.20	-3.27
$A' = \dfrac{A}{3.42}$	$+1$	-1.734	$+0.421$	-0.643	-0.956
B	$+1.95$	$+4.68$	-2.73	$+3.00$	$+6.90$
$-1.95A'$	-1.95	$+3.38$	-0.82	$+1.25$	$+1.86$
$B - 1.95A'$		$+8.06$	-3.55	$+4.25$	$+8.76$
$B' = \dfrac{1}{8.06}(B - 1.95A')$		$+1$	-0.440	$+0.528$	$+1.087$
C	$+2.14$	$+3.64$	$+2.92$	$+13.58$	$+22.28$
$-2.14A'$	-2.14	$+3.71$	-0.90	$+1.38$	$+2.05$
$-7.35B'$			-7.35	$+3.24$	-3.88	-7.99
$C - 2.14A' - 7.35B'$			$+5.26$	$+11.08$	$+16.34$
$C' = \dfrac{1}{5.26}(C - 2.14A' - 7.35B')$			$+1$	$+2.11$	$+3.11$

$$z = 2.11,$$
$$y = 0.528 + 0.440z = 1.46,$$
$$x = -0.643 - 0.421z + 1.734y = 0.99.$$

The answers can be checked by substituting the values of x, y, z in equation C:

$$2.14x + 3.62y + 2.92z = 2.12 + 5.31 + 6.17 = 13.60.$$

This result is close enough to the correct value 13.58.

The Doolittle method was explained here with a system of three simultaneous equations with three unknown quantities. It can be used for any number of equations. If the system consists of only two equations with two unknowns, the third step in the computation is not required. If more than three equations are given, the procedure can be continued to the point where only one equation with one unknown is obtained.

Exercises

Solve the following systems of simultaneous equations (slide-rule precision):

1. $2.57x + 4.73y = 7.21$,
    $7.91x + 10.84y = 14.8$.

2. $0.715x + 0.672y = -2.19$,
    $0.832x - 0.548y = -0.873$.

3. $60.2r + 76.5s = 1213$,
    $23.8r - 35.2s = 153.7$.

4. $1.43e_1 - 1.38e_2 = 4.51$,
    $4.52e_1 - 0.327e_2 = -21.6$.

5. $3.47R_1 - 1.25R_2 = -8.29$,
    $5.72R_1 + 1.19R_3 = -13.5$,
    $5.61R_2 - 3.52R_3 = 38.1$.

6. $213a + 325b - 507c = 314$,
    $128a - 96.3b - 194c = 486$,
    $387a - 193b - 215c = -207$.

7. $3.15x - 0.912y + 0.553z = 1.37$,
    $-0.923x + 3.68y - 0.376z = 1.89$,
    $0.542x - 0.375y + 2.78z = 2.71$.

8. $1.05x + 0.231y - 0.195z = 1.07$,
    $0.245x + 1.09y + 0.343z = 1.63$,
    $-0.192x + 0.338y + 0.982z = 1.28$.

13–3. Graphical Representation and Discussion of a System of Two Linear Equations

The problem of finding the solution of a system of simultaneous linear equations may sometimes present difficulties. For a system with two unknowns these difficulties will be discussed in this section. The graphical method explained in Sec. 3–13 is very helpful in obtaining a clear understanding of the different possibilities.

The graph corresponding to the linear equation $ax + by = c$ consists of all points whose coordinates x and y satisfy this equation. It was stated in Chapter 3 that the graph of a linear equation is always a straight line. This line, as described in Sec. 3–12, can be constructed after two of its points are found.

In order to find the coordinates of a point on the line, an arbitrary value can be assumed for one of the two variables, and the corresponding value of the other variable can then be computed from the given equation. In particular, the x intercept, that is, the abscissa of the point where the straight line meets the x axis, is found by substituting $y = 0$ in the given equation; the y intercept is found by substituting $x = 0$.

Example 1.   Construct the straight line L corresponding to the equation

$$3x + 4y = 6.$$

Points on this line can be found by substituting particular values for one of the variables.   Thus, for $x = 3$, the corresponding value of y is

$$y = \frac{6 - 3x}{4} = \frac{6 - 9}{4} = -\frac{3}{4},$$

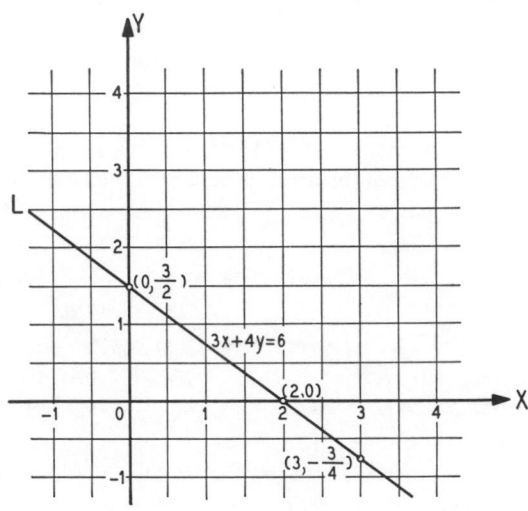

Fig. 13–1

and the point $(3, -\frac{3}{4})$ is on the line L.   In this way any number of points on the line can be found.

The x intercept of the line is the point on the line whose ordinate has the value $y = 0$ and is found, therefore, from the equation

$$3x + 4 \cdot 0 = 6,$$

$$3x = 6,$$

$$x = 2.$$

The y intercept is found by substituting $x = 0$, which gives

$$3 \cdot 0 + 4y = 6, \qquad y = \tfrac{3}{2}.$$

The graph of this straight line is plotted in Fig. 13–1.

This graphical representation can be used to solve a system of two simultaneous equations with two unknowns.   Each of the two equations can be regarded as an equation between two variables, and the two corresponding straight lines can be constructed.   The coordinates of a point where the two lines intersect each other satisfy both equations and are, therefore, a solution of the given system of equations.

Example 2. Solve the following system of equations:

$$4x - 3y = 3,$$

$$3x + 4y = 6.$$

The corresponding straight lines are plotted in Fig. 13–2. They intersect at the point with coordinates $x = 1.2$, $y = 0.6$. This pair of values satisfies, therefore, both equations and is the solution of the given system.

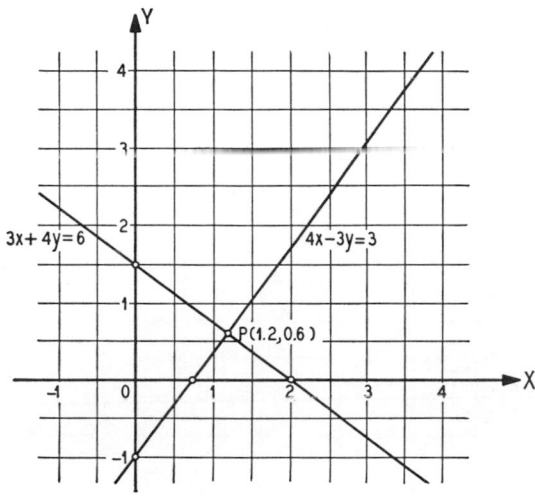

FIG. 13–2

Two straight lines have no point in common if they are parallel. Thus a system of two simultaneous linear equations has no solution if the corresponding straight lines are parallel. Consider, for example, the system of equations

$$3x + 4y = 6,$$

$$6x + 8y = 3.$$
(1)

The graphs of the two equations are plotted in Fig. 13–3. Since the two straight lines are parallel and therefore have no point of intersection, it can be concluded that the given system of equations has no solution.

The fact that there is no solution in this case can be shown without using a graphical representation in the following way. We start with the hypothesis that the given system of equations has a solution and then show that this hypothesis leads to an impossible conclusion and that, therefore, it must be discarded. Thus, if it is assumed that the system has solutions, two numbers x and y could be found such that

$$3x + 4y = 6,$$

$$6x + 8y = 3$$

are satisfied simultaneously. Multiplying the first equation by 2 and subtracting it from the second equation, we have $0 = -9$, which is obviously impossible. The hypothesis, therefore, that the given equations have a solution is wrong because it leads to a false statement.

A system of simultaneous equations that have no common solutions is called **inconsistent.** Correspondingly, the expression **consistent** is used for any system of simultaneous equations that have common solutions. Thus, the system of example 2 is consistent; the system 1 is inconsistent.

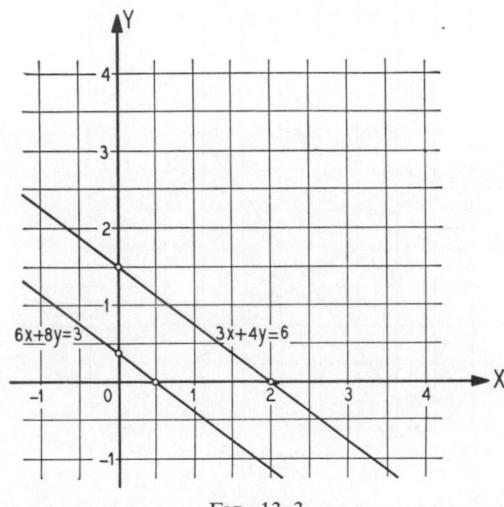

FIG. 13–3

In addition to the two cases discussed above, there is a third possibility, namely, that the two lines may be identical.   Since in this case the same straight line corresponds to both equations, the coordinates of every point on this line satisfy both equations, and therefore infinitely many solutions can be found.

Example 3.   Solve the system

$$3x + 4y = 6,$$

$$5.19x + 6.92y = 10.38.$$

Both equations correspond to the same straight line, which is plotted in Fig. 13–1, and the coordinates of each point of this line satisfy both equations. Multiplying the first equation by 1.73, we obtain the second equation, whence the given equations are equivalent.

The discussion of a system of two simultaneous linear equations with two unknowns given in this section can be now summarized in the following three cases.

1. *The two straight lines corresponding to the equations of the system are not parallel.* In this case the system is **consistent** and has just one solution, and the equations are said to be **independent.**

2. *The two straight lines corresponding to the equations of the system are parallel and different from each other.* In this case the system is **inconsistent** and has no solution.

3. *The two straight lines corresponding to the two equations of the system coincide.* This system is **consistent** and has infinitely many solutions. In this case the two equations are always equivalent. They are therefore called **dependent** on one another, and the system itself is called **dependent.** Every solution of one equation is also a solution of the other.

Exercises

Construct the graphs corresponding to the following systems of equations, and classify the systems as independent, inconsistent, or dependent. If the lines intersect, estimate the coordinates of the point of intersection from the graph.

1. $3x + 5y = 12,$
   $x - 2y = 5.$

2. $x + y = 0,$
   $x - y = 0,$

3. $x + y = 3,$
   $x + y = 7.$

4. $2r_1 - 2r_2 = 5,$
   $3r_1 - 3r_2 = 11.$

5. $4u = 6v,$
   $10u = 15v.$

6. $3A - 2B = 0,$
   $3A + 2B = 6.$

7. $4x + 2y = 0,$
   $6x + 3y = 0.$

8. $u - v = 0,$
   $4u + 3v = 0.$

9. $5x - 7y = 10,$
   $10x - 14y = 7.$

10. $3x + 5y = 12,$
    $2x - y = 3.$

11. $2e_1 - e_2 = 3,$
    $8e_1 - 4e_2 = 12.$

12. $2.4r - 4.8s = 7.2,$
    $0.9r - 1.8s = 2.7.$

13. $3m - 2n - 6 = 0,$
    $4n - 6m - 9 = 0.$

14. $5E_1 + 2E_2 - 10 = 0,$
    $10E_1 + 4E_2 + 17 = 0.$

15. $3.8M - 1.4N = 3.6,$
    $5.7M - 2.1N = 5.4.$

16. $I_1 + 2I_2 - 3 = 0,$
    $2I_1 - I_2 + \frac{1}{2} = 0.$

13–4. Solutions of Systems of Two Linear Equations by Determinants

The graphical method of Sec. 13–3 is often inconvenient and inaccurate, and it cannot easily be extended to systems with three or more unknowns. It is therefore desirable to have a method that permits us to examine a system of linear equations by computations.

We shall start with a system of two equations with two unknowns. In order to make statements that will be true for all such systems, the equations will be written with letters as coefficients:

$$a_1x + b_1y = c_1,$$
$$a_2x + b_2y = c_2. \tag{1}$$

In order to eliminate y, the first equation is multiplied by b_2, the second by b_1, giving

$$a_1 b_2 x + b_1 b_2 y = c_1 b_2,$$

$$a_2 b_1 x + b_2 b_1 y = c_2 b_1.$$

By subtracting the second equation from the first it is found that

$$(a_1 b_2 - a_2 b_1)x = c_1 b_2 - c_2 b_1$$

or

$$x = \frac{c_1 b_2 - c_2 b_1}{a_1 b_2 - a_2 b_1}.$$

Similarly, it can be shown that

$$y = \frac{a_1 c_2 - a_2 c_1}{a_1 b_2 - a_2 b_1}.$$

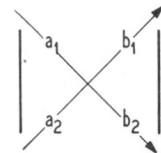

Fig. 13–4

Thus, two formulas have been found by which the solutions of the system 1 can be computed. They look complicated, but a closer examination will show that it is very easy to write them down. The two formulas have identical denominators, given by the expression

$$a_1 b_2 - a_2 b_1.$$

Expressions of this kind occur frequently in different branches of mathematics. It is, therefore, useful to have a name for them and to examine their properties.

The expression $a_1 b_2 - a_2 b_1$ is called a **determinant of the second order,** *and it is denoted by the symbol*

$$\begin{vmatrix} a_1 & b_1 \\ a_2 & b_2 \end{vmatrix}.$$

The four numbers a_1, b_1, a_2, b_2 are arranged in two horizontal lines called **rows** and two vertical lines called **columns.** Each number is called an **element** of the determinant. Thus, the first row consists of the elements a_1, b_1; the second of the elements a_2, b_2. The first column consists of the elements a_1, a_2; the second of the elements b_1, b_2. The four elements form a square with two diagonals, one from the upper left to the lower right corner, the other from the lower left to the upper right corner, as indicated in Fig. 13–4.

A determinant of the second order is computed by taking the product $a_1 b_2$ of the elements in the first diagonal and subtracting the product $a_2 b_1$ of the elements in the other diagonal.

Example 1. Compute the determinant

$$\begin{vmatrix} 7 & 2 \\ 4 & 3 \end{vmatrix}.$$

According to the rule given above the product $7 \cdot 3$ of the elements in the diagonal going from the left down has to be formed, and the product $4 \cdot 2$ of the elements in the diagonal from the left up has to be subtracted, so that

$$\begin{vmatrix} 7 & 2 \\ 4 & 3 \end{vmatrix} = 7 \cdot 3 - 4 \cdot 2 = 21 - 8 = 13.$$

Example 2. Similarly, it is found that

$$\begin{vmatrix} -2 & -5 \\ 4 & 10 \end{vmatrix} = (-2) \cdot 10 - 4 \cdot (-5) = -20 + 20 = 0.$$

Example 3. The computation is simplified if there is an element of the determinant which is zero,

$$\begin{vmatrix} 3 & 9 \\ 0 & -4 \end{vmatrix} = 3 \cdot (-4) - 0 \cdot 9 = -12.$$

The formulas for the solution of a system of two linear equations

$$x = \frac{c_1 b_2 - c_2 b_1}{a_1 b_2 - a_2 b_1}, \qquad y = \frac{a_1 c_2 - a_2 c_1}{a_1 b_2 - a_2 b_1}$$

can be written using the determinant notation:

$$x = \frac{\begin{vmatrix} c_1 & b_1 \\ c_2 & b_2 \end{vmatrix}}{\begin{vmatrix} a_1 & b_1 \\ a_2 & b_2 \end{vmatrix}}, \qquad y = \frac{\begin{vmatrix} a_1 & c_1 \\ a_2 & c_2 \end{vmatrix}}{\begin{vmatrix} a_1 & b_1 \\ a_2 & b_2 \end{vmatrix}}.$$

This result can be stated in the following rule.

The values of x and y that satisfy the system

$$a_1 x + b_1 y = c_1, \tag{1}$$

$$a_2 x + b_2 y = c_2$$

are two fractions, each one having the denominator

$$\begin{vmatrix} a_1 & b_1 \\ a_2 & b_2 \end{vmatrix} \tag{2}$$

which is the determinant formed by the coefficients of x and y in the given equations 1. The numerator in the expressions for x is

$$\begin{vmatrix} c_1 & b_1 \\ c_2 & b_2 \end{vmatrix}$$

which is the determinant obtained by replacing the elements of the first column of (2) by the right members c_1 and c_2 of equations 1. The numerator in the expression for y is obtained similarly by replacing the elements of the second column of (2) by the right members of the given equations 1.

Example 4. Using determinants, find the solution of

$$7x - 5y = 9,$$
$$3x + 4y = 10.$$

The denominator of x and y is

$$\begin{vmatrix} 7 & -5 \\ 3 & 4 \end{vmatrix} = 28 - 3 \cdot (-5) = 28 + 15 = 43.$$

The numerator of x is the determinant obtained when the elements 7 and 3 of the first column of the above determinant are replaced by 9 and 10. This numerator has the value

$$\begin{vmatrix} 9 & -5 \\ 10 & 4 \end{vmatrix} = 36 - 10(-5) = 36 + 50 = 86.$$

Similarly, the numerator of y is equal to

$$\begin{vmatrix} 7 & 9 \\ 3 & 10 \end{vmatrix} = 7 \cdot 10 - 3 \cdot 9 = 70 - 27 = 43,$$

and therefore

$$x = \tfrac{86}{43} = 2, \qquad y = \tfrac{43}{43} = 1.$$

Exercises

Solve the following systems of equations by determinants:

1. $5x - 3y = -1,$
     $2x + y = 4.$

2. $x + y = 1,$
     $x - y = -5.$

3. $2x - y = 13,$
     $3x + 5y = 0.$

4. $7E_1 + 2E_2 = 10,$
     $3E_1 + 9E_2 = 7.$

5. $3x + 7y = 0,$
     $x - 5y = 0.$

6. $4x + 3y = 3,$
     $3x - 6y = -1.$

7. $17z_1 + 3z_2 = 20,$
     $12z_1 - z_2 = 11.$

8. $5m - 5n = 20,$
     $2m + n = 29.$

9. $5a - b = 0.3,$
     $a + b = 0.3.$

10. $3u + 5v - 4 = 0,$
     $7u + 6v - 9 = 0.$

11. $10P_1 - 5P_2 + 19 = 0,$
     $3P_1 + 2P_2 - 12 = 0.$

12. $x + 1 = 5(3y - 5),$
     $5(x - y) = 4(x + 1).$

13. $3z_1 - 2z_2 = 7,$
$3z_1 + z_2 = 2.$

14. $2I_1 = 26 - 3I_2,$
$4I_2 - 96 = 5I_1.$

15. $771r - 183s = 0,$
$453r + 217s = 0.$

16. $\frac{2}{5}R - \frac{4}{3}L - 1 = 0,$
$\frac{1}{2}R - \frac{1}{3}L - \frac{3}{4} = 0.$

17. $p - 18 = \dfrac{11 - q}{3},$

$29 - 2p = \dfrac{q - 13}{4}.$

18. $\frac{1}{3}(x - y) + \frac{1}{3}(x + y) = -1,$
$\frac{1}{2}(x + y) - \frac{1}{3}(x - y) = 1.$

19. $\dfrac{x + 1}{2} - \dfrac{2y - 3}{3} = \dfrac{4}{3},$

$\dfrac{x + y}{4} - \dfrac{x + 1}{2} = -\dfrac{1}{2}.$

13–5. Discussion of a System of Two Linear Equations by Means of Determinants

The solutions of a system of two linear equations are, according to the rule of the preceding section, fractions with the denominator $\begin{vmatrix} a_1 & b_1 \\ a_2 & b_2 \end{vmatrix}$.

These fractions can always be computed if this determinant is not zero. Therefore, the following theorem is true:

Two simultaneous linear equations

$$a_1x + b_1y = c_1,$$
$$a_2x + b_2y = c_2,$$
(1)

are independent, which means that they may be solved and that there is only one pair of values in the solution, if the determinant of the system

$$\begin{vmatrix} a_1 & b_1 \\ a_2 & b_2 \end{vmatrix} \neq 0.$$
(2)

If the determinant 2 is equal to zero, then the equations in 1 are either inconsistent or dependent. We shall state here without proof that, *when the three determinants*

$$\begin{vmatrix} a_1 & b_1 \\ a_2 & b_2 \end{vmatrix}, \quad \begin{vmatrix} c_1 & b_1 \\ c_2 & b_2 \end{vmatrix}, \quad \begin{vmatrix} a_1 & c_1 \\ a_2 & c_2 \end{vmatrix}$$

all have the value zero, the system is dependent. If the first is zero but the other two are not zero, then the system is inconsistent. A few examples will show how to apply these criteria.

Example 1.   Consider the system of equations

$$3x - 7y = 10,$$
$$5x + 2y = 20.$$

Find the value of the determinant:

$$\begin{vmatrix} 3 & -7 \\ 5 & 2 \end{vmatrix} = 6 - 5(-7) = 6 + 35 = 41.$$

Since this value is different from zero, we conclude, without solving the equations, that the system is independent and has only one solution.

There are many cases in which it is sufficient to know that the system is independent and has only one solution and in which it is not required to compute the values of the unknowns.

Example 2.   The system of equations

$$6x + 8y = 10,$$
$$9x + 12y = 12$$

is not independent, for

$$D = \begin{vmatrix} 6 & 8 \\ 9 & 12 \end{vmatrix} = 72 - 72 = 0.$$

In order to decide whether this system is inconsistent or dependent, two other determinants must be computed.   These determinants are obtained by replacing, respectively, the first and the second column of D by the right members of the given equations.   We have then

$$D_1 = \begin{vmatrix} 10 & 8 \\ 12 & 12 \end{vmatrix} = 120 - 96 = 24,$$

$$D_2 = \begin{vmatrix} 6 & 10 \\ 9 & 12 \end{vmatrix} = 72 - 90 = -18.$$

Since D_1 and D_2 are different from zero, the given system of equations is inconsistent and has, therefore, no solution.   The graphs of the two equations are, in this case, two parallel lines.

Example 3.   Consider the system of equations

$$6x + 8y = 10,$$
$$9x + 12y = 15.$$

The three determinants whose values we need are:

$$D = \begin{vmatrix} 6 & 8 \\ 9 & 12 \end{vmatrix} = 72 - 72 = 0,$$

$$D_1 = \begin{vmatrix} 10 & 8 \\ 15 & 12 \end{vmatrix} = 120 - 120 = 0,$$

$$D_2 = \begin{vmatrix} 6 & 10 \\ 9 & 15 \end{vmatrix} = 90 - 90 = 0.$$

Hence it follows that the given system is dependent and that every solution of the first equation is also a solution of the second. The two equations are equivalent, for the second equation is obtained by multiplying the first one by $\frac{3}{2}$.

Example 4. Given the following system of two simultaneous equations with the two unknowns x and y:

$$kx + 8y = 10,$$
$$9x + 2ky = 15.$$

Without solving the system, find the values of k for which the equations are independent, inconsistent, or dependent.

The first step is to compute the determinant of the coefficients of x and y. This gives

$$D = \begin{vmatrix} k & 8 \\ 9 & 2k \end{vmatrix} = 2k^2 - 72.$$

The system is independent for all values of k for which $2k^2 - 72 \neq 0$, that is, for which $k \neq \pm 6$. For example, if $k = 4$, then

$$D = 2k^2 - 72 = 2 \cdot 16 - 72 = -40 \neq 0,$$

and the system obtained by substituting $k = 4$,

$$4x + 8y = 10,$$
$$9x + 8y = 15,$$

has a solution which can be easily computed.

The system is not independent if the determinant

$$D = 2k^2 - 72 = 0,$$

or when

$$k = \pm 6.$$

In order to decide whether in this case the system is inconsistent or dependent the two other determinants

$$D_1 = \begin{vmatrix} 10 & 8 \\ 15 & 2k \end{vmatrix} = 20k - 120,$$

$$D_2 = \begin{vmatrix} k & 10 \\ 9 & 15 \end{vmatrix} = 15k - 90$$

must be computed for $k = \pm 6$. The result is the following:

Case 1. For $k = +6$: $D_1 = 120 - 120 = 0$, $D_2 = 90 - 90 = 0$. The given system is in this case *dependent*.

Case 2. For $k = -6$: $D_1 = -120 - 120 = -240 \neq 0$, $D_2 = -90 - 90 = -180 \neq 0$. The given system is in this case *inconsistent*.

The results may be summarized as follows:

When $k \neq \pm 6$, the system is independent.

When $k = +6$, the system is dependent.

When $k = -6$, the system is inconsistent.

Exercises

1–16. Using determinants, decide whether the systems of equations in the exercises of Sec. 13–3 are independent, inconsistent, or dependent, and compute the values of the unknowns when the equations are independent.

Find the values of k for which the following systems are independent, dependent, and inconsistent:

17. $kx + 15y = 20$,
  $5x + 3ky = 20$.

18. $(k + 3)u + (k + 4)v = 1$,
  $(k + 2)u + (k + 5)v = 3$.

19. $(1 - k)x + 5y = 7$,
  $5x + (25 - k)y = 0$.

20. $(3 - k)x + 4y = 8$,
  $3x + (7 - k)y = 12$.

21. $5x + 8y = k$,
  $10x + ky = 2k$.

22. $kR_1 - R_2 = k$,
  $R_1 + kR_2 = 1$.

23. $kx + y = -m$,
  $kx - y = m$,
  where $m \neq 0$.

24. $\dfrac{A}{4 - k} + \dfrac{B}{6 - k} = 1$,

  $\dfrac{x}{5 - k} + \dfrac{y}{7 - k} = 1$.

13–6. Systems of Two Linear Homogeneous Equations

A linear equation is called **homogeneous** if its right member is zero. Thus, $3x + 4y = 0$ and $a_1x + b_1y = 0$ are homogeneous equations. There are many problems leading to the solution of a system of two homogeneous equations, and, therefore, such systems will be studied in this section.

The graph of a linear equation without the absolute term as, for example, $3x + 4y = 0$ is a straight line passing through the origin of the coordinate system. This is so because the equation is satisfied when $x = 0$ and $y = 0$ are substituted. A second point of this line can be easily found by choosing an arbitrary value for x and computing the corresponding value of y. Thus, when $x = 4$, $y = -3$. The graph of $3x + 4y = 0$ is plotted in Fig. 13–5.

Suppose a second homogeneous equation be given by $7x - 5y = 0$. The straight line corresponding to this equation also passes through the origin. To obtain a second point on it, the value 5 is substituted for x, giving $y = 7$. From Fig. 13–5, it is obvious that the two straight lines have only the origin as a common point; therefore, the coordinates of this point, $x = 0$, $y = 0$, are the only solution of the system

$$3x + 4y = 0,$$
$$7x - 5y = 0.$$

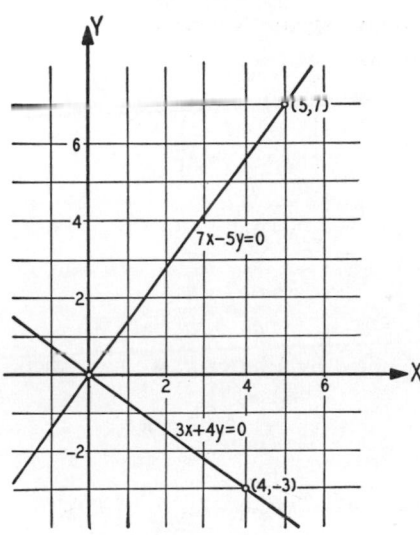

FIG. 13—5

A similar discussion would show that any two linear homogeneous equations are represented by two straight lines which intersect at the origin, and that, therefore, the coordinates $x = 0$, $y = 0$ are the only solution of the system formed by the two equations. But there is one exception. It may happen that *the two equations are represented by the same straight line.* The coordinates of every point on this line satisfy both equations, so that there are infinitely many solutions besides $x = 0$, $y = 0$. This exceptional case is very important for the application of the theory, because we are often required to investigate whether or not there are solutions different from zero. A test by which the two cases may be distinguished will be found by the rules of Sec. 13–5.

In general, consider the system

$$a_1x + b_1y = 0,$$

$$a_2x + b_2y = 0.$$

(1)

The determinant of the coefficients of x and y is given by

$$\begin{vmatrix} a_1 & b_1 \\ a_2 & b_2 \end{vmatrix} = a_1 b_2 - a_2 b_1.$$

In Sec. 13–5 it was stated that the system has only one solution if this determinant is different from zero. This solution is, in this case, $x = 0$, $y = 0$. This gives the theorem:

When the determinant of the system 1 of two homogeneous linear equations is not zero, then the only solution is given by $x = 0$, $y = 0$.

Consider now the case when

$$\begin{vmatrix} a_1 & b_1 \\ a_2 & b_2 \end{vmatrix} = 0,$$

and hence

$$a_1 b_2 - a_2 b_1 = 0, \quad \text{or} \quad a_1 b_2 = a_2 b_1,$$

so that

$$\frac{a_1}{a_2} = \frac{b_1}{b_2}.$$

If this quotient is denoted by a letter k, we have then,

$$\frac{a_1}{a_2} = \frac{b_1}{b_2} = k$$

and

$$a_1 = k a_2, \qquad b_1 = k b_2,$$

or

$$a_1 x + b_1 y = k(a_2 x + b_2 y).$$

The first equation $a_1 x + b_1 y = 0$ is, therefore, equivalent to the second equation $a_2 x + b_2 y = 0$, from which it is derived by multiplication by the factor k. A solution of one of the two equations satisfies, therefore, the second equation, and the lines representing the two equations are identical. Thus, the second part of the theorem can be stated as follows.

A system 1 of two homogeneous linear equations has infinitely many solutions if its determinant is zero.

A few examples will show how to apply the rules of this section.

Example 1. The system

$$3.7x - 4.9y = 0,$$
$$1.3x + 8.5y = 0$$

has only the solution $x = 0$, $y = 0$, because its determinant

$$\begin{vmatrix} 3.7 & -4.9 \\ 1.3 & 8.5 \end{vmatrix} = 3.7 \cdot 8.5 + 1.3 \cdot 4.9 \neq 0.$$

Example 2.   For what values of k has the system

$$(7 - k)x + 2y = 0,$$
$$2x + (4 - k)y = 0$$

only the solution $x = 0$, $y = 0$?

The determinant of the system is

$$\begin{vmatrix} 7 - k & 2 \\ 2 & 4 - k \end{vmatrix} = (7 - k)(4 - k) - 2^2$$

$$= 28 - 11k + k^2 - 4$$
$$= k^2 - 11k + 24.$$

The system has solutions different from $x = 0$, $y = 0$ only if this determinant is zero, or only if

$$k^2 - 11k + 24 = 0.$$

The solution of this quadratic equation gives the two values

$$k_1 = 8, \qquad k_2 = 3.$$

Hence, it can be stated that the given system has only the solution $x = 0$, $y = 0$, if a value is chosen for k different from 8 or 3.

If $k = 8$ is substituted, then the given system becomes

$$-x + 2y = 0,$$
$$2x - 4y = 0.$$

These two equations are equivalent, the second being the product of the first equation and -2.   Each pair of numbers satisfying the first equation is a solution of the system.   For example,

$$x = 2, \quad y = 1, \qquad \text{or} \qquad x = 10, \quad y = 5, \qquad \text{or} \qquad x = 400, \quad y = 200,$$

satisfy both equations.

The substitution $k = 3$ into the given system yields

$$4x + 2y = 0,$$
$$2x + y = 0.$$

Again, both equations are equivalent, the first being twice the second.

Exercises

Investigate whether the following systems have solutions different from $x = 0$, $y = 0$:

1. $3x + 5y = 0,$
   $2x + 7y = 0.$

2. $x + y = 0,$
   $x - y = 0.$

3. $16x - 8y = 0,$
   $6x - 3y = 0.$

4. $14e_1 + 7e_2 = 0,$
   $-10e_1 - 5e_2 = 0.$

5. $-4R_1 + 9R_2 = 0,$
   $5R_1 - R_2 = 0.$

6. $1.8u = 1.5v,$
   $4.2u = 3.5v.$

7. $2.1a - 5.9b,$
   $4.6a = 3.7b.$

8. $0.22x - 0.14y = 0,$
   $-0.33x + 0.21y = 0.$

9. $\frac{1}{2}m + \frac{1}{3}n = 0,$
   $\frac{2}{3}m + \frac{3}{4}n = 0.$

Find values for k such that the following systems have solutions different from $x = 0$, $y = 0$:

10. $2x + y = 0$,   **11.** $4x - 3y = 0$,   **12.** $2u - kv = 0$,

$16x + (k + 4)y = 0$. $(7 - k)x - 5y = 0$. $ku - 18v = 0$.

13. $y - kx = 0$,   **14.** $3A + 4B = 0$,   **15.** $11x + 3y = kx$,

$x - ky = 0$. $12A + k^2 B = 0$. $3x + 11y - ky$.

16. $(k - 11)E_1 - 7E_2 = 0$,   **17.** $(2k - 13)x = -(15 + 4k)y$,

$2E_1 + 6E_2 = kE_2$. $11x + 6y = 0$.

13–7. Determinants of the Third Order

In the previous sections determinants of the second order were defined and used for the discussion of systems of two linear equations with two unknowns. Determinants of higher order must be introduced if systems of linear equations with more than two unknowns are to be examined.

A determinant of the third order is a system of nine numbers which are arranged in three rows and three columns. It is denoted by enclosing the system of nine numbers by two vertical lines, like

$$\begin{vmatrix} 3 & 5 & 11 \\ 9 & -3 & 8 \\ 1 & 5 & 7 \end{vmatrix} \quad \text{or} \quad \begin{vmatrix} a_1 & b_1 & c_1 \\ a_2 & b_2 & c_2 \\ a_3 & b_3 & c_3 \end{vmatrix}.$$

Each of the nine numbers is called an **element** of the determinant. Every element is located in a certain row and in a certain column. For example, in the last determinant the element c_2 is in the second row and in the third column; this can be also expressed by saying that the row number of c_2 is 2 and its column number is 3.

An element is called **even** if the sum of its row and column numbers is even; it is called an **odd** element if this sum is odd. Thus the element c_2 is odd because $2 + 3 = 5$. The element a_1 is even because it is in the first row and first column and the sum $1 + 1 = 2$ is an even number. *Throughout this section the terms even and odd will be used exclusively in the sense just described.*

The computation of a determinant of the third order can be reduced to the computation of determinants of the second order. How this can be done will be shown in what follows.

If in a determinant of the third order one row and one column are omitted, four elements remain from which a determinant of the second order can be formed.   Thus, if in

$$\begin{vmatrix} 3 & 5 & 11 \\ 9 & -3 & 8 \\ 1 & 5 & 7 \end{vmatrix}$$

the second row and the third column are crossed out,

$$\begin{vmatrix} 3 & 5 & 1\!\!1 \\ \cancel{9} & \cancel{-3} & \cancel{8} \\ 1 & 5 & \cancel{7} \end{vmatrix},$$

the remaining four elements

$$\begin{vmatrix} 3 & 5 \\ 1 & 5 \end{vmatrix}$$

form a determinant of the second order.

To each element of the third-order determinant, there corresponds a second-order determinant consisting of the elements that remain if the row and the column of the given element are crossed out.   This procedure is used in the following definition of a **cofactor.**

The cofactor of a given element in a determinant of the third order is defined in the following way.

(a) *The row and the column of the given element are crossed out.*

(b) *The second-order determinant of the four remaining elements is formed.*

(c) *This determinant is the cofactor of the given element if this element is even.   For an odd element, the negative value of this determinant is the cofactor of the given element.*

Example 1.   Find the cofactors of the elements of the first row of the determinant

$$D = \begin{vmatrix} a_1 & b_1 & c_1 \\ a_2 & b_2 & c_2 \\ a_3 & b_3 & c_3 \end{vmatrix}.$$

In order to find the cofactor of a_1, the first line and the first column are omitted, which gives

$$\begin{vmatrix} a_1 & b_1 & c_1 \\ a_2 & b_2 & c_2 \\ a_3 & b_3 & c_3 \end{vmatrix}.$$

The four remaining elements form a determinant of the second order

$$\begin{vmatrix} b_2 & c_2 \\ b_3 & c_3 \end{vmatrix}.$$

Since the element a_1 is even, its cofactor is equal to this determinant.

The element b_1 is odd. Its cofactor, therefore, is the negative value of the determinant obtained after crossing out the first row and second column in D. Thus the cofactor of b_1 is

$$- \begin{vmatrix} a_2 & c_2 \\ a_3 & c_3 \end{vmatrix}.$$

Similarly, the cofactor of c_1 is

$$\begin{vmatrix} a_2 & b_2 \\ a_3 & b_3 \end{vmatrix}.$$

Example 2. Compute the cofactor of the element -3 in the determinant

$$\begin{vmatrix} 3 & 5 & 11 \\ 9 & -3 & 8 \\ 1 & 5 & 7 \end{vmatrix}.$$

This element, being in the second row and second column, is even. Its cofactor therefore, is

$$\begin{vmatrix} 3 & 11 \\ 1 & 7 \end{vmatrix} = 21 - 11 = 10.$$

The definition of a determinant of the third order can now be stated as follows:

In order to compute a determinant of the third order, each element of the first row is multiplied by its cofactor, and the three products are added.

When this definition is applied in order to find the value of

$$D = \begin{vmatrix} a_1 & b_1 & c_1 \\ a_2 & b_2 & c_2 \\ a_3 & b_3 & c_3 \end{vmatrix}, \tag{1}$$

we obtain

$$D = a_1 \begin{vmatrix} b_2 & c_2 \\ b_3 & c_3 \end{vmatrix} - b_1 \begin{vmatrix} a_2 & c_2 \\ a_3 & c_3 \end{vmatrix} + c_1 \begin{vmatrix} a_2 & b_2 \\ a_3 & b_3 \end{vmatrix},$$

where the values of the cofactors are taken from example 1.

If the three second-order determinants of the last expression are computed, we have

$$D = a_1(b_2c_3 - b_3c_2) - b_1(a_2c_3 - a_3c_2) + c_1(a_2b_3 - a_3b_2)$$

$$= a_1b_2c_3 - a_1b_3c_2 - b_1a_2c_3 + b_1a_3c_2 + c_1a_2b_3 - c_1a_3b_2.$$

Rearranging the last sum by writing first the positive terms, we have the following expression for the determinant of the third order given in equation 1:

$$D = a_1b_2c_3 + b_1c_2a_3 + c_1a_2b_3 - a_3b_2c_1 - b_3c_2a_1 - c_3a_2b_1.$$

This expression, which looks complicated, can be found by the following simple rule, which applies *only* to determinants of third order.

To the right of the given determinant, the first two columns are repeated so that the following array is obtained:

$$\begin{vmatrix} a_1 & b_1 & c_1 \\ a_2 & b_2 & c_2 \\ a_3 & b_3 & c_3 \end{vmatrix} \begin{matrix} a_1 & b_1 \\ a_2 & b_2 \\ a_3 & b_3 \end{matrix}.$$

Then multiply the elements in the diagonal of the original square which runs from the upper left corner downward and the elements in the two lines parallel to this diagonal indicated in Fig. 13–6. These three products are the positive terms of D. In order to obtain the negative terms, multiply the elements in the diagonal of the original square, running from the lower left corner upward, and the elements in the two lines parallel to this diagonal as shown in Fig. 13–6.

An alternative rule which gives the same result is indicated in Fig. 13–7.

Example 3. Compute the value of the determinant

$$D = \begin{vmatrix} 3 & 5 & 11 \\ 9 & -3 & 8 \\ 1 & 5 & 7 \end{vmatrix}.$$

By the rule given above we write

$$
\begin{vmatrix} 3 & 5 & 11 \\ 9 & -3 & 8 \\ 1 & 5 & 7 \end{vmatrix} \begin{matrix} 3 & 5 \\ 9 & -3 \\ 1 & 5 \end{matrix} .
$$

Hence we have for our determinant:

$$D = 3 \cdot (-3) \cdot 7 + 5 \cdot 8 \cdot 1 + 11 \cdot 9 \cdot 5 - 1 \cdot (-3) \cdot 11 - 5 \cdot 8 \cdot 3 - 7 \cdot 9 \cdot 5$$
$$= -63 + 40 + 495 + 33 - 120 - 315 = 568 - 498 = 70.$$

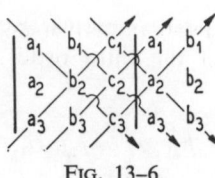

Fig. 13–6

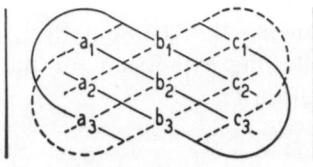

Fig. 13–7

Example 4. Compute the value of the determinant

$$
D = \begin{vmatrix} 10 & 0 & -5 \\ 1 & 6 & 0 \\ 0 & 9 & 2 \end{vmatrix}.
$$

As in the previous example we write

$$
\begin{vmatrix} 10 & 0 & -5 \\ 1 & 6 & 0 \\ 0 & 9 & 2 \end{vmatrix} \begin{matrix} 10 & 0 \\ 1 & 6 \\ 0 & 9 \end{matrix} .
$$

which gives

$$D = 10 \cdot 6 \cdot 2 + 0 + (-5) \cdot 1 \cdot 9 - 0 - 0 - 0 = 120 - 45 = 75.$$

When the elements are complicated numbers, the computation of determinants may be very troublesome. Such computations can be greatly simplified by several rules relating to determinants that are given below without proof. The same rules hold also for determinants of the second order and for determinants of higher orders.

The value of a determinant of the third order was originally defined as the sum of the products of the elements of the first row by their cofactors. The same value is obtained if the elements of the first row are replaced by the elements of another row or of a column.   Thus, the first of the theorems to be stated is:

1. *The value of a determinant is obtained if the elements of any row or any column are multiplied by their respective cofactors and these products are added.*

2. *If only one element in a row or in a column is different from zero, the determinant is the product of this element and its cofactor.*

This statement is an immediate consequence of the first theorem.

Example 5.   Using the last result we have

$$\begin{vmatrix} 3 & 4 & 2 \\ 0 & 0 & 4 \\ 2 & 1 & 3 \end{vmatrix} = -4 \begin{vmatrix} 3 & 4 \\ 2 & 1 \end{vmatrix} = -4(3 - 8) = 20.$$

The following two statements are also consequences of the first theorem.

3. *A determinant is zero if all elements of one row or all elements of one column are zero.*

4. *The value of the determinant changes its sign if two rows or two columns are interchanged.*

This last theorem will be illustrated in the next two examples.

Example 6.   Consider the determinant

$$\begin{vmatrix} 1 & 4 \\ 2 & 5 \end{vmatrix} = 5 - 8 = -3.$$

Interchanging the two rows, we obtain

$$\begin{vmatrix} 2 & 5 \\ 1 & 4 \end{vmatrix} = 8 - 5 = 3.$$

Example 7.   Consider the determinant

$$\begin{vmatrix} 1 & 2 & -1 \\ 3 & 1 & 4 \\ 2 & 3 & 1 \end{vmatrix} = 1 \cdot 1 \cdot 1 + 2 \cdot 4 \cdot 2 + (-1) \cdot 3 \cdot 3 - 2 \cdot 1 \cdot (-1)$$
$$- 3 \cdot 4 \cdot 1 - 1 \cdot 3 \cdot 2$$
$$= 1 + 16 - 9 + 2 - 12 - 6 = -8.$$

The determinant obtained by interchanging the first and third columns is given by

$$\begin{vmatrix} -1 & 2 & 1 \\ 4 & 1 & 3 \\ 1 & 3 & 2 \end{vmatrix} = (-1) \cdot 1 \cdot 2 + 2 \cdot 3 \cdot 1 + 1 \cdot 4 \cdot 3 - 1 \cdot 1 \cdot 1$$
$$- 3 \cdot 3 \cdot (-1) - 2 \cdot 4 \cdot 2$$
$$= -2 + 6 + 12 - 1 + 9 - 16 = +8.$$

5. *A determinant is zero if the elements of one row or of one column are equal to the corresponding elements of another row or column, respectively.*

Example 8. To illustrate the last theorem, consider the following determinant having the elements of the first column equal to the corresponding elements of the third column:

$$\begin{vmatrix} 3 & 1 & 3 \\ 2 & 3 & 2 \\ 1 & 2 & 1 \end{vmatrix} = 3 \cdot 3 \cdot 1 + 1 \cdot 2 \cdot 1 + 3 \cdot 2 \cdot 2 - 1 \cdot 3 \cdot 3 - 2 \cdot 2 \cdot 3 \\ - 1 \cdot 2 \cdot 1$$

$$= 9 + 2 + 12 - 9 - 12 - 2 = 0.$$

6. *A determinant is multiplied or divided by a number k if all elements of one row or of one column are multiplied or divided, respectively, by this number k.*

It follows from this theorem that any factor common to all the elements of a row or column may be removed and written before the determinant.

Example 9.

$$\begin{vmatrix} 24 & 32 & 16 \\ 5 & 1 & 3 \\ 3 & 4 & 7 \end{vmatrix} = \begin{vmatrix} 8 \cdot 3 & 8 \cdot 4 & 8 \cdot 2 \\ 5 & 1 & 3 \\ 3 & 4 & 7 \end{vmatrix} = 8 \begin{vmatrix} 3 & 4 & 2 \\ 5 & 1 & 3 \\ 3 & 4 & 7 \end{vmatrix}.$$

7. *If to the elements of any column we add or subtract k times the corresponding elements of any other column, the value of the determinant remains unchanged. The same rule holds for rows.*

Example 10. If the elements of the second column of the determinant

$$\begin{vmatrix} 3 & 4 & 10 \\ 1 & 2 & 6 \\ 2 & 6 & 15 \end{vmatrix}$$

are multiplied by 3 and then subtracted from the third column, the result is

$$\begin{vmatrix} 3 & 4 & -2 \\ 1 & 2 & 0 \\ 2 & 6 & -3 \end{vmatrix}.$$

In order to check the given theorem, both determinants can be computed:

$$\begin{vmatrix} 3 & 4 & 10 \\ 1 & 2 & 6 \\ 2 & 6 & 15 \end{vmatrix} = 3 \cdot 2 \cdot 15 + 4 \cdot 6 \cdot 2 + 10 \cdot 1 \cdot 6 - 2 \cdot 2 \cdot 10 \\ - 6 \cdot 6 \cdot 3 - 15 \cdot 1 \cdot 4$$

$$= 90 + 48 + 60 - 40 - 108 - 60 = -10.$$

$$\begin{vmatrix} 3 & 4 & -2 \\ 1 & 2 & 0 \\ 2 & 6 & -3 \end{vmatrix} = -3 \cdot 2 \cdot 3 + 4 \cdot 0 \cdot 2 + (-2) \cdot 1 \cdot 6 - 2 \cdot 2 \cdot (-2) \\ - 6 \cdot 0 \cdot 3 - (-3) \cdot 1 \cdot 4$$

$$= -18 - 12 + 8 + 12 = -10.$$

The computation of a determinant can often be greatly simplified by the application of these rules. Theorem 7 is especially useful in decreasing the numerical value of elements of the determinant or, as in example 10, in reducing an element to zero.

A few examples will show how these theorems are used for the computation of determinants.

Example 11. Find the value of the determinant

$$D = \begin{vmatrix} 32 & 31 & 15 \\ 40 & 44 & 23 \\ 35 & 34 & 18 \end{vmatrix}.$$

By Theorem 7 the value of D is not changed if the second column is subtracted from the first. This operation diminishes the terms of the first column and makes

$$D = \begin{vmatrix} 1 & 31 & 15 \\ -4 & 44 & 23 \\ 1 & 34 & 18 \end{vmatrix}.$$

In order to continue with the simplification of D, the first row is subtracted from the third, making one element zero. The result is

$$D = \begin{vmatrix} 1 & 31 & 15 \\ -4 & 44 & 23 \\ 0 & 3 & 3 \end{vmatrix}.$$

All the elements of the third row are divisible by 3. Hence the application of Theorem 6 gives

$$D = 3 \begin{vmatrix} 1 & 31 & 15 \\ -4 & 44 & 23 \\ 0 & 1 & 1 \end{vmatrix}.$$

Subtraction of the third column from the second gives

$$D = 3 \begin{vmatrix} 1 & 16 & 15 \\ -4 & 21 & 23 \\ 0 & 0 & 1 \end{vmatrix}.$$

The last determinant can be easily computed because of the zeros in the third row. We have finally

$$D = 3[1 \cdot 21 - (-4) \cdot 16] = 3 \cdot 85 = 255.$$

Example 12. Compute the determinant

$$D = \begin{vmatrix} 13 & 19 & 11 \\ 7 & 9 & 5 \\ 10 & 14 & 8 \end{vmatrix}.$$

The elements of the third row are divisible by 2; therefore

$$D = 2 \begin{vmatrix} 13 & 19 & 11 \\ 7 & 9 & 5 \\ 5 & 7 & 4 \end{vmatrix}.$$

The terms of D can be diminished in different ways. For example, in the last determinant the elements of the third row may be multiplied by 2 and subtracted from the first, and then the elements of the third row may be subtracted from the second. Both operations, when performed simultaneously, give

$$D = 2 \begin{vmatrix} 3 & 5 & 3 \\ 2 & 2 & 1 \\ 5 & 7 & 4 \end{vmatrix}.$$

Subtraction of the second row from the third in the last determinant gives

$$D = 2 \begin{vmatrix} 3 & 5 & 3 \\ 2 & 2 & 1 \\ 3 & 5 & 3 \end{vmatrix}.$$

In this determinant, the elements of the first and third row are, respectively, equal; therefore, according to Theorem 5,
$$D = 0.$$

Example 13. Compute the determinant

$$D = \begin{vmatrix} 1 & 1 & 1 \\ a & b & c \\ a^2 & b^2 & c^2 \end{vmatrix}.$$

If the first column is subtracted from the second and from the third column, we obtain

$$D = \begin{vmatrix} 1 & 0 & 0 \\ a & b-a & c-a \\ a^2 & b^2-a^2 & c^2-a^2 \end{vmatrix}$$

$$= \begin{vmatrix} 1 & 0 & 0 \\ a & b-a & c-a \\ a^2 & (b-a)(b+a) & (c-a)(c+a) \end{vmatrix}.$$

All the terms of the second column are divisible by $b - a$, and all the terms of the third column by $c - a$. Applying Theorem 6, we get,

$$D = (b-a)(c-a) \begin{vmatrix} 1 & 0 & 0 \\ a & 1 & 1 \\ a^2 & b+a & c+a \end{vmatrix}.$$

To simplify the last determinant, the second column is subtracted from the third, yielding

$$D = (b-a)(c-a) \begin{vmatrix} 1 & 0 & 0 \\ a & 1 & 0 \\ a^2 & b+a & c-b \end{vmatrix},$$

and therefore

$$D = (b-a)(c-a)(c-b).$$

Exercises

Evaluate each of the following determinants:

1. $\begin{vmatrix} 2 & 0 & 3 \\ 5 & 0 & 9 \\ 7 & 0 & 13 \end{vmatrix}$.

2. $\begin{vmatrix} 1 & 0 & 0 \\ 0 & 1 & 0 \\ 0 & 0 & 1 \end{vmatrix}$.

3. $\begin{vmatrix} 0 & 0 & 4 \\ 2 & 0 & 1 \\ -1 & 5 & 6 \end{vmatrix}$.

4. $\begin{vmatrix} 3 & 0 & 0 \\ 0 & 2 & 0 \\ 0 & 0 & -1 \end{vmatrix}$.

5. $\begin{vmatrix} 2 & 0 & 0 \\ 0 & 0 & -4 \\ 0 & 1 & 0 \end{vmatrix}$.

6. $\begin{vmatrix} 1 & 3 & 2 \\ 4 & -1 & 3 \\ 5 & 2 & 1 \end{vmatrix}$.

7. $\begin{vmatrix} 2 & 0 & -1 \\ 0 & 5 & 3 \\ 1 & -1 & 4 \end{vmatrix}$.

8. $\begin{vmatrix} 3 & 7 & 9 \\ 0 & 0 & 0 \\ -1 & 2 & 5 \end{vmatrix}$.

9. $\begin{vmatrix} 3 & 0 & 0 \\ 0 & -2 & 0 \\ 0 & 0 & 4 \end{vmatrix}$.

10. $\begin{vmatrix} 0 & 1 & 0 \\ 1 & 0 & 0 \\ 0 & 0 & -1 \end{vmatrix}$.

11. $\begin{vmatrix} 0 & 0 & 1 \\ 0 & 1 & 0 \\ -1 & 0 & 0 \end{vmatrix}$.

12. $\begin{vmatrix} 0 & 0 & 3 \\ 0 & -1 & 0 \\ 2 & 0 & 0 \end{vmatrix}$.

13. $\begin{vmatrix} 1 & 2 & 4 \\ 0 & -3 & 5 \\ 0 & 0 & 3 \end{vmatrix}$.

14. $\begin{vmatrix} -2 & 0 & 0 \\ 5 & -4 & 0 \\ 7 & 8 & 3 \end{vmatrix}$.

15. $\begin{vmatrix} 3 & 5 & 4 \\ 6 & 10 & 8 \\ -1 & 0 & 5 \end{vmatrix}$.

16. $\begin{vmatrix} -2 & 4 & -1 \\ 5 & -10 & 3 \\ 3 & -6 & 11 \end{vmatrix}$.

17. $\begin{vmatrix} 4 & -5 & 1 \\ 6 & \frac{1}{2} & 3 \\ 2 & \frac{1}{3} & -1 \end{vmatrix}$.

18. $\begin{vmatrix} \frac{1}{4} & 1 & 0 \\ 3 & \frac{1}{2} & -2 \\ -\frac{1}{3} & -5 & 1 \end{vmatrix}$.

19. $\begin{vmatrix} 0 & -p & -q \\ p & 0 & -r \\ q & r & 0 \end{vmatrix}.$ 20. $\begin{vmatrix} x_1 & x_2 & x_3 \\ y_1 & y_2 & y_3 \\ z_1 & z_2 & z_3 \end{vmatrix}.$ 21. $\begin{vmatrix} x & a & a \\ a & x & a \\ a & a & x \end{vmatrix}.$

22. $\begin{vmatrix} 1 & -c & b \\ c & 1 & -a \\ -b & a & 1 \end{vmatrix}.$ 23. $\begin{vmatrix} -21 & 9 & -6 \\ 7 & -1 & 2 \\ 14 & -4 & 8 \end{vmatrix}.$ 24. $\begin{vmatrix} 150 & -30 & 20 \\ 5 & 3 & -4 \\ -35 & 21 & 14 \end{vmatrix}.$

25. $\begin{vmatrix} 33 & 22 & -55 \\ -18 & 4 & 10 \\ 12 & -8 & -20 \end{vmatrix}.$ 26. $\begin{vmatrix} 26 & 39 & 65 \\ 10 & 15 & 25 \\ -2 & -3 & -5 \end{vmatrix}.$ 27. $\begin{vmatrix} 58 & -21 & 19 \\ 0 & 0 & 0 \\ 11 & 17 & -23 \end{vmatrix}.$

28. $\begin{vmatrix} 9 & 5 & -7 \\ 11 & -3 & 4 \\ 12 & -1 & 3 \end{vmatrix}.$ 29. $\begin{vmatrix} 69 & -73 & 54 \\ 28 & 31 & -19 \\ 35 & 41 & 43 \end{vmatrix}.$ 30. $\begin{vmatrix} 34 & 25 & 39 \\ 47 & 41 & 52 \\ 68 & 75 & 60 \end{vmatrix}.$

31. $\begin{vmatrix} 2 & 4+x & 1 \\ -1 & 3 & 4-x \\ 0 & 5 & 3 \end{vmatrix}.$ 32. $\begin{vmatrix} 0 & 5 & 1 \\ 3+x & 2 & -1 \\ 3 & 3-x & 6 \end{vmatrix}.$ 33. $\begin{vmatrix} 1 & 1 & 1 \\ a_1 & a_2 & a_3 \\ x_1 & x_2 & x_3 \end{vmatrix}.$

34. $\begin{vmatrix} a & a & a \\ y & b & c \\ y^2 & b^2 & c^2 \end{vmatrix}.$ 35. $\begin{vmatrix} 1 & 1 & 1 \\ a_1 & a_2 & a_3 \\ a_1^2 & a_2^2 & a_3^2 \end{vmatrix}.$ 36. $\begin{vmatrix} x^2 & x & 1 \\ y^2 & y & 1 \\ z^2 & z & 1 \end{vmatrix}.$

37. $\begin{vmatrix} x & y & 1 \\ x_1 & y_1 & 1 \\ x_2 & y_2 & 1 \end{vmatrix}.$ 38. $\begin{vmatrix} k-1 & 2 & 3 \\ 1 & k-1 & 2 \\ -1 & -2 & k-1 \end{vmatrix}.$ 39. $\begin{vmatrix} a & b & c \\ a^2 & b^2 & c^2 \\ a^3 & b^3 & c^3 \end{vmatrix}.$

40. $\begin{vmatrix} a & b & c \\ d & e & f \\ a+2d & b+2e & c+2f \end{vmatrix}.$

13–8. Systems of Linear Equations with Three Unknowns

Determinants of the third order can be used to obtain the solution of a system of three simultaneous linear equations just as determinants of the second order were used for solving a system of two simultaneous equations. Since space does not permit the development in this book of the general theory of systems of three or more equations, only three of the more important problems concerning such systems will be discussed.   These are:

1. Is the given system independent?
2. When does a system of homogeneous equations have a solution in which not all the unknowns are equal to zero?
3. What is the solution of a system of two homogeneous equations in three unknowns?

Each one of these three problems is solved by a theorem.   The first two of the three theorems will be stated without proof; a brief proof of the third will be given.

A system of three linear equations with three unknown quantities is called **independent** if it can be solved, that is, if there is only one set of three values (one for each unknown) satisfying all the given equations. This definition of independence is analogous to the one that was given in the discussion of systems of two equations (Sec. 13–3).

The first of the problems listed above is solved by the following theorem:

1. *The system of equations*

$$a_1x + b_1y + c_1z = d_1,$$
$$a_2x + b_2y + c_2z = d_2,$$
$$a_3x + b_3y + c_3z = d_3,$$

is independent if the determinant

$$\begin{vmatrix} a_1 & b_1 & c_1 \\ a_2 & b_2 & c_2 \\ a_3 & b_3 & c_3 \end{vmatrix}$$

which is formed from the coefficients of the unknowns is not zero.

If this determinant is zero, the system is either inconsistent or dependent.   In the first case there is no solution, and in the second case there are infinitely many solutions.

Example 1.   Consider the system of equations

$$x + y + z = 3,$$
$$2x - y + 3z = 1,$$
$$4x + y + 5z = 5.$$

This system is not independent, because the determinant of the coefficients

$$\begin{vmatrix} 1 & 1 & 1 \\ 2 & -1 & 3 \\ 4 & 1 & 5 \end{vmatrix} = -5 + 12 + 2 + 4 - 3 - 10 = 0.$$

By theorems that are not given here it could be shown that this system of equations is inconsistent. This fact, however, can be shown directly by multiplying the first equation by 2, the second by 1, the third by -1, and adding the resulting equations.

$$\begin{array}{rcr} 2x + 2y + 2z = & 6 \\ 2x - y + 3z = & 1 \\ -4x - y - 5z = & -5 \\ \hline 0 = & 2 \end{array}$$

This, obviously, is impossible.

Methods for solving an independent system of three linear equations with three unknowns were given in Sec. 13–1 and in Sec. 13–2.

The second problem deals with a system of equations of the form

$$a_1 x + b_1 y + c_1 z = 0,$$
$$a_2 x + b_2 y + c_2 z = 0,$$
$$a_3 x + b_3 y + c_3 z = 0.$$

In these equations the constant term is missing, and they are called homogeneous. It is obvious that these equations are satisfied if

$$x = 0, \qquad y = 0, \qquad z = 0.$$

In many problems it is important to know whether or not there are other solutions such that not all the unknowns are zero. The answer to this question is given by the following theorem.

2. *The system of three homogeneous equations with three unknowns*

$$a_1 x + b_1 y + c_1 z = 0,$$
$$a_2 x + b_2 y + c_2 z = 0,$$
$$a_3 x + b_3 y + c_3 z = 0$$

has only one solution,

$$x = 0, \qquad y = 0, \qquad y = 0,$$

if the determinant of the system

$$D = \begin{vmatrix} a_1 & b_1 & c_1 \\ a_2 & b_2 & c_2 \\ a_3 & b_3 & c_3 \end{vmatrix}$$

is not zero. It has infinitely many solutions if this determinant D is equal to zero.

For this problem we may go further and give some simple rules for finding the solutions of the given system of equations.

When $D = 0$, formulas for all the solutions can be found by multiplying the cofactors of the elements of the first row of D by an arbitrary number t to give solutions as follows:

$$x = t \begin{vmatrix} b_2 & c_2 \\ b_3 & c_3 \end{vmatrix} = t(b_2c_3 - b_3c_2),$$

$$y = -t \begin{vmatrix} a_2 & c_2 \\ a_3 & c_3 \end{vmatrix} = t(a_3c_2 - a_2c_3), \tag{1}$$

$$z = t \begin{vmatrix} a_2 & b_2 \\ a_3 & b_3 \end{vmatrix} = t(a_2b_3 - a_3b_2),$$

provided at least one of these three cofactors is not zero.

Similar results, involving the cofactors of the elements of each of the other two rows may also be stated.

If all the cofactors of the elements of the first row of D are zero, then the three formulas in (1) yield only the known solution $x = 0, y = 0, z = 0$. When this happens, we may turn to the similar formulas in terms of the cofactors of the elements of either the second or the third row of D, provided at least one of these cofactors is not zero.   In this way it is always possible to find formulas for the solutions, provided at least one of the nine cofactors of the elements of D is not zero.

If all nine cofactors of the elements of D are zero, it can be shown that all three of the original homogeneous equations are equivalent, that is, all three of the left members are constants times the same expression. Thus, in this case any three values of x, y, and z that satisfy one of these equations also satisfy each of the other two.

Example 2.   Solve the system of equations

$$x + y + z = 0,$$
$$2x - y + 3z = 0,$$
$$4x + y + 5z = 0.$$

The determinant of this system as computed in example 1 is

$$\begin{vmatrix} 1 & 1 & 1 \\ 2 & -1 & 3 \\ 4 & 1 & 5 \end{vmatrix} = 0. \tag{1}$$

According to Theorem 2, the given system has infinitely many solutions. In order to find them, we compute the cofactors of the elements of the first row of the above determinant:

$$\begin{vmatrix} -1 & 3 \\ 1 & 5 \end{vmatrix} = -5 - 3 = -8, \qquad -\begin{vmatrix} 2 & 3 \\ 4 & 5 \end{vmatrix} = -(10 - 12) = 2,$$

$$\begin{vmatrix} 2 & -1 \\ 4 & 1 \end{vmatrix} = 2 + 4 = 6.$$

All the solutions of the system are now given by

$$x = -8t, \qquad y = 2t, \qquad z = 6t, \tag{2}$$

where t is an arbitrary number. This solution can be checked by substituting the expressions for x, y, and z in the given system of equations. Thus we obtain

$$x + y + z = -8t + 2t + 6t = 0,$$
$$2x - y + 3z = -16t - 2t + 18t = 0,$$
$$4x + y + 5z = -32t + 2t + 30t = 0.$$

A simple solution is obtained by choosing, for example, $t = \frac{1}{2}$, so that

$$x = -4, \qquad y = 1, \qquad z = 3.$$

The same solution is obtained if the cofactors of another row of the determinant in (1) are used. Computing the cofactors of the third row we obtain

$$\begin{vmatrix} 1 & 1 \\ -1 & 3 \end{vmatrix} = 3 + 1 = 4, \qquad -\begin{vmatrix} 1 & 1 \\ 2 & 3 \end{vmatrix} = -(3 - 2) = -1,$$

$$\begin{vmatrix} 1 & 1 \\ 2 & -1 \end{vmatrix} = -1 - 2 = -3.$$

Hence, if u is an arbitrary number, we have as the solution

$$x = 4u, \qquad y = -u, \qquad z = -3u. \tag{3}$$

These expressions give exactly the same values for x, y, z as those obtained in equations 2 for, when $u = -2t$, we see that equations 3 are the same as equations 2.

Example 3. Solve the system of equations

$$x + 2y + z = 0,$$
$$x + y + 2z = 0,$$
$$3x + 3y + 6z = 0.$$

The student may easily verify that all three of the cofactors of the elements of the first row of the determinant of this system are zero. The solution $x = 3t$, $y = -t$, $z = -t$, where t is arbitrary, may be obtained from the cofactors of the elements of the second row. The student should derive and check this solution.

Example 4. Solve the system of equations

$$3x - y + 2z = 0,$$
$$-6x + 2y - 4z = 0,$$
$$15x - 5y + 10z = 0.$$

The student may easily verify that all nine of the cofactors of the elements of the determinant of this system are zero, and therefore that any three values of x, y, and z satisfying one equation also satisfy the other two equations.

The student should then notice that the second left member is -2 times the first and the third is 5 times the first. It follows that the three equations are equivalent.

The third question asked at the beginning of this section is answered by the following theorem:

3. *The solutions of the system of two simultaneous linear equations with three unknowns*

$$a_1x + b_1y + c_1z = 0,$$

$$a_2x + b_2y + c_2z = 0$$

are given by the expressions

$$x = t \begin{vmatrix} b_1 & c_1 \\ b_2 & c_2 \end{vmatrix}, \quad y = t \begin{vmatrix} c_1 & a_1 \\ c_2 & a_2 \end{vmatrix}, \quad z = t \begin{vmatrix} a_1 & b_1 \\ a_2 & b_2 \end{vmatrix},$$

t being an arbitrary number, provided at least one of the three determinants is not zero.

This result is easy to derive. For example, suppose that the determinant

$$d = \begin{vmatrix} a_1 & b_1 \\ a_2 & b_2 \end{vmatrix} \neq 0,$$

and consider the given equations in the form

$$a_1x + b_1y = -c_1z,$$

$$a_2x + b_2y = -c_2z.$$

As we saw in Sec. 13–4, these equations are satisfied by

$$x = \frac{\begin{vmatrix} -c_1z & b_1 \\ -c_2z & b_2 \end{vmatrix}}{d} = \frac{z\begin{vmatrix} b_1 & c_1 \\ b_2 & c_2 \end{vmatrix}}{d}, \quad y = \frac{\begin{vmatrix} a_1 & -c_1z \\ a_2 & -c_2z \end{vmatrix}}{d} = \frac{z\begin{vmatrix} c_1 & a_1 \\ c_2 & a_2 \end{vmatrix}}{d}.$$

If we set $z = dt$, we obtain the desired result.

The student may easily furnish a discussion of the situation that occurs when all three determinants of the solution stated in Theorem 3 are zero.

Example 5. Solve the system of equations

$$3x - 4y + 5z = 0,$$
$$2x + y + z = 0.$$

The three determinants, used in Theorem 3, are in this case

$$\begin{vmatrix} -4 & 5 \\ 1 & 1 \end{vmatrix} = -4 - 5 = -9, \qquad \begin{vmatrix} 5 & 3 \\ 1 & 2 \end{vmatrix} = 10 - 3 = 7,$$

$$\begin{vmatrix} 3 & -4 \\ 2 & 1 \end{vmatrix} = 3 + 8 = 11.$$

Hence all the solutions of the given system are given by

$$x = -9t, \qquad y = 7t, \qquad z = 11t,$$

where t is arbitrary. If, for example, the value $t = 1$ is chosen, then

$$x = -9, \qquad y = 7, \qquad z = 11.$$

These values are easily checked to be solutions by substitution in the given equations. The result is

$$3(-9) - 4 \cdot 7 + 5 \cdot 11 = -27 - 28 + 55 = 0,$$
$$2(-9) + 7 + 11 = -18 + 7 + 11 = 0.$$

The definition of determinants of the third order by means of the cofactors of the elements of a row or column can be extended to define determinants of any order. It can be shown that the theorems of Sec. 13–7 hold for determinants of any order. Likewise, theorems about systems of linear equations like those given above hold for systems with any number of unknowns.

Exercises

Decide whether or not the following systems are independent:

1. $x + y = 2,$
   $x + z = 4,$
   $y + z = -1.$

2. $x - 2y = 3,$
   $y + 3z = 5,$
   $x + 6z = 0.$

3. $\quad 2x + y - 1 = 0,$
   $\quad x + z - 2 = 0,$
   $4x + 3y - 2z = -1.$

4. $\quad u + 2v = 6,$
   $2u - v - 2 = 0,$
   $u + v + w = 0.$

5. $\quad 2R_1 - 3R_2 + R_3 = 5,$
   $\quad R_1 + 5R_2 + 4R_3 = 1,$
   $7R_1 + 9R_2 + 14R_3 = 8.$

6. $\quad 2i_1 - 3i_2 = -2,$
   $\quad i_1 + i_2 - i_3 = 1,$
   $\quad 3i_1 - i_3 = 0.$

7. $-3x + y - 5z = 7,$
   $\quad x - y - 3 = 0,$
   $\quad 2x + 4z = 1.$

8. $\quad p + q - r = 0,$
   $\quad 2p - q + 3r = 1,$
   $13p - q + 7r = 4.$

Solve the following systems of homogeneous equations:

9. $x + y = 0,$
$x + z = 0,$
$y - z = 0.$

10. $2x + y = 0,$
$x - z = 0,$
$5x + 2y - z = 0.$

11. $r_1 + r_2 = 0,$
$2r_1 - r_3 = 0,$
$r_2 + 4r_3 = 0.$

12. $2I_1 - I_2 + 4I_3 = 0,$
$I_1 + 3I_2 - 2I_3 = 0,$
$I_1 - 11I_2 + 14I_3 = 0.$

13. $M + N + 3P = 0,$
$M + 2N + 2P = 0,$
$M + 5N - P = 0.$

14. $2A - B + C = 0,$
$A + 2B - C = 0,$
$3A + B - 2C = 0.$

15. $5E_1 + 2E_2 - 7E_3 = 0,$
$E_1 + E_2 - 2E_3 = 0,$
$E_2 - E_3 = 0.$

16. $2x - y + 3z = 0,$
$8x - 9y + 7z = 0,$
$-x + 3y + z = 0.$

17. $-3x + 6y + 5z = 0,$
$x - 2y + 4z = 0,$
$-5x + 10y - 20z = 0.$

18. $2x - y + z = 0,$
$5x + 2y - 2z = 0,$
$6z - 6y - 15x = 0.$

19. $2x - y + 3z = 0,$
$9z + 6x - 3y = 0,$
$10x + 15z - 5y = 0.$

20. $2x - 7y + 3z = 0,$
$14y - 4x - 6z = 0,$
$12z + 8x - 28y = 0.$

21. $3x + 4y - 2z = 0,$
$x - 7y + z = 0.$

22. $e_1 - 3e_2 - 4e_3 = 0,$
$2e_1 + e_2 + 5e_3 = 0.$

23. $u + 2v = 0,$
$2u - v + 3w = 0.$

24. $3A_x + 2A_y - A_z = 0,$
$5A_x - A_y + 4A_z = 0.$

25. $i_1 - 3i_2 = 0,$
$i_2 - i_3 = 0.$

26. $2A - 7C = 0,$
$B + 3C = 0.$

27. $5x - 2y + 3z = 0,$
$4y - 10x = 6z - 0.$

28. $2y - 3x + 5z = 0,$
$6x   10z - 4y - 0.$

29. $x + y + 2z = 0,$
$3x + 3y + 4z = 0.$

30. $2x - y + 3z = 0,$
$4x - 3y + 6z = 0.$

Solve the following equations for $A, B, C,$ and give the particular solution for which the arbitrary constant t, used in the third theorem of this section, is equal to 1:

31. $A + B + C = 0,$
$2A - 3B + C = 0.$

32. $A - 2B + 3C = 0,$
$3A + B - C = 0.$

33. $m^2A + mB + C = 0,$
$A - B - C = 0.$

34. $Ax_1 + By_1 + C = 0,$
$Ax_2 + By_2 + C = 0.$

Find, if possible, values of k such that the following systems of equations have other solutions besides $x = 0$, $y = 0$, $z = 0$:

35. $kx + 2y + z = 0$,
$\quad\;\; x + y + z = 0$,
$\quad\;\; kx + 4y - z = 0$.

36. $2x + (k + 1)y - z = 0$,
$\quad\;\; x - (k - 2)y + 2z = 0$,
$\quad\quad\;\; 3x + y + z = 0$.

37. $kx + z = 0$,
$\quad\;\; y + 2z = 0$,
$\quad\;\; y + kz = 0$.

38. $\quad\;\; kx + 8y = 0$,
$\quad\quad\;\; 2x + ky = 0$,
$\quad\;\; 3x + 2y + z = 0$.

39. $5x + 7y - 9z = 0$,
$\quad\quad\; y - kz = 0$,
$\quad\quad\; ky + z = 0$.

40. $(k - 1)x + y + 3z = 0$,
$\quad\;\; kx - 2y + z = 0$,
$\quad (k + 1)x + y + z = 0$.

41. Verify that the solutions given in Theorem 3 of this section are correct by substituting them into the given equations. *Hint*: After substituting, note that the left member is the expansion of a third order determinant. The verification can be completed by applying an appropriate theorem from Sec. 13–7 to this determinant.

42. Prove Theorem 2 of this section by substituting the specified solutions into each of the given equations. Use the hint given in exercise 41.

13–9. Engineering Applications of Simultaneous Equations

Simultaneous linear equations arise in many engineering problems, especially in electric networks in electrical engineering and in the theory of structures in civil engineering. We shall consider in this section a system of simultaneous linear equations arising in connection with an electric network.

In an automobile, airplane, tank, or tractor, an electric generator charges a storage battery, and both of these in turn supply the current for the ignition system and lights. Figure 13–8 shows a pictorial diagram and a schematic diagram of such a circuit. The voltage of the generator is E_1 and the voltage of the battery is E_2. The internal resistance of the generator is R_1 and the internal resistance of the battery is R_2. The load will be considered simply as a resistance R_3. The currents flowing in the branch circuits are I_1, I_2, and I_3 as shown. Their direction of flow as indicated by the arrows is chosen arbitrarily. If we should later find that in the solution of a particular problem a current comes out with a negative value, then the current flows in a direction opposite to that made in the original assumption. The current junction points are marked a and b.

Kirchhoff's laws state that

1. The algebraic sum of the currents at a junction is equal to zero.

2. In an electric circuit the algebraic sum of the voltage rises and voltage drops is zero.

The algebraic sign given to any voltage rise is positive if the direction of travel is from negative to positive.   The sign associated with a voltage drop across a resistance is negative if the direction of travel is in the assumed direction of current flow.

From the circuit given above, by these laws and Ohm's law (see Sec. 2–18) we can write the circuit equations.   If we regard the voltage and resistance values as known, then there are three unknown currents, the current through the generator, the current through the battery, and the current through the load.   By Kirchhoff's first law, we see that

$$I_3 = I_2 - I_1. \tag{1}$$

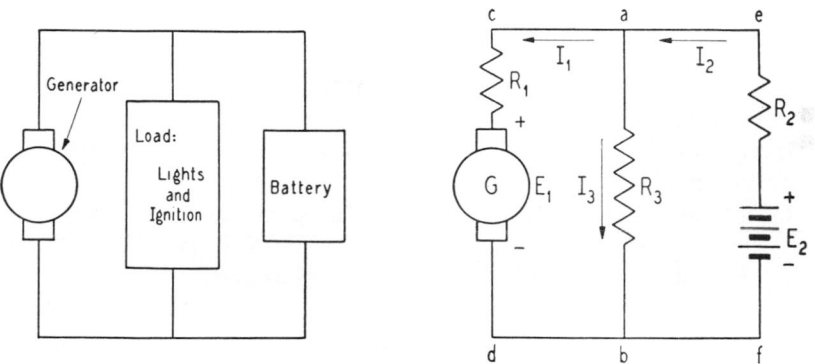

FIG. 13–8

Adding the voltage rises and drops in the circuit *acdb* by going around the circuit in a counterclockwise direction, we get

$$-I_1R_1 - E_1 + I_3R_3 = 0. \tag{2}$$

Similarly, for the circuit *abfe* we obtain

$$-I_3R_3 + E_2 - I_2R_2 = 0. \tag{3}$$

We have then the system of equations:

$$I_1 - I_2 + I_3 = 0, \tag{1}$$

$$-R_1I_1 + R_3I_3 = E_1, \tag{2}$$

$$R_2I_2 + R_3I_3 = E_2, \tag{3}$$

which can be solved by the methods of this chapter for I_1, I_2, and I_3.

Exercises

1. In the circuit in Fig. 13–8 let the generator voltage be 9.1 volts and the voltage of the battery 7.2 volts. The internal resistance of the generator is 0.5 ohm and that of the battery 0.4 ohm. The load has a resistance of 0.4 ohm. What is the current delivered by the generator and the battery through the load?

2. In a submarine that has its power plant arranged as shown in Fig. 13–8, the generator voltage is 120.5 volts and the battery voltage is 120.1 volts. The internal resistance of the generator is 0.0008 ohm and that of the battery 0.0006 ohm. The load resistance is 0.75 ohm. What are the values of the generator, battery, and load currents?

3. Reverse the direction of the arrows for I_1 and I_2 in the above circuit, and write the three simultaneous equations.

4. Rewrite the simultaneous equations for the circuit in Sec. 13–9, assuming that the internal resistance of the battery is zero and that the resistance of the load is twice that of the generator.

Progress Report

This chapter was devoted to a group of topics connected with systems of linear equations.

1. The solution of simple systems of linear equations by elimination was considered.

2. A systematic elimination procedure called the Doolittle method for solving linear systems with several digits in the coefficients was considered.

3. For systems of two linear equations we considered:

    (*a*) Solution by graphical means.

    (*b*) Solution by means of determinants.

    (*c*) The classification of such systems as independent, inconsistent, and dependent by means of determinants and graphs.

    (*d*) The solution of a homogeneous system.

4. Determinants of the third order were defined, and seven fundamental properties of these determinants were stated. The definition is easily extended to determinants of higher order, and the properties also hold for such determinants.

5. Determinants were used to classify systems of three linear equations in three unknowns as independent and dependent and to solve homogeneous systems of two and three linear equations.

6. Finally, a brief example of how a linear system of equations can arise in an electrical application was given.

14

Quadratic Equations and Equations of Higher Degree

In many applications it is necessary to know how to operate with polynomials and how to find the roots of polynomial equations in one unknown. In this chapter we shall learn to handle operations with polynomials and how to find the real roots, both rational and irrational, of polynomials in one unknown.

14–1. Polynomials

An expression of the form

$$7x^3 + 2x^2 - 5x^4 + 9 - x \qquad (1)$$

is called a **polynomial** or, more precisely, a **polynomial in x.** The exponent of the highest power of x (in this case 4) is called the **degree** of the polynomial. It is usually convenient to arrange the terms of the polynomial according to descending or ascending exponents of x. The polynomial 1, if arranged according to descending exponents of x, can be written as

$$-5x^4 + 7x^3 + 2x^2 - x + 9.$$

It is a polynomial of the fourth degree. The numbers -5, 7, 2, -1, 9 are called the **coefficients** of the polynomial, -5 being the coefficient of the highest power of x. The number 9 is called the **absolute term** of the polynomial. As examples: $10 + 30t - 16t^2$ is a polynomial of second degree in t; the absolute term is 10. $Q^3 - 2Q$ is a polynomial of third degree in Q; the absolute term is zero.

When speaking about polynomials in general, we denote the coefficients by letters. Thus, an arbitrary polynomial of the fourth degree can be written in the form

$$Ax^4 + Bx^3 + Cx^2 + Dx + E.$$

It is sometimes useful to denote all the coefficients by the same letter and to distinguish them by subscripts. Thus, a polynomial of the nth degree can be written as

$$A_0x^n + A_1x^{n-1} + A_2x^{n-2} + \cdots + A_{n-1}x + A_n.$$

Polynomials of the first degree, namely,

$$A_0 x + A_1,$$

are also called **linear.** These were studied in Sec. 3–12. Polynomials of the second degree, namely,

$$A_0 x^2 + A_1 x + A_2,$$

are called **quadratic.** The next few paragraphs are devoted to the study of quadratic polynomials and equations connected with them.

14–2. Quadratic Polynomials and Functions

An arbitrary quadratic polynomial in x will be written, in the following paragraphs, in the form

$$A x^2 + B x + C,$$

A being the coefficient of the highest or quadratic term, B the coefficient of the linear term, and C the absolute term.

If x is regarded as variable, the function

$$y = A x^2 + B x + C$$

is called a quadratic function of x. For convenience we shall sometimes use the functional notation discussed in Chapter 3 and write

$$y = f(x) = A x^2 + B x + C.$$

In order to obtain a good idea of the behavior of the quadratic function, we shall construct tables of values for several examples of quadratic functions and plot them.

Example 1. Constructing a table of values and a graph for the quadratic function

$$y = \tfrac{1}{2} x^2 - x + 2,$$

we obtain

x	-3	-2	-1	0	1	2	3	4
y	9.5	6	3.5	2	1.5	2	3.5	6

The corresponding graph is plotted in Fig. 14–1.

Example 2. Similarly we obtain for the quadratic function

$$y = 5 x^2 + 2 x - 4,$$

x	-3	-2	-1	0	1	2	3
y	35	12	-1	-4	3	20	47

The function is plotted in Fig. 14–2.

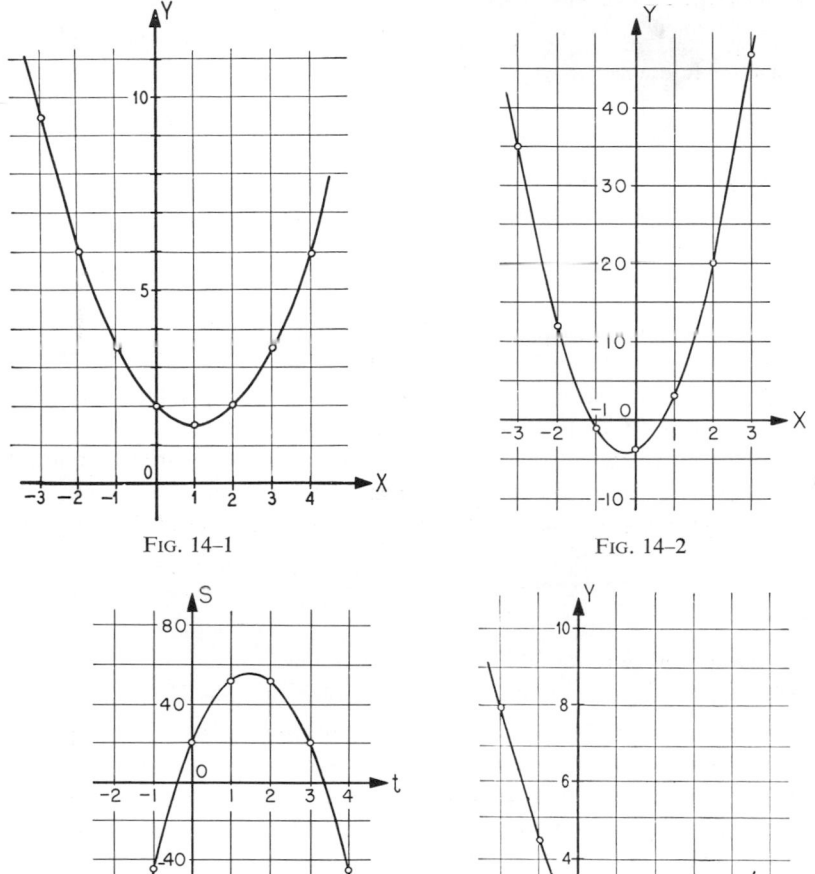

FIG. 14–1

FIG. 14–2

FIG. 14–3

FIG. 14–4

Example 3.   In a like manner we obtain for the quadratic function
$$s = 20 + 48t - 16t^2$$
the following table and the graph in Fig. 14–3.

t	-2	-1	0	1	2	3	4
s	-140	-44	20	52	52	20	-44

Example 4.   For the quadratic function

$$y = \tfrac{1}{2}x^2 - 2x + 2$$

we have the following values and the graph in Fig. 14–4.

x	-2	-1	0	1	2	3	4
y	8	4.5	2	0.5	0	0.5	2

The curves obtained in Fig. 14–1, 14–2, 14–3, and 14–4 by plotting the graphs of quadratic functions are called **parabolas.**   In general, *the graph of the quadratic function*

$$y = Ax^2 + Bx + C$$

is called a parabola.   Inspecting the graphs plotted above, we can see that a parabola has the following characteristics:

1. If the coefficient A of the second-degree term is positive, the graph has a lowest point, called the **vertex** of the parabola.

2. If the coefficient A of the second-degree term is negative, the graph has a highest point, also called the **vertex** of the parabola.

It is sometimes convenient to know that the coordinates of the vertex are

$$x = -\frac{B}{2A}, y = -\frac{B^2 - 4AC}{4A}.$$

14–3. Graphical Solution of Quadratic Equations

An equation whose left-hand side is a polynomial and whose right-hand side is zero is said to be an equation in the **polynomial form.**   An equation that can be **reduced** to the polynomial form is called an **algebraic equation.** For example, equations involving radicals of the unknown are algebraic equations since, as shown in Sec. 8–10, they can be reduced to the polynomial form.   Throughout this chapter, the term algebraic equation will refer to equations in the polynomial form.

In particular, an equation that can be reduced to $Ax^2 + Bx + C = 0$ is called a **quadratic equation.**   The problem of finding its roots can be put in this way: *Find the values of x such that the function*

$$y = Ax^2 + Bx + C$$

assumes the value zero. This can be done by plotting the graph of the function and estimating the abscissas of the points where the graph crossed the x axis.   To illustrate this we use the examples of the preceding section.

Example 1.   Find the roots of $\tfrac{1}{2}x^2 - x + 2 = 0$.

From Fig. 14–1 the student will see immediately that the graph of the function $y = \tfrac{1}{2}x^2 - x + 2$ has no intersection with the x axis.   The given equation, therefore, has no real roots.

Example 2. Find the roots of $5x^2 + 2x - 4 = 0$.

Figure 14–2 shows that the graph has two x intercepts. The given equation, therefore, has two roots; by measuring the x intercepts, the roots are found to be approximately -1.1 and 0.7.

Example 3. Find the roots of $20 + 48t - 16t^2 = 0$.

From Fig. 14–3 it is seen that the graph of the corresponding quadratic has two intercepts with the t axis. Hence the corresponding equation has two roots; they are approximately -0.4 and 3.4.

Example 4. Find the roots of $\frac{1}{2}x^2 - 2x + 2 = 0$.

Inspection of the corresponding graph (Fig. 14–4) shows that there is only one point where the curve meets the x axis. The corresponding abscissa is 2; the given equation has therefore only one root $x = 2$.

If the graph of the curve in Fig. 14–4 were slightly lowered (see Fig. 14–5), one

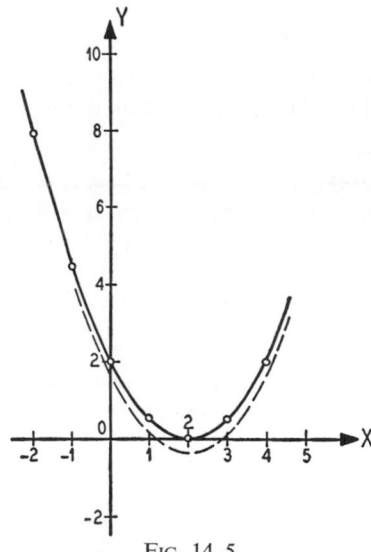

FIG. 14–5

would obtain two roots which are very near to each other. The single root $x = 2$ of this example can therefore be thought of as replacing two roots of a more general case and is called, therefore, a *double root.*

Exercises

1. On the same coordinate system plot the graphs of

     (a) x^2,   (b) $-x^2$,   (c) $3x^2$,   (d) $-2x^2$,   (e) $\frac{1}{3}x^2$.

2. On the same coordinate system plot the graphs of

     (a) $x^2 + 2x$,   (b) $x^2 - 4x$,   (c) $-x^2 + 6$.

3. On the same coordinate system plot the graphs of

     (a) $x^2 + 2x + 1$,   (b) $-x^2 + 2x - 5$,   (c) $3x^2 - 6x + 1$.

4. On the same coordinate system plot the graphs of

(a) $-2x^2 + 9$, (b) $(x + 4)^2 - 9$, (c) $-(x - 1)^2 + 2$.

Estimate graphically the roots, if any exist, of the following equations.

5. $x^2 + 3x - 7 = 0$.   **6.** $4v^2 - 4v + 1 = 0$.   **7.** $2s^2 + s + 5 = 0$.

8. $-3t^2 + 4t + 5 = 0$.   **9.** $u^2 - 6u + 7 = 0$.   **10.** $-5z^2 + z - 1 = 0$.

11. Given $u = 3w^2 - 6w + 5$, find the values of w such that

(a) $u = 0$,   (b) $u = 1$,   (c) $u = -1$,   (d) $u = 5$,

(e) $u = 2$,   (f) $u = -4$,   (g) $u = 6$,   (h) $u = -5$.

14–4. Solution by Factoring

The graphical method described in Sec. 14–3 gives only approximate answers.   If more accurate values of the roots are required, other methods must be used.

One method, which can be used only occasionally, uses factoring of the quadratic polynomial $Ax^2 + Bx + C$. This method has been described in Sec. 2–13. The student should review this section before proceeding.

Example 1.   Consider the equation $x^2 + x - 6 = 0$.   The polynomial on the left side of this equation can easily be factored by writing

$$x^2 + x - 6 = (x + 3)(x - 2).$$

Thus the given equation can be written as

$$x^2 + x - 6 = (x + 3)(x - 2) = 0.$$

As pointed out in Sec. 2–13, the product of two factors can be zero only if at least one of the factors is zero.   It follows then that this equation can be satisfied if either

$$x + 3 = 0 \text{ or } x - 2 = 0.$$

The first possibility gives $x = -3$; the second $x = 2$. The equation has, therefore, the two roots -3 and 2.

The method of factoring can always be applied, if there is no absolute term in the given equation.

Example 2.   To solve $40t - 16t^2 = 0$, we factor the left-hand member and obtain

$$t(40 - 16t) = 0.$$

This equation is satisfied when

$$t = 0 \text{ or } t = 2.5.$$

14–5. Pure Quadratic Equations

A quadratic equation whose linear term is missing can be solved very easily.   If, in the general quadratic $Ax^2 + Bx + C = 0$, we have $B = 0$, the equation takes the form sometimes referred to as a **pure quadratic equation**:

$$Ax^2 + C = 0. \tag{1}$$

To solve this equation we have at once

$$x^2 = -\frac{C}{A} \quad \text{and} \quad x = \pm\sqrt{-\frac{C}{A}}.$$

Thus, equation 1 has two real roots when A and C have opposite signs, and two imaginary roots when A and C have like signs.

Example 1. To solve the equation

$$3x^2 - 75 = 0,$$

we have at once

$$x^2 = 25, \qquad x = \pm 5.$$

Example 2. For the equation $3x^2 + 10 = 0$ we have

$$x^2 = -\frac{10}{3}, \qquad x = \pm\sqrt{-\frac{10}{3}} = \pm j\frac{1}{3}\sqrt{30} = \pm j1.826.$$

Example 3. Compute to three significant digits a positive value of ω such that

$$3 \times 10^{-3}\omega - \frac{1}{2 \times 10^{-6}\omega} = 0.$$

Multiplication by the denominator, which is permitted since $\omega \neq 0$, yields

$$6 \times 10^{-9}\omega^2 = 1$$

or

$$\omega^2 = \tfrac{1}{6} \times 10^9 = \tfrac{10}{6} \times 10^8 = 1.67 \times 10^8, \quad \text{and} \quad \omega = 1.29 \times 10^4.$$

Exercises

Solve the following equations, using the slide rule for the numerical computations:

1. $3.92t^2 = 19.7$.   **2.** $15E^2 - 43 = 0$.   **3.** $7R^2 = 29$.

4. $9.12Z^2 + 37.5 = 0$.   **5.** $\pi L^2 = 15.8$.   **6.** $10^4 v^2 = 7 \cdot 10^{12}$.

7. $11M^2 + 17 = -2M^2$.   **8.** $7.2\omega^2 - 3.4\omega = -2.3\omega^2$.

9. $3.25u^2 - 1.31u = 4.73u^2$.   **10.** $18.2m + 3.11m^2 = -7.56m^2$.

11. $E_1^2 - 6E_1 - 7 = 0$.   **12.** $r^2 + 2r - 15 = 0$.

13. $2X^2 + X - 1 = 0$.   **14.** $12s^2 - 11s = 5$.

15. $h(h - 3) = 3(5 - h) + 7$.   **16.** $(3L - 2)^2 - (L - 6)^2 = -40$.

17. $2(N^2 - 16) = 8N + 32$.   **18.** $7R + 3 = -2(R^2 + 1)$.

19. $\dfrac{7}{a^2} + \dfrac{5}{a} - 2 = 0$.   **20.** $\dfrac{17}{6(y^2 - 4)} - \dfrac{1}{y^2 - 4} = \dfrac{2}{3}$.

21. $\dfrac{u}{u + 4} - 6 = \dfrac{u}{4 - u}$.   **22.** $\dfrac{2p}{3} - 2 = \dfrac{2}{p} + \dfrac{5}{3}$.

23. The kinetic energy of a moving electron is given by the formula $KE = \frac{1}{2}mv^2$ where the kinetic energy is given in ergs, the mass m in grams, and the velocity v in centimeters per second. If the mass of the electron is 9.03×10^{-28} gram and the kinetic energy in a given case is 3.2×10^{-10} erg, what is the velocity?

24. The volume of a cylinder is $0.785d^2h$ where d is the diameter and h the height. What is the diameter of a steel rod 5 in. long that contains a volume of 4.2 cu in.?

25. The energy W in joules stored in a capacitor of capacitance C farads connected to a supply circuit of E volts is given by the equation $W = CE^2/2$. What voltage is required to store 0.05 joule in a capacitor of 10 microfarads?

26. The heat, measured in calories, developed in a d-c circuit during t sec is $H = 0.24I^2Rt$, where R is the resistance in ohms of the circuit and I the current in amperes. (*a*) Solve this equation for I. (*b*) What current is required to develop 60,000 calories of heat in 10 min if the resistance is 30 ohms?

27. A general formula for the brake horsepower of a single-cylinder gasoline engine is $P = d^2ln/18{,}000$ where d is the diameter of the cylinder in inches, l the stroke in inches, and n the number of revolutions per minute. Calculate the diameter of the cylinder of an engine that has a stroke of 5 in. and is to deliver 20 hp at a speed of 1300 rpm.

28. The energy stored in a magnetic field is given by the expression $W = LI^2/2$ where W is the energy in watts, L the inductance in henrys, and I the current in amperes. Compute the current when $W = 30$ watts and $L = 4.6$ henrys.

29. The centrifugal force acting on a revolving body is given by the expression $F = 0.00034Wrn^2$ where F is the force in pounds, W is weight in pounds of the body, r is the radius in feet, and n is the number of revolutions per minute the body makes. A 30-lb weight revolves at a radius of 4.5 ft. At what speed will the centrifugal force on a radial member be equal to 4000 lb?

14–6. Quadratic Formula

In practical problems the methods of the two preceding paragraphs can be used only very occasionally. We shall, therefore, develop in the present section a method that can always be applied.

Consider the general quadratic equation

$$Ax^2 + Bx + C = 0, \tag{1}$$

where we may suppose that $A \neq 0$, since otherwise we obtain a linear and not a quadratic equation. When this equation is divided by A, the following equivalent equation is obtained:

$$x^2 + \frac{B}{A}x + \frac{C}{A} = 0,$$

or

$$x^2 + \frac{B}{A}x = -\frac{C}{A}. \tag{2}$$

If $B^2/4A^2$ is added to both sides of equation 2, the left member becomes the square of a linear expression, since

$$\left(x + \frac{B}{2A}\right)^2 = x^2 + 2x\frac{B}{2A} + \left(\frac{B}{2A}\right)^2 = x^2 + \frac{B}{A}x + \frac{B^2}{4A^2}.$$

From equation 2 we obtain, therefore,

$$\left(x+\frac{B}{2A}\right)^2 = x^2 + \frac{B}{A}x + \frac{B^2}{4A^2} = -\frac{C}{A} + \frac{B^2}{4A^2},$$

or

$$\left(x+\frac{B}{2A}\right)^2 = \frac{B^2 - 4AC}{4A^2}.$$

Hence by computing the square root

$$x + \frac{B}{2A} = \pm\sqrt{\frac{B^2 - 4AC}{4A^2}} = \pm\frac{\sqrt{B^2 - 4AC}}{2A},$$

and therefore

$$x = \frac{-B \pm \sqrt{B^2 - 4AC}}{2A} \tag{3}$$

gives us the formula for the two roots of the general quadratic equation 1. The expression in (3) is called the **quadratic formula.**

For practical applications when using the quadratic formula it is convenient to apply it by the following steps:

  1. Find the values of A, B, C.
  2. Compute the quantity $D = B^2 - 4AC$. *This is called the* **discriminant** *of the quadratic equation.*
  3. Compute the roots by the quadratic formula

$$x = \frac{-B \pm \sqrt{D}}{2A}.$$

There are three essentially different cases of the quadratic equation according to the value of D, which are enumerated in the following table:

Case 1. $D > 0$. $\sqrt{D}$ is real.   There are *two real and unequal roots.*

Case 2. $D = 0$. $\sqrt{D} = 0$.   There is only *one real root* called a *double root.*

Case 3. $D < 0$. $\sqrt{D}$ is an imaginary number.   There are *two unequal complex roots*, which are conjugates.

In order to illustrate the various cases, we apply the quadratic formula to the functions plotted in Sec. 14–2.

Example 1. To solve the quadratic equation

$$\tfrac{1}{2}x^2 - x + 2 = 0$$

we follow the steps outlined above.

Step 1. $A = \frac{1}{2}, \; B = -1, \; C = 2.$

Step 2. $D = B^2 - 4AC = (-1)^2 - 4 \cdot \frac{1}{2} \cdot 2 = -3.$

Step 3. $x = \dfrac{-B \pm \sqrt{D}}{2A} = \dfrac{1 \pm \sqrt{-3}}{2 \cdot \frac{1}{2}} = 1 \pm j\sqrt{3}.$

Since $D = -3 < 0$, this quadratic equation belongs to case 3.   The roots are conjugate complex numbers.

The student should compare example 1 of Sec. 14–2 and example 1 of Sec. 14–3.

Example 2.   For the equation

$$5x^2 + 2x - 4 = 0$$

we have

$$A = 5, \qquad B = 2, \qquad C = -4,$$

$$D = 2^2 - 4 \cdot 5 \cdot (-4) = 4 + 80 = 84.$$

$$\sqrt{D} = \sqrt{84} = 9.17.$$

$$x = \frac{-2 \pm 9.17}{10}.$$

$$x_1 = \frac{-2 + 9.17}{10} = \frac{7.17}{10} = 0.717.$$

$$x_2 = \frac{-2 - 9.17}{10} = \frac{-11.17}{10} = -1.117.$$

Since $D = 84 > 0$, the equation belongs to case 1 and has two real roots 0.717 and -1.117.   This should be compared with the answers obtained in example 2 of Sec. 14–3.

Example 3.   Solve the equation discussed in example 3 of Sec. 14–3:

$$20 + 48t - 16t^2 = 0.$$

It is always useful to investigate whether a given equation can be simplified. In this case, the whole equation may be divided by 4.   If we simplify the equation and rearrange it, we have

$$-4t^2 + 12t + 5 = 0.$$

To solve this equation we write,

$$A = -4, \qquad B = 12, \qquad C = 5,$$

$$D = 12^2 - 4 \cdot (-4) \cdot 5 = 144 + 80 = 224 > 0, \quad \text{(Case 1)}$$

$$\sqrt{D} = \sqrt{224} = 14.97.$$

$$t_1 = \frac{-12 + 14.97}{-8} = -\frac{2.97}{8} = -0.37.$$

$$t_2 = \frac{-12 - 14.97}{-8} = \frac{26.97}{8} = 3.37.$$

Example 4. To solve the equation

$$\tfrac{1}{2}x^2 - 2x + 2 = 0$$

we write

$$A = \tfrac{1}{2}, \qquad B = -2, \qquad C = 2,$$

$$D = 4 - 4 \cdot \tfrac{1}{2} \cdot 2 = 0.$$

$$x = \frac{2 \pm 0}{2 \cdot \tfrac{1}{2}} = 2.$$

Since $D = 0$, by case 2, the equation has a double root $x = 2$. Again the student should compare this result with example 4 of Sec. 14–3.

Example 5. Solve the equation

$$x + \frac{1}{x} = k,$$

and discuss the character of the roots.

By clearing fractions, the given equation may be written in the form

$$x^2 - kx + 1 = 0.$$

To solve this equation we write

$$A = 1, \qquad B = -k, \qquad C = 1,$$

$$D = k^2 - 4.$$

Hence the values of the roots of the above equation found by the quadratic formula are

$$x = \frac{k \pm \sqrt{k^2 - 4}}{2}.$$

To discuss the character of the roots for different values of k, we must examine carefully the discriminant $D = k^2 - 4$.

Case 1. When

$$D = k^2 - 4 > 0,$$

the roots are real and different. Then we can write $k^2 > 4$ or $k > +2$ and $k < -2$.

Case 2. When

$$D = k^2 - 4 = 0 \quad \text{or} \quad k = \pm 2,$$

the roots are equal and real.

Case 3. When

$$D = k^2 - 4 < 0 \quad \text{or} \quad k^2 < 4,$$

the roots are conjugate complex numbers. Then we may write $-2 < k < +2$.

When the notation $|k|$ for the absolute value of k is used, the result can be stated in the following way.

The equation $x + \dfrac{1}{x} = k$ has

(*a*) Two different real roots if $|k| > 2$,
(*b*) A double root if $|k| = 2$,
(*c*) Two complex roots if $|k| < 2$.

Exercises

Solve the following quadratic equations by any of the methods discussed in this chapter. Give all real roots in decimal form correct to three significant digits.

1. $x^2 - 8x + 13 = 0.$

2. $y^2 + 5y - 3 = 0.$

3. $z^{2\cdot} + 7z + 1 = 0.$

4. $s^2 - 12s + 8 = 0.$

5. $r^2 - r - 4 = 0.$

6. $u^2 - 2u - 3 = 0.$

7. $d^2 + 10d = 5.$

8. $v^2 + v - 1 = 0.$

9. $h^2 - 5h + 8 = 0.$

10. $w^2 + 6w + 2 = 0.$

11. $t^2 + 6t + 10 = 0.$

12. $R^2 + R = -2.$

13. $2E^2 + 3E + 3 = 0.$

14. $-m^2 + 7m - 15 = 0.$

15. $2p(p - 3) = 1.$

16. $4R_x(R_x - 1) = 1.$

17. $7 + 8P = -3P^2.$

18. $Q(7Q + 1) = -3.$

19. $t^2 + \sqrt{3}t + 1 = 0.$

20. $-s^2 + \sqrt{2}s + 1 = 0.$

21. $0.6L_x{}^2 = 1.6L_x - 1.4.$

22. $0.6M_x + 0.05 = -0.2M_x{}^2.$

23. $0.4M^2 - 0.12 = 0.16M.$

24. $0.16x + 0.08 = -0.06x^2.$

25. $\frac{1}{2}N^2 + \frac{5}{8} = \frac{1}{3}N.$

26. $3k = \frac{5}{6}k^2 - \frac{1}{2}.$

27. $\frac{3}{8}S^2 = 2S + \frac{1}{6}.$

28. $2L_1{}^2 - 6\frac{3}{5}L_1 = -4.$

29. $2x^2 - \sqrt{3}x + \sqrt{2} = 0.$

30. $-\sqrt{3}y^2 + 5y + \sqrt{2} = 0.$

31. $2.73w^2 - 3.28w + 1.82 = 0.$

32. $26.7k^2 + 36.5k - 15.2 = 0.$

33. $\dfrac{e_1}{e_1 - 2} - (e_1 + 1) = \dfrac{1 - e_1}{2}.$

34. $\dfrac{2}{Z} - 2 = \dfrac{3}{1 - Z}.$

35. $5u + 7 = \dfrac{6}{2u - 3}.$

36. $\dfrac{1}{m - 3} + \dfrac{1}{m - 2} - \dfrac{1}{2} = 0.$

37. $\dfrac{2 + R_1}{3R_1 - 2} + \dfrac{R_1 - 4}{2R_1} = \dfrac{2}{3}.$

38. $\dfrac{E - 3}{E - 2} = 4 + \dfrac{E + 2}{1 - E}.$

39. $\dfrac{2}{P_y} + \dfrac{3}{P_y + 1} = \dfrac{1}{1 - P_y}.$

40. $\dfrac{1}{R + 1} + \dfrac{1}{R - 1} = \dfrac{1}{R + 3}.$

41. $7.6 \times 10^8 m^2 + 2.6 \times 10^4 m = 4.3.$

42. $4 \times 10^{-5}\omega - \dfrac{1}{5 \times 10^{-12}\omega} = 6 \times 10^{-8}.$

Discuss the character of the roots of the following equations without computing the roots:

43. $5x^2 - 24x + 17 = 0.$

44. $8z^2 + 15z + 14 = 0.$

45. $49u^2 + 25 = 70u.$

46. $66v + 121 = -9v^2.$

47. $t^2 + 2\sqrt{2}t + 2 = 0.$ **48.** $6\sqrt{3}s^2 - 6s - \sqrt{3} = 0.$

49. $3.4E^2 + 0.47 = 2.3E.$ **50.** $7.2R^2 + 6.7R + 3.9 = 0.$

51. $4 \times 10^{-6}L_x^2 - 7 \times 10^{-5}L_x + 6 \times 10^{-3} = 0.$

52. $7.1 \times 10^{-8}\omega^2 + 8.7 \times 10^{-12}\omega - 3.7 \times 10^{-10} = 0.$

Determine k (real) so that the following equations have double roots:

53. $3kx^2 + k + 1 = 12x.$ **54.** $x^2 - kz = 4k - 9.$

55. $\dfrac{9}{k}y^2 + 6y + 8 = 0.$ **56.** $y^2 + 2ky = 2k + 1.$

57. $z^2 - 5z = -\dfrac{25}{k}.$ **58.** $u^2 + 1 = \dfrac{3}{k}u.$

59. $(k+1)v^2 + 9k = 20v + 2.$ **60.** $z^2 - kz = k - z.$

Solve the following equations for x, and discuss the character of the roots:

61. $kx^2 + 4x + 4 = 0.$ **62.** $mx^2 + 2x - 1 = 0.$

63. $x^2 + 6x + p + 5 = 0.$ **64.** $x^2 - 4x = k + 2.$

65. $(x+2)(x+1) = m.$ **66.** $(x-3)(x+2) = p.$

67. $x + \dfrac{1}{x} = -k.$ **68.** $x - \dfrac{1}{x} = k.$

Solve the equations in exercises 69 through 84, and check the results. These equations involve radicals, and, before solving them, the student should review Sec. 8–10, where such equations were discussed.

69. $\sqrt{x} = x - 12.$ **70.** $\sqrt{x} + 20 = x.$

71. $\sqrt{y^2 - 5} + 2 = 0.$ **72.** $z - 3\sqrt{z} - 10 = 0.$

73. $s_1 + 4\sqrt{s_1} - 21 = 0.$ **74.** $\sqrt{t^2 + 5} + 1 = 2t.$

75. $\sqrt{R} + \sqrt{2R - 5} = 2.$ **76.** $\sqrt{3w - 2} - 6 = -\sqrt{5w \times 6}.$

77. $2\sqrt{3E + 10} = 1 + \sqrt{5 - 2E}.$ **78.** $\sqrt{L} + \sqrt{2L + 6} = 3.$

79. $\sqrt{2x_1 + 1} + 3\sqrt{x_1} = 11.$ **80.** $\sqrt{2N - 5} - 1 = \frac{1}{2}\sqrt{N + 9}.$

81. $\sqrt{\sqrt{2v - 2} + \sqrt{v + 1}} = 2.$ **82.** $\sqrt{\sqrt{r + 4} - \sqrt{2r + 1}} = 1.$

83. $\sqrt{10u - 1} - \sqrt{3u + 1} = \sqrt{2u - 1}.$

84. $\sqrt{2M + 4} + \sqrt{M - 1} = \sqrt{5M + 7}.$

85. Find two consecutive integers whose product is 272.
86. Find two consecutive odd integers whose product is 483.
87. Find two consecutive even integers whose product is 224.
88. Find three consecutive positive integers the sum of whose squares is equal to 194.
89. Find two consecutive positive integers such that the difference of their cubes is equal to 91.
90. In an isosceles right triangle the hypotenuse is 2 in. longer than each of the other sides. Find the lengths of the sides of this triangle.

91. The length and breadth of a rectangular piece of sheet metal are 30.2 in. and 21.5 in.   Both have to be increased by the same amount so that the area may be doubled. What are the dimensions of the new rectangular piece of sheet metal?

92. An open box containing 520 cu in. is made by cutting out a 5-in. square from each corner of a square piece of tin and turning up the sides.   Find the dimensions of the original piece of tin.

93. A long sheet of metal is to be made into a gutter by turning strips up vertically along the two sides.   If the sheet of metal is 33.5 in. wide, how much should be turned up to give the rectangular cross section of the gutter an area of 98.6 sq in.?

94. Find the dimensions of a rectangular lot whose area is 7500 sq ft, if the fence enclosing it is 370 ft long.

95. A motorboat takes 3 hr to travel 12 miles downstream and back on a river which flows at a rate of 3 mph.   Find the speed at which the boat would travel in still water.

96. After traveling 60 miles at a certain speed, a motorist increases his speed by 5 mph and travels 48 miles farther.   If he took 4 hr to cover the whole distance of 108 miles, find his speed during the first 60 miles.

97. It would take a flywheel 3 min less to make 10,000 revolutions if its speed were 12 rpm faster.   Find the speed of the flywheel.

98. An object shot upwards with a speed of 400 ft per sec will be, after t sec, at a height of approximately

$$s = 400t - 16t^2$$

if air resistance is neglected.   When will the object be 2000 ft above the starting point?

99. The emf of a storage battery is 6.2 volts, and its internal resistance is 0.02 ohm. What current will the battery deliver to an electromagnet that requires 200 watts?   Use the formulas $I(R_i + R_e) = E$ and $I^2 R_e = P$, where I is the current in amperes, R_i the internal resistance of the battery in ohms, R_e the resistance of the electromagnet in ohms, and P the power in watts.

14–7. Properties of the Roots of a Quadratic Equation

The roots of the quadratic equation

$$Ax^2 + Bx + C = 0 \qquad (1)$$

were found by the previous section to be given by the formulas

$$x_1 = \frac{-B + \sqrt{D}}{2A}, \qquad x_2 = \frac{-B - \sqrt{D}}{2A},$$

where

$$D = B^2 - 4AC.$$

From these formulas we have at once

$$x_1 + x_2 = \frac{-B + \sqrt{D}}{2A} + \frac{-B - \sqrt{D}}{2A} = \frac{-2B}{2A} = -\frac{B}{A},$$

$$x_1 x_2 = \frac{-B + \sqrt{D}}{2A} \cdot \frac{-B - \sqrt{D}}{2A} = \frac{B^2 - (\sqrt{D})^2}{4A^2} = \frac{B^2 - D}{4A^2}$$

$$= \frac{B^2 - (B^2 - 4AC)}{4A^2} = \frac{4AC}{4A^2} = \frac{C}{A}.$$

Hence it follows that when x_1 and x_2 are the roots of equation 1

$$x_1 + x_2 = -\frac{B}{A}, \qquad x_1 x_2 = \frac{C}{A}. \tag{2}$$

In the special case when $A = 1$, we may state that the sum of the roots is equal to the negative value of the coefficient of x and the product of the roots equal to the absolute term; i.e., for the equation $x^2 + Bx + C = 0$ we have $x_1 + x_2 = -B$, $x_1 x_2 = C$.

The sum and the product of the roots of a quadratic equation can be found without solving the equation and are given by formulas 2. This will be illustrated by the following examples.

Example 1. Consider the equation

$$3x^2 - 7x - 20 = 0$$

where $A = 3$, $B = -7$, $C = -20$. Using formulas 2, we have

$$x_1 + x_2 = \tfrac{7}{3} \quad \text{and} \quad x_1 x_2 = -\tfrac{20}{3}.$$

Thus without solving the above equation we know that the sum of its roots is $\tfrac{7}{3}$ and their product $-\tfrac{20}{3}$. This can be verified by solving the equation, for $D = 49 - 4 \cdot 3(-20) = 289$ or $\sqrt{D} = 17$, and hence

$$x_1 = \frac{7 + 17}{6} = 4 \quad \text{and} \quad x_2 = \frac{7 - 17}{6} = -\frac{5}{3},$$

so that

$$x_1 + x_2 = 4 - \tfrac{5}{3} = \tfrac{7}{3} \quad \text{and} \quad x_1 x_2 = 4(-\tfrac{5}{3}) = -\tfrac{20}{3}.$$

The formulas for the sum and the product of the roots can be used to show that every quadratic polynomial can be factored. This can be stated as follows:

Let $Ax^2 + Bx + C$ be a given polynomial, and let x_1, x_2 be the roots of $Ax^2 + Bx + C = 0$. Then $Ax^2 + Bx + C = A(x - x_1)(x - x_2)$.

This statement is very easily proved if the multiplication of the right member of the last equation is performed as follows:

$$A(x - x_1)(x - x_2) = A[x^2 - (x_1 + x_2)x + x_1 x_2]$$

$$= A\left[x^2 - \left(-\frac{B}{A}\right)x + \frac{C}{A}\right]$$

$$= Ax^2 + Bx + C.$$

Example 2. The roots of the equation

$$3x^2 - 7x - 20 = 0$$

have been found to be

$$x_1 = 4, \qquad x_2 = -\tfrac{5}{3}.$$

We can now factor the corresponding quadratic polynomial by writing

$$3x^2 - 7x - 20 = 3(x - 4)(x + \tfrac{5}{3}) = (x - 4)(3x + 5).$$

Example 3. Factor the polynomial $x^2 - x - 1$. The roots of the corresponding equation $x^2 - x - 1 = 0$ are found to be

$$x = \frac{1 \pm \sqrt{5}}{2} = \frac{1 \pm 2.236}{2},$$

whence

$$x_1 = \frac{3.236}{2} = 1.618, \qquad x_2 = \frac{-1.236}{2} = -0.618,$$

and therefore

$$x^2 - x - 1 = (x - 1.618)(x + 0.618).$$

Example 4. Factor the expression $3x^2 - 5xy + 2y^2$.

First solve the equation

$$3x^2 - 5xy + 2y^2 = 0$$

for x. Using the quadratic formula, we have in this case

$$A = 3, \qquad B = -5y, \qquad C = 2y^2,$$
$$D = B^2 - 4AC = 25y^2 - 24y^2 = y^2,$$
$$x = \frac{5y \pm y}{6},$$

and the roots are given by

$$x_1 = y, \qquad x_2 = \tfrac{2}{3}y.$$

Therefore

$$3x^2 - 5xy + 2y^2 = 3(x - x_1)(x - x_2) = 3(x - y)(x - \tfrac{2}{3}y) = (x - y)(3x - 2y).$$

Exercises

Without solving the following equations, find the sum and the product of their roots:

1. $x^2 + 5x = 7$.
2. $3y^2 = 8y - 1$.
3. $2z = 4 - 5z^2$.
4. $u^2 - 6 = 3(u + 1)$.
5. $(v - 1)(2v + 3) - 5 = 0$.
6. $(3t + 1)(1 - t) = 12$.
7. $5(x_1^2 - 2) = 3(1 - x_1)$.
8. $(2R_1 - 3) = 9(R_1^2 + 1)$.
9. $(s - 2)(s + 5) = 2s^2 + 3s$.
10. $(E + 4)(2E - 1) = E^2 - 4$.

Form quadratic equations, with integral coefficients, having the following numbers as roots:

11. $2, 3$.
12. $4, 7$.
13. $3, -1$.
14. $-4, -5$.

15. $0, 7$.
16. $\tfrac{1}{2}, \tfrac{1}{3}$.
17. $\tfrac{1}{4}, -\tfrac{1}{3}$.
18. $-0.3, 1.5$.

19. $\pm\sqrt{3}$.
20. $\pm\sqrt{5}$.
21. $\pm j$.
22. $\pm 2j$.

23. $1 \pm \sqrt{3}$.
24. $\dfrac{3 \pm \sqrt{5}}{2}$.
25. $2 \pm 3j$.
26. $\dfrac{4 \pm 5j}{2}$.

Factor the following expressions by solving quadratic equations.   Express irrationals in decimal form with three significant digits.

27. $y^2 + y - 6$.

28. $6R^2 + 13R - 5$.

29. $z^2 + 3z - 5$.

30. $2t^2 - 7t + 1$.

31. $21x^2 + 11xy - 2y^2$.

32. $12p^2 + 23pq - 24q^2$.

33. $m^2 - 3mn + n^2$.

34. $E_1{}^2 + 9E_1E_2 - 2E_2{}^2$.

35. $3R_1{}^2 + 12R_1R_2 + 5R_2{}^2$.

36. $x^2 - 2x + 2$.

37. $z^2 - 4z + 13$.

38. $E^2 + 3E + 11$.

14–8. Equations Reducible to Quadratics

There are many complicated equations that can be reduced by a simple substitution to quadratic equations.   A few such examples are given below.

Example 1.   Solve the fourth-degree equation $x^4 - x^2 - 12 = 0$.

When x^2 is replaced by y this equation becomes

$$y^2 - y - 12 = 0$$

which is a quadratic.   This equation has the two roots:

$$y_1 = 4 \quad \text{and} \quad y_2 = -3,$$

from which it follows that

$$x^2 = 4 \quad \text{or} \quad x^2 = -3.$$

The first possibility yields

$$x = \pm\sqrt{4},$$

and the second

$$x = \pm j\sqrt{3}.$$

The given equation has the four roots $2, -2, j\sqrt{3}$, and $-j\sqrt{3}$.   On substitution it can be verified that these numbers satisfy the given equation.

Example 2.   Solve the equation $12\dfrac{x^2 + 1}{x^2 - 1} + 25\dfrac{x^2 - 1}{x^2 + 1} = 35$.

Let

$$\frac{x^2 + 1}{x^2 - 1} = y; \qquad \text{then} \qquad \frac{x^2 - 1}{x^2 + 1} = \frac{1}{y},$$

and the given equation is reduced to

$$12y + 25\frac{1}{y} = 35,$$

or the quadratic form:

$$12y^2 - 35y + 25 = 0.$$

The solutions of this equation are

$$y = \frac{35 \pm \sqrt{35^2 - 4 \cdot 12 \cdot 25}}{24} = \frac{35 \pm 5}{24};$$

$$y_1 = \frac{5}{3}, \qquad y_2 = \frac{5}{4}.$$

There are, therefore, two possibilities:

$$\frac{x^2 + 1}{x^2 - 1} = \frac{5}{3}, \qquad\qquad \frac{x^2 + 1}{x^2 - 1} = \frac{5}{4},$$

$$3x^2 + 3 = 5x^2 - 5, \qquad\qquad 4x^2 + 4 = 5x^2 - 5,$$

$$x^2 = 4, \qquad\qquad x^2 = 9,$$

$$x_1 = +2, \qquad\qquad x_3 = +3,$$

$$x_2 = -2. \qquad\qquad x_4 = -3.$$

The numbers 2, -2, 3, and -3 are the roots of the given equation, since, on substitution, they are seen to satisfy the given equation.

Example 3. Find all the angles between $0°$ and $360°$ so that $25 \sin^2 x + 30 \sin x - 7 = 0$.

The substitution $y = \sin x$ reduces the given equation to

$$25y^2 + 30y - 7 = 0,$$

the solution of which is given by

$$y = \frac{-30 \pm \sqrt{900 + 700}}{50} = \frac{-30 \pm 40}{50},$$

$$y_1 = 0.2, \qquad y_2 = -1.4.$$

We have therefore two possibilities:

$$\sin x = 0.2, \qquad\qquad \sin x = -1.4,$$

$$x_1 = 11.5°,$$

$$x_2 = 180° - 11.5°$$

$$= 168.5°.$$

which gives no solution, since the absolute value of $\sin x$ cannot be greater than 1.

The two angles $11.5°$ and $168.5°$ are the required roots of the given equation.

Example 4. Find all the angles between $0°$ and $360°$ so that $\tan x + \cot x = 3$.

Let $\tan x = y$, then $\cot x = 1/y$, and the given equation is reduced to quadratic form:

$$y + \frac{1}{y} = 3 \quad \text{or} \quad y^2 - 3y + 1 = 0.$$

The solutions of this equation are

$$y = \frac{3 \pm \sqrt{9 - 4}}{2} = \frac{3 \pm \sqrt{5}}{2},$$

$$y_1 = \frac{3 + 2.236}{2} = \frac{5.236}{2} = 2.618,$$

$$y_2 = \frac{3 - 2.236}{2} = \frac{0.764}{2} = 0.382.$$

We have therefore two possibilities:

$$\tan x = 2.618, \qquad\qquad \tan x = 0.382,$$
$$x_1 = 69.1°, \qquad\qquad x_3 = 20.9°,$$

or   or

$$x_2 = 180° + 69.1° \qquad\qquad x_4 = 180° + 20.9°$$
$$= 249.1°. \qquad\qquad\qquad = 200.9°.$$

The four angles 20.9°, 69.1°, 200.9°, and 249.1° are the required roots of the given equation.

Exercises

Solve the following equations. Give all irrational answers in decimal form, correct to three significant digits.

1. $s^4 - s^2 - 2 = 0.$

2. $L_y{}^4 - L_y{}^2 - 20 = 0.$

3. $M_x{}^4 + M_x{}^2 = 2.$

4. $Z_1{}^4 - 4Z_1{}^2 - 45 = 0.$

5. $R_x{}^4 + 4R_x{}^2 = 21.$

6. $E_1{}^4 + E_1{}^2 = 6.$

7. $W_z{}^4 + 7W_z{}^2 + 10 = 0.$

8. $P_1{}^4 + 9P_1{}^2 + 18 = 0.$

9. $z - 3z^{\frac{1}{2}} = 4.$

10. $t + 5t^{\frac{1}{2}} - 1 = 0.$

11. $x^{\frac{2}{3}} + x^{\frac{1}{3}} - 2 = 0.$

12. $w^{\frac{2}{3}} - 2w^{\frac{1}{3}} = 3.$

13. $\left(\dfrac{2}{L} - 3L\right)^2 + 6 = 5\left(\dfrac{2}{L} - 3L\right).$

14. $2\left(E + \dfrac{1}{E}\right)^2 + \left(E + \dfrac{1}{E}\right) = 10.$

15. $\left(R + \dfrac{1}{R}\right)^2 + \left(R + \dfrac{1}{R}\right) = 6.$

16. $\left(m + \dfrac{1}{m}\right)^2 = 4 + 3\left(m + \dfrac{1}{m}\right).$

17. $(p + 3)^4 = 4 - 3(p + 3)^2.$

18. $2y^{-2} + 5y^{-1} = 3.$

19. $T^{-2} - T^{-1} = 12.$

20. $M^{-2} - M^{-1} - 20 = 0.$

21. $(2v + v^2)^2 - 6 = (2v + v^2).$

22. $3^\alpha + 3^{-\alpha} = 10.$

23. $3^{2y} - 10 \cdot 3^y + 9 = 0.$

24. $5^x + 5^{-x} = 5.2.$

25. $\dfrac{z^2 - 6}{z} - 5\dfrac{z}{z^2 - 6} = 4.$

26. $\dfrac{2u}{1 - u^2} + \dfrac{1 - u^2}{2u} + \dfrac{25}{12} = 0.$

27. $2\sin^2 \alpha + \sin \alpha = 1.$

28. $\tan x + \cot x = 4.$

29. $\tan \beta = 3 \cot \beta + 2.$

30. $7\cos^2 \omega - 9 \cos \omega + 1 = 0.$

31. $\tan^2 x + 2 \tan x = 1.$

32. $6.2 \sin^2 x + 5.3 \sin x = 2.5.$

Solve for x:

33. $\dfrac{e^x - e^{-x}}{2} = u.$

34. $\dfrac{e^x + e^{-x}}{2} = u.$

35. $\dfrac{e^x - e^{-x}}{e^x + e^{-x}} = v.$

14–9. Polynomials and Algebraic Equations with Two Unknowns

A function of two variables x and y is called a polynomial in x and y if it is a sum of terms of the form

$$ax^p y^q$$

where p and q are positive integers or zero. The sum $p + q$ is called the **degree of the term,** and the highest degree of any term of the sum is called the **degree of the polynomial.**

Example 1. $3x^2 + 2xy - y^2 + 5x - 4y + 6$ is a polynomial of the second degree, or a quadratic polynomial in x and y.
$P^4 + Q^4$ is a polynomial of the fourth degree in P and Q.
$P + 2P^2Q + Q^2 - P^3$ is a polynomial of the third degree in P and Q.
$uv - 2u + 3v$ is a polynomial of second degree in u and v.
$st^2 - 1$ is a polynomial of the third degree in s and t.

An equation between two quantities x and y is called an **algebraic equation of nth degree** if it can be reduced to $P(x, y) = 0$ where $P(x, y)$ is a polynomial of nth degree.

A pair of particular values for x and y, satisfying the equation $P(x, y) = 0$, is called a **pair of solutions** or briefly a **solution** of the given equation. There are infinitely many solutions of such an equation. They can be found by solving the equation for one of the variables; for instance, we can solve for y and then compute the values of y which correspond to arbitrary values of x. This is illustrated in the following example.

Example 2. Let the given equation be $P(x, y) = x^3 - 3x^2 y - 10 = 0$.
Solving for y in terms of x, we obtain

$$y = -\frac{10 - x^3}{3x^2}.$$

By using this expression, the value of y corresponding to an arbitrary value of x can be found. For example:

When $x = 1$, then $y = -3$;
When $x = 2$, then $y = -\frac{1}{6}$;
When $x = -1$, then $y = -\frac{11}{3}$.

When two equations involving x and y are given simultaneously, the problem is to find numerical values of x and y satisfying both equations; or we can say simply that our problem is to *solve the simultaneous system of equations.* For example, the system of simultaneous equations

$$x^3 - y^2 = 23,$$
$$x^2 - y^3 = 1$$

has the solution $x = 3$, $y = 2$. This solution has been obtained by inspection and can be checked by direct substitution since $3^3 - 2^2 = 23$ and $3^2 - 2^3 = 1$. We do not know, however, whether or not there are other solutions. The general solution of the above system of equations is rather difficult.

When the equations are not linear, it is often very difficult to find solutions of a system of simultaneous equations in two unknowns. A few cases where the solution can be found are discussed in the next section.

14–10. Systems of Equations Involving Quadratics

Some systems of two simultaneous equations can be solved by eliminating one of the variables. To do this, solve one of the equations for one of the unknowns in terms of the other and substitute this solution into the second equation. Thus an equation with only one unknown is obtained. This procedure is illustrated by the following example.

Example 1. Solve the following system of equations:

$$y^2 = 4x + 20,$$
$$xy = 1.$$

Substituting $1/x$ in place of y in the first of the given equations, we obtain

$$\frac{1}{x^2} = 4x + 20$$

or

$$4x^3 + 20x^2 - 1 = 0.$$

The last equation is of the third degree and cannot be solved by the methods developed so far. Thus we are unable to solve the given system of equations.

This example shows that in general it is very difficult to find solutions of a system of simultaneous equations in two unknowns. There are, however, a few types where the solution can be found. In the remaining portion of this section we will discuss systems of simultaneous equations which can be reduced to quadratics in one unknown.

Type A. One equation of the system is linear; the other is of the second degree. In this case the linear equation is solved for one of the unknowns and the result substituted in the second equation. The following example illustrates this method.

Example 2. Solve the system of equations

$$x^2 + 2y^2 = 6, \tag{1}$$
$$2x + 3y = 7. \tag{2}$$

Solving the linear equation 2, we obtain

$$x = \frac{7 - 3y}{2} \tag{3}$$

which when substituted in equation 1 yields

$$\left(\frac{7 - 3y}{2}\right)^2 + 2y^2 = 6,$$

$$49 - 42y + 9y^2 + 8y^2 = 24,$$

$$17y^2 - 42y + 25 = 0, \tag{4}$$

a quadratic in one unknown y.   Solving equation 4 with the aid of the quadratic formula, we obtain

$$y = \frac{42 \pm \sqrt{42^2 - 4 \times 17 \times 25}}{34} = \frac{42 \pm 8}{34},$$

$$y_1 = \frac{50}{34} = \frac{25}{17}, \qquad y_2 = \frac{34}{34} = 1.$$

The corresponding values of x are found by substituting y_1 and y_2 in equation 3, so that

$$x_1 = \frac{7 - 3 \times \frac{25}{17}}{2} = \frac{7 - \frac{75}{17}}{2} = \frac{119 - 75}{34} = \frac{22}{17},$$

$$x_2 = \frac{7 - 3}{2} = 2.$$

The given system of equations has, therefore, two sets of solutions:

$$x_1 = \tfrac{22}{17}, \quad y_1 = \tfrac{25}{17},$$

and

$$x_2 = 2, \quad y_2 = 1.$$

These solutions should be checked by substituting them in the given equations.

Type B.   *Both equations are linear in x^2 and y^2, that is, when both equations are of the form $ax^2 + by^2 = c$.*   Such a system of equations can always be solved by the methods used to solve linear equations (see Sec. 13–1). This will now be illustrated.

Example 3.   Solve

$$x^2 + 2y^2 = 17,$$

$$2x^2 - 3y^2 = 6.$$

Multiplying the first equation by 2 and subtracting the second equation from the product, we obtain

$$7y^2 = 28 \quad \text{or} \quad y^2 = 4.$$

Therefore

$$y = \pm 2.$$

Using the first of the given equations, we obtain

$$x = \pm 3.$$

The system has four sets of solutions, because each of the y values can be combined with each of the x values.   The solutions are:

$$x_1 = 3, \qquad y_1 = 2,$$
$$x_2 = 3, \qquad y_2 = -2,$$
$$x_3 = -3, \qquad y_3 = 2,$$
$$x_4 = -3, \qquad y_4 = -2.$$

They can be checked by substituting in the original equations.

Exercises

Solve the following systems of equations, and check the solutions:

1. $x + y = 9,$
$xy = 20.$

2. $2A - B = -7,$
$AB = 4.$

3. $m + y = 5.83,$
$xy = 8.27.$

4. $r^2 - 4s = 0,$
$r + s = 3.$

5. $u + 2v = 6,$
$uv + 8 = 0.$

6. $M - 2N = 7,$
$MN = 4.$

7. $x^2 + y^2 = 1,$
$x + y = 1.$

8. $I_1^2 + I_2^2 = 16,$
$I_1 + I_2 = 4.$

9. $m^2 - 4n = 0,$
$m - n = 3.$

10. $4R_1^2 - R_2^2 + 16 = 0,$
$2R_1 - R_2 - 2 = 0.$

11. $p^2 + 4q^2 = 4,$
$9p^2 - q^2 = 9.$

12. $z_1^2 + 4z_2^2 = 25,$
$z_1^2 - 2z_2^2 = 1.$

13. $4h^2 + 3k^2 = 19,$
$2h^2 - 5k^2 = 3.$

14. $7x^2 - 8y^2 = 36,$
$11x^2 - 5y^2 + 4 = 0.$

15. $e_1^2 - 3e_2^2 = 4,$
$2e_1 - 3e_2 = 1.$

16. $4L_1^2 + L_2^2 = 100,$
$8L_1 + 3L_2 = 50.$

17. $E_1^2 - E_2^2 = 15,$
$E_1 E_2 = 4.$

18. $3.74x^2 - 5.31y^2 = 9.72,$
$2.83x + 8.76y = 25.1$

19. $10u^2 - v^2 = 25,$
$50u^2 - 7v^2 = 125.$

20. $r^2 + rs + s^2 = 0,$
$r + s + 1 = 0.$

21. $2ab + 4a + 3b = 1,$
$2a + b = -3.$

22. $m^2 - mn - 3n^2 + 2m + n = 0,$
$$4m + 5n = 0.$$

Solve the following systems for x and y:

23. $x^2 + y^2 = 2k^2,$
$xy = k^2.$

24. $3x - y = -2c,$
$xy = c^2.$

25. $x^2 + y^2 = r^2,$
$y = mx + r\sqrt{1 + m^2}.$

26. $\dfrac{x^2}{a^2} + \dfrac{y^2}{b^2} = 1,$
$$\dfrac{x}{a} + \dfrac{y}{b} = 1.$$

Solve the following systems for x and y, and find the values of k for which there is only one set of solutions:

27. $xy = k,$
$x + y = 1.$

28. $y^2 = 4x,$
$y - 3x = k.$

29. $x^2 + y^2 = 9,$
$y = x + k.$

30. $x^2 + y^2 = 25,$
$3x + 4y = k.$

14–11. Problems from Science and Engineering

Many problems arising in science and engineering involve formulas that have quadratic forms. In working with such formulas one uses the methods developed in this chapter.

Example 1. The angle θ subtended in t sec by a particle having an initial angular velocity of ω_0 radians per second and a uniform angular acceleration of α radians per second per second, is given by the formula

$$\theta = \omega_0 t + \tfrac{1}{2}\alpha t^2.$$

Solve this equation for t.

Rearranging the above equation gives us

$$\alpha t^2 + 2\omega_0 t - 2\theta = 0.$$

This is a quadratic equation in t, which can be solved by the quadratic formula 3 of Sec. 14–6:

$$t = \frac{-2\omega_0 \pm \sqrt{4\omega_0{}^2 + 8\alpha\theta}}{2\alpha} = \frac{-\omega_0 \pm \sqrt{\omega_0{}^2 + 2\alpha\theta}}{\alpha}$$

Since t is positive, only the upper sign is used, and we have

$$t = \frac{-\omega_0 + \sqrt{\omega_0{}^2 + 2\alpha\theta}}{\alpha}.$$

Example 2. Two resistors have to be selected which when combined in parallel have a resistance of 1.82 ohms and when combined in series have a resistance of 10 ohms.

Let R_1 and R_2 be the resistances of the two resistors to be selected. In the theory of electricity it is shown that the series resistance R_s is given by

$$R_s = R_1 + R_2$$

and the parallel resistance R_p by the equation

$$\frac{1}{R_p} = \frac{1}{R_1} + \frac{1}{R_2}.$$

Therefore, to determine R_1 and R_2, the following system of simultaneous equations has to be solved:

$$R_1 + R_2 = 10, \tag{1}$$

$$\frac{1}{R_1} + \frac{1}{R_2} = \frac{1}{1.82} = 0.55. \tag{2}$$

Multiplying equation 2 by the product $R_1 R_2$ of the denominators gives

$$R_2 + R_1 = 0.55 R_1 R_2,$$

and, making use of equation 1, we obtain

$$0.55 R_1 R_2 = 10.$$

From equation 1,

$$R_2 = 10 - R_1,$$

and therefore

$$0.55R_1(10 - R_1) = 10,$$

or

$$5.5R_1 - 0.55R_1{}^2 = 10,$$

or

$$0.55R_1{}^2 - 5.5R_1 + 10 = 0.$$

This is a quadratic equation which can be solved by the quadratic formula, in which

$$A = 0.55, \quad B = -5.5, \quad C = 10,$$
$$D = B^2 - 4AC = 30.25 - 22 = 8.25.$$
$$\sqrt{D} = 2.87.$$

$$R_1 = \frac{5.5 \pm 2.87}{1.1},$$

$$R_2 = 10 - R_1 = 10 - \frac{5.5 \pm 2.87}{1.1} = \frac{5.5 \mp 2.87}{1.1}.$$

If the upper signs are used,

$$R_1 = \frac{8.37}{1.1} = 7.6 \text{ ohms},$$

$$R_2 = \frac{2.63}{1.1} = 2.4 \text{ ohms}.$$

The lower signs give the same values except that the order is changed.

Exercises

In the formulas below, consider all of the quantities as given except the quantity to be solved for. A hint as to the meaning of each formula is given. If the reader is interested in a description of the exact physical meaning of the formula and the units involved, an appropriate textbook should be consulted. The results obtained should be simplified as much as possible.

Given	Solve for	Description
1. $I_d = \dfrac{3m}{20}(r^2 + 4h^2)$	r, h	Moment of inertia
2. $J_0 = \dfrac{A(a^2 + b^2)}{4}$	a, b	Polar moment of inertia
3. $k_0 = \sqrt{\dfrac{b^2 + h^2}{12}}$	b, h	Radius of gyration
4. $C_s = V \sqrt[5]{\dfrac{\rho}{PN^2}}$	N	Propeller characteristics
5. $F = \dfrac{m(v_1{}^2 - v_2{}^2)}{2s}$	v_1, v_2	Kinetics
6. $T = \dfrac{J_m(\omega_1{}^2 - \omega_2{}^2)}{2\theta}$	ω_1, ω_2	Torque

7. $\dfrac{U_2{}^2 - U_1{}^2}{64.34} = J(H_1 - H_2)$ U_1, U_2 Energy equation for adiabatic flow

8. $P_i = E_m I + r_m I^2$ I Power input to motor armature

9. $P_0 = E_g I - r_g I^2$ I Power output of generator armature

10. $c = a + bt + ft^2$ t Specific heat as a function of temperature

11. $s = v_0 t + \frac{1}{2} a t^2$ t Distance traversed by a particle

12. $S = \dfrac{`3a - 2b}{a} \cdot \dfrac{wb^2}{t^2}$ b Stress in elliptical plates

13. $x_c = \dfrac{J_0 + A x_0{}^2}{A x_0}$ x_0 Distance to center of pressure

14. $l = \dfrac{J_m}{x_0 m} + x_0$ x_0 Distance to center of percussion

15. $v_f = \dfrac{1}{2\pi} \sqrt{\dfrac{f}{m} - \dfrac{R^2}{m^2}}$ m Oscillations in a vibrating system

16. $I_1 = I_2 + a(b + c)^2$ b, c Design of a built-up column

17. $K = \dfrac{x^2}{(1 - x)v}$ x Ionization constant of acid

18. $y = \sqrt{s^2 - m^2} - m$ m Theory of suspended cable

19. $P^2 = \dfrac{2W^2 h}{L} + 2WP$ P, W Stress in a machine

20. $W_L = (V_G I_1 - R_1 I_1{}^2) \dfrac{h}{1000}$ I_1 Energy in a transmission line

21. $G = \dfrac{1.06 a_t}{\sqrt{T_1}} \sqrt{p_2(p_1 - p_2)}$ p_2 Flow of air

22. $G = 0.0292 a_t \sqrt{p_2(p_1 - p_2)}$ p_2 Flow of steam

23. If L is the inductance in henrys and C the capacitance in farads of an a-c circuit, the resonance frequency f in cycles per second satisfies the relation

$$2\pi f L - \frac{1}{2\pi f C} = 0.$$

Solve this equation for f, and compute f if $L = 2 \times 10^{-3}$ henry and $C = 300\ \mu\mu f$. ($1\mu\mu f = 10^{-12}$ farad).

24. For a simple beam loaded and supported in a certain way, the *bending moment* at any distance of x ft from one end is $M = 20x - x^2$. For what value of x will $M = 70$?

25. If a square wooden column x in. on each side is to carry a certain load, the smallest safe value of x is a root of $x^4 - 125x^2 - 10{,}368 = 0$. Find that root.

26. The alternating component i_p of a vacuum-tube plate current of a variable-mu tube can be expressed approximately in terms of grid voltage e_g by the equation $i_p = c_1 e_g + c_2 e_g{}^2$ where c_1 and c_2 are constants.   Solve this equation for e_g.

27. In an electric circuit comprising a given heater in series with a resistance, the following relationship exists, $I^2 R + IE = 600$, where R is the resistance in ohms and E is the applied voltage.   If the resistance has a value of 2 ohms and the applied voltage is 110, what is the current I in amperes that flows in the circuit?

28. The formula

$$h = v_0 t - 16 t^2$$

gives the height h in feet that is reached in t sec by a body projected vertically upward with an initial velocity of v_0 ft per sec.   Solve for t.   Also find t for $v_0 = 79.5$ ft per sec and $h = 94.6$ ft.

29. The height h of a projectile t sec after firing is given by

$$h = k v_0 t - \tfrac{1}{2} g t^2,$$

where k depends on the angle of elevation of the cannon, v_0 is the initial velocity, and g the acceleration of gravity.   Solve for t.

30. The formula

$$s = v_0 t + \frac{g}{2} t^2$$

gives the distance s in feet passed over in t sec by a falling body whose initial velocity was v_0 ft per sec.   If $g = 32$ ft per sec per sec and $v_0 = 80$ ft per sec, how long will it take the object to fall 800 ft?

31. In a vacuum tube with cylindrical electrodes, the amplification constant μ depends on the radius r_g of the grid, the radius r_p of the plate, and on a constant K according to the equation $\mu = K\left(r_g - \dfrac{r_g{}^2}{r_p} \right)$.   Solve this equation for r_g.

32. The formula for the surface S of a right circular cone of altitude h and base radius r is given by $S = \pi r \sqrt{r^2 + h^2}$.   Solve for r^2.

33. The reactance of an electric circuit is

$$X = 2\pi f L - \frac{1}{2\pi f C}.$$

Compute f, if $L = 30 \times 10^{-6}$, $C = 200 \times 10^{-12}$, and $X = 150$.

34. The formula for the reactance necessary to tune a transmission line is

$$X_l = \frac{z_0 R_e}{\pm \sqrt{R_e(z_0 - R_e)}}.$$

Solve this for z_0 in terms of the other quantities.

35. If a lens of focal length f is used to produce an image of an object, the following relation exists between the distances of object and image from the lens.

$$\frac{1}{a} + \frac{1}{b} = \frac{1}{f}.$$

Compute a and b if $f = 5$ in., and $a + b = 2$ ft.

14–12. Operations with Polynomials in One Variable

In order to solve problems involving algebraic equations, a certain amount of skill in dealing with polynomials is necessary. The student will acquire this skill in the present section.

In almost all problems involving polynomials it is advantageous to arrange all the polynomials according to descending powers of the variable, that is, starting with the term of highest degree.

Addition and subtraction. If polynomials have to be combined by addition and subtraction, all the terms of like degree are collected. One starts with the terms of highest degree and then proceeds to terms of the next highest degree and so on until all terms of like degree are collected.

Example 1.

$$5(2x^3 - 4x^2 + 7x + 9) - 7(x^3 + x) + 5(x^2 + 7)$$
$$= (5 \cdot 2 - 7)x^3 + [5(-4) + 5]x^2 + (5 \cdot 7 - 7)x + 5 \cdot 9 + 5 \cdot 7$$
$$= 3x^3 - 15x^2 + 28x + 80.$$

Multiplication. To multiply two polynomials, first arrange them according to descending powers of the variable. Then multiply all of the terms of the first polynomial successively by the first term, second term, etc., of the second factor. When all the terms of like degree are collected the work is finished.

Example 2. To find the product $(2x^3 - 3x^2 - x + 5)(3x^2 + 2x - 4)$ we arrange our work in the following way:

$$(2x^3 - 3x^2 - x + 5)(3x^2 + 2x - 4)$$

The product of the first factor by $3x^2$:	$6x^5 - 9x^4 - 3x^3 + 15x^2$
The product of the first factor by $2x$:	$4x^4 - 6x^3 - 2x^2 + 10x$
The product of the first factor by -4:	$- 8x^3 + 12x^2 + 4x - 20$
Sum of all the terms:	$6x^5 - 5x^4 - 17x^3 + 25x^2 + 14x - 20$

Example 3. Similarly, to find the product $(x^4 + 3x^2 - 2)(x^2 + 5x + 1)$ we write

$$(x^4 + 3x^2 - 2)(x^2 + 5x + 1)$$

$$
\begin{array}{llll}
x^6 & + 3x^4 & - 2x^2 & \\
+ 5x^5 & + 15x^3 & - 10x & \\
+ x^4 & + 3x^2 & - 2 &
\end{array}
$$

$$x^6 + 5x^5 + 4x^4 + 15x^3 + x^2 - 10x - 2$$

Division. Each division may be regarded as a repeated subtraction, the divisor being subtracted from the dividend as many times as possible. Thus the statement: "If 20 is divided by 3, the quotient is 6 and the remainder is 2" means that 3 can be subtracted 6 times from 20 without getting a negative difference.   This is written in the form,

$$20 - 3 \cdot 6 = 2, \tag{1}$$

or

$$20 = 3 \cdot 6 + 2. \tag{2}$$

In this example, 20 is the dividend, 3 is the divisor, 6 is the quotient, and 2 is the remainder.   Equation 1 is a special case of the relation

$$\textbf{dividend} - \textbf{divisor} \times \textbf{quotient} = \textbf{remainder} \tag{3}$$

which holds whenever one number is divided by another.

Using relation 3, we now give the following definition for the division of two polynomials.

Dividing a polynomial $F(x)$ by a polynomial $G(x)$ means finding a third polynomial $Q(x)$ such that the difference

$$F(x) - G(x) \cdot Q(x) = R(x) \tag{4}$$

where $R(x)$ is a polynomial whose degree is smaller than the degree of $G(x)$.

Relation 3 may be also written as

$$\textbf{dividend} = \textbf{divisor} \times \textbf{quotient} + \textbf{remainder} \tag{5}$$

of which equation 2 is a special case.   If we use formula 5, the above definition for division of polynomials takes on the following form:

If two polynomials $F(x)$ and $G(x)$ are given, a third polynomial $Q(x)$, called the **quotient,** *can be found such that*

$$F(x) = G(x) \cdot Q(x) + R(x), \tag{6}$$

where the polynomial $R(x)$, called **remainder,** *is of lower degree than $G(x)$.*

The process of division of polynomials is then the same as long division in arithmetic.   This process consists in finding a series of multiples of the divisor and subtracting these from the dividend until the last remainder is smaller than the divisor.   The mechanics of division of polynomials will become clear from the following illustrations.

Example 4.   Divide the polynomial $6x^4 + x^3 + 5x^2 + 9x + 4$ by $3x^2 + 2x - 1$.

The two given polynomials are already arranged according to the descending powers of x.   If this were not so, they would have to be arranged in this way.

To start the division, the highest term of the dividend, $6x^4$, is divided by the highest term of the divisor, $3x^2$.   The quotient is $2x^2$.   The divisor is now multiplied by $2x^2$ and the product subtracted from the dividend.   The remainder found in this way is used as a dividend for the next step, and this process is repeated until the highest term of the last remainder is of lower degree than the divisor.

The computations described above can be arranged in the following way:

$$\text{(Dividend) } 6x^4 + x^3 + 5x^2 + 9x + 4 \,\big|\, 3x^2 + 2x - 1 \text{ (Divisor)}$$

$$ 2x^2 - x + 3 \text{ (Quotient)}$$

$$6x^4 + 4x^3 - 2x^2$$

$$- 3x^3 + 7x^2 + 9x$$
$$- 3x^3 - 2x^2 + x$$

$$9x^2 + 8x + 4$$
$$9x^2 + 6x - 3$$

$$2x + 7 \text{ (Remainder)}$$

The quotient is $2x^2 - x + 3$ and the remainder $2x + 7$. The result can be checked by the fundamental relation 6 given above. We obtain

$$6x^4 + x^3 + 5x^2 + 9x + 4 = (3x^2 + 2x - 1)(2x^2 - x + 3) + 2x + 7.$$

The result of the operations on the right side of this equation is equal to the left side, and therefore the division was performed without mistake.

Example 5. Divide $2x^3 - x + 5$ by $x - 2$.

The dividend does not contain a term of the second degree. In such a case we write the dividend with space left free for any missing terms. The required division may be arranged in the following way:

$$2x^3 - x + 5 \,\big|\, x - 2$$

$$ 2x^2 + 4x + 7$$

$$2x^3 - 4x^2$$

$$4x^2 - x$$
$$4x^2 - 8x$$

$$7x + 5$$
$$7x - 14$$

$$19$$

A check for this computation is obtained by testing the relation

$$2x^3 - x + 5 = (x - 2)(2x^2 + 4x + 7) + 19.$$

It should also be noted that the divisor $x - 2$ is of degree one; hence the remainder, being of lower degree, has no terms containing x and is therefore a constant, in this case 19.

From relation 6 it follows *that the dividend is divisible by the quotient or that the quotient is a factor of the dividend, if the remainder is zero.* Since in this case $R(x) = 0$, we have

$$F(x) = G(x) \cdot Q(x). \tag{7}$$

Example 6.   Divide $x^3 + 8x^2 + 3x - 18$ by $x + 2$.

$$
\begin{array}{r|l}
x^3 + 8x^2 + 3x - 18 & x + 2 \\
\cline{2-2}
& x^2 + 6x - 9
\end{array}
$$

$x^3 + 2x^2$

$\underline{}$

$6x^2 + 3x$
$6x^2 + 12x$

$\underline{}$

$-9x - 18$
$-9x - 18$

$\underline{}$

0

A check on these computations is obtained by showing that

$$x^3 + 8x^2 + 3x - 18 = (x + 2)(x^2 + 6x - 9).$$

Exercises

Perform the indicated operations and simplify.

1. $(x^4 + x^2 - 4) + 3(x^3 + x^2 - x + 5) - 2(7 - x + 3x^2 - x^4) - 5(x + 3x^2 - 2)$.
2. $4(a^5 - 2a^3 + a - 1) - 6(a + a^2 - 3 + a^4) + 5(a^4 - a^2 + 2a^5 - 11 - 3a)$.
3. $(2y^2 - 3y + 1)(y^2 + 2y - 5) - 7(y^5 + 3y - 2 + y^4) + 2y^3 - 5y + 2$.
4. $(6u^5 - 3u^3 + 4u^2 + 2)(u^2 - 1 + u) - (5u^2 - 7u + 1)(2u^2 - 3 + u)$.
5. $(a^2 - 3ab + 2b^2)(a^2 + ab - b^2) + (a - b)(a^3 + 2a^2b - ab^2 + b^3)$.
6. $(x - 2)^3 + 2(x + 3)^2(3x - 1) - (2x^2 - x + 5)^2 + (7x^4 - 2x^3 + 1 - x^2 + x)$.

In the following exercises find the quotient and the remainder.   Check the results by using relation 6.

7. $(x^4 + 2x^3 - 6x^2 - 5x + 7) \div (x - 1)$.
8. $(2x^5 + 4x^4 - 7x^3 + x^2 - 3x + 2) \div (x + 1)$.
9. $(r^5 - 3r^4 + 4r^3 + 9r^2 - 5r - 2) \div (r + 2)$.
10. $(m^4 + 4m^3 - m^2 + 5m - 2) \div (m^2 + 3m - 1)$.
11. $(y^5 - 2y^4 + 3y^3 - 6y^2 + y - 2) \div (y^2 - y + 1)$.
12. $(12r^3 + 2r^2 - 5r - 3) \div (3r + 2)$.
13. $(2E^4 - 3E^3 - 2E^2 + 5E - 2) \div (E^2 - 2E + 1)$.
14. $(6R^4 - 3R^3 - 5R^2 + 24R - 16) \div (3R^2 + 3R - 4)$.
15. $(Z^4 + 3Z^2 - 4) \div (Z^2 - 1)$.   16. $(u^4 - 5u^2 - 1) \div (u^2 + 2)$.
17. $(x^3 - 1) \div (x - 1)$.   18. $(v^3 - 1) \div (v + 1)$.
19. $(y^3 + 1) \div (y - 1)$.   20. $(m^3 + 1) \div (m + 1)$.
21. $(a^4 - 1) \div (a - 1)$.   22. $(a^4 - 1) \div (a + 1)$.
23. $(b^4 + 1) \div (b + 1)$.   24. $(e^4 + 1) \div (e - 1)$.
25. $(x^3 - y^3) \div (x - y)$.   26. $(x^3 + y^3) \div (x + y)$.
27. $(27x^3 + 8y^3) \div (3x + 2y)$.   28. $(E_1^3 + 8E_2^3) \div (E_1 + 2E_2)$.

14–13. Remainder Theorem and Factor Theorem

Let us restate relation 6 of the section just preceding, namely,

$$F(x) = G(x) \cdot Q(x) + R(x) \qquad (1)$$

where $R(x)$ is of lower degree than $G(x)$. Now, if the divisor $G(x)$ is of the first degree, then it follows that $R(x)$ is a constant. Hence, when the dividend $F(x)$ is a given polynomial and the divisor $G(x) = x - r$ (a linear expression), then equation 1 becomes

$$F(x) = (x - r) \, Q(x) + R, \qquad (2)$$

where $Q(x)$ is the quotient of the division $\dfrac{F(x)}{x - r}$ and the *constant* R is the remainder.

Equation 2 is an identity, true for all values of the variable x. If an equation is true for all values of x, it remains true if a particular value is substituted for x. Substituting $x = r$ on both sides of equation 2, we obtain

$$F(r) = R.$$

Hence equation 2 can be written in the form

$$F(x) = (x - r) \cdot Q(x) + F(r). \qquad (3)$$

The result given in equation 3 states symbolically the following theorem:

If a polynomial $F(x)$ is divided by $x - r$, the remainder is $F(r)$; that is, the remainder is the number obtained by replacing x by r in $F(x)$. This result is known as the **remainder theorem.**

Example 1. In Example 5 of Sec. 14–12 it was seen that when $F(x) = 2x^3 - x + 5$ is divided by $x - 2$ the remainder is 19. This can now be verified by the remainder theorem, for, replacing x by 2 in $F(x) = 2x^3 - x + 5$, we obtain

$$R = F(2) = 2 \cdot 8 - 2 + 5 = 19.$$

Example 2. What is the remainder when $x^{59} - 1$ is divided by $x + 1$?

In this case $F(x) = x^{59} - 1$ and $r = -1$; hence, by the remainder theorem,

$$R = F(r) = F(-1) = (-1)^{59} - 1 = -1 - 1 = -2.$$

The remainder theorem has enabled us to find the remainder $R = -2$ without performing the extremely long division of $x^{59} - 1$ by $x + 1$.

If, in particular, $F(r) = 0$, then we obtain from equation 3

$$F(x) = (x - r) \cdot Q(x). \qquad (4)$$

This result can be stated in the following theorem.

If the polynomial F(x) is equal to zero when x = r, then (x − r) is a factor of F(x), and, conversely, if (x − r) is a factor of F(x), then F(r) = 0. This result is known as the **factor theorem.** This theorem enables us to find out whether or not a polynomial $F(x)$ is divisible by a linear factor $x − r$ without actually carrying out the division.

Example 3. The polynomial $F(x) = x^3 − 8$ vanishes for $x = 2$. Hence, by the factor theorem, $F(x)$ is divisible by $x − 2$; in fact,

$$x^3 − 8 = (x − 2)(x^2 \mid 2x + 4).$$

Example 4. Decide whether or not $x + 3$ is a factor of $x^3 − 3x + 9$.

In this case $F(x) = x^3 − 3x + 9$ and $x − r = x + 3$; therefore $r = −3$ and

$$F(−3) = (−3)^3 − 3(−3) + 9 = −27 + 9 + 9 = −9 \neq 0,$$

which shows that $x + 3$ is not a factor of $x^3 − 3x + 9$.

Exercises

Perform the following divisions, and check the remainder by using the remainder theorem.

1. $(x^3 − 2x^2 + 3x − 1) ÷ (x − 2)$.   **2.** $(y^4 + 8) ÷ (y + 1)$.

3. $(R^4 − 2R^2 + 3R − 2) ÷ (R + 2)$.   **4.** $(2L^4 − L + 1) ÷ (L + 3)$.

5. $(Z^4 − 2Z^2 + 3Z − 5) ÷ (Z − 2)$.   **6.** $(x^5 − 32) ÷ (x − 2)$.

7. $(x^5 − 3x^4 + x^2 − 7) ÷ (x + 1)$.   **8.** $(3E^4 + 5) ÷ (E − 5)$.

9. $(y^5 − 2y^4 + 3y^3 + y^2 − 5y − 18) ÷ (y − 1)$.

10. $(2v^5 − 3v^4 + v^3 + 5v^2 − 4v + 5) ÷ (v + 1)$.

Find the remainders of the following divisions without actually dividing:

11. $(3x^{31} − 25) ÷ (x − 1)$.   **12.** $(5x^{53} + 2) ÷ (x + 1)$.

13. $(4y^3 − 3y^2 − y + 7) ÷ (y + 2)$.   **14.** $(z^{25} − 5) ÷ (z − 1)$.

15. $(5E^4 − 2E^3 + E + 8) ÷ (E + 3)$.   **16.** $(\sin^5 \theta + 4) ÷ (\sin \theta − 1)$.

17. $(8v^9 + 3v^4 − 5v + 2) ÷ (v − 1)$.   **18.** $(64u^8 + 32u^4 − 1) ÷ (u + \frac{1}{2})$.

19. $(6 \cos^7 \alpha + 3 \cos^3 \alpha − \cos \alpha − 25) ÷ (\cos \alpha − 2)$.

20. $(12x^7 − 8x^5 + 7x^4 − 3x^3 − 2x^2 − 6) : (x − 1)$.

Decide whether $x − 2$ is a factor of the following polynomials:

21. $x^2 + 8x − 1$.   **22.** $x^3 − 7x + 6$.

23. $x^4 − 2x^3 + x^2 + x − 6$.   **24.** $x^4 + 10x^2 + 16$.

25. $3x^5 + 7x − 8$.   **26.** $x^4 − 16$.

27. $x^6 − 2x^5 + x − 2$.   **28.** $x^6 + x^5 − 2x^4 − x^2 + x − 12$.

Decide whether $x + 3$ is a factor of the following polynomials:

29. $x^3 + 27$.   **30.** $x^2 + 9x − 27$.

31. $5x^3 + 7x^2 + 3x − 1$.   **32.** $x^9 + 3x^8 − 2x − 6$.

33. $x^5 − 2x^3 − 21$.   **34.** $x^4 − 3x^3 + 7x + 3$.

35. $x^3 + 2x^2 + 2x + 15$.   **36.** $x^{12} − 81^3$.

37. Is $\cos^3 A + 1$ divisible by $\cos A - 1$?

38. Is $8 \sin^2 \theta - 3 \sin \theta - \frac{1}{2}$ divisible by $\sin \theta - \frac{1}{2}$?

39. Find the value or values of m so that $2x^3 - 3x^2 - mx + 20$ is divisible by $x - 1$.

40. Find the value or values of k so that $4y^3 - k^2y^2 - 4ky + 64$ is divisible by $y + 1$.

41. Find the value or values of m so that $z + m$ will divide into $2z^2 + z - 5$ with a remainder equal to -2.

42. Find the value or values of p so that $y - 2$ will divide $y^3 - 3y^2 - 6p^2y + 2p + 7$ with a remainder equal to 1.

14–14. The Fundamental Theorem of Algebra

Let $F(x)$ be any polynomial of degree n in the variable x. Then *the equation*

$$F(x) = 0$$

is called an algebraic equation of nth degree. Every number, real or imaginary, which when substituted for x satisfies this equation is called a root of $F(x) = 0$.

Thus, for example,

$$2x^3 + x^2 - 8x - 4 = 0$$

is an algebraic equation of the third degree. The left-hand side vanishes for $x = 2$. Therefore, $x = 2$ is a root of the given equation.

Every algebraic equation has at least one root. This statement, even though it may seem obvious to the student, is one of the most important theorems in algebra and is therefore known as the **fundamental theorem of algebra.** This theorem was proved the first time by the mathematician Gauss in 1799. A proof may be found in advanced books on algebra.

Obviously this theorem is true only when the complex numbers are admitted as roots. For example, the equation $x^2 + 1 = 0$ has only the roots $x = \pm j$.

14–15. Number of Roots of an Algebraic Equation

We can now restate the factor theorem given in Sec. 14–13 in the following way:

If r is a root of the algebraic equation $F(x) = 0$, then $(x - r)$ is a factor of $F(x)$.

The linear expression $x - r$ is called the **root factor** corresponding to the root $x = r$.

We shall now use the factor theorem to determine the number of roots of an algebraic equation.

Let $F(x)$ be a given polynomial. By the fundamental theorem of algebra stated in Sec. 14–14, the equation

$$F(x) = 0$$

has at least one root which we may call $x = r_1$. The polynomial $F(x)$ is then divisible by $x - r_1$ and can be factored, so that

$$F(x) = (x - r_1)\, F_1(x),$$

where $F_1(x)$ is found by dividing $F(x)$ by $x - r_1$. The degree of $F_1(x)$ is one less than that of $F(x)$.

The same procedure can be repeated with $F_1(x)$. The equation $F_1(x) = 0$ has at least one root $x = r_2$, and the polynomial $F_1(x)$ can be factored, so that

$$F_1(x) = (x - r_2\,) F_2(x),$$

and therefore

$$F(x) = (x - r_1)(x - r_2)\, F_2(x),$$

where the degree of $F_2(x)$ is two less than the degree of $F(x)$.

This procedure can be repeated until the degree of one of the quotients $F_1(x)$, $F_2(x)$, $F_3(x)$, $\cdots$ is zero, that is, until a quotient is a constant A. If $F(x)$ is a polynomial of degree n, we have then

$$F(x) = A(x - r_1)(x - r_2) \cdots (x - r_n). \tag{1}$$

This argument proves the following theorem:

Every polynomial of the nth degree can be factored in n linear factors.

The equation $F(x) = 0$ when written in factored form

$$F(x) = A(x - r_1)(x - r_2) \cdots (x - r_n) = 0$$

is easily seen to be satisfied when $x = r_1$, $x = r_2$, $\cdots$, $x = r_n$. The equation $F(x) = 0$, therefore, has n different roots, provided that all the numbers r_1, r_2, $\cdots$, r_n are different. If some of these numbers are equal, then the corresponding root is said to be a **multiple root.** For example, when $r_1 = r_2$, then the root r_1 is said to have multiplicity two or to be a **double root;** when $r_1 = r_2 = r_3$, then the root r_1 is said to have multiplicity three or to be a **triple root,** and so on. This leads to the following theorem:

Given an algebraic equation of degree n:

$$F(x) - 0.$$

The left-hand member of this equation is the product of n linear factors, and the equation has exactly n roots if each root is counted as often as its multiplicity indicates.

Example. Consider the equation $x^5 + 2x^4 - 2x^3 - 4x^2 + x + 2 = 0$.

The left-hand side can be factored and written in the form

$$(x - 1)^2(x + 1)^2(x + 2) = 0,$$

so that it is the product of five linear root factors. The corresponding roots are, therefore,

$$r_1 = r_2 = 1, \qquad r_3 = r_4 = -1, \qquad r_5 = -2.$$

The equation has three different roots, $1, -1, -2$. Each of the first two of these roots is a double root so that the total number of roots is five.

14–16. Reduced Equation

If $x = r$ is a root of the algebraic equation $F(x) = 0$, then this equation can be written in the form

$$(x - r_1) F_1(x) = 0.$$

Here the left-hand member can be zero only if either $x - r_1 = 0$ or $F_1(x) = 0$. The first possibility gives $x = r_1$, the root already known. All the other roots of $F(x) = 0$ are, therefore, roots of the **reduced equation**

$$F_1(x) = 0.$$

The degree of this reduced equation is one less than the degree of the original equation $F(x) = 0$.

The problem of finding the roots of the equation $F(x) = 0$ which is of degree n is therefore reduced to finding the roots of the equation $F_1(x) = 0$, which is simpler, being of degree $n - 1$.

Example. Solve the equation $x^3 + x^2 - 8x + 6 = 0$, given that one of the roots is $r_1 = 1$.

That this equation has the root $r_1 = 1$ can easily be checked by substitution in the given equation for $1 + 1 - 8 \cdot 1 + 6 = 0$. To find the reduced equation, divide $(x^3 + x^2 - 8x + 6)$ by $(x - 1)$ as follows:

$$
\begin{array}{r|l}
x^3 + x^2 - 8x + 6 & \,x - 1 \\
\hline
 & x^2 + 2x - 6 \\
\end{array}
$$

$$x^3 - x^2$$

$$
\begin{array}{l}
2x^2 - 8x \\
2x^2 - 2x \\
\end{array}
$$

$$
\begin{array}{l}
- 6x + 6 \\
- 6x + 6 \\
\end{array}
$$

$$0$$

The reduced equation is

$$x^2 + 2x - 6 = 0,$$

and its roots are

$$x = \frac{-2 \pm \sqrt{4 + 24}}{2} = \frac{-2 \pm \sqrt{28}}{2} = -1 \pm \sqrt{7}.$$

The given equation has, therefore, the following three roots:

$$x_1 = 1,$$

$$x_2 = -1 + \sqrt{7} = -1 + 2.646 = 1.646,$$

$$x_3 = -1 - \sqrt{7} = -1 - 2.646 = -3.646.$$

Exercises

In the following equation one root is given.   Find all the other roots.   Express all irrational roots in decimal form, correct to three significant figures.

1. $x^3 - 6x^2 + 11x - 6 = 0$; $r_1 = 1$.
2. $x^3 - 3x^2 - 7x + 6 = 0$; $r_1 = -2$.
3. $y^3 - y^2 - 8y + 12 = 0$; $r_1 = 2$.
4. $x^3 - x^2 + x - 1 = 0$; $r_1 = 1$.
5. $x^3 - 10x^2 + 31x - 30 = 0$; $r_1 = 5$.
6. $z^3 - 5z^2 - 17z + 21 = 0$; $r_1 = 7$.
7. $x^3 + x^2 - 17x - 20 = 0$; $r_1 = -4$.
8. $v^3 + 6v^2 + v + 6 = 0$; $r_1 = -6$.
9. $2z^3 + 3z^2 - 5z - 18 = 0$; $r_1 = 2$.
10. $3x^3 - 14x^2 + 14x + 3 = 0$; $r_1 = 3$.
11. $2x^3 + 3x^2 - 4x + 1 = 0$; $r_1 = \frac{1}{2}$.
12. $5u^3 + 3u^2 - 13u + 2 = 0$; $r_1 = -2$.
13. $3 \tan^3 \theta - 20 \tan^2 \theta + 36 \tan \theta - 16 = 0$; $\tan \theta = 4$.
14. $2 \sin^3 \alpha - 3 \sin^2 \alpha + 1 = 0$; $\sin \alpha = -\frac{1}{2}$.
15. $6 \cos^3 \theta + \cos^2 \theta - 4 \cos \theta + 1 = 0$; $\cos \theta = \frac{1}{3}$.

In the following equations two roots are given.   Find all the other roots.

16. $x^4 + 4x^3 - x - 4 = 0$; $r_1 = 1, r_2 = -4$.
17. $3x^4 - 4x^3 + 3x^2 - 6x + 4 = 0$; $r_1 = 1$ is a double root.
18. $y^4 + 4y^3 - 6y^2 - 36y - 27 = 0$; $r_1 = -3$ is a double root.
19. $z^4 - z^3 - 5z^2 + 15z - 18 = 0$; $r_1 = 2, r_2 = -3$.
20. $y^4 + 2y^3 - 2y^2 + 2y - 3 = 0$; $r_1 = -3, r_2 = 1$.
21. $u^4 - 4u^3 + 3u^2 + 4u - 4 = 0$; $r_1 = 2, r_2 = -1$.
22. $z^4 + 2z^3 + z + 2 = 0$; $r_1 = -1, r_2 = -2$.
23. $v^4 + v^3 - 2v^2 + 4v - 24 = 0$; $r_1 = 2, r_2 = -3$.
24. $x^4 + 4x^3 - 20x^2 + 17x - 2 = 0$; $r_1 = \frac{1}{3}, r_2 = \frac{2}{3}$.
25. $\tan^4 \theta - 4 \tan^3 \theta - 28 \tan^2 \theta + 64 \tan \theta + 192 = 0$; $\tan \theta = 4, \tan \theta = -2$.

14–17. Rational Roots

We shall first consider an algebraic equation in which the coefficient of the highest power of the unknown is unity and all the other coefficients are integers.   For example,

$$x^3 + 2x^2 - 3x - 6 = 0 \tag{1}$$

is such an equation.   It can be proved that the real roots of such an equation are either irrational numbers or integers (they cannot be fractions).   In the present section we shall explain the method for finding the integral roots which such an equation may have.   In this connection the following theorem is true:

For an algebraic equation in which the coefficient of the highest power of the unknown is unity and all other coefficients are integers, any existing integral roots are factors of the absolute term.

Applying this result to equation 1, we see that if integral roots exist they have to be factors of the absolute term 6.   In order to find these roots, the factors of 6 have to be tested.   All the possible integral factors of 6 are:

$$-1, +1, -2, +2, -3, +3, -6, +6.$$

By substituting in the given equation it is found that of these numbers only -2 satisfies equation 1 and is, therefore, a root of this equation. Since (1) is an equation of the third degree and has therefore three roots, the other two roots are either irrational or complex. These can be determined by dividing $x^3 + 2x^2 - 3x - 6$ by $x + 2$ and solving the resulting reduced equation. Carrying out the proper computations, we get

$$x^3 + 2x^2 - 3x - 6 = (x + 2)(x^2 - 3);$$

so that the three roots of the equation 1 are

$$r_1 = -2, \qquad r_2 = \sqrt{3}, \qquad r_3 = -\sqrt{3}.$$

This example illustrates the following rule:

In order to find all existing integral roots of an algebraic equation with integral coefficients and the highest coefficient 1, test all factors of the absolute term.

If the highest coefficient is not unity, but all the coefficients are integers, the equation may have rational roots which are fractions. The method for finding them will be shown in the following example:

Let

$$2x^3 - 5x^2 + 7x - 6 = 0 \tag{2}$$

be the given equation. The coefficient of the highest power of x is 2, while all the other coefficients are integers. Introduce a new unknown quantity y by the substitution

$$x = \frac{y}{2}, \tag{3}$$

where the denominator on the right-hand side is equal to the coefficient of the highest power of x in the given equation. The result of the replacing x by $y/2$ in the given equation 2 is

$$2\frac{y^3}{8} - 5\frac{y^2}{4} + 7\frac{y}{2} - 6 = 0,$$

from which

$$y^3 - 5y^2 + 14y - 24 = 0. \tag{4}$$

This is an equation with integral coefficients and the highest coefficient equal to 1. Hence, by the method discussed previously, each integral root of this equation is a factor of 24. Testing the various factors of 24 we find that $y = 3$ is a root of equation 4, and therefore by equation 3, $x = y/2 = \frac{3}{2}$ is a root of the original equation 2.

Once an integer or fractional root has been found it can be used, by the method of Sec. 14–16, to form a reduced equation of lower degree. Thus, in the last example the polynomial on the left-hand side of equation 2 can be divided by $x - \frac{3}{2}$ and the resulting quotient (a quadratic) solved for the other two roots of the given equation. An alternative method which will avoid working with fractions is to divide the polynomial on the left-hand side of equation 4 by $y - 3$ and thus obtain the other two roots of equation 4. Then by equation 3 we shall have the corresponding roots of equation 2.

Example. Find all the roots of the equation
$$5x^3 + 28x^2 + 10x - 3 = 0. \tag{5}$$
Since the coefficient of the highest power of x is 5, we substitute $x = y/5$ in equation 5 and obtain
$$5 \cdot \frac{y^3}{125} + 28 \cdot \frac{y^2}{25} + 10 \cdot \frac{y}{5} - 3 = 0, \tag{6}$$
$$y^3 + 28y^2 + 50y - 75 = 0.$$

Each integral root of this equation is a factor of 75. By testing the different factors of 75, it is found that -3 is a root of the given equation. To find the other roots we find the reduced equation. This is accomplished by the following long division:

$$
\begin{array}{r|l}
y^3 + 28y^2 + 50y - 75 & \underline{y + 3} \\
& y^2 + 25y - 25 \\[2pt]
\underline{y^3 + \ 3y^2} & \\[2pt]
25y^2 + 50y & \\
25y^2 + 75y & \\[2pt]
\overline{-25y - 75} & \\
-25y - 75 & \\[2pt]
\overline{0} &
\end{array}
$$

The reduced equation is therefore
$$y^2 + 25y - 25 = 0,$$
and its roots are found by the quadratic formula as follows:
$$D = 25^2 - 4(-25) = 625 + 100 = 725.$$
$$\sqrt{D} = \sqrt{725} = \sqrt{25 \cdot 29} = 5\sqrt{29}.$$
$$y_2 = \frac{-25 + 5\sqrt{29}}{2}, \qquad y_3 = \frac{-25 - 5\sqrt{29}}{2},$$

which together with $y_1 = -3$ give us the three roots of equation 6.
Using the relation $x = y/5$, we find the three roots of equation 5:
$$x_1 = \frac{y_1}{5} = -\frac{3}{5} = -0.600,$$
$$x_2 = \frac{y_2}{5} = \frac{-5 + \sqrt{29}}{2} = \frac{-5 + 5.385}{2} = 0.192,$$
$$x_3 = \frac{y_3}{5} = \frac{-5 - \sqrt{29}}{2} = \frac{-5 - 5.385}{2} = -5.192.$$

Exercises

Find the rational roots of the following equations:

1. $x^3 - x^2 - 4x + 4 = 0$.
2. $x^3 - x^2 - 5x + 2 = 0$.
3. $x^3 + 2x^2 + 3x + 2 = 0$.
4. $x^3 - 4x^2 + 7x - 12 = 0$.
5. $y^3 + 2y^2 = 5y + 6$.
6. $z^3 + 14z + 5 = 8z^2$.
7. $2x^3 - 9x^2 + 7x + 6 = 0$.
8. $12v^3 + 4v^2 - 3v = 1$.
9. $2u^3 - 3u^2 + 5u - 2 = 0$.
10. $3R^3 + 4R^2 - 7R + 2 - 0$.
11. $L^4 - L^3 + 2L^2 = 4L + 8$.
12. $x^4 + 2x^3 + x + 2 = 0$.
13. $x^4 + x^3 - 7x^2 - 13x = 6$.
14. $2y^4 - 7y^3 + 5y^2 = 7y - 3$.
15. $2x^4 + x^3 + 5x^2 + 3x - 1 = 0$.
16. $3x^4 + 2x^3 + 11x^2 + 8x - 4 = 0$.
17. $\tan^4 \theta - 2\tan^3 \theta - 13\tan^2 \theta + 14\tan \theta + 24 = 0$.
18. $x^5 - 8x^4 + 24x^3 - 34x^2 + 23x - 6 = 0$.
19. $2\sin^4 \alpha + \sin^3 \alpha - 3\sin^2 \alpha - \sin \alpha + 1 = 0$.
20. $3M^5 + M^4 + 12M^3 + 4M^2 - 15M - 5 = 0$.

Find all the roots of the following equations. Express all irrational roots in decimal form, correct to three significant figures.

21. $x^3 - 3x^2 - 7x + 6 = 0$.
22. $x^3 - 8x - 3 = 0$.
23. $z^3 + 2z^2 - 2z + 3 = 0$.
24. $y^3 + 2y^2 = y + 2$.
25. $v^3 = 2(v - 2)$.
26. $L_x{}^3 - 7L_x - 6 = 0$.
27. $2x^3 + 3x^2 + 1 = 4x$.
28. $2u^3 + 5u^2 - u = 1$.
29. $4s^3 - 5s - 2 = 0$.
30. $3R^3 + 5R^2 - R = 2$.
31. $z^4 + 3z^3 - 6z^2 = 18z - 20$.
32. $L^4 - 2L^3 = 12L^2 + 15L$.
33. $E^4 + 3E^3 + 4E^2 + 3E + 1 = 0$.
34. $2y^4 - 13y^3 + 12y^2 + 17y = 10$.
35. $\tan^4 \theta + 3\tan^3 \theta - 12\tan^2 \theta = 4(5\tan \theta - 12)$.
36. $\cot^4 \theta - 9\cot^3 \theta + 21\cot^2 \theta + \cot \theta = 30$.
37. $w^5 - 8w^4 + 15w^3 + 20w^2 - 76w + 48 = 0$.
38. $E_1{}^6 - 3E_1{}^5 + 6E_1{}^3 - 3E_1{}^2 - 3E_1 + 2 = 0$.

14–18. Graphical Solution of Equations

After having worked with algebraic equations, we could hardly expect that all the roots of such an equation are rational, or even in many cases that an algebraic equation has any rational roots at all. The method for finding rational roots, explained in Sec. 14–17, can therefore be used only occasionally. If we are to find all the real roots of an algebraic equation, we are still faced with the problem of finding the irrational roots. A method for finding the approximate values of the irrational roots will be developed in the remaining sections of this chapter. The student should be reminded that an algebraic equation may also have roots that are complex numbers. Finding the complex roots of an equation of a degree higher than the quadratic is difficult and will therefore not be described in this book.   This, however, is not a serious omission in view of the fact that in the applications of mathematics one is usually interested only in the real roots.

In a quadratic equation, the irrational roots can be found by the quadratic formula (explained in Sec. 14–6).   This fact may suggest the idea of finding corresponding formulas for higher-degree equations.   Formulas of this kind which permit the computation of the roots are known for equations of the third and fourth degree, but they are too complicated to be used generally for solving such equations.   That no such formulas exist for equations of the fifth or higher degree was proved by the Norwegian mathematician Abel in 1827.

Even though formulas for obtaining irrational roots of an algebraic equation are difficult or impossible to derive, there are methods for finding approximate values of these irrational roots to any desired degree of accuracy.   Some of these methods will be explained in the remainder of this chapter.   We shall start in the present section with the graphical solution of equations.

The first method for the graphical solution of equations has been already explained in Sec. 3–9.   In order to locate approximately the real roots of the equation

$$F(x) = 0,$$

plot the graph of the function

$$y = F(x),$$

and find its x intercepts, the abscissas of the points where the graph meets the x axis.

Example 1.   Locate approximately the roots of the equation $x^3 - 3x + 1 = 0$.
In order to plot the graph of

$$y = x^3 - 3x + 1$$

the following table of corresponding values is computed,

x	-3	-2	-1	0	1	2	3
y	-17	-1	3	1	-1	3	19

and the graph is plotted on cross-section paper (Fig. 14–6).   From the graph it can be concluded that the given equation has three real roots, approximately

$$x_1 = -1.9, \qquad x_2 = 0.3, \qquad x_3 = 1.5.$$

To plot the graph of the function $y = F(x)$ is often troublesome. Another method, therefore, will be given here, which can be used occasionally and which requires less work.   The method will be explained in connection with the equation of the previous example,

$$x^3 - 3x + 1 = 0. \tag{1}$$

Write this equation in the form

$$x^3 = 3x - 1,$$

and plot the graphs of the two functions

$$y = x^3, \qquad y = 3x - 1.$$

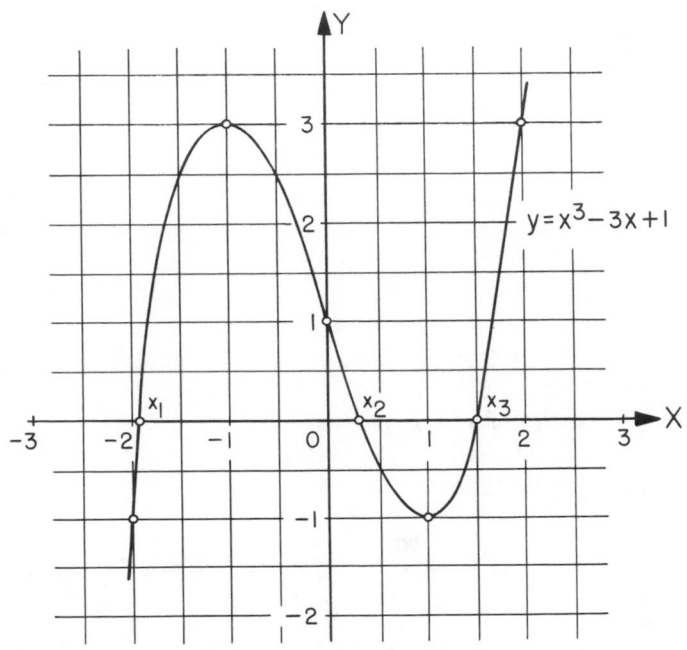

FIG. 14–6

These graphs can be easily plotted. Moreover, if other cubic equations have to be solved, the graph of $y = x^3$ can always be used. The graph of $y = x^3$ has been discussed in Sec. 8–13, while the graph of $y = 3x - 1$ is a straight line (Fig. 14–7).

If the two graphs intersect in a point P, the two functions have the same values at P, namely,

$$y = x^3 = 3x - 1.$$

The abscissa x of P, therefore, satisfies the given equation 1. From Fig. 14–7 it can be inferred that there are three points of intersection, P_1, P_2, P_3 whose abscissas are approximately

$$x_1 = -1.9, \qquad x_2 = 0.3, \qquad x_3 = 1.5.$$

This method can always be used when a few terms of the given equation $F(x) = 0$ can be transposed so that both sides of the equation can be plotted with less effort than the graph of $F(x)$. The method can also be applied to nonalgebraic equations.

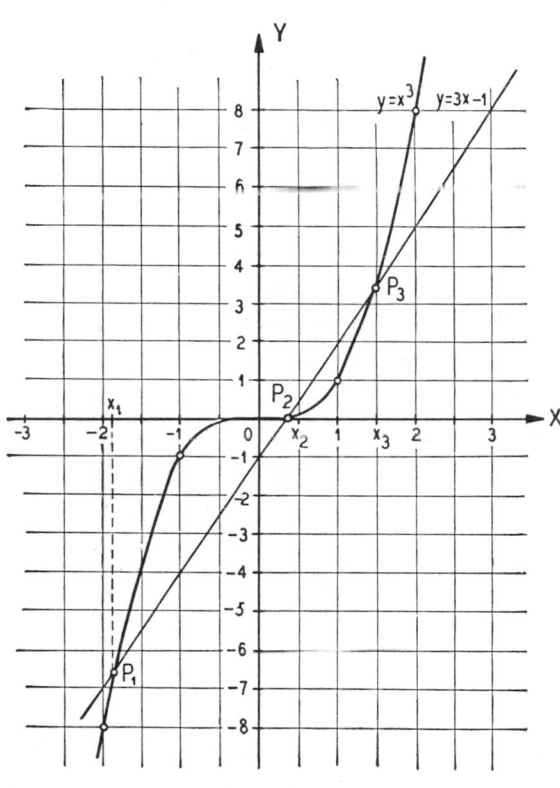

FIG. 14–7

Example 2. Solve approximately the equation $x^x = 10$.
Computing the logarithms of both sides of the equation, we have

$$x \log x = 1 \quad \text{or} \quad \log x - \frac{1}{x}.$$

The graphs of the two functions $y = \log x$ and $y = 1/x$ are easier to plot than the graph of $x^x = 10$.
Figure 14–8 shows that the two graphs intersect at a point whose abcissa is roughly 2.5. Hence, $x_0 = 2.5$ is an approximate value satisfying the equation $x^x = 10$. In fact, $2.5^{2.5} = 9.88$.

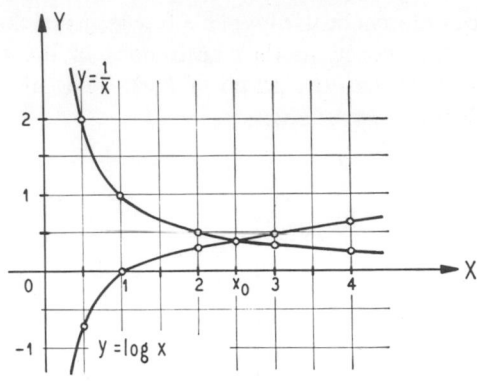

Fig. 14–8

Exercises

Find by the methods of this section approximate values for the real roots of the following equations:

1. $x^3 + 2x + 20 = 0.$ 2. $x^3 - 8x + 2 = 0.$ 3. $x^3 - 7x - 7 = 0.$

4. $x^3 = 6x + 2.$ 5. $x^3 - 6x + 12 = 0.$ 6. $3x^3 - 4x = 1.$

7. $x^3 + 2x = 8.$ 8. $x^3 - 4x + 2 = 0.$ 9. $x^3 - 3x = 1.$

10. $x^3 + 4x^2 = 7.$ 11. $x^3 + 3x^2 = 10.$ 1₃. $x^3 - 7x + 7 = 0.$

13. $x^4 + 10x - 100 = 0.$ 14. $x^4 + 8x - 12 = 0.$ 15. $x^4 - (x - 5) = 0.$

16. $3x^4 - 4x^2 + 8 = 0.$ 17. $\log x = 2x - 9.$ 18. $\log x = 3x - 9.$

19. $2x - \log x = 8.$ 20. $e^x + x = 3.$ 21. $5e^{-x} + x = 5.$

22. $2^x - 4x = 0.$ 23. $x^2 - 2^x = 0.$ 24. $e^x + x = 0.$

25. $e^x - 4x = 0.$

In the following equations the unknown is an angle measured in radians,

26. $x = \cos x.$ 27. $x + \sin 2x = 1.2.$ 28. $x + \cos x = 0.$

29. $\sin x = x - 2.$ 30. $x^2 - 3 \sin x = 0.$ 31. $\cos 2x - x = 0.$

32. The angle of a circular sector whose area is bisected by its chord satisfies the equation $\theta = 2 \sin \theta.$ Find an approximate value for θ.

14–19. Method of Repeated Linear Interpolation

Various methods have been developed which permit us to find the irrational roots of an equation with greater precision than is possible by the graphical methods of Sec. 14–18.   The **method of trial and error** or the **method of linear interpolation** will be discussed in this section.   This method is simple and applies to all cases in which the graph of the function $y = F(x)$ intersects the x axis so that y is positive on one side and negative on the other side of the point of intersection of the curve with the x axis, as in Fig. 14–9. This method cannot be used when the curve touches the x axis as in Fig. 14–10. This latter case, however, occurs seldom and will not be discussed here.

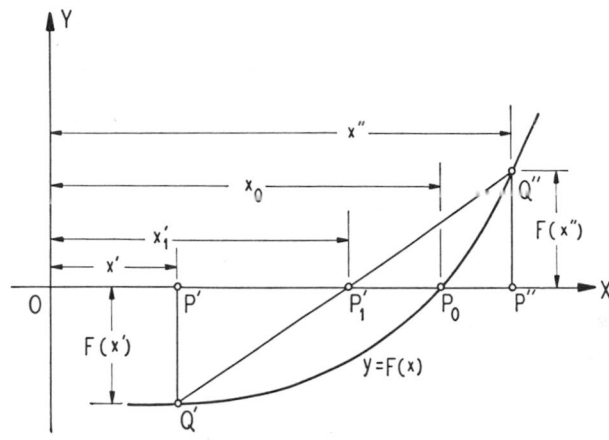

FIG. 14–9

The method of repeated linear interpolation is based on the following two statements:

1. If x' and x'' are two values sufficiently close to each other (Fig. 14–9) so that $F(x')$ and $F(x'')$ have different signs, then the graph of $y = F(x)$ intersects the x axis at a point P_0 whose abscissa x_0 is between x' and x''. This statement means that there is a value $x = x_0$ between x' and x'' such $F(x_0) = 0$.

2. Let Q' and Q'' (Fig. 14–9) be the two points on the graph of $y = F(x)$ with abscissas x' and x'', respectively. Also let P' and P'' be the projections on the x axis of the points Q' and Q'', respectively. Now, if $F(x')$ and $F(x'')$ have different signs

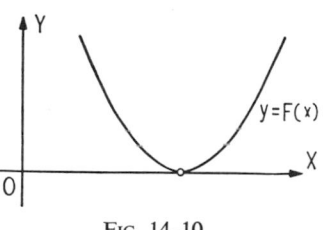

FIG. 14–10

and the points Q' and Q'' are sufficiently close to each other, then the straight line $Q'Q''$ crosses the x axis at the point P_1' which, as a rule, will be nearer to the point P_0 than either of the points P' and P''. Thus the line $Q'Q''$ becomes a close approximation of the section of curve of $y = F(x)$ between the points Q' and Q''.

Our object now is to find the abscissa x_1' of the point P_1'. This value x_1' will give us the first approximation to the abscissa x_0 of the point P_0, which is the root in question.

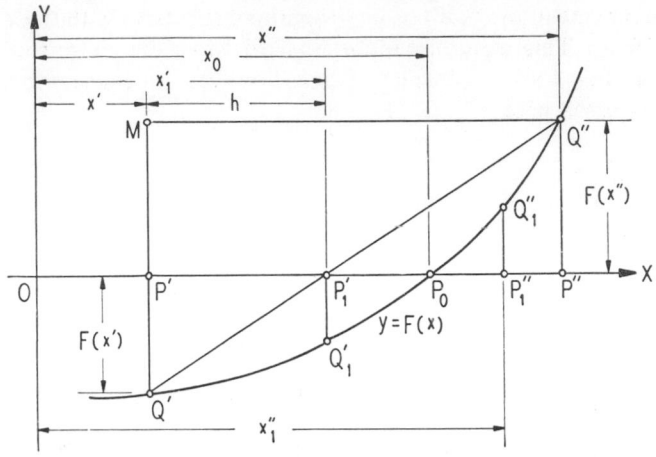

Fig. 14–11

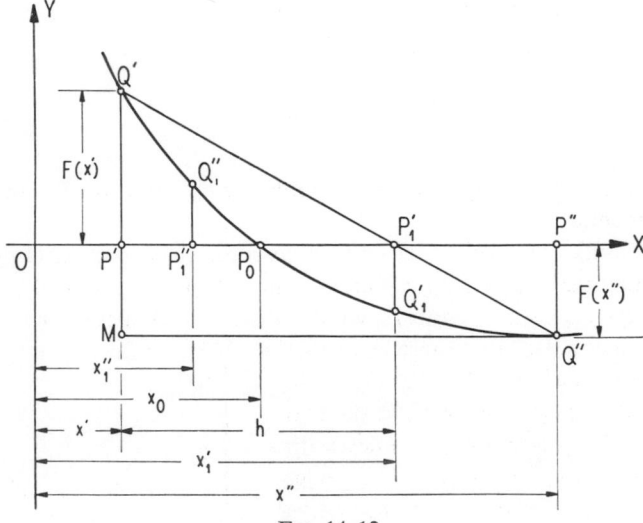

Fig. 14–12

The abscissa x_1' of the point P_1' can be found very easily from the fact that the two triangles $P'P_1'Q'$ and $MQ''Q'$ are similar (see Figs. 14–11 and 14–12). Therefore it follows that

$$\frac{P'P_1'}{MQ''} = \frac{P'Q'}{MQ'}.$$

Now, since we have $P'P_1' = h$ and $MQ'' = |x'' - x'|$,

$$h = \frac{P'Q'}{MQ'}|x'' - x'|.$$

In the last expression $P'Q'$ is the absolute value of $F(x')$ and MQ' is the sum of the absolute values of $F(x')$ and $F(x'')$, so that

$$h = \frac{|F(x')|}{|F(x')| + |F(x'')|} |x'' - x'| \tag{1}$$

and

$$x_1' = x' + h. \tag{2}$$

We can now summarize our work.

To find a first approximation to a real root of the equation

$$F(x) = 0,$$

proceed as follows:

1. *Find two values x' and x'' sufficiently close to each other that $F(x')$ and $F(x'')$ have different signs.*

2. *Having found the values of $F(x')$ and $F(x'')$, use equation 1 to compute h.*

3. *Using equation 2, find the value of $x' + h$ which gives us the first approximation to the required root.*

This procedure of replacing a curve by a straight line in order to find an approximation to its x intercept is the process of **linear interpolation.** In the examples below it will be explained how a second value x_1'' can be found such that $F(x_1')$ and $F(x_1'')$ have different signs. By using the values x_1' and x_1'' instead of the original values x' and x'' the whole procedure can be repeated in order to obtain a closer approximation to the root. Thus, by successive repetition of this process, the root can be found with any desired degree of precision.

Example 1.   Find correct to three significant digits, the greater positive root of the equation

$$F(x) = x^3 - 3x + 1 = 0.$$

This equation has already been investigated in Sec. 14–18.   It was seen there that the greater positive root is roughly $x' = 1.5$ and that the corresponding graph (Fig. 14–6) crosses the x axis at this point from negative to positive values of $F(x)$.

Step 1.   Compute

$$F(x') = F(1.5) = 1.5^3 - 3(1.5) + 1 = 3.375 - 4.5 + 1 = -0.125.$$

From the graph it can be inferred that x' is to the left of the unknown root.   A second value x'' to the right of this root is found by trials, increasing x' by one or more units of its rightmost place.   Trying the value $x = 1.6$, we obtain

$$F(1.6) = 1.6^3 - 3(1.6) + 1 = 4.096 - 4.8 + 1 = 0.296.$$

Since this value of the function is positive, we choose $x'' = 1.6$.   The required root lies between $x' = 1.5$ and $x'' = 1.6$.   These two values can now be used to find a better approximation by linear interpolation.   For this purpose it is always advisable to plot a graph, as shown in Fig. 14–13, which is large enough so that all the necessary arguments can be made in connection with the figure.

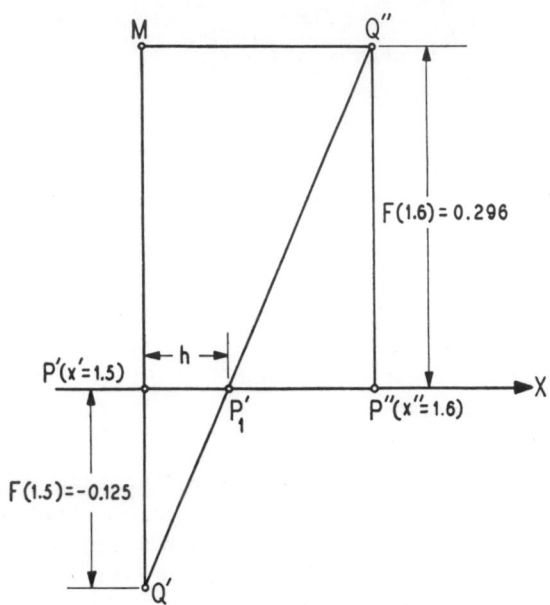

FIG. 14–13

Substituting in equation 1, we obtain

$$h = \frac{0.125}{0.125 + 0.296}(1.6 - 1.5) = \frac{0.125}{0.421}(0.1) = 0.03.$$

It is generally sufficient to compute only one digit of h. The new approximate value for the root is the abscissa of P_1',

$$x_1' = x' + h = 1.5 + 0.03 = 1.53.$$

In order to test the precision of this value and to obtain a closer approximation, the whole procedure is now repeated, starting with the new value $x_1' = 1.53$.

Step 2. Compute

$$F(1.53) = 1.53^3 - 3(1.53) + 1 = 3.582 - 4.59 + 1 = -0.008.$$

The value $x_1' = 1.53$, therefore, is smaller than the correct root value. In order to find a value greater than the root, increase the value 1.53 and compute the corresponding value of $F(x)$.

$$F(1.54) = 1.54^3 - 3(1.54) + 1 = 3.652 - 4.62 + 1 = 0.032.$$

We, therefore, take $x_1'' = 1.54$, and x_1' and x_1'' are two values that can be used for the linear interpolation. The corresponding part of the graph is plotted on a much enlarged scale in Fig. 14–14.

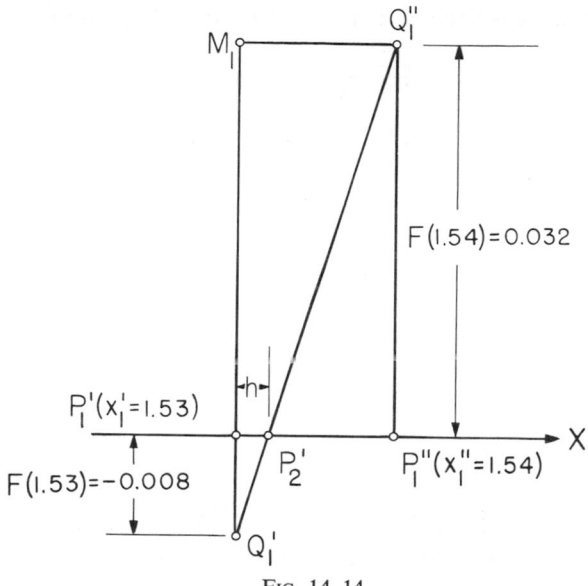

FIG. 14–14

Computing h as before, we find it to be

$$h = \frac{0.008}{0.008 + 0.032}(0.01) = \frac{8}{40}(0.01) = 0.002,$$

and therefore

$$x_2' = 1.53 + 0.002 = 1.532.$$

Since the value $x_2' = 1.532$ is a better approximation than the previous value $x_1' = 1.53$, it can, therefore, be concluded that $x = 1.53$ is the value of the root to three significant digits.

If a greater precision is required, the procedure can be repeated as often as necessary.

Example 2. Find the negative root of the equation

$$F(x) = 2x^4 - x - 5 - 0.$$

We first construct a table of values for the function

$$y = 2x^4 - x - 5,$$

obtaining

x	-2	-1	0	1	2
y	29	-2	-5	-4	25

There is a negative root between -2 and -1, for, as the table shows, $F(-2)=29$ and $F(-1) = -2$ have opposite signs. In order to find closer approximations of this root, the process of linear interpolation is applied starting with the values

$$x' = -2, \qquad x'' = -1.$$

Step 1.   As in the previous example, we have from Fig. 14–15

$$h = \frac{29}{29 + 2}(-1 + 2) = 0.9$$

whence

$$x_1' = -2 + 0.9 = -1.1.$$

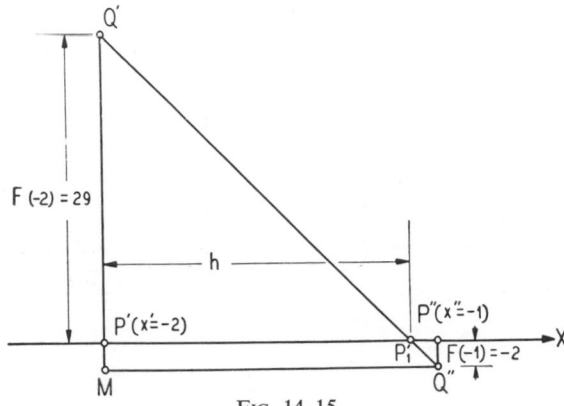

Fig. 14–15

Step 2.   From the table of functional values it can be concluded that the function is positive to the left and negative to the right of the required root.   The value $x_1' = -1.1$ is too large, because

$$F(-1.1) = 2(-1.1)^4 - (-1.1) - 5 = 2(1.46) + 1.1 - 5 = -0.98$$

is negative.   In order to have a value x_1'' on the other side of the root, a smaller value is tested:

$$F(-1.2) = 2(-1.2)^4 - (-1.2) - 5 = 2(2.07) + 1.2 - 5 = 0.34.$$

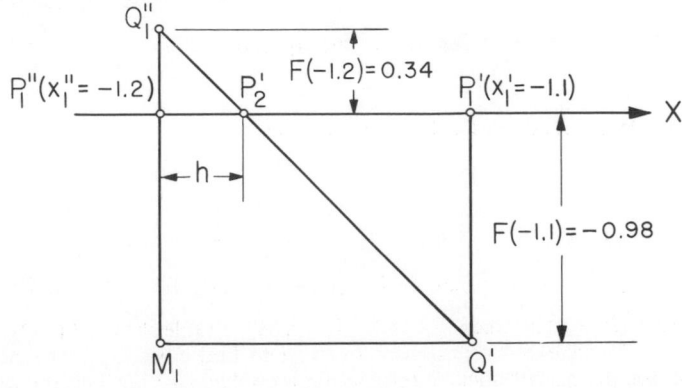

Fig. 14–16

The unknown root, therefore, is located between -1.2 and -1.1, and these two values are used for the next linear interpolation (Fig. 14–16):

$$h = \frac{0.34}{0.34 + 0.98}(0.1) = \frac{34}{132}(0.1) = 0.03$$

whence

$$x_2' = -1.2 + 0.03 = -1.17.$$

If the accuracy of the digit 7 needs to be tested, the whole procedure can be repeated.

The fact that $F(x)$ is a polynomial is not essential for the application of the method of linear interpolation. The same procedure, therefore, can be applied to nonalgebraic equations.

Exercises

Find the real roots of the following equations correct to three significant figures:

1. $x^3 + 4x - 7 = 0.$
2. $u^3 - 4u + 2 = 0.$
3. $r^3 - r + 1 = 0.$
4. $x^3 + 2x^2 - 7 = 0.$
5. $v^3 - 3v^2 + 3 = 0.$
6. $r^3 + 6r^2 - r = 10.$
7. $z^3 + 2z^2 + 8z - 1 = 0.$
8. $s^3 - 6s^2 + 13s = 7.$
9. $p^4 + p^3 - p - 2 = 0.$
10. $x^4 + x^2 = x + 3.$
11. $y^4 - 3y^3 + 8y - 6 = 0.$
12. $x^4 + x^3 - 4x^2 - 16 = 0.$
13. Find the cube root of 40.
14. Find the cube root of 7.
15. Find two fourth roots of 6.
16. Find two fourth roots of 20.

17. A sphere of ice, 1 ft in diameter, floating in water, sinks to a depth of h ft, given by the equation $2h^3 - 3h^2 + 0.9 = 0$. Find the depth correct to three significant digits.

18. The values of $\sin \theta$ and $\sin \dfrac{\theta}{3}$ satisfy the following equation:

$$\sin \theta = 3 \sin \frac{\theta}{3} - 4 \sin^3 \frac{\theta}{3}.$$

If $\sin \theta = 0.7$, find $\sin \dfrac{\theta}{3}$ correct to two significant digits.

19. The diameter d in inches of the bolts needed to couple the flanges of certain cylindrical shafts is the positive root of the equation $d^4 + 800d^2 - 18d - 360 = 0$. Find d to two decimal places.

20. Find the radius of a sphere whose volume is 1 cu ft. Give the answer in inches correct to three significant figures.

21. The base of a box is a square and the height is 1 ft less than the side of the base. Find the dimensions of the box if the volume is 5 cu ft. Give your result correct to three significant figures.

22. An open tank is to be made from a rectangle of sheet metal, 20 by 30 in., by cutting out equal squares from each corner and turning up the sides. What size squares should be cut out in order that the volume of the tank should be 900 cu in.? Give your result correct to two significant digits.

23. When the edge of a cube is increased by 1 in. its volume is doubled. Find the edge of the cube correct to three significant figures.

24. What is the thickness of a hollow spherical vessel whose outer diameter is 10 in. and which can hold 400 cu in. ?   Give your result correct to three significant figures.

Progress Report

This chapter was devoted to the study of polynomials and polynomial (or algebraic) equations.   In particular, the following topics were discussed:

1. Quadratic polynomials and their graphs.

2. Graphic and algebraic methods of solving quadratic equations, properties of the roots of quadratic equations, equations reducible to quadratic equations, systems of equations involving quadratic expressions, and problems from applications involving quadratic expressions.

3. Algebraic operations with polynomials in one variable.

4. The remainder and factor theorems for algebraic equations.

5. The fundamental theorem of algebra and the number of roots of an algebraic equation.

6. Finding the rational roots of an algebraic equation.

7. The graphical solution of algebraic equations.

8. The method of linear interpolation for approximating the irrational roots of an algebraic equation.

15

The Straight Line

In plane geometry problems were solved by construction and by geo-metrical reasonings. However, by introducing a coordinate system as described in Chapter 3, it is possible to apply algebraic processes to the solution of geometrical problems. Since the methods of algebra are more direct and usually easier than geometrical reasonings, much is gained in the solving of problems in this way. Thus by means of a coordinate system algebra and geometry are united; the resulting theory is called **analytic geometry,** first introduced by René Descartes (1596–1650), a French mathematician.

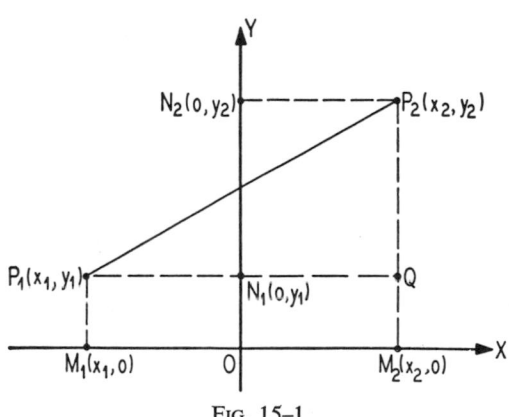

FIG. 15–1

15–1. The Distance between Two Points

Let $P_1(x_1, y_1)$ and $P_2(x_2, y_2)$ be any two points in the plane, as shown in Fig. 15–1. They determine the segment joining them, which will be denoted by P_1P_2. *We shall also let P_1P_2 represent the length of the segment, a positive number.*

The perpendiculars from P_1 and P_2 to the x axis cut that axis at $M_1(x_1, 0)$ and $M_2(x_2, 0)$, respectively. Since M_2 is to the right of M_1, $x_2 > x_1$, and $M_1M_2 = x_2 - x_1$. Similarly, if the perpendiculars from P_1 and P_2 to the y axis cut that axis at $N_1(0, y_1)$ and $N_2(0, y_2)$ respectively, then $N_1N_2 = y_2 - y_1$. Both of these statements depend, of course, on how Fig. 15–1 has been drawn. If it were drawn differently, the order of the subscripts might be changed.

From Fig. 15–1 and the Pythagorean theorem we have

$$(P_1P_2)^2 = (P_1Q)^2 + (QP_2)^2$$
$$= (M_1M_2)^2 + (N_1N_2)^2$$
$$= (x_2 - x_1)^2 + (y_2 - y_1)^2.$$

If we denote the distance P_1P_2 by d we have that

$$d = \sqrt{(x_2 - x_1)^2 + (y_2 - y_1)^2}. \tag{1}$$

Since the order of subscripts in equation 1 does not matter, the formula is valid for any two points in the plane.

Example. Find the distance between the points $(-2, 5)$ and $(7, -3)$. Let P_2 be $(-2, 5)$ and P_1 be $(7, -3)$. Then

$$d = \sqrt{(-2 - 7)^2 + (5 + 3)^2} = \sqrt{81 + 64} = \sqrt{145}.$$

Exercises

Find the distance between the points of each of the following pairs:

1. $(4, 2), (1, -3)$.　　　　　　　　**2.** $(5, 1), (-2, -2)$.

3. $(-4, 7), (-8, 3)$.　　　　　　　**4.** $(6, 9), (-2, 6)$.

5. $(6, 8), (-6, -8)$.　　　　　　　**6.** $(2, 3), (2, -5)$.

7. $(7, 4), (6, 2)$.　　　　　　　　**8.** $(7, 4), (-6, -2)$.

9. $(8, 5), (-4, -3)$.　　　　　　**10.** $(13, 7), (-5, 9)$.

Prove by using the distance formula that the three points given in each exercise form the vertices of a right triangle.

11. $(0, 0), (5, 3), (2, 8)$.　　　　**12.** $(0, 6), (6, 0), (-8, -2)$.

13. $(-3, -2), (1, 0), (-1, 4)$.　　**14.** $(-5, -1), (-3, -5), (-7, -7)$.

15. $(-5, -3), (9, 5), (-9, 4)$.　　**16.** $(-2, 1), (12, 9), (-6, 8)$.

Prove by using the distance formula that the three points given in each exercise form the vertices of an isosceles triangle.

17. $(4, -4), (3, 4), (10, 0)$.　　　**18.** $(-1, 0), (0, -1), (2, 2)$.

19. $(-6, 3), (3, -3), (9, 6)$.　　　**20.** $(0, 0), (6, 2), (5, -5)$.

21. $(-5, 5), (-2, 2), (-6, 1)$.　　**22.** $(2, -3), (8, 6), (-7, -9)$.

Prove by using the distance formula that the four points given in each exercise form the vertices of a parallelogram. By finding the lengths of the diagonals, find out whether or not the parallelogram is a rectangle.

23. $(3, 3), (3, 0), (-3, -3), (-3, 0).$

24. $(-1, -1), (2, -4), (6, 0), (3, 3).$

25. $(-6, 3), (-6, 1), (-2, 3), (-2, 5).$

26. $(-5, 1), (-2, -2), (1, 7), (4, 4).$

27. $(-1, 3), (-3, 7), (-5, 3), (-3, -1).$

28. Show that the points $(1, -1), (4, 2), (7, -1),$ and $(4, -4)$ form the vertices of a rhombus.   Show that the figure is a square.

29. Show that the points $(-4, 2), (3, 1),$ and $(3, -5)$ are equidistant from $(-1, -2).$ What are the center and radius of the circle determined by the first three points given?

30. Find a point on the y axis equidistant from $(4, 2)$ and $(-1, -1).$

31. Find a point on the x axis equidistant from $(3, 2)$ and $(1, -4).$

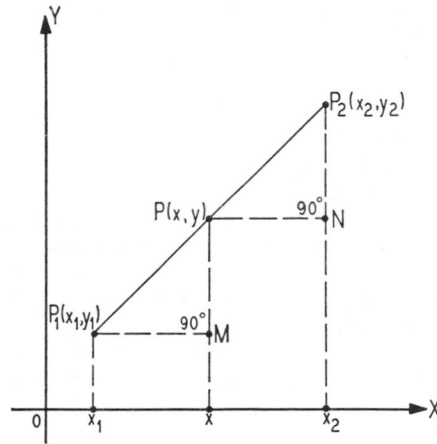

FIG. 15–2

15–2. The Midpoint of a Segment

Let $P(x, y)$ be the midpoint of the segment joining $P_1(x_1, y_1)$ and $P_2(x_2, y_2)$ as shown in Fig. 15–2.   With the constructions shown in this figure, it can be demonstrated easily that triangle P_1MP is congruent to triangle PNP_2.   Then $P_1M = PN$.   Obviously $P_1M = x - x_1$ and $PN = x_2 - x$.   Then

$$x - x_1 = x_2 - x.$$

whence

$$x = \frac{x_1 + x_2}{2}. \tag{1}$$

In like manner

$$y = \frac{y_1 + y_2}{2}. \tag{2}$$

This discussion, of course, depends on the way the figure was drawn, but the results 1 and 2 are the same no matter what figure is drawn, as the reader may easily verify.

Example. Find the midpoint of the segment joining (1, 2) and (−5, 4).

By formula 1, $x = \dfrac{1 - 5}{2} = -2$, and, by formula 2, $y = \dfrac{2 + 4}{2} = 3$. Thus, the midpoint is (−2, 3).

Exercises

Find the coordinates of the midpoint of the segment joining:

1. (3, 7), (5, −1).   **2.** (3, −7), (−5, 1).   **3.** (−4, −2), (−8, 3).

4. (6, −2), (3, 9).   **5.** (7, 0), (−5, 4).   **6.** (−3, −6), (−8, 5).

7. Find the coordinates of the midpoints of the sides of a triangle whose vertices are (4, 2), (−3, −4), and (7, −1).

8. Find the coordinates of the midpoints of the sides of a triangle whose vertices are (−2, 3), (1, −2), and (−6, −3).

9. The midpoint of a segment is (−1, 2), and one extremity is (5, 6). Find the coordinates of the other extremity.

10. The midpoint of a segment is (9, −6), and one extremity is (4, −8). Find the coordinates of the other extremity.

11. The points (4, 6), (8, −2), (0, 8), and (−6, 6) form the vertices of a quadrilateral. Prove that the segments joining the midpoints of opposite sides bisect each other.

12. The points (0, 3), (5, −3), (1, −7), and (4, 0) form the vertices of a quadrilateral. Prove that the segments joining the midpoints of opposite sides bisect each other.

13. For the quadrilateral of exercise 11, show that the segment joining the midpoints of a pair of opposite sides bisects the segment joining the midpoints of the diagonals.

14. For the quadrilateral of exercise 12, show that the segment joining the midpoints of a pair of opposite sides bisects the segment joining the midpoints of the diagonals.

15–3. The Inclination and Slope of a Line

Let $P_1(x_1, y_1)$ and $P_2(x_2, y_2)$ be two points in the plane. By the line P_1P_2 we mean the line determined by P_1 and P_2 and extending infinitely far in both directions, and by the segment P_1P_2 we mean the portion of the line included between P_1 and P_2.

The smallest angle through which the positive half of the x axis must be rotated counterclockwise to bring it parallel to a given line P_1P_2 is called the angle of inclination of P_1P_2. In Fig. 15–3, θ is the angle of inclination.

Any line P_1P_2 will either cut the x axis or be parallel to it. If it cuts the x axis, the angle of inclination is less than 180°; if it is parallel to the x axis, the angle of inclination is taken as 0°.

Another description of the direction of a line is given by its slope m, which is defined by the relation

$$m = \tan \theta, \tag{1}$$

where θ is the angle of inclination of the given line.

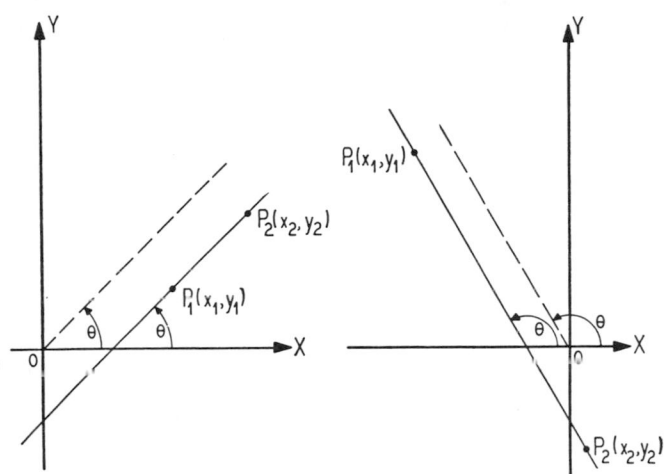

FIG. 15–3

Figure 15–4 shows two possible positions of the line P_1P_2, with θ between $0°$ and $90°$ in a, and with θ between $90°$ and $180°$ in b. We see then that

$$|m| = |\tan\theta| = \frac{MP_2}{P_1M} = \left|\frac{y_2 - y_1}{x_2 - x_1}\right|.$$

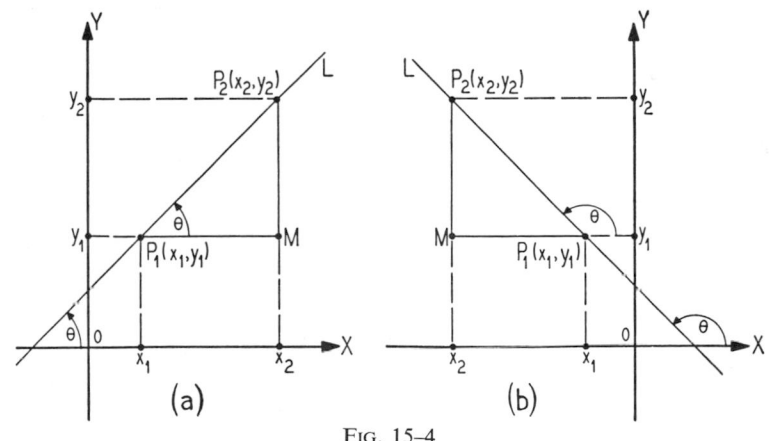

FIG. 15–4

Now, when θ is between $0°$ and $90°$, $m = \tan\theta$ is a positive number, and we see that in Fig. 15–4a the expression $\dfrac{y_2 - y_1}{x_2 - x_1}$ is positive. When θ is between $90°$ and $180°$, $m = \tan\theta$ is negative, and $\dfrac{y_2 - y_1}{x_2 - x_1}$ for Fig. 15–4b is negative. Hence we state simply that

$$m = \frac{y_2 - y_1}{x_2 - x_1}.\tag{2}$$

Since $\dfrac{y_2 - y_1}{x_2 - x_1} = \dfrac{y_1 - y_2}{x_1 - x_2}$, the order in which the points are chosen in

equation 2 is immaterial, and hence equation 2 is valid for any line P_1P_2. However, when P_1P_2 is parallel to the y axis, $x_1 = x_2$ and the denominator of equation 2 is zero. Thus equation 2 has no meaning, and we say that the line has no slope. Since the angle of inclination of a line parallel to the y axis is 90°, and tan 90° $= \infty$, we can say that *the slope of a line parallel to the y axis is infinite.* Also, *the slope of a line parallel to the x axis is zero.*

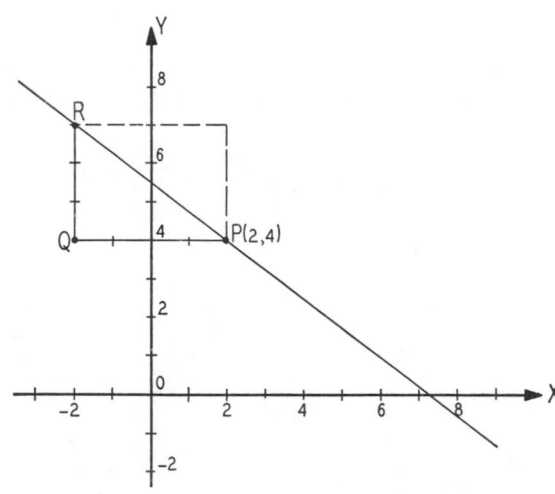

FIG. 15–5

Example 1. Find the slope and angle of inclination of the line through (1, 0) and (4, −2). By formula 2,

$$m = \frac{0 - (-2)}{1 - 4} = \frac{2}{-3} = -\frac{2}{3} = -0.6667$$

whence $\theta = 146.3°$.

Example 2. Draw the line through $P(2, 4)$ with slope $-\frac{3}{4}$.

From $P(2, 4)$ lay off a horizontal segment PQ four units to the left. Then from Q lay off a vertical segment upward three units as shown in Fig. 15–5. Then the line RP is the required line. The same result could have been achieved by making the construction indicated by the dotted lines in Fig. 15–5.

Exercises

Find the slope of the line through each pair of points. Find also the angle of inclination of each line, accurate to the nearest tenth of a degree.

1. $(1, 4), (5, 8)$.

2. $(4, 2), (8, -2)$.

3. $(-2, -1), (-4, 7)$.

4. $(9, -5), (3, 0)$.

5. $(3, -5), (8, -1)$.

6. $(5, 2), (-4, -6)$.

7. $(-2, 7), (-14, -9)$.

8. $(-3, 12), (8, -7)$.

If three points P_1, P_2, and P_3 are given, and the slope of $P_1 P_2$ is equal to the slope of $P_1 P_3$, then these two lines coincide. Use this fact to show that the points in each of the following sets lie on a straight line :

9. $(0, 0), (2, 3), (6, 9)$.

10. $(0, 0), (-1, 3), (-3, 9)$.

11. $(4, 6), (0, 4), (-8, 0)$.

12. $(-5, 2), (-1, 1), (3, 0)$.

13. $(-5, -3), (-1, -1), (5, 2)$.

14. $(-4, 2), (-1, -2), (2, -6)$.

15. $(-1, 2), (0, 0), (2, -4)$.

16. $(-3, 3), (3, 2), (9, 1)$.

Determine k so that the three points in each of the following sets lie on the same line :

17. $(1, 2), (4, 4), (k, 6)$.

18. $(2, 7), (3, 5), (k, 1)$.

19. $(1, 1), (-3, -1), (3, k)$.

20. $(4, 2), (2, -1), (-2, k)$.

21. $(-5, 5), (-2, k), (1, -7)$.

22. $(-3, 1), (-2, k), (0, -2)$.

23. $(-9, -9), (k, -3), (-3, 3)$.

24. $(-5, 4), (1, 3), (k, k)$.

Construct the line through the given point with the given slope.

25. $P(3, 5), m = \frac{1}{3}$.

26. $P(4, -2), m = \frac{4}{5}$.

27. $P(-1, 3), m = 2$.

28. $P(-5, 3), m = 4$.

29. $P(-2, -4), m = -\frac{1}{4}$.

30. $P(6, 1), m = -\frac{1}{6}$.

31. $P(3, 5), m = -5$.

32. $P(2, 4), m = -3$.

15–4. Parallel and Perpendicular Lines

When two lines L_1 and L_2 are parallel, they have the same angles of inclination θ_1 and θ_2, respectively. Then, since $\theta_1 = \theta_2$, $\tan \theta_1 = \tan \theta_2$, and

$$m_1 = m_2. \tag{1}$$

It follows that, *if two lines are parallel, their slopes are equal. Conversely, if the slopes of two lines are equal, the lines are parallel.*

Let L_1 and L_2 be lines with angles of inclination θ_1 and θ_2, respectively, and let L_1 be perpendicular to L_2. Now either $\theta_2 = \theta_1 + 90°$ as shown in Fig. 15–6 or $\theta_1 = \theta_2 + 90°$. In the first case, if we denote the slopes of L_1 and L_2 by m_1 and m_2, respectively,

$$m_2 = \tan \theta_2 = \tan (\theta_1 + 90°) = -\cot \theta_1 = -\frac{1}{\tan \theta_1} = -\frac{1}{m_1}.$$

In the second case,

$$m_1 = \tan \theta_1 = \tan (\theta_2 + 90°) = -\cot \theta_2 = -\frac{1}{\tan \theta_2} = -\frac{1}{m_2}.$$

Hence, in either case,

$$m_1 m_2 = -1 \quad \text{or} \quad m_1 = -\frac{1}{m_2}. \tag{2}$$

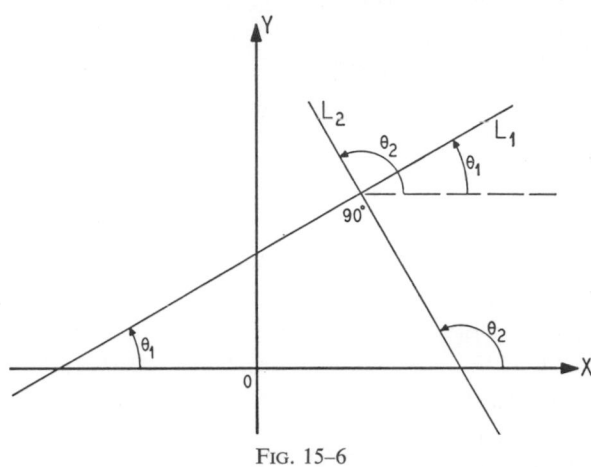

FIG. 15–6

That is to say, *if two lines are perpendicular, their slopes are negative reciprocals. Conversely, if two lines have slopes that are negative reciprocals, the lines are perpendicular.*

Let the reader draw a figure illustrating the case when $\theta_1 = \theta_2 + 90°$.

Exercises

The three points given in each exercise form the vertices of a triangle. Find the slopes of the sides of the triangle, and, using these data, show that each is a right triangle.

1. (1, 2), (5, 4), (3, 8).　　　　　　**2.** (0, 0), (2, 2), (3, −3).

3. (−1, 1), (1, −1), (5, 3).　　　　**4.** (2, 8), (3, 0), (−1, 2).

5. (2, −2), (−6, −3), (−4, 1).　　**6.** (4, −6), (−3, −10), (−4, 8).

7. Show that the line through (1, −5) and (3, 1) is parallel to the line through (1, 0) and (2, 3).

8. Show that the line through (−1, −2) and (6, 3) is parallel to the line through (−4, −5) and (3, 0).

9. Show that the segment joining (3, 0) and (0, 4) is parallel to and half as long as the segment joining (6, 0) and (0, 8).

10. Show that the line through (3, 1) and (−5, −3) is perpendicular to the line through (0, 3) and (4, −5).

11. Show that the line through (−1, −4) and (3, 2) is perpendicular to the line through (−7, −2) and (2, −8).

12. Prove that the points (1, 2), (5, 3), (3, −6), and (7, −5) form the vertices of a rectangle.

13. Using only the idea of slope, show that the points (−4, −1), (1, 2), (4, −3), and (−1, −6) form the vertices of a square.

14. Prove that the points $(0, -3), (-1, -7), (-3, -2)$, and $(-2, 2)$ form the vertices of a quadrilateral whose opposite sides are parallel.

Using the slope, show that the four points given in each exercise below form the vertices of a parallelogram. Which of these parallelograms are rectangles?

15. $(-1, 0), (-1, 3), (4, 3), (4, 0)$.   **16.** $(1, 4), (2, -1), (-8, -3), (-9, 2)$.

17. $(1, 2), (-2, -2), (-5, 2), (-2, 6)$.   **18.** $(6, -4), (-1, -10), (-8, -3), (-1, 3)$.

19. $(3, 5), (9, -1), (4, -6), (-2, 0)$.   **20.** $(10, -7), (9, -8), (-3, 4), (-2, 5)$.

21. The points $(1, 3), (-3, 3), (-1, -5)$, and $(7, -7)$ form the vertices of a quadrilateral. Prove that the figure formed by joining in order the midpoints of the sides of the quadrilateral is a parallelogram.

22. The points $(4, 3), (-6, -5), (-2, -9)$, and $(4, -7)$ form the vertices of a quadrilateral. Prove that the figure formed by joining in order the midpoints of the sides of the quadrilateral is a parallelogram.

23. Show that the points $(6, 2), (-2, 6), (-6, -2)$, and $(2, -6)$ form the vertices of a square. Then prove that the midpoints of the sides of this square form the vertices of a square.

24. Show that the points $(3, 2), (-3, 0), (-1, -6)$, and $(5, -4)$ form the vertices of a square. Then prove that the midpoints of the sides of this square form the vertices of a square.

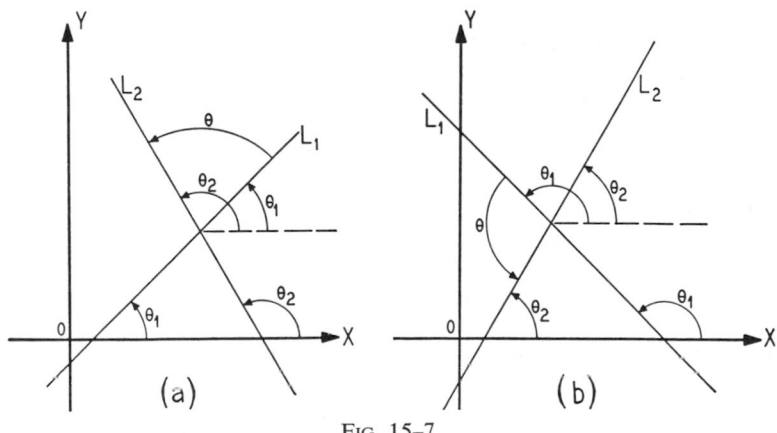

FIG. 15–7

15–5. The Angle between Two Intersecting Lines

Let L_1 and L_2 be two intersecting lines with angles of inclination θ_1 and θ_2, respectively. Then, either $\theta_1 < \theta_2$ as in Fig. 15–7a or $\theta_1 > \theta_2$ as in Fig. 15–7b. In order to specify one angle between L_1 and L_2, we adopt the following definition.

The angle θ from L_1 to L_2 is the positive angle less than 180° through which L_1 must be rotated in counterclockwise direction in order to coincide with L_2.

Now in Fig. 15–7a, $\theta = \theta_2 - \theta_1$ and

$$\tan \theta = \tan (\theta_2 - \theta_1) = \frac{\tan \theta_2 - \tan \theta_1}{1 + \tan \theta_1 \tan \theta_2}. \tag{1}$$

Since $\tan \theta_1$ is the slope m_1 of L_1, and $\tan \theta_2$ is the slope m_2 of L_2, we may write equation 1 as

$$\tan \theta = \frac{m_2 - m_1}{1 + m_1 m_2}. \tag{2}$$

In the case of Fig. 15–7b, $\theta_1 - \theta_2 + \theta = 180°$ whence $\theta = 180° + \theta_2 - \theta_1$. Then

$$\tan \theta = \tan [180° + (\theta_2 - \theta_1)] = \tan (\theta_2 - \theta_1) \tag{3}$$

by the reduction formulas in Sec. 4–11, and equation 2 holds as in the previous case.

Hence, *if L_1 and L_2 are two intersecting lines with slopes m_1 and m_2, respectively, the angle θ from L_1 to L_2 is given by*

$$\boldsymbol{\tan \theta} = \frac{\boldsymbol{m_2 - m_1}}{\boldsymbol{1 + m_1 m_2}}. \tag{3}$$

Formula 3 requires that neither m_1 nor m_2 be infinite. But this exceptional case occurs when L_1 or L_2 are parallel to the y axis, and θ can then be found without using formula 3, as will be shown in the example that follows.

Example 1.   Find the interior angles of the triangle whose vertices are $A(-3, 5)$, $B(3, 3)$, and $C(3, -2)$.
   The slope of AB is $\dfrac{5 - 3}{-3 - 3} = -\dfrac{1}{3}$, and the slope of AC is $-\dfrac{7}{6}$.
   The angle A is the angle from AC to AB, whence from formula 3

$$\tan A = \frac{\text{slope } AB - \text{slope } AC}{1 + (\text{slope } AB)(\text{slope } AC)}$$

$$= \frac{-\frac{1}{3} - (-\frac{7}{6})}{1 + (-\frac{1}{3})(-\frac{7}{6})} = \frac{3}{5} = 0.6000.$$

From Table 3,

$$A = 31.0°.$$

In this case BC is parallel to the y axis so that formula 3 cannot be used to find angles C and B. Now, if θ_1 and θ_2 denote the angles of inclination of AC and AB, respectively, from Fig. 15–8 it is obvious that

$$C = \theta_1 - 90°, \qquad B = 270° - \theta_2.$$

Now since

$$\tan \theta_1 = -\tfrac{7}{6} = -1.1667 \quad \text{and} \quad \tan \theta_2 = -\tfrac{1}{3} = -0.3333,$$

we have

$$\theta_1 = 130.6° \quad \text{and} \quad \theta_2 = 161.6°.$$

Hence

$$C = 40.6° \quad \text{and} \quad B = 108.4°.$$

As a check we have

$$31.0° + 40.6° + 108.4° = 180°.$$

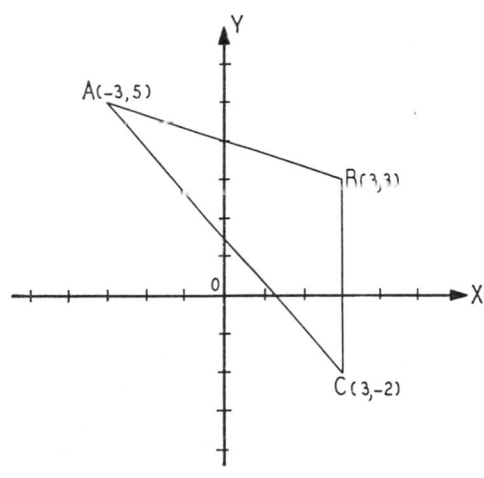

FIG. 15–8

Example 2. If L_1 has the slope $\frac{1}{3}$, find the slope of a line L_2 such that the angle from L_1 to L_2 is 45°.

If the slopes of L_1 and L_2 are m_1 and m_2, then $m_1 = \frac{1}{3}$, and

$$\tan 45° = \frac{m_2 - \frac{1}{3}}{1 + \frac{1}{3}m_2},$$

whence

$$1 = \frac{3m_2 - 1}{3 + m_2},$$

$$3 + m_2 = 3m_2 - 1,$$

$$m_2 = 2.$$

Exercises

Find the angle from L_1 to L_2 when the following data are given:

1. L_1 has the slope 1, L_2 has the slope 4.
2. L_1 has the slope 2, L_2 has the slope 3.
3. L_1 has the slope $\frac{1}{4}$, L_2 has the slope 2.
4. L_1 has the slope $-\frac{2}{3}$, L_2 has the slope 4.
5. L_1 has the slope $\frac{6}{5}$, L_2 has the slope -1.
6. L_1 has the slope 3, L_2 has the slope $-\frac{1}{2}$.
7. L_1 passes through $(1, 1)$ and $(4, 3)$; L_2 passes through $(0, 5)$ and $(2, -1)$.

8. L_1 passes through $(0, 0)$ and $(5, 3)$; L_2 passes through $(-2, 3)$ and $(4, 1)$.

9. L_1 passes through $(0, 3)$ and $(8, -4)$; L_2 passes through $(7, 2)$ and $(0, -4)$.

10. L_1 passes through $(-1, 3)$ and $(-4, -4)$; L_2 passes through $(-2, -7)$ and $(0, -5)$.

Find the interior angles of each of the triangles whose vertices are given.

11. $(-4, -5)$, $(-1, 1)$, $(0, -3)$.   **12.** $(-2, 2)$, $(-1, -1)$, $(1, 2)$.

13. $(-2, -1)$, $(4, -3)$, $(1, 2)$.   **14.** $(-3, 0)$, $(1, -1)$, $(0, 5)$.

15. $(-4, 4)$, $(-3, 1)$, $(6, 2)$.   **16.** $(2, 1)$, $(3, -2)$, $(9, 1)$.

17. $(-4, 0)$, $(4, -3)$, $(4, 3)$.   **18.** $(-2, 2)$, $(-4, 0)$, $(1, -3)$.

In each of the following problems find m_1, the slope of L_1, when m_2, the slope of L_2, and θ, the angle from L_1 to L_2, are given:

19. $m_2 = 3$, $\theta = 45°$.   **20.** $m_2 = \frac{1}{4}$, $\theta = 60°$.   **21.** $m_2 = 0$, $\theta = 135°$.

22. $m_2 = -1$, $\theta = 150°$.   **23.** $m_2 = -\frac{1}{2}$, $\theta = 60°$.   **24.** $m_2 = 5$, $\theta = 30°$.

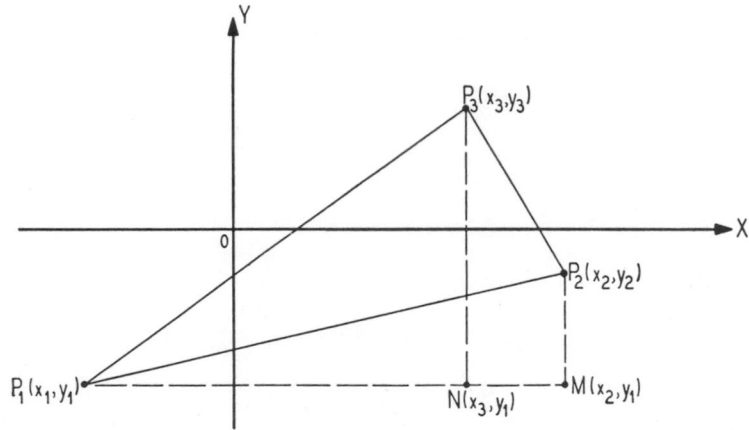

FIG. 15–9

15–6. The Area of a Triangle

Consider the triangle shown in Fig. 15–9 with vertices $P_1(x_1, y_1)$, $P_2(x_2, y_2)$ and $P_3(x_3, y_3)$. Start with P_1, which is the lowest vertex, and locate the points $M(x_2, y_1)$ and $N(x_3, y_1)$, and construct the dotted segments P_1N, NM, P_3N, and P_2M. Now the area $P_1P_2P_3$ equals the area P_1P_3N plus the area NP_3P_2M minus the area P_1P_2M. Since the area of a triangle is one-half the base times the altitude and the area of a trapezoid is one-half the sum of the parallel sides times the distance between those sides, we have

$$\text{Area } P_1P_3N = \tfrac{1}{2}(x_3 - x_1)(y_3 - y_1),$$

$$\text{Area } NP_3P_2M = \tfrac{1}{2}(x_2 - x_3)(y_2 - y_1 + y_3 - y_1),$$

$$\text{Area } P_1P_2M = \tfrac{1}{2}(x_2 - x_1)(y_2 - y_1).$$

Then

$$\text{Area } P_1P_2P_3 = \tfrac{1}{2}[(x_3 - x_1)(y_3 - y_1) + (x_2 - x_3)(y_2 + y_3 - 2y_1)$$
$$- (x_2 - x_1)(y_2 - y_1)]$$
$$= \tfrac{1}{2}[x_1(y_2 - y_3) + x_2(y_3 - y_1) + x_3(y_1 - y_2)]$$

which we may write as

$$\frac{1}{2}\left[x_1 \begin{vmatrix} y_2 & 1 \\ y_3 & 1 \end{vmatrix} - x_2 \begin{vmatrix} y_1 & 1 \\ y_3 & 1 \end{vmatrix} + x_3 \begin{vmatrix} y_1 & 1 \\ y_2 & 1 \end{vmatrix} \right].$$

Since this is obviously the expansion of a third-order determinant, we have

$$\text{Area } P_1P_2P_3 = \tfrac{1}{2} \begin{vmatrix} x_1 & y_1 & 1 \\ x_2 & y_2 & 1 \\ x_3 & y_3 & 1 \end{vmatrix}. \tag{1}$$

The proof given here has been based on Fig. 15–9. However, similar proofs cover all other cases, and the results obtained are the same as formula 1 with the possible difference of a sign. Thus for any triangle we may state the following:

The area A of a triangle given by its three vertices $P_1(x_1, y_1)$, $P_2(x_2, y_2)$, and $P_3(x_3, y_3)$ is given by the formula

$$A = \pm\tfrac{1}{2} \begin{vmatrix} x_1 & y_1 & 1 \\ x_2 & y_2 & 1 \\ x_3 & y_3 & 1 \end{vmatrix} \tag{2}$$

where the sign is chosen so that A is positive.

Example. Find the area of the triangle whose vertices are (3, 5), (−2, 2), and (1, −2).

By formula 2 we have

$$\frac{1}{2}\begin{vmatrix} 3 & 5 & 1 \\ -2 & 2 & 1 \\ 1 & -2 & 1 \end{vmatrix} = \frac{1}{2}\begin{vmatrix} 2 & 7 & 0 \\ -3 & 4 & 0 \\ 1 & -2 & 1 \end{vmatrix} = \frac{1}{2}\begin{vmatrix} 2 & 7 \\ -3 & 4 \end{vmatrix}$$
$$= \tfrac{1}{2}(8 + 21) = \tfrac{29}{2} = 14.5.$$

It follows that the area of the triangle is 14.5.

In this example the points were chosen for insertion in formula 2 in the order in which they are met by starting with (3, 5) and traversing the perimeter of the triangle in the counterclockwise direction (shown by the arrows in Fig. 15–10). It can be shown in general that whenever the points are chosen in this counterclockwise order the determinant in formula 2 is positive, and, further, that this is true, regardless of which point is used as a starting point. If the points are chosen in the opposite,

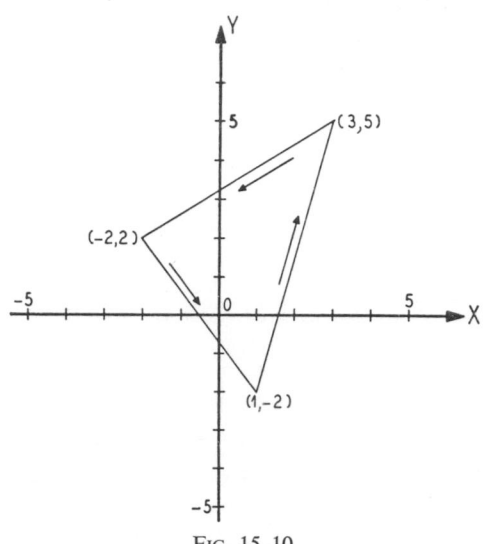

FIG. 15–10

or clockwise, order the determinant in formula 2 will be negative, regardless of which point is used as a starting point. The reader may easily verify that these general statements are correct for this example.

Exercises

Find the areas of the triangles whose vertices are given.

1. $(-2, 2)$, $(-2, -3)$, $(2, 1)$.   **2.** $(3, 5)$, $(5, 1)$, $(1, 3)$.

3. $(1, 1)$, $(7, 5)$, $(8, 6)$.   **4.** $(-7, -7)$, $(-4, -1)$, $(-1, -4)$.

5. $(-4, -1)$, $(2, 3)$, $(3, -4)$.   **6.** $(3, -2)$, $(7, 7)$, $(-1, 5)$.

7. $(0, 0)$, $(4, 2)$, $(5, -4)$.   **8.** $(-3, 0)$, $(0, -3)$, $(0, 3)$.

9. $(-3, -3)$, $(-1, 1)$, $(1, 5)$.   **10.** $(-1, 2)$, $(2, -3)$, $(5, -8)$.

Prove that the points in each exercise form the vertices of a parallelogram, and then find its area.

11. $(5, 1)$, $(1, 1)$, $(-1, 5)$, $(7, -3)$.   **12.** $(-4, 7)$, $(4, -1)$, $(-4, -1)$, $(4, 7)$.

13. $(-5, -6)$, $(2, 4)$, $(-4, 2)$, $(1, -4)$.   **14.** $(1, 5)$, $(-3, -1)$, $(5, 3)$, $(1, -3)$.

15. $(-3, -3)$, $(-4, -2)$, $(8, 1)$, $(9, 0)$.   **16.** $(0, 3)$, $(0, 7)$, $(1, 2)$, $(1, -2)$.

Form a quadrilateral by joining the points in the order given.   Then find the area of the quadrilateral.

17. $(1, 5), (0, 6), (-1, -1), (3, -3)$.   **18.** $(1, 3), (-5, 0), (-1, -9), (5, -1)$.

19. $(6, 0), (4, 3), (-2, 2), (-4, -4)$.   **20.** $(0, 2), (-5, -4), (9, -3), (3, 1)$.

21. $(0, 7), (-4, -1), (-3, -5), (3, -5)$.   **22.** $(1, 1), (4, 0), (2, 2), (-4, 0)$.

23. Find the area of the polygon formed by joining in order the points $(6, -4)$, $(4, -1), (0, 1), (-2, -4), (2, -1)$.

24. Find the area of the polygon formed by joining in order the points $(2, 3), (-3, 0)$, $(-7, 2), (-4, -3), (1, -1)$.

25. Find the area of the polygon formed by joining in order the points $(5, 1), (3, 5)$, $(0, 7), (-2, -1), (3, -3), (8, -1)$.

15-7. The Equation of a Straight Line

By definition, the equation of a straight line is an equation in x and y of the form $f(x, y) = 0$ which is satisfied by the coordinates of every point on the line and which is not satisfied by the coordinates of any point not on the line.

A line may be determined by giving its slope and a point through which it passes or by giving two points through which it passes.   In the next few sections we shall find the equations of lines given by these geometrical conditions.   When these geometrical conditions are stated in general literal notation, the resulting equations are called **standard forms.**   These standard forms will enable us to show that the points satisfying a first-degree equation form a straight line, and conversely.

15-8. The Point-Slope Form

Let L be a line passing through the point $P_1(x_1, y_1)$ and having a slope m. If $P(x, y)$ is any point on the line different from P_1, then the slope of PP_1 is the same as the slope of L, and we have

$$m = \frac{y - y_1}{x - x_1}, \qquad (1)$$

and, hence,

$$y - y_1 = m(x - x_1). \qquad (2)$$

Since equation 1 holds for any point on L different from $P_1(x_1, y_1)$, equation 2 does also.   In addition equation 2 is satisfied by (x_1, y_1), whence it holds for every point on L.

On the other hand, if (x, y) is not on L, equation 1 and therefore equation 2 does not hold.   Thus equation 2 is satisfied by the coordinates of the points on L and by the coordinates of no other points, and therefore it is the equation of L.

The line passing through $P_1(x_1, y_1)$ with slope m has the equation

$$y - y_1 = m(x - x_1). \qquad (2)$$

This equation is called the **point-slope** form. It enables us to write very readily the equation of a line given by its slope and a point through which it passes. Also, any equation of first degree that can be put into form 2 is an equation of a line whose slope is m and which passes through the point (x_1, y_1).

Example 1. Find an equation of the line through $(2, -3)$ with slope $\frac{1}{2}$. By equation 2 we have

$$y + 3 = \tfrac{1}{2}(x - 2)$$

which can be written

$$x - 2y - 8 = 0.$$

This is an equation of the given line.

Example 2. Show that $2x - y + 2 = 0$ is an equation of a straight line, and find the slope of the line.

The equation can be written as

$$y = 2x + 2$$

or

$$y - 0 = 2[x - (-1)].$$

It follows that the given equation is the equation of a straight line, and that the line passes through $(-1, 0)$ and has slope 2.

15–9. The Slope-Intercept Form

If a line intersects the x axis, the abscissa of the point of intersection is called the **x intercept** of the line. Similarly the **y intercept** is the ordinate of the point of intersection of the line and the y axis.

As a special case of the point-slope form, let the point be given by the y intercept b. Then the line passes through $(0, b)$, and has slope m and has the equation

$$y - b = m(x - 0),$$

or

$$y = mx + b.$$

Hence, *the line with slope m and y intercept b has the equation*

$$y = mx + b. \tag{1}$$

This equation is called the **slope-intercept** form. Since a line with a finite slope is not parallel to the y axis, it intersects the y axis, and therefore has a y intercept. Thus any line not parallel to the y axis has an equation of form 1. Also any first-degree equation whose y coefficient is not zero can be put in form 1 by solving for y. Thus we see that, *if a first-degree equation is solved for y, the coefficient of x is the slope of the line given by the equation.*

Example 1. Find the slope of the line whose equation is

$$3x - 2y + 6 = 0.$$

Solving the equation for y, we obtain

$$2y = 3x + 6, \qquad y = \tfrac{3}{2}x + 3$$

whence the line has slope $\tfrac{3}{2}$ and y intercept 3.

Example 2. Find the equation of the straight line through (3, 4) and perpendicular to $3x - 2y + 6 = 0$.

The equation of the line can be written as $y = \tfrac{3}{2}x + 3$, whence its slope is $\tfrac{3}{2}$. Then the slope of any perpendicular line is $-\tfrac{2}{3}$. Using the point-slope form, the required line has the equation

$$(y - 4) = -\tfrac{2}{3}(x - 3)$$

or

$$2x + 3y - 18 = 0.$$

15–10 Lines Parallel to the Axes

Since a line parallel to the y axis has no finite slope, neither of the preceding forms of the equation of a line apply. However, it is obvious that any line parallel to the y axis has an equation of the form

$$x = a \qquad (1)$$

where a is a constant and equal to the abscissa of any point on the line. Thus the line through (5, −3) and parallel to the y axis is $x = 5$.

Any line parallel to the x axis has slope zero and hence has the form

$$y = b \qquad (2)$$

where b is the y intercept of the line. Thus b is the ordinate of any point on the line. For example, the line through (5, −3) and parallel to the x axis is $y = -3$.

15–11. The General Linear Equation

As we have just seen, every line not parallel to the y axis has an equation of the form $y = mx + b$. A line parallel to the y axis has an equation of the form $x = a$. Hence the following theorem holds:

Every straight line has an equation of first degree.

The converse theorem is also true.

Every equation of first degree is satisfied by the coordinates of the points on a straight line.

To prove this statement we start with the general equation of first degree:

$$Ax + By + C = 0.$$

Since equation 1 would be trivial if A and B were both zero, we assume that A and B are not both zero. We shall consider two cases in connection with equation 1, (*a*) when $B = 0$, and (*b*) when $B \neq 0$.

(a) If $B = 0$, then $A \neq 0$ by hypothesis, and we have $Ax + C = 0$ which can be written as

$$x = -\frac{C}{A},$$

which is the equation of a line parallel to the y axis.

(b) If $B \neq 0$, we can solve equation 1 for y and obtain

$$y = -\frac{A}{B}x - \frac{C}{B}.$$

This is the equation of a line with slope $-A/B$ and y intercept $-C/B$.

Exercises

Find an equation of the line passing through the given point with the given slope.

1. $(4, 3), m = \frac{1}{4}$.   **2.** $(5, 7), m = \frac{3}{5}$.   **3.** $(2, 6), m = -\frac{1}{4}$.

4. $(4, -3), m = -\frac{1}{3}$.   **5.** $(-5, -1), m = -5$.   **6.** $(-2, 2), m = 2$.

7. $(7, 9), m = 3$.   **8.** $(6, 1), m = -4$.   **9.** $(-7, -5), m = 0$.

10. $(-8, -8), m = -\frac{3}{4}$.

Find the slope-intercept form of the equation of the line whose slope m and y intercept b are given.

11. $m = \frac{1}{3}, b = 2$.   **12.** $m = \frac{2}{5}, b = -3$.   **13.** $m = -3, b = 5$.

14. $m = -\frac{3}{4}, b = 6$.   **15.** $m = -\frac{2}{5}, b = -1$.   **16.** $m = 6, b = -3$.

17. $m = 8, b = 0$.   **18.** $m = -5, b = -2$.   **19.** $m = -4, b = -2$.

20. $m = -1, b = 4$.

Find an equation of the line passing through the given point with the given angle of inclination θ.

21. $(3, 3), \theta = 45°$.   **22.** $(2, -5), \theta = 135°$.   **23.** $(-1, 4), \theta = 30°$.

24. $(-3, 4), \theta = 60°$.   **25.** $(0, 2), \theta = 120°$.   **26.** $(-7, -1), \theta = 150°$.

Find the slope-intercept form of the equation of the line whose y intercept b and angle of inclination θ are given.

27. $b = 2, \theta = 45°$.   **28,** $b = 7, \theta = 135°$.   **29.** $b = -3, \theta = 30°$.

30. $b = -5, \theta = 60°$.   **31.** $b = 0, \theta = 150°$.   **32.** $b = 4, \theta = 120°$.

Find the slope and y intercept of each of the following lines:

33. $x + y = 4$.   **34.** $4x + 3y = 24$.   **35.** $5x + 4y - 20 = 0$.

36. $3x + 7y - 21 = 0$.   **37.** $x - 2y + 8 = 0$.   **38.** $y - 6x + 12 = 0$.

39. $4x - y + 4 = 0$.   **40.** $9x + 7y + 18 = 0$.   **41.** $3x - 5y - 1 = 0$.

42. $y - 3x - 7 = 0$.   **43.** $5x - 5y - 14 = 0$.   **44.** $5y - 6 = 0$.

Find an equation of the line through the given point (a) parallel and (b) perpendicular to the given line.

45. $(3, 1), x - y = 4$.   **46.** $(0, 6), x + 3y = 2$.   **47.** $(5, -4), 2x + y = 1$.

48. $(-2, 5), 2x - 5y = 10.$ **49.** $(-2, -2), 3x - 2y = 12.$ **50.** $(7, -4), 4x - y = 8.$

51. $(-4, 3), x = 2y + 6.$ **52.** $(-3, -6), y = 3x + 5.$

53. Find the equation of the y axis.

54. Find the equation of the x axis.

55. Find the equation of the line through $(7, -3)$ parallel to the y axis.

56. Find the equation of the line through $(7, -3)$ parallel to the x axis.

57. Find the equation of the line through $(2, -5)$ perpendicular to the y axis.

58. Find the equation of the line through $(2, -5)$ perpendicular to the x axis.

Find the angle from L_1 to L_2.

59. L_1: $3x - 3y = 4,$ **60.** L_1: $3x - 2y = 5,$ **61.** L_1: $4x - 2y = 7,$

L_2: $2x - y = 6.$ L_2: $5x - y = 8.$ L_2: $3x + y = 2.$

62. L_1: $2x - 3y = 12,$ **63.** L_1: $4x + 5y = 20,$ **64.** L_1: $x + 2y = 6,$

L_2: $5x + 3y = 15.$ L_2: $5x - 4y = 10.$ L_2: $2x + y = 4.$

15–12. The Two-Point Form

If two points $P_1(x_1, y_1)$ and $P_2(x_2, y_2)$ are given on a line not parallel to the y axis, the slope of the line is

$$m = \frac{y_2 - y_1}{x_2 - x_1}.$$

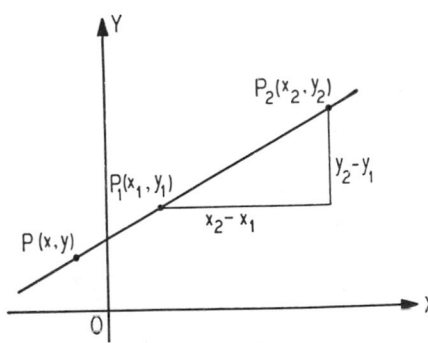

FIG. 15–11

Substituting this for m in the point-slope form, we have

$$y - y_1 = \frac{y_2 - y_1}{x_2 - x_1}(x - x_1). \tag{1}$$

An equation of the straight line passing through $P_1(x_1, y_1)$ and $P_2(x_2, y_2)$ is given by (1). This is called the **two-point form** of the equation of a line. Writing equation 1 as

$$(y - y_1)(x_2 - x_1) = (y_2 - y_1)(x - x_1),$$

we have a second form of the equation of the line. This form has no exceptions since it is also valid in the case $x_1 = x_2$.

The reader may easily verify that equation 1 can be written as a determinant

$$\begin{vmatrix} x & y & 1 \\ x_1 & y_1 & 1 \\ x_2 & y_2 & 1 \end{vmatrix} = 0. \tag{2}$$

This determinant is not so useful as equation 1 for actually finding an equation of a line, but it is sometimes convenient in theoretical discussions.

Example. Find an equation of the line through $(3, 4)$ and $(-1, 2)$. From equation 1 we have

$$(y - 4) = \frac{(2 - 4)}{(-1 - 3)}(x - 3)$$

or

$$x - 2y + 5 = 0.$$

Exercises

Find an equation of the line passing through each of the following pairs of points:

1. $(0, 0)$, $(3, 4)$.   **2.** $(2, 1)$, $(5, 2)$.   **3.** $(2, -1)$, $(3, -2)$.

4. $(4, 3)$, $(-1, 9)$.   **5.** $(0, 2)$, $(3, 0)$.   **6.** $(-4, 0)$, $(5, 5)$.

7. $(1, -3)$, $(-2, 3)$.   **8.** $(-5, -1)$, $(4, 4)$.   **9.** $(-6, -2)$, $(-1, 3)$.

10. $(-3, 2)$, $(1, 5)$.   **11.** $(2, -2)$, $(-2, 2)$.   **12.** $(0, 0)$, $(-4, 3)$.

13. $(9, -3)$, $(1, -2)$.   **14.** $(7, -4)$, $(-3, -4)$.   **15.** $(3, 1)$, $(5, 5)$.

Find the equations of the medians of the triangles whose vertices are as follows:

16. $(1, 1)$, $(3, 7)$, $(7, 4)$.   **17.** $(1, -2)$, $(3, 6)$, $(7, -4)$.

18. $(4, 0)$, $(-1, 5)$, $(-2, -3)$.   **19.** $(5, -1)$, $(3, -3)$, $(-2, 7)$.

20. $(1, 3)$, $(-1, -3)$, $(-3, -2)$.   **21.** $(8, 2)$, $(-8, -4)$, $(-2, -6)$.

22. Find the equation of the line through $(2, 3)$ parallel to the line through $(5, 5)$ and $(1, -7)$.

23. Find the equation of the line through $(2, -5)$ parallel to the line through $(-3, 2)$ and $(-1, -2)$.

24. Find the equation of the line through $(2, 3)$ perpendicular to the line through $(5, 5)$ and $(1, -7)$.

25. Find the equation of the line through $(2, -5)$ perpendicular to the line through $(5, 5)$ and $(1, -7)$.

26. Show that $(1, 3)$, $(5, 5)$, and $(7, 6)$ lie on a straight line.

27. Show that $(5, -1)$, $(2, -2)$, and $(-7, -5)$ lie on a straight line.

28. Show that $(9, 8)$, $(7, 5)$, and $(-3, -10)$ lie on a straight line.

15–13. The Intercept Form

One of the easiest ways of plotting the line corresponding to a given equation is to find its intercepts on the axes. We shall now use the intercepts of a line to find its equation.

Let a be the x intercept and b the y intercept (Fig. 15–12), neither of which is zero. Since the line thus contains the points $(a, 0)$ and $(0, b)$, its equation, found from the two-point form, is

$$y - 0 = -\frac{b}{a}(x - a),$$

or

$$\frac{x}{a} + \frac{y}{b} = 1. \tag{1}$$

This is called the **intercept form** of the equation of a line.

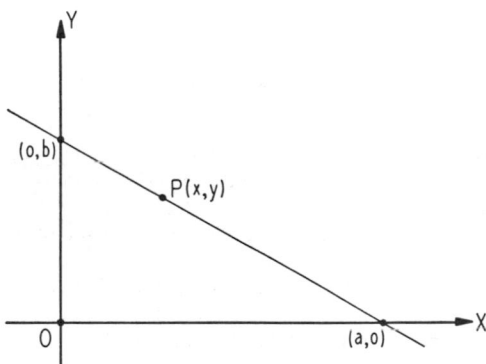

FIG. 15–12

An equation of the form $Ax + By + C = 0$ where A, B, and C are not 0, can be reduced to the intercept form after dividing by $-C$, obtaining

$$\left(-\frac{A}{C}\right)x + \left(-\frac{B}{C}\right)y = 1$$

whence

$$a = -\frac{C}{A}, \qquad b = -\frac{C}{B}.$$

If A, B, or C is 0, the equation cannot be written in intercept form. (Why?)

Example 1. Find the equation of the line with x intercept 5 and y intercept -6.
From equation 1, $a = 5$ and $b = -6$, whence the equation of the line is

$$\frac{x}{5} + \frac{y}{-6} = 1,$$

which we may write as

$$6x - 5y = 30.$$

Example 2. Find the intercepts of the line $2x - 3y = 6$.
Dividing by 6 to make the right member unity, we obtain

$$\frac{x}{3} + \frac{y}{-2} = 1,$$

whence the x intercept is 3 and the y intercept is -2.

Exercises

Find an equation of each of the lines determined by the following data:

1. x intercept 5, y intercept 3.   2. x intercept 4, y intercept -1.

3. x intercept -2, y intercept 8.   4. x intercept -5, y intercept -6.

5. x intercept -3, y intercept -5.   6. x intercept -2, y intercept 7.

7. x intercept 4, y intercept 1.   8. x intercept -1, y intercept 4.

9. x intercept 8, y intercept -3.   10. x intercept 5, y intercept 5.

Find the intercepts of the following lines by reducing each of them to form 1:

11. $3x - 5y - 15 = 0$.   12. $3x + 7y - 21 = 0$.   13. $2x - 5y = 10$.

14. $5x + 4y = 10$.   15. $4x + 3y = 24$.   16. $x - 6y = 9$.

17. $4x - 12y = 18$.   18. $6x - 8y = 12$.   19. $3x - y + 60 = 0$.

20. $2y - 5x - 10 = 0$.   21. $7y + x - 9 = 0$.   22. $y - x - 4 = 0$.

23. Find the equation of the line passing through $(5, 3)$ having equal intercepts.
24. Find the equations of the lines passing through $(2, 4)$ and forming with the axes a triangle of area 18.
25. Find the equations of the lines passing through $(3, 4)$ and forming with the axes a triangle of area 25.
26. Find the equation of the line passing through $(7, 5)$ having equal intercepts.
27. Find the equation of the line passing through $(6, 14)$ having equal intercepts.
28. Find the equations of the lines passing through $(4, 4)$ and forming with the axes a triangle of area 36.

15–14. Distance from a Point to a Line

Let a given line L have the equation $Ax + By + C = 0$, and let a given point P_1 have coordinates (x_1, y_1). Then the distance d from P_1 to L (see Fig. 15–13) can be shown to be given by the formula

$$d = \left| \frac{Ax_1 + By_1 + C}{\sqrt{A^2 + B^2}} \right|, \tag{1}$$

the absolute value being indicated because length is a positive number by definition. The derivation of this formula is left as an exercise for the student; however, exercise 14 below gives an outline of the steps in the derivation.

Example 1. Find the distance from the point $(-2, 4)$ to the line $2x - 3y - 5 = 0$.
Substituting in equation 1, we obtain

$$d = \left| \frac{2(-2) - 3(4) - 5}{\sqrt{4 + 9}} \right| = \left| \frac{-21}{\sqrt{13}} \right| = \frac{21}{\sqrt{13}} = 5.82.$$

Example 2. Find the length of the altitude drawn from the vertex A in the triangle whose vertices are $A(1, 4)$, $B(-2, -3)$, $C(3, -1)$.

In order to find, by using (1), the distance from the line BC to the point A, we must have the equation of BC. Using equation 1 in Sec. 15–12, we get that the equation of BC is $2x - 5y - 11 = 0$. Hence substituting in equation 1 of this section we get

$$\text{Altitude} = \left| \frac{2(1) - 5(4) - 11}{\sqrt{4 + 25}} \right| = \left| \frac{-29}{\sqrt{29}} \right| = \frac{29}{\sqrt{29}} = \sqrt{29} = 5.39.$$

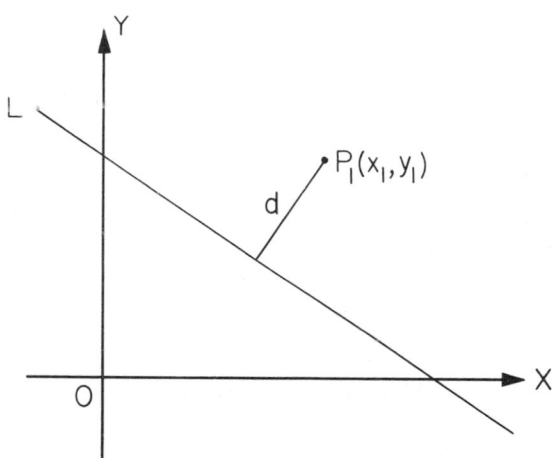

FIG. 5–13

Exercises

In each of the following examples find the distance of the given point from the given line:

1. $(1, 2)$, $2x - y + 10 = 0$.
2. $(0, 0)$, $7x + 3y - 5 = 0$.
3. $(-1, -3)$, $x + 3y - 2 = 0$.
4. $(-3, 5)$, $2x - y = 0$.
5. $(0, 2)$, $4x + y - 10 = 0$.
6. $(3, 0)$, $x - y - 1 = 0$.

7. Find the lengths of the three altitudes of the triangle whose vertices are the points $(0, 0)$, $(2, 4)$, $(-3, 3)$.

8. Find the lengths of the three altitudes of the triangle whose vertices are the points $(1, 2)$, $(-2, 4)$, $(-3, -6)$.

9. Find the distance between the two parallel lines whose equations are $3x + y + 5 = 0$ and $6x + 2y - 7 = 0$.

10. Find the distance between the two parallel lines whose equations are $x - 2y = 0$ and $3x - 6y + 10 = 0$.

11. How far is the point $(3, -2)$ from the line that passes through the origin with an angle of inclination of $45°$?

12. How far is the point $(1, 1)$ from the line that passes through the point $(-1, -2)$ and has slope 3?

13. Using the results of this section and the formula of Sec. 15–1, find the areas of the triangles whose vertices are:

(*a*) (0, 3), (5, 1), (2, 4).   (*b*) (0, 0), (8, −3), (2, 4).

(*c*) (−4, 3), (1, 2), (1, −3).   (*d*) (5, 5), (−10, −3), (4, −2).

Check your work by using equation 2 of Sec. 15–6.

14. Derive formula 1 of this section. This derivation may be carried out as follows: Consider the line L and the point P_1 in the positions shown in Fig. 15–13. First draw the perpendicular from the origin to L, and find its length; this length is the distance from the origin to L. Second, show that the equation of the line L_1 through P_1 parallel to L is $A(x - x_1) + B(y - y_1) = 0$. Third, find the distance from the origin to L_1 by means of the result derived in the first step. Fourth, the formula for the distance d can now be found as the difference of the distances found in the first and third steps. The final step is to verify that the formula holds, no matter what positions are taken by P_1 and L with respect to each other and to the coordinate axes.

15–15. Intersection of Two Lines

The coordinates of the point of intersection of two nonparallel lines satisfy the equations of both lines. The coordinates of this point of intersection are the values of x and y obtained by solving the two equations simultaneously. This has been done algebraically and graphically in Chapters 3 and 13.

Miscellaneous Exercises

1. Find the equation of the line through (4, −1) parallel to the line through (−2, −2) and (−1, 4).

2. Find the coordinates of the vertices of the triangle the equations of whose sides are $2x - 3y + 8 = 0$, $x + 8y + 4 = 0$, and $5x + 2y - 18 = 0$. Also find the lengths of the three altitudes of this triangle.

3. A diagonal of a square joins the vertices (3, 2) and (4, 9). Find the coordinates of the other two vertices.

4. Find the distance from the origin to the point of intersection of the lines $x - 2y + 7 = 0$ and $2x - y - 1 = 0$.

5. Find the distance from the point (3, −3) to the point of intersection of the lines $7x + y - 16 = 0$ and $3x + y - 2 = 0$.

6. Write the equations of the sides of the triangle whose vertices are (8, 8), (−8, 8), and (0, 0).

7. Show that the lines $x - 4y + 13 = 0$, $2x + 3y - 18 = 0$, and $7x - 2y - 13 = 0$ intersect in a common point.

8. Find the equation of the line perpendicular to $4x - y - 13 = 0$ and bisecting the segment joining (−3, 1) and (1, 11).

9. Find the equation of the perpendicular bisector of the segment whose ends are (−3, −5) and (−1, 7).

10. Find the equation of the line through the point (−6, 1) bisecting the segment whose ends are (−2, 5) and (8, 0).

11. Find the point of intersection of the two diagonals of the rectangle whose vertices are (5, −2), (5, 4), (−7, 4), and (−7, −2).

12. Find the area of the triangle whose sides are formed by the lines $3x - 2y = 0$, $2x + 3y = 0$, and $5x + y - 13 = 0$.

13. Find the area of the triangle whose sides are formed by the lines $x + 3y - 12 = 0$, $3x - y + 14 = 0$, and $x - y - 4 = 0$.

14. Show that the segments joining the two points $(7, 3)$ and $(6, -14)$ to the origin subtend a right angle at the origin.

15. Find the equation of the line whose intercepts are twice those of the line $3x + 5y - 15 = 0$.

16. Find the equation of the line through the point $(-1, -2)$ and parallel to the line $4x - y + 3 = 0$.

17. Find the coordinates of the foot of the perpendicular from the origin to the line whose equation is $x + 3y - 10 = 0$.

18. Find the coordinates of the foot of the perpendicular from the point $(-3, -2)$ to the line whose equation is $x - 5y + 19 = 0$.

19. Find the equation of the line through the origin whose slope is twice that of the line $5x - y - 3 = 0$.

20. Determine p so that the line $4x - py = 8$ shall be parallel to the line $4x - 5y = 6$.

21. Find the form of formula 2 in Sec. 15–6 when one of the vertices, say (x_1, y_1), is at the origin.

22. Find the point on the x axis that is equidistant from the points $(5, 3)$ and $(-3, 5)$.

23. The vertices of a triangle are $(3, -2)$, $(5, 2)$, and $(-5, -2)$.   Find:

(a) The equations of the sides.

(b) The equations of the lines through the vertices parallel to the opposite sides.

(c) The equations of the perpendicular bisectors of the sides.

(d) The equations of the lines through the vertices perpendicular to the opposite sides.

(e) The area of the triangle.

(f) The lengths of the three altitudes.

24. For the triangle whose sides are $7x - 4y - 15 = 0$, $2x - 9y + 35 = 0$, $x + y + 1 = 0$ find:

(a) The coordinates of the vertices.

(b) The equations of the lines through the vertices perpendicular to the opposite sides.

(c) The equations of the lines through the vertices parallel to the opposite sides.

(d) The equations of the perpendicular bisectors of the sides.

(e) The area of the triangle.

(f) The lengths of the three altitudes.

25. Find the equation of the line that makes equal intercepts on the axes and passes through the point $(-4, 3)$.

26. Find the equation of the line whose y intercept is four times its x intercept and which passes through the point $(5, 4)$.

27. Find the slopes of the following straight lines:

$$(a)\ \begin{vmatrix} x & y & 1 \\ 3 & 7 & 1 \\ -2 & 4 & 1 \end{vmatrix} = 0, \qquad (b)\ \begin{vmatrix} x & y & 1 \\ 0 & 6 & 1 \\ -5 & 0 & 1 \end{vmatrix} = 0.$$

28. A linear relation exists between the Fahrenheit and the centigrade temperature scales.   The temperature of $32°$ F is equivalent to $0°$ C and a temperature of $212°$ F is equivalent to $100°$ C.   Find (a) the equation of the line that shows the relation between the scales, and (b) the slope of the line, if the centigrade scale is taken along the horizontal axis and the Fahrenheit scale along the vertical axis.

29. The deflection d in inches of a spring is a linear function of the loading w in pounds. Two observations were made in the laboratory on the performance of a spring mounted in the chassis of an automobile. At loadings in addition to the weight of the automobile of 400 and 550 lb the deflections of the spring were 0.70 and 0.85 in., respectively. Find the formula for d. What is the initial deflection of the spring due to the weight of the automobile chassis?

30. The velocity v_t in feet per second after a time t in seconds of a uniformly accelerated body which has an initial velocity of v_0 ft per sec is given by the equation $v_t = v_0 + \alpha t$, where α is the acceleration in feet per second per second. In a specific case a body with an initial velocity of 18 ft per sec was uniformly accelerated at the rate of 3 ft per sec per sec. In another case the initial velocity was 36 ft per sec, and the acceleration was 6 ft per sec per sec. At what time are the two velocities equal?

31. The resistance of Nichrome wire changes linearly with temperature. Two readings were taken in the laboratory. At a temperature of $100°$ C the resistance of the sample was 24.8 ohms; at $200°$ C the resistance was 32.4 ohms. What is the resistance of the sample at $0°$ C?

32. The voltage E of a storage-battery cell varies with the concentration Z in grams of sulfuric acid per liter of electrolyte throughout a given range according to the equation $E = 1.850 + 0.00057Z$. Determine the voltage increase when the acid concentration changes from 1.1 to 1.5 kilograms per liter of electrolyte. What is the slope of the line given by the equation?

Progress Report

This chapter was devoted to the study of those parts of geometry that involve the straight line. In this study we have translated certain geometrical concepts into algebra. In doing so, a geometry-algebra dictionary has been created in which the geometrical concepts of point, distance, angle, line, etc., are written as algebraic symbols, expressions, and equations. Anyone who works in the field of mathematics and its applications should be well acquainted with the results of this chapter. We shall therefore conclude with a summary of the principal results in the form of a geometry-algebra dictionary.

GEOMETRY-ALGEBRA DICTIONARY

Geometry	Algebra
A point	(x, y)
Distance between two points	$\sqrt{(x_2 - x_1)^2 + (y_2 - y_1)^2}$
Midpoint of a line segment	$\left(\dfrac{x_1 + x_2}{2}, \dfrac{y_1 + y_2}{2}\right)$
Slope of a line segment	$\dfrac{y_2 - y_1}{x_2 - x_1} = \tan \theta = m$
Parallel lines	Slopes are equals: $m_1 = m_2$
Perpendicular lines	Slopes are negative reciprocals: $m_2 = -\dfrac{1}{m_1}$

Geometry	Algebra
Angle between two intersecting lines	$\tan\theta = \dfrac{m_2 - m_1}{1 + m_1 m_2}$
Line parallel to the x axis	$y = b$
Line parallel to the y axis	$x = a$
Line through a given point and with a given slope	$y - y_1 = m(x - x_1)$
Line with a given slope and a given y intercept	$y = mx + b$
Line through two given points	$\dfrac{y - y_1}{x - x_1} = \dfrac{y_2 - y_1}{x_2 - x_1}$
Line with given intercepts	$\dfrac{x}{a} + \dfrac{y}{b} = 1$
Any straight line	$Ax + By + C = 0$
Point of intersection of two lines	Solution of two equations
Distance from a point to a line	$\left\| \dfrac{Ax_1 + By_1 + C}{\sqrt{A^2 + B^2}} \right\|$

16

Circles and Loci

In this·chapter the method of studying geometry by means of algebra will be applied to the circle. A standard form for the equation of a circle similar to the forms derived in the last chapter for the line will be derived and used in solving a variety of problems involving circles.

A circle may be defined as a set of points at a given distance from a fixed point called the center. A translation of this requirement into algebraic terms leads to an equation which is satisfied by the coordinates of the points which lie on the circle. The algebraic method provides in this way a powerful tool for studying sets of points specified by geometrical properties of various kinds, as we shall see in this chapter. We shall discuss not only how geometric properties are translated into algebraic terms but also how the geometric representations of various algebraic equations can be arrived at systematically.

The chapter closes with a discussion of how graphs are drawn on a polar coordinate system, a system of coordinates associated with the circle.

16–1. The Circle

From his study of geometry the student knows that a circle is a figure consisting of all the points in a plane that are at a constant distance from a fixed point. The fixed point is called the center of the circle, and the constant distance is called the radius. To obtain an algebraic representation of the circle we shall translate this geometrical definition of the circle into algebra.

Let $C(h, k)$ be the fixed point, and let r be the radius (Fig. 16–1). If $P(x, y)$ is any point of the circle, then from the formula for the distance between two points we have

$$\sqrt{(x - h)^2 + (y - k)^2} = r, \tag{1}$$

which becomes, after squaring,

$$(x - h)^2 + (y - k)^2 = r^2. \tag{2}$$

Thus, if a point P is on the circle, its coordinates satisfy equation 2. Equation 2 is called the **standard form** of the equation of the circle with center at (h, k) and radius r.

If the center of the circle is at the origin, then h and k are both zero, and equation 2 reduces to

$$x^2 + y^2 = r^2. \tag{3}$$

Example 1.   Find the equation of the circle with center at $(-2, 1)$ and radius 3. We have

$$h = -2, \qquad k = 1, \qquad r = 3.$$

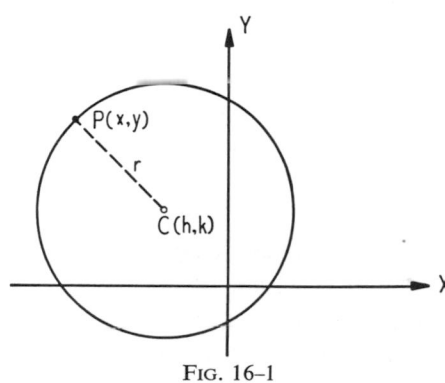

FIG. 16–1

Therefore the equation of the required circle is

$$(x + 2)^2 + (y - 1)^2 = 9,$$

or

$$x^2 + y^2 + 4x - 2y - 4 = 0.$$

Example 2.   Find the equation of the circle with center $(2, -1)$ which is tangent to the line $3x - 2y - 12 = 0$.

The radius is equal to the distance of the center $(2, -1)$ from the line.   By equation 1 of Sec. 15–14, the radius is therefore

$$r = \left| \frac{3 \cdot 2 - 2(-1) - 12}{\sqrt{9 + 4}} \right| = \frac{4}{\sqrt{13}}.$$

Therefore the equation of the required circle is

$$(x - 2)^2 + (y + 1)^2 = \tfrac{16}{13},$$

or

$$13x^2 + 13y^2 - 52x + 26y + 49 = 0.$$

Exercises

Find the equations of the circles satisfying the given conditions.

1. Center at $(4, 3)$, radius $= 6$.

2. Center at $(1, 2)$, radius $= 5$.

3. Center at $(0, 0)$, radius $= 7$.

4. Center at $(1, 0)$, radius $= 4$.

5. Center at $(0, 3)$, radius $= 8$.

6. Center at $(-1, 4)$, radius $= 3$.

7. Center at $(2, -5)$, radius $= \sqrt{11}$.

8. Center at $(-3, -2)$, radius $= 1$.

9. Center at (a, a), radius $= a$.

10. Center at (a, a), radius $= b$.

11. Center at $(a, -a)$, radius $= a$.

12. Center at $(-a, a)$, radius $= a$.

13. Center at $(0, a)$, radius $= 2a$.

14. Center at (c, c), radius $= c\sqrt{2}$.

15. Center at $(3, 0)$, and passing through the origin.

16. Center at $(2, -1)$, and passing through the origin.

17. Center at (a, b), and passing through the origin.

18. Center at $(3, 2)$, and tangent to the x axis.

19. Center at $(-2, 4)$, and tangent to the y axis.

20. Center at $(1, 1)$, and tangent to both axes.

21. Center at (a, a), and tangent to both axes.

22. Center at $(0, 0)$, and tangent to the line $x + 2y - 10 = 0$.

23. Center at $(1, 2)$, and tangent to the line $2x - y + 6 = 0$.

24. Center at $(-5, -3)$, and tangent to the line $x = y$.

25. Center at $(-1, 3)$, and tangent to the line $y = \frac{1}{2}x$.

26. Center at $(-3, 4)$, and passing through the point $(2, 1)$.

27. Center at $(2, -5)$, and passing through the point $(3, 0)$.

28. Center at $(-1, -2)$, and passing through the point $(3, 4)$.

29. Diameter the segment from $(4, 2)$ to $(2, 6)$.

30. Diameter the segment from $(-1, 3)$ to $(4, 0)$.

31. Diameter the segment from $(0, 0)$ to $(2a, 2a)$.

32. Touching the x axis at the origin and radius $= 3$ (two solutions).

33. Touching the y axis at the origin and radius $= 2$ (two solutions).

34. Touching both axes, radius $= 1$ (four solutions).

35. Center on the line $3x + 5y = 16$ and tangent to both axes (two solutions).

36. Center at the point of intersection of the lines $2x + 3y = 12$ and $2x - 3y = 18$, and passing through the point $(-1, 5)$.

37. Center at the point of intersection of the lines $x - 2y - 4 = 0$ and $3x + 2y - 4 = 0$, and passing through the point $(-2, 4)$.

38. Center at the point of intersection of the lines $4x - y = 3$ and $2x + y = 3$, and radius $= 5$.

39. Center on the x axis and passing through $(6, 4)$ and $(2, -2)$.

40. Center on the y axis and passing through $(4, 0)$ and $(-2, -6)$.

41. Center on the line $x - y + 1 = 0$ and passing through $(2, 1)$ and $(4, 3)$.

16–2. Reduction of the General Equation of the Circle to Standard Form

The standard form of the equation of the circle, given in equation 2 of the previous section, when expanded is an equation of the form

$$x^2 + y^2 + Dx + Ey + F = 0, \tag{1}$$

where D, E, and F are constants.

The question now arises: Is every equation of the form 1 the equation of a circle? To answer this question we complete the squares in x and y, so that equation 1 becomes

$$x^2 + Dx + \frac{D^2}{4} + y^2 + Ey + \frac{E^2}{4} = \frac{D^2}{4} + \frac{E^2}{4} - F$$

or

$$\left(x + \frac{D}{2}\right)^2 + \left(y + \frac{E}{2}\right)^2 = \frac{1}{4}(D^2 + E^2 - 4F). \qquad (2)$$

This equation has the standard form of a circle with center at $(-D/2, -E/2)$ and radius $\frac{1}{2}\sqrt{D^2 + E^2 - 4F}$, provided the right member of equation 2 is positive. If the right member of equation 2 is negative, no real numbers x, y can satisfy the equation, since the sum of the squares of two real numbers cannot equal a negative number. Therefore in this case there is no circle. If the right number of equation 2 is zero, the center is the only point whose coordinates satisfy the equation, and hence the circle is a single point.

Consider now the equation

$$Ax^2 + Ay^2 + Dx + Ey + F = 0. \qquad (3)$$

When this equation is divided by A, it reduces to the form of equation 1. But division of both members of an equation by a constant does not affect the geometrical figure it represents. Therefore equation 3 is also a general equation of a circle.

Example. Find the center and the radius, and draw the circle

$$x^2 + y^2 - 2x - 4y - 20 = 0.$$

The equation may be written

$$(x^2 - 2x) + (y^2 - 4y) = 20.$$

Completing the square in each set of parentheses, we have

$$(x^2 - 2x + 1) + (y^2 - 4y + 4) = 20 + 1 + 4$$

or

$$(x - 1)^2 + (y - 2)^2 = 25.$$

From this equation it follows that the center of the circle is at the point $(1, 2)$ and the radius is 5 (Fig. 16–2).

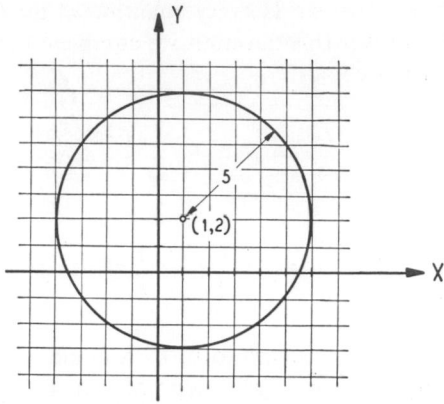

FIG. 16–2.   The graph of $x^2 + y^2 - 2x - 4y - 20 = 0$.

Exercises

By reducing each equation to standard form, determine whether or not it is the equation of a circle. If it is, find the center and radius of the circle, and draw the circle.

1. $x^2 + y^2 + 4x - 2y + 5 = 0$.   **2.** $x^2 + y^2 - 8x + 4y = 44$.

3. $x^2 + y^2 + 16y + 28 = 0$.   **4.** $x^2 + y^2 + 25 = 0$.

5. $x^2 + y^2 + 1 = 0$.   **6.** $x^2 + y^2 - 8x = 9$.

7. $x^2 + y^2 + 2x + 4y + 5 = 0$.   **8.** $x^2 + y^2 - 4x + 4 = 0$.

9. $2x^2 + 2y^2 - 6x - 10y = 1$.   **10.** $x^2 + y^2 - 6x - 8y + 16 = 0$.

11. $x^2 + y^2 + 2x + 6y + 11 = 0$.   **12.** $3x^2 + 3y^2 + 8x - 4y + 15 = 0$.

13. $x^2 + y^2 + 6x = 8y - 9$.   **14.** $36x^2 + 36y^2 - 36x + 24y + 13 = 0$.

15. $3x^2 + 3y^2 + 5x + 12y = 0$.   **16.** $x^2 + y^2 = 4x - 6y - 23$.

17. $4x^2 + 4y^2 = 8x - 16y + 5$.   **18.** $x^2 + y^2 - ax - ay = 0$.

19. If the circle $x^2 + y^2 - 2x + 4y + 6 + k = 0$ passes through the origin, find the value of k.

20. Find the value of a for which the circle $x^2 + y^2 + 2x - 4y + a = 0$ reduces to a point.

16–3. The Circle Satisfying Three Conditions

It was shown in the preceding section that every circle has an equation of the form

$$x^2 + y^2 + Dx + Ey + F = 0, \tag{1}$$

where D, E, and F are constants. Thus, if a circle is given by certain conditions, its equation can be found by determining the proper values of D, E, and F in equation 1.

It is known that only one circle can be drawn through any three given points which do not lie on a straight line. Hence three points (x_1, y_1),

(x_2, y_2), and (x_3, y_3) not on a straight line determine a circle. The values of D, E, and F which give the equation of this circle can be found by sub-stituting the coordinates of these points in equation 1, thereby obtaining three linear equations from which the values of D, E, and F can be found. This process will be illustrated in the following example.

Example 1. Find the equation of the circle through the three points $(1, -3)$ $(1, 7)$, and $(4, -2)$.

The equation of the circle has the form 1. Since the coordinates of each of the three points must satisfy the equation of the circle, the three equations obtained by substituting the coordinates of these points in equation 1 must hold simultaneously.

$$1 + 9 + D \quad 3E + F - 0,$$
$$1 + 49 + D + 7E + F = 0,$$
$$16 + 4 + 4D - 2E + F = 0.$$

Simplifying these equations and then solving for D, E, and F, we find

$$D = -2, \quad E = -4, \quad F = -20.$$

Hence the required equation is

$$x^2 + y^2 - 2x - 4y - 20 = 0.$$

The student should find the center and radius of this circle by reducing its equa-tion to the standard form. He should then sketch the circle on coordinate paper and verify the fact that the three given points lie on the circle. This process will give a graphical check on the accuracy of our work.

More generally, a circle can frequently be determined when three suitably chosen elementary facts are stated about it. For example, we may state that a circle is tangent to the x axis, tangent to the y axis, and passes through the point $(1, 2)$. In geometry we would find such a circle by devising a ruler and compass construction for it. However, in analytic geometry we can find the circle by algebraic methods. To find the circle by algebraic methods we may write the equation of the circle either in the form 1 or in the form

$$(x - h)^2 + (y - k)^2 = r^2, \tag{2}$$

where the constants D, E, and F in equation 1 or h, k, and r in equation 2 are to be determined from the given conditions. To see how this is done in an example we shall now find the circle satisfying the conditions given at the beginning of this paragraph.

Example 2. Find the equation of the circle that is tangent to both coordinate axes and passes through the point $(1, 2)$.

Assume the equation of the circle to be in the form 2. Since the point $(1, 2)$ lies on the circle it follows that

$$(1 - h)^2 + (2 - k)^2 = r^2. \tag{3}$$

Also, since the circle is tangent to the coordinate axes, its center (h, k) must be at a distance r (the radius) from each axis. It follows that $h = r$ and $k = r$. Substituting these values into equation 3 and simplifying, we obtain

$$r^2 - 6r + 5 = 0,$$

whence $r = 1$ and 5. It follows that there are two solutions:

$$h = k = r = 1,$$
$$h = k = r = 5.$$

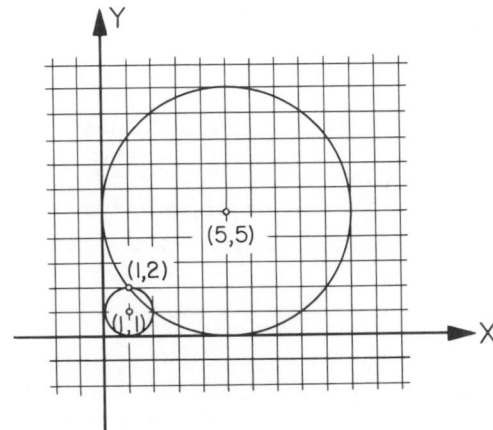

FIG. 16–3

Therefore there are two circles satisfying the given conditions:

$$(x - 1)^2 + (y - 1)^2 = 1,$$
$$(x - 5)^2 + (y - 5)^2 = 25.$$

These circles are drawn in Fig. 16–3, which gives a graphical check on the correctness of our work.

In problems of the kind considered in this section the student should always draw the circle (or circles) and observe whether or not the given conditions are satisfied.

Exercises

Find the equation of the circle passing through the given three points. Draw the corresponding figure.

1. $(1, 0)$, $(0, -1)$, $(0, 0)$.

2. $(0, 0)$, $(0, 8)$, $(6, 0)$.

3. $(0, 0)$, $(4, 8)$, $(9, 9)$.

4. $(0, 0)$, $(8, 0)$, $(-2, -6)$.

5. $(1, 0)$, $(5, 0)$, $(6, 4)$.

6. $(0, 4)$, $(1, 0)$, $(-2, 0)$.

7. $(4, -1)$, $(-5, 2)$, $(3, 6)$.

8. $(2, 2)$, $(2, 4)$, $(10, 2)$.

9. $(2, 2)$, $(-5, 1)$, $(4, -2)$.

10. $(-2, 5)$, $(8, 5)$, $(-6, 9)$.

Find the equation of the circle (or circles) determined by the given data. Draw the corresponding figure.

11. Through the points (3, 5) and (5, −1), and center on the x axis.

12. Through the points (0, 2) and (3, 5), and center on the y axis.

13. Through the points (2, 6) and (8, 4), and center on the line $2x + y + 4 = 0$.

14. Through the points (−1, 3) and (3, 1), and center on the line $3x + y = 5$.

15. Through the point (−3, 1), radius = 2, and center on the line $2x − 3y + 3 = 0$.

16. Through the point (4, 2), radius = 2, center on the line $x = y$.

17. Through the point (−2, 4), center on the line $y = 2x$, and tangent to the x axis.

18. Through the point (4, −2), center on the line $x = 2y$, and tangent to the y axis.

19. Through the point (1, 2) and tangent to both coordinate axes.

20. Through the points (1, 2) and (2, 1) and radius = $1/\sqrt{2}$.

21. Through the points (4, 1) and (6, 3) and radius = $\sqrt{10}$.

22. Through the point (2, −4) and tangent to both coordinate axes.

23. Tangent to both coordinate axes, and center on the line $3x − 2y = 2$.

24. Tangent to both coordinate axes, and center on the line $2x − y − 1 = 0$.

25. Circumscribing the triangle formed by the coordinate axes and the line $2x+3y=6$.

26. Circumscribing the triangle formed by the coordinate axes and the line $2x−y+4=0$.

27. Circumscribing the triangle formed by the lines $x + y = 1$, $2x + y + 3 = 0$, and $x − y + 2 = 0$.

28. Center at the point (−1, 2) and tangent to the line $2x − y + 3 = 0$.

29. Center at the point (−3, −4) and tangent to the line $3x + 4y = 20$.

30. Through the points (6, 4) and (3, 7) and tangent to the y axis.

16–4. Combinations of Curves

Consider the equations

$$x - y = 0, \tag{1}$$

$$x + 2y - 6 = 0, \tag{2}$$

which represent, respectively, the lines L_1 and L_2 of Fig. 16–4. If we form the product of the left members of equations 1 and 2 and set it equal to zero, we obtain

$$(x - y)(x + 2y - 6) = 0, \tag{3}$$

which may be written in the form

$$x^2 + xy - 2y^2 - 6x + 6y = 0. \tag{4}$$

Clearly any coordinates (x, y) that make the left member of equation 1 equal to zero also make the left members of equations 3 and 4 equal to zero. In other words, any point on the graph of equation 1 also lies on the graph of equation 4; similarly any point on the graph of equation 2 also lies on the graph of equation 4. These remarks show that the graph of equation 4 contains both of the lines L_1 and L_2. Does the graph of equation 4 consist of any points beside those on either L_1 or L_2? To answer this question consider the factored form of equation 4 given in equation 3. The product in the left member of equation 3 is zero only if at least one of the factors is zero; thus coordinates (x, y) which make the left member of equation 3 equal to zero must make at least one of the factors equal to zero; i.e., they must satisfy at least one of equations 1 and 2. Therefore any point on the graph of equation 4 lies on either L_1 or L_2 (or possibly both).

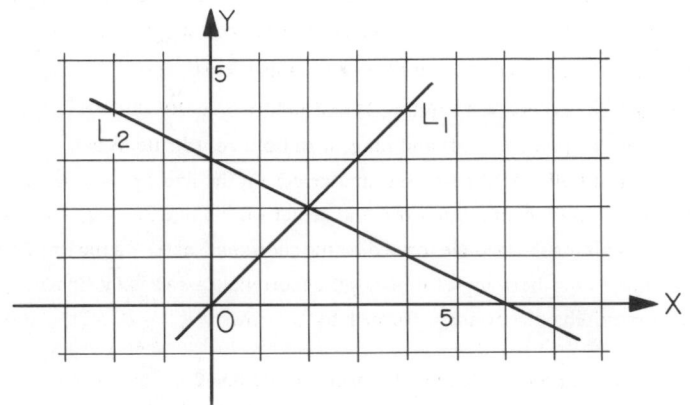

FIG. 16–4

We have now shown that the graph of equation 4 contains L_1 and L_2 and that any point on the graph of equation 4 lies on either L_1 or L_2. It follows that the graph of equation 4 consists precisely of the lines L_1 and L_2.

In like manner we may arrive at a general proposition:

If the graph of $f(x, y) = 0$ is a curve C_1 and the graph of $g(x, y) = 0$ is a curve C_2, then the equation

$$f(x, y) \cdot g(x, y) = 0$$

has as its graph a curve consisting of the points on the curves C_1 and C_2 and no other points.

The student should prove this proposition by an argument similar to the one given above. It is clear that this argument can be extended to cover any number of factors, and therefore that the proposition can be extended to cover any number of factors.

Example 1. The equation

$$x^4 - y^4 - x^2 + y^2 = 0,$$

which may be written in the factored form

$$(x - y)(x + y)(x^2 + y^2 - 1) = 0,$$

has as its graph a curve consisting of the lines $x - y = 0$ and $x + y = 0$, and the circle $x^2 + y^2 = 1$.

Let us now consider the following problem: If the graph of $f(x, y) = 0$ is a curve C_1 and the graph of $g(x, y) = 0$ is a curve C_2, what can be said about the graph of

$$f(x, y) + k \cdot g(x, y) = 0, \tag{5}$$

where k is any constant? If (x, y) is a point on both C_1 and C_2 (i.e., a point of intersection of C_1 and C_2), then its coordinates make both $f(x, y)$ and $g(x, y)$ equal to zero, and therefore satisfy equation 5. Therefore we can make this statement: *The graph of equation 5 passes through every point of intersection of C_1 and C_2.* For different values of k we may expect to obtain different curves, but this statement holds for each of these curves.

In particular, let C_1 and C_2 be two intersecting straight lines whose equations are, respectively,

$$a_1 x + b_1 y + c_1 = 0, \qquad a_2 x + b_2 y + c_2 = 0, \tag{6}$$

and consider the expression

$$(a_1 x + b_1 y + c_1) + k \cdot (a_2 x + b_2 y + c_2) = 0. \tag{7}$$

It is clear that, for a given value of k, this is the equation of a straight line which passes through the point of intersection of C_1 and C_2. Each value of k gives such a line; therefore corresponding to the set of all possible values of k there is a set of lines, each one passing through the point of intersection of C_1 and C_2. This set of lines specified by equation 7 is called *the system of lines through the point of intersection of C_1 and C_2.* Equation 7 enables us to find a member of this system without finding the coordinates of the point of intersection of the two given lines, as the following example shows.

Example 2. Find the equation of the line through the point $(1, 2)$ and the intersection of $2x + y - 7 = 0$ and $3x + 4y + 6 = 0$.

The equation of the system of lines through the point of intersection of the two given lines is

$$(2x + y - 7) + k \cdot (3x + 4y + 6) = 0. \tag{8}$$

Since we seek the line of the system that passes through $(1, 2)$, we substitute $x = 1$ and $y = 2$ in equation 8 and obtain $k = \frac{3}{17}$. When this value of k is substituted in equation 8 and the result simplified, we obtain $43x + 29y - 101 = 0$ as the equation of the required line.

The student should solve this problem by finding the coordinates of the point of intersection of the two given lines and then finding the equation of the line through this point and the given point. He should also draw the graph.

Example 3. Find the equation of the line that passes through the point of intersection of the lines $3x + 7y - 1 = 0$ and $3x + y - 4 = 0$ and that is perpendicular to the line $x + 2y - 3 = 0$.

The system of lines through the point of intersection of the given lines is given by the equation

$$(3x + 7y - 1) + k \cdot (3x + y - 4) = 0,$$

which may be written in the form

$$(3 + 3k)x + (7 + k)y + (-1 - 4) = 0. \tag{9}$$

The slope of any line of this system is $-\dfrac{3 + 3k}{7 + k}$, and the slope of $x + 2y - 3 = 0$ is $-\frac{1}{2}$. Since we must find the member of the system that is perpendicular to this given line, we must find the value of k that satisfies the equation

$$-\frac{3 + 3k}{7 + k} = 2.$$

This value of k is $-\frac{17}{5}$, whence the equation of the required line is given by

$$(3x + 7y - 1) - \tfrac{17}{5}(3x + y - 4) = 0,$$

which becomes, on simplification, $4x - 2y - 7 = 0$.

The student should verify this result by solving the problem by finding the point of intersection of the given lines. He should also draw the graph.

Let

$$x^2 + y^2 + D_1x + E_1y + F_1 = 0, \tag{10}$$

$$x^2 + y^2 + D_2x + E_2y + F_2 = 0 \tag{11}$$

be the equations of two intersecting circles, and consider the equation

$$(x^2 + y^2 + D_1x + E_1y + F_1) + k \cdot (x^2 + y^2 + D_2x + E_2y + F_2) = 0. \tag{12}$$

When $k = -1$ it is clear that equation 12 is the equation of a line passing through the points of intersection of these circles; the chord common to these two circles lies on this line. This line is therefore sometimes referred to as the **common chord**, although it extends outside the circles. When $k \neq -1$, it is clear that equation 12 is the equation of a circle passing through the points of intersection of equations 10 and 11.

Example 4. Find the circle that passes through the origin and the points of intersection of the circles $x^2 + y^2 - 4x - 6y - 3 = 0$ and $x^2 + y^2 + 4x - 2y - 4 = 0$.

The required circle is a member of the system given by

$$(x^2 + y^2 - 4x - 6y - 3) + k \cdot (x^2 + y^2 + 4x - 2y - 4) = 0. \tag{13}$$

Since the required circle passes through the origin, we must find the value of k that will enable the coordinates of the origin to satisfy equation 13. Thus we substitute $x = y = 0$ in equation 13 and obtain $-3 - 4k = 0$, or $k = -\frac{3}{4}$. Substituting this value of k in equation 13 and simplifying, we obtain $x^2 + y^2 - 28x - 18y = 0$ as the equation of the required circle.

The student should draw graphs of these circles to verify that the correct solution has been obtained. He should also solve the problem by finding the points of intersection of the given circles.

Example 5. Find the common chord of the circles given in the preceding example.

If we set $k = -1$ in equation 13, we obtain $8x + 4y - 1 = 0$ as the equation of the common chord. The student should verify that this is the correct result by drawing the line on the graph for the previous problem and by finding the equation of the line from the two points of intersection of the given circles.

Exercises

In each of the following problems write a single equation which represents a curve consisting of the given curves and draw the corresponding graph:

1. $x - 3 = 0$, $x + 4y + 3 = 0$.   2. $y = 1$, $2x - y + 1 = 0$.

3. $x = 0$, $y = 5$, $x - 3y - 1 = 0$.   4. $2x - y = 1$, $2x + y = 1$.

5. $x = y$, $x = -y$, $x^2 + y^2 = 4$.

6. $x + y - 1 = 0$, $x - y = 1$, $x^2 + y^2 = 1$.

7. $2x - y - 1 = 0$, $x^2 + y^2 - x + y = 0$.

8. $x - y = 0$, $x + y = 0$, $x^2 + y^2 - 2x + 4y = 0$.

Sketch the graphs of each of the following equations after factoring their left-hand members:

9. $x^2 - 3x = 0$.   10. $x^2 - 4xy = 0$.   11. $x^2 - 9 = 0$.

12. $y^2 - 1 = 0$.   13. $9x^2 - 4y^2 = 0$.   14. $x^2 - x - 2 = 0$.

15. $x^3y - xy^3 = 0$.   16. $x^2 + xy - 2y^2 = 0$.   17. $x^2 - y^2 + 2x - 2y = 0$.

18. $xy - 3x - 2y + 6 = 0$.   19. $x^3 - y^3 + xy^2 - x^2y - x + y = 0$.

20. $x^3 + y^3 + xy^2 + xy^2 - x - y = 0$.

Draw a graph corresponding to each of the following exercises:

21. Find the equation of the line through the point $(2, 1)$ and the point of intersection of $3x + 4y - 1 = 0$ and $4x - y - 3 = 0$.

22. Find the equation of the line through the point $(4, -2)$ and the point of intersection of $x + 2y - 5 = 0$ and $3x - 4y + 5 = 0$.

23. Find the equation of the line through the point of intersection of $x - y + 4 = 0$ and $2x - 3y + 6 = 0$ and having an x intercept equal to 2.

24. Find the equation of the line through the point of intersection of $4x - 3y - 9 = 0$ and $2x + y + 5 = 0$ and having a slope of 3.

25. Find the equation of the line that passes through the point of intersection of $x + y - 2 = 0$ and $2x - y - 8 = 0$ and is parallel to the line $5x - y = 0$.

26. Find the equation of the line that passes through the point of intersection of $3x + y - 5 = 0$ and $x + 2y - 3 = 0$ and has a slope equal to $\frac{1}{2}$.

27. Find the equation of the line passing through the intersection of $x + 2y - 5 = 0$ and $3x - 4y + 5 = 0$ and perpendicular to the line $2x - y = 6$.

28. Find the equation of the line passing through the intersection of $2x + y = 5$ and $3y = 4x + 7$ and perpendicular to the line $2x - y = 5$.

29. Find the equation of the line that passes through the point of intersection of $y = 2(2x - 1)$ and $x + y = -3$ and is perpendicular to the line $y + 2x = 5$.

30. Find the equation of the line that passes through the point of intersection of $5x + 7 = 4y$ and $7x = 8 - 3y$ and is perpendicular to the y axis.

31. Find the equation of the line passing through the point of intersection of $x + 2y = 5$ and $3x + 5 = 4y$ and forming with the axes a triangle of area 4.

32. Describe the system of lines whose equation is $(2x+3y-1)+k(2x+3y+7)=0$.

33. What does equation 7 of Sec. 16–4 represent when the two lines given in equation 6 are parallel?

34. Find the equation of the circle that passes through the origin and the points of intersection of the circles $x^2 + y^2 = 25$ and $x^2 + y^2 - 8x + 6y + 11 = 0$.

35. Find the equation of the common chord of the circles given in the preceding exercise.

36. Find the equation of the circle that passes through the origin and the points of intersection of the circles $x^2 + y^2 = 4x + 6y + 3$ and $x^2 + y^2 + 4x - 2y = 4$.

37. Find the equation of the common chord of the circles given in the preceding exercise.

38. Find the equation of the circle that passes through the point $(3, 2)$ and through the points of intersection of the circles $x^2 + y^2 = 4$ and $x^2 + y^2 + 4x = 0$.

39. Find the equation of the common chord of the circles given in the preceding exercise.

40. Find the equation of the circle that passes through the point $(1, -3)$ and through the points of intersection of the circles $x^2 + y^2 + 3x - 5y - 2 = 0$ and $x^2 + y^2 - 8x - 4y - 1 = 0$.

41. Find the equation of the common chord and the points of intersection of the circles $x^2 + y^2 = 100$ and $x^2 + y^2 + 4x - 8y - 220 = 0$.

42. Find the equation of the common chord and the points of intersection of the circles $x^2 + y^2 + 3x - 5y = 0$ and $x^2 + y^2 + x - 4y + 1 = 0$.

43. Find the equations of the three common chords of the three circles $x^2 + y^2 = 9$, $x^2 + y^2 - 6x - 8y + 9 = 0$, and $x^2 + y^2 - 4x + 6y - 23 = 0$, taken in pairs, and show that these lines meet in a point.

44. Find the equations of the three common chords of the three circles $x^2 + y^2 + 6x - 7 = 0$, $x^2 + y^2 - 6x - 2y - 15 = 0$, and $x^2 + y^2 + 8y - 33 = 0$, taken in pairs, and show that these lines meet in a point.

16–5. Loci

If S is a set of points such that (a) every point that satisfies a given condition is a point of S, and (b) every point that does not satisfy the given condition is not in S, then S is called the **locus** of a point satisfying the given condition. For example, a circle is the locus of a point satisfying the condition that each point of the locus lies a given distance (called the radius) from a fixed point (called the center).

We have seen that by means of coordinates analytic geometry unites algebra, which deals with numbers and equations, and geometry, which deals with points and loci. This union gives rise to two definitions:

1. An **equation of a locus** is an equation that is satisfied by the coordinates of every point on the locus and is not satisfied by any point not on the locus.

2. The **locus of an equation** is a figure consisting of all the points whose coordinates satisfy the equation.

In Chapter 15 we defined a line as the locus of a point P satisfying this condition: If P_1 is a fixed point and m is a given slope, then the segment PP_1 has slope m (see Sec. 15–8). From this definition we found an equation of the locus. Later, in Sec. 15–11, we found that the locus of an equation of the form $Ax + By + C = 0$ is a straight line. Earlier in this chapter we found the form of the equation of the circle, which may be defined as a locus, and this result enabled us to determine the locus of an equation of the form $x^2 + y^2 + Dx + Ey + F = 0$.

Thus we have already considered some examples of two of the principal problems of analytic geometry:

1. Given a geometrical locus, to find a corresponding algebraic equation.

2. Given an algebraic equation, to find the corresponding geometrical locus.

We shall consider these two problems further in the next two sections.

16–6. The Equation of a Locus

Plane geometry studies the problems of determining the configurations of loci. However, plane geometry has no general methods for finding these configurations: Each problem must be approached by methods peculiar to the problem itself.

In contrast to plane geometry, analytic geometry provides a general method for solving locus problems. We merely translate the given geometrical conditions, describing the locus, into algebra. Two examples of this method have already been given. In Sec. 15–8 we found the equation of a line through a given point by defining it as the locus of a point satisfying the condition that the line segment joining it to the given point has a constant slope. In Sec. 16–1 the circle was considered as the locus of a point whose distance from the center is constant. In both cases we translated the given conditions into algebra and simplified the resulting expressions algebraically.

In the examples that follow we shall show how we can apply the method of analytic geometry to other locus problems.

Example 1. Find the equation of the locus of a point such that the sum of the squares of its distances from (0, 0) and (4, 6) is 76.

Let P_1 and P_2 be, respectively, the given points (0, 0) and (4, 6), and let $P(x, y)$

be any point on the required locus (Fig. 16–5). Our requirement is that
$$(PP_1)^2 + (PP_2)^2 = 76$$
which becomes, when stated algebraically,
$$x^2 + y^2 + (x - 4)^2 + (y - 6)^2 = 76.$$

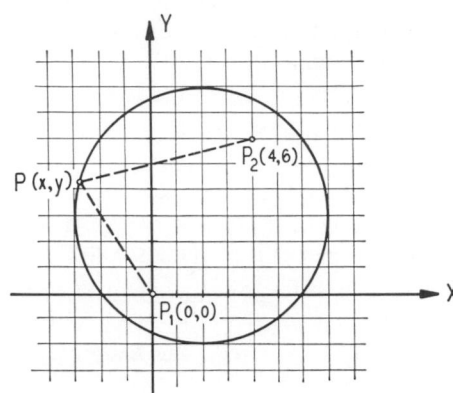

Fig. 16–5

Simplifying yields
$$2x^2 + 2y^2 - 8x - 12y = 24,$$
$$x^2 + y^2 - 4x - 6y = 12,$$
$$(x^2 - 4x + 4) + (y^2 - 6y + 9) = 12 + 9 + 4,$$
$$(x - 2)^2 + (y - 3)^2 = 25.$$
This is the equation of a circle with center at (2, 3) and radius 5.

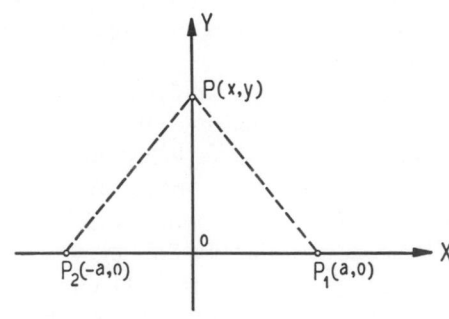

Fig. 16–6

Example 2. Find the equation of the locus of a point equidistant from two given points.

We construct a figure (Fig. 16–6) and introduce a coordinate system so that the given points have the coordinates $P_1(a, 0)$ and $P_2(-a, 0)$. This choice of co-ordinates will simplify our work. Now geometrically the requirement is that
$$P_2P = P_1P.$$

When written algebraically, this requirement becomes

$$\sqrt{(x + a)^2 + (y - 0)^2} = \sqrt{(x - a)^2 + (y - 0)^2}.$$

Simplifying, we get

$$x^2 + 2ax + a^2 + y^2 = x^2 - 2ax + a^2 + y^2,$$

$$4ax = 0,$$

and, since $a \neq 0$, we obtain finally

$$x = 0.$$

This is the equation of the y axis, which is the perpendicular bisector of P_1P_2.

When a geometrical locus is specified, the work of finding the corresponding equation frequently proceeds by the following steps, as the examples in this section illustrate.

1. A figure is drawn, showing the given data for a representative point $P(x, y)$ on the locus.

2. If a coordinate system is not given, one is introduced. An appropriate selection will often simplify the later work.

3. The description of the locus is written down as an equation involving the geometrical quantities given in the specification of the locus.

4. The equation set down in the previous step is now expressed in terms of the coordinates (x, y) of P, and then simplified algebraically.

5. Since it is possible that the algebraic simplification process will yield an equation that is satisfied by the coordinates of points not on the locus (just as simplified equations in one variable may yield extraneous roots), it is important to verify that no points not on the locus satisfy the simplified equation. This frequently overlooked, but important, step can be made by observing that the geometrical description of the locus can be derived from the simplified equation by reversing the steps used in deriving this equation. The student should carry out this step for the two examples of this section.

Exercises

Find an equation of the locus of a point that satisfies each condition described below. Whenever possible, identify the locus.

1. Four units to the right of the y axis.
2. Six units to the left of the y axis.
3. Five units above the x axis.
4. Ten units below the x axis.
5. Its distance from the origin is 5.
6. Its distance from $(1, 2)$ is 7.
7. Its distance from $(-2, -3)$ is 8.
8. Equidistant from $(0, 2)$ and $(4, 6)$.
9. Equidistant from $(2, 4)$ and $(8, 4)$.
10. Equidistant from $(-1, 3)$ and $(2, -5)$.
11. Equidistant from $(2, -3)$ and $(-1, 6)$.
12. The sum of the squares of its distances from $(-3, 0)$ and $(3, 0)$ is 10.

13. The sum of the squares of its distances from $(2, 0)$ and $(6, 0)$ is 16.
14. The sum of the squares of its distances from two fixed points is a constant k.
15. Its distance from $(0, 0)$ is twice its distance from the line $y = 4$.
16. Its distance from $(0, 0)$ is half its distance from the line $y + 6 = 0$.
17. Its distance from $(0, 0)$ is half its distance from the line $x = 3$.
18. Its distance from $(0, 2)$ is three times its distance from the line $x + 4 = 0$.
19. The difference of the squares of its distances from $(-2, 0)$ and $(2, 0)$ is 3.
20. The difference of the squares of its distances from $(0, 0)$ and $(3, 5)$ is 12.

16–7. The Locus of an Equation

In this section we shall consider the problem of how to find the geometrical locus corresponding to a given equation.   Problems of this kind have already been studied in various places in this book: For example, simple algebraic curves were discussed in Sec. 3–9, Sec. 8–13, and Sec. 14–2, trigonometric functions in Chapter 5, exponential functions in Sec. 8–14, and logarithmic curves in Sec. 9–4.   The reader should review this material before going on.

In the present section we shall consider several examples of more complicated algebraic curves, sometimes called *higher plane curves*.   We shall discuss several important properties that are frequently found in such curves and that are helpful in drawing them.

Symmetry.   Two points P_1 and P_2 are said to be symmetric with respect to a given line l when the given line l is the perpendicular bisector of the line P_1P_2.   In particular we are interested in symmetry with respect to the coordinate axes.   In this case we have the following rules:

Rule 1.   If an equation is unchanged by the substitution of $-x$ for x, its locus is symmetric with respect to the y axis.

Rule 2.   If an equation is unchanged by the substitution of $-y$ for y, its locus is symmetric with respect to the x axis.

The proof of the above rules is left for the reader.

Two points P_1 and P_2 are said to be symmetric with respect to a third point Q if the point Q is the midpoint of the segment P_1P_2.   In particular we are interested in symmetry with respect to the origin of the coordinate system.   In this case we have the following rule:

Rule 3.   If an equation is unchanged by the substitution of $-x$ for x and of $-y$ for y, its locus is symmetric with respect to the origin.

Extent.   In trying to plot the locus of an equation, it sometimes happens that for some values of one coordinate the other may be imaginary and hence the points having these coordinates cannot be plotted.   Thus there may be intervals in which there is no curve.   These intervals can often be determined by inspection after solving the equation for y in terms of x, and for x in terms of y.

Asymptotes.   We recall from our study of the graph of $y = \tan x$

(Sec. 5–4) that tan x approaches infinity as x approaches $\pi/2$. The line $x = \pi/2$ in Fig. 5–6 is called an *asymptote*, a line whose distance from the curve of $y = \tan x$ decreases as we move out on the curve in such a way that, as a point on the curve recedes sufficiently far from the origin along the infinite branch of the curve, the distance of the point from the line comes as near to zero as we please without actually becoming zero. The location of asymptotes, when there are any, is an important aid in drawing a curve.

The following examples will illustrate how we can use the properties listed above in plotting the loci of equations.

Example 1. Discuss and plot the graph of the equation $x^2 y^2 + x^2 - y^2 = 0$.

If we replace x by $-x$ the equation is left unchanged. Also if we replace y by $-y$ the equation is left unchanged. Hence the curve is symmetric with respect to the y axis and with respect to the x axis. Clearly the curve is also symmetric with respect to the origin.

Solving the given equation for y, we obtain

$$y = \pm \frac{x}{\sqrt{1 - x^2}},$$

which shows that for all values of x for which $x^2 > 1$ the ordinate y is imaginary. There are, accordingly, no points on the curve for which $x > 1$ and $x < -1$, and the extent in x is from -1 to $+1$. Also if x is made to approach $+1$ or -1 the numerical value of y will increase indefinitely. The lines $x = 1$ and $x = -1$ are therefore asymptotes.

By the aid of a few computed points the curve can now be sketched as shown in Fig. 16–7.

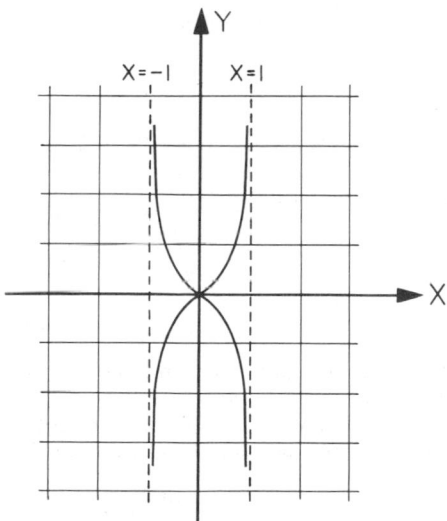

FIG. 16–7. The graph of $x^2 y^2 + x^2 - y^2 = 0$

Example 2. Discuss and plot the graph of the equation $x^2y + y - 1 = 0$.

If we replace x by $-x$ the equation is left unchanged. Hence the curve is symmetric with respect to the y axis.

Solving for x, we have

$$x = \pm \sqrt{\frac{1-y}{y}}.$$

Hence x is imaginary if y is negative or greater than 1, so that the extent in y is from 0 to 1. Also, if y is made to approach 0, the numerical value of x will increase indefinitely. The line $y = 0$ or the x axis is an asymptote.

Making a table of values we obtain

x	0	1	2	3
y	1	0.5	0.2	0.1

This curve, plotted in Fig. 16–8, is known in mathematical literature as a *witch of Agnesi.*

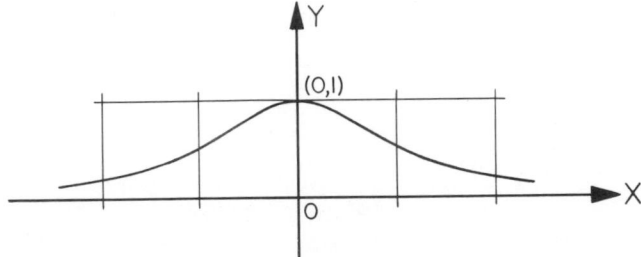

FIG. 16–8. The graph of $x^2y + y - 1 = 0$

Example 3. Discuss and plot the graph of the equation

$$y = \frac{16}{x^2 - 4}. \tag{1}$$

If we replace x by $-x$ the equation is left unchanged. Hence the curve is symmetric with respect to the y axis.

Solving the equation for x, we have

$$x = \pm 2 \sqrt{\frac{4+y}{y}}. \tag{2}$$

Values for which the expression $\dfrac{4+y}{y}$ is negative must be excluded. Since a fraction is negative when its numerator and denominator have opposite signs, it follows that there is no curve in the interval $-4 < y \le 0$. From equation 2 it also follows that $y = 0$ or the x axis is an asymptote.

Factoring the denominator, we have from equation 1

$$y = \frac{16}{(x-2)(x+2)},$$

which shows that the lines $x = 2$ and $x = -2$ are asymptotes.

By the aid of a few computed points the curve of equation 1 can now be sketched, and the result is given in Fig. 16–9.

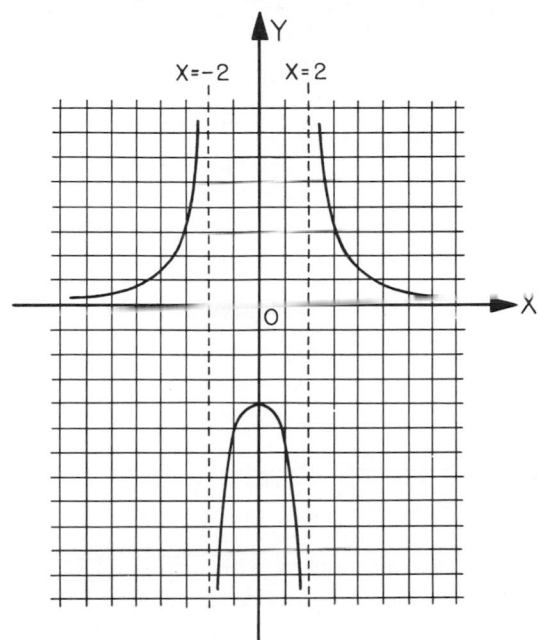

FIG. 16–9. The graph of $x^2 y - 4y = 16$

Exercises

Discuss and plot the graphs of the following equations:

1. $xy = 1$.
2. $xy + 2 = 0$.
3. $y = 2x^2$.

4. $y = 3x^2 - 5$.
5. $x = 2y^2 + 1$.
6. $x + 2 = 3y^2$.

7. $x^2 - y^2 = 1$.
8. $x^2 - y^2 = 4$.
9. $9x^2 + y^2 = 36$.

10. $4x^2 + 9y^2 = 36$.
11. $x^2 y^2 = 144$.
12. $x^2 y = 24$.

13. $x^2 y = 12$.
14. $xy^2 = 1$.
15. $xy = 2y + 4$.

16. $xy - 3y + 2 = 0$.
17. $xy + x^2 - 12 = 0$.
18. $xy - 6 = y^2$.

19. $(x - 3)^2 y - x = 0$.
20. $(x^2 - 4)y = 3x$.
21. $x^2 y + y - 2 = 0$.

22. $x^2 y - x^2 + 9 = 0$.
23. $x^2 y - x^3 = 8$.
24. $x^4 + y^4 = 1$.

25. $y^2(1 + x^2) = 1$.
26. $x^2 y^2 = x^2 + y^2$.
27. $y^2 = (x-1)(x-2)(x-3)$.

28. $y^2 = x(x-1)(x-2)$.
29. $y = \dfrac{8 - x}{x - 2}$.
30. $y = \dfrac{x^2}{x^2 - 25}$.

31. $y = \dfrac{x}{x^2 - 4}$.
32. $y^2 = \dfrac{x}{x^2 - 4}$.

16–8. Polar Coordinates

In Chapter 3 we introduced the **rectangular coordinate** system, a device for locating points in the plane by setting up a correspondence between pairs of real numbers (x, y) and the points in a plane. In the subsequent chapters this device was used frequently in making geometrical or graphical interpretations of algebraic relations.

The **polar coordinate system,** a second method for locating points in the plane, is often useful. Let O be a fixed point in the plane, and let OA be a fixed line extending from O infinitely far in one direction only (Fig. 16–10), and let the unit of length be chosen along OA as shown. Now, any

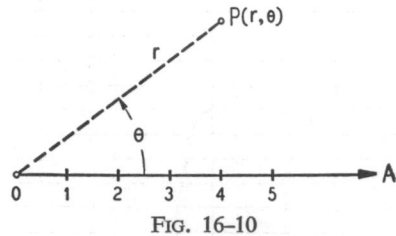

FIG. 16–10

point P in the plane can be located with respect to O and OA by giving θ, any angle with initial side OA and terminal side OP, and r, the length OP measured on the terminal side of θ. These two values are customarily written (r, θ) and are called a set of **polar coordinates** of the point P. The point O is called the **pole** of the system; the line OA is called the **polar axis** of the system.

We shall adopt the following conventions in measuring r and θ:

1. If θ is measured counterclockwise it is positive; if θ is measured clockwise it is negative.

2. If r is measured from O to P along the terminal side of θ, r is positive; if r is measured along the extension through O of the terminal side of θ, then r is negative.

Example. Fig. 16–11 shows the points $P_1(5, 30°)$, $P_2(-5, 60°)$, $P_3(3, -210°)$, and $P_4(-4, -90°)$.

For a given pair of coordinates (r, θ) only one point in the plane can be located. However, for a given point in the plane many pairs of coordinates can be found. For example, the point P_1 in Fig. 16–11 located by the coordinates $(5, 30°)$ can also be located by the coordinates $(-5, 210°)$, $(5, -330°)$, $(5, 390°)$, $(-5, -150°)$, $(5, 750°)$, etc. For a given point, it is easily seen that there are infinitely many pairs of coordinates. Thus, unlike rectangular coordinates, a system of polar coordinates does not establish a one-to-one correspondence between points in the plane and pairs of values (r, θ).

The student will recall that a complex number could be given in polar form by giving its magnitude and angle.   In this way the position of the point on the complex plane corresponding to the number is given by specifying its distance from the origin and the angle the line joining it to the origin makes with the positive x axis.   Hence the polar form of the complex numbers and the polar coordinate system are based on the same idea.

To facilitate the plotting of points and curves in polar coordinates, especially designed polar coordinate paper should be used.

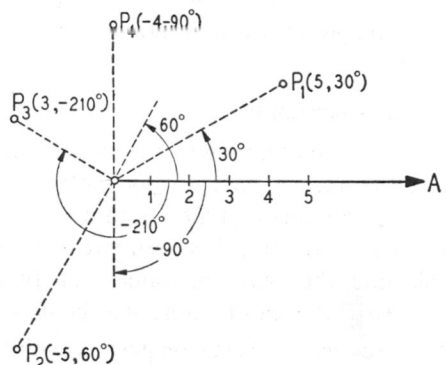

FIG. 16–11

Exercises

Using either a scale and a protractor or polar coordinate paper, plot the following points:

1. (2, 30°), (3, 120°), (1, 215°), (3.2, 311°), (1.5, −45°), (2.5, −90°), (1.8, 180°), (1.8, −180°), (2, $\pi/6$), (1.7, 2π), (3, 2$\pi/3$), (5, 13$\pi/8$), (−2, 3$\pi/5$), (3, −$\pi/2$).

2. (3, 170°), (8, 430°), (1.3, −220°), (−3, 65°), (−8, 250°), (−8, −50°), (−6.3, −470°), (6.3, −470°), (2, 15$\pi/4$), (−2, 15$\pi/4$), (2, −15$\pi/4$), (−2, −15$\pi/4$).

3. (3, 70°), (3, 430°), (3, −290°), (3, 650°), (−3, 250°), (−3, −110°), (−3, 610°).

4. (1.3, $\pi/6$), (1.3, 13$\pi/6$), (−1.3, 7$\pi/6$), (−1.3, −5$\pi/6$), (1.3, −11$\pi/6$).

5. (0, 70°), (0, 0°), (0, 10°), (0, 412°), (0, −80°), (0, −360°), (0, 720°).

6. (2, 70°), (2, 0°), (2, 120°), (2, 180°), (2, 90°), (2, 270°), (2, −60°), (−2, 45°).

Find five other pairs of coordinates for each of the following points:

7. (2, 90°), (3, 210°).   **8.** (−3, 180°), (3, 135°).

9. (1, 120°), (1, −120°).   **10.** (5, 30°), (5, 270°).

11. What curve will be obtained from all points (r, θ) where r satisfies the equation $r = a$, a being a constant?   Where r satisfies the equation $r = -a$?

12. What locus or curve will be obtained from all points (r, θ) where θ satisfies the equation

$$(a)\ \theta = \frac{\pi}{3}, \quad (b)\ \theta = \frac{4\pi}{3}, \quad (c)\ \theta = -\frac{2\pi}{3}, \quad (d)\ \theta = 60°\ ?$$

13. Show that the points (r, θ) and $(r, -\theta)$ are symmetric with respect to the polar axis.

14. Show that the points (r, θ) and $(-r, \theta)$ are symmetric with respect to the pole.

15. Show that the points (r, θ) and $(r, \theta + 180°)$ are symmetric with respect to the pole.

16. Use the law of cosines (Sec. 11–2) to show that the distance between (r_1, θ_1) and (r_2, θ_2) is

$$\sqrt{r_1{}^2 + r_2{}^2 - 2r_1r_2 \cos(\theta_2 - \theta_1)}.$$

17. Use the formula of the preceding exercise to find the distance between $(2, 45°)$ and $(7, 60°)$.

16–9. Graphs in Polar Coordinates

In Chapter 3 it was seen that an algebraic relation between two variables x and y can be represented pictorially by plotting the pairs of values satisfying the relation on rectangular coordinates. Likewise, we shall now see that an algebraic relation between two variables r and θ can be represented by plotting the pairs of values satisfying the relation on polar coordinates. How this can be done will be illustrated by examples.

Example 1. Plot the curve $r = 5 \sin \theta$ on polar coordinates.

To construct a table of values from which to plot this function, we may substitute chosen values for θ and find r, or we may substitute chosen values for r and find θ. In this example the former course presents least difficulty. The choice of values to be substituted for θ is, of course, at our discretion, but usually the principal angles are the most convenient. If the principal angles do not give enough points to determine the general character of the curve, more values of θ should be plotted, using tables of natural functions.

For this example we have the table:

θ	$r = 5 \sin \theta$	θ	$r = 5 \sin \theta$
0°	0.00	150°	2.50
30°	2.50	180°	0.00
60°	4.33	210°	−2.50
90°	5.00	etc.	etc.
120°	4.33		

By analyzing the function $r = 5 \sin \theta$, the general behavior of the curve between the plotted points can be determined. For example, as θ increases between 0° and 90°, the function $5 \sin \theta$ increases from 0 to 5. As θ passes from 90° to 180°, $5 \sin \theta$ decreases from 5 to 0. Now, as θ takes on values from 180° to 360°, $r = 5 \sin \theta$ takes on negative values, giving points already found when θ was less than 180° as the student may verify by extending the table. For example, when $\theta = 30°$ we find the point $(2.5, 30°)$ and for $\theta = 210°$ we obtain the same point located by $(-2.5, 210°)$. Thus, $r = 5 \sin \theta$ gives the circle shown in Fig. 16–12.

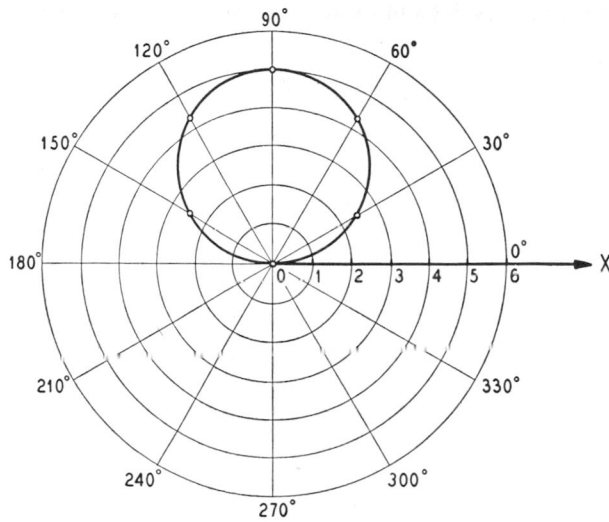

FIG. 16–12. The graph of $r = 5 \sin \theta$

In some cases the negative values of r do not give points already found. This is illustrated in what follows.

Example 2. Plot the function $r = 2 \cos 3\theta$.

Substituting values for θ we obtain the following table:

θ	3θ	$r = 2 \cos 3\theta$	θ	3θ	$r = 2 \cos 3\theta$
0°	0°	2.00	140°	420°	1.00
20°	60°	1.00	150°	450°	0.00
30°	90°	0.00	160°	480°	−1.00
40°	120°	−1.00	180°	540°	−2.00
60°	180°	−2.00	200°	600°	−1.00
80°	240°	−1.00	210°	630°	0.00
90°	270°	0.00	220°	660°	1.00
100°	300°	1.00	240°	720°	2.00
120°	360°	2.00	etc.	etc.	etc.

Note that, although the angle 3θ is used to evaluate r, the angle θ is the angle used in plotting. Hence the curve represented by $r = 2 \cos 3\theta$ is the three-leaved rose of Fig. 16–13. As in example 1, the curve retraces itself as θ passes 180°.

Curves of this general type plotted on polar coordinate paper are obtained in plotting the field strength of radio energy directed by two or more antennas suitably placed and excited. Advanced treatises on radio waves and their propagation show many forms of field patterns, of which Fig. 16–13 may be considered a special example. In another engineering application, the horizontal distribution of the intensity of light from an incandescent lamp or fixture is plotted on polar coordinate paper.

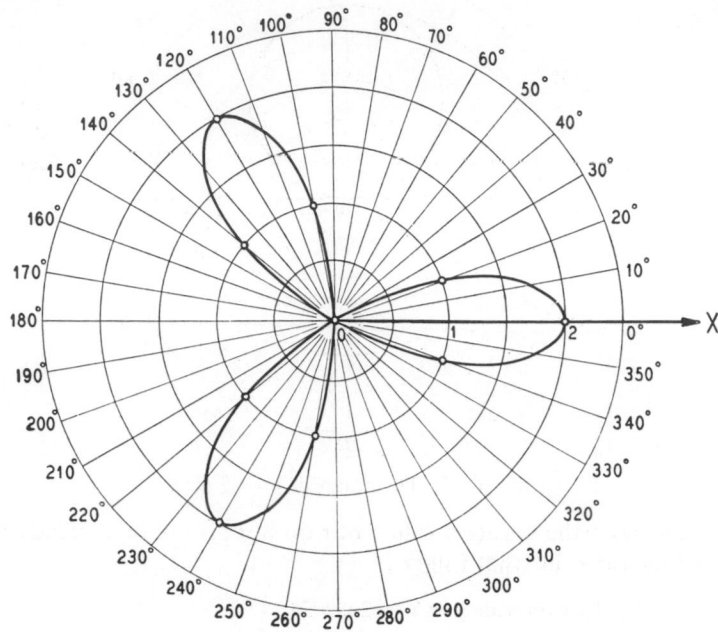

FIG. 16–13. The graph of $r = 2 \cos 3\theta$

Exercises

Using polar coordinate paper plot the curves representing the following functions:

1. $r = 3$.   **2.** $r = -2$.   **3.** $\theta = 30°$.   **4.** $\theta = -60°$.

5. $r \sin \theta = 1$.   **6.** $r \cos \theta = 1$.   **7.** $r = \sin \theta$.   **8.** $r = \cos \theta$.

9. $r = 3 \sin \theta$.   **10.** $r = 4 \cos \theta$.   **11.** $r = 1 + \sin \theta$.   **12.** $r = 1 - \sin \theta$.

13. $r = 1 + \cos \theta$.   **14.** $r = 4 + \cos \theta$.   **15.** $r = 4 \sin^2 \theta$.   **16.** $r = 1 + 2 \sin \theta$.

17. $r^2 = \cos 2\theta$.   **18.** $r = 1 + 2 \cos \theta$.   **19.** $r = 2\theta$.   **20.** $r = 3\theta$.

21. $r = \dfrac{2}{1 + \cos \theta}$.   **22.** $r = \dfrac{6}{2 - \cos \theta}$.   **23.** $r = \sin 3\theta$.   **24.** $r = \cos 3\theta$.

25. $r = 3 \sin 2\theta$.   **26.** $r = 2 \cos 2\theta$.

27. Using Fig. 16–14, show that any equation of the type

$$r = a \cos \theta$$

is a circle of diameter a.

28. Show that any equation of the type

$$r = a \sin \theta$$

is a circle of diameter a.

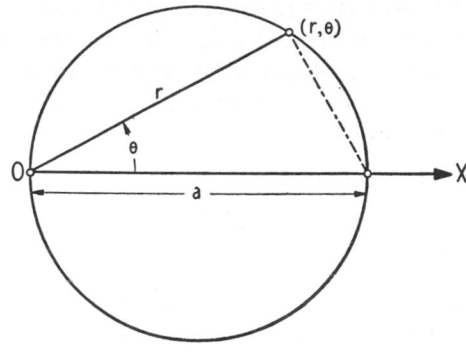

FIG. 16–14

16–10. Relations between Polar Coordinates and Rectangular Coordinates

Occasionally it is advantageous to be able to change from rectangular coordinates to polar coordinates, or vice versa. To obtain a relation between the two coordinate systems, let the pole and polar axis in a polar coordinate system coincide with the origin and positive x axis of a rectangular coordinate system (Fig. 16–15). Also choose the same unit of

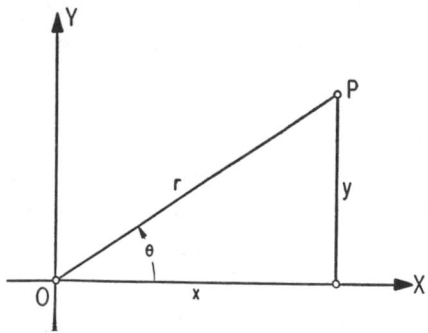

FIG. 16–15

length for both systems. Now consider any point P in the plane. If we denote the polar coordinates of P by (r, θ) and the rectangular coordinates of P by (x, y), then obviously

$$x = r \cos \theta, \qquad y = r \sin \theta, \qquad (1)$$

$$r^2 = x^2 + y^2, \qquad \theta = \arctan \frac{y}{x}. \qquad (2)$$

Since $\arctan \frac{y}{x}$ is a multiple-valued function, a correct value of θ can be determined by noting the quadrant in which P lies, or the signs of y and x.

By using the formulas 1 it is possible to transform a relation given in rectangular coordinates to polar coordinates; by using formulas 2 it is possible to transform a relation given in polar coordinates to rectangular coordinates. The curve that represents the given relation is the same in both coordinate systems.

Example 1. Transform the equation

$$r \sin \theta = 4 \tag{3}$$

to rectangular coordinates.

If equation 3 is plotted in polar coordinates, it can be shown to be a straight line. However, this fact is made obvious if equation 3 is transformed to rectangular coordinates, for, from equations 1, $r \sin \theta = y$, whence equation 3 becomes

$$y = 4.$$

Example 2. Transform the equation

$$x^2 - 4x + y^2 = 0 \tag{4}$$

to polar coordinates.

If we substitute $x^2 + y^2 = r^2$ from equations 2 and $x = r \cos \theta$ from equations 1, the equation becomes

$$r^2 - 4r \cos \theta = 0.$$

Factoring, we obtain

$$r = 0 \quad \text{or} \quad r - 4 \cos \theta = 0.$$

But, since $r = 4 \cos \theta$ gives the value $r = 0$ when $\theta = 90°$, we can express equation 4 in polar form simply as

$$r = 4 \cos \theta.$$

Exercises

Transform the following equations into rectangular coordinates:

1. $r = 4$. **2.** $r = a$. **3.** $\theta = 0$. **4.** $\theta = \dfrac{\pi}{4}$.

5. $r \sin \theta = 6$. **6.** $r \cos \theta = 5$. **7.** $r = 0$. **8.** $\sin \theta = \frac{1}{2}$.

9. $r = 8 \sec \theta$. **10.** $r = 6 \csc \theta$. **11.** $r \sin 2\theta = 1$. **12.** $r = 2 \sin \theta$.

13. $r = \dfrac{1}{1 - \cos \theta}$. **14.** $r = \dfrac{12}{2 + \cos \theta}$. **15.** $r = \cos \theta - \sin \theta$. **16.** $r = 1 - \cos \theta$.

17. $r = 2 \cos \theta + 3 \sin \theta$. **18.** $r^2 = \sin 2\theta$.

19. $r = \tan \theta$. **20.** $r^2 = 2 \cos 2\theta$.

Transform the following equations into polar coordinates:

21. $y = 4$. **22.** $x = y$. **23.** $x^2 + y^2 = 1$.

24. $x^2 + y^2 = a^2$. **25.** $x^2 + y^2 = 4x$. **26.** $x^2 + y^2 + 2y = 0$.

27. $x + y = 0$. **28.** $x + y = 1$. **29.** $x - y = 3$.

30. $2x - y = 3$.   **31.** $x^2 - 8x + y^2 = 0$.   **32.** $x^2 + 4y + y^2 = 0$.

33. $x^2 - y^2 = 4$.   **34.** $x^2 - y^2 = 9$.   **35.** $y = x^2$.

36. $x = y^2$.   **37.** $y^2 = 4ax$.   **38.** $x^2 = 4ay$.

39. $x^2 - x + y^2 + y = 0$.   **40.** $(x^2 + y^2)^2 = y(x^2 - y^2)$.

Progress Report

In this chapter the following topics were considered:
1. The circle:
    (a) The standard form of the equation of a circle.
    (b) The reduction of the general equation of a circle to the standard form.
    (c) The determination of the equations of circles satisfying various given conditions.
2. Combinations of curves:
    (a) Graphs of equations of the form $f(x, y) \cdot g(x, y) = 0$.
    (b) Systems of curves through the points of intersection of two given curves and the selection of a member of the system satisfying a given condition.
3. Loci:
    (a) The definition of a locus.
    (b) Determining the equation of a locus.
    (c) Sketching the locus of an equation using information on the symmetry, the extent, and the asymptotes of the curve.
4. Polar coordinates:
    (a) The definition of the polar coordinate system.
    (b) Sketching graphs on a polar coordinate system.
    (c) The relations between the polar coordinate system and the rectangular system.

The principal results of Chapter 15 were summarized at its conclusion in the form of a geometry-algebra dictionary. The reader should now complete this progress report by extending that earlier dictionary to include the principal results of this chapter.

17

Equations of the Second Degree
The Conics

For the study of the calculus, a familiarity with the graphs of certain second-degree equations is important. In order to establish this familiarity we shall discuss in this chapter the graphs of such equations and some of their analytic and geometrical properties. We shall in this way obtain a partial answer to the question: What is the locus of the general second-degree equation in two variables?

The equations of conic sections occur frequently in physical problems. Planetary motion is elliptical, and recent theories of electron motion suggest that electrons travel about the nucleus of an atom in ellipses. The properties of a parabola are utilized in ultrahigh-frequency techniques; if the source of radio waves is placed at the focus of a paraboloid, the radiations into space will be along parallel, straight lines. This is analogous to the light radiation from a parabolic reflector of an automobile headlight. It is known that a cable supporting a uniform, horizontal load (as cables of a suspension bridge do) assumes the shape of a parabola. But, if the cable bears no load, the configuration is that of the catenary, and a solution of this problem hinges on certain functions associated with the hyperbola. These problems are beyond the scope of this book, but the comprehension of such practical applications depends on the fundamental knowledge of the properties of conic sections.

17-1. The Conic Sections

The curves of intersection of a plane and a right circular cone are called **conic sections** or **conics.** If the plane cuts across one nappe of the cone, the curve of intersection is an **ellipse,** as in Fig. 17–1. If the plane is parallel to a straight line in the surface of the cone, the curve of intersection is a **parabola,** as in Fig. 17–2. If the plane cuts both nappes of the cone, as in Fig. 17–3, the curve of intersection is a **hyperbola.** These definitions explain the origin of the term **conic,** which is applied to these curves.　A study of these curves can be made from these definitions, but we shall find it more convenient to start from other definitions which will be given in succeeding sections.

We have already made a study of the equation of the first degree, and we found that its locus was a straight line.　The conic sections are of

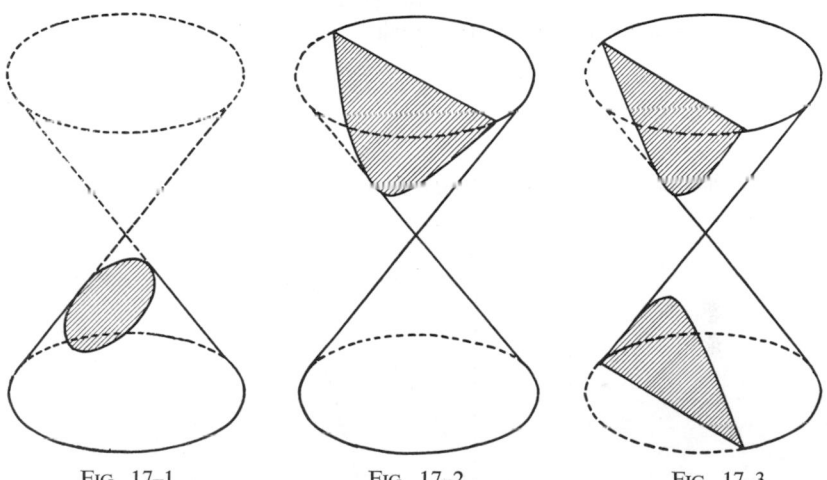

FIG. 17–1　　　　　　　FIG. 17–2　　　　　　　FIG. 17–3

interest because, as we shall see, their equations are equations of second degree.　Further, it can be shown that the locus of the general second-degree equation in two variables,

$$Ax^2 + Bxy + Cy^2 + Dx + Ey + F = 0$$

is a conic, if we also include in this term certain degenerate loci.

17–2. The Parabola

The geometrical definition of the parabola which we shall use is the following.

The parabola is the locus of a point in a plane which satisfies the condition that its distance from a fixed point is equal to its distance from a fixed line. The fixed point is called the **focus,** the fixed line the **directrix.**　The line perpendicular to the directrix through the focus is the **axis** of the parabola. The intersection of the parabola and the axis is the **vertex.**

The equation of a parabola is particularly simple if we choose as focus the point $(p, 0)$, where p is positive, and as directrix the line $x = -p$. Then from Fig. 17–4, the definition requires that

$$KP - FP. \tag{1}$$

Obviously,

$$KP = x + p, \tag{2}$$

and

$$FP = \sqrt{(x - p)^2 + (y - 0)^2}. \tag{3}$$

Using equations 2 and 3 in equation 1, we obtain

$$x + p = \sqrt{(x - p)^2 + y^2}.$$

Squaring both sides of the equation yields

$$(x + p)^2 = (x - p)^2 + y^2$$

which reduces to

$$y^2 = 4px, \tag{4}$$

which is *the standard form of the equation of the parabola.*

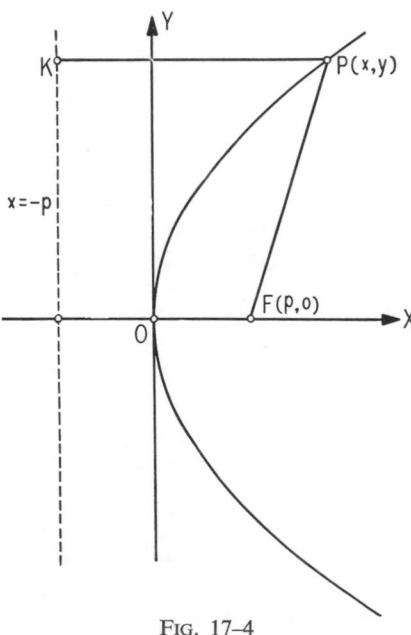

FIG. 17–4

The chord of a parabola drawn through the focus perpendicular to the axis is called the **latus rectum.** When $x = p$, then $y^2 = 4p^2$, and $y = \pm 2p$; hence the *length of the latus rectum is 4p* (Fig. 17–5a).

It is left as an exercise to the student to show that, with the vertex at $(0, 0)$, focus at $(-p, 0)$, and directrix $x = p$, the equation of the parabola is

$$y^2 = -4px. \tag{5}$$

This information is summarized in Fig. 17–5b. The student may also verify the information summarized in Fig. 17–6a and in Fig. 17–6b.

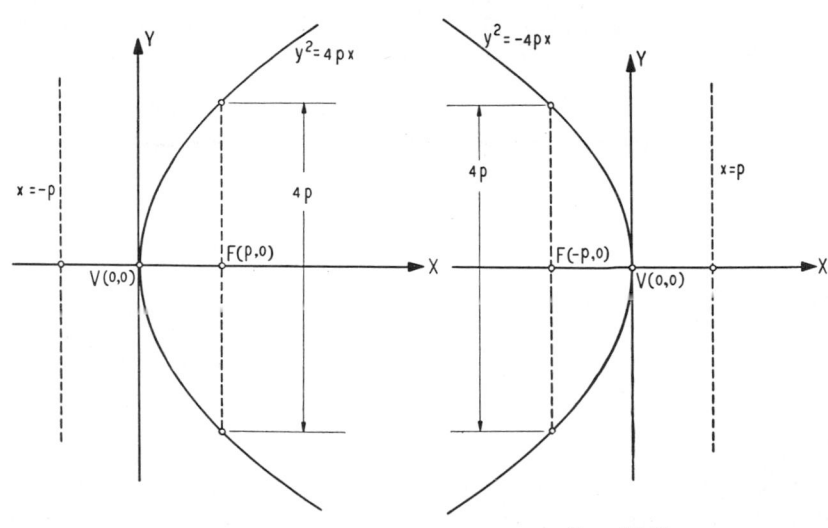

Fig. 17–5a   Fig. 17–5b

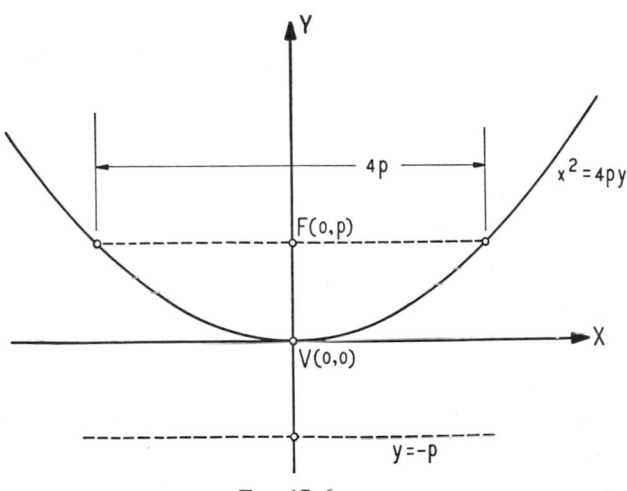

Fig. 17–6a

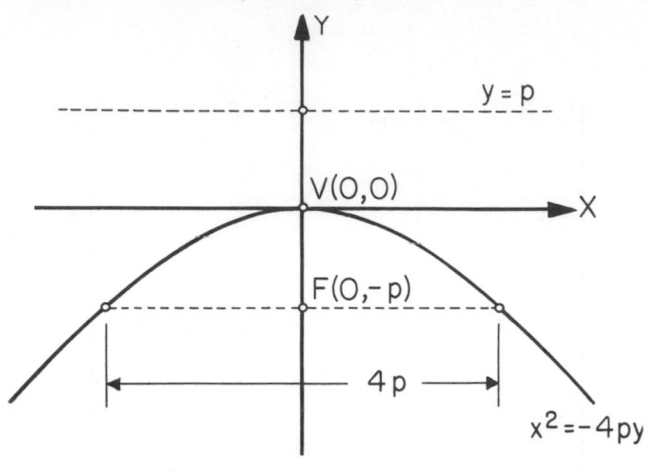

FIG. 17–6*b*

A parabola can be sketched readily by locating the vertex V and the focus F and drawing the latus rectum, since the general shape of the figure is indicated clearly in Fig. 17–5 and Fig. 17–6. The procedure is to locate these parts and then to sketch in the curve freehand.

Exercises

For each of the following parabolas, find the coordinates of vertex and focus, the equation of the directrix, and the length of latus rectum.   Sketch the curve.

1. $y^2 = 4x$.
2. $y^2 = -4x$.
3. $x^2 = 4y$.
4. $x^2 = -4y$.
5. $y^2 = 16x$.
6. $y^2 = 6x$.
7. $y^2 = -8x$.
8. $y^2 = -6x$.
9. $x^2 = 8y$.
10. $x^2 = -8y$.
11. $x^2 = 6y$.
12. $x^2 = -6y$.
13. $y^2 = 32x$.
14. $x^2 = 64y$.
15. $x^2 = -12y$.
16. $y^2 = 3x$.
17. $y^2 = x$.
18. $x^2 = -y$.
19. $5x^2 = 16y$.
20. $y^2 = -5x$.
21. $3y^2 = -10x$.

Find the equations of the following parabolas, and sketch:

22. Vertex at $(0, 0)$, focus at $(4, 0)$.
23. Vertex at $(0, 0)$, focus at $(2, 0)$.
24. Vertex at $(0, 0)$, focus at $(-6, 0)$.
25. Vertex at $(0, 0)$, focus at $(0, 4)$.
26. Vertex at $(0, 0)$, focus at $(0, 8)$.
27. Vertex at $(0, 0)$, focus at $(0, -8)$.
28. Vertex at $(0, 0)$, focus at $(3, 0)$.
29. Vertex at $(0, 0)$, focus at $(0, 5)$.
30. Vertex at $(0, 0)$ focus at $(0, -1)$.
31. Vertex at $(0, 0)$, directrix, $x - 4 = 0$.
32. Vertex at $(0, 0)$, directrix, $y + 8 = 0$.
33. Vertex at $(0, 0)$, directrix, $3y - 8 = 0$.
34. Vertex at $(0, 0)$, directrix, $2x + 7 = 0$.
35. Focus at $(4, 0)$, directrix, $x + 4 = 0$.

36. Focus at (6, 0), directrix, $x + 6 = 0$.

37. Focus at (0, 4), directrix, $y + 4 = 0$.

38. Focus at (0, −6), directrix, $y − 6 = 0$.

39. Find the equation of the parabola that has its axis coinciding with the x axis, its vertex at the origin, and that passes through the point (2, 4).

40. Find the equation of the parabola that has its axis coinciding with the x axis, its vertex at the origin, and that passes through the point (6, 2).

41. Find the equation of the parabola that has its axis coinciding with the y axis, its vertex at the origin, and that passes through the point (3, 5).

42. Find the equation of the parabola that has its axis coinciding with the y axis, its vertex at the origin, and that passes through (−4, −2).

43. Find the equation of the parabola that has its axis coinciding with the y axis, its vertex at the origin, and that passes through (1, −4).

44. Using the definition of a parabola, derive the equation of the parabola whose vertex is at (4, 0) and whose directrix is the y axis.

45. Using the definition of a parabola, derive the equation of the parabola whose vertex is at (0, 8) and whose directrix is the x axis.

17–3. The Ellipse

The geometrical definition of the ellipse which we shall use is the following.

An ellipse is the locus of a point in the plane which satisfies the condition that the sum of its distances from two fixed points is a constant, greater than the distance between the two points. The student may devise very easily, on the basis of this definition, a mechanical method of constructing an ellipse.

The two fixed points are the **foci** of the ellipse; the midpoint of the segment joining them is the **center** of the ellipse. We shall choose the coordinate system so that the center is at the origin and the foci are $F_1(c, 0)$ and $F_2(−c, 0)$ where c is positive, as in Fig. 17–7. For convenience we shall call the constant length sum $2a$, which must then be greater than $2c$ by the definition of the ellipse. Our requirement is that

$$F_1P + F_2P = 2a. \tag{1}$$

Using the distance formula, equation 1 can be written as

$$\sqrt{(x − c)^2 + y^2} + \sqrt{(x + c)^2 + y^2} = 2a$$

or

$$\sqrt{(x − c)^2 + y^2} = 2a − \sqrt{(x + c)^2 + y^2},$$

which becomes, when both sides of the equation are squared,

$$(x − c)^2 + y^2 = 4a^2 − 4a\sqrt{(x + c)^2 + y^2} + (x + c)^2 + y^2.$$

Simplifying, we obtain

$$a\sqrt{(x + c)^2 + y^2} = a^2 + cx.$$

Squaring again, we have

$$a^2(x + c)^2 + a^2 y^2 = a^4 + 2a^2 cx + c^2 x^2$$

or

$$(a^2 - c^2)x^2 + a^2 y^2 = a^2(a^2 - c^2). \tag{2}$$

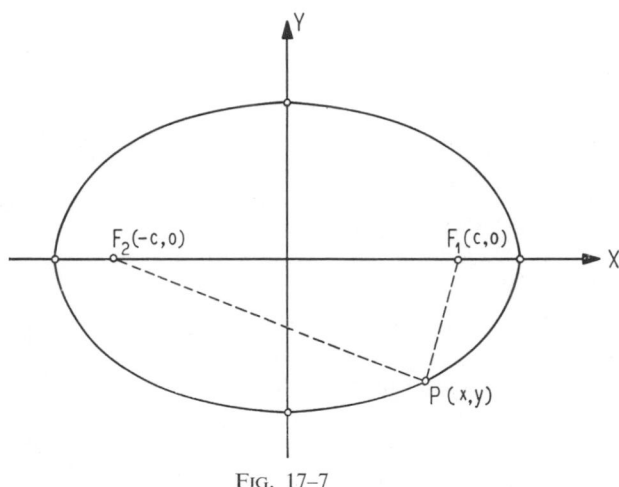

FIG. 17–7

Since $a > c$, then $a^2 - c^2$ is positive.   Setting

$$b^2 = a^2 - c^2 \tag{3}$$

in equation 2, we have

$$b^2 x^2 + a^2 y^2 = a^2 b^2. \tag{4}$$

Dividing both sides of equation 4 by $a^2 b^2$, we obtain *the standard form of the equation of the ellipse*

$$\frac{x^2}{a^2} + \frac{y^2}{b^2} = 1, \tag{5}$$

where a^2 is greater than b^2.

When $x = 0$, $y = \pm b$, and, when $y = 0$, $x = \pm a$.   The locus is plotted as shown in Fig. 17–8.

The segment $V_1 V_2$ is the **major axis,** the segment $Q_1 Q_2$ is the **minor axis.** V_1 and V_2 are the **vertices** of the ellipse.   Thus the **semimajor** axis has length a, and the **semiminor** axis has length b.

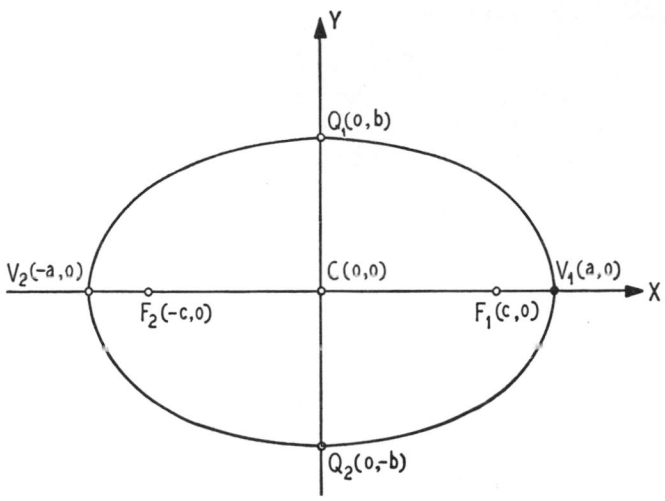

FIG. 17–8. The graph of $\dfrac{x^2}{a^2} + \dfrac{y^2}{b^2} = 1$.

The student may verify as an exercise that, if we start with foci $F_1(0, c)$ and $F_2(0, -c)$, and proceed as above, we obtain

$$\frac{y^2}{a^2} + \frac{x^2}{b^2} = 1, \tag{6}$$

where a^2 *is greater than* b^2. The locus is plotted in Fig. 17–9. In this case the major axis and the vertices are along the y axis.

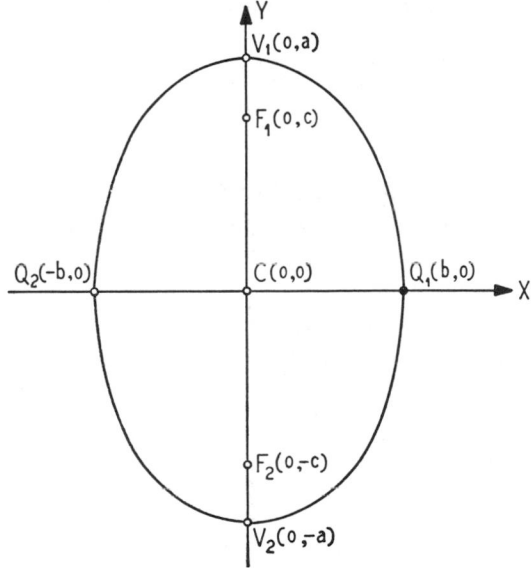

FIG. 17–9. The graph of $\dfrac{y^2}{a^2} + \dfrac{x^2}{b^2} = 1$

Example 1. For the ellipse

$$16x^2 + 25y^2 = 400,$$

find (*a*) the coordinates of the vertices, (*b*) the coordinates of the foci, (*c*) the length of the semiaxes. Sketch the ellipse, showing all these data on the figure.

Dividing both sides of the equation by 400, we obtain the standard form

$$\frac{x^2}{25} + \frac{y^2}{16} = 1.$$

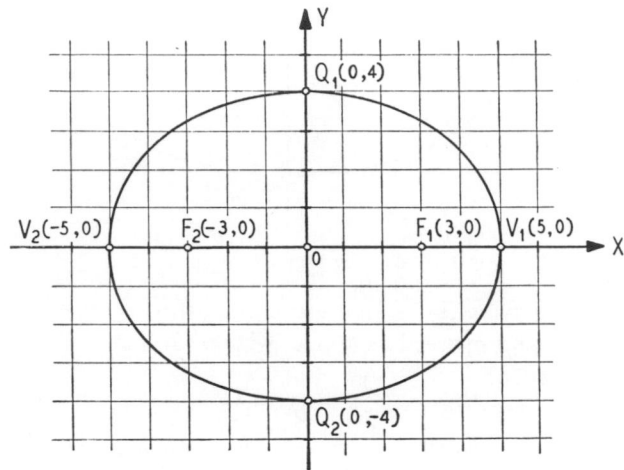

Fig. 17–10. The graph of $\dfrac{x^2}{25} + \dfrac{y^2}{16} = 1$

From this equation and the relation $b^2 = a^2 - c^2$ we obtain the following information:

Semimajor axis: $a = 5$.
Semiminor axis: $b = 4$.
Coordinates of the vertices: $(5, 0)$, $(-5, 0)$.
Coordinates of the foci: $(3, 0)$, $(-3, 0)$.

The graph is shown in Fig. 17–10.

Example 2. Find the equation of the ellipse whose foci are $(0, 4)$ and $(0, -4)$ and whose vertices are $(0, 5)$ and $(0, -5)$.

The midpoint of the line joining the foci is the origin, and hence the origin is the center of the ellipse. The major axis is along the y axis. Since $a = 5$ and $c = 4$, from $b^2 = a^2 - c^2$ we have $b^2 = 25 - 16 = 9$. Thus the standard form of the equation of the ellipse is

$$\frac{x^2}{9} + \frac{y^2}{25} = 1.$$

The ellipse is sketched in Fig. 17–11.

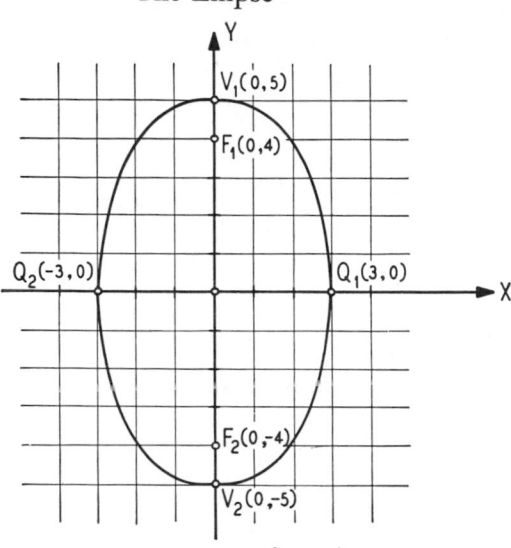

FIG. 17–11. The graph of $\dfrac{x^2}{9} + \dfrac{y^2}{25} = 1$

Exercises

For the following ellipses find (a) the coordinates of the vertices, (b) the coordinates of the foci, and (c) the lengths of the semiaxes, and sketch, showing all these data on the figure.

1. $4x^2 + 25y^2 = 100.$　　　**2.** $25x^2 + 4y^2 = 100.$　　　**3.** $9x^2 + 25y^2 = 900.$

4. $25x^2 + 9y^2 = 900.$　　　**5.** $4x^2 + 9y^2 = 36.$　　　**6.** $9x^2 + 4y^2 = 36.$

7. $4x^2 + y^2 = 16.$　　　**8.** $x^2 + 4y^2 = 16.$　　　**9.** $36x^2 + y^2 = 36.$

10. $9x^2 + 16y^2 = 144.$　　**11.** $x^2 + 49y^2 = 196.$　　**12.** $25x^2 + 36y^2 = 900.$

13. $36x^2 + 25y^2 - 400 = 0.$　　　　**14.** $49x^2 + 36y^2 - 900 = 0.$

15. $3x^2 + 4y^2 - 9 = 0.$　　　　　**16.** $4x^2 + 25y^2 - 625 = 0.$

17. $\dfrac{x^2}{4} + \dfrac{y^2}{25} - 1 = 0.$　　　　**18.** $\dfrac{x^2}{36} + \dfrac{y^2}{16} = 4.$

Find the equation of each of the ellipses satisfying the following conditions.　Sketch the curves.

19. Foci $(4, 0)$ and $(-4, 0)$; vertices $(10, 0)$ and $(-10, 0)$.

20. Foci $(0, 4)$ and $(0, -4)$; vertices $(0, 10)$ and $(0, -10)$.

21. Foci $(-3, 0)$ and $(3, 0)$; vertices $(-5, 0)$ and $(5, 0)$.

22. Foci $(0, 4)$ and $(0, -4)$; length of major axis 10.

23. Foci $(-3, 0)$ and $(3, 0)$; length of minor axis 8.

24. Foci $(8, 0)$ and $(-8, 0)$; length of semimajor axis 16.

25. Semimajor axis on the y axis and equal to 7; semiminor axis 2; center at $(0, 0)$.

26. Semimajor axis on the x axis and equal to 4; semiminor axis 1; center at $(0, 0)$.

27. Vertices (15, 0) and (−15, 0); length of minor axis 10.

28. Vertices (0, 12) and (0, −12); length of minor axis 8.

29. Vertices (0, 5) and (0, −5); length of semiminor axis 1.

30. Vertices (17, 0) and (−17, 0); length of semiminor axis 10.

31. Find the equation of the ellipse whose vertices are the points (−3, 0), (3, 0) and which passes through the point (2, 1).   Sketch the curve.

32. Find the equation of the ellipse whose vertices are the points (0, −4), (0, 4) and which passes through the point (2, 3).   Sketch the curve.

33. Find the equation of the ellipse whose vertices are the points (0, 6) and (0, −6) and which passes through the point (4, 0).   Sketch the curve.

34. Find the equation of the ellipse whose vertices are the points (−8, 0), (8, 0) and which passes through the point (6, 4).   Sketch the curve.

35. Find the equation of the ellipse whose foci are the points (−3, 0), (3, 0) and whose semimajor axis has length 5.   Sketch the curve.

36. Find the equation of the ellipse whose vertices are (10, 0) and (−10, 0) and which passes through $(5, \sqrt{6})$.   Sketch the curve.

37. Find the equation of the ellipse whose foci are the points (−3, 0), (3, 0) and whose major axis is twice the minor axis.   Sketch the curve.

38. Find the equation of the locus of a point that moves so that the sum of its distances from (0, 4) and (0, −4) is 10.   Sketch the curve.

39. Find the equation of the locus of a point that moves so that the sum of its distances from (8, 0) and (−8, 0) is 20.   Sketch the curve.

40. Find the equation of the locus of a point that moves so that the sum of its distances from (−2, 1) and (6, 1) is 10.   Sketch the curve.

41. Find the equation of the locus of a point that moves so that the sum of its distances from the points (2, 0) and (−4, 0) is equal to 10.   Sketch the curve.

17–4. The Hyperbola

The geometrical definition of the hyperbola which we shall use is the following.

The hyperbola is the locus of a point in a plane which satisfies the condition that the difference of its distances in either order from two fixed points is a positive constant, less than the distance between the two fixed points.

The two fixed points are the **foci** of the hyperbola; the midpoint of the segment joining them is the **center** of the hyperbola.   We shall choose the coordinate system so that the center is at the origin and the foci are $F_1(c, 0)$ and $F_2(−c, 0)$ where c is positive, as shown in Fig. 17–12.   For convenience we shall call the positive constant length difference $2a$, which by the definition of the hyperbola is less than $2c$.   Then we require, from Fig. 17–12, that either of these two requirements be fulfilled:

$$F_2P - F_1P = 2a, \tag{1}$$

$$F_1P - F_2P = 2a. \tag{2}$$

If equation 1 is satisfied, since $2a$ is positive, we get the right branch of the hyperbola as shown in Fig. 17–12; if equation 2 is satisfied, we get the left branch of the hyperbola. Using the distance formula to write equations 1 and 2 algebraically, we obtain, respectively,

$$\sqrt{(x+c)^2+y^2} - \sqrt{(x-c)^2+y^2} = 2a, \qquad (3)$$

$$\sqrt{(x-c)^2+y^2} - \sqrt{(x+c)^2+y^2} = 2a. \qquad (4)$$

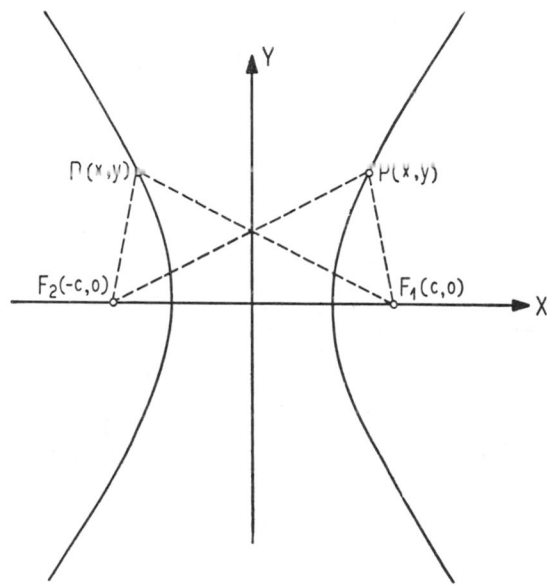

FIG. 17–12

The student may simplify both equations 3 and 4 by the method used to simplify the corresponding equation in the case of the ellipse. Both simplifications result in the equation

$$(c^2 - a^2)x^2 - a^2y^2 = a^2(c^2 - a^2). \qquad (5)$$

Since $c > a$ by definition, $c^2 - a^2 > 0$, and we may write

$$b^2 = c^2 - a^2. \qquad (6)$$

Making this substitution in equation 5, we obtain

$$b^2x^2 - a^2y^2 = a^2b^2. \qquad (7)$$

Dividing both members of equation 7 by a^2b^2, we obtain the result

$$\frac{x^2}{a^2} - \frac{y^2}{b^2} = 1, \qquad (8)$$

the *standard form* of the equation of this hyperbola. In this case, unlike that of the ellipse, b^2 can have any size relative to a^2. The student may verify that equation 8 is equivalent to equations 3 and 4 jointly.

When $y = 0$, then $x = \pm a$, and the points $V_1(a, 0)$ and $V_2(-a, 0)$ are the **vertices** of the hyperbola. The segment V_1V_2 is the **transverse axis,** whence the **semitransverse axis** has length a. When $x = 0$, y has no real values, whence the locus does not cross the y axis. The segment B_1B_2, where B_1 and B_2 have the coordinates $B_1(0, b)$, $B_2(0, -b)$, is called the **conjugate axis.** The **semiconjugate axis** has length b.

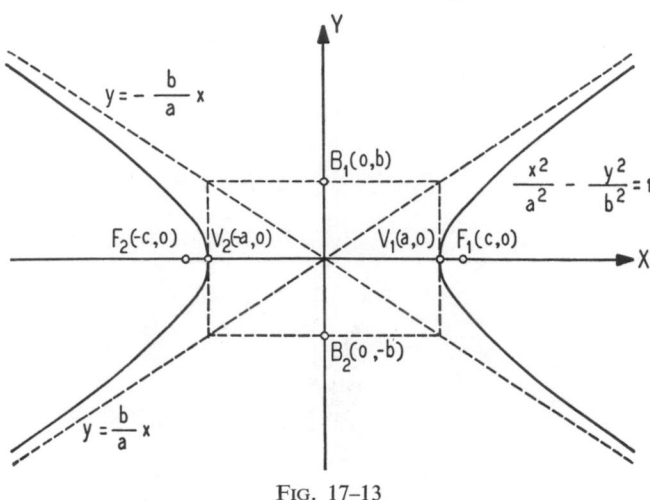

Fig. 17–13

If a fixed straight line is so related to an infinite branch of a curve that, as a point on the curve recedes indefinitely along the infinite branch, the distance of the point from the line comes as near to zero as we please, but never equals zero, then the line is called an **asymptote** of the curve (see Sec. 16–7).

The hyperbola of equation 8 has as asymptotes the lines

$$y = \frac{b}{a}x, \qquad y = -\frac{b}{a}x. \tag{9}$$

Proof of this fact is beyond the scope of this book.

To sketch the hyperbola, draw the rectangle with sides perpendicular to the x axis through the points V_1 and V_2, and with sides perpendicular to the y axis through the points B_1 and B_2, as shown in Fig. 17–13. The straight lines through the center of the hyperbola and the opposite corners of the rectangle are the asymptotes 9, as can be readily seen. After these preliminary constructions, the hyperbola can be sketched easily, as shown in Fig. 17–13.

The student may verify as an exercise that, if we start with the foci $F_1(0, c)$ and $F_2(0, -c)$ and proceed as above, we obtain the *standard form*

$$\frac{y^2}{a^2} - \frac{x^2}{b^2} = 1 \tag{10}$$

where b^2 is given by equation 6. In this case the asymptotes are

$$y = \frac{a}{b}x, \qquad y = -\frac{a}{b}x, \tag{11}$$

and the curve is sketched as shown in Fig. 17–14.

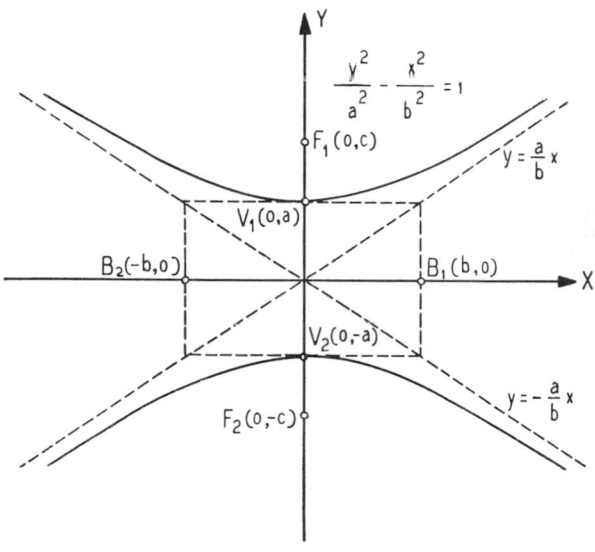

FIG. 17–14

Example. Sketch the curve of
$$16y^2 - 9x^2 = 144,$$
showing on the graph (*a*) the coordinates of the foci, (*b*) the coordinates of the vertices, and (*c*) the asymptotes.

Dividing the equation by 144, we obtain the standard form

$$\frac{y^2}{9} - \frac{x^2}{16} = 1.$$

Thus the semitransverse axis $a = 3$, and the semiconjugate axis $= 4$. The vertices are $V_1(0, 3)$ and $V_2(0, -3)$; the ends of the semiconjugate axes are $B_1(4, 0)$ and $B_2(-4, 0)$. After locating these points, we sketch the rectangular box as shown by Fig. 17–15. Then the asymptotes are drawn through the corners of the box. They are seen to be the lines

$$y = \tfrac{3}{4}x, \qquad y = -\tfrac{3}{4}x.$$

From equation 6, $c^2 = a^2 + b^2 = 9 + 16 = 25$, whence $c = 5$, and the foci are at $F_1(0, 5)$ and $F_2(0, -5)$. Finally, the curve can be sketched as shown in Fig. 17–15.

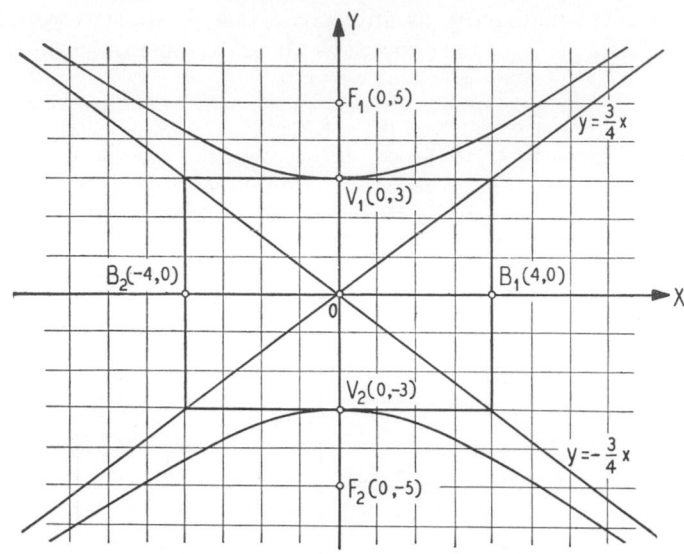

FIG. 17–15. The graph of $16y^2 - 9x^2 = 144$

Exercises

For the following hyperbolas with center at the origin, find (a) coordinates of the vertices, (b) coordinates of the foci, (c) lengths of the semiaxes, and (d) equations of the asymptotes, and sketch, showing all this information on the figure.

1. $x^2 - y^2 = 1$. **2.** $y^2 - x^2 = 1$. **3.** $x^2 - y^2 = 4$.

4. $y^2 - x^2 = 4$. **5.** $y^2 - x^2 = 9$. **6.** $x^2 - y^2 = 9$.

7. $y^2 - 4x^2 = 36$. **8.** $x^2 - 9y^2 = 36$. **9.** $x^2 - 4y^2 = 36$.

10. $25x^2 = 100 + y^2$. **11.** $9y^2 = 225 + 25x^2$. **12.** $9x^2 = 4y^2 + 36$.

13. $x^2 = y^2 + 64$. **14.** $y^2 = x^2 + 25$. **15.** $x^2 - 25y^2 = 100$.

16. $25x^2 - 144y^2 = 3600$. **17.** $9x^2 - 4y^2 - 144 = 0$. **18.** $x^2 - 2y^2 = 36$.

19. $4y^2 - x^2 = 25$. **20.** $4x^2 - 9y^2 = 25$. **21.** $4y^2 - 9x^2 - 25 = 0$.

22. $3x^2 - y^2 - 48 = 0$. **23.** $4x^2 - 5y^2 = 100$. **24.** $x^2 - 2y^2 + 18 = 0$.

Find the equations of the hyperbolas given by the following data. Sketch a graph of each.

25. Foci at $(5, 0)$ and $(-5, 0)$; vertices at $(3, 0)$ and $(-3, 0)$.

26. Foci at $(0, 5)$ and $(0, -5)$; vertices at $(0, 4)$ and $(0, -4)$.

27. Foci at $(0, 13)$ and $(0, -13)$; vertices at $(0, 5)$ and $(0, -5)$.

28. Foci at $(17, 0)$ and $(-17, 0)$; length of transverse axis 30.

29. Foci at $(0, 17)$ and $(0, -17)$; length of transverse axis 16.

30. Vertices at $(0, 3)$ and $(0, -3)$; length of conjugate axis 8.

31. Vertices at $(15, 0)$ and $(-15, 0)$; length of conjugate axis 16.

32. Foci at $(0, 10)$ and $(0, -10)$; length of conjugate axis 16.

33. Center at origin; transverse axis along the x axis and of length 8; length of conjugate axis 6.

34. Vertices at $(4, 0)$ and $(-4, 0)$; passes through $(5, 4)$.

35. Find the equation of the hyperbola with center at the origin and transverse axis on the x axis which passes through the points $(-6, 2)$ and $(5, 1)$. Sketch the hyperbola.

36. Find the equation of the hyperbola with center at the origin and transverse axis on the x axis which passes through the points $(4, 0)$ and $(8, -6)$. Sketch the hyperbola.

37. Find the equation of the hyperbola with center at the origin and transverse axis on the y axis which passes through the points $(5, 8)$ and $(1, -4)$. Sketch the hyperbola.

38. Find the equation of the hyperbola with center at the origin and transverse axis on the y axis which passes through the points $(5, 10)$ and $(\ 2, 5)$. Sketch the hyperbola.

39. Find the equation of the hyperbola whose vertices are $(4, 0)$ and $(-4, 0)$ and which passes through $(5, 4)$. Sketch the hyperbola.

40 A hyperbola with center at the origin has its transverse axis along the x axis of length 24 and conjugate axis of length 10. Find the equation of the hyperbola, and sketch its graph.

41. Find the equation of the hyperbola whose foci are $(0, 6)$ and $(0, -6)$ and whose transverse axis is twice its conjugate axis.

42. Find the equation of the hyperbola whose foci are $(3, 0)$ and $(-3, 0)$ and whose transverse axis is twice its conjugate axis.

17–5. Change of Coordinate Axes

In analytic geometry the solution of a problem can often be simplified by the use of a new set of axes, different from the set employed in the statement of the problem. In fact, we shall devote the next few sections to reducing by a change of axes a number of apparently difficult problems to ones which we have just solved.

If the new axes are parallel, respectively, to the old ones and the positive directions are the same, the transformation is called a **translation of axes.** If the origin remains unchanged and the new axes are obtained by revolving the old ones about the origin through some given angle, then the transformation is called a **rotation of axes.** It is readily seen that, in general, any change of axes in the plane can be accomplished by a translation and then a rotation or vice versa.

17–6. Translation of Coordinate Axes

Let OX and OY of Fig. 17–16 be the original axes and let $O'X'$ and $O'Y'$ be the new axes, parallel, respectively, to the old and having the same positive directions. Let the coordinates of O' referred to the original axes be (h, k).

Let P be any point in the plane, and let its coordinates referred to the old axes be (x, y), and let its coordinates referred to the new axes be (x', y'). From Fig. 17–16 we see that

$$x = NP = NN' + N'P = h + x',$$
$$y = MP = MM' + M'P = k + y',$$

whence

$$x = x' + h,$$

$$y = y' + k.$$

(1)

The above discussion depends on the picture as arranged in Fig. 17–16. The student should verify that the same formulas hold for all translations and all positions of P.

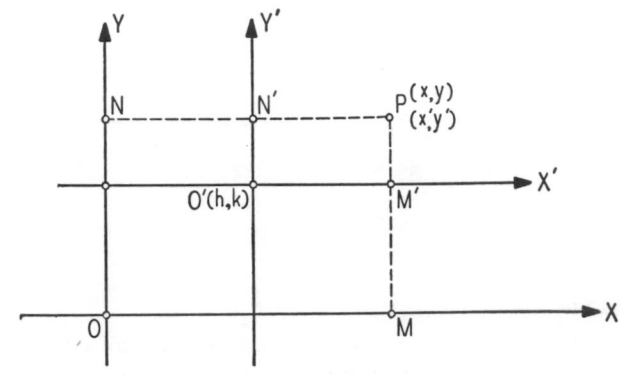

FIG. 17–16

Example. The equation of a curve is

$$4x^2 + y^2 - 24x + 4y + 24 = 0.$$

Find the equation of the same curve with respect to a new coordinate system obtained by translating the origin of the axes to the point $(3, -2)$, and sketch the curve.

We have from equations 1

$$x = x' + 3,$$

$$y = y' - 2.$$

(2)

If we make this substitution, the given equation becomes

$$4(x' + 3)^2 + (y' - 2)^2 - 24(x' + 3) + 4(y' - 2) + 24 = 0.$$

Simplifying, we obtain

$$4x'^2 + y'^2 = 16,$$

which we can write in standard form as

$$\frac{x'^2}{4} + \frac{y'^2}{16} = 1.$$

This is the equation of an ellipse. By drawing its graph with respect to the new axes we obtain the graph of the original equation with respect to the original axes. With respect to the new axes we have immediately the following information:

Vertices: V_1: $x' = 0$, $y' = 4$; V_2: $x' = 0$, $y' = -4$.

Ends of minor axis: Q_1: $x' = 2$, $y' = 0$; Q_2: $x' = -2$, $y' = 0$.

Foci: F_1: $x' = 0$, $y' = 2\sqrt{3}$; F_2: $x' = 0$, $y' = -2\sqrt{3}$.

From equations 2 we have immediately the information:

Vertices: V_1: $x = 3$, $y = 2$; V_2: $x = 3$, $y = -6$.

Ends of minor axis: Q_1: $x = 5$, $y = -2$; Q_2: $x = 1$, $y = -2$.

Foci: F_1: $x = 3$, $y = 2\sqrt{3} - 2$; F_2: $x = 3$, $y = -2\sqrt{3} - 2$.

The graph is drawn in Fig. 17–17.

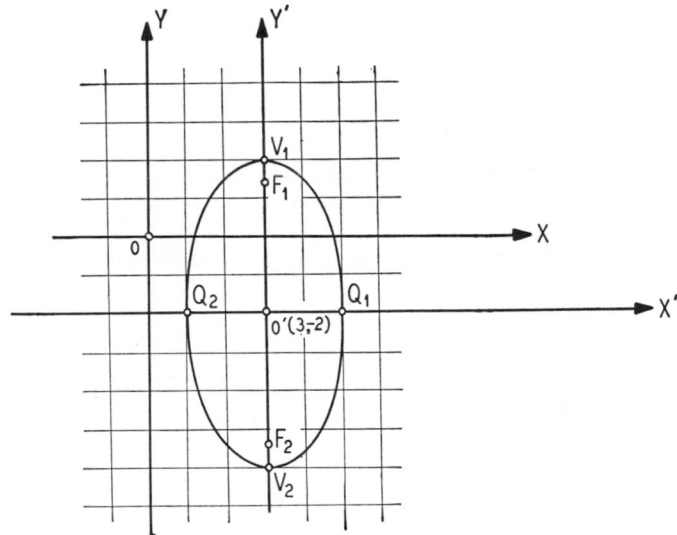

FIG. 17–17. The graph of $4x^2 + y^2 - 24x + 4y + 24 = 0$

Exercises

Find the coordinates of the following points after translating the axes to the new origin indicated. Draw a figure for each exercise to verify the results.

1. New origin $O'(2, 4)$. Points: $(2, 1)$, $(0, 2)$, $(3, 5)$, $(-1, 2)$.
2. New origin $O'(1, 3)$. Points: $(2, -1)$, $(0, 0)$, $(-1, 5)$, $(4, -3)$.
3. New origin $O'(-1, 2)$. Points: $(0, 0)$, $(2, 4)$, $(4, 2)$, $(3, -5)$.
4. New origin $O'(3, -4)$. Points: $(0, 0)$, $(-1, 3)$; $(5, -2)$, $(-1, -2)$.
5. New origin $O'(-1, -5)$. Points: $(1, 1)$, $(0, 0)$, $(-1, 3)$, $(2, -3)$.
6. New origin $O'(8, -4)$. Points: $(0, 0)$, $(6, -2)$, $(-5, 3)$, $(-7, -5)$.

7. Find the coordinates of the point $(4, -3)$ when the origin is translated to

(*a*) $(4, 6)$;   (*b*) $(-2, 3)$;   (*c*) $(5, -4)$;   (*d*) $(-3, -2)$;   (*e*) $(0, -7)$.

Transform each of the following equations so that the axes are translated to the new origin O'.   Plot a graph showing both pairs of axes and the curve.

8. $2x - y = 0$; $O'(6, 2)$.   **9.** $x - 2y + 5 = 0$; $O'(-1, 2)$.

10. $4x + 3y = 10$; $O'(2, -3)$.   **11.** $3y - x = 5$; $O'(1, 0)$.

12. $x + 2y = -5$; $O'(2, -1)$.   **13.** $7x + y - 1 = 0$; $O'(0, -3)$.

14. $x^2 + y^2 - 4x - 6y - 3 = 0$; $O'(2, 3)$.

15. $x^2 + y^2 - 8x = 0$; $O'(4, 0)$.

16. $x^2 + y^2 + 4x - 8y - 5 = 0$; $O'(-2, 4)$.

17. $3x^2 + 3y^2 + 5x + 12y = 0$; $O'(-\frac{5}{6}, -2)$.

18. $y^2 - 4x - 12y + 44 = 0$; $O'(2, 6)$.

19. $x^2 - 4x - 8y - 20 = 0$; $O'(2, -3)$.

20. $4x^2 + y^2 - 8x + 4y - 8 = 0$; $O'(1, -2)$.

21. $4x^2 + 9y^2 + 8x - 18y - 3 = 0$; $O'(-1, 1)$.

22. $9x^2 - 4y^2 - 24y - 72 = 0$; $O'(0, -3)$.

23. $x^2 - y^2 - 6x - 8y - 23 = 0$; $O'(3, -4)$.

24. $y^2 - 4x^2 + 8x + 4y - 16 = 0$; $O'(1, -2)$.

25. $(x - h)^2 + (y - k)^2 = r^2$; $O'(h, k)$.

17–7. Application of Translation of Axes to the Equations of the Conics

In the preceding sections the equations of the conics were found to be the following (the vertices of the parabolas and the centers of the other conics at the origin):

Parabola: $y^2 = 4px$, $y^2 = -4px$, $x^2 = 4py$, $x^2 = -4py$. (1)

Ellipse: $\dfrac{x^2}{a^2} + \dfrac{y^2}{b^2} = 1$, $\dfrac{y^2}{a^2} + \dfrac{x^2}{b^2} = 1$. (2)

Hyperbola: $\dfrac{x^2}{a^2} - \dfrac{y^2}{b^2} = 1$, $\dfrac{y^2}{a^2} - \dfrac{x^2}{b^2} = 1$. (3)

We shall now consider the question: *What equations can be reduced by a translation of axes to the forms* 1, 2, *or* 3?

For example, we want to know what equation can be reduced to

$$y'^2 = 4px'$$ (4)

by the substitution

$$x = x' + h, \quad y = y' + k.$$ (5)

The answer can be found by making in equation 4 the inverse substitution

$$x' = x - h, \qquad y' = y - k.$$ (6)

This substitution gives the result

$$(y - k)^2 = 4p(x - h). \tag{7}$$

Thus, any equation of the form 7 can be reduced to the form 4 by the substitution 5. It follows that the locus of equation 7 is a parabola with vertex at the point (h, k) and axis parallel to the x axis.

In like manner we can establish the following results:
The graph of each of the equations

$$(y - k)^2 = \pm 4p(x - h), \qquad (x - h)^2 = \pm 4p(y - k) \tag{8}$$

is a parabola with **vertex** (h, k) *and axis parallel to a coordinate axis.*
The graph of each of the equations

$$\frac{(x - h)^2}{a^2} + \frac{(y - k)^2}{b^2} = 1, \qquad \frac{(x - h)^2}{b^2} + \frac{(y - k)^2}{a^2} = 1 \tag{9}$$

is an **ellipse** *with* **center** (h, k) *and axes parallel to the coordinate axes.*
The graph of each of the equations

$$\frac{(x - h)^2}{a^2} - \frac{(y - k)^2}{b^2} = 1, \qquad \frac{(y - k)^2}{a^2} - \frac{(x - h)^2}{b^2} = 1 \tag{10}$$

is a **hyperbola** *with* **center** (h, k) *and axes parallel to the coordinate axes.*
The converse of each statement also holds.

Equations 8, 9, and 10 are termed **standard forms.**

Thus, knowing the vertex of a parabola, or the center of an ellipse or hyperbola, we can make use of all the facts developed thus far in the chapter, provided the axes of the conics are parallel to the coordinate axes. An example will illustrate how this can be done.

Example. Find an equation of the ellipse whose foci are the points $(2, -2)$, $(2, 6)$ and the length of whose minor axis is 6.

The center is midway between the foci, at $(2, 2)$; and hence, in equations 9, $h = 2$, $k = 2$. The distance between foci is 8, whence $c = 4$. Also, $2b = 6$, or $b = 3$. Since $a^2 = b^2 + c^2$, $a = 5$. Hence the desired equation is

$$\frac{(x - 2)^2}{9} + \frac{(y - 2)^2}{25} = 1$$

or

$$25x^2 + 9y^2 - 100x - 36y = 89.$$

We plot the locus as shown in Fig. 17–18. We see that the major axis is parallel to the y axis, the vertices are $(2, 7)$ and $(2, -3)$, and the ends of the minor axis are $(-1, 2)$ and $(5, 2)$.

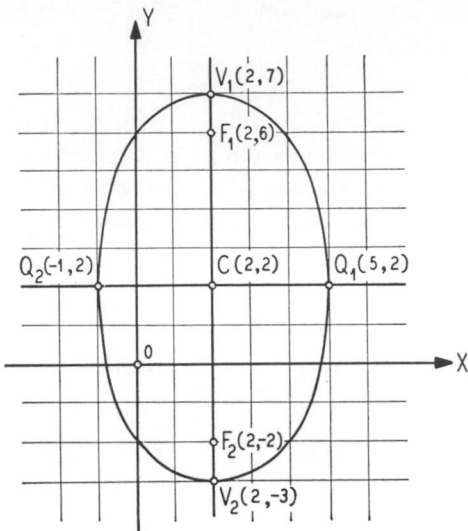

FIG. 17–18. The graph of $25x^2 + 9y^2 - 100x - 36y = 89$

Exercises

Sketch the following parabolas. Find the coordinates of vertex, focus, ends of latus rectum, and length of latus rectum. Show this information on your figure.

1. $(y - 2)^2 = 8(x - 3)$.
2. $(x - 5)^2 = -20(y + 2)$.

3. $(x + 1)^2 = 12(y + 3)$.
4. $(y + 3)^2 = 16(x - 4)$.

5. $(x + 8)^2 = -2(y - 20)$.
6. $(y - 3)^2 = 8(x + 3)$.

7. $(y + 3)^2 = 8(x - 3)$.
8. $(y + 6)^2 = -2x$.

9. $(x - \frac{1}{4})^2 = \frac{1}{2}y$.
10. $(y - 1)^2 = 2(x + \frac{1}{2})$.

11. $(x - 8)^2 = -2(y + 10)$.
12. $(x - 20)^2 = 100(y + 4)$.

Sketch the following ellipses, showing on the figure the coordinates of the center, foci, vertices, and ends of minor axis:

13. $(x - 2)^2 + 4(y + 2)^2 = 16$.
14. $25(x - 2)^2 + 36(y + 1)^2 = 900$

15. $4(x - 2)^2 + (y + 2)^2 = 16$.
16. $16(x + 1)^2 + 9(y - 1)^2 = 576$.

17. $9(x + 1)^2 + 25y^2 = 225$.
18. $9x^2 + 16(y + 2)^2 = 144$.

19. $\frac{4}{9}(x + \frac{3}{2})^2 + 4(y - 1)^2 = 1$.
20. $\dfrac{(x + 1)^2}{3} + 3(y + 2)^2 = 3$.

21. $81(x + 2)^2 + 100y^2 = 8100$.
22. $36(x - 1)^2 + 25(y - 2)^2 = 400$.

23. $2(x - a)^2 + y^2 = 2a^2$.
24. $4(x - a)^2 + (y + 2a)^2 = 4a^2$.

Sketch the following hyperbolas, showing on the figure the coordinates of the center, vertices, foci, ends of conjugate axis, and equations of asymptotes:

25. $9(x - 1)^2 - 16(y + 2)^2 = 144$.
26. $9(x - 2)^2 - 4(y - 4)^2 = 36$.

27. $4(x - 3)^2 - 9(y - 1)^2 = 36$.
28. $(x + 2)^2 - (y - 2)^2 = 64$.

29. $25(x + 2)^2 - 144(y - 3)^2 = 3600$.   **30.** $25(x - 3)^2 - y^2 = 100$.

31. $16(y + 1)^2 - x^2 = 16$.   **32.** $9(x + 2)^2 - 4(y - 1)^2 = -36$.

33. $y^2 - 4(x + 4)^2 = 36$.   **34.** $x^2 - (y + 4)^2 = -64$.

35. $(y + 7)^2 - (x - 3)^2 = 16$.   **36.** $4(y + 1)^2 - 9x^2 = 25$.

Find an equation of the conic section determined by the following conditions, and sketch, supplying on the figure the standard data indicated in the directions for the preceding exercises:

37. Parabola, vertex $(2, 2)$, focus $(2, -6)$.

38. Parabola, vertex $(-2, 4)$, focus $(-6, 4)$.

39. Ellipse, major axis 10, foci $(3, -2)$ and $(-3, -2)$.

40. Ellipse, minor axis 6, focus $(2, 4)$, vertex $(2, 6)$.

41. Ellipse, major axis 10, foci $(3, 2)$ and $(1, 2)$.

42. Ellipse, minor axis 6, focus $(-2, -2)$, vertex $(0, -2)$.

43. Ellipse, minor axis 6, focus $(-3, -1)$, vertex $(-3, 1)$.

44. Hyperbola, center $(2, 1)$, focus $(7, 1)$, vertex $(5, 1)$.

45. Hyperbola, center $(-1, 4)$, vertex $(-1, 8)$, conjugate axis 6.

46. Hyperbola, center $(2, 3)$, vertex $(2, 7)$, conjugate axis 6.

47. Hyperbola, foci $(-3, -2)$ and $(7, -2)$, conjugate axis 6.

48. Hyperbola, foci $(0, -3)$ and $(10, -3)$, conjugate axis 8.

17–8. Reduction of Equations of Second Degree to Standard Forms

If we carry out the multiplications indicated in the standard forms 8, 9, and 10 of Sec. 17–7, we see that all these equations have the general form

$$Ax^2 + Cy^2 + Dx + Ey + F = 0. \qquad (1)$$

From equations 8, 9 and 10 of Sec. 17–7 we see that

1. The standard form 8 of the equation of a parabola yields an equation of the form 1 with either $A = 0$ or $C = 0$.

2. The standard form 9 of the equation of an ellipse (which includes the standard form of the equation of a circle as a special case) yields an equation of the form 1 with A and C having the same sign.

3. The standard form 10 of the equation of a hyperbola yields an equation of the form 1 with A and C having opposite signs.

Since these standard forms yield equations of the form 1, it is clear that, conversely, there are equations of the form 1 that can be transformed into these standard forms. How this can be done is illustrated in the following examples.

Example 1. Reduce $x^2 + 6x + 8y = 7$ to standard form.
There is a squared term in x but not in y; we therefore try to write the equation in one of the standard forms of the parabola. We obtain

$$x^2 + 6x = -8y + 7.$$

Adding 9 to both sides to complete the square in the left member, we obtain

$$x^2 + 6x + 9 = -8y + 16,$$

$$(x + 3)^2 = -8(y - 2),$$

which is one of the standard forms 8 of Sec. 17–7. Thus the locus is a parabola with vertex at $(-3, 2)$ and axis parallel to the y axis.

Example 2. Reduce $24x^2 + 49y^2 - 96x + 294y - 639 = 0$ to standard form.

There are squared terms in x and y, and their coefficients have the same signs; we therefore try to reduce the equation to one of the forms of the ellipse. Thus,

$$24(x^2 - 4x) + 49(y^2 + 6y) = 639,$$

$$24(x^2 - 4x + 4) + 49(y^2 + 6y + 9) = 639 + 96 + 441,$$

$$24(x - 2)^2 + 49(y + 3)^2 = 1176,$$

which gives as the desired result:

$$\frac{(x - 2)^2}{49} + \frac{(y + 3)^2}{24} = 1.$$

Thus we have an ellipse with center at $(2, -3)$.

Example 3. Reduce $36x^2 - 25y^2 + 216x + 100y - 676 = 0$ to standard form.

There are squared terms in x and y, and their coefficients have opposite signs, and so we reduce the equation to one of the forms of the hyperbola.

$$36(x^2 + 6x) - 25(y^2 - 4y) = 676,$$

$$36(x^2 + 6x + 9) - 25(y^2 - 4y + 4) = 676 + 324 - 100,$$

$$36(x + 3)^2 - 25(y - 2)^2 = 900,$$

giving the desired result

$$\frac{(x + 3)^2}{25} - \frac{(y - 2)^2}{36} = 1.$$

This is a hyperbola with center at $(-3, 2)$.

Since these examples show that some equations of the form 1 can be reduced to the standard forms of equations of conics, it is natural to ask this question: Can equation 1 always be reduced to a standard form of the equation of a conic? This question can be answered positively if we extend our definition of conic section to include certain degenerate loci such as a point considered as a degenerate circle or ellipse, two intersecting lines considered as a degenerate hyperbola, a single line or two parallel lines considered as a degenerate parabola, and certain "imaginary" loci (including, for example, the "imaginary" circle that arises when the right member of equation 2 of Sec. 16–2 is negative). To establish this result one performs a reduction on equation 1 similar to the reductions used in the examples of this section and studies the various possibilities that can occur.   Although this analysis is not difficult, it will not be presented here; the student may wish to carry it out as an exercise.

It will be shown in Sec. 17–10 that the general equation of second degree,

$$Ax^2 + Bxy + Cy^2 + Dx + Ey + F = 0, \tag{2}$$

can be reduced to the form 1 by a rotation of axes.   It follows, therefore, that *the locus of equation 2 is a conic if we include in this term the degenerate loci.*   A more detailed analysis of these matters is beyond the scope of this book.   A textbook in analytic geometry may be consulted for such an analysis; for example, Chapter XII of *Analytic Geometry* by D. R. Curtiss and E. J. Moulton gives an excellent treatment that is unusually complete.

Exercises

Reduce each of the following equations to standard form, and sketch, showing on the graph all the standard information as indicated in the directions for the exercises following Sec. 17–7:

1. $y^2 + 8x + 8 = 0$.
2. $x^2 - 4x - 4y = 0$.
3. $x^2 - x - y = 3$.
4. $y^2 - 4x + 4y = 0$.
5. $y^2 + 16x - 2y - 95 = 0$.
6. $x^2 - 20x + y - 20 = 0$.
7. $4x^2 + y^2 = 4x$.
8. $16x^2 + y^2 + 16y = 105$.
9. $9x^2 + 4y^2 = 24y$.
10. $x^2 + 4y^2 + 8x = 8y + 5$.
11. $x^2 + 2y^2 - 2x - 4y = 1$.
12. $2x^2 + y^2 + 8x + 4y = 0$.
13. $x^2 - 2y^2 + 4x + 4y + 4 = 0$.
14. $x^2 - 2y^2 + 2x = 0$.
15. $x^2 - y^2 + 20x = 0$.
16. $x^2 - 4y^2 - 4x - 8y = 4$.
17. $x^2 - 4y^2 + 2x + 8y - 11 = 0$.
18. $4x^2 - 4y^2 + 4x - 8y - 1 = 0$.
19. $x^2 + 4y^2 - 16x + 16y + 76 = 0$.
20. $2x^2 - 24x + 3y + 78 = 0$.
21. $x^2 - 4x - 2y + 10 = 0$.
22. $x^2 - y^2 + 2x - 2y - 2 = 0$.
23. $4x^2 + 9y^2 - 8x - 36y + 4 = 0$.
24. $y^2 + 2y - 4x + 5 = 0$.
25. $16x^2 - 9y^2 - 32x - 36y - 164 = 0$.
26. $y^2 - 12x - 6y = 51$.
27. $x^2 + 4y^2 - 2x - 24y + 21 = 0$.
28. $4x^2 + 5y^2 + 16x - 20y + 31 = 0$.
29. $2x^2 - 4y^2 + 12x + 16y = 7$.
30. $x^2 + 2x - 7y + 29 = 0$.

17–9. Rotation of Axes

It was observed in Sec. 17–5 that any change of rectangular coordinate axes in the plane can be accomplished by a translation and a rotation. Changes of axes by means of translations have already been considered and found to be useful in reducing equations of the form $Ax^2 + Cy^2 + Dx + Ey + F = 0$ to the standard forms for the conics.   This section will consider changes of axes by means of rotations, and the next section will consider how rotations can be used to eliminate the xy terms in second-degree equations.

Let OX and OY (see Fig. 17–19) be the original axes, let OX' and OY' be the new axes, and let θ be the angle of rotation. Let P be any point

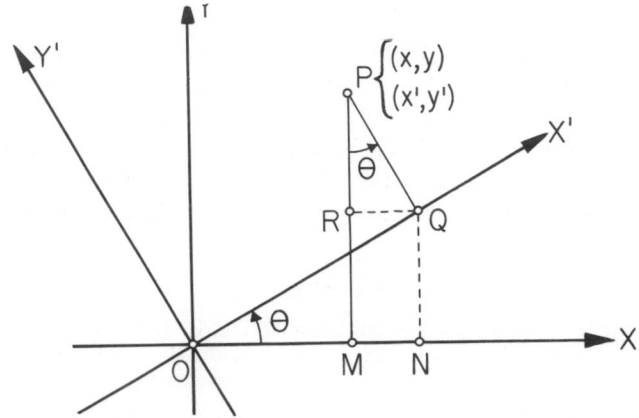

FIG. 17–19

in the plane with coordinates (x, y) and (x', y') with reference to the old and new axes, respectively. We wish to find the relation between the old coordinates (x, y) and the new coordinates (x', y'). From Fig. 17–19 we see that

$$x = OM, \qquad x' = OQ,$$

$$y = MP, \qquad y' = QP.$$

Since the sides of angle RPQ are perpendicular to the sides of angle NOQ, it follows that angle $RPQ = \theta$. Hence

$$RQ = y' \sin \theta, \qquad ON = x' \cos \theta,$$

$$RP = y' \cos \theta, \qquad NQ = x' \sin \theta.$$

Since $x = OM = ON - RQ$ and $y = MP = NQ + RP$, we have

$$x = x' \cos \theta - y' \sin \theta,$$

$$y = x' \sin \theta + y' \cos \theta. \tag{1}$$

The student should verify that these formulas hold, regardless of the position of the point P.

The substitutions given in equations 1 can be used to transform an equation with variables x and y into a new equation with variables x' and y'.

If the opposite transformation is desired, the formulas obtained by solving the formulas in (1) for x' and y' must be used:

$$x' = \quad x \cos \theta + y \sin \theta,$$
$$y' = - x \sin \theta + y \cos \theta. \tag{2}$$

Example 1. Find the equation of the locus $xy = 1$ when referred to new axes obtained from the given axes by rotating them through an angle of 45°.

Since $\theta = 45°$, $\sin \theta = 1/\sqrt{2}$, $\cos \theta = 1/\sqrt{2}$, whence from equations 1

$$x = \frac{1}{\sqrt{2}} x' - \frac{1}{\sqrt{2}} y',$$

$$y = \frac{1}{\sqrt{2}} x' + \frac{1}{\sqrt{2}} y'. \tag{3}$$

Using the substitutions given in equations 3 in the equation $xy = 1$ and simplifying, we obtain $x'^2 - y'^2 = 2$, which is the equation of a hyperbola. The graphical representation is given in Fig. 17–20.

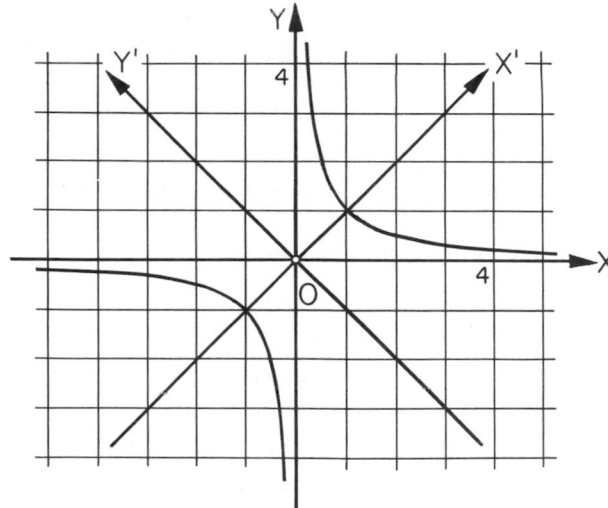

FIG. 17–20. The graph of $xy = 1$

Example 2. Find the coordinates of the point $(4, 2)$ after rotating the axes through 30°.

Here $\theta = 30°$, $\sin \theta = \frac{1}{2}$, $\cos \theta = \frac{1}{2}\sqrt{3}$. Hence, after substituting in equations 2, we have

$$x' = \tfrac{1}{2}\sqrt{3}x + \tfrac{1}{2}y, \qquad y' = -\tfrac{1}{2}x + \tfrac{1}{2}\sqrt{3}y.$$

Since $x = 4$ and $y = 2$, we have

$$x' = \tfrac{1}{2}\sqrt{3} \cdot 4 + \tfrac{1}{2} \cdot 2 = 2\sqrt{3} + 1 = 4.46,$$

$$y' = -\tfrac{1}{2} \cdot 4 + \tfrac{1}{2}\sqrt{3} \cdot 2 = \sqrt{3} - 2 = -0.27.$$

Let the reader draw the figure.

Exercises

1. Transform the equation $xy = 2$ by rotating the axes through an angle of 45°.
2. Transform the equation $xy = 5$ by rotating the axes through an angle of 45°.
3. Transform the equation $xy = k$ by rotating the axes through an angle of 45°.
4. Transform the equation $x^2 - y^2 = 4$ by rotating the axes through an angle of 45°.
5. Transform the equation $x^2 - y^2 = k$ by rotating the axes through an angle of 45°.
6. Find the coordinates of

    (a) $(2, 3)$; (b) $(5, -1)$; (c) $(-4, 0)$; (d) $(-6, 2)$; (e) $(-2, -4)$

when the axes are rotated through an angle of 30°.

7. Find the coordinates of the points of exercise 6 when the angle of rotation is 60°.
8. Show that the equation of the circle $x^2 + y^2 = r^2$ is unchanged by rotating the axes through any angle θ.
9. Transform the equation $x^2 + 4xy + y^2 - 10 = 0$ by rotating the axes through an angle of 45°.
10. Transform the equation $6x^2 - 24xy - y^2 + 30 = 0$ by rotating the axes through an angle $\theta = \arccos \tfrac{3}{5}$.
11. Transform the equation $5x^2 - 6xy + 5y^2 = 8$ by rotating the axes through an angle of 135°.
12. Show that the equation $(x^2 + y^2)^2 = ky(3x^2 - y^2)$ is left unchanged by rotating the axes through an angle of 120°.
13. Transform the equation $xy + 2x - y = 6$ by translating the axes so the new origin is set at the point $(1, -2)$, and then rotate the axes through 45°.

17–10. Simplification of Equations by Rotation

In Sec. 17–8 it was stated that the locus of the equation

$$Ax^2 + Cy^2 + Dx + Ey + F = 0, \tag{1}$$

is a conic section, provided we include in this term certain degenerate loci. We shall now show that the most general equation of the second degree

$$Ax^2 + Bxy + Cy^2 + Dx + Ey + F = 0, \quad \text{where} \quad B \neq 0, \tag{2}$$

can be reduced to the form of equation 1 by a rotation of axes. In other words, we shall show that, if the xy term is present, as in equation 2, it can be removed by rotating the axes through a properly chosen angle θ.

If we rotate the axes through an angle θ by means of formulas 1 of Sec. 17–9, the given equation 2 becomes

$$A(x' \cos \theta - y' \sin \theta)^2 + B(x' \cos \theta - y' \sin \theta)(x' \sin \theta + y' \cos \theta)$$
$$+ C(x' \sin \theta + y' \cos \theta)^2 + D(x' \cos \theta - y' \sin \theta)$$
$$+ E(x' \sin \theta + y' \cos \theta) + F = 0. \tag{3}$$

The coefficient of $x'y'$ in equation 3 is

$$B(\cos^2 \theta - \sin^2 \theta) - (A - C) \cdot 2 \sin \theta \cos \theta,$$

or

$$B \cos 2\theta - (A - C) \sin 2\theta.$$

We wish to choose θ so that the coefficient of $x'y'$ is zero, that is, so that

$$B \cos 2\theta - (A - C) \sin 2\theta = 0,$$

or

$$\cot 2\theta = \frac{A - C}{B}. \tag{4}$$

It follows that, if an angle of rotation θ is chosen that satisfies equation 4, the coefficient of $x'y'$ in equation 3 will be zero.

While the right member of equation 4 may be any positive or negative number, the properties of the cotangent make it possible to choose a value of 2θ lying between $0°$ and $180°$ for any value of the right member, a fact that may be verified by an inspection of a cotangent curve (see Fig. 5–8). Therefore, for any set of values of A, B, and C the angle θ can be chosen between $0°$ and $90°$, and we may state this result:

For any equation of the form 2 there is an angle θ between $0°$ and $90°$ such that the substitution of formulas 1 of Sec. 17–9 will transform the equation into one containing no $x'y'$ term.

Example 1. By rotating the coordinate axes, transform the equation

$$x^2 - 4xy + y^2 + 3 = 0 \tag{5}$$

into one that contains no term in $x'y'$. Identify and sketch the locus.

The given equation is of the form 2, where $A = 1$, $B = -4$, and $C = 1$.

From equation 4 we obtain $\cot 2\theta = \dfrac{1 - 1}{-4} = 0$. This relation is satisfied by

$2\theta = 90°$, or $\theta = 45°$. Using this value in equations 1 of Sec. 17–9, we obtain

$$x = \frac{1}{\sqrt{2}} x' - \frac{1}{\sqrt{2}} y' = \frac{1}{\sqrt{2}} (x' - y'),$$

$$y = \frac{1}{\sqrt{2}} x' + \frac{1}{\sqrt{2}} y' = \frac{1}{\sqrt{2}} (x' + y'),$$

as the equations of rotation. Substituting these expressions in equation 5, and simplifying, we get $x'^2 - 3y'^2 = 3$, or

$$\frac{x'^2}{(\sqrt{3})^2} \quad \frac{y'^2}{1^2} - 1, \tag{6}$$

which is the desired equation.

The curve is a hyperbola, and its graph (see Fig. 17–21) is obtained by plotting equation 6 with respect to the new axes OX' and OY'.

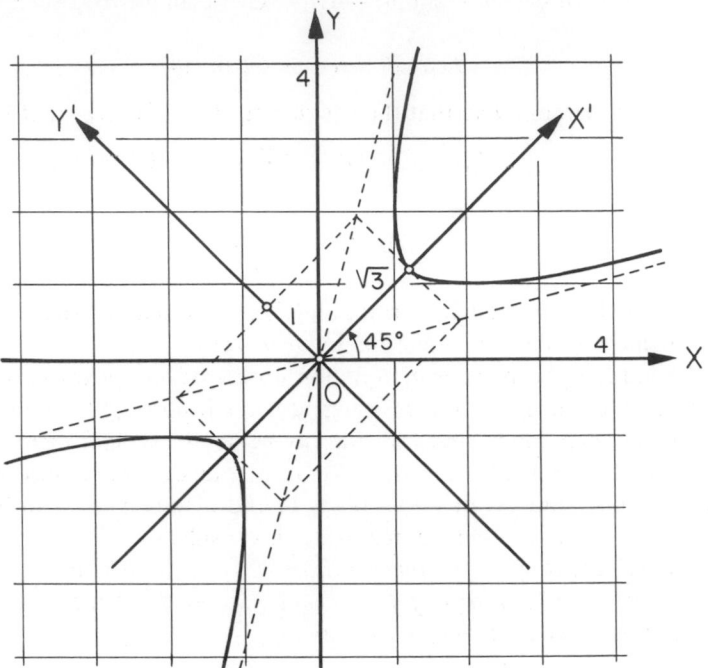

FIG. 17–21. The graph of $x^2 - 4xy + y^2 + 3 = 0$

Example 2. By rotating the coordinate axes, transform the equation

$$8x^2 + 4xy + 5y^2 - 36 = 0 \tag{7}$$

into one which contains no term in $x'y'$. Identify and sketch the locus.

The given equation is of the form 2, where $A = 8$, $B = 4$, and $C = 5$. Hence, using equation 4, we get

$$\cot 2\theta = \frac{8 - 5}{4} = \frac{3}{4}.$$

It is better not to try to determine the angle θ from tables; instead we shall find $\sin \theta$ and $\cos \theta$ by using formulas from trigonometry.

Since $\cot 2\theta = \frac{3}{4}$, we have at once $\cos 2\theta = \frac{3}{5}$. Substituting in the half-angle formulas of trigonometry (see equations 7 in Sec. 10–2), we get

$$\sin \theta = \sqrt{\frac{1 - \cos 2\theta}{2}} = \frac{1}{\sqrt{5}},$$

$$\cos \theta = \sqrt{\frac{1 + \cos 2\theta}{2}} = \frac{2}{\sqrt{5}}.$$

Using these values of $\sin \theta$ and $\cos \theta$ in equations 1 of Sec. 17–9, we get

$$x = \frac{2}{\sqrt{5}} x' - \frac{1}{\sqrt{5}} y' = \frac{1}{\sqrt{5}} (2x' - y'),$$

$$y = \frac{1}{\sqrt{5}} x' + \frac{2}{\sqrt{5}} y' = \frac{1}{\sqrt{5}} (x' + 2y').$$

Substituting these expressions in equation 7 and simplifying, we get

$$9x'^2 + 4y'^2 = 36,$$

or

$$\frac{x'^2}{4} + \frac{y'^2}{9} = 1, \tag{8}$$

which is the desired equation.

The curve is an ellipse and its graph (see Fig. 17–22) is obtained by plotting equation 8 with respect to the new axes OX' and OY'.

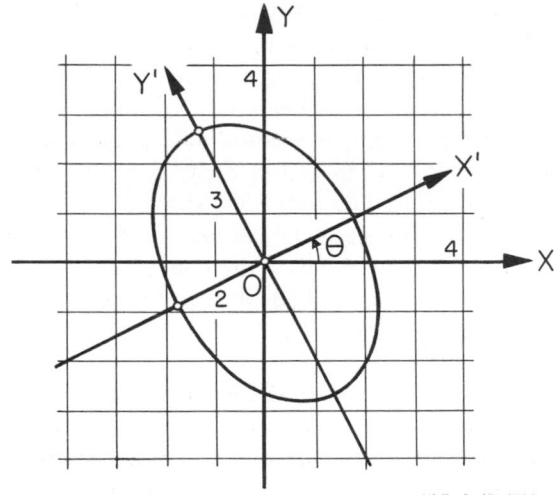

FIG. 17–22. The graph of $8x^2 + 4xy + 5y^2 - 36 = 0$

Exercises

In each of the following equations, rotate the axes through a positive acute angle so as to remove the xy term. Determine in each case the new equation, and use it to draw the locus of the original equation.

1. $3x^2 + 4xy - 4 = 0.$

2. $x^2 - xy + y^2 - 3 = 0.$

3. $9x^2 - 24xy + 41y^2 = 0.$

4. $x^2 - 4xy + 4y^2 - 8 = 0.$

5. $11x^2 + 6xy + 3y^2 = 20.$

6. $3x^2 - 10xy + 3y^2 = -32.$

7. $2x^2 - 12xy = 3y^2 - 42.$

8. $5x^2 + 6xy + 5y^2 - 8 = 0.$

9. $7x^2 + 48xy = 7y^2 - 25.$

10. $16x^2 - 24xy + 9y^2 = 30(3x + 4y).$

11. $7x^2 + 8xy + y^2 + 81 = 0.$

12. $41x^2 - 84xy + 76y^2 - 208 = 0.$

13. $52x^2 + 73y^2 = 72xy - 8x + 294y + 1167.$

14. $13x^2 + 12xy = 3y^2 - 10(4x + 3y + 1).$

15. $16x^2 + 24xy + 9y^2 = 50(20 - 3x + 4y).$

16. $8x^2 + 12xy + 17y^2 - 28x - 46y + 17 = 0.$

17. $x^2 + 24xy - 6y^2 + 4x + 48y + 34 = 0.$

18. $9x^2 - 12xy + 4y^2 - 20x - 30y - 50 = 0.$

19. $x^2 + 5xy + y^2 - 70x - 70y = 0.$

20. $25x^2 - 14xy + 25y^2 + 142x - 178y + 121 = 0.$

Progress Report

In this chapter we considered the graphs of the parabola, ellipse, and hyperbola. The equations of these loci were derived, how to sketch their graphs simply and quickly was explained, and various related items were discussed as follows:

1. For the parabola: Focus, directrix, axis, vertex, latus rectum.

2. For the ellipse: Foci, center, major and minor axes, vertices.

3. For the hyperbola: Foci, center, transverse and conjugate axes, vertices, asymptotes.

Since any change of rectangular coordinate axes in the plane can be thought of as a suitable combination of a translation and a rotation, these two changes of coordinate axes were studied:

1. The translation: After deriving the equations of the transformation, we applied them to the equations of the conics to derive new standard forms for the conics, and we applied them to the reduction of equations of the form $Ax^2 + Cy^2 + Dx + Ey + F = 0$ to the standard forms for the conics.

2. The rotation: After deriving the equations of the transformation, we used them to eliminate the xy terms from equations of second degree.

A conclusion of this chapter is that the locus of a second-degree equation is a conic, if we include in this term certain degenerate loci.

The reader should conclude this progress report on Chapter 17 by adding its principal results to the geometry-algebra dictionary begun in Chapter 15 and continued at the end of Chapter 16.

18————————————

Elements of
Solid Analytic Geometry

Just as plane analytic geometry is the application of the methods of algebra to the study of geometrical figures in the plane, so solid analytic geometry is the application of the methods of algebra to the study of figures in three-dimensional space.   Frequently, in their work, scientists and engineers are concerned with geometrical problems in three-dimensional space.   Therefore, one of the purposes of this chapter is to train the student in the visualization of spatial relations.

18–1. Directed Lines

Consider any line and a point at position A on that line (Fig. 18–1). This point can move away from A and along the line in either of two

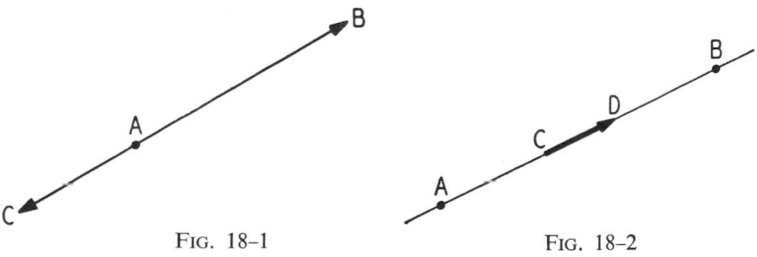

Fig. 18–1   Fig. 18–2

directions.   Suppose we move the point along the line to position B. Then the line segment AB is called a **directed line segment** with **initial point** A and **terminal point** B.   In Chapter 6 a directed line segment was called a **vector,** and we shall use the terms interchangeably in this chapter also.

The directed line segment or vector AC, obtained by moving a point from position A to position C, represents the other possiblity for direction of motion along the line.   The vectors AB and AC are said to have **opposite** directions.

A directed line segment which lies on a line may be thought of as giving its direction to the line.   Thus *a line is said to be* **directed** *if there is given a vector or directed segment that lies on the line, the* **direction of the line** *being the same as the direction of the line segment.*   For example, in Fig. 18–2, the line passing through A and B has the direction left to right because the vector CD lying on it has that direction.

The direction of a number line is always chosen in the direction of increasing values of the numbers.   Thus the usual arrows on the axes in a rectangular coordinate system indicate the directions of these lines.

18–2. Rectangular Coordinates

In Chapter 3 we adopted the convention of locating points in a plane by relating each point to a pair of real numbers called the coordinates of the point.   A simple extension of this idea can be used to locate points in space.

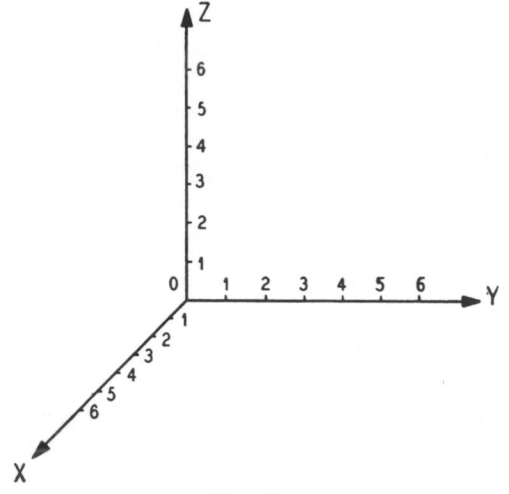

FIG. 18–3

Consider three lines in space which intersect at a point O and which are such that each line is perpendicular to the other two (Fig. 18–3).   These three lines are called the x **axis,** y **axis,** and z **axis,** and on each a unit length and a positive direction are chosen as shown in Fig. 18–3.   Usually the same unit is chosen for all three axes.   The point O at which the axes intersect is called the **origin.**   The plane determined by the x axis and the y axis is called the xy **plane** or the xy **coordinate plane;** the plane determined by the x axis and the z axis is the xz **coordinate plane;** the plane determined by the y axis and the z axis is the yz **coordinate plane.**

Now let P be any point in space (Fig. 18–4).   Pass through P a plane parallel to the xy plane, a second plane parallel to the xz plane, and a third plane parallel to the yz plane.   These planes cut the x axis, y axis, and z axis at A, B, and C, respectively (Fig. 18–4).   If we denote the number on the line OX at A by x, the number on the line OY at B by y, and the number on the line OZ by z, then the numbers (x, y, z) written in that order are called the **coordinates of the point.**   Thus OA has length x, OB has length y, and OC has length z, as shown in Fig. 18–4.   In this way we can associate a triple of numbers (x, y, z) with any point in space.   It is obvious

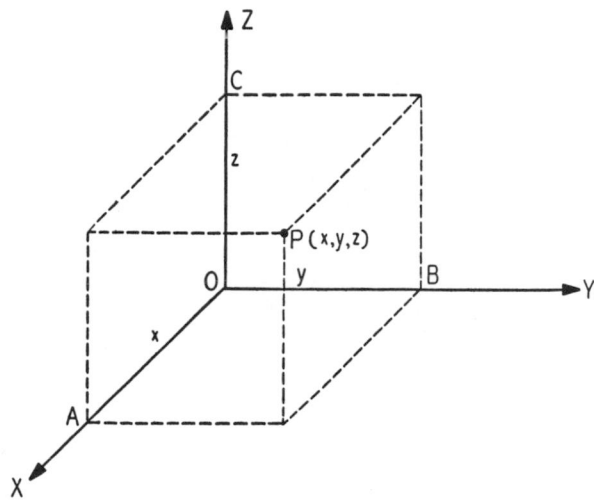

Fig. 18–4

that for the point in Fig. 18–4 all three coordinates are positive numbers. For a point on the other side of the yz plane, the x coordinate is negative; for a point on the other side of the xz plane, the y coordinate would be negative, etc.

Each of the eight parts into which the coordinate planes divide space is called an **octant.**   It is not customary to assign a numerical order to the octants.   However, the octant in which all three coordinates are positive is usually called the **first octant.**

The point P with coordinates (x, y, z) can be located by reversing the above process.   The point P can also be located as follows. Find the point A corresponding to x on the x axis, the point B corresponding to y on the y axis.   In the xy plane find D, the intersection of the perpendicular to OX at A with the perpendicular to OY at B.   Then erect a perpendicular to the xy plane at D, and on this perpendicular measure the length z, up if z is positive, down if z is negative.   Such a construction is carried out in Fig. 18–5.

That a point P has coordinates (x, y, z) is denoted by $P(x, y, z)$.   Thus $P(0, 0, 0)$ is the origin.

In this way to every point P in space there corresponds one triple of numbers (x, y, z), and to every triple of numbers (x, y, z) there corresponds one point.

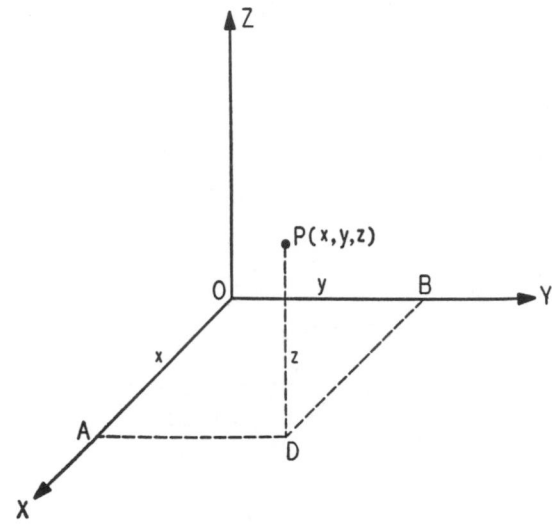

FIG. 18–5

Example. Locate on a sketch the points $P_1(3, 5, 4)$, $P_2(-5, 7, 4)$, and $P_3(4, -4, -2)$.

These points are located in Fig. 18–6.

In drawing the graphs called for in this chapter the student is urged to follow this simple principle: *Lines that are parallel in space should be drawn parallel on the graph.*   The figures shown in this chapter follow this principle.   While it does not meet, as the reader can see, every demand of artistic perspective, it will amply meet the student's mathematical needs.

Exercises

Locate the following points on a sketch:

1. $P_1(1, 2, 3)$, $P_2(-2, 4, 5)$, $P_3(0, 1, -3)$.
2. $P_1(4, 3, -5)$, $P_2(-4, -5, 2)$, $P_3(5, -3, 2)$.
3. $P_1(4, 0, 0)$, $P_2(4, -4, 0)$, $P_3(0, 0, 4)$.
4. $P_1(5, 5, 5)$, $P_2(-5, 5, 5)$, $P_3(-5, -5, 5)$.
5. $P_1(0, 0, -3)$, $P_2(0, 0, 3)$, $P_3(3, -3, -3)$.
6. $P_1(-2, 0, 0)$, $P_2(0, -2, 0)$, $P_3(0, 0, -2)$.
7. $P_1(3, -4, -2)$, $P_2(3, 2, -4)$, $P_3(-3, -4, -2)$.

8. Plot on one figure the following eight points: $(2, 3, 4)$, $(2, 3, -4)$, $(2, -3, 4)$, $(-2, 3, 4)$, $(2, -3, -4)$, $(-2, 3, -4)$, $(-2, -3, 4)$, $(-2, -3, -4)$.   What are the relative positions of these points?

9. Draw a rectangular box in the first octant with one vertex at the origin if the lengths of the sides are 3, 5, and 7, and if three of the faces lie in the coordinate planes.

10. Describe the positions of the points whose coordinates satisfy the following conditions:
(a) $x = 0, y = 0$;   (b) $x = 0, z = 0$;   (c) $y = 0, z = 0$;   (d) $x = 0$;   (e) $y = 0$;
(f) $z = 0$;   (g) $x = 1$;   (h) $y = 2$;   (i) $x = 3, y = 1$;   (j) $x = y$;   (k) $x = z$;
(l) $x = y, z = 0$;   (m) $x = z, y = 0$;   (n) $y = z, x = 0$.

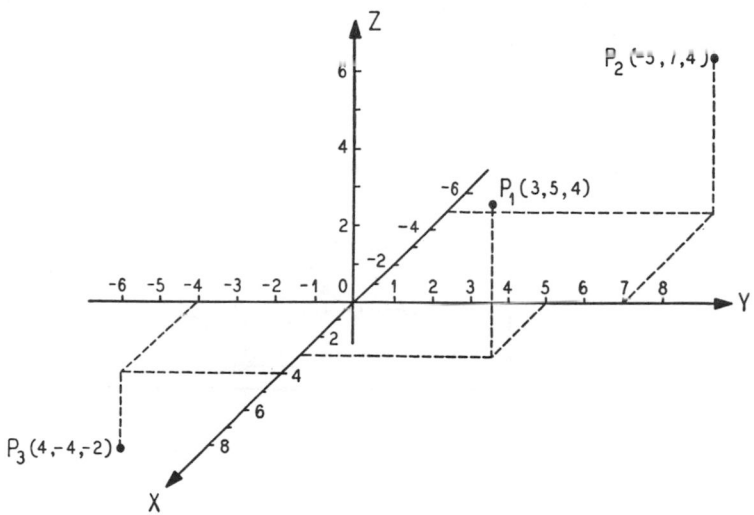

FIG. 18–6

18–3. The Distance between Two Points

Let $P_1(x_1, y_1, z_1)$ and $P_2(x_2, y_2, z_2)$ be any two distinct points.   Through each of the points pass three planes, one parallel to each coordinate plane. These planes intersect and form a rectangular parallelepiped which has $P_1 P_2$ as a diagonal (Fig. 18–7).   It is easily seen that the dimensions of the rectangular parallelepiped are

$$P_1 E = x_2 - x_1, \qquad EF = y_2 - y_1, \qquad FP_2 = z_2 - z_1. \qquad (1)$$

Since the triangles $P_1 EF$ and $P_1 FP_2$ are right triangles,

$$(P_1 F)^2 = (P_1 E)^2 + (EF)^2$$

and

$$(P_1 P_2)^2 = (P_2 F)^2 + (P_1 F)^2,$$

whence

$$(P_1 P_2)^2 = (P_2 F)^2 + (P_1 E)^2 + (EF)^2. \qquad (2)$$

Combining the results of equations 1 and 2, we have that the distance d between P_1 and P_2 is given by

$$d = \sqrt{(x_2 - x_1)^2 + (y_2 - y_1)^2 + (z_2 - z_1)^2}. \qquad (3)$$

Note that the order of subscripts does not matter. If P_2 is at the origin, equation 3 becomes

$$d = \sqrt{x_1^2 + y_1^2 + z_1^2}. \qquad (4)$$

The proof of equation 3 given above depends on the way Fig. 18–7 is drawn. However, the student should satisfy himself that the result is the same, no matter what the relative positions of P_1 and P_2 are.

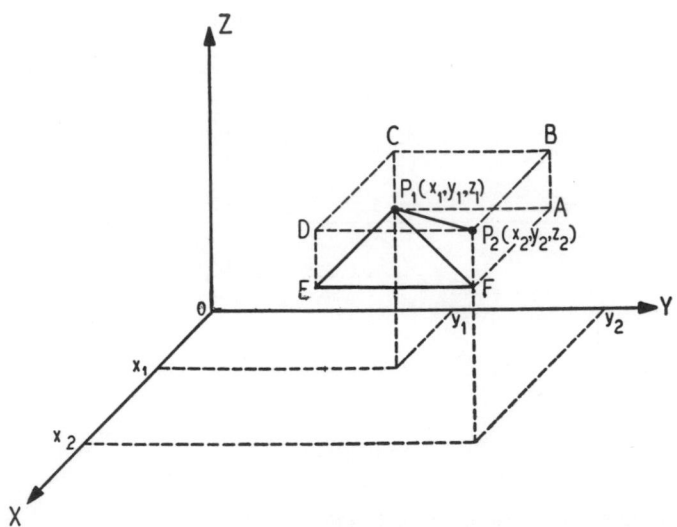

FIG. 18–7

Exercises

Find the distance between the points of each pair.

1. $(4, 1, 1)$ and $(0, 0, 0)$.   2. $(1, 2, 3)$ and $(4, -1, 6)$.

3. $(5, 1, 2)$ and $(3, 0, -3)$.   4. $(0, 0, 0)$ and $(6, 3, 2)$.

5. $(0, 0, -10)$ and $(10, 0, 0)$.   6. $(1, 1, 1)$ and $(-1, -1, -1)$.

7. $(6, -3, -2)$ and $(5, -9, 3)$.   8. $(-6, -3, -1)$ and $(3, 4, 5)$.

9. Show that $(5, 4, -2)$, $(7, 3, 4)$ and $(1, 2, -1)$ are vertices of a right triangle, and find its area.

10. Show that $(-8, 4, 10)$, $(-2, 10, 4)$ and $(-6, 6, 0)$ are vertices of an isosceles triangle.

11. Show that $(-4, 1, -3)$, $(2, 0, 4)$ and $(3, 7, -2)$ are vertices of an equilateral triangle.

12. Show that $(2, 2, 2)$, $(-1, 3, 6)$, $(1, 0, -1)$ and $(-2, 1, 3)$ are vertices of a quadrilateral whose opposite sides are equal in length.

13. Show that the points $(4, 3, 1)$, $(3, 1, 2)$ and $(1, -3, 4)$ lie on a line.

14. Find the equation of the locus of a point whose distance from the origin is equal to 6. What locus is defined by this equation?

15. What is the locus of the equation $x^2 + y^2 + z^2 = 9$?

16. Find the equation of the locus of a point whose distance from the point $(-7, 3, -2)$ is equal to 5. What locus is defined by this equation?

17. What locus is defined by the equation $(x + 1)^2 + (y - 5)^2 + (z + 3)^2 = 49$?

18. Find the equation of the locus of a point whose distances from $(-1, 2, 3)$ and $(4, 6, 8)$ are equal. What locus is defined by this equation?

19. Find the equation of the locus of a point whose distances from $(-2, 1, -2)$ and $(2, -2, 3)$ are equal. What locus is defined by this equation?

20. Find the equation of a plane that bisects perpendicularly the line segment joining the points $(-2, 3, 2)$ and $(6, 5, -6)$.

18–4. The Angle between Two Lines

Since two lines in space may not meet, we must define what we mean by the angle between them.

Let L_1 and L_2 be two lines in space, and assume that directions have been assigned to these two lines. Choose a rectangular coordinate system in the space. Now, parallel to L_1, draw a directed segment l_1 which has its

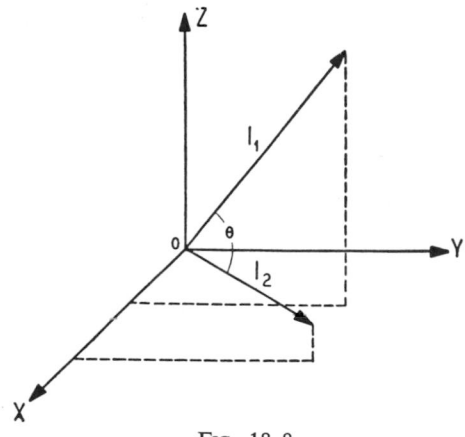

FIG. 18–8

initial point at the origin and the same direction as L_1; parallel to L_2 draw a directed segment l_2 which has its initial point at the origin and the same direction as L_2 (Fig. 18–8). These two segments determine a plane, and in this plane they determine many angles. We shall define the angle between L_1 and L_2 as the angle θ between $0°$ and $180°$ determined by l_1 and l_2. Summarizing, we have this definition:

The **angle θ between two directed lines L_1 and L_2** *is the angle between* *0° and 180° determined by the two directed segments l_1 and l_2 with initial points at the origin, which are parallel to, and have the same direction as, L_1 and L_2, respectively.*

18–5. Direction Cosines

In plane analytic geometry the direction of a line with respect to the coordinate axes is given by the angle of inclination, or its tangent, called

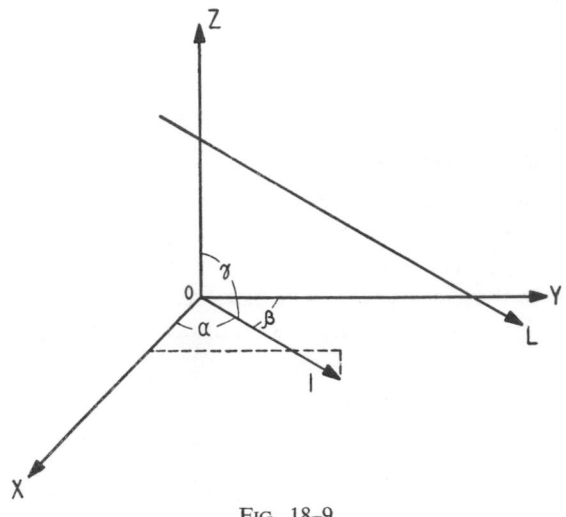

Fig. 18–9

the slope of the line.   In solid analytic geometry we shall give the direction of a line with respect to the axes by three angles or by the cosines of these angles.

Let L be a directed line in space.   As usual, we shall assume the co-ordinate axes to be directed as indicated by the arrows in Fig. 18–9.   Let l be a directed segment with initial point at the origin parallel to L and with the same direction as L.   Let α be the angle between l and the x axis, β the angle between l and the y axis, and γ the angle between l and the z axis (Fig. 18–9).   These are then by definition the angles that L makes with the axes.   The angles α, β, and γ are called the **direction angles** of L. The cosines of these angles, $\cos \alpha$, $\cos \beta$, and $\cos \gamma$, are called the **direction cosines** of L.

Obviously, *two lines that are parallel and have the same direction have the same direction angles and the same direction cosines.*   Let L_1 be a line parallel to L but with the opposite direction.   Then the parallel directed segment l_1 with initial point at the origin has the direction opposite to that of l. Since l and l_1 are parallel, l_1 is then an extension of l through the origin. Then the direction angles α_1, β_1, and γ_1 of L_1 are

$$\alpha_1 = 180° - \alpha, \qquad \beta_1 = 180° - \beta, \qquad \gamma_1 = 180° - \gamma.$$

Since $\cos(180° - \theta) = -\cos\theta$, it follows that

$$\cos\alpha_1 = -\cos\alpha, \qquad \cos\beta_1 = -\cos\beta, \qquad \cos\gamma_1 = -\cos\gamma.$$

Thus *the direction cosines of two parallel lines that have opposite directions are opposite in sign.*

If the coordinates of two points on a line are given and its direction is known, the direction cosines of the line can be determined. Let L be a directed line on which the two points $P_1(x_1, y_1, z_1)$ and $P_2(x_2, y_2, z_2)$ are given, such that P_1 is the initial point of the directed segment P_1P_2 which has the same direction as L (Fig. 18–10). For convenience, let us suppose

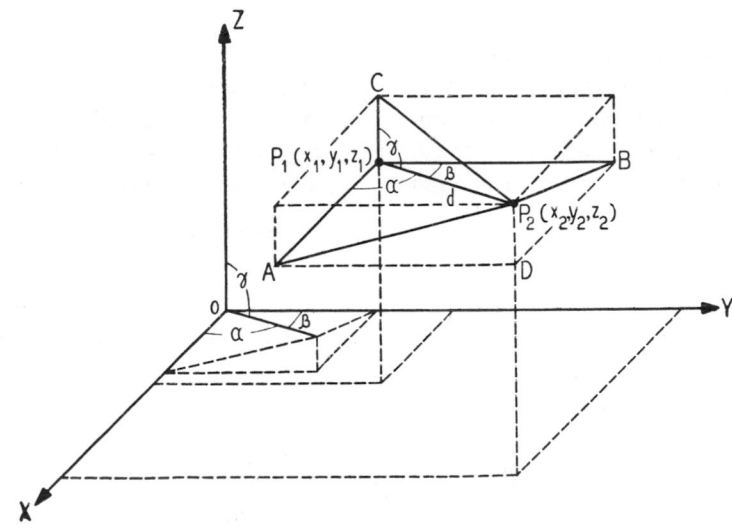

FIG. 18–10

as in Fig. 18–10 that the direction angles α, β, and γ of L are all between $0°$ and $90°$. The sides of the rectangle AP_1BD are parallel to the x and y axes, and P_1C is parallel to the z axis. It is evident that angle AP_1P_2 is equal to α, angle BP_1P_2 is equal to β, and angle CP_1P_2 is equal to γ, as shown. Now the triangles AP_1P_2, BP_1P_2, CP_1P_2 are right triangles. Since

$$d = P_1P_2 = \sqrt{(x_2 - x_1)^2 + (y_2 - y_1)^2 + (z_2 - z_1)^2}$$

and

$$P_1A = x_2 - x_1, \quad P_1B = y_2 - y_1, \quad P_1C = z_2 - z_1,$$

it is obvious that

$$\cos \alpha = \frac{x_2 - x_1}{d}, \qquad \cos \beta = \frac{y_2 - y_1}{d}, \qquad \cos \gamma = \frac{z_2 - z_1}{d}. \quad (1)$$

If α is greater than $90°$, then $\cos \alpha$ is negative. It is easily seen that in this case $\left| \cos \alpha \right| = \left| \dfrac{x_2 - x_1}{d} \right|$, but if $\alpha > 90°$ then $x_2 < x_1$; $x_2 - x_1$ is negative; and $\cos \alpha = \dfrac{x_2 - x_1}{d}$. In this way it can be shown in general that

If L is a directed line passing through the points $P_1(x_1, y_1, z_1)$ and $P_2(x_2, y_2, z_2)$, and if the vector with initial point P_1 and terminal point P_2 has the same direction as L, then the direction cosines of L are

$$\frac{x_2 - x_1}{d}, \qquad \frac{y_2 - y_1}{d}, \qquad \frac{z_2 - z_1}{d}, \qquad (2)$$

where d is the length of the segment P_1P_2.

Since the direction angles are by definition between $0°$ and $180°$, the cosines 2 determine the direction angles uniquely (see Sec. 10–6).

If the direction of L is reversed, the new direction is that of the directed segment P_2P_1 with P_2 as initial point. By the rule the direction cosines are

$$\frac{x_1 - x_2}{d}, \qquad \frac{y_1 - y_2}{d}, \qquad \frac{z_1 - z_2}{d},$$

the negatives of those given in expressions 2. This verifies the remark made previously that two parallel lines that have opposite directions have direction cosines that are opposite in sign.

From expressions 2 we have that

$$\cos^2 \alpha + \cos^2 \beta + \cos^2 \gamma = \frac{(x_2 - x_1)^2 + (y_2 - y_1)^2 + (z_2 - z_1)^2}{d^2} = 1.$$

Thus, if α, β, and γ are the direction angles of a line,

$$\cos^2 \alpha + \cos^2 \beta + \cos^2 \gamma = 1. \qquad (3)$$

From equation 3 we see that, if any two direction angles or direction cosines are given, then the third direction cosine is determined except for sign.

Example. Find the direction cosines of the line passing through the points $P_1(1, 4, -2)$ and $P_2(-1, 0, 4)$, the direction of the line being given by the directed segment P_1P_2 which has P_1 as its initial point.

The distance d between the two points is

$$d = \sqrt{(-1 - 1)^2 + (0 - 4)^2 + (4 - [-2])^2}$$

$$= \sqrt{4 + 16 + 36} = \sqrt{56} = 2\sqrt{14}.$$

Then by the rule the direction cosines are

$$\cos \alpha = \frac{-1-1}{2\sqrt{14}} = -\frac{\sqrt{14}}{14} = -0.267,$$

$$\cos \beta = \frac{0-4}{2\sqrt{14}} = -\frac{\sqrt{14}}{7} = -0.535,$$

$$\cos \gamma = \frac{4-(-2)}{2\sqrt{14}} = \frac{3\sqrt{14}}{14} = 0.802,$$

and the direction angles are

$$\alpha = 105.5°, \qquad \beta = 122.3°, \qquad \gamma = 36.7°.$$

As a check on the accuracy of the direction cosines we may verify that

$$\cos^2 \alpha + \cos^2 \beta + \cos^2 \gamma = \left(-\frac{\sqrt{14}}{14}\right)^2 + \left(-\frac{\sqrt{14}}{7}\right)^2 + \left(\frac{3\sqrt{14}}{14}\right)^2$$

$$= \frac{14+56+126}{196} = \frac{196}{196} = 1.$$

Exercises

Find the direction cosines of the line passing through the given points P_1 and P_2, the direction of the line being given by the directed segment P_1P_2 which has P_1 as its initial point. Find the direction angles to the nearest tenth of a degree, and check the direction cosines by showing that the sum of their squares is 1.

1. $P_1(0, 0, 0)$, $P_2(6, -9, 2)$.
2. $P_1(0, 0, 0)$, $P_2(0, 3, 0)$.
3. $P_1(0, 0, 5)$, $P_2(0, 0, 0)$.
4. $P_1(0, -1, 0)$, $P_2(0, 0, 0)$.
5. $P_1(0, 0, 2)$, $P_2(2, 0, 0)$.
6. $P_1(0, -5, 0,)$ $P_2(-5, 0, 0)$.
7. $P_1(1, 5, 2)$, $P_2(7, 2, 4)$.
8. $P_1(2, 8, 3)$, $P_2(4, 1, 6)$.
9. $P_1(1, 2, 3)$, $P_2(4, 5, 8)$.
10. $P_1(-1, 0, 5)$, $P_2(5, 1, -2)$.
11. $P_1(2, 8, 3)$, $P_2(4, 1, 6)$.
12. $P_1(-4, 3, 1)$, $P_2(0, 1, 3)$.

Using relation 3 of this section, find the direction cosines and direction angles not specified by the following data.

13. $\alpha = 60°$, $\gamma = 90°$.
14. $\alpha = 30°$, $\beta = 90°$.
15. $\alpha = 60°$, $\beta = 120°$.
16. $\beta = 45°$, $\gamma = 120°$.
17. $\beta = 135°$, $\gamma = 90°$.
18. $\alpha = 45°$, $\beta = 45°$.
19. $\alpha = 135°$, $\beta = 45°$.
20. $\alpha = \beta = \gamma$.

21. What are the direction angles and direction cosines of the x axis?

22. What are the direction angles and direction cosines of a line parallel to the y axis but with the opposite direction?

23. What are the direction angles and direction cosines of the z axis?

24. Using direction cosines, show that the points $(1, 4, -2)$, $(5, 8, 1)$, and $(-7, -4, -8)$ lie on a line.

25. Show that the points $(5, 3, 9)$, $(3, -1, 7)$, and $(7, 7, 11)$ lie on a line.

26. Show that the points $(-3, -1, 5)$, $(3, -4, 7)$, $(0, 5, -1)$, and $(6, 2, 1)$ are the vertices of a parallelogram.

27. Using direction cosines, show that the points $(3, 3, 3)$, $(1, 2, -1)$, $(4, 1, 1)$, and $(6, 2, 5)$ are the vertices of a parallelogram.

18–6. A Formula for the Angle between Two Lines

Let L_1 and L_2 be two directed lines with direction angles α_1, β_1, γ_1, and α_2, β_2, γ_2, respectively, and let θ be the angle between L_1 and L_2. Let l_1 and l_2 be directed segments with initial points at the origin, parallel to and with the same directions as L_1 and L_2, respectively (Fig. 18–11). Then

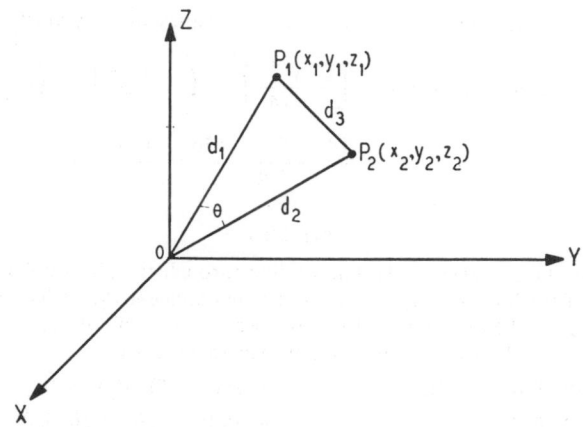

FIG. 18–11

by definition the angle θ between l_1 and l_2 is the angle between L_1 and L_2. By definition the direction angles α_1, β_1, and γ_1 of L_1 are the direction angles of l_1; similarly α_2, β_2, and γ_2 are the direction angles of l_2. Let $P_1(x_1, y_1, z_1)$ be the terminal point of l_1, $P_2(x_2, y_2, z_2)$ be the terminal point of l_2. Denote the length of l_1 by d_1, the length of l_2 by d_2. Denote the length of the segment P_1P_2 by d_3. Applying the law of cosines (see Sec. 11–2) to the triangle P_1OP_2, we have that

$$\cos \theta = \frac{d_1{}^2 + d_2{}^2 - d_3{}^2}{2d_1d_2}. \tag{1}$$

Since

$$d_1{}^2 = x_1{}^2 + y_1{}^2 + z_1{}^2,$$

$$d_2{}^2 = x_2{}^2 + y_2{}^2 + z_2{}^2,$$

$$d_3{}^2 = (x_2 - x_1)^2 + (y_2 - y_1)^2 + (z_2 - z_1)^2,$$

from equation 1 we have that

$$\cos\theta = \frac{x_1 x_2 + y_1 y_2 + z_1 z_2}{d_1 d_2}. \qquad (2)$$

This expression can be written in the form

$$\cos\theta = \left(\frac{x_1}{d_1}\right)\left(\frac{x_2}{d_2}\right) + \left(\frac{y_1}{d_1}\right)\left(\frac{y_2}{d_2}\right) + \left(\frac{z_1}{d_1}\right)\left(\frac{z_2}{d_2}\right). \qquad (3)$$

Now

$$\cos\alpha_1 = \frac{x_1}{d_1}, \qquad \cos\alpha_2 = \frac{x_2}{d_2};$$

$$\cos\beta_1 = \frac{y_1}{d_1}, \qquad \cos\beta_2 = \frac{y_2}{d_2};$$

$$\cos\gamma_1 = \frac{z_1}{d_1}, \qquad \cos\gamma_2 = \frac{z_2}{d_2}.$$

Using these formulas in equation 3, we have this result:

If the direction angles of L_1 are α_1, β_1, γ_1 and the direction angles of L_2 are α_2, β_2, γ_2, then the angle θ between L_1 and L_2 is given by

$$\cos\theta = \cos\alpha_1 \cos\alpha_2 + \cos\beta_1 \cos\beta_2 + \cos\gamma_1 \cos\gamma_2. \qquad (4)$$

Since θ by definition lies between $0°$ and $180°$, equation 4 determines θ uniquely (see Sec. 10–6).

If L_1 and L_2 are parallel, $\theta = 0°$ and

$$\cos\alpha_1 \cos\alpha_2 + \cos\beta_1 \cos\beta_2 + \cos\gamma_1 \cos\gamma_2 = 1. \qquad (5)$$

If L_1 and L_2 are perpendicular, $\theta = 90°$ and

$$\cos\alpha_1 \cos\alpha_2 + \cos\beta_1 \cos\beta_2 + \cos\gamma_1 \cos\gamma_2 = 0. \qquad (6)$$

18–7. Direction Numbers

Any three numbers a, b, c that are proportional to the direction cosines of a directed line and also have their respective signs are called a set of **direction numbers** of the line. Thus, *if L is a line with direction angles α, β, γ, then any three numbers a, b, c such that*

$$a = k\cos\alpha, \qquad b = k\cos\beta, \qquad c = k\cos\gamma, \qquad (1)$$

where k is positive, form a set of direction numbers for L.

If $P_1(x_1, y_1, z_1)$ and $P_2(x_2, y_2, z_2)$ are two points on L, and L is directed from P_1 to P_2, then from equations 1 of Sec. 18–5 we have that

$$x_2 - x_1 = d\cos\alpha, \qquad y_2 - y_1 = d\cos\beta, \qquad z_2 - z_1 = d\cos\gamma$$

where d is positive. Thus *the differences $x_2 - x_1$, $y_2 - y_1$, $z_2 - z_1$ form a set of direction numbers for L.*

Example 1. For the example of Sec. 18–5, the numbers -2, -4, 6 form a set of direction numbers, for they are $2\sqrt{14}$ times the direction cosines. Obviously -4, -8, 12, and -1, -2, 3 are also sets of direction numbers for this line. On the other hand 2, 4, -6, and 8, 16, -24 are sets of direction numbers for a parallel but oppositely directed line.

If a set of direction numbers a, b, c is given, it is possible to find the direction cosines. From equations 1 and the identity 3 of Sec. 18–5,

$$a^2 + b^2 + c^2 = k^2(\cos^2 \alpha + \cos^2 \beta + \cos^2 \gamma) = k^2.$$

Since by definition k is positive, we have

$$k = \sqrt{a^2 + b^2 + c^2}.$$

Thus finally we have

$$\cos \alpha = \frac{a}{\sqrt{a^2 + b^2 + c^2}}, \qquad \cos \beta = \frac{b}{\sqrt{a^2 + b^2 + c^2}},$$

$$\cos \gamma = \frac{c}{\sqrt{a^2 + b^2 + c^2}}. \tag{2}$$

Obviously, *two lines are parallel if their direction numbers a_1, b_1, c_1 and a_2, b_2, c_2 are proportional,* that is, if

$$a_1 = ka_2, \qquad b_1 = kb_2, \qquad c_1 = kc_2, \tag{3}$$

where k is a positive or negative constant. If none of the direction numbers are zero, equations 3 can be written

$$\frac{a_1}{a_2} = \frac{b_1}{b_2} = \frac{c_1}{c_2}. \tag{4}$$

If L_1 and L_2 are two lines with direction numbers a_1, b_1, c_1 and a_2, b_2, c_2, respectively, by Sec. 18–6 and equations 2 above, *the angle θ between L_1 and L_2 is given by*

$$\cos \theta = \frac{a_1 a_2 + b_1 b_2 + c_1 c_2}{\sqrt{a_1^2 + b_1^2 + c_1^2}\,\sqrt{a_2^2 + b_2^2 + c_2^2}}. \tag{5}$$

The two lines are perpendicular if $\cos \theta = 0$, *whence the lines are perpendicular if*

$$a_1 a_2 + b_1 b_2 + c_1 c_2 = 0, \tag{6}$$

and conversely.

Example 2. Given a line L_1 with direction numbers 1, −2, 2, and a line L_2 with direction numbers 3, 1, −2. Find the angle between the lines.

By equation 5

$$\cos \theta = \frac{(1 \cdot 3) - (2 \cdot 1) - (2 \cdot 2)}{\sqrt{1^2 + (-2)^2 + 2^2} \sqrt{3^2 + 1^2 + (-2)^2}}$$

$$= \frac{-3}{\sqrt{9}\sqrt{14}} = -\frac{\sqrt{14}}{14} = -0.267,$$

whence

$$\theta = 105.5°.$$

Exercises

Find a set of direction numbers for the line passing through the given points P_1 and P_2, the direction of the line being given by the directed segment P_1P_2 which has P_1 as its initial point.

1. $P_1(4, 7, 3)$, $P_2(-5, -3, 4)$.	**2.** $P_1(0, -1, 2)$, $P_2(1, -5, 0)$.

3. $P_1(3, -1, 5)$, $P_2(7, 0, -3)$.	**4.** $P_1(0, 7, 9)$ $P_2(1, -1, 0)$.

5. $P_1(7, 0, 5)$, $P_2(-1, 5, 3)$.	**6.** $P_1(-2, -1, 9)$, $P_2(-3, -7, 0)$.

Find the direction cosines of the lines with the following direction numbers. Find the direction angles to the nearest tenth of a degree.

7. 2, 0, 0.	**8.** 0, 4, 0.	**9.** 1, 1, 1.	**10.** 3, 4, 0.

11. 1, −2, 2.	**12.** 0, 5, −12.	**13.** 1, −4, 8.	**14.** 8, −1, −4.

15. 1, −2, 3.	**16.** 2, 3, 4.

Find to the nearest tenth of a degree the angle between the lines of each pair, if the direction numbers are as follows:

17. L_1: 0, 0, 3; L_2: 5, 12, 0.	**18.** L_1: 1, 0, 1; L_2: 0, 2, 0.

19. L_1: 4, 1, 5; L_2: 0, 0, 2.	**20.** L_1: 1, 1, 1; L_2: −2, 1, 2.

21. L_1: 1, 2, 3; L_2: 0, 1, 1.	**22.** L_1: 2, 3, −4; L_2: −1, 2, 3.

23. Given six lines by two points through which each passes as follows:

$$L_1 : (1, 2, -2), (0, 0, 0); \quad L_2 : (0, 1, -4), (4, 7, 4);$$
$$L_3 : (5, 1, 2), (4, -1, 4); \quad L_4 : (2, 3, 4), (0, 0, 0);$$
$$L_5 : (2, 1, -5), (0, -2, -9); \quad L_6 : (4, -2, 3), (1, -8, 9).$$

Find the groups of parallel lines in this set.

24. Find all groups of mutually perpendicular lines among the six lines of exercise 23.

25. Given six lines by two points through which each passes as follows:

$$L_1 : (3, 1, -1), (0, 0, 0); \quad L_2 : (5, 1, 4), (3, 4, 1);$$
$$L_3 : (3, 4, 5), (3, 2, 3); \quad L_4 : (-1, 5, 4), (-1, 2, 1);$$
$$L_5 : (0, 0, 0), (0, 1, 1); \quad L_6 : (2, -3, 3), (0, 0, 0).$$

Find the groups of parallel lines in this set.

26. Find all groups of mutually perpendicular lines among the six lines of exercise 25.

27. Given six lines whose direction numbers are: L_1: 1, 2, −1; L_2: −7, 4, 1; L_3: 1, 1, 3; L_4: $\frac{1}{2}$, −$\frac{2}{7}$, −$\frac{1}{14}$; L_5: −2, −4, 2; L_6: 1, 1, 2. Find all groups of mutually perpendicular lines.

28. Among the six lines of exercise 27, find all groups of parallel lines.

29. Show that the triangle with vertices $(6, -10, 0)$ $(1, 0, -5)$, $(6, 10, 10)$ is a right triangle.

30. Show that the points $(3, -1, -1)$, $(1, 2, -2)$, and $(5, 5, -1)$ are the vertices of a right triangle. What is the perimeter of the triangle? Find the two acute angles.

31. Show that the line through $(5, 1, -2)$ and $(-4, -5, 13)$ is perpendicular to the line segment joining $(-5, 2, 0)$ and $(9, -4, 6)$.

32. Show that the points $(1, -1, 3)$, $(5, -13, 11)$, and $(2, -4, 5)$ lie on a straight line.

33. Show that the points $(2, 3, 7)$, $(1, 3, 2)$, and $(0, 3, -3)$ lie on a straight line.

34. Show that the points $(2, 1, -1)$, $(4, 3, -2)$, $(6, 2, 0)$, and $(4, 0, 1)$ are the vertices of a square.

35. Show that the quadrilateral with vertices at $(2, 3, 4)$, $(3, 7, 3)$, $(5, 0, 1)$, $(6, 4, 0)$ is a parallelogram. Find the angles of the parallelogram.

36. Show that the quadrilateral with vertices $(0, 0, 0)$, $(0, 3, 3)$, $(3\sqrt{2}, 3, 3)$, $(3\sqrt{2}, 0, 0)$ is a square.

37. Show that the points $(0, 1, 3)$, $(-1, -1, -1)$, $(-2, 2, -1)$, and $(-1, 4, 3)$ form the vertices of a parallelogram.

38. Show that the points $(7, -5, 4)$, $(3, -1, 2)$, $(5, 3, 6)$, and $(9, -1, 8)$ are vertices of a square, and find its area.

39. Show that the points $(7, 2, -1)$, $(1, 5, 1)$, $(9, -4, 2)$, and $(3, -1, 4)$ are the vertices of a parallelogram, and find the lengths of its diagonals.

40. Show that the triangle determined by the points $(0, -2, 3)$, $(1, 0, 2)$, and $(2, -3, 0)$ is isosceles by showing that two angles are equal.

41. Show that the triangle determined by the points $(12, 3, 4)$, $(4, 12, 3)$, and $(3, 4, 12)$ is equilateral by finding the angles of the triangle.

42. Show that the triangle determined by $(16, -4, 6)$, $(4, 8, 6)$, and $(4, -4, -6)$ is equilateral by finding the angles of the triangle.

18–8. Simple Loci in Space: Surfaces

A surface in space is frequently specified as the locus of a point satisfying a single condition. For example, the locus of a point that satisfies the condition that it is a distance r from a fixed point is a sphere with center at the fixed point and radius r. To find the equation of such a locus, the geometrical condition must be expressed as an equation involving the variable coordinates x, y, z of the typical point satisfying the given condition. The procedure is what we would expect from our knowledge of plane analytic geometry.

Example 1. Find the equation of the locus of a point that satisfies the condition that it is 5 units from the origin.

If $P(x, y, z)$ is the point, then $OP = \sqrt{x^2 + y^2 + z^2}$, and since $OP = 5$ the equation of the sphere is

$$x^2 + y^2 + z^2 = 25.$$

Example 2. Find the equation of the locus of a point which satisfies the condition that it is 4 times as far from the xy plane as from the yz plane.

Let the point P have coordinates (x, y, z). The distance from the xy plane is $|z|$, and the distance from the yz plane is $|x|$.   Hence the requirement is that $|z| = 4|x|$, whence the equations of the locus are

$$z = 4x, \quad \text{or} \quad z = -4x,$$

which can be written together as $z^2 = 16x^2$.

In some simple cases the locus of an equation is easily described. For example, $x = 5$ is the equation of the plane perpendicular to the x axis at the point $x = 5$; $x^2 + y^2 = 25$ is the equation of a right circular cylinder whose axis is the z axis.   The sections of this chapter that follow will show how the loci of first-degree equations and some second-degree equations can be described.

Exercises

Find the equation of the locus in space of the point that satisfies the following condition.   Identify the locus.

1. The point is the same distance from the xy plane as from the xz plane.
2. The point is the same distance from the yz plane as from the xz plane.
3. The point is the same distance from the xy plane as from the yz plane.
4. The point is twice as far from the xy plane as from the xz plane.
5. The point is three times as far from the xy plane as from the yz plane.
6. The point is 4 units from the xy plane.
7. The point is 3 units from the yz plane.
8. The point is 5 units from the xz plane.
9. The point is 8 units from the origin.
10. The point is 10 units from the origin.
11. The point is 6 units from $(0, 0, 1)$.
12. The point is 2 units from $(1, 0, -2)$.
13. The point is 4 units from $(-1, 2, 3)$.
14. The point is 2 units from $(-2, 0, 5)$.
15. The point is equidistant from the origin and the point $(0, 0, 6)$.
16. The point is equidistant from $(2, 2, 0)$ and $(-2, -2, 0)$.
17. The point is equidistant from $(0, 3, 1)$ and $(1, -2, -4)$.
18. The point is equidistant from $(3, 1, -4)$ and $(-2, 4, 6)$.
19. The point is 5 units from the x axis.
20. The point is 4 units from the y axis.
21. The point is 6 units from the z axis.

Describe the surface that is the locus of each of the following equations:

22. $x = 0$.
23. $y = 0$.
24. $z = 0$.
25. $x = 5$.
26. $y = -6$.
27. $z = 1$.
28. $x = y$.
29. $x = z$.
30. $y - z = 0$.
31. $y = 3x$.
32. $y - 2x = 0$.
33. $x^2 = 1$.
34. $y^2 = 9$.
35. $x^2 + y^2 = 1$.
36. $y^2 + z^2 = 25$.
37. $x^2 + y^2 + z^2 = 16$.
38. $x^2 + y^2 + z^2 = 25$.
39. $(x - 1)^2 + (y - 2)^2 + (z - 4)^2 = 9$.
40. $(x + 2)^2 + (y - 5)^2 + (z + 1)^2 = 100$.
41. Find the equation of a sphere with center $(2, -2, 3)$ and passing through $(7, -3, 5)$.

42. Find the equation of a sphere with its center at $(4, -3, 2)$ and tangent to the plane $x = -2$.

43. Find the equation of the locus of a point, the sum of whose distances from $(2, 0, 0)$ and $(-2, 0, 0)$ is 8.

44. Find the equation of the locus of a point whose distance from $(-1, 2, -2)$ is equal to its distance from the xy plane.

18–9. The Plane

A line that is perpendicular to a plane is called a **normal** to the plane. Since all normals to a plane are parallel, their direction numbers are proportional (Sec. 18–7).

Let RS (Fig. 18–12) be any plane, and let $P_0(x_0, y_0, z_0)$ be a fixed point of the plane. Let L be a directed normal to the plane at P_0, and let A, B, C

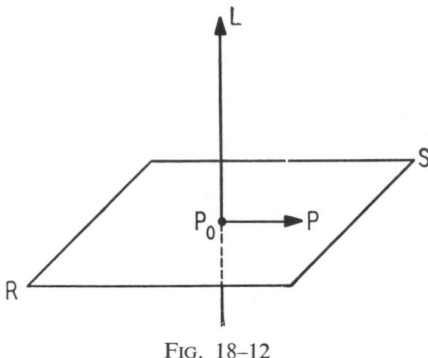

FIG. 18–12

be a set of direction numbers for L. Now the plane may be defined as the locus of points $P(x, y, z)$ such that P_0P is perpendicular to L. The numbers $x - x_0$, $y - y_0$, and $z - z_0$ form a set of direction numbers for P_0P (Sec. 18–7). By equation 6 of Sec. 18–7, P_0P is perpendicular to L if and only if

$$A(x - x_0) + B(y - y_0) + C(z - z_0) = 0. \tag{1}$$

Hence the equation of a plane passing through a point $P_0(x_0, y_0, z_0)$ and perpendicular to a line with direction numbers A, B, and C is given by equation 1. The direction numbers A, B, C are called the direction numbers for the plane.

Since any plane can be specified in this way, we see from equation 1 that *every plane is represented by an equation of first degree.*

We shall now prove the converse of this theorem: *Every equation of the first degree represents a plane.*

The general equation of first degree is

$$Ax + By + Cz + D = 0, \tag{2}$$

where not all three of the constants A, B, and C are zero. Let x_0, y_0, z_0 be a set of values satisfying equation 2. Then

$$Ax_0 + By_0 + Cz_0 + D = 0. \tag{3}$$

Subtracting equation 3 from equation 2, we obtain

$$A(x - x_0) + B(y - y_0) + C(z - z_0) = 0, \tag{4}$$

which is the equation of a plane containing the point (x_0, y_0, z_0) and normal to a line with direction numbers A, B, C.

As a consequence of these results, equation 2 is called the **general equation of the plane,** and equation 1 is called **the general equation of the plane containing** (x_0, y_0, z_0).

Since two planes are parallel if and only if they have parallel normals, *two planes are parallel if and only if the coefficients of x, y, and z in their equations are proportional.*

Similarly, *two planes*

$$A_1x + B_1y + C_1z + D_1 = 0,$$
$$A_2x + B_2y + C_2z + D_2 = 0$$

are perpendicular if and only if

$$A_1A_2 + B_1B_2 + C_1C_2 = 0. \tag{5}$$

Example 1. Find the equation of the plane which passes through $(2, -1, \frac{3}{2})$ and whose normal has the direction numbers $\frac{1}{2}, -2, 3$.

Using equation 1, we have

$$\tfrac{1}{2}(x - 2) - 2(y + 1) + 3(z - \tfrac{3}{2}) = 0,$$

which becomes, on simplifying,

$$x - 4y + 6z - 15 = 0.$$

Example 2. Show that the plane $4x + y = 0$ is perpendicular to the plane of example 1.

Since 4, 1, 0 is a set of direction numbers for this plane, and 1, -4, 6 is a set for the plane of example 1, by equation 5,

$$(4 \cdot 1) + (1 \cdot [-4]) + (0 \cdot 6) = 0$$

shows that the planes are perpendicular.

18–10. The Plane Determined by Three Points

The following example illustrates how to find the equation of a plane determined by three points.

Example. Find the equation of the plane that passes through the points $P_1(7, 1, 2)$, $P_2(-3, -3, 1)$, and $P_3(5, -1, 1)$.

Since P_1 lies in the plane, the equation of the plane has the form

$$A(x - 7) + B(y - 1) + C(z - 2) = 0. \tag{1}$$

Substituting the coordinates of P_2 in equation 1, we have

$$-10A - 4B - C = 0, \tag{2}$$

and, substituting the coordinates of P_3 in equation 1, we have
$$-2A - 2B - C = 0. \tag{3}$$
Solving equations 2 and 3 simultaneously for A and B in terms of C (Sec. 13–8), we obtain $A = \frac{1}{6}C$ and $B = -\frac{2}{3}C$, whence equation 1 becomes
$$\tfrac{1}{6}C(x - 7) - \tfrac{2}{3}C\,(y - 1) + C(z - 2) = 0.$$
This reduces to
$$x - 4y + 6z - 15 = 0,$$
the required equation.

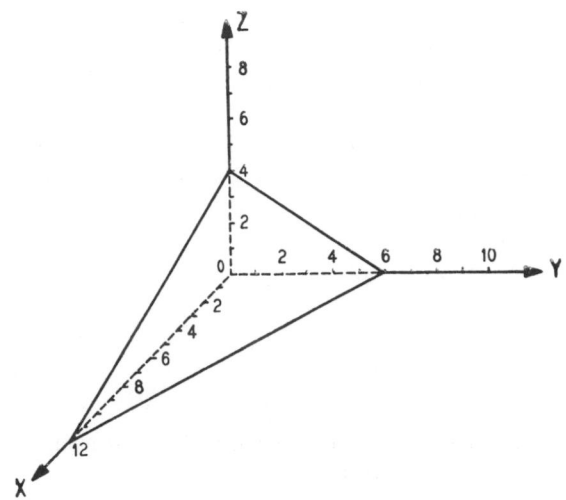

Fig. 18–13. The plane $x + 2y + 3z - 12 = 0$

18–11. Sketching a Plane

The x coordinate of the point at which a plane crosses the x axis is called the **x intercept** of the plane; the **y intercept** and **z intercept** are defined similarly. To find the x intercept of a plane find the value of x when $y = z = 0$ in the equation of the plane; to find the y intercept of a plane find the value of y when $x = z = 0$ in the equation of the plane, etc.

When a plane is not parallel to any of the coordinate axes it can be sketched conveniently by finding its intercepts on the axes as indicated in the following example. A plane parallel to an axis can be sketched easily; the modification required is left as an exercise for the reader.

Example. Sketch the plane $x + 2y + 3z - 12 = 0$.

The x intercept is 12, the y intercept is 6, and the z intercept is 4. Since the intersection of two planes is a straight line, the given plane and the xy plane intersect in a straight line passing through the x intercept and y intercept of the given plane, called the **trace** of the given plane in the xy plane. It is convenient to draw the portion of this trace between the intercepts; similarly in the other coordinate planes. Then the triangular section as shown in Fig. 18–13 makes a convenient sketch of the plane.

Exercises

Find the equation of the plane which passes through the given point and whose normal has the given direction numbers.

1. $(0, 0, 0)$; direction numbers $1, 2, 1$.

2. $(0, 1, 0)$; direction numbers $0, 3, 1$.

3. $(3, 2, -1)$; direction numbers $0, 1, 0$.

4. $(1, -2, 3)$; direction numbers $0, 1, 1$.

5. $(1, 2, -4)$; direction numbers $-2, -1, 3$.

6. $(-1, 2, \frac{1}{2})$; direction numbers $1, \frac{1}{2}, -2$.

7. $(2, \frac{1}{3}, 0)$; direction numbers $-2, 1, \frac{1}{2}$.

8. $(1, 3, -\frac{2}{3})$; direction numbers $\frac{1}{4}, \frac{3}{4}, -1$.

For each of the following planes (a) find the intercepts, (b) find a set of direction numbers for the normal to the plane, (c) sketch the plane.

9. $x + y + 3z = 6$.

10. $2x - 3y + 4z = 12$.

11. $3x + 2y + 4z + 12 = 0$.

12. $3x - 5y + z = 15$.

13. $5x - y + 4z = 10$.

14. $7x - 3y + 2z - 14 = 0$.

15. $3x + 2y = 6$.

16. $3x - 5z + 15 = 0$.

17. $4y - z - 12 = 0$.

18. $2x - 3y = 15 - z$.

Determine whether the planes are parallel, perpendicular, or neither.

19. $x - 2y + 3z = 8$, $2x - 4y + 6z - 11 = 0$.

20. $3x + y + 2z = 1$, $4x + 2y - 7z = 3$.

21. $3x + 4y - z - 1 = 0$, $2z - 6x - 8y = 5$.

22. $3x - 2y - z = 5$, $2x + y + 4z - 1 = 0$.

23. $x + 2y - z = 1$, $4x + 3y = 5$.

24. $x - 2y = 7 - 3z$, $4x - 1 = y + 2z$.

25. $5x - 7y = 15 - 2z$, $15x = 21y - 6z$.

26. $5x = 7y + 1$, $x + y = z + 5$.

Determine the equation of the plane passing through the given three points.

27. $(0, 0, 0)$, $(2, -1, 4)$, $(3, 7, -1)$.

28. $(0, 0, 0)$, $(0, 4, -1)$, $(4, 1, 4)$.

29. $(2, 1, 6)$, $(3, 2, 4)$, $(3, 1, 4)$.

30. $(0, 4, -5)$, $(4, 2, 1)$, $(-1, -2, 2)$.

31. $(a, 0, 0)$, $(0, b, 0)$, $(0, 0, c)$.

32. Find the equation of the plane through $(1, -2, 3)$, and perpendicular to a line whose direction numbers are $2, -1, 5$.

33. Find the equation of the plane through $(2, 0, 5)$, and perpendicular to a line whose direction numbers are $1, 3, -1$.

34. Find the equation of the plane through $(2, 1, 3)$ perpendicular to the segment joining this point and $(-1, -2, 6)$.

35. Find the equation of the plane through $(3, -2, 4)$ perpendicular to the segment joining this point and $(0, 4, -2)$.

36. Find the equation of the plane that passes through $(5, 1, 2)$ and is parallel to the plane $3x - 2y + z - 12 = 0$.

37. Find the equation of the plane that passes through $(2, 0, -5)$ and is parallel to the plane $x + 2y + 3z = 6$.

38. Find the equation of the plane that passes through $(3, 1, -2)$ and is parallel to $5x - y + z = 15$.

39. Find the equation of the plane that passes through $(1, 2, -1)$ and $(2, 4, -2)$ and is perpendicular to $2x + 3y - z + 4 = 0$.

40. Find the equation of the plane that passes through $(4, -1, -2)$ and $(7, 3, 1)$ and is perpendicular to $x - 3y + 4z -.2 = 0$.

41. Find the equation of the plane that passes through $(1, 2, 3)$ and is perpendicular to each of the planes $x - y + 2z = 3$ and $2x - y - 3z = 0$.

42. Find the equation of the plane that passes through $(1, 0, 2)$ and is perpendicular to each of the two planes $2x - y + 4z = 0$ and $3x + 2y - z = 10$.

43. Assuming that the angle θ between the two planes $A_1x + B_1y + C_1z + D_1 = 0$ and $A_2x + B_2y + C_2z + D_2 = 0$ is the same as the angle between the normals to the planes, show that

$$\cos \theta = \pm \frac{A_1A_2 + B_1B_2 + C_1C_2}{\sqrt{A_1^2 + B_1^2 + C_1^2}\sqrt{A_2^2 + B_2^2 + C_2^2}}.$$

44. Find the cosine of the smallest angle between the planes $4x + y - z = 6$ and $3x - 2y + z = 7$.

45. Find the acute angle between the planes $2x - y + z = 8$ and $x + y + 2z - 11 = 0$.

46. Find the acute angle between the planes $x + 2y - z = 1$ and $x - 2y = 2z + 5$.

47. Find the angle which the plane $x + y + z = 5$ makes with each of the coordinate planes.

18–12. Sketching a Surface

The surfaces corresponding to the equations discussed in the previous sections were easily identified. Frequently, however, the surface corresponding to an equation is not so easily identified but can be discovered by a simple analysis. We shall show how such an analysis can be carried out in a few simple cases.

A surface cannot be plotted satisfactorily by plotting points and joining them, as can a plane curve. However, a study of the traces, intercepts, and plane sections is sufficient to reveal the nature of the surfaces we shall consider.

1. *Intercepts.* To find the intercepts of the surface on a given axis, let the variables measured along the other two axes be zero, and then solve the equation of the surface for the third variable.

2. *Traces.* The trace of a surface on a coordinate plane is the curve of intersection of the surface and the plane. The equation of a trace of a surface on a coordinate plane is found by substituting zero for the variable measured perpendicular to that plane.

3. *Plane Sections.* The curve in which a plane cuts a surface is called

a plane section of the surface. Thus the traces of a surface in the co-ordinate planes are plane sections. If $x = a$ is substituted in the equation of the surface, the resulting equation in the variables y and z is the equation of the plane section, a curve lying in the plane $x = a$. Similar statements hold for planes parallel to the other two coordinate planes. By sketching a few plane sections the shape of a surface can be indicated very easily.

Example 1. Sketch the portion of the surface

$$\frac{x^2}{16} + \frac{y^2}{25} + \frac{z^2}{9} = 1$$

that lies in the first octant.

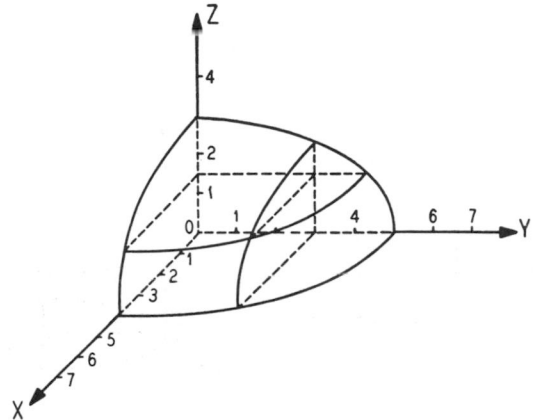

FIG. 18–14

Setting y and z equal to zero, we find the x intercepts to be $+4$ and -4. Similarly the y intercepts are $+5$ and -5, and the z intercepts are $+3$ and -3. The traces in the xy plane, xz plane, and yz plane are, respectively,

$$\frac{x^2}{16} + \frac{y^2}{25} = 1, \qquad \frac{x^2}{16} + \frac{z^2}{9} = 1, \qquad \frac{y^2}{25} + \frac{z^2}{9} = 1,$$

which are ellipses and can be sketched very easily, as shown in Fig. 18–14. The plane section in the plane $z = \frac{3}{2}$ is

$$\frac{x^2}{12} + \frac{y^2}{\left(\frac{7\,5}{4}\right)} = 1,$$

an ellipse as shown in the figure. The plane section in the plane $y = 3$ is

$$\frac{x^2}{\left(\frac{2\,5\,6}{2\,5}\right)} + \frac{z^2}{\left(\frac{1\,4\,4}{2\,5}\right)} = 1$$

as shown.

Consider a curve and a fixed line in space; if a line parallel to the given fixed line is passed through each point of the curve, then the set of lines so constructed forms a surface that is called a **cylindrical surface,** or **cylinder.** The curve is called the **directrix** and the lines are called the

rulings of the cylindrical surface. Of particular interest are the cylindrical surfaces that arise when the directrix is a plane curve and the rulings are perpendicular to the plane of the curve: When the directrix is a circle, the surface is called a **right circular cylinder,** and, when the directrix is a parabola, an ellipse, or a hyperbola, the surface is called a **parabolic, elliptic,** or **hyperbolic cylinder,** respectively.

Consider a surface with equation $f(x, y) = 0$. The trace of this surface in the xy plane is the curve given by the equation $f(x, y) = 0$, and clearly, if $(x_0, y_0, 0)$ is a point on this trace, then (x_0, y_0, z), where z is any number, lies on the surface. Now the points (x_0, y_0, z) for all z form a line through

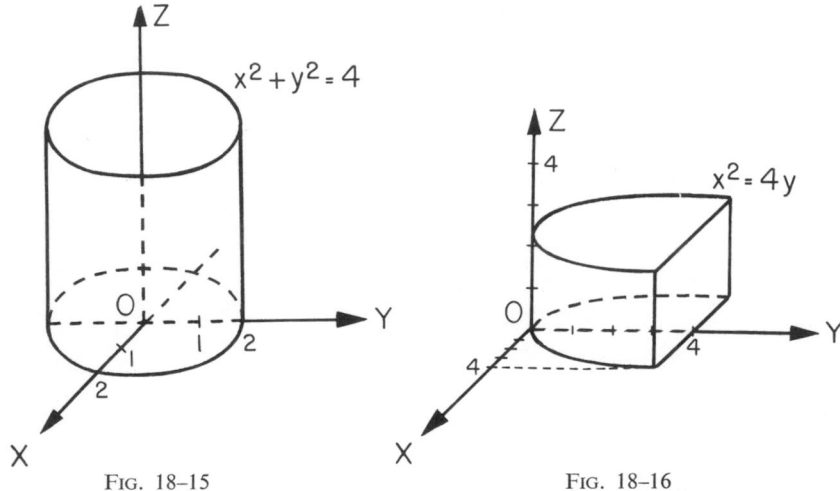

Fig. 18–15   Fig. 18–16

$(x_0, y_0, 0)$ parallel to the z axis. It follows that *the surface $f(x, y) = 0$ is a cylindrical surface with rulings parallel to the z axis and directrix given by the curve $f(x, y) = 0$ in the xy plane.* For example, the surface $x^2 + y^2 = 4$ is a right circular cylinder, as sketched in Fig. 18–15.

Conversely, it is easily established that, *if a cylindrical surface has a directrix which is a curve $f(x, y) = 0$ in the xy plane and rulings parallel to the z axis, then the equation of the surface is $f(x, y) = 0$.* For example, the parabolic cylinder with directrix the curve $x^2 = 4y$ in the xy plane and rulings parallel to the z axis, as shown in Fig. 18–16, has as its equation $x^2 = 4y$.

More generally, the following statement can be made: *The locus of an equation $F(x, y, z) = 0$ in which one of the variables x, y, z is missing is a cylinder whose rulings are parallel to the axis of the missing variable and whose directrix is the curve given by $F(x, y, z) = 0$ in the coordinate plane of the two variables that are present, and conversely.*

Exercises

Sketch the portion of each of the following surfaces that lies in the first octant:

1. $x^2 + y^2 = 1.$

2. $x^2 + z^2 = 4.$

3. $y^2 + z^2 = 9.$

4. $x^2 + z^2 - 16 = 0.$

5. $y^2 = 2x.$

6. $y^2 = 6x.$

7. $z^2 = 4y.$

8. $z^2 = 2y.$

9. $x^2 = 4y.$

10. $x^2 = 2y.$

11. $z^2 = 8x.$

12. $x^2 - 4z.$

13. $4x^2 + 9y^2 = 36.$

14. $4y^2 + 9z^2 = 36.$

15. $25x^2 + 9z^2 = 225.$

16. $4x^2 - 9y^2 = 36.$

17. $4y^2 - 9z^2 = 36.$

18. $x^2 - 4y^2 = 1.$

19. $4y = 16 - x^2.$

20. $y^2 + 4x - 8 = 0.$

21. $y^2 + z^2 - 4z = 0.$

22. $4x + y^2 = 12.$

23. $x^2 + y^2 + z^2 = 9.$

24. $x^2 + y^2 + z^2 = 16.$

25. $\dfrac{x^2}{1} + \dfrac{y^2}{4} + \dfrac{z^2}{9} = 1.$

26. $\dfrac{x^2}{4} + \dfrac{y^2}{16} + \dfrac{z^2}{36} = 1.$

27. $\dfrac{x^2}{4} + \dfrac{y^2}{25} + \dfrac{z^2}{36} = 1.$

28. $\dfrac{x^2}{1} + \dfrac{y^2}{36} + \dfrac{z^2}{4} = 1.$

29. $x^2 + y^2 = 4z.$

30. $x^2 + y^2 = 12z.$

31. $x^2 + y^2 - z^2 = 1.$

32. $x^2 + y^2 - 2z^2 = 1.$

33. $x^2 - 16y^2 = 0.$

34. $x^2 - 25z^2 = 0.$

35. $y^2 - 9z^2 = 0.$

36. $4x^2 - 25y^2 = 0.$

18–13. Curves in Space

A curve in space may be specified as the locus of a point satisfying two conditions as follows: The locus of the point satisfying one condition is a surface, the locus of the point satisfying the other condition is another surface, and then the curve is the intersection of these two surfaces. It follows that a curve in space can be specified by giving the equations of two surfaces whose intersection is the curve. Thus a straight line can be specified by giving the equations of two planes which intersect in the line; how this can be done systematically will be discussed in the next section.

Example. Sketch the intersection in the first octant of the surfaces $y^2 + z^2 = 4$ and $x^2 = -8(z - 2).$

The surface $y^2 + z^2 = 4$ is a right circular cylinder of radius 2 whose axis is the x axis. The surface $x^2 = -8(z - 2)$ is a cylinder such that any plane section parallel to the xz plane is a parabola. The surfaces and curve are shown in Fig. 18–17.

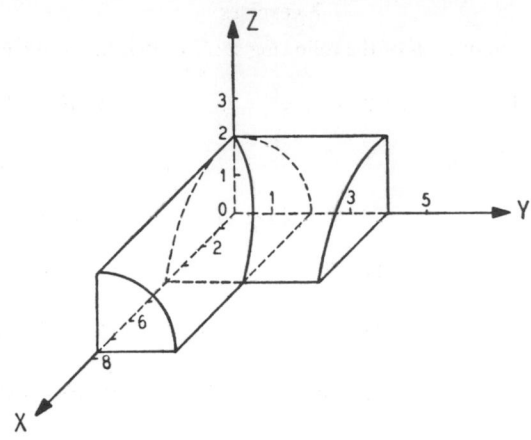

FIG. 18–17. The curve of intersection of the surfaces $y^2 + z^2 = 4$
and $x^2 = -8 \, (z - 2)$

Exercises

Sketch the curve of intersection in the first octant of the following pairs of surfaces :

1. $x^2 + y^2 = 16$; $x + z = 4$.

2. $x^2 + y^2 = 9$; $x + z = 3$.

3. $x^2 + z^2 = 25$; $x + y = 5$.

4. $x^2 + z^2 = 1$; $y + z = 1$.

5. $x^2 + y^2 = 1$; $x^2 + z^2 = 1$.

6. $x^2 + y^2 = 25$; $y^2 + z^2 = 25$.

7. $x^2 + y^2 + z^2 = 9$; $x - y = 0$.

8. $x^2 + y^2 + z^2 = 1$; $y = 2x$.

9. $x^2 + y^2 + z^2 = 36$; $x^2 + y^2 = 9$.

10. $x^2 + y^2 + z^2 = 25$; $x^2 + z^2 = 4$.

11. $x + y = 4$; $y + z = 4$.

12. $x + y = 4$; $x + y + z = 6$.

13. $\dfrac{x^2}{16} + \dfrac{y^2}{9} + \dfrac{z^2}{25} = 1$; $x = y$.

14. $\dfrac{x^2}{4} + \dfrac{y^2}{1} + \dfrac{z^2}{9} = 1$; $x^2 + z^2 = 9$.

18–14. The Equations of a Line

In the preceding section we saw that a straight line in space may be specified by giving two planes which intersect in the line. Thus two equations,

$$A_1 x + B_1 y + C_1 z + D_1 = 0,$$

$$A_2 x + B_2 y + C_2 z + D_2 = 0,$$

(1)

of two intersecting planes may be considered as equations of their line of intersection.

Certain standard forms of the equations of a straight line are useful; these forms will now be discussed.

Let L be a line passing through the point $P_1(x_1, y_1, z_1)$ and with direction given by the direction angles α, β, γ. If $P(x, y, z)$ is another point on the line a distance d in the positive direction from P_1, then from equations 1 of Sec. 18–5 we have

$$\cos \alpha = \frac{x - x_1}{d}, \qquad \cos \beta = \frac{y - y_1}{d}, \qquad \cos \gamma = \frac{z - z_1}{d}.$$

After solving each of these equations for d we may write

$$\frac{x - x_1}{\cos \alpha} = \frac{y - y_1}{\cos \beta} = \frac{z - z_1}{\cos \gamma}, \tag{2}$$

the *standard form of the equations of a straight line through the point* (x_1, y_1, z_1) *and with direction angles* α, β, γ. Replacing the direction cosines in equation 2 with direction numbers, we obtain

$$\frac{x - x_1}{a} = \frac{y - y_1}{b} = \frac{z - z_1}{c}. \tag{3}$$

Equations 2 and 3 are called **symmetric equations** of the line.

If $P_1(x_1, y_1, z_1)$ and $P_2(x_2, y_2, z_2)$ are two distinct points on a line L, then the three numbers $x_2 - x_1$, $y_2 - y_1$, $z_2 - z_1$ constitute a set of direction numbers for L. Thus from equation 3 we obtain

$$\frac{x - x_1}{x_2 - x_1} = \frac{y - y_1}{y_2 - y_1} = \frac{z - z_1}{z_2 - z_1}, \tag{4}$$

known as the *two-point form of the equations of the line.*

It should be noted that there are three possible equations in each of the relations 2, 3, and 4, and that each of these equations is the equation of a plane parallel to a coordinate axis. For example, equation 3 yields

$$\frac{x - x_1}{a} = \frac{y - y_1}{b}, \qquad \frac{x - x_1}{a} = \frac{z - z_1}{c}, \qquad \frac{y - y_1}{b} = \frac{z - z_1}{c}, \tag{5}$$

any two of these planes determining the line, which of course also lies in the third plane. A choice of two of the standard planes given in equations 5 makes the line particularly easy to sketch by the methods of the preceding section.

The standard forms 2, 3, and 4 cannot be used if any of the denominators are zero. However, if $a = 0$, for example, the line may be specified by means of the planes

$$x - x_1 = 0, \qquad \frac{y - y_1}{b} = \frac{z - z_1}{c}. \tag{6}$$

If, for example, $a = b = 0$, the line may be specified by means of the planes

$$x - x_1 = 0, \qquad y - y_1 = 0. \tag{7}$$

The reader should verify that equations 6 and 7 are correct.

In the first paragraph of this section it was observed that the equations of two intersecting planes in the general form 1 may be considered as equations of the line of intersection of the two planes. However, all of the standard forms considered so far have specified lines by means of special planes which are parallel to the coordinate axes. Thus we may ask how to find a standard form for the equations of a line when the line is given by the equations of two planes of the form 1. One obvious answer to this question is to find two points on the line from the given equations and then to use the two-point form 4. However, the standard form can also be found directly, as example 4 below shows.

Example 1.   Find the equations of the line through $(1, -2, 3)$ with direction angles $\alpha = 60°$, $\beta = 45°$, $\gamma = 60°$.
From the standard form 2 we obtain

$$\frac{x - 1}{\frac{1}{2}} = \frac{y + 2}{\frac{\sqrt{2}}{2}} = \frac{z - 3}{\frac{1}{2}},$$

which may be written as

$$\frac{x - 1}{1} = \frac{y + 2}{\sqrt{2}} = \frac{z - 3}{1}.$$

Since the sum of the squares of the direction cosines must equal 1 (see Sec. 18–5), three direction angles cannot be chosen arbitrarily. The student should verify that direction angles given in this example satisfy the requirement that the sum of the squares of their cosines is 1.

Example 2.   Find equations of the line L passing through the points $(1, 4, -2)$ and $(5, -1, 3)$.
From the two-point form 4 we obtain

$$\frac{x - 1}{4} = \frac{y - 4}{-5} = \frac{z + 2}{5}.$$

Alternatively, if the other given point is used in the numerator, we obtain

$$\frac{x - 5}{4} = \frac{y + 1}{-5} = \frac{z - 3}{5}.$$

However, it is interesting to note that both of these standard forms yield the same three planes,

$$5x + 4y - 21 = 0, \qquad 5x - 4z - 13 = 0, \qquad y + z - 2 = 0.$$

The portion of the line L that lies in the first octant is shown in Fig. 18–18. Since the sketch shows the plane parallel to the x axis and the plane parallel to the z axis, the student should verify that the given line lies in the third plane parallel to the y axis.

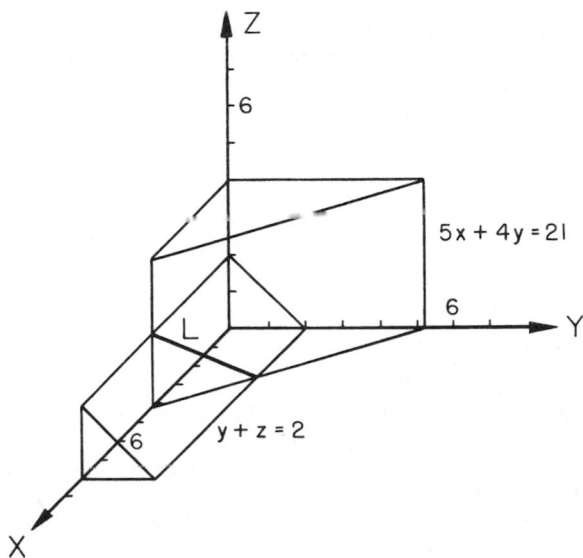

FIG. 18–18

Example 3.   Find equations of the line passing through $(-1, 3, 4)$ and perpendicular to the plane $7x + 2y - 5z = 6$.

The required line is parallel to the normal to the given plane, and therefore $7, 2, -5$ are direction numbers for it.   From equations 3 we have

$$\frac{x + 1}{7} = \frac{y - 3}{2} = \frac{z - 4}{-5}$$

as the equations of the required line.

Example 4.   Find a standard form for the line specified by the planes $x + y + z = 12$ and $x + 2y + 3z = 24$.

Eliminating first z and then y between the two equations, we obtain

$$2x + y = 12, \qquad x - z = 0,$$

which can be written as

$$x = \frac{y - 12}{-2}, \qquad x = z,$$

which yield

$$\frac{x - 0}{1} = \frac{y - 12}{-2} = \frac{z - 0}{1} \tag{8}$$

as a standard form for the equation of the line.   We note that the line passes through $(0, 12, 0)$ and has direction numbers $1, -2, 1$.

Since the point (4, 4, 4) lies in both of the given planes it also lies on the line. Then the points (4, 4, 4) and (0, 12, 0) on the line can be used with the two-point standard form to obtain

$$\frac{x-0}{4} = \frac{y-12}{-8} = \frac{z-0}{4},$$

which is equivalent to the standard form obtained earlier.

In this example we have shown only the algebraic technique for obtaining equations 8 from the equations of the given planes. The student should supply the arguments that show that the line specified by the given planes is the line given by equations 8, and conversely.

Exercises

Find equations of the lines determined by the following pairs of points:

1. (4, 2, 3) and (1, 3, 5).   **2.** (−1, 2, 4) and (0, −1, 3).

3. (1, 0, −5) and (−2, 6, 0).   **4.** (4, 2, 0) and (−1, 1, 1).

5. (3, 4, 5) and (−2, 5, −1).   **6.** (4, 0, 0) and (0, 3, −1).

7. (0, 0, −3) and (−1, 2, 0).   **8.** (−7, 3, 5) and (0, 5, −3).

9. Find equations of the line that passes through the point $(2, 3, -4)$ and has direction angles $\alpha = 45°$, $\beta = 60°$, and $\gamma = 60°$.

10. Find equations of the line that passes through the point $(2, -5, 4)$ and has direction numbers $1, 2,$ and 5.

11. Find equations of the line that passes through the point $(2, 0, 1)$ and has direction numbers $-1, 3,$ and 8.

12. Find equations of the line passing through the point $(3, -1, 5)$ and parallel to the line through the points $(5, -3, 2)$ and $(4, 0, 1)$.

13. Find equations of the line through the origin and perpendicular to the plane $2x + 4y + 7z - 10 = 0$.

14. Find equations of the line through the point $(2, -1, 0)$ and perpendicular to the plane $5x - 2y + 6z = 1$.

15. Find equations of the lines through the point $(1, 4, 3)$ with $\cos \beta = \frac{4}{5}$ and $\cos \gamma = \frac{3}{5}$.

16. Find equations of the line that passes through the point $(2, 1, -3)$ and is parallel to the line $\dfrac{x-6}{3} = \dfrac{y+7}{1} = \dfrac{z}{-5}$.

17. Find the points in which the line determined by $2x + y - 3z = 1$ and $2x - 3y + 5z = 3$ meet the coordinate planes. What are the direction angles of this line?

18. Show that the lines $\dfrac{x}{6} = \dfrac{y-1}{-2} = \dfrac{z+4}{-4}$ and $\dfrac{x+1}{4} = \dfrac{y}{6} = \dfrac{z-1}{3}$ are perpendicular to each other.

19. Find the point at which the line determined by $x = z + 2$, $y + 3z = 1$ meets the plane $x - 2y = 7$.

20. Find the point at which the line determined by $x + 2y + 4z = 2$ and $2x + 3y - 2z + 3 = 0$ meets the plane $2x - y + 4z + 8 = 0$. Find a standard form for the equations of the given line.

21. Show that the points $(2, -3, 1)$, $(8, 11, -9)$ and $(5, 4, -4)$ lie in a straight line by finding a standard form for the equations of the line.

22. Find the point of intersection of the line determined by $2x + y = 5$ and $3x + z = 14$, and the line determined by $x - 4y = 7$ and $5x + 4z = 35$.

23. Show that the line $\dfrac{x + 6}{3} = \dfrac{y - 4}{-2} = \dfrac{z - 1}{5}$ is perpendicular to the plane $3x - 2y + 5z + 59 = 0$, and find the point at which it pierces that plane.

24. Find equations of the line through $(1, -3, 2)$ with direction numbers $0, 1, -2$.

25. Find equations of the line through $(-3, 4, 5)$ with direction numbers $-1, 0, 2$.

26. Find equations of the line through $(3, 6, -7)$ with direction numbers $0, 0, -1$.

27. Find equations of the line through $(1, -2, 3)$ with direction numbers $0, 1, 0$.

28. Find a standard form for the equations of the line given by $3x + 2y - 2z + 2 = 0$ and $6x + 7y - 6z = 3$.

29. Find a standard form for the equations of the line given by $x - 2y + z = 15$ and $2x + y - 3z = 8$.

30. Show that the line given by $3x - y = 0$, $5y + 3z = 0$ is parallel to the line given by $2x + y + z = 0$, $4x - 3y - z = 2$.

31. Show that the line given by $x - 3y + 7 = 0$, $x - 2z = 4$ lies in the plane $2x - 3y - 2z + 3 = 0$.

32. Show that the line of exercise 31 is perpendicular to the plane $6x + 2y + 3z = 15$.

33. If we let t equal the common value of the three ratios in equations 3 above, then we obtain parametric equations for the line through (x_1, y_1, z_1) with direction numbers a, b, c:

$$x = x_1 + at,$$
$$y = y_1 + bt,$$
$$z = z_1 + ct.$$

How may t be interpreted if a, b, c are the direction cosines of the line?

34. Find parametric equations for the line through $(1, 6, -4)$ and $(2, -1, 3)$.

35. Find parametric equations for the line through $(3, -2, 4)$ and perpendicular to the plane $3x - y + 2z = 15$.

Progress Report

It is convenient to summarize this chapter on solid analytic geometry by continuing the geometry-algebra dictionary begun in the earlier chapters on plane analytic geometry.

GEOMETRY-ALGEBRA DICTIONARY

Geometry	Algebra
Point P in space	$P(x, y, z)$
Distance between two points	$\sqrt{(x_2 - x_1)^2 + (y_2 - y_1)^2 + (z_1 - z_2)^2}$
Direction angles of a line	α, β, γ
Direction cosines of a line	$\cos \alpha, \cos \beta, \cos \gamma$
Direction cosines of the segment $P_1 P_2$ of length d	$\dfrac{x_2 - x_1}{d}, \dfrac{y_2 - y_1}{d}, \dfrac{z_2 - z_1}{d}$
Angle θ between L_1 and L_2	$\cos \theta = \cos \alpha_1 \cos \alpha_2 + \cos \beta_1 \cos \beta_2 + \cos \gamma_1 \cos \gamma_2$
Direction numbers of a line	$a = k \cos \alpha, b = k \cos \beta, c = k \cos \gamma$

Geometry	Algebra
Lines L_1 and L_2 are parallel	$\dfrac{a_1}{a_2} = \dfrac{b_1}{b_2} = \dfrac{c_1}{c_2}$
Angle θ between L_1 and L_2	$\cos \theta = \dfrac{a_1 a_2 + b_1 b_2 + c_1 c_2}{\sqrt{a_1^2 + b_1^2 + c_1^2}\ \sqrt{a_2^2 + b_2^2 + c_2^2}}$
Lines L_1 and L_2 are perpendicular	$a_1 a_2 + b_1 b_2 + c_1 c_2 = 0$
Direction numbers of $P_1 P_2$	$x_2 - x_1,\ y_2 - y_1,\ z_2 - z_1$
Surface	$f(x, y, z) = 0$
Plane	$Ax + By + Cz + D = 0$
Plane with direction numbers A, B, C through (x_0, y_0, z_0)	$A(x - x_0) + B(y - y_0) + C(z - z_0) = 0$
Two planes are perpendicular	$A_1 A_2 + B_1 B_2 + C_1 C_2 = 0$
Line through P_1 with direction angles α, β, γ	$\dfrac{x - x_1}{\cos \alpha} = \dfrac{y - y_1}{\cos \beta} = \dfrac{z - z_1}{\cos \gamma}$
Line through P_1 with direction numbers a, b, c	$\dfrac{x - x_1}{a} = \dfrac{y - y_1}{b} = \dfrac{z - z_1}{c}$
Line through P_1 and P_2	$\dfrac{x - x_1}{x_2 - x_1} = \dfrac{y - y_1}{y_2 - y_1} = \dfrac{z - z_1}{z_2 - z_1}$

In addition to establishing and applying the results given in the table above, the chapter considered methods for sketching planes, surfaces (including cylindrical surfaces), curves, and lines in space.

19———

The Elements
of Differential Calculus

The concept of a function was introduced in Chapter 3 and was used throughout the book. In almost every chapter, new functions were examined, and new methods were developed which permitted us to investigate properties of these functions and to solve problems in which they were involved. There still remains one extremely important type of problem for which no adequate method of solution has been developed so far. These problems can be exemplified by the following questions. What is the best way of describing the speed of a car or the cooling of a hot object? How does the change of the plate current of a tube depend on the change of the grid voltage? There are instances of innumerable problems in which the rate of change of different quantities has to be investigated. For such investigations new ideas have been introduced and new methods have been developed. The system of all these concepts and procedures, called **differential calculus,** is the main topic of this chapter.

19–1. Increments and Δ Notation

The pressure P on the surface of a structure is a function of the velocity V of the wind. The following table gives the results of a wind-tunnel experiment on an exposed surface having an area of one square foot; in the experiment P was measured in pounds for different values of V measured in miles per hour.

V	5	6	7	8	9	10	11	12	13	14	15
P	0.88	1.26	1.72	2.24	2.84	3.50	4.26	5.04	5.92	6.86	7.88

Assume that the experiment starts with a wind velocity of 12 mph and that this velocity is then successively changed. Each increase or decrease of the velocity is called an increment and is denoted by the symbol ΔV, read: **increment of V** or simply **delta V.** Thus the increment of V is 2 mph if V is changed from 12 to 14 mph; and, if the velocity is changed from 12 mph to 7, the increment is -5 mph. In the first case we write $\Delta V = 2$ mph; in the second case $\Delta V = -5$ mph.

If V changes from 12 to 14, the pressure changes from 5.04 to 6.86 lb. The change of the pressure is then 1.82 lb and is called *the increment of the pressure which corresponds to the increment of 2 mph of the velocity.* The symbol ΔP is used to denote the increment of the pressure. The experiment has shown that P increases 1.82 lb if the velocity is increased by 2 mph from 12 to 14 mph. These two values $\Delta V = 2$ mph and $\Delta P = 1.82$ lb are called *corresponding increments for the interval from $V = 12$ mph to $V = 14$ mph.*

Example 1. What are the corresponding increments in the above experiment if V changes from 13 to 6 mph?

In this case $\Delta V = 6 - 13 = -7$ mph, and, from the table, the corresponding $\Delta P = 1.26 - 5.92 = -4.66$ lb.

Example 2. From a table of logarithms find the increment Δy of $y = \log x$ corresponding to $\Delta x = 0.3$ at the point $x = 2$.

From the table of logarithms, $\log 2 = 0.3010$ and $\log 2.3 = 0.3617$; hence

$$\Delta x = 0.3, \qquad \Delta y = \log 2.3 - \log 2 = 0.3617 - 0.3010 = 0.0607.$$

The functions used in the preceding examples were given by tables. But a function can be given also by means of a formula or by a graph. The concept of **corresponding increments** is applicable to a function given in any form and can be explained generally in the following way.

Assume that a function $y = f(x)$ is given. In order to compute the increment Δy corresponding to an increment Δx, compute the value $f(x + \Delta x)$. The increment of the function is thus

$$\Delta y = \Delta f(x) = f(x + \Delta x) - f(x).$$

Since $y = f(x)$, it follows that

$$f(x + \Delta x) = y + \Delta y.$$

Example 3. Compute the corresponding increments of x and the function

$$y = f(x) = 3x^2 - 5$$

if x changes from 2 to 2.5.

Following the above explanation, we have

$$x = 2, \qquad \Delta x = 2.5 - 2 = 0.5,$$

$$f(x) = f(2) = 3 \cdot 4 - 5 = 7,$$

$$f(x + \Delta x) = f(2.5) = 3 \cdot 6.25 - 5 = 13.75,$$

$$\Delta y = f(x + \Delta x) - f(x) = 13.75 - 7 = 6.75.$$

Example 4. From the formula $d = \dfrac{140}{n}$ compute the corresponding increments of d and n if n changes from 40 to 45.

In this case we have

$$n = 40, \qquad \Delta n = 5,$$

$$d = f(n) = f(40) = \frac{140}{40} = 3.50,$$

$$f(n + \Delta n) = f(45) = \frac{140}{45} = 3.11,$$

$$\Delta d = f(n + \Delta n) - f(n) = 3.11 - 3.50 = -0.39.$$

Corresponding increments are found very easily if the function $y = f(x)$ is given by a graph. Let P and Q be two points on the graph (Fig. 19–1),

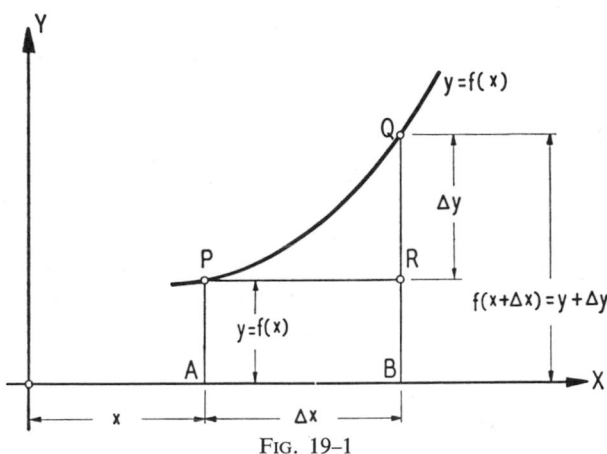

FIG. 19–1

x the abscissa of P, and Δx the increment of x if we proceed from P to Q. From the figure it can be seen that $\Delta x = PR$ is the difference of the abscissas of P and Q. The corresponding increment of $f(x)$ is, as seen from the figure,

$$\Delta y = RQ = BQ - BR = f(x + \Delta x) - f(x).$$

Example 5. The degree of magnetization or magnetic flux density in a piece of iron depends on the magnetizing force. For a certain kind of steel, the flux density B, measured in gausses, is plotted in Fig. 19–2 against the magnetizing force H, measured in oersteds. What is the increment ΔB if H increases from 0.4 to 0.8, and what is the value of ΔB if H increases from 1.0 to 1.4? In both cases $\Delta H = 0.4$. The graph shows that the increment of B in the first case is

$$\Delta B = R_1 Q_1 = 4000,$$

and the increment in the second case is

$$\Delta B = R_2 Q_2 = 1000$$

approximately.

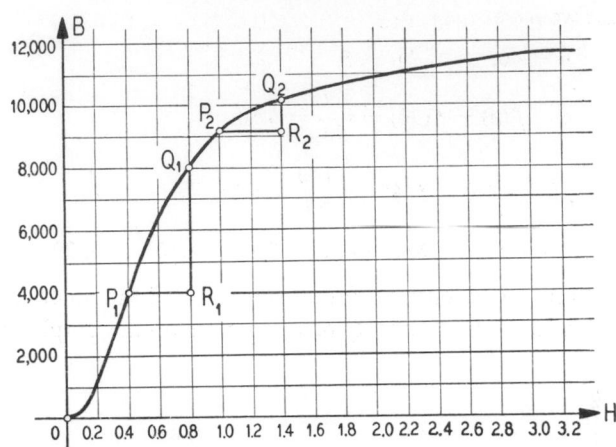

FIG. 19–2

Exercises

1. Using the table of exercise 15 of Sec. 3–8, find the corresponding increments if t changes from (*a*) 2 to 3 min, (*b*) 3 to 4 min, (*c*) 6 to 7 min, (*f*) 8 to 9 min.

2. Using the table of exercise 20 of Sec. 3–8, find the corresponding increments if θ changes from (*a*) 10° to 20°, (*b*) 30° to 40°, (*c*) 70° to 80°, (*d*) 80° to 90°.

Compute $\Delta \log x$ for $\Delta x = 10, 5, 1, 0.5, 0.1$ at a point where

3. $x = 40$.   **4.** $x = 60$.   **5.** $x = 24$.

6. $x = 78$.   **7.** $x = 320$.   **8.** $x = 122$.

Compute $\Delta \sin \theta$ for $\Delta\theta = 2°, 1°, 0.5°, 0.1°$ at a point where

9. $\theta = 6°$.   **10.** $\theta = 14°$.   **11.** $\theta = 25°$.

12. $\theta = 42°$.   **13.** $\theta = 63°$.   **14.** $\theta = 87°$.

Compute $\Delta \tan \theta$ for $\Delta\theta = 2°, 1°, 0.5°, 0.1°$ at a point where

15. $\theta = 5°$.   **16.** $\theta = 22°$.   **17.** $\theta = 44°$.

18. $\theta = 58°$.   **19.** $\theta = 75°$.   **20.** $\theta = 84°$.

Compute $\Delta \cos \theta$ for $\Delta\theta = 1°, 0.6°, 0.2°, 0.1°$ at a point where

21. $\theta = 10°$.   **22.** $\theta = 24°$.   **23.** $\theta = 36°$.

24. $\theta = 52°$.   **25.** $\theta = 61.4°$.   **26.** $\theta = 83.8°$.

Using a table of squares and square roots, compute the increments indicated in exercises 27–30 for $\Delta x = 5, 1, 0.1$.

27. $\Delta(x^2)$ at $x = 10$, $x = 20$, $x = 40$, $x = 80$.

28. $\Delta(\sqrt{x})$ at $x = 8$, $x = 16$, $x = 31$, $x = 63$.

29. $\Delta(\frac{1}{3}x^2 + 5\sqrt{x})$ at $x = 10$, $x = 20$, $x = 30$.

30. $\Delta(x^2 - 2x + 3\sqrt{x})$ at $x = 4$, $x = 8$, $x = 16$.

Compute the increments of the functions given in exercises 31–44.

31. $y = 3x + 7$ at $x = 4$ for $\Delta x = -1, 1, 0.5, 0.1$.

32. $y = 5x - 2$ at $x = 6$ for $\Delta x = -2, 2, 1, 0.1$.

33. $y = x^2 + 2x - 3$ at $x = 2$ for $\Delta x = 1, 0.5, 0.1$.

34. $y = 6x^2 - 8x + 5$ at $x = 3$ for $\Delta x = 0.6, 0.2, 0.1$.

35. $y = \dfrac{1}{x}$ at $x = 40$ for $\Delta x = 2, 1, -0.5$.

36. $y = \dfrac{1}{x}$ at $x = 4$ for $\Delta x = 2, 1, -0.5$.

37. $d = \dfrac{140}{n}$ at $n = 20$ for $\Delta n = -4, 2, 1$.

38. $I = \dfrac{220}{R}$ at $R = 40$ for $\Delta R = -5, -2, -1$.

39. $M = \dfrac{300}{\sqrt{70 + 3D}}$ at $D = 10$ for $\Delta D = 3, 1, 0.3$.

40. $I = \dfrac{440}{\sqrt{R^2 + 15}}$ at $R = 7$ for $\Delta R = 5, 2, 1$.

41. $A = \pi r^2$ at $r = 3.75$ for $\Delta r = 1, 0.5, 0.1$.

42. $V = \frac{4}{3}\pi r^3$ at $r = 6$ for $\Delta r = 3, 1, 0.1$.

43. $P = I^2 R$ at $R = 3$, $I = 18$ for $\Delta I = 2, 0.5, 0.1$.

44. $s = 1 + 0.00014(t - 60)$ at $t = 72$ for $\Delta t = 2, 1, 0.5$.

45. The power in an a-c circuit is $P = EI \cos \theta$.   Compute ΔP if $E = 220$ volts, $I = 12$ amperes, and if θ changes from $10°$ to $14°$.

46. The minimum distance D in feet in which an automobile with perfect brakes can be stopped on a level road is given by the formula

$$D = \frac{v^2}{2g\mu},$$

where v is the speed of the car in feet per second.   Compute ΔD if $g = 32.2$ ft per sec per sec, $\mu = 0.8$, and the speed of an automobile is increased from 50 to 55 mph.

47. The maximum torque T to be applied to a solid circular shaft for a design stress s is given by the formula

$$T = \frac{\pi s D^3}{16},$$

where D is the diameter of the shaft.   Compute ΔT if $s = 8000$ lb per sq in., $D = 3.00$ in., $\Delta D = 0.50$ in.

48. The speed of sound in a gas at any temperature T is given by the expression

$$v = 916.3k \sqrt{\frac{450 + T}{482}},$$

where k is a constant depending on the nature of the gas.   Compute Δv for $k = 1.2$, $T = 32°$, $\Delta T = 12°$.

49. The angle of bank β for a proper turn of an airplane is given by the formula

$$\tan \beta = \frac{v^2}{gR},$$

where v is the speed of the airplane, g is the acceleration due to gravity, and R is the radius of the turn.   Compute $\Delta\beta$ if $g = 32.2$ ft per sec per sec, $R = 4000$ ft, $v = 400$ ft per sec, $\Delta v = 100$ ft per sec.

50. The magnetization B of a certain kind of steel as a function of the magnetizing force H is given in Fig. 19–2. Find ΔB for $\Delta H = 0.2$ at $H = 0.2$, at $H = 0.4$, at $H = 0.8$, and at $H = 2.6$.

51. The distance s from the starting point of an object thrown upward with an initial velocity of 100 ft per sec as a function of the elapsed time t is given in Fig. 3–9.   Find Δs for $\Delta t = 0.5$ at $t = 0$, at $t = 2$, at $t = 4$, and at $t = 5.5$.

19–2. The Average Rate of Change

It is obvious that the increment of a function depends on the corresponding increment of the independent variable and on the point where the increment begins.   The table in Sec. 19–1 for corresponding values of pressure and wind velocity, which is repeated here, illustrates this fact very clearly.

V	5	6	7	8	9	10	11	12	13	14	15
P	0.88	1.26	1.72	2.24	2.84	3.50	4.26	5.04	5.92	6.86	7.88

If the increments of P corresponding to $\Delta V = 1, 2, 3$ are computed, first starting with $V = 6$, and then with $V = 12$, the following corresponding increments are obtained:

$$V = 6, \qquad \Delta V = 1, \qquad \Delta P = 0.46,$$
$$\Delta V = 2, \qquad \Delta P = 0.98,$$
$$\Delta V = 3, \qquad \Delta P = 1.58,$$
$$V = 12, \qquad \Delta V = 1, \qquad \Delta P = 0.88,$$
$$\Delta V = 2, \qquad \Delta P = 1.82,$$
$$\Delta V = 3, \qquad \Delta P = 2.84.$$

The increment ΔP increases in both cases if the corresponding ΔV increases, but the increase of ΔP is much greater at the point $V = 12$ than at the point $V = 6$.   When V increases from $V = 6$ to $V = 8$ by $\Delta V = 2$, the corresponding $\Delta P = 0.98$; when V increases from $V = 12$ to $V = 14$, we have again $\Delta V = 2$, but the corresponding $\Delta P = 1.82$. We say then that P increases faster in the interval from $V = 12$ to 14 than in the interval from $V = 6$ to 8.

In order to compare the behavior of P as a function of V in both intervals, the ratio $\Delta P / \Delta V$ of corresponding increments can be used.   The value of this ratio for the interval $V = 6$ to $V = 8$ is

$$\frac{\Delta P}{\Delta V} = \frac{0.98}{2} = 0.49;$$

the corresponding ratio computed for the interval $V = 12$ to $V = 14$ is

$$\frac{\Delta P}{\Delta V} = \frac{1.82}{2} = 0.91.$$

The ratio $\Delta P / \Delta V$ is called **average rate of change** of P in the interval from V to $V + \Delta V$. Thus 0.91 is the average rate of change of P in the interval from $V = 12$ to $V = 14$, and 0.49 is the average rate of change of P if V changes from 6 to 8.

The idea of an average rate of change is used very frequently. A few examples will illustrate the application of this concept.

Example 1. A car starts at noon, and at 2 P.M. is a distance of 50 miles, and at 3 P.M. a distance of 140 miles, from the starting point. Within the 3 hr from 2 P.M. to 5 P.M. the car covered a distance of $140 - 50 = 90$ miles at an **average rate** of $\frac{90}{3} = 30$ mph.

Example 2. On a certain day, the temperature at 7 A.M. was 30°, and at 1 P.M. it was 60°. The increment of the temperature was 30° during 6 hr, and the temperature changed at an average rate of $\frac{30°}{6} = 5°$ per hour.

Example 3. A city had a population of 80,000 people in 1920 and 120,000 people in 1940. The population increased by 40,000 people in 20 years. The average rate of increase was $\frac{40,000}{20} = 2000$ people per year.

In the above examples, it has not been stated that the observed quantities—distance, temperature, population—change uniformly. There is no reason why the population of the city in example 3 should have increased by exactly 2000 every year. The meaning of the result obtained in example 3 is that the change from 80,000 to 120,000 could have been produced by a uniform increase of 2000 people per year.

The average rate of change in the above examples was computed as the ratio of the increment of an observed quantity and the corresponding increment of time. The observed quantity was regarded as a function of the time. This concept of an average rate of change can be defined generally in the following way.

The average rate of change of a function $y = f(x)$ in the interval from x to $x + \Delta x$ is defined as the ratio of the corresponding increments,

$$\frac{\Delta y}{\Delta x} = \frac{\Delta f(x)}{\Delta x} = \frac{f(x + \Delta x) - f(x)}{\Delta x}.$$

It is easy to see that the average rates in the above examples were computed in this way. Consider for instance example 1. The distance s from the starting point may be regarded as a function of the time t, denoted by the symbol $s = f(t)$. We have then that $f(2) = 50$ and

$f(5) = 140$. The increment of t is $\Delta t = 5 - 2 = 3$, and the increment of s is $\Delta s = f(2 + \Delta t) - f(2) = f(5) - f(2) = 140 - 50 = 90$. Hence

$$\frac{\Delta s}{\Delta t} = \frac{90}{3} = 30 \text{ mph}.$$

A very striking geometrical interpretation of the average rate of change can be given if the graph of the function is plotted. The function $y = f(x)$ may be represented by the graph of Fig. 19–3. The increment Δx is given

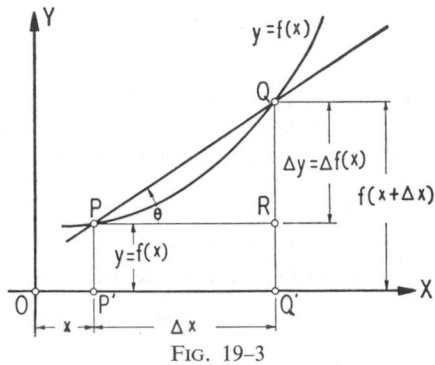

FIG. 19–3

by the segment $P'Q' = PR$. The corresponding increment $\Delta y = \Delta f(x)$ is given by the segment RQ. Hence it follows that the average rate of change in the interval $P'Q'$ is given by

$$\frac{\Delta y}{\Delta x} = \frac{RQ}{PR} = \tan \theta$$

where θ is the angle of inclination of the line PQ. Thus $\Delta y/\Delta x$ is the slope of the line PQ, and we may state that the *average rate of change of the function $y = f(x)$ in the interval from x to $x + \Delta x$ is equal to the slope of the straight line PQ connecting the points on the graph that correspond to the values x and $x + \Delta x$*. Such a straight line through two points of a curve is called a **secant** of the curve.

The average rate of change is computed very easily when the function is given by a formula. The following examples will show the procedure.

Example 4. Compute the average rate of change of the function $y = f(x) = 3x^2 - 2$ if x changes from 1 to 3.

Following the explanations given above, we have

$$\Delta x = 3 - 1 = 2,$$
$$f(x) = f(1) = 1,$$
$$f(x + \Delta x) = f(1 + 2) = f(3) = 25,$$
$$\Delta y = \Delta f(x) = f(x + \Delta x) - f(x) = 25 - 1 = 24,$$
$$\frac{\Delta y}{\Delta x} = \frac{f(x + \Delta x) - f(x)}{\Delta x} = \frac{24}{2} = 12.$$

Example 5. The plate current in milliamperes of a particular triode tube is a function of the grid potential, in volts, approximately given by the formula.

$$I_p = f(E_g) = 0.4(16 + E_g)^{\frac{3}{2}}.$$

Compute the average rate of change of the current when the grid potential changes from -12 to -8 volts.

We have, in this case,

$$\Delta E_g = 4 \text{ volts},$$

$$I_p = f(-12) = 0.4(16 - 12)^{\frac{3}{2}} = 3.20 \text{ milliamperes},$$

$$I_p + \Delta I_p = f(-8) = 0.4(16 - 8)^{\frac{3}{2}} = 9.05 \text{ milliamperes},$$

$$\Delta I_p = f(-8) - f(-12) = 5.85 \text{ milliamperes},$$

$$\frac{\Delta I_p}{\Delta E_g} = \frac{5.85}{4} = 1.46 \text{ milliamperes per volt}.$$

Exercises

1. A physics student performs an experiment on the boiling point of water and observes that the boiling temperature is $150°$ C at a pressure of 356.9 cm of mercury, and $200°$ C at a pressure of 1165.0 cm of mercury. Compute the average rate of increase of the boiling temperature of water with pressure for this temperature interval.

2. The electromotive force of a standard Clark cell is 1.4267 volts at $20°$ C and 1.4202 volts at $25°$ C. Compute the average rate of change of the voltage for this temperature interval.

3. Using Fig. 19–2, find the average rate of increase of B if H changes from 0.4 to 0.8, if H changes from 1.0 to 1.4, and if H changes from 2.2 to 2.6.

4. Using Fig. 3–6, find the average rate of change of s if t changes from 0.5 to 2.5 sec, if t changes from 0.5 to 2.0 sec, and if t changes from 0.5 to 1.0 sec.

In exercises 5–18 compute the average rate of change of each of the functions for the indicated intervals.

5. $y = \log x$; (*a*) from $x = 10$ to $x = 14$, (*b*) from $x = 30$ to $x = 40$, (*c*) from $x = 600$ to $x = 608$.

6. $y = \log x$; (*a*) from $x = 5$ to $x = 6$, (*b*) from $x = 5$ to $x = 5.1$, (*c*) from $x = 5$ to $x = 5.01$.

7. $y = \sin \theta$; (*a*) from $\theta = 16°$ to $\theta = 18°$, (*b*) from $\theta = 43°$ to $\theta = 44°$, (*c*) from $\theta = 61°$ to $\theta = 60.5°$.

8. $y = \tan \theta$; (*a*) from $\theta = 2°$ to $\theta = 4°$, (*b*) from $\theta = 47°$ to $\theta = 48°$, (*c*) from $\theta = 87°$ to $\theta = 88°$.

9. $y = \cot \theta$; (*a*) from $\theta = 1°$ to $\theta = 2°$, (*b*) from $\theta = 1°$ to $\theta = 1.5°$, (*c*) from $\theta = 1°$ to $\theta = 1.1°$.

10. $y = \cos \theta$; (*a*) from $\theta = 10°$ to $\theta = 11°$, (*b*) from $\theta = 36.1°$ to $\theta = 36.2°$, (*c*) from $\theta = 71.8°$ to $\theta = 71.6°$.

11. $y = x^2$; (a) from $x = 5$ to $x = 6$, (b) from $x = 5$ to $x = 5.1$, (c) from $x = 5$ to $x = 5.01$.

12. $y = x^2$; (a) from $x = 6$ to $x = 6.06$, (b) from $x = 6$ to $x = 6.03$, (c) from $x = 6.03$ to $x = 6.06$.

13. $s = 16t^2$; (a) from $t = 0$ to $t = 1$, (b) from $t = 3$ to $t = 5$, (c) from $t = 8$ to $t = 8.1$.

14. $C = \dfrac{v^2}{4}$; (a) from $v = 3$ to $v = 4$, (b) from $v = 4$ to $v = 3$, (c) from $v = 9.6$ to $v = 9.8$.

15. $u = \sqrt{3w}$; (a) from $w = 2$ to $w = 4$, (b) from $w = 3$ to $w = 3.5$, (c) from $w = 3$ to $w = 2.8$.

16. $d = 0.5\sqrt{2n}$; (a) from $n = 0$ to $n = 1$, (b) from $n = 2$ to $n = 2.2$, (c) from $n = 2$ to $n = 2.02$.

17. $i = 10 \sin 120\pi t$; (a) from $t = 0$ to $t = 0.001$, (b) from $t = 0$ to $t = \dfrac{1}{240}$, (c) from $t = 1$ to $t = 1.001$.

18. $d = 6 \sin \dfrac{\pi}{3} t$; (a) from $t = 6$ to $t = 7$, (b) from $t = 6$ to $t = 7.5$, (c) from $t = 3$ to $t = 3.1$.

19. Compute the average rate of increase in volume during inflation of a spherical balloon if the inside diameter increases from 6 to 6.6 ft.

20. The plate current I_p in amperes of a certain radio tube is given by the formula $I_p = 5.2 \cdot 10^{-6} E_p^{\frac{3}{2}}$, where E_p is the plate potential in volts. Compute the average change of the plate current if E_p changes from 60 to 80 volts.

21. A bolt in tension can carry a load F in pounds given by the formula $F = 5000 A^{\frac{3}{2}}$, where A in square inches is the area at the root of the thread of the bolt. Compute the average rate of change in the load a bolt can carry if A is changed from 0.40 to 0.60 sq in.

22. The quantity Q in cubic feet of water discharged per second through a circular opening of 1 sq ft area at the bottom of a tank is given by the formula $Q = 4.96\sqrt{h}$, where h in feet is the distance from the opening to the free surface of the water. Compute the average rate of change in Q if h changes from 8 to 8.1 ft.

19–3. The Instantaneous Rate of Change

There are many rate problems in practical life that are not satisfactorily solved by merely computing an average rate of change of a function. If an automobile accident happens, the driver cannot shake off his responsibility by proving that he drove at an average rate of 20 mph during the last 2 hr. The important questions are: What was his rate at the moment of the accident? Did he drive at this very moment at a rate of 15 or 60 mph? The interest is concentrated on his **instantaneous rate** or his speed at the instant of the accident.

The precise meaning of this concept of an instantaneous rate needs to be investigated. It was easy to define an average rate of change of a function, but difficulties arise if one tries to use the same approach to the idea of an instantaneous rate. There is no increment of time for the moment of the accident and no corresponding distance covered by the automobile. The actual experiments that have been designed in order to find the speed of a car, or of a bullet, measure the average rate during a very short time. It is assumed that during this short time the motion of the object is uniform so that the average rate during a short time interval can be used in order to characterize the motion at any instant during this interval. This is a somewhat vague description of how to arrive at the idea of an instantaneous rate. The problem of this section is to clarify and to give a correct definition of this concept.

Consider the rate of a car that moves with a constant speed of 30 mph. The average rate for 2 hr is $\dfrac{60}{2} = 30$ mph, the rate for 0.5 hr is $\dfrac{15}{0.5} = 30$ mph, the average rate for 1 min is $\dfrac{\frac{1}{2}}{\frac{1}{60}} = 30$ mph. It is obvious that the rate can be computed for any interval of time and that the value 30 mph will always result. If the time interval is small, the distance covered by the car during this interval is small, but the ratio of this distance Δs and the time Δt is the same. For example, if $\Delta t = 1$ sec $= \dfrac{1}{3600}$ of an hour, the corresponding $\Delta s = \dfrac{30}{3600} = \dfrac{1}{120}$ mile and $\dfrac{\Delta s}{\Delta t} = \dfrac{3600}{120} = 30$ mph.

In more complicated cases, the rates are not constant, but observations show that generally they are nearly constant if small intervals are investigated. This experience has been encountered so often that it is used, almost instinctively, in many applications. Assume, for instance, that an experiment shows an increment of $\Delta I_p = 2$ milliamperes of the plate current of a tube if the grid voltage changes from $E_g = -9$ to $E_g = -8$ volts. The average rate of change in this interval is $\dfrac{\Delta I_p}{\Delta E_g} = \dfrac{2}{1} = 2$ milliamperes per volt. It can therefore be assumed that the increment of the current within this interval is approximately proportional to the increment of E_g, so that $\Delta I_p = 1, \dfrac{1}{2}, \dfrac{1}{5}$ milliampere, when $\Delta E_g = \dfrac{1}{2}, \dfrac{1}{4}, \dfrac{1}{10}$ volt, respectively, and that $\dfrac{\Delta I_p}{\Delta E_g} = 2$ for all corresponding increments of I_p and E_g within the observed interval.

The whole situation can more easily be examined, using a function for which many values are given. The function $y = \log x$ may be selected for this purpose, and the average rates of change will be computed for different intervals starting with $x = 1.5$. The following table contains the required data.

x	$\log x$	$\Delta x = x - 1.5$	$\Delta y = \log x - \log 1.5$	$\Delta y / \Delta x$
1.3	0.1139	−0.2	−0.0622	0.311
1.4	0.1461	−0.1	−0.0300	0.300
1.5	0.1761	0	0	
1.6	0.2041	0.1	0.0280	0.280
1.7	0.2304	0.2	0.0543	0.272

It can be observed that the average rates of change of $y = \log x$ for segments to the left of $x = 1.5$ are greater than the average rates for segments to the right of this point, and that the values of these rates are, for the investigated segments, rather close to each other. We may expect that for still smaller intervals, starting with the same point $x = 1.5$, the rates of change will be between 0.300 and 0.280. A four-place table of logarithms is not sufficient to study the behavior of these rates for smaller intervals and, therefore, a seven-place logarithmic table has been used to obtain the following values.

x	$\log x$	$\Delta x = x - 1.500$	$\Delta y = \log x - \log 1.500$	$\Delta y / \Delta x$
1.498	0.1755118	−0.002	−0.0005795	0.2898
1.499	0.1758016	−0.001	−0.0002897	0.2897
1.500	0.1760913	0	0	
1.501	0.1763807	0.001	0.0002894	0.2894
1.502	0.1766699	0.002	0.0005786	0.2893

It may be observed that the values of the average rates of change for these small intervals are very close to each other, and it may be expected that, if still smaller intervals of both sides of $x = 1.5$ are investigated, the corresponding average rates of change will be between 0.2897 and 0.2894. We say that the behavior of $y = \log x$ in the neighborhood of $x = 1.500$ is characterized by a number between 0.2894 and 0.2897, which could be found with any precision if logarithmic tables with enough decimals were available. This number is called the **instantaneous rate of change** or the **derivative** of the function $y = \log x$ with respect to x at the point $x = 1.5$. Its correct value with seven decimals is 0.2895297. This number means that the average rate of change of the function $y = \log x$ for intervals adjacent to $x = 1.5$ approaches this value if smaller and smaller intervals are investigated.

A good approximation for this instantaneous rate of change or derivative at a given point may be found by computing the average rates of change for two small segments of equal length adjacent to the given point at both sides of it and by taking half the sum of the two values. In this way we obtain, from the first of the two tables given above:

$$\frac{\Delta \log x}{\Delta x} = 0.300 \text{ in the interval from } x = 1.4 \text{ to } x = 1.5,$$

$$\frac{\Delta \log x}{\Delta x} = 0.280 \text{ in the interval from } x = 1.5 \text{ to } x = 1.6.$$

Taking half the sum of these two values, we get

$$\tfrac{1}{2}(0.300 + 0.280) = 0.290$$

as the value of the instantaneous rate of change, correct to three decimals.
The following examples will illustrate this new concept of an instantaneous rate of change.

Example 1. If the angle x is expressed in radians, compute approximately the derivative of the function $y = \sin x$ at the point $x = 1.2$.

From Table 5 in the appendix we have the following table of values:

x	$\sin x$	$\Delta x = x - 1.2$	$\Delta y = \sin x - \sin 1.2$	$\Delta y / \Delta x$
1.10	0.8912	−0.10	−0.0408	0.41
1.19	0.9284	−0.01	−0.0036	0.36
1.20	0.9320			
1.21	0.9356	0.01	0.0036	0.36
1.30	0.9636	0.10	0.0316	0.32

Hence the derivative or instantaneous rate of change of $y = \sin x$, at $x = 1.2$, is, with two decimals, 0.36.

Example 2. Using the table of Sec. 19–2, compute an approximate value for the instantaneous rate of change of P with respect to V at the point $V = 11$ mph.

From the given table we obtain the following values:

V	P	ΔV	ΔP	$\Delta P / \Delta V$
10	3.50	−1	−0.76	0.76
11	4.26	0	0	
12	5.04	+1	0.78	0.78

An approximate value for the instantaneous rate of change of P with respect to V at the point $V = 11$, is given by

$$\tfrac{1}{2}(0.76 + 0.78) = 0.77 \text{ lb per mph.}$$

Exercises

Compute, with the precision attainable by the tables in the appendix, approximate values for the derivatives of the functions in exercises 1–14 at the indicated points:

1. $y = \log x$ at (a) $x = 3$, (b) $x = 9$, (c) $x = 0.4$.
2. $y = \log x$ at (a) $x = 4$, (b) $x = 17$, (c) $x = 6.6$.
3. $y = \sqrt{x}$ at (a) $x = 3$, (b) $x = 12$, (c) $x = 6.72$.

4. $y = \sqrt{x}$ at (a) $x = 1$, (b) $x = 0.8$, (c) $x = 3.48$.

5. $y = x^2$ at (a) $x = 5$, (b) $x = 5.1$, (c) $x = 5.01$.

6. $y = x^2$ at (a) $x = 620$, (b) $x = 62$, (c) $x = 6.2$.

7. $y = \sin x$ (x in radians) at (a) $x = 0.4$, (b) $x = 0.9$, (c) $x = 0$.

8. $y = \sin x$ (x in radians) at (a) $x = 1.3$, (b) $x = 0.52$, (c) $x = \dfrac{\pi}{2}$.

9. $y = \cos x$ (x in radians) at (a) $x = 0.22$, (b) $x - 0.8$, (c) $x = 1.44$.

10. $y = \tan x$ (x in radians) at (a) $x = 0.10$, (b) $x = 0.20$, (c) $x = 0.40$.

11. $y = \log \cos \theta$ at (a) $\theta = 9.9°$, (b) $\theta = 29.1°$, (c) $\theta = 64.7°$.

12. $y = \log \tan \theta$ at (a) $\theta = 2.9°$, (b) $\theta = 39.7°$, (c) $\theta = 76.9°$.

13. $y = \log \sin x$ (x in radians) at (a) $x = 0.3$, (b) $x = 0.7$, (c) $x = 1.2$.

14. $y = \log \cot x$ (x in radians) at (a) $x = 0.5$, (b) $x = 0.92$, (c) $x = 1.35$.

15. Compute, using the table of exercise 13 of Sec. 3–8, an approximate value for the instantaneous rate of change of S with respect to P at (a) $P = 1$ per cent, (b) $P = 3$ per cent, (c) $P = 6$ per cent.

16. Compute, using the table of exercise 15 of Sec. 3–8, an approximate value for the instantaneous rate of change of T with respect to t at (a) $t = 2$ min, (b) $t = 5$ min, (c) $t = 8$ min.

17. Compute, using the table of exercise 20 of Sec. 3–8, an approximate value for the instantaneous rate of change of L with respect to θ at (a) $\theta = 20°$, (b) $\theta = 40°$, (c) $\theta = 70°$.

18. Compute, using the table of exercise 22 of Sec. 3–8, an approximate value for the instantaneous rate of change of I with respect to E at (a) $E = 20$ volts, (b) $E = 50$ volts, (c) $E = 100$ volts.

19. Compute, using the table of exercise 12 of Sec. 3–8, an approximate value for the instantaneous rate of change of P with respect to S at (a) $S = 600$ rpm, (b) $S = 1000$ rpm, (c) $S = 1400$ rpm.

20. Compute, using the table of exercise 16 of Sec. 3–8, an approximate value for the instantaneous rate of change of e_0 with respect to F at (a) $F = 1406$ kilocycles per second, (b) $F = 1402$ kilocycles per second, (c) $F = 1394$ kilocycles per second.

21. Compute, using the table of exercise 14 of Sec. 3–8, an approximate value for the instantaneous rate of change of S with respect to T at (a) $T = -20°$, (b) $T = 0°$, (c) $T = 60°$.

22. Compute, using the table of exercise 17 of Sec. 3–8, an approximate value for the instantaneous rate of change of D with respect to H at (a) $H = 1$ ft, (b) $H = 3.5$ ft, (c) $H = 8$ ft.

23. Compute, using the table of exercise 21 of Sec. 3–8, an approximate value for the instantaneous rate of change of v with respect to p at (a) $p = 16$ lb per sq in., (b) $p = 25$ lb per sq in., (c) $p = 50$ lb per sq in.

24. Compute, using the table of exercise 23 of Sec. 3–8, an approximate value for the instantaneous rate of change of d with respect to E at (a) $E = 40$ volts, (b) $E = 0$ volts, (c) $E = -40$ volts.

19–4. Limits

The explanation of the instantaneous rate of change and the method of computing an approximate value of this rate were based on the observation that the average rate of change approaches a certain value if the interval for which it is computed becomes very small. Situations of this kind, where a quantity approaches a certain value, occur in different branches of mathematics and its applications and are especially important for the subject of this chapter.

Consider, as an example, the sequence of numbers

$$0, \tfrac{1}{2}, \tfrac{2}{3}, \tfrac{3}{4}, \tfrac{4}{5}, \tfrac{5}{6}, \ldots$$

which can be continued indefinitely. The numbers obviously approach one. In order to state this fact, we say that 1 is the **limit** of the numbers of the given sequence. Every number of the given sequence can be found by the formula $a_n = \dfrac{n-1}{n}$, for when n is replaced by particular values we then obtain

$$a_1 = \frac{1-1}{1} = 0, a_2 = \frac{2-1}{2} = \frac{1}{2}, a_3 = \frac{3-1}{3} = \frac{2}{3}, a_4 = \frac{4-1}{4} = \frac{3}{4}, \text{ etc.}$$

From the expression $a_n = \dfrac{n-1}{n} = 1 - \dfrac{1}{n}$ it can be inferred that a_n approaches the value 1 when n increases indefinitely, for $1/n$ becomes as small as we want if n increases. In order to state the fact that the numbers of the above sequence have the limit 1, the following notation is used:

$$\lim a_n = \lim \frac{n-1}{n} = 1, \quad \text{if} \quad n \to \infty,$$

or more briefly

$$\lim_{n \to \infty} \frac{n-1}{n} = 1,$$

which reads as follows: *the limit of* $\dfrac{n-1}{n}$ *is equal to 1 if n increases indefinitely*, or if n becomes *indefinitely great*, or if n *approaches infinity*.

An illustration of the concept of limit is given by the following engineering application.

Example 1. The combined resistance of two resistances R_1 and R_2, connected in parallel, is given by the formula

$$R = \frac{R_1 R_2}{R_1 + R_2}.$$

What is the value of the limit of R as R_2 approaches infinity?

The general method for finding the limit, in this and similar examples, is to divide numerator and denominator of R by R_2, yielding

$$R = \frac{R_1}{\dfrac{R_1}{R_2} + 1}.$$

The fraction R_1/R_2 approaches zero, when R_2 increases indefinitely, and, therefore, R approaches the value $\dfrac{R_1}{1} = R_1$, so that we have finally

$$\lim R = R_1, \quad \text{if} \quad R_2 \to \infty.$$

The concept of limit refers to many similar situations. Two more examples will illustrate this.

Example 2. Examine the values of the function

$$y = \frac{\log x}{x - 1}$$

when x assumes values which approach 1.

First observe that no value of y is obtained when $x = 1$ because the numerator and the denominator of the given function become zero when the value $x = 1$ is substituted.   We shall therefore investigate the value of y in the neighborhood of $x = 1$.   This is done in the following table, where the values of y are computed for different values of x approaching 1.

x	$\log x$	$x - 1$	$\dfrac{\log x}{x - 1}$
0.50	$0.6990 - 1 = -0.3010$	-0.50	0.602
0.80	$0.9031 - 1 = -0.0969$	-0.20	0.484
0.90	$0.9542 - 1 = -0.0458$	-0.10	0.458
0.99	$0.9956 - 1 = -0.0044$	-0.01	0.44
1	0	0	
1.01	0.0043	0.01	0.43
1.10	0.0414	0.10	0.414
1.20	0.0792	0.20	0.396
1.50	0.1761	0.50	0.352

From this table it is seen that, when x approaches 1, the function $\dfrac{\log x}{x - 1}$ approaches the value M, which lies between 0.43 and 0.44.   By using logarithmic tables with more decimals this value M could be found with greater precision. Thus we could get $M = 0.434$ correct to three significant digits.

Using the notation that has been introduced before, we write

$$\lim_{x \to 1} \frac{\log x}{x - 1} = 0.434,$$

which reads: *the limit of* $\dfrac{\log x}{x - 1}$ *is 0.434 if x approaches 1.*

The situation appearing in the above example occurs very often and can be described as follows.   *The limit of a fraction has to be computed if numerator and denominator become very small.*   Indeed, the concept of instantaneous rate is based on such a situation, as will be shown in the following example.

Example 3. Compute the instantaneous rate of change of the function $y = \sqrt{x}$ at the point $x = 1$.

When $x = 1$, then $y = 1$; and, when $x = 1 + \Delta x$, then $y = \sqrt{1 + \Delta x}$. Hence the average rate of change of $\sqrt{x}$ in the interval from 1 to $1 + \Delta x$ is given by the expression

$$\frac{\sqrt{1 + \Delta x} - 1}{\Delta x},$$

and the instantaneous rate of change is the limit of this expression as Δx approaches zero. But when $\Delta x = 0$ both the numerator and the denominator of the last expression are equal to zero, and we must investigate its value in the neighborhood of $\Delta x = 0$. This is done in the following table.

Δx	$\sqrt{1 + \Delta x} - 1$	$\dfrac{\sqrt{1 + \Delta x} - 1}{\Delta x}$
-0.1	-0.05132	0.5132
-0.01	-0.005013	0.5013
-0.001	-0.0005001	0.5001
0	0	
0.001	0.0004999	0.4999
0.01	0.004988	0.4988
0.1	0.04881	0.4881

We can infer, from this table, that the value sought lies between 0.5001 and 0.4999 and that it is 0.5. Hence

$$\lim \frac{\sqrt{1 + \Delta x} - 1}{\Delta x} = 0.5, \quad \text{if} \quad \Delta x \to 0,$$

and 0.5 is the value of the instantaneous rate of change, or the derivative, of the function $\sqrt{x}$ at the point $x = 1$.

Exercises

Compute the following limits.

1. $\lim \left(6 + \dfrac{1}{n} \right)$, $n \to \infty$.

2. $\lim \left(3 - \dfrac{1}{n} \right)$, $n \to \infty$.

3. $\lim \left(0.001 - \dfrac{2}{n} \right)$, $n \to \infty$.

4. $\lim \left(\dfrac{3}{n} + 2\pi \right)$, $n \to \infty$.

5. $\lim \left(\dfrac{5}{n^2} + 8 \right)$, $n \to \infty$.

6. $\lim \left(27 - \dfrac{9}{x^3} \right)$, $x \to \infty$.

7. $\lim \dfrac{4n + 1}{2n + 1}$, $n \to \infty$. (*Hint.* Divide numerator and denominator by n.)

8. $\lim \dfrac{9n - 1}{3n - 1}$, $n \to \infty$.

9. $\lim \dfrac{3 + 4n}{7 + 5n}$, $n \to \infty$.

10. $\lim \dfrac{n^2 + 1}{n^2 - 1}$, $n \to \infty$.

11. $\lim \dfrac{5 + 2n^2}{1 - n^2}$, $n \to \infty$.

12. $\lim \dfrac{6 - 3x^2}{x^2 - 2x}$, $x \to \infty$.

13. $\lim \dfrac{2u + u^2}{7 - 3u^2}$, $u \to \infty$.

14. $\lim 2^{-t}$, $t \to \infty$.

15. $\lim 5.4^{-t}$, $t \to \infty$.

16. $\lim (8 + 2 \cdot 6.1^{-12t})$, $t \to \infty$.

17. $\lim (30 - 10 \cdot 1.5^{-2t})$, $t \to \infty$.

In the following exercises assume E, I, R, L, C to be positive quantities, $e = 2.718$, and compute the limits if $t \to \infty$.

18. $\lim (10 - Ie^{-Rt/L})$.

19. $\lim (60 + Ee^{-t/RC})$.

20. $\lim E(1 - e^{-t/RC})$.

21. $\lim [4I + I(1 - e^{-t/RC})]$.

22. $\lim \dfrac{7E}{3R} (21 - e^{-Rt/L})$.

23. $\lim \dfrac{E}{377L} (e^{-10t} - e^{-20t})$.

24. $\lim \dfrac{E^2}{R} (e^{-t/RC} - e^{-2t/RC})$.

25. $\lim [(Ie^{-10t} \cdot \sin 60t + 10)]$.

26. $\lim (4 - 10e^{-2t} \cos 377t)$.

27. $\lim Ee^{-5t}(3 \sin 240t + 5 \cos 240t)$.

The angle θ in the following exercises is expressed in radians. Find the results with the precision that can be attained if the tables in the appendix are used.

28. $\lim \dfrac{\sin \theta}{\theta}$, $\theta \to 0$.

29. $\lim \dfrac{\sin \theta}{\tan \theta}$, $\theta \to 0$.

30. $\lim \dfrac{1 - \cos \theta}{\theta^2}$, $\theta \to 0$.

31. $\lim \theta \cot \theta$, $\theta \to 0$.

32. The combined reactance X_C of two capacitors C_1 and C_2, connected in series and operated at a frequency f, is given by the formula

$$X_C = \frac{2\pi f C_1 C_2}{C_1 + C_2}.$$

Find the limit of X_C as (a) C_1 approaches infinity, (b) C_1 approaches zero.

33. The capacitance C of a circuit consisting of three capacitors, C_1, C_2, and C_3, in series, is given by the equation

$$C = \frac{C_1 C_2 C_3}{C_1 C_2 + C_2 C_3 + C_1 C_3}.$$

Find the limit of C as C_2 approaches infinity.

19–5. Geometrical Investigation of the Instantaneous Rate of Change

A geometrical interpretation of the average rate of change as the slope of a straight line was studied in Sec. 19–2. This will be used in this section in discussing the instantaneous rate of change.

Assume that a function is given by the graph of Fig. 19–4. The average rate of change of the function for the segment QQ_1 is equal to the slope of the secant PP_1. If different points, P_1, P_2, P_3, $\cdots$ to the right of P are chosen, different secants PP_1, PP_2, PP_3, $\cdots$ are obtained. Similarly, different points P_1', P_2', P_3', $\cdots$ to the left of P may be used in order to construct the secants PP_1', PP_2', PP_3', $\cdots$. From the figure it is seen that there is a line PT that separates the secants meeting the curve to the right of P from the secants meeting the curve to the left of P. This line is called the **tangent of the curve** at P. The secants PP_1, PP_2, PP_3, $\cdots$ approach the tangent if the points P_1, P_2, P_3, $\cdots$ approach P, or, which is the same, if the lengths of the segments QQ_1, QQ_2, QQ_3, $\cdots$ approach zero.

The slopes of the secants PP_1, PP_2, PP_3, $\cdots$ approach the slope of the

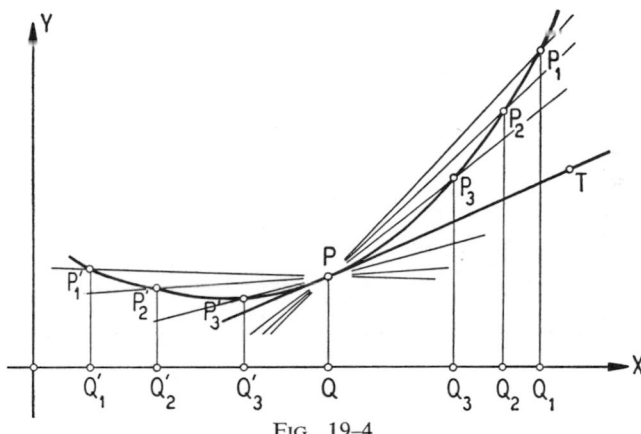

FIG. 19–4

tangent PT when the points P_1, P_2, P_3, $\cdots$ approach P. We may say that *the slope of the tangent at P is the limit of the slopes of the secants through P whose second point of intersection with the graph approaches P.*

The slopes of the secants PP_1, PP_2, PP_3, $\cdots$ are the average rates of change of the function for the intervals QQ_1, QQ_2, QQ_3, $\cdots$, respectively. These average rates approach the instantaneous rate of change at P. But we have seen before that the slopes of the secants approach the slope of the tangent. We can state, therefore, that *the instantaneous rate of change at P is equal to the slope of the tangent at P.*

In order to find the slope of the tangent, consider an arbitrary point T on the tangent (Fig. 19–5). The increment of the coordinates, if one passes from P to T, may be denoted by dx and dy. We have, then, for the slope of the tangent PT,

$$\tan \theta = \frac{dy}{dx}.$$

But it was stated above that the instantaneous rate of change at the point P, which may be denoted by m, is equal to the slope of the tangent PT, and therefore

$$m = \frac{dy}{dx}.$$

Since the instantaneous rate of change was defined as the limit of the average rate of change for the interval Δx, as Δx approaches zero, we have then

$$m = \lim \frac{\Delta y}{\Delta x}, \qquad \Delta x \to 0.$$

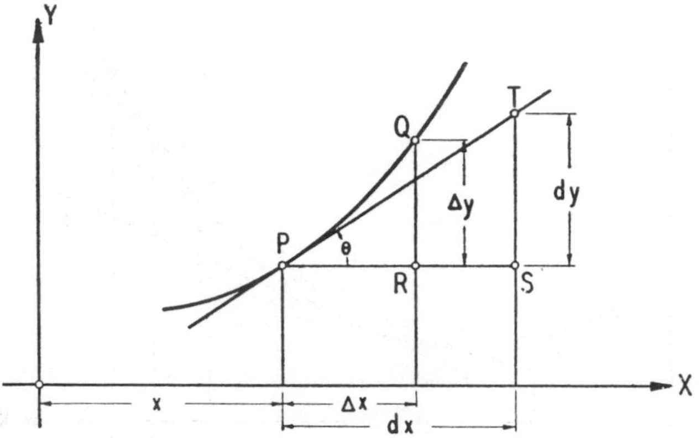

Fig. 19–5

To summarize the contents of this and the preceding section, *the derivative, or instantaneous rate of change, of a function $y = f(x)$ at a particular point P is the value of* $\lim \dfrac{\Delta y}{\Delta x}$ *as Δx approaches zero. This value is equal to the slope of the tangent at P. If dx and dy are the increments of x and y from P to an arbitrary point on the tangent, then the derivative at P is given by*

$$\lim_{\Delta x \to 0} \frac{\Delta y}{\Delta x} = \frac{dy}{dx}.$$

19–6. Derivative of a Function Given by a Formula

It has been explained, in the preceding sections of this chapter, how to estimate the **derivative** (this will be used in preference to **instantaneous rate of change**) of a function that is defined by a table or by a graph. The next step is to show how to find the derivative of a function that is defined by a formula.

Consider, for example, the function

$$y = f(x) = 3x^2 - 5. \tag{1}$$

The problem is to find the derivative of this function at some particular point, for example, $x = 2$. According to the discussion in the preceding sections, our first step is to find the average rate of change of the function in the interval from $x = 2$ to $x = 2 + \Delta x$. We thus obtain the following:

$$f(2) = 3 \cdot 4 - 5 = 7,$$

$$f(2 + \Delta x) = 3(2 + \Delta x)^2 - 5 = 7 + 12 \cdot \Delta x + 3 \cdot (\Delta x)^2,$$

$$\Delta y = f(2 + \Delta x) - f(2) = 12 \cdot \Delta x + 3 \cdot (\Delta x)^2,$$

$$\frac{\Delta y}{\Delta x} = 12 + 3 \cdot \Delta x.$$

The next step is to find the limit of the average rate of change $\dfrac{\Delta y}{\Delta x}$ as $\Delta x \to 0$. We have then

$$\lim \frac{\Delta y}{\Delta x} = \lim (12 + 3 \cdot \Delta x) = 12, \quad \text{as} \quad \Delta x \to 0.$$

Hence 12 is the value of the derivative of the given function $y = 3x^2 - 5$ at the point $x = 2$.

The same method can be used if the derivative of this function 1 is wanted for other values of x. In order to avoid the repetition of the same steps, we shall derive a formula that gives the derivative of function 1 for any desired value of x. This can be accomplished if in the preceding computations the number 2 is replaced by the symbol x. We thus obtain

$$f(x) = 3x^2 - 5,$$

$$f(x + \Delta x) = 3(x + \Delta x)^2 - 5 = 3x^2 + 6x \cdot \Delta x + 3 \cdot (\Delta x)^2 - 5.$$

Hence $\Delta f(x)$, which is the increment of $f(x)$ corresponding to the increment Δx of x, is given by

$$\Delta f(x) = f(x + \Delta x) - f(x) = 6x \cdot \Delta x + 3 \cdot (\Delta x)^2.$$

Thus the average rate of change of $f(x)$ for the interval from x to $x + \Delta x$ is given by

$$\frac{\Delta f(x)}{\Delta x} = \frac{f(x + \Delta x) - f(x)}{\Delta x} = \frac{6x \cdot \Delta x + 3 \cdot (\Delta x)^2}{\Delta x} = 6x + 3 \cdot \Delta x,$$

whence

$$\lim \frac{\Delta f(x)}{\Delta x} = \lim (6x + 3 \cdot \Delta x) = 6x, \quad \text{as} \quad \Delta x > 0.$$

The result $6x$ gives the derivative of the function $y = 3x^2 - 5$ at any point x. Substituting $x = 2$ yields the value 12, which is the same as the answer previously obtained.

It is necessary to have a convenient notation for derivatives. In the preceding section it has been shown that the derivative is equal to the ratio dy/dx, where dx and dy are the corresponding increments if one passes from the point $P(x, y)$ to an arbitrary point on the tangent at P. This ratio dy/dx will be used systematically in denoting the derivative. Thus the following are the notations used to denote the derivative of the function 1 at an arbitrary point x:

$$\text{Derivative} = \lim_{\Delta x \to 0} \frac{\Delta f(x)}{\Delta x} = \frac{dy}{dx} = \frac{df(x)}{dx} = \frac{d(3x^2 - 5)}{dx} = \frac{d}{dx}(3x^2 - 5),$$

and we have

$$\frac{d(3x^2 - 5)}{dx} = 6x,$$

which is read: *the derivative of $3x^2 - 5$ with respect to x is equal to $6x$* or simply $d(3x^2 - 5)$ *over dx is $6x$.*

Sometimes it is convenient to have a still shorter notation for the derivative of $y = f(x)$, and we write

$$\frac{dy}{dx} = y' = f'(x),$$

which is read: *y prime*, or *f prime of x.*

The derivative of a function, when computed for an arbitrary value x is another function of this variable x. The operation which consists in finding the derivative is called **differentiation.** The following examples will illustrate the concepts of this section.

Example 1. The force F in dynes required to produce work of 12,000 ergs is given by

$$F = \frac{12,000}{s},$$

where s in centimeters is the space through which the force acts. Compute the derivative dF/ds.

The force F is here a function $f(s)$ of the independent variable s. If s increases by Δs, the corresponding increment of F is given by

$$\Delta F = f(s + \Delta s) - f(s) = \frac{12,000}{s + \Delta s} - \frac{12,000}{s} = -\frac{12,000 \cdot \Delta s}{(s + \Delta s)s},$$

and hence the average rate of change is

$$\frac{\Delta F}{\Delta s} = -\frac{12,000}{(s + \Delta s)s}.$$

The derivative is the limit of this expression as $\Delta s \to 0$, and hence we get

$$\frac{dF}{ds} = \lim_{\Delta s \to 0} \frac{-12,000}{(s + \Delta s)s} = -\frac{12,000}{s^2} \text{ dynes per centimeter.}$$

If it is given that $s = 40$ cm, we then obtain

$$F = \frac{12{,}000}{s} = 300 \text{ dynes,}$$

$$\frac{dF}{ds} = -\frac{12{,}000}{s^2} = -\frac{12{,}000}{1600} = -7.5 \text{ dynes per centimeter.}$$

Example 2.   The velocity of a moving object is equal to the instantaneous rate of change with respect to the time of the distance between the object and its starting point.   If an object is thrown vertically upwards with an initial speed of 100 ft per sec, its distance from the starting point after t sec is given by the expression

$$s = 100t - 16t^2.$$

What is its velocity after t sec?   After 3 sec?
    Proceeding as before, we have the following:

$$s = f(t) = 100t - 16t^2,$$
$$s + \Delta s = f(t + \Delta t) = 100(t + \Delta t) - 16(t + \Delta t)^2$$
$$= 100t + 100 \cdot \Delta t - 16t^2 - 32t \cdot \Delta t - 16 \cdot (\Delta t)^2,$$
$$\Delta s = f(t + \Delta t) - f(t) = 100 \cdot \Delta t - 32t \cdot \Delta t - 16 \cdot (\Delta t)^2.$$

Hence the average velocity during the time interval Δt is given by

$$\frac{\Delta s}{\Delta t} = \frac{100 \cdot \Delta t - 32t \cdot \Delta t - 16 \cdot (\Delta t)^2}{\Delta t} = 100 - 32t - 16 \cdot \Delta t,$$

whence

$$\frac{ds}{dt} = \lim_{\Delta t \to 0} (100 - 32t - 16 \cdot \Delta t) = 100 - 32t.$$

We have, finally, that the velocity of the object after t sec is $(100 - 32t)$ ft per sec.   After 3 sec the velocity is $100 - 32 \cdot 3 = 4$ ft per sec.

Exercises

Compute the derivatives of the following functions at the indicated points:

1. $f(x) = 5x^2$ at (a) $x = 4$, (b) $x = -2$, (c) $x = 0$.

2. $f(x) = 7x^2$ at (a) $x = 3$, (b) $x = 0$, (c) $x = 0.1$.

3. $f(x) = x^3$ at (a) $x = 2$, (b) $x = -3$, (c) $x = 1$.

4. $f(x) = 2x^3$ at (a) $x = 3$, (b) $x = -1$, (c) $x = 2$.

5. $f(x) = 2x^2 + 1$ at (a) $x = 10$, (b) $x = 1$, (c) $x = -1$.

6. $f(x) = 3x^2 - 2$ at (a) $x = 6$, (b) $x = -3$, (c) $x = 0$.

7. $f(x) = 6x^2 - 0.2$ at (a) $x = 2$, (b) $x = 9$, (c) $x = 1.5$.

8. $f(x) = 0.5x^2 - 0.1$ at (a) $x = 0.2$, (b) $x = 0.8$, (c) $x = -0.1$.

9. $f(x) = x^3 + 2x$ at (a) $x = 1$, (b) $x = -6$, (c) $x = 2$.

10. $f(x) = 2x^3 + x$ at (a) $x = 3$, (b) $x = 0$, (c) $x = -1$.

Compute the derivatives of the following functions for arbitrary values of the independent variable:

11. $P = 24t^2$.

12. $W = 7.6v^2$.

13. $A = \pi r^2$, (π = constant).   **14.** $V = \frac{4}{3}\pi r^3$, (π = constant).

15. $F = 9m + 12$.   **16.** $H = 2T + 14$.

17. $s = 16t^2 - 8$.   **18.** $s = 16t^2 + 8$.

19. $s = 18t + 3.6t^2$.   **20.** $s = 22t^2 - 6.1t$.

21. $s = 2t^3 - t^2$.   **22.** $s = 2t^2 - t^3$.

23. $X_C = \dfrac{1}{200C}$.   **24.** $P = \dfrac{640}{M}$.

25. $I = \dfrac{10}{1 + R}$.   **26.** $E = \dfrac{80}{I + 2}$.

27. $Q = 2w^2 - 3w + \dfrac{1}{w}$.   **28.** $s = \dfrac{4}{t} + 6t - 3t^2$.

29. The centrifugal force F in dynes of a certain mass moving uniformly in a circular path is given by the expression

$$F = \frac{200{,}000\pi^2}{T^2},$$

where T in seconds is the time of one revolution. Compute the derivative dF/dT for $T = 10$ sec.

30. The refractive index μ of a prism for a wavelength λ is given by Cauchy's formula

$$\mu = A + \frac{B}{\lambda^2},$$

where A and B are constants. Compute the derivative $d\mu/d\lambda$.

19–7. Differential Calculus

The computation of derivatives as described in the preceding section is troublesome, especially if derivatives of more complicated functions are needed. *The differential calculus is a system of formulas and rules which permits one to find, in a comparatively simple way, the derivatives of practically all functions used in the various applications of mathematics.* The method for computing derivatives is easily understood if one observes that functions defined by complicated formulas are a combination of a small number of simple functions which are connected in a more or less complicated way. Thus, for example, the computation of the function

$$y = \sqrt{x^3 + 5x^2}$$

can be resolved into the following steps:

1. Compute the value of x^2.
2. Compute the value of $5u$ where $u = x^2$.
3. Compute the value of x^3.
4. Add the results of steps 2 and 3.
5. Compute the function $\sqrt{v}$ where $v = x^3 + 5x^2$.

It is obvious that the given function is a combination of the simple functions $y = x^2$, $y = x^3$, $y = \sqrt{x}$, which are special cases of $y = x^n$.

The fundamental result of the differential calculus is that the derivatives of all such functions which are built up from simpler functions can be computed if

1. The derivatives of a small number of elementary functions are known.

2. Rules are known that permit one to compute the derivative of a function that is a combination of functions whose derivatives are already known.

The purpose of this chapter is to serve as an introduction to differential calculus. Therefore, not all the formulas and rules will be given here, but only those that are needed to obtain a general idea of the calculus and to solve simple problems.

19–8. The Derivatives of x^n, sin x, and cos x

It will be sufficient for our needs to know the derivatives of the functions x^n, sin x, and cos x. These derivatives are given without proof by the following formulas:

$$\frac{dx^n}{dx} = nx^{n-1}. \tag{1}$$

$$\frac{d \sin x}{dx} = \cos x. \tag{2}$$

$$\frac{d \cos x}{dx} = -\sin x. \tag{3}$$

The formula given in (1) is true for all possible values of n. The following are a few important special cases of formula 1:

$n = 0$, $\dfrac{dx^0}{dx} = \dfrac{d1}{dx} = 0$,

$n = 1$, $\dfrac{dx}{dx} = 1$,

$n = 2$, $\dfrac{dx^2}{dx} = 2x$,

$n = \dfrac{1}{2}$, $\dfrac{dx^{\frac{1}{2}}}{dx} = \dfrac{d\sqrt{x}}{dx} = \dfrac{1}{2} x^{\frac{1}{2}-1} = \dfrac{1}{2} x^{-\frac{1}{2}} = \dfrac{1}{2\sqrt{x}}$,

$n = -1$, $\dfrac{dx^{-1}}{dx} = \dfrac{d\left(\dfrac{1}{x}\right)}{dx} = -1 \cdot x^{-1-1} = -x^{-2} = -\dfrac{1}{x^2}$.

It must be pointed out that formulas 2 and 3 are true only when x is measured in radians. When the degree measure of x is used these formulas have to be replaced by more complicated ones. This is one reason why the radian measure of angles is important in many problems.

19–9. Rules for Computing Derivatives

The second group of fundamental theorems in calculus consists of a set of rules that permits one to find the derivatives of functions formed by combining simpler functions with known derivatives. These rules are given without proof in this section.

Rule 1. The derivative of the product of a constant and a function

$$y = af(x)$$

is given by the formula

$$\frac{dy}{dx} = \frac{d[af(x)]}{dx} = a\frac{df(x)}{dx}.$$

Example 1. Find the derivative of $y = 7x^3$.
Using the above rule, we have

$$\frac{d(7x^3)}{dx} = 7\frac{dx^3}{dx} = 7 \cdot 3x^2 = 21x^2.$$

Example 2. Find the derivative of a constant function $y = C$.
Using rule 1, and formula 1 of Sec. 19–8, we obtain

$$\frac{dC}{dx} = \frac{d(C \cdot 1)}{dx} = C\frac{d1}{dx} = 0.$$

Hence the derivative of a constant is zero.

Rule 2. The derivative of the sum of two functions

$$y = f(x) + g(x)$$

is given by the formula

$$\frac{dy}{dx} = \frac{d[f(x) + g(x)]}{dx} = \frac{df(x)}{dx} + \frac{dg(x)}{dx}.$$

Example 3. $\dfrac{d(x^3 + x^5)}{dx} = \dfrac{dx^3}{dx} + \dfrac{dx^5}{dx} = 3x^2 + 5x^4.$

Example 4. Find the derivative of $y = 3x^3 - 2x + 5 - 3\sqrt{x} + \frac{1}{2}\sqrt{x}.$

Since rule 2 can be extended to a sum consisting of any number of terms, we have

$$\frac{dy}{dx} = 3\frac{dx^3}{dx} - 2\frac{dx}{dx} + \frac{d5}{dx} - 3\frac{dx^{\frac{1}{2}}}{dx} + \frac{1}{2}\frac{dx^{-\frac{1}{2}}}{dx}$$

$$= 3 \cdot 3x^2 - 2 + 0 - 3 \cdot \frac{1}{2}x^{\frac{1}{2}-1} + \frac{1}{2} \cdot \left(-\frac{1}{2}\right)x^{-\frac{1}{2}-1}$$

$$= 9x^2 - 2 - \frac{3}{2\sqrt{x}} - \frac{1}{4x\sqrt{x}}.$$

Before an expression is differentiated, radicals have to be replaced by fractional exponents, as shown in the last example, in order to apply formula 1 of Sec. 19–8.

Example 5.   Find the derivative of $E = 10 \sin t + 5 \cos t$.
Using the various formulas and rules, we obtain

$$\frac{dE}{dt} = \frac{d(10 \sin t)}{dt} + \frac{d(5 \cos t)}{dt} = 10 \frac{d \sin t}{dt} + 5 \frac{d \cos t}{dt} = 10 \cos t - 5 \sin t.$$

Rule 3. The derivative of the product of two functions

$$y = f(x)\, g(x)$$

is given by the formula

$$\frac{dy}{dx} = \frac{d[f(x)g(x)]}{dx} = g(x) \frac{df(x)}{dx} + f(x) \frac{dg(x)}{dx}.$$

Example 6.   Find the derivative of the function $y = (3 + 5x)\sqrt[3]{x}$.
Using rule 3, we have

$$\frac{dy}{dx} = \frac{d[(3 + 5x)\sqrt[3]{x}]}{dx} = \sqrt[3]{x}\,\frac{d(3 + 5x)}{dx} + (3 + 5x)\frac{dx^{\frac{1}{3}}}{dx}$$

$$= \sqrt[3]{x} \cdot (0 + 5 \cdot 1) + (3 + 5x) \cdot \frac{1}{3} x^{\frac{1}{3}-1}$$

$$= 5\sqrt[3]{x} + \frac{1}{3}(3 + 5x)\,\frac{1}{\sqrt[3]{x^2}}.$$

The result can be simplified by rationalizing the second term:

$$\frac{dy}{dx} = 5\sqrt[3]{x} + \frac{1}{3}(3 + 5x)\,\frac{\sqrt[3]{x}}{x} = 5\sqrt[3]{x} + \frac{\sqrt[3]{x}}{x} + \frac{5}{3}\sqrt[3]{x}$$

$$= \frac{20}{3}\sqrt[3]{x} + \frac{\sqrt[3]{x}}{x} = \left(\frac{20}{3} + \frac{1}{x}\right)\sqrt[3]{x}.$$

Example 7.   Find the derivative of $r = 5\theta \sin \theta$.
We obtain

$$\frac{dr}{d\theta} = 5\,\frac{d(\theta \sin \theta)}{d\theta} = 5 \sin \theta\,\frac{d\theta}{d\theta} + 5\theta\,\frac{d \sin \theta}{d\theta}$$

$$= 5 \sin \theta + 5\theta \cos \theta = 5(\sin \theta + \theta \cos \theta).$$

Rule 4. The derivative of the quotient of two functions

$$y = \frac{f(x)}{g(x)}$$

is given by the formula

$$\frac{dy}{dx} = \frac{d\left[\dfrac{f(x)}{g(x)}\right]}{dx} = \frac{g(x)\dfrac{df(x)}{dx} - f(x)\dfrac{dg(x)}{dx}}{[g(x)]^2}.$$

Example 8.   Find the derivative of $y = \dfrac{2 + 3x}{1 + x^2}$.

Using the last rule, we have

$$\frac{dy}{dx} = \frac{d\dfrac{2 + 3x}{1 + x^2}}{dx} = \frac{(1 + x^2)\dfrac{d(2 + 3x)}{dx} - (2 + 3x)\dfrac{d(1 + x^2)}{dx}}{(1 + x^2)^2}$$

$$= \frac{(1 + x^2)(0 + 3 \cdot 1) - (2 + 3x)(0 + 2x)}{(1 + x^2)^2}$$

$$= \frac{3 + 3x^2 - 4x - 6x^2}{(1 + x^2)^2} = \frac{3 - 4x - 3x^2}{(1 + x^2)^2}.$$

Example 9.   Find the derivative of $\tan \theta$.

Using the relation $\tan \theta = \dfrac{\sin \theta}{\cos \theta}$, we have

$$\frac{d \tan \theta}{d\theta} = \frac{d\left(\dfrac{\sin \theta}{\cos \theta}\right)}{d\theta} = \frac{\cos \theta \dfrac{d \sin \theta}{d\theta} - \sin \theta \dfrac{d \cos \theta}{d\theta}}{\cos^2 \theta}$$

$$= \frac{\cos \theta \cdot \cos \theta - \sin \theta \cdot (-\sin \theta)}{\cos^2 \theta}$$

$$= \frac{\cos^2 \theta + \sin^2 \theta}{\cos^2 \theta} = \frac{1}{\cos^2 \theta} = \sec^2 \theta.$$

Rule 5. If u is a function of x, then the formulas of Sec. 19–8 are replaced by the following :

$$\frac{du^n}{dx} = nu^{n-1} \frac{du}{dx},$$

$$\frac{d \sin u}{dx} = \cos u \cdot \frac{du}{dx},$$

$$\frac{d \cos u}{dx} = -\sin u \cdot \frac{du}{dx}.$$

Example 10.   Compute $\dfrac{d\sqrt{1 + x^2}}{dx}$.

Substituting $u = 1 + x^2$, we obtain

$$\frac{d\sqrt{1 + x^2}}{dx} = \frac{du^{\frac{1}{2}}}{dx} = \frac{1}{2}u^{\frac{1}{2}-1}\frac{du}{dx} = \frac{1}{2\sqrt{1 + x^2}}\frac{d(1 + x^2)}{dx}$$

$$= \frac{1}{2\sqrt{1 + x^2}} \cdot 2x = \frac{x}{\sqrt{1 + x^2}}.$$

Example 11. Compute $\dfrac{d \cos 5t}{dt}$.

Substituting $u = 5t$, we obtain

$$\frac{d \cos 5t}{dt} = \frac{d \cos u}{dt} = -\sin u \cdot \frac{du}{dt} = -\sin u \cdot 5 = -5 \sin 5t.$$

Example 12. Compute dp/dt where $p = P_m \sin (\omega t + \theta)$, P_m, ω, and θ being constants.

According to the rules developed in this section we obtain

$$\frac{dp}{dt} - P_m \frac{d \sin (\omega t + \theta)}{dt} = P_m \frac{d \sin u}{dt},$$

where $u = \omega t + \theta$.   Hence

$$\frac{dp}{dt} = P_m \cos u \cdot \frac{du}{dt} = P_m \cos (\omega t + \theta) \cdot (\omega + 0) = \omega P_m \cos (\omega t + \theta).$$

Exercises

Find the derivatives of the following functions:

1. $y = x^4$.
2. $y = x^7$.
3. $y = x^9$.

4. $y = x^{\frac{4}{3}}$.
5. $y = x^{\frac{6}{5}}$.
6. $y = x^{\frac{3}{2}}$.

7. $y = x^{\frac{5}{2}}$.
8. $y = x^{-2}$.
9. $y = x^{-6}$.

10. $y = x^{-\frac{4}{5}}$.
11. $y = x^{-\frac{1}{2}}$.
12. $y = x^{-\frac{1}{3}}$.

13. $y = 9x^3$.
14. $y = 4x^4$.
15. $y = \dfrac{x^2}{3}$.

16. $y = \dfrac{3x^4}{2}$.
17. $s = \dfrac{7}{5t^2}$.
18. $s = \dfrac{8}{3t^3}$.

19. $s = 6\sqrt[3]{t}$.
20. $u = 4\sqrt{w}$.
21. $v = 5u + 2$.

22. $y = 4x^2 + 9$.
23. $y = 3x^3 - 3$.
24. $y = 12x^2 - 1$.

25. $y = 12x^2 + 1$.
26. $u = t^2 + 2t - 3$.
27. $u = 4t^2 - t - 2$.

28. $v = u^4 - 2u^2 + 4$.
29. $s = 7t^3 + 5t^2 - t$.
30. $u = \dfrac{8}{v^2} - \dfrac{3}{v} + 1$.

31. $y = 4 - \dfrac{2}{x} - \dfrac{1}{x^2}$.
32. $q = \dfrac{2 \cdot 5}{t^2} - \dfrac{1 \cdot 4}{t}$.
33. $q = 8 - \dfrac{3 \cdot 2}{t} + \dfrac{8 \cdot 3}{t^2}$.

34. $s = \sqrt{t} + 3\sqrt[3]{t}$.
35. $y = 3\sqrt{x} - 5\sqrt[3]{x}$.
36. $y = \sqrt[3]{2x} + \sqrt{5x}$.

37. $y = \sin x + \cos x$.
38. $y = 1 - \cos x$.
39. $y = 4 \sin x - \cos x$.

40. $u = 2t^2 - 5 \sin t$.
41. $r = 6 \cos \theta + 5\theta$.
42. $s = 4t^3 + 7 \cos t$.

43. $y = 3(1 + x)^4$.
44. $y = 6(2 - x)^3$.
45. $u = \sqrt{t^2 + 5}$.

46. $v = 2\sqrt{3 - w^3}$.　　　**47.** $s = \dfrac{2}{4 + t}$.　　　**48.** $y = \dfrac{10}{x^3 - 4}$.

49. $r = 0.5 \cos 3\theta$.　　　**50.** $r = 6 \sin 2\theta$.　　　**51.** $I = 40 \sin 377t + 8$.

52. $r = 18 \cos \frac{14}{3} \theta$.　　　　　**53.** $E = 5 \sin (680t + 28)$.

54. $I = 10 \sin (150t + 12)$.　　　　　**55.** $s = 4 \sin^2 t$.

56. $z = 7 \cos^2 \omega$.　　　**57.** $y = \sin x \cos x$.　　　**58.** $y = x^2 \cos x$.

59. $r = x^3 \cos 3x$.　　　**60.** $u = 7v \sin 5v$.　　　**61.** $s = (1 + t)^4 \sqrt{t}$.

62. $r = (2 + x)^3 \sqrt{x}$.　　　　　**63.** $u = 10(1 + v)^2(5 + 7v)^3$.

64. $s = 6(3t + 2)^3(t + 4)^2$.　　　　　**65.** $y = 16(3 + x^2)^3(3 - x^2)^3$.

66. $y = \dfrac{x + 1}{x + 4}$.　　　**67.** $s = \dfrac{8t + 3}{5t - 2}$.　　　**68.** $u = \dfrac{3v - 4}{6 - 2v}$.

69. $r = \dfrac{\sin \theta}{\theta^2}$.　　　**70.** $y = \dfrac{\cos x}{1 + x}$.　　　**71.** $y = \dfrac{1 + x}{\cos x}$.

72. $y = \cot x$.　　　**73.** $r = \sec \theta$.　　　**74.** $y = \csc x$.

75. $y = \dfrac{\sin x}{\cos 2x}$.　　　**76.** $y = \dfrac{\sin 2x}{\cos x}$.　　　**77.** $s = \dfrac{1 - t^2}{1 + t^2}$.

78. $z = \dfrac{3\omega^2 + 1}{\omega^2 - 1}$.　　　**79.** $u = \dfrac{v^4 + 1}{v^2 + 1}$.　　　**80.** $v = \dfrac{3w^3 + 6}{2 - w}$.

81. $y = \dfrac{(4x^2 + 3)^3}{2x}$.　　　**82.** $s = \dfrac{(3t^2 - 6)^3}{5t^2}$.　　　**83.** $u = \dfrac{1}{\sqrt{1 + v}}$.

84. $y = \dfrac{3}{\sqrt{x^2 + 1}}$.　　　**85.** $z = \sqrt{4 + \dfrac{1}{30\omega^2}}$.　　　**86.** $r = \sqrt{60 - \dfrac{2}{5\theta^3}}$.

Compute the derivatives of the following functions of t, assuming that all the other symbols in the right members denote constants:

87. $s = \dfrac{a - bt}{m + nt}$.　　　　　**88.** $v = \sqrt{12 + F^2 \sin^2 t}$.

89. $w = 2k\sqrt{m + U \cos^2 t}$.　　　　　**90.** $e = RI \sin (\omega t + \alpha)$.

91. $F = F_0 + M \sin \omega t + N \cos \omega t$.　　　**92.** $w = EI \sin^2 \omega t$.

93. $e = E_0 + E_1 \sin (\omega t + \alpha_1) + E_2 \sin (\omega t + \alpha_2)$.

94. $z = a_0 + a_1 \sin t + a_2 \sin^2 t + a_3 \sin^3 t$.

19–10. Applications of the Differential Calculus

There are many problems in geometry and engineering that can be solved by the methods of the preceding sections. A few of them will be discussed in this section.

Example 1.   Compute the slope of the tangent of the curve given by the equation $y = \frac{1}{6}(x^2 + 4x - 11)$ at the point P with the abscissa $x = 1$ (Fig. 19–6).

The required slope is, according to Sec. 19–5, equal to the derivative of y at the point $x = 1$.   This derivative is found to be

$$\frac{dy}{dx} = \frac{d\frac{1}{6}(x^2 + 4x - 11)}{dx} = \frac{1}{3}x + \frac{2}{3},$$

and its value for $x = 1$ is 1.

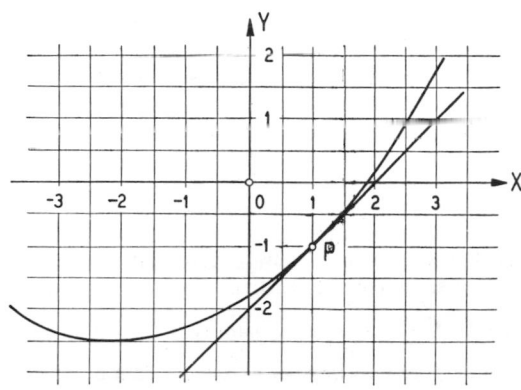

FIG. 19–6

The equation of the tangent line to the given curve at the point $x = 1$, $y = \frac{1}{6}(1 + 4 - 11) = -1$, can now be found, for this is the equation of a straight line passing through the point $(1, -1)$ and having the slope 1.   Hence, by using the point slope form of the equation of a straight line (Sec. 15–8), the equation of the tangent is found to be

$$y + 1 = x - 1,$$
$$y = x - 2.$$

Example 2.   The motion of a particle P along a straight line is completely described if the distance s between a point of reference O on this line and the particle (Fig. 19–7) is given as a function $s = f(t)$ of the time t.

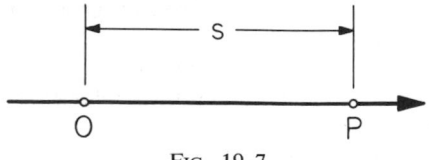

FIG. 19–7

The velocity v of the particle at the time t is defined as the instantaneous rate of change of s with respect to t and, therefore, can be found by differentiating the function $f(t)$ with respect to t,

$$v = \frac{ds}{dt} = \frac{df(t)}{dt}.$$

In Sec. 5–16 we discussed the *harmonic* motion of a point described by the equation $s = A \sin \omega t$. The velocity of the moving point at time t is given by

$$v = \frac{ds}{dt} = \frac{d(A \sin \omega t)}{dt} = A\omega \cos \omega t.$$

Now the velocity of the moving particle is itself a function of the time. The rate of change of this function with respect to time is called **acceleration**, often denoted by the letter a. For the harmonic motion we have

$$a = \frac{dv}{dt} = \frac{d(A\omega \cos \omega t)}{dt} = -A\omega^2 \sin \omega t.$$

The acceleration is found by computing the derivative $\dfrac{ds}{dt} = \dfrac{df(t)}{dt} = v$ and by differentiating this result another time, so that

$$a = \frac{dv}{dt} = \frac{d\left(\dfrac{df(t)}{dt}\right)}{dt}.$$

This value, which has been obtained by differentiating the result of a first differentiation, is called the **second derivative** of $s = f(t)$ with respect to t. It is generally denoted by $\dfrac{d^2s}{dt^2} = \dfrac{d^2f(t)}{dt^2}$ or simply $s'' = f''(t)$ (read: s two prime, or s double prime, f two prime of t or f double prime of t). In this connection, $f'(t) = \dfrac{df(t)}{dt}$ is often called the **first derivative** of $f(t)$.

The result of the preceding discussion is summarized in the following statement: *The velocity of a particle moving along a straight line is the first derivative of the distance s with respect to time, and the acceleration is the second derivative of the distance s with respect to time, where s is measured from a fixed point of reference to the moving particle.*

Example 3. The force F in pounds required to produce linear motion in a mechanical system, which has a mass M in slugs, a damping constant k_d in pounds per foot per foot, and a spring constant k_s in pounds per foot, through a distance s in feet is given by

$$F = M\frac{d^2s}{dt^2} + k_d\frac{ds}{dt} + k_s s.$$

This equation can be used to compute F when s is given. In a vibrating system s may be given by the equation

$$s = S_m \sin \omega t.$$

The force required to maintain vibration is then obtained as follows:

$$F = M\frac{d^2s}{dt^2} + k_d\frac{ds}{dt} + k_s s$$

$$= M\frac{d\left(\dfrac{d(S_m \sin \omega t)}{dt}\right)}{dt} + k_d\frac{d(S_m \sin \omega t)}{dt} + k_s S_m \sin \omega t$$

$$= -MS_m\omega^2 \sin \omega t + k_d S_m\omega \cos \omega t + k_s S_m \sin \omega t$$

$$= S_m(k_s - M\omega^2)\sin \omega t + k_d\omega \cos \omega t.$$

Exercises

Find the equations of the tangents of the following curves at the indicated points:

1. $y = 2x^2$, (a) $x = 1$, (b) $x = -2$, (c) $x = 0$.

2. $y = -x^2$, (a) $x = 3$, (b) $x = 0$, (c) $x = -3$.

3. $y = 0.5x^2 - x + 2$, (a) $x = 4$, (b) $x = 1$, (c) $x = -2$.

4. $y = 20 + 48x - 16x^2$, (a) $x = 1$, (b) $x = 2$, (c) $x = 0$.

5. $y = \sqrt{25 - x^2}$, (a) $x = 3$, (b) $x = 4$, (c) $x = 0$.

6. $y = \sqrt{100 - x^2}$, (a) $x = -6$, (b) $x = 0$, (c) $x = 8$.

7. $y = \sin x$, (a) $x = 0$, (b) $x = \dfrac{\pi}{4}$, (c) $x = \dfrac{\pi}{2}$.

8. $y = \cos x$, (a) $x = 0$, (b) $x = \dfrac{\pi}{6}$, (c) $x = \dfrac{\pi}{2}$.

9. $y = \sin x + 3 \cos x$, (a) $x = 0$, (b) $x = \dfrac{\pi}{6}$, (c) $x = \dfrac{\pi}{2}$.

10. $y = 2 \cos x - \sin 2x$, (a) $x = 0$, (b) $x = \dfrac{\pi}{4}$, (c) $x = \dfrac{\pi}{2}$.

Find the velocity and the acceleration at the time t of a particle moving along a straight line according to the following equations:

11. $s = 0.003t^3$.

12. $s = 8t^4 - 3t^2 + 26$.

13. $s = v_0 t + \tfrac{1}{2}gt^2$.

14. $s = \dfrac{2}{t^2 + 1} + \dfrac{3}{t + 1}$.

15. $s = 7t^2 \sin 3t$.

16. $s = 4t^2 \cos t - 2t \sin 2t$.

17. $s = 2t^3 \cos 2t + 3t$.

18. $s = A_0 + A_1 \sin \omega t + A_2 \sin 3\omega t$.

The electromotive force E, across a circuit of resistance R ohms and inductance L henrys, when a current I amperes flows in the circuit, is given by the equation

$$E = RI + L\frac{dI}{dt} ..$$

Find the electromotive force across a circuit that has a resistance of 20 ohms and an inductance of 0.3 henry if the current

19. $I = 8$ amperes.

20. $I = 0.8$ ampere.

21. $I = (0.2t + 3)$ amperes.

22. $I = (0.5t + 9.2)$ amperes.

23. $I = (5t^2 - 2t + 1)$ amperes.

24. $I = (3.2 + 8.6t - 5.4t^2)$ amperes.

25. $I = 10 \sin (377t + 0.1)$ amperes.

26. $I = (6 \sin 377t + 4 \cos 377t)$ amperes.

A mechanical system has a mass of 30 slugs, a damping constant of 420 lb per ft per ft, and a spring constant of 3750 lb per ft. Find the force required to cause a displacement of

27. $s = 0.12$ ft.

28. $s = 1.25$ ft.

29. $s = (0.36t + 0.04)$ ft.

30. $s = (0.80t + 0.24)$ ft.

31. $s = (0.08t^2 + 0.12t - 0.16)$ ft.

32. $s = (0.06t^2 - 0.20t + 0.08)$ ft.

33. $s = 0.1 \sin (12.6t + 0.2)$ ft.

34. $s = 0.5 \cos (9.4t + 0.6)$ ft.

19–11. Maxima and Minima

There are many practical problems in which it is necessary to obtain a quantity as great or as small as possible. Examples of such problems are given in the following statements: Find the sides of the largest rectangular lot that can be fenced with 200 ft of wire. Find the dimensions of a tin can of given volume if as little sheet metal as possible should be used for its construction. Find the value of the resistance that must be in series with a source of electricity if the power output should be as large as possible. The differential calculus permits us to develop a method for dealing with these and similar problems. All problems of this kind, when expressed in mathematical language, can be stated in the following way: *Find the values of x for which a given function $y = f(x)$ assumes its greatest or smallest value, and compute this value.* The first of the problems mentioned above will be used to illustrate the subject of the present section.

Our problem is to find the sides of the largest rectangular lot that can be fenced with 200 ft of wire. Let x and x_1 denote the unknown sides of the lot. The sum of the four sides (perimeter) of this lot is given by

$$x + x + x_1 + x_1,$$

and hence we have the equation

$$2x + 2x_1 = 200,$$

or

$$x_1 = (100 - x) \text{ ft.}$$

Now the area of this lot is given by

$$y = xx_1 = x(100 - x) \text{ sq ft.}$$

We have to find a value of x for which y has the greatest value. For this purpose the following table of corresponding values of x and y is constructed:

x	$x_1 = 100 - x$	$y = x(100 - x)$
0	100	0
10	90	900
20	80	1600
30	70	2100
40	60	2400
50	50	2500
60	40	2400
70	30	2100
80	20	1600
90	10	900
100	0	0

From this table and from the corresponding graph (Fig. 19–8), it can be seen that the greatest value or **maximum** of the area is 2500 sq ft and that the lot having this area is a square whose sides are 50 ft in length.

The next step is to solve the above problem without constructing a table or plotting a graph. From Fig. 19–8 it is obvious that the tangent at P_0, the highest point of the graph, is parallel to the x axis and, therefore, has slope zero. From Sec. 19–5 it is known that the slope of the tangent is equal to the derivative of the function of the graph. Hence it can be

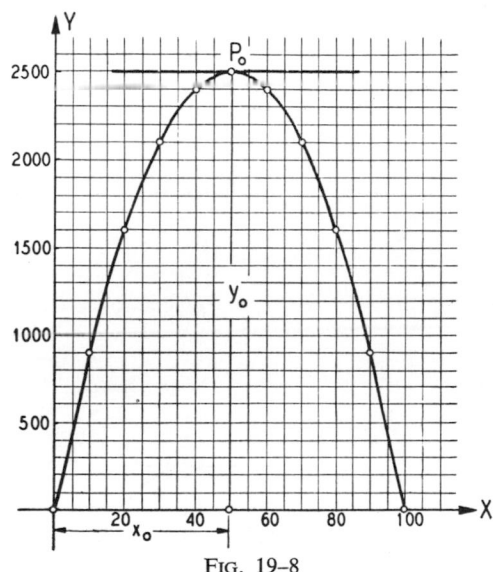

FIG. 19–8

stated that the derivative of this function, computed at the point P_0, is zero. Now the equation of the graph of Fig. 19–8 is given by

$$y = x(100 - x) = 100x - x^2,$$

and hence its derivative is

$$\frac{dy}{dx} = 100 - 2x.$$

This derivative is zero when computed for $x = x_0$, the abscissa of P_0. Hence the following equation for x_0 is obtained:

$$100 - 2x_0 = 0,$$

and therefore

$$x_0 = 50,$$

$$y_0 = x_0(100 - x_0) = 2500.$$

These values are the same as those previously obtained by constructing a table of corresponding values of the sides and the area of the lot.

The same method can be used to find the smallest value or **minimum** of a function given by $y = f(x)$. Assume that the graph of this function is given in Fig. 19–9 and P is the lowest point of the graph. Again it

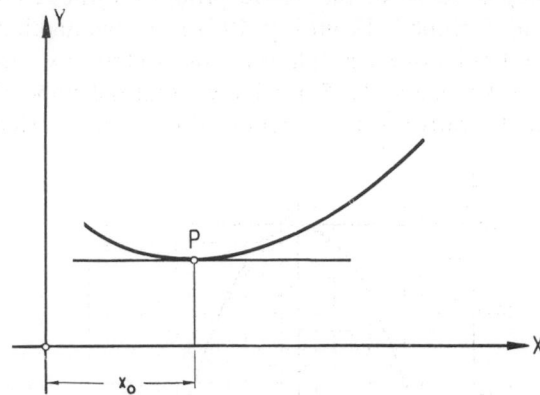

FIG. 19–9

can be stated that $f'(x) = 0$ for $x = x_0$, because the tangent at P has slope zero.

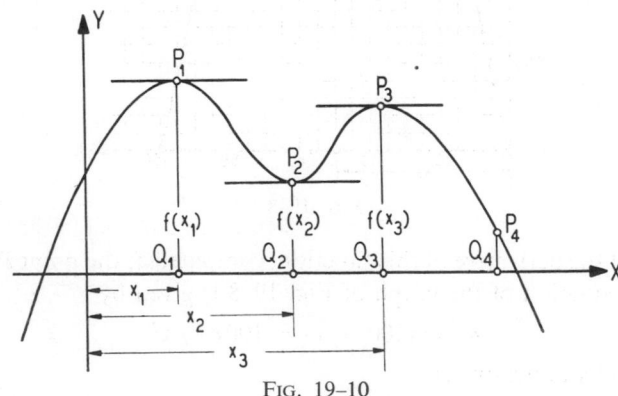

FIG. 19–10

The foregoing discussion shows that the problem of finding the maximum or minimum of a function $y = f(x)$ can be reduced to the problem of solving the equation

$$\frac{df(x)}{dx} = 0.$$

Assume that the graph of some function $y = f(x)$ is plotted in Fig. 19–10. The roots of the equation

$$\frac{df(x)}{dx} = 0$$

are the abscissas of the points P_1, P_2, P_3 where the tangent is parallel to the
x axis.   The figure shows that $Q_1P_1 = f(x_1)$ is the maximum of $y = f(x)$,
but that $Q_2P_2 = f(x_2)$ is not the *smallest* value of y.   The ordinate Q_2P_2
is only *smaller* than the ordinates of the other points in the neighborhood
of P_2, but there are points whose ordinates are smaller than Q_2P_2.   For
example, P_4 is such a point since Q_4P_4 is smaller than Q_2P_2.   At points
like P_2 the given function is said to have a **relative minimum.**   Similarly the
ordinate Q_3P_3 is a **relative maximum** of the function.   There are many
problems in which it is important to find such a relative minimum or a
relative maximum.   *In the following discussion, the expressions maximum
and minimum are used to denote a relative maximum or relative minimum.*
The value $f(x_1) = Q_1P_1$ is then called an **absolute maximum.**

The solution of the equation $\dfrac{df(x)}{dx} = 0$ yields the abscissas of the

points P_1, P_2, P_3, but it does not settle the question whether, at any one
of these points, the function has a maximum or a minimum.   This
question can be answered by plotting the graph, although it is sometimes
clear from the nature of the problem whether we have a maximum or
minimum.   A better method for determining whether a point is a maximum or a minimum is given by the following rule:

*The function $y = f(x)$ has a maximum or a minimum for $x = x_0$ if, at
this point, the first derivative $f'(x_0) = 0$ and the second derivative $f''(x_0)$
is, respectively, negative or positive.*

The proof of this rule is beyond the scope of this book.

The results of this section can now be summarized in the following
statement:

*In order to find the maxima or minima of a function $y = f(x)$, compute
the first derivative $f'(x)$ and the second derivative $f''(x)$.   Then solve the
equation $f'(x) = 0$.   If $x = x_0$ is a particular root of this equation, the
value $f(x_0)$ is a maximum if $f''(x_0) < 0$, and a minimum if $f''(x_0) > 0$.*

A few examples will illustrate the application of the material developed
in this section.

Example 1.   Find the maxima and minima of the function

$$y = f(x) = \frac{x^3}{4} - \frac{3x^2}{4} + 2.$$

According to the rule given above we find the first and second derivatives:

$$f'(x) = \frac{3x^2}{4} - \frac{3x}{2}, \qquad f''(x) = \frac{3}{2}x - \frac{3}{2}.$$

Solving the equation

$$f'(x) = \frac{3x^2}{4} - \frac{3x}{2} = 0,$$

we obtain

$$\tfrac{3}{4}x(x-2) = 0,$$

$$x_1 = 0, \qquad x_2 = 2.$$

Computing $f''(x)$ for these values, we get

$$f''(0) = -\tfrac{3}{2} < 0, \qquad f''(2) = \tfrac{3}{2} > 0.$$

It can be stated, therefore, that the value $f(0) = 2$ is a maximum and the value $(2) = 1$ is a minimum of the given function. These results are illustrated by the graph of Fig. 19–11.

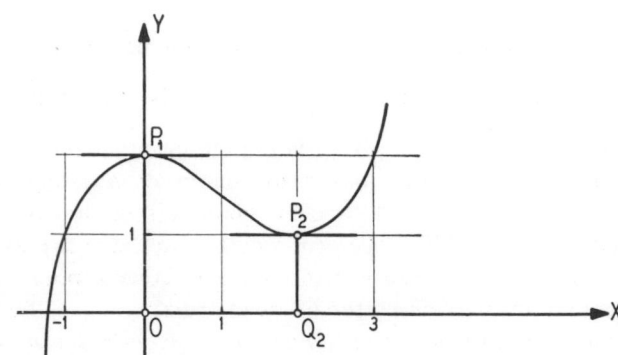

FIG. 19–11

Example 2. What is the resistance R of a load connected in series with a battery of electromotive force E and internal resistance R_i if the power output is to be as great as possible?

The power output P is given by the formula

$$P = I^2 R.$$

By Ohm's law we have

$$I = \frac{E}{R_i + R}.$$

We obtain, therefore,

$$P = \frac{E^2 R}{(R_i + R)^2}.$$

In order to find the value of R for which P is a maximum, we have to compute $\dfrac{dP}{dR}$ and $\dfrac{d^2P}{dR^2}$ from the last formula. This computation is troublesome in this case and can be simplified by observing that $1/P$ has a minimum when P has a maximum. Hence, if we substitute $\dfrac{1}{P} = u$ in the expression for P, we obtain

$$u = \frac{(R_i + R)^2}{E^2 R} = \frac{R_i^2 + 2R_i R + R^2}{E^2 R} = \frac{1}{E^2}\left(\frac{R_i^2}{R} + 2R_i + R\right).$$

Finding the first and second derivatives, we have

$$\frac{du}{dR} = \frac{1}{E^2}\left(-\frac{R_i^{\,2}}{R^2} + 1\right),$$

$$\frac{d^2u}{dR^2} = \frac{1}{E^2}\left(\frac{2R_i^{\,2}}{R^3}\right) = \frac{2R_i^{\,2}}{E^2R^3}.$$

Equating the first derivative du/dR to zero, we obtain

$$\frac{1}{E^2}\left(-\frac{R_i^{\,2}}{R^2} + 1\right) = 0,$$

and hence

$$R = R_i \quad \text{or} \quad R = -R_i.$$

Since the negative value $R = -R_i$ has no physical meaning in our problem, we discard it and use the value $R = R_i$. When this value $R = R_i$ is substituted in the second derivative $\dfrac{d^2u}{dR^2} = \dfrac{2R_i^{\,2}}{E^2R^3}$, the result is positive, so that the corresponding value of u is a minimum and hence the value $P = \dfrac{1}{u}$ is a maximum. The power output, therefore, is a maximum if the load resistance R is equal to the internal resistance R_i. Substituting $R = R_i$ in the expression for P, we get that the maximum power output is given by

$$P = \frac{E^2R_i}{4R_i^{\,2}} = \frac{E^2}{4R_i}.$$

Exercises

Find the maxima and minima of the following functions:

1. $y = 6x^2 - 24x$.

2. $y = 4x^2 - 32x$.

3. $y = 2x - 8x^2$.

4. $y = 3x - 9x^2$.

5. $u = 3v^2 + 2v - 1$.

6. $s = 2t^2 - 3t + 7$.

7. $q = r^2 - 5r + 3$.

8. $v = 5w^2 + 2w - 4$,

9. $s = 20 + 48t - 16t^2$.

10. $u = 6 + 5v - 2v^2$.

11. $u = t^3 - 3t + 1$.

12. $q = 3x^3 - x + 2$.

13. $y = 4.5x^3 - x$.

14. $y = 12x^3 - 4x$.

15. $q = r + \dfrac{1}{r}$.

16. $v = 4w^2 + \dfrac{1}{w}$.

17. $s = \dfrac{1}{1 + t^2}$.

18. $y = 9r + \dfrac{1}{r} - 6$.

19. $m = 4N^3 + 3N^2 - 90N + 144$.

20. $s = t^4 - 4t^3 - 8t^2 + 1$.

Find the maxima and minima within the indicated intervals of the functions given in exercises 21–26.

21. $y = 2 \sin x, -\pi \leq x \leq \pi$.

22. $r = 5 \cos \theta, 0 \leq \theta \leq 2\pi$.

23. $r = 3 \cos \theta + 2, -\dfrac{\pi}{2} \leq \theta \leq \dfrac{3\pi}{2}$.

24. $y = \sin x + \cos x, 0 \leq x \leq \dfrac{\pi}{2}$.

25. $s = \sin t + \sin 2t, 0 \leq t \leq \pi$.

26. $s = \sin 2t - \sin t, 0 \leq t \leq \pi$.

27. The volume V of water at any temperature θ between $0°$ and $30°$ C may be determined approximately from the formula

$$V = V_0(1 + a\theta + b\theta^2).$$

where V_0 is the volume at $0°$ C, and a and b are constants having values of -6.2×10^{-5} and 8.1×10^{-6}, respectively. Find the temperature where the volume of water is a minimum, that is, where its density is a maximum.

28. An open box is formed from a square sheet of tin, the side of which is 24 in., by cutting off equal squares at each corner and turning up the sides. What size of squares should be cut out in order to obtain a box of maximum content?

29. Solve the problem of exercise 28 if, instead of a square piece of tin, a rectangular sheet 15 by 24 in. is given.

30. A cylindrical can open at one end is to be made so that its volume is 42 cu in. and its surface is as small as possible. Find the diameter and the height of the can.

31. A closed cylindrical can is to be made so that its volume is 128 cu in. Find its dimensions if the total surface is to be a minimum.

32. Find the side and height of a rectangular parallelepiped with a square base and 125 cu in. in volume if the surface is to be a minimum.

33. A water tank is to be constructed having a square base, an open top, and vertical walls; it is to contain 864 cu yd. Find the side of the base and the height if the sum of the areas of the bottom and the walls is to be a minimum.

34. The power output P of a battery of electromotive force E and internal resistance R_i is given by $P = EI - R_iI^2$, where I is the current delivered by the battery. Find the current for which the power output is a maximum.

35. The time, in seconds, needed for an object to slide down an inclined plane, if friction and air resistance are neglected, is given by the formula

$$t = \frac{1}{4}\sqrt{\frac{2L}{\sin 2\theta}},$$

where θ is the angle between the inclined plane and its base, and L is the length of the base measured in feet. Find the value of θ that makes t a minimum.

36. The power developed in watts in an a-c circuit is given by the expression

$$P = \frac{E^2R}{R^2 + X^2},$$

where E is the impressed voltage, R the resistance in ohms, and X the reactance in ohms. If E and X have constant values, compute the value of R which makes P a maximum.

37. A chemist folds a piece of circular filter paper into the shape of a right circular cone having a slant height of 6 cm. Find the radius of the base for which the volume of the cone is as large as possible and compute this maximum.

19–12. Increments and Differentials

The computation of an increment of a function $y = f(x)$ according to the formula

$$\Delta y = f(x + \Delta x) - f(x)$$

given in Sec. 19–1 is often troublesome. In many cases the computation can be greatly simplified if the increment Δx is small and only an approximate value of Δy is required, conditions that are satisfied in many practical problems.

Assume that P and P_1 are the two points on the graph of $y = f(x)$, Fig. 19–12, which correspond to the abscissas x and $x + \Delta x$. The increment Δy of the ordinate is given by the segment $P_1 Q$. Now construct

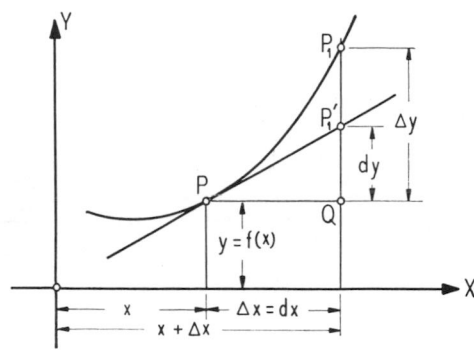

FIG. 19–12

the tangent to the curve at P, and let P_1' be the point on the tangent that has the same abscissa as P_1. In Sec. 19–5, the term differential was used for each of the increments connected with a point on the tangent, and the notations dx and dy were adopted. Accordingly, $dx = PQ$, $dy = QP_1'$, and $\Delta x = dx$.

The differential dy is easier to compute than the increment Δy because, according to Sec. 19–5, dy/dx is the derivative $f'(x)$ computed at the point P, and therefore

$$dy = f'(x)\, dx.$$

From Fig. 19–12 it is obvious that dy is an approximation for Δy when $\Delta x = dx$ is small, because the points P_1 and P_1' approach each other if Q approaches P. This discussion permits us to state the following rule:

The increment of $y = f(x)$ for a small increment of x, denoted by $\Delta x = dx$, is given approximately by the formula for the differential

$$\boldsymbol{dy = f'(x)\, dx.}$$

A few examples will illustrate the use of the differential dy as an approximation to the increment Δy.

Example 1. Compute the differential of $u = \sqrt{1 + v^2}$.
Since $u = f(v)$, and

$$\frac{du}{dv} = f'(v) = \frac{v}{\sqrt{1 + v^2}},$$

it follows that

$$du = \frac{v}{\sqrt{1 + v^2}} dv.$$

Example 2. What is the increase of the volume V of a sphere if its radius r is increased from 24 to 24.2 in.? The volume of a sphere is given by the formula

$$V = \frac{4\pi r^3}{3}.$$

Using the differential dV as an approximation to ΔV in order to simplify the computation, we obtain

$$\frac{dV}{dr} = 4\pi r^2, \qquad dV = 4\pi r^2 \, dr.$$

Since $r = 24$ and $dr = 0.2$, we obtain

$$dV = 4\pi(24)^2(0.2) = 1450 \text{ cu in.,}$$

the computation being carried out by the slide rule.

In order to investigate how close this value is to the correct value ΔV, it is necessary to compute V for $r = 24$ in. and $r = 24.2$ in. and to subtract the first value from the second. The results obtained by use of the slide rule are not precise enough for this purpose. Use of logarithms gives the following results:

$$V = \frac{4\pi}{3} (24)^3 = 57{,}900 \text{ cu in.,}$$

$$V + \Delta V = \frac{4\pi}{3} (24.2)^3 = 59{,}360 \text{ cu in.,}$$

$$\Delta V = 1460 \text{ cu in.}$$

The error made by using $dV = 1450$ instead of $\Delta V = 1460$ is less than 1 per cent.

Example 3. The period t in seconds for a complete vibration of a simple pendulum of length L in centimeters is given by the formula

$$t = 2\pi \sqrt{\frac{L}{g}},$$

where g in centimeters per second per second is the acceleration due to gravity. What is the approximate change of t if L is increased by 1 per cent?

In this problem, t is regarded as a function of L, and the value of dt is to be computed for $dL = 0.01L$. We compute first the derivative

$$\frac{dt}{dL} = \frac{d\,2\pi\sqrt{\dfrac{L}{g}}}{dL} = \frac{2\pi}{\sqrt{g}}\frac{dL^{\frac{1}{2}}}{dL} = \frac{2\pi}{\sqrt{g}}\left(\frac{1}{2}\right)L^{-\frac{1}{2}} = \frac{\pi}{\sqrt{gL}} = \frac{t}{2L}.$$

Then

$$dt = \frac{t}{2L}\cdot dL = \frac{t}{2}\cdot\frac{dL}{L} = \frac{0.01}{2}t = 0.005t.$$

This result shows that the period increases approximately $\frac{1}{2}$ per cent if the length of the pendulum is increased by 1 per cent.

Example 4. Compute approximately $\sin 30.1°$, using the known value $\sin 30° = 0.5$.
Since

$$\Delta \sin x = \sin(x + \Delta x) - \sin x$$

or

$$\sin(x + \Delta x) = \sin x + \Delta \sin x \tag{1}$$

we can compute $\sin 30.1°$ by computing equation 1 for $x = 30°$ and $\Delta x = 0.1°$. As an approximation to the term $\Delta \sin x$ we shall use the differential

$$d \sin x = \cos x\, dx. \tag{2}$$

We stressed in Sec. 19–8 that in the formulas for the differentiation of trigonometric functions it is assumed that x is measured in radians. Hence

$$dx = 0.1° = (0.1)\frac{\pi}{180}\text{ radians} = 0.00175\text{ radians},$$

and therefore, if $x = 30° = \frac{\pi}{6}$ radians, we have, by equation 2,

$$d \sin x = \left(\cos\frac{\pi}{6}\right)(0.00175) = (0.8660)(0.00175) = 0.0015.$$

By equation 1 we have that

$$\sin 30.1° = \sin 30° + 0.0015 = 0.5000 + 0.0015 = 0.5015.$$

The approximation in this case is so good that all four digits of the result are correct, as the student may verify from the tables of trigonometric functions.

Exercises

Compute the differentials of the following functions:

1. $y = 10x - 5$. 　　　　**2.** $y = 4x^2 + 6$. 　　　　**3.** $s = 80t - 16t^2$.

4. $q = 2r^2 - 3r + 7$. 　　**5.** $u = 5x^3 - 2x^2 + x$. 　**6.** $v = 4 - 2w - 5w^2$.

7. $y = \dfrac{1}{x}$. 　　　　　**8.** $r = \dfrac{1}{\theta} + \dfrac{1}{\theta^2}$. 　　　**9.** $u = \dfrac{1}{v^2 + 1}$.

10. $y = \sqrt{x} + \dfrac{1}{\sqrt{x}}.$ **11.** $y = \dfrac{1}{\sqrt{1 - x^2}}.$ **12.** $s = \dfrac{1}{\sqrt{2 + 5t}}.$

13. $u = \dfrac{1 - v^2}{1 + v^2}.$ **14.** $q = \dfrac{2r}{3r + 4}.$ **15.** $v = \dfrac{u^2 + 1}{u - 2}.$

16. $y = 2x \cos x.$ **17.** $r = 3\theta \sin \theta.$ **18.** $r = \dfrac{5x}{\sin x}.$

19. Compute approximately the volume of material in a hollow sphere of internal radius $r = 18$ in. and thickness $a = 0.2$ in.

20. How much does the volume of a cube with 4-ft sides increase if each side is increased by 1 in.?

21. Compute approximately the volume of material in a cylindrical shell having a length of 7 ft, an internal radius of 2.6 in., and a wall thickness of 0.1 in.

22. A platinum right circular cylinder has a length of 2 in. and a diameter of 0.378 in. Compute approximately the volume of the platinum that is removed if the cylinder is machined to a diameter of 0.375 in.

23. Find the approximate decrease in the volume of a spherical balloon having a diameter of 20.0 ft if the diameter is decreased to 19.9 ft.

24. The kinetic energy W in foot-pounds of a mass m measured in slugs that moves with a speed of v in feet per second is given by the formula

$$W = \tfrac{1}{2}mv^2.$$

Compute the approximate change in the kinetic energy of an automobile having a mass of 70 slugs if its speed is decreased from 80 to 78 ft per sec.

25. A projectile, fired vertically downward with an initial velocity v_0 in feet per second from a stationary blimp, travels a distance s ft below its starting point in a time t sec given by the formula

$$s = v_0 t + \tfrac{1}{2}gt^2,$$

where g is the acceleration due to gravity. If $v_0 = 200$ ft per sec and $g = 32.2$ ft per sec per sec, compute approximately how far a projectile will travel between the time $t = 4.0$ sec and $t = 4.1$ sec.

26. The impedance of an a-c circuit, measured in ohms, is given by the expression

$$Z = \sqrt{R^2 + 4\pi^2 f^2 L^2},$$

where R is the resistance in ohms, f the frequency in cycles per second, and L the inductance in henrys. Compute approximately the increase of Z if $R = 8$ ohms, $L = 0.3$ henry, $f = 60.0$ cycles per second, and if f increases to 60.4 cycles per second.

27. The speed, measured in centimeters per second, of a liquid that flows through an orifice in the wall of a container is given by the formula $v = \sqrt{2gh}$, where g in centimeters per second per second is the acceleration due to gravity, and h in centimeters is the distance from the orifice to the free surface of the liquid. Compute the approximate change of v if h is decreased by 2 per cent.

28. The frequency f in cycles per second of the electric oscillations that can be produced by a circuit of inductance of L henrys and capacitance of C farads is

$$f = \dfrac{1}{2\pi\sqrt{LC}}.$$

What is the approximate change of f if C is increased by 1 per cent?

Compute, approximately, the following quantities:

29. $\sqrt{9.2}$, starting from the value $\sqrt{9} = 3$.

30. $\sqrt{25.4}$, starting from the value $\sqrt{25} = 5$.

31. $\sqrt[5]{245}$, using the value $\sqrt[5]{243} = 3$.

32. $(4.02)^{\frac{3}{2}}$, using the value $4^{\frac{3}{2}} = 8$.

33. $\sin 45.3°$.   **34.** $\cos 60.1°$.   **35.** $\tan 45.2°$.   **36.** $\tan 45.4°$.

Progress Report

This chapter began with a discussion of related increments of a function and the average rate of change of a function. This discussion led to the preliminary consideration of the instantaneous rate of change of a function. The mathematical idea of a limit was next introduced, on an intuitive basis, in order to provide a foundation for further consideration of the instantaneous rate of change. After a geometrical investigation of the instantaneous rate of change of a function, the derivative of a function was defined and the differential calculus was defined as the study of derivatives.

The remainder of the chapter was devoted to the following topics from the differential calculus:

1. How to find the derivative of a function given by a simple formula by means of the definition of the derivative.

2. The derivatives of x^n, $\sin x$, and $\cos x$.

3. Computing derivatives of a constant times a function, the sum of two functions, the product of two functions, the quotient of two functions, and a function of a function.

4. How to compute the derivatives of a wide variety of elementary functions involving various combinations of x^n, $\sin x$, and $\cos x$.

5. Using derivatives to find the tangent to a curve at a point and to compute the velocity and acceleration of a point moving on a line.

6. How to find the maximum and minimum points of a function by means of its derivatives.

7. The approximation to the increment of a function by means of the differential.

20

The Elements
of Integral Calculus

Newton and Leibnitz made one of the greatest discoveries of mathematics when they found that there is a close connection between computing an area and finding a function whose derivative is given, and that

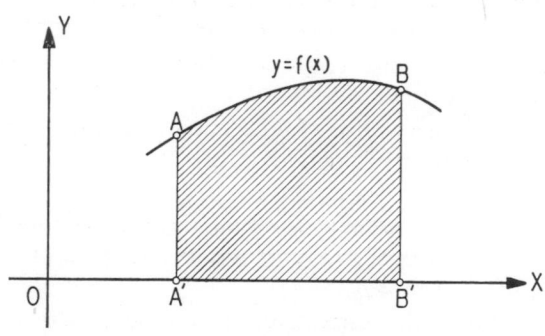

Fig. 20–1

the first problem can be reduced to the second. The consequences of this discovery together with the solution of the second problem are of the utmost importance and possess innumerable applications in engineering and the physical sciences.

20–1. The Approximate Computation of an Area

The problem of computing an area is often presented in the following way. The graph of a function $y = f(x)$ is plotted (Fig. 20–1), and the shaded area bounded by the graph AB and the segments AA', BB', and $A'B'$ is to be found. In order to obtain an approximate value for this area, the segment $A'B'$ is divided into a number of equal parts. For example, in Fig. 20–2 the segment $A'B'$ is divided into 5 equal parts. The length of each of these parts is denoted by Δx. Now in each of these parts select a point arbitrarily. In this way we select P_1, P_2, $\cdots$, whose abscissas are denoted by x_1, x_2, $\cdots$. The points on the curve with these abscissas are denoted by Q_1, Q_2, $\cdots$, whence $P_1Q_1 = f(x_1)$, $P_2Q_2 = f(x_2)$, $\cdots$. Now, as shown in the figure, rectangles are constructed whose bases are the segments of length Δx on the x axis and whose altitudes are the segments P_1Q_1, P_2Q_2, $\cdots$. The area of a rectangle

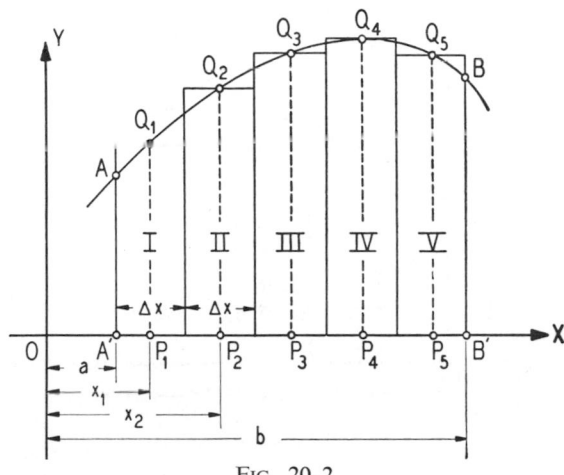

FIG. 20–2

is the product of the length of its base and its altitude, and therefore,

$$\text{Area of rectangle I} = (P_1Q_1)\,\Delta x = f(x_1) \cdot \Delta x,$$
$$\text{Area of rectangle II} = (P_2Q_2)\,\Delta x = f(x_2) \cdot \Delta x,$$

$$\cdots$$

The sum of these areas

$$\text{I} + \text{II} + \cdots = (P_1Q_1)\,\Delta x + (P_2Q_2)\,\Delta x + \cdots$$
$$= f(x_1) \cdot \Delta x + f(x_2) \cdot \Delta x + \cdots$$

is obviously an approximation for the area $ABB'A'$. The error made by using the sum of the rectangles instead of the correct, but unknown value of the area $ABB'A'$ becomes smaller if the number of segments into which $A'B'$ is divided is increased. Thus the area can be found with any desired degree of precision if the number of segments is large enough or, what means the same thing, if the bases Δx are small enough.

The sum

$$f(x_1) \cdot \Delta x + f(x_2) \cdot \Delta x + \cdots$$

which is used above can be denoted symbolically by

$$Sf(x) \cdot \Delta x,$$

where the symbol S denotes that a series of similar terms $f(x_1) \cdot \Delta x$, $f(x_2) \cdot \Delta x, \cdots$ are to be added. In order to indicate that the area can be found as precisely as we desire by taking Δx small enough and computing these sums, the symbol dx is used instead of Δx, and the sum is written with the symbol $\int$ instead of S:

$$\text{Area } ABB'A' = \int f(x)\, dx.$$

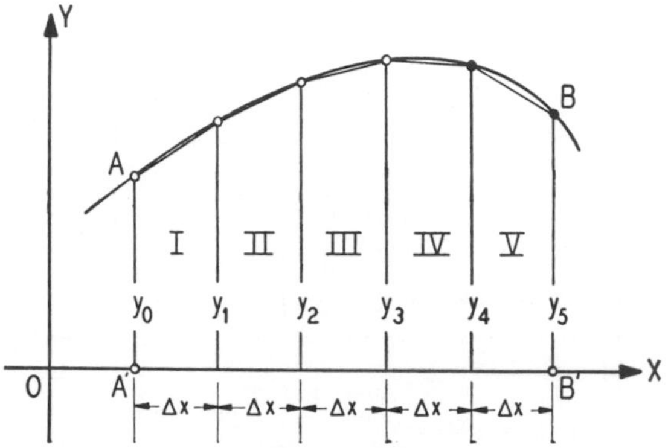

Fig. 20–3

In order to indicate that the area bounded by the lines $x = a$ and $x = b$ is to be computed, we write

$$\text{Area} = \int_a^b f(x)\, dx.$$

The right member of this equation is called **the definite integral of $f(x)\, dx$ between the limits a and b** or **from a to b**. This discussion has shown that the definite integral is an area; the symbol $\int$ (read "the integral of") indicates the method of computing the area.

For practical computation it is advantageous to replace the rectangles of Fig. 20–2 by trapezoids as shown in Fig. 20–3. The area of a trapezoid is the product of one-half the sum of the two parallel sides times the distance between them. If the notations of Fig. 20–3 are used, these areas are

$$\text{I} = \tfrac{1}{2}(y_0 + y_1)\, \Delta x,$$
$$\text{II} = \tfrac{1}{2}(y_1 + y_2)\, \Delta x,$$
$$\text{III} = \tfrac{1}{2}(y_2 + y_3)\, \Delta x,$$
$$\text{IV} = \tfrac{1}{2}(y_3 + y_4)\, \Delta x,$$
$$\text{V} = \tfrac{1}{2}(y_4 + y_5)\, \Delta x,$$

whence

$$I + II + III + IV + V$$

$$= \frac{\Delta x}{2}\left[(y_0 + y_1) + (y_1 + y_2) + (y_2 + y_3) + (y_3 + y_4) + (y_4 + y_5)\right]$$

$$= \frac{\Delta x}{2}\left(y_0 + y_5 + 2y_1 + 2y_2 + 2y_3 + 2y_4\right)$$

$$- \Delta x \left(\frac{y_0 + y_5}{2} + y_1 + y_2 + y_3 + y_4\right).$$

When the interval $A'D'$ is divided into any number of equal parts, a similar formula can be found. The general result of this discussion is:

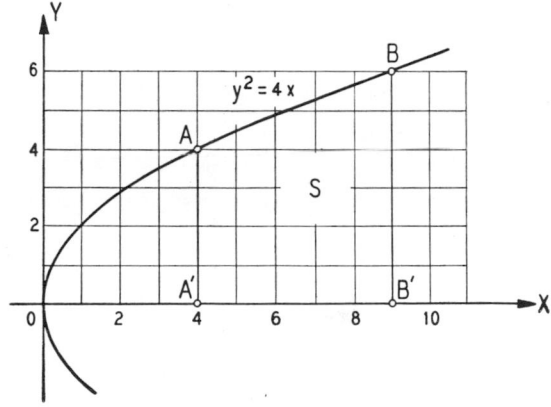

$$y^2 = 4x$$

FIG. 20–4

In order to find an approximate value for the area $ABB'A'$ bounded by a curve, the axis of the abscissas, and the two ordinates, divide the segment $A'B'$ into n equal parts, the length of each part being denoted by Δx. Then find the ordinates $y_0, y_1, y_2, \cdots, y_n$ corresponding to the division points of $A'B'$. The formula

$$\Delta x \left(\frac{y_0 + y_n}{2} + y_1 + y_2 + \cdots y_{n-1}\right)$$

gives an approximation for the area $ABB'A'$. This formula is known as the trapezoidal formula.

A few examples will show how this formula is applied.

Example 1. Compute the area S bounded by the graph of $y^2 = 4x$, the ordinates corresponding to $x = 4$ and $x = 9$, and the x axis, as shown in Fig. 20–4.

If the segment $A'B'$ is divided into five equal parts each of length $\Delta x = 1$, the following table can be constructed giving corresponding values of x and $y = \sqrt{4x}$, the values of x being the points of division of the segment $A'B'$.

x	y
4	4.000
5	4.472
6	4.899
7	5.292
8	5.657
9	6.000

The trapezoidal formula gives as an approximate value of the area

$$S = \int_4^9 \sqrt{4x}\, dx = 1 \left(\frac{4.000 + 6.000}{2} + 4.472 + 4.899 + 5.292 + 5.657 \right)$$

$$= 5.00 + 20.230 = 25.320.$$

Since the correct value is 25.333, the value obtained by this computation is correct to about four significant figures, a very good approximation.

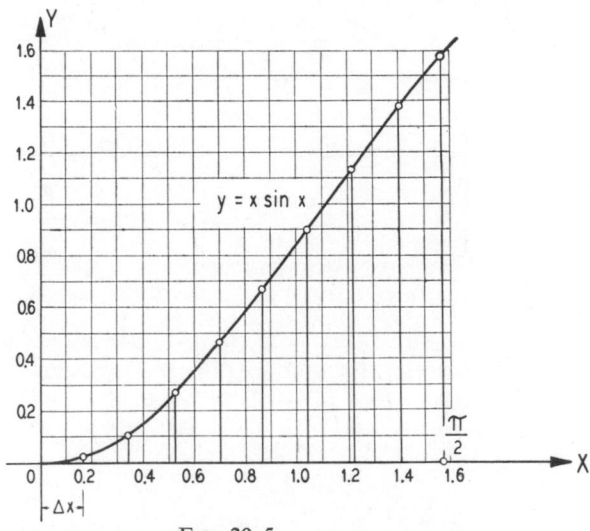

FIG. 20–5

Example 2. Compute the area denoted by the integral $\int_0^{\pi/2} x \sin x\, dx$.

The graph of $y = x \sin x$ is plotted in Fig. 20–5. The area to be computed is included by the curve, the x axis from 0 to $\pi/2$, and the line $x = \pi/2$. The interval from 0 to $\pi/2$ is divided in 9 equal segments of the length $\Delta x = \pi/18 = 0.1745$, corresponding to $10°$. In order to apply the trapezoidal formula, the following table is computed, the computations being performed with the slide rule:

Number	x		sin x	$y = x \sin x$
0	$0°$		0.0000	0.000
1	$10° = \dfrac{\pi}{18} = 0.175$		0.1736	0.030
2	$20° = 2\dfrac{\pi}{18} = 0.349$		0.3420	0.119
3	$30° = 3\dfrac{\pi}{18} = 0.524$		0.5000	0.262
4	$40° = 4\dfrac{\pi}{18} = 0.698$		0.6428	0.449
5	$50° = 5\dfrac{\pi}{18} = 0.873$		0.7660	0.668
6	$60° = 6\dfrac{\pi}{18} = 1.047$		0.8660	0.907
7	$70° = 7\dfrac{\pi}{18} = 1.222$		0.9397	1.148
8	$80° = 8\dfrac{\pi}{18} = 1.396$		0.9848	1.375
9	$90° = \dfrac{\pi}{2} = 1.571$		1.0000	1.571

By the trapezoidal formula, we add the average of the first and last y values to the sum of the others.

$$\frac{y_0 + y_9}{2} + y_1 + y_2 + \cdots + y_8 = 0.786 + 4.958 = 5.744.$$

The area is this sum multiplied by $\Delta x = 0.1745$.   The result is

$$\int_0^{\pi/2} x \sin x \, dx = S = 0.1745 \cdot 5.744 = 1.000.$$

Exercises

Using the trapezoidal rule, find to three significant figures the area bounded by the given curve, the given two vertical lines, and the x axis.   Use the given value of Δx.

1. $y = x^2$, $x = 0$, $x = 4$; $\Delta x = 1$.   2. $y = x^2$, $x = 1$, $x = 5$; $\Delta x = 1$.

3. $y = x^2$, $x = 0$, $x = 4$; $\Delta x = \frac{1}{2}$.   4. $y = x^2$, $x = 1$, $x = 5$; $\Delta x = \frac{1}{2}$.

5. $y = x^3$, $x = 1$, $x = 5$; $\Delta x = 1$.   6. $y = x^4$, $x = 0$, $x = 6$; $\Delta x = 1$.

7. $y = \sqrt{x}$, $x = 4$, $x = 10$; $\Delta x = 1$.   8. $y = \sqrt{x}$, $x = 0$, $x = 20$; $\Delta x = 2$.

9. $y = \sqrt[3]{x}$, $x = 1$, $x = 11$; $\Delta x = 2$.   **10.** $y = \sqrt[3]{x^2}$, $x = 1$, $x = 11$; $\Delta x = 2$.

11. $y = \dfrac{1}{x}$, $x = 1$, $x = 3$; $\Delta x = \frac{1}{3}$.   **12.** $y = \dfrac{2}{x^2}$, $x = 2$, $x = 6$; $\Delta x = \frac{1}{2}$.

13. $y = \sin x$, $x = 0$, $x = \dfrac{\pi}{2}$; $\Delta x = \dfrac{\pi}{10}$.   **14.** $y = \sin x$, $x = 0$, $x = \pi$; $\Delta x = \dfrac{\pi}{12}$.

15. $y = \cos x$, $x = 0$, $x = \dfrac{\pi}{2}$; $\Delta x = \dfrac{\pi}{18}$.   **16.** $y - \cos x$, $x = 0$, $x - \dfrac{\pi}{2}$; $\Delta x = \dfrac{\pi}{20}$.

17. $y = \tan x$, $x = 0$, $x = \dfrac{\pi}{3}$; $\Delta x = \dfrac{\pi}{18}$.   **18.** $y = \frac{1}{3}\tan x$, $x = 0$, $x = 1$; $\Delta x = 0.2$.

19. $y = \log x$, $x = 1$, $x = 2.4$; $\Delta x = 0.2$.   **20.** $y = 3^x$, $x = 0$, $x = 4$; $\Delta x = 0.5$.

Compute the area denoted by the given integral to three significant figures, using the given value of Δx.

21. $\displaystyle\int_0^6 x^3\, dx$, $\Delta x = 1$.   **22.** $\displaystyle\int_0^4 4x^4\, dx$, $\Delta x = 1$.

23. $\displaystyle\int_2^{10} (x^2 - 2x)\, dx$, $\Delta x = 1$.   **24.** $\displaystyle\int_0^7 (2x^3 + 4x)\, dx$, $\Delta x = 1$.

25. $\displaystyle\int_1^3 x^{-\frac{1}{2}}\, dx$, $\Delta x = \frac{1}{4}$.   **26.** $\displaystyle\int_1^4 \frac{1}{x}\, dx$, $\Delta x = 0.5$.

27. $\displaystyle\int_0^4 (x+1)^2\, dx$, $\Delta x = \frac{1}{2}$.   **28.** $\displaystyle\int_0^6 (1-2x)^2\, dx$, $\Delta x = \frac{1}{2}$.

29. $\displaystyle\int_0^5 \sqrt{25 - x^2}\, dx$, $\Delta x = 1$.   **30.** $\displaystyle\int_0^3 \sqrt{36 - x^2}\, dx$, $\Delta x = \frac{1}{2}$.

31. $\displaystyle\int_0^{\pi/4} \tan x\, dx$, $\Delta x = \dfrac{\pi}{12}$.   **32.** $\displaystyle\int_3^5 \log x\, dx$, $\Delta x = 0.4$.

33. $\displaystyle\int_0^{\pi/2} x\cos x\, dx$, $\Delta x = \dfrac{\pi}{20}$.   **34.** $\displaystyle\int_0^{\pi/2} x^2\cos x\, dx$, $\Delta x = \dfrac{\pi}{12}$.

35. $\displaystyle\int_0^{\pi/2} x^2\sin x\, dx$, $\Delta x = \dfrac{\pi}{12}$.   **36.** $\displaystyle\int_0^{\pi} \sin^2 x\, dx$, $\Delta x = \dfrac{\pi}{6}$.

37. Compute the value of $\displaystyle\int_0^8 x^3\, dx$, using the trapezoidal rule and $\Delta x = 1$. Then compute the same integral using $\Delta x = \frac{1}{2}$. If the correct value of the integral is 1024, by how much does each approximation differ from the correct value, and what is the per cent error in each case?

38. Find the area in the first quadrant bounded by the graph of the equation $y = 6x^2 - 2x^3$ and the x axis, using the trapezoidal rule and $\Delta x = \frac{1}{2}$. Then compute the same area using $\Delta x = \frac{1}{3}$. If the correct value of the area is $\frac{27}{2}$, what is the per cent error in each case?

39. The contour of a rivet head given by the equation $y = \sqrt{2.25 - x^2}$ is a semicircle with center at the origin and lying above the x axis. Compute the area of the semicircle by the trapezoidal rule, using $\Delta x = 0.25$, and find the amount by which this approximation differs from the correct result.

40. An engineer designs a cam that has an elliptical contour given by the equation

$$y = \pm \sqrt{\frac{36 - 9x^2}{4}}$$

with center at the origin. Compute the area of the ellipse using the trapezoidal rule and $\Delta x = \frac{1}{4}$.

41. The average displacement from the normal position of the prongs of a tuning fork that vibrate with a peak displacement of D_m cm is $a D_m$ cm, where

$$a = \frac{1}{\pi} \int_0^\pi \sin x \, dx.$$

Compute the value of a correct to three significant figures.

42. The effective value of a sinusoidal alternating current with a peak value of I_m amperes is $k I_m$ amperes, where

$$k = \sqrt{\frac{1}{\pi} \int_0^\pi \sin^2 x \, dx}.$$

Compute the value of k correct to three significant figures.

20–2. The Definite Integral and the Inverse Operation of Differentiation

The method given in the preceding section for computing approximate values of an area made use of a table of values of the function corresponding to equally spaced values of the independent variable, but did not make use of any further properties of the function defining the area. When a function is defined by a mathematical formula, it is desirable to find the area under the curve precisely, without the use of the approximation method of Sec. 20–1. That such a precise result can be achieved by using the properties of the function defining the area is one of the most important discoveries of mathematics. How this precise result can be obtained will be discussed in this section.

The problem considered in Section 20–1 was: Given a function $y = f(x)$, compute the area bounded by an arc of the graph of $y = f(x)$ the two lines $x = a$ and $x = b$, and the x axis. Now we shall investigate how the area S changes if the value of b changes, or, using functional language, we shall investigate the area S as a function of b. If we write x instead of b we may state the problem as follows. Investigate the area S under the curve $y = f(x)$ and above the x axis between the abscissas a and x (see Fig. 20–6). Since this area depends on x, it is a function of x, which we shall denote by $S(x)$.

We shall now compute the derivative of $S(x)$ with respect to x. The increase $\Delta S(x)$ corresponding to the increase Δx of x is the shaded area $BCC'B'$. This area is equal to the area of a rectangle with base $B'C'$

and altitude PP', where P is a point on the curve between B and C. Thus

$$\Delta S(x) = \text{area } BCC'B' = (PP')(B'C') = (PP') \cdot \Delta x.$$

The average rate of change of $S(x)$ as x increases from x to $x + \Delta x$ is

$$\frac{\Delta S(x)}{\Delta x} = PP'.$$

The derivative $\dfrac{dS(x)}{dx}$ is the limit of this expression as Δx approaches zero. From Fig. 20–6 it can be inferred that PP' approaches BB' as Δx

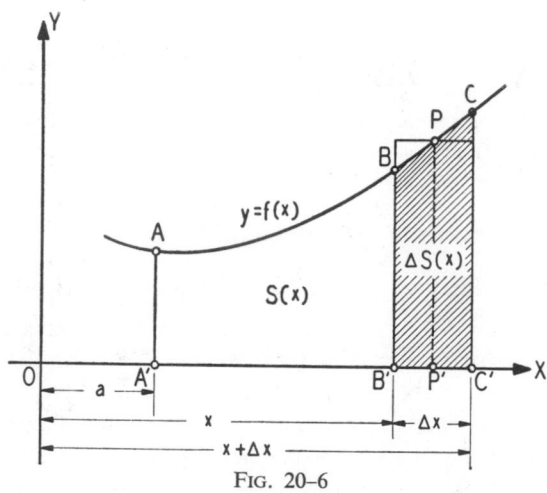

Fig. 20–6

approaches zero, and therefore C approaches B. We have, then, that

$$\frac{dS(x)}{dx} = \lim_{\Delta x \to 0} \frac{\Delta S(x)}{\Delta x} = \lim_{\Delta x \to 0} PP' = BB'.$$

The length of BB' is the value of the function $y = f(x)$ for the value of x indicated in the figure. Hence we have the fundamental formula

$$\frac{dS(x)}{dx} = f(x).$$

The meaning of this formula can be stated as follows.

If the area bounded by the curve $y = f(x)$, the x axis, and the ordinates corresponding to the abscissas a and x is denoted by $S(x)$, then the derivative of $S(x)$ with respect to x is the given function $f(x)$.

The problem of finding a formula for $S(x)$ is thus the problem of finding a function whose derivative is the given function $f(x)$. This statement shows that the problem of finding an area is closely connected with the problem of *finding a function whose derivative is a given function of the independent variable.*

The operation of finding a function whose derivative is known is called **integration.** Integration is the inverse operation of differentiation, just as subtraction is the inverse operation of addition and division is the inverse operation of multiplication.

The operation of differentiation applied to a given function yields a completely determined result, the derivative of the given function. The inverse operation does not yield a unique result, because it can be

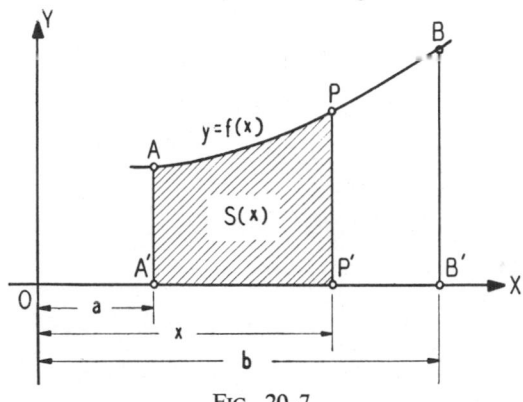

FIG. 20–7

shown that there are infinitely many functions having the same derivative. For example, the functions $2x^2$, $2x^2 + 10$, $2x^2 - 5$ have the same derivative $4x$, because the constant terms $+10$ and -5 have the derivative zero. All the functions that are obtained by adding an arbitrary constant C to a function $F(x)$ have the same derivative as $F(x)$ because

$$\frac{d[F(x) + C]}{dx} = \frac{dF(x)}{dx}.$$

Conversely, it can be shown that *all functions that have the same derivative as $F(x)$ can be obtained from the formula $F(x) + C$.* This fact can be expressed as follows.

Theorem A. If $F(x)$ is a function with $\dfrac{dF(x)}{dx} = f(x)$, then every function of x having the same derivative $f(x)$ can be written as $F(x) + C$, where C is a constant.

The result of this theorem will now be applied to the area problem. Consider the shaded area $S(x)$ under the curve $y = f(x)$ as shown in Fig. 20–7. We have already seen that

$$\frac{dS(x)}{dx} = f(x).$$

Suppose that we succeed in finding a function $F(x)$ such that $\dfrac{dF(x)}{dx} = f(x)$.

Then according to Theorem A, there is a constant C such that

$$S(x) = F(x) + C.$$

When $x = a$ the area is obviously zero, so that $S(a) = 0$.   Therefore,

$$S(a) = F(a) + C = 0$$

or

$$C = -F(a).$$

Hence the formula for $S(x)$ becomes

$$S(x) = F(x) - F(a).$$

This formula permits us to find the area for any value of x after $F(x)$ has been found.   Substituting $x = b$, we obtain, in particular, the area $ABB'A'$, which was previously denoted by

$$\int_a^b f(x)\, dx.$$

Therefore

$$\int_a^b f(x)\, dx = S(b) = F(b) - F(a).$$

We have established then this important result:

Theorem B.   If $f(x)$ is given and $F(x)$ is a function such that

$$\frac{dF(x)}{dx} = f(x),$$

then

$$\int_a^b f(x)\, dx = F(b) - F(a).$$

Example 1.   In example 1 of Sec. 20–1, the area under the curve $y = \sqrt{4x}$ and above the x axis lying between $x = 4$ and $x = 9$ was computed.   This area can be denoted by

$$\int_4^9 \sqrt{4x}\, dx.$$

According to the rule just discussed, this definite integral can be computed if a function $F(x)$ can be found such that $\dfrac{dF(x)}{dx} = \sqrt{4x} = 2\sqrt{x}$.   The function

$$F(x) = \tfrac{4}{3}x^{\frac{3}{2}}$$

is such a function, for

$$\frac{d(\tfrac{4}{3}x^{\frac{3}{2}})}{dx} = \frac{4}{3} \cdot \frac{3}{2} x^{\frac{1}{2}} = 2\sqrt{x}.$$

According to Theorem B the value of the area is

$$\int_4^9 \sqrt{4x}\, dx = F(9) - F(4) = \tfrac{4}{3}(9^{\frac{3}{2}}) - \tfrac{4}{3}(4^{\frac{3}{2}}) = \tfrac{4}{3}(27 - 8) = 25.33.$$

Example 2. The area computed in example 2 of Sec. 20–1 was denoted by

$$\int_0^{\pi/2} x \sin x\, dx.$$

We find that for the function

$$F(x) - \sin x - x \cos x,$$

the derivative

$$\frac{dF(x)}{dx} - \cos x - (\cos x - x \sin x) = x \sin x,$$

and therefore

$$\int_0^{\pi/2} x \sin x\, dx = F\left(\frac{\pi}{2}\right) - F(0)$$

$$= \left(\sin \frac{\pi}{2} - \frac{\pi}{2} \cos \frac{\pi}{2}\right) - (\sin 0 - 0 \cdot \cos 0)$$

$$= (1 - 0) - (0 - 0) = 1.$$

So far, as in the preceding examples, we are able to find the function $F(x)$ only by trial and error, by testing various choices to see whether the derivative of the chosen $F(x)$ is $f(x)$. In the next section a few methods for finding $F(x)$ when $f(x)$ is given will be developed, using the definitions that follow in this section. Because of the connection between the problem of finding the function $F(x)$ and the area problem whose solution was denoted by the symbol $\int_a^b f(x)\, dx$, called a definite integral, the following definitions are introduced.

Each function $F(x)$ such that $\dfrac{dF(x)}{dx} = f(x)$, *or such that* $dF(x) = f(x)\, dx$, *is called a* **particular integral** *of* $f(x)\, dx$.

Since every other function with the same derivative $f(x)$ can be expressed in the form $F(x) + C$, where C is a constant, we state the following definition.

If $F(x)$ is a particular integral of $f(x)\, dx$, the expression $F(x) + C$ is called the **general or indefinite integral** *of* **$f(x)\, dx$,** *and is denoted by the symbol* $\int f(x)\, dx$. *C is called the* **constant of integration.**

The symbol $\int f(x)\, dx$ which is defined by the relation

$$\int f(x)\, dx = F(x) + C$$

is read as *the indefinite integral of f(x) dx,* or briefly, *the integral of f(x) dx.*

The abbreviation $\left[F(x) \right]_a^b$ is often used for the expression $F(b) - F(a)$,

whence the fundamental relation of Theorem *B* can be written as

$$\int_a^b f(x) \, dx = \left[F(x) \right]_a^b = F(b) - F(a). \qquad (1)$$

The integral in this expression is called the **definite integral of f(x) from a to b,** as we saw in Sec. 20–1.

The operation of finding an integral is called **integration;** the integral of $f(x) \, dx$ is found by **integrating.** As an illustration of how these notations are used, the results of examples 1 and 2 of this section will be restated here.

Example 3. The problem of example 1 was to compute the value of $\int_4^9 \sqrt{4x} \, dx$. We found that $\frac{4}{3}x^{\frac{3}{2}}$ was a function with the derivative $\sqrt{4x}$. We state this fact by saying that $\frac{4}{3}x^{\frac{3}{2}}$ is a particular integral of $\sqrt{4x} \, dx$. Then the general or indefinite integral of $\sqrt{4x} \, dx$ is

$$\int \sqrt{4x} \, dx = \frac{4}{3} x^{\frac{3}{2}} + C.$$

In other words, $\frac{4}{3}x^{\frac{3}{2}} + C$ is the most general function whose derivative is $\sqrt{4x}$. Then by relation 1,

$$\int_4^9 \sqrt{4x} \, dx = \left[\frac{4}{3} x^{\frac{3}{2}} \right]_4^9 = \frac{4}{3} \cdot 27 - \frac{4}{3} \cdot 8 = 25.33.$$

Example 4. The problem of example 2 was to compute the value of $\int_0^{\pi/2} x \sin x \, dx$. We found that

$$F(x) = \sin x - x \cos x$$

was a particular integral of $f(x) \, dx$, so that

$$\int x \sin x \, dx = \sin x - x \cos x + C,$$

and

$$\int_0^{\pi/2} x \sin x \, dx = \left[\sin x - x \cos x \right]_0^{\pi/2} = 1.$$

Example 5. Prove the statement that

$$\int x^2 \cos x \, dx = 2x \cos x + (x^2 - 2) \sin x + C.$$

This equation states that the derivative of the right member equals $x^2 \cos x$, and therefore the equation can be checked by computing this derivative.

$$\frac{d[2x \cos x + (x^2 - 2) \sin x + C]}{dx} = \frac{d(2x \cos x)}{dx} + \frac{d[(x^2 - 2) \sin x]}{dx}$$

$$= 2(\cos x - x \sin x) + [2x \sin x + (x^2 - 2) \cos x]$$

$$= 2 \cos x - 2x \sin x + 2x \sin x + x^2 \cos x - 2 \cos x$$

$$= x^2 \cos x.$$

Exercises

Prove each of the following statements:

1. $\displaystyle\int x^2 \, dx = \frac{1}{3} x^3 + C.$

2. $\displaystyle\int 3x^3 \, dx = \frac{3}{4} x^4 + C.$

3. $\displaystyle\int x^{\frac{1}{2}} \, dx = \frac{2}{3} x^{\frac{3}{2}} + C.$

4. $\displaystyle\int ax^n \, dx = \frac{a}{n+1} x^{n+1} + C, \ n = 1, 2, 3, \cdots .$

5. $\displaystyle\int (2x^2 - 3x + 5) \, dx = \frac{2}{3} x^3 - \frac{3}{2} x^2 + 5x + C.$

6. $\displaystyle\int (x^5 + 4x^3 + x) \, dx = \frac{1}{6} x^6 + x^4 + \frac{1}{2} x^2 + C.$

7. $\displaystyle\int (1 + x^n) \, dx = x + \frac{x^{n+1}}{n+1} + C, \ n = 1, 2, 3, \cdots .$

8. $\displaystyle\int (1 + x)^n \, dx = \frac{(1 + x)^{n+1}}{n+1} + C, n = 1, 2, 3, \cdots .$

9. $\displaystyle\int \frac{dx}{(1 + x)^2} = \frac{-1}{1 + x} + C.$

10. $\displaystyle\int \frac{x \, dx}{\sqrt{1 + x^2}} = \sqrt{1 + x^2} + C.$

11. $\displaystyle\int \sin^2 x \, dx = \frac{1}{2}(x - \sin x \cos x) + C.$

12. $\displaystyle\int \cos^2 x \, dx = \frac{1}{2}(x + \sin x \cos x) + C.$

13. $\displaystyle\int \cos^3 x \, dx = \frac{1}{3} \sin x(\cos^2 x + 2) + C.$

14. $\int \sin^3 x \, dx = -\dfrac{1}{3} \cos x(\sin^2 x + 2) + C.$

15. $\int \sin (ax + b) \, dx = -\dfrac{1}{a} \cos (ax + b) + C.$

16. $\int \cos (ax + b) \, dx = \dfrac{1}{a} \sin (ax + b) + C.$

Using the statements of exercises 1–16, compute the following definite integrals giving your results correct to three significant figures. By drawing the corresponding graph, interpret each integral as an area.

17. $\int_0^3 x^2 \, dx.$

18. $\int_3^6 x^2 \, dx.$

19. $\int_{-1}^5 x^2 \, dx.$

20. $\int_{-2}^4 x^2 \, dx.$

21. $\int_2^4 3x^3 \, dx.$

22. $\int_3^5 3x^3 \, dx.$

23. $\int_0^2 3x^3 \, dx.$

24. $\int_3^6 3x^3 \, dx.$

25. $\int_0^3 5x^2 \, dx.$

26. $\int_2^5 4x^2 \, dx.$

27. $\int_1^2 x^7 \, dx.$

28. $\int_{-2}^3 2x^6 \, dx.$

29. $\int_0^1 x^{\frac{1}{2}} \, dx.$

30. $\int_1^2 x^{\frac{1}{2}} \, dx.$

31. $\int_1^3 (2x^2 - 3x + 5) \, dx.$

32. $\int_2^3 (2x^2 - 3x + 5) \, dx.$

33. $\int_1^2 (x^5 + 4x^3 + x) \, dx.$

34. $\int_0^3 (x^5 + 4x^3 + x) \, dx.$

35. $\int_{-1}^1 (1 + x^2) \, dx.$

36. $\int_3^6 (1 + x^2) \, dx.$

37. $\int_2^3 (1 + x^4) \, dx.$

38. $\int_{-1}^5 (1 + x^3) \, dx.$

39. $\int_0^4 (1 + x)^2 \, dx.$

40. $\int_1^3 (1 + x)^3 \, dx.$

41. $\int_0^3 \dfrac{dx}{(1 + x)^2}.$

42. $\int_1^2 \dfrac{dx}{(1 + x)^2}.$

43. $\int_0^2 \dfrac{x \, dx}{\sqrt{1 + x^2}}.$

44. $\int_0^5 \dfrac{x \, dx}{\sqrt{1 + x^2}}.$

45. $\int_0^\pi \sin^2 x \, dx.$

46. $\int_\pi^{2\pi} \cos^2 x \, dx.$

47. $\int_{\pi/2}^{2\pi} \cos^2 x \, dx.$

48. $\int_{\pi/2}^{3\pi/2} \sin^2 x \, dx.$

49. $\int_0^{\pi/4} \sin^2 x \, dx.$

50. $\int_0^{\pi/2} \cos^3 x \, dx.$

51. $\int_{-\pi/2}^{\pi/2} \cos^3 x \, dx.$

52. $\int_0^\pi \sin^3 x \, dx.$

53. $\int_0^{\pi/4} \sin^3 x \, dx.$

54. $\int_0^{\pi/2} \sin 2x \, dx.$

55. $\int_{3\pi/4}^{5\pi/4} \cos 2x \, dx.$

56. $\displaystyle\int_0^{\pi/8} \cos 2x \, dx.$ **57.** $\displaystyle\int_0^{\pi/8} \sin 2x \, dx.$ **58.** $\displaystyle\int_{\pi/6}^{\pi/3} \sin 3x \, dx.$

59. $\displaystyle\int_0^{\pi/8} \sin\left(4x + \frac{\pi}{2}\right) dx.$ **60.** $\displaystyle\int_{-\pi/6}^{\pi/12} \cos\left(3x + \frac{\pi}{4}\right) dx.$

61. Find the area bounded by the graph of $y = 2x^2 - 3x + 5$, the ordinates corresponding to $x = -5$ and $x = 9$, and the x axis.

62. The effective value of a sinusoidal alternating current with a peak value of I_m amperes is kI_m amperes, where

$$k = \sqrt{\frac{1}{\pi}\int_0^\pi \sin^2 x \, dx}.$$

Compute the value of k correct to three significant figures. Compare this result with that obtained in exercise 42 of Sec. 20–1.

20–3. The Integral Calculus

From the discussion in Sec. 20–2 it is obvious that the most important problem now is to find the indefinite integral of a given function. This problem is much more complicated than the inverse problem: to find the derivative of a given function. The system of all the methods that have been developed for obtaining integrals of given functions is called **integral calculus.** A few of the more elementary formulas and rules of this calculus will be discussed in this section.

Each formula of the differential calculus can be used to obtain a corresponding formula of the integral calculus. Thus the formula $\dfrac{dx^n}{dx} = nx^{n-1}$ can be used to prove that

$$\int x^n \, dx = \frac{x^{n+1}}{n+1} + C. \tag{1}$$

This formula is proved by computing the derivative of the right member:

$$\frac{d\left(\dfrac{x^{n+1}}{n+1} + C\right)}{dx} = \frac{1}{n+1} \cdot \frac{dx^{n+1}}{dx} = \frac{1}{n+1} \cdot (n+1)\, x^n = x^n.$$

A particular case is obtained for $n = 0$,

$$\int dx = x + C. \tag{2}$$

In a similar way the formulas

$$\int \sin x \, dx = -\cos x + C, \tag{3}$$

$$\int \cos x \, dx = \sin x + C \tag{4}$$

can be proved.

The following rules correspond to the analogous rules of differential calculus.

A constant factor may be written before or after the sign of integration.

$$\int Cf(x) \, dx = C \int f(x) \, dx. \tag{5}$$

Example 1. $\displaystyle\int 4x^2 \, dx = 4 \int x^2 \, dx = \frac{4}{3} x^3 + C.$

The integral of a sum or of a difference is equal to the sum or the difference of the integrals of each term:

$$\int [f(x) \pm g(x)] \, dx = \int f(x) \, dx \pm \int g(x) \, dx. \tag{6}$$

Example 2. Integrate $\displaystyle\int (3x + 5\sqrt{x}) \, dx.$

$$\int (3x + 5\sqrt{x}) \, dx = \int 3x \, dx + \int 5\sqrt{x} \, dx \qquad \text{By (6)}$$

$$= 3 \int x \, dx + 5 \int x^{\frac{1}{2}} \, dx \qquad \text{By (5)}$$

$$= 3 \left(\frac{x^2}{2} \right) + 5 \cdot \frac{x^{\frac{3}{2}}}{\frac{3}{2}} + C \qquad \text{By (1)}$$

$$= \tfrac{3}{2} x^2 + \tfrac{10}{3} x\sqrt{x} + C.$$

Example 3. Integrate $\displaystyle\int (2 \sin x + 5 \cos x) \, dx.$

$$\int (2 \sin x + 5 \cos x) \, dx = \int 2 \sin x \, dx + \int 5 \cos x \, dx \qquad \text{By (6)}$$

$$= 2 \int \sin x \, dx + 5 \int \cos x \, dx \qquad \text{By (5)}$$

$$= -2 \cos x + 5 \sin x + C. \quad \text{By (3) and (4)}$$

Many more complicated integration problems, to which the formulas given above do not apply directly, can be reduced to simpler problems where the formulas do apply by the introduction of a new variable. The following examples will illustrate how this can be done.

Example 4. Find $\displaystyle\int x\sqrt{1 + 2x} \, dx.$

The formulas and rules developed so far do not apply to this problem. A simplification can be obtained by introducing a new variable t such that

$$1 + 2x = t \quad \text{or} \quad x = \frac{t - 1}{2}.$$

It is important that not only the new variable t is to be substituted for x in the function $x\sqrt{1 + 2x}$, but also that the differential dx is to be replaced by an expression containing only t and dt according to the formula

$$dx = \frac{dx}{dt} dt.$$

In our case we obtain

$$dx = \tfrac{1}{2}dt,$$

and therefore

$$\int x\sqrt{1 + 2x}\, dx = \int \frac{t - 1}{2} \cdot \sqrt{t} \cdot \frac{1}{2} dt$$

$$= \frac{1}{4}\int (t^{\frac{3}{2}} - t^{\frac{1}{2}})\, dt$$

$$= \frac{1}{4}\left[\frac{t^{\frac{5}{2}}}{\frac{5}{2}} - \frac{t^{\frac{3}{2}}}{\frac{3}{2}}\right] + C$$

$$= \tfrac{1}{4}[\tfrac{2}{5}t^{2}\sqrt{t} - \tfrac{2}{3}t\sqrt{t}] + C$$

$$= \tfrac{1}{30}\sqrt{t}(3t^{2} - 5t) + C.$$

The value of the integral has now been found as a function of t. Substituting for t its original value in terms of x, we obtain

$$\int x\sqrt{1 + 2x}\, dx = \tfrac{1}{30}\sqrt{1 + 2x}[3(1 + 2x)^{2} - 5(1 + 2x)] + C$$

$$= \tfrac{1}{15}\sqrt{1 + 2x}[6x^{2} + x - 1] + C.$$

The result can be checked by differentiating, because the derivative of the result has to be $x\sqrt{1 + 2x}$.

Example 5. Find $\int \cos 4x\, dx$.

The function $\cos 4x$ suggests the introduction of a new variable t such that

$$t = 4x,$$
$$dt = 4dx,$$
$$dx = \tfrac{1}{4}dt.$$

This substitution gives

$$\int \cos 4x\, dx = \int (\cos t)\frac{dt}{4} = \frac{1}{4}\int \cos t\, dt = \frac{1}{4}\sin t + C = \frac{1}{4}\sin 4x + C.$$

Example 6. Find $\int \cos^{2} x \sin x\, dx$.

By the substitution $t = \cos x$, $dt = -\sin x\, dx$ we obtain

$$\int \cos^{2} x \sin x\, dx = -\int t^{2}\, dt = -\frac{t^{3}}{3} + C = -\frac{\cos^{3} x}{3} + C.$$

Example 7. Compute $\int \sin(\omega t + \theta)\, dt$, where ω and θ are constants.

The substitution

$$u = \omega t + \theta$$

gives

$$du = \omega\, dt \quad \text{or} \quad dt = \frac{1}{\omega}\, du.$$

Hence

$$\int \sin(\omega t + \theta)\, dt = \int (\sin u)\,\frac{1}{\omega}\, du = -\frac{1}{\omega}\cos u + C = -\frac{1}{\omega}\cos(\omega t + \theta) + C.$$

Example 8. Find $\int_0^1 x\sqrt{1 + 2x}\, dx$.

The computation of this definite integral proceeds in two steps:
(*a*) First, an indefinite integral is found. This was done in this case in example 4 above, where we found

$$\int x\sqrt{1 + 2x}\, dx = \tfrac{1}{15}\sqrt{1 + 2x}(6x^2 + x - 1) + C.$$

(*b*) Second, the indefinite integral is used with the limits 0 and 1 to compute the desired result:

$$\int_0^1 x\sqrt{1 + 2x}\, dx = \left[\tfrac{1}{15}\sqrt{1 + 2x}(6x^2 + x - 1)\right]_0^1$$

$$= [\tfrac{1}{15}\sqrt{3}(6)] - [\tfrac{1}{15}\sqrt{1}(-1)] = \tfrac{1}{15}(6\sqrt{3} + 1).$$

Example 9. Find $\int_0^{\pi/2} \sin 2x\, dx$.

If we let $2x = t$, then $dx = \tfrac{1}{2}dt$, and

$$\int \sin 2x\, dx = \frac{1}{2}\int \sin t\, dt = -\frac{1}{2}\cos t + C = -\frac{1}{2}\cos 2x + C.$$

Using this result, we obtain

$$\int_0^{\pi/2} \sin 2x\, dx = \left[-\frac{1}{2}\cos 2x\right]_0^{\pi/2} = -\frac{1}{2}(-1) + \frac{1}{2}(1) = 1.$$

In this introductory chapter on integration we have considered only definite integrals $\int_a^b f(x)\, dx$ for which $f(x) \geq 0$ between a and b and for which $a < b$. Textbooks on the calculus show how the definitions are extended for functions $f(x)$ that are negative and for values of a and b for which $b \geq a$. These extended definitions require some adjustment of the interpretation of $\int_a^b f(x)\, dx$ as an area. The problems that are faced in making this extension are not difficult, and therefore the student may wish to investigate this problem.

Exercises

Find the following indefinite integrals:

1. $\displaystyle\int x^4\,dx.$ **2.** $\displaystyle\int x^6\,dx.$ **3.** $\displaystyle\int 4r^3\,dr.$

4. $\displaystyle\int 9x^2\,dx.$ **5.** $\displaystyle\int \frac{dx}{x^2}.$ **6.** $\displaystyle\int \frac{5\,du}{u^3}.$

7. $\displaystyle\int \sqrt[3]{t^2}\,dt.$ **8.** $\displaystyle\int r\sqrt{r}\,dr.$ **9.** $\displaystyle\int (x^2 - 2x + 1)\,dx.$

10. $\displaystyle\int (x^3 - 6x^2 + 9x - 4)\,dx.$ **11.** $\displaystyle\int (a^2 + bt - t^2)\,dt.$

12. $\displaystyle\int \left(\sqrt{5r} - 3r^2 + \frac{6}{r^2}\right)dr.$ **13.** $\displaystyle\int \left(\frac{a}{x^6} + \frac{b}{x^4} + \frac{c}{x^2}\right)dx.$

14. $\displaystyle\int (x^{\frac{2}{3}} - 4x^{\frac{3}{4}})\,dx.$ **15.** $\displaystyle\int (x - 1)^3\,dx.$

16. $\displaystyle\int (x^2 + 3)^2\,dx.$ **17.** $\displaystyle\int \sqrt{3x + 2}\,dx.$

18. $\displaystyle\int \frac{dx}{(x - 1)^3}.$ **19.** $\displaystyle\int \frac{dx}{(5x + 8)^4}.$

20. $\displaystyle\int \sqrt[5]{2x + 9}\,dx.$ **21.** $\displaystyle\int (3\sin x - 4\cos x)\,dx.$

22. $\displaystyle\int \left(\frac{\sin x}{2} + \frac{\cos x}{3}\right)dx.$ **23.** $\displaystyle\int \left(\frac{\sin \theta}{6} + 10\right)d\theta.$

24. $\displaystyle\int \left(\frac{5}{3}\cos \theta - \frac{1}{2}\theta^4\right)d\theta.$ **25.** $\displaystyle\int x(x + 1)^2\,dx.$

26. $\displaystyle\int x^2(3x - 2)^2\,dx.$ **27.** $\displaystyle\int \frac{x\,dx}{\sqrt{x^2 + 2}}.$

28. $\displaystyle\int \frac{x^2\,dx}{\sqrt{3 + x^3}}.$ **29.** $\displaystyle\int x\sqrt{x^2 - 5}\,dx.$

30. $\displaystyle\int \frac{2x^3\,dx}{(x^4 + 2)^2}.$ **31.** $\displaystyle\int \frac{8r\,dr}{(2r^2 + 1)^3}.$

32. $\displaystyle\int (6t - 5)^8\,dt.$ **33.** $\displaystyle\int \sin 3x\,dx.$

34. $\displaystyle\int 4\cos 2\pi x\,dx.$ **35.** $\displaystyle\int (2\sin 2\theta + 3\cos 2\theta)\,d\theta.$

36. $\displaystyle\int \sin\left(377t + \frac{\pi}{6}\right) dt.$

37. $\displaystyle\int [3\cos(2t+4) + 5\cos(4t-6)]dt.$

38. $\displaystyle\int (2 - 4\sin\pi t + 6\cos 2\pi t)\, dt.$

39. $\displaystyle\int x\sqrt{4x - 2}\; dx.$

40. $\displaystyle\int \frac{x\,dx}{\sqrt{x+5}}.$

41. $\displaystyle\int \frac{(t^3 + 1)\,dt}{\sqrt{t^4 + 4t}}.$

42. $\displaystyle\int 4\sin^3\theta\cos\theta\, d\theta.$

Find the following definite integrals:

43. $\displaystyle\int_0^5 x^3\, dx.$

44. $\displaystyle\int_2^4 18x^2\, dx.$

45. $\displaystyle\int_{-1}^2 (3x^2 - 5x + 7)\, dx.$

46. $\displaystyle\int_0^3 (x+4)^2\, dx.$

47. $\displaystyle\int_1^4 \left(t + 2\sqrt{t} - 3\frac{1}{\sqrt{t}}\right) dt.$

48. $\displaystyle\int_\pi^{2\pi} \sin\theta\, d\theta.$

49. $\displaystyle\int_0^{\pi/2} (\sin x + \cos x)\, dx.$

50. $\displaystyle\int_{-2}^2 (6 - x^2)\, dx.$

51. $\displaystyle\int_0^m (3m^3 - x^3)\, dx,\; m > 0.$

52. $\displaystyle\int_0^6 \frac{x\,dx}{\sqrt{x^2 + 13}}.$

53. $\displaystyle\int_1^2 x^3(x^4 + 2)^2\, dx.$

54. $\displaystyle\int_{-2}^3 \frac{(3x^2 - 7)\,dx}{\sqrt{x^3 - 7x + 9}}.$

55. $\displaystyle\int_0^{\pi/\omega} \sin\omega t\, dt.$

56. $\displaystyle\int_0^{\pi/2} 2\sin^3\theta\cos\theta\, d\theta.$

57. $\displaystyle\int_{-\pi/2}^{\pi/2} \sin^2\theta\cos^3\theta\, d\theta.$

58. $\displaystyle\int_2^8 \frac{dr}{r\sqrt{2r}}.$

59. A *cycloid* is a curve generated by a point P on the circumference of a circle which rolls along a straight line. One complete revolution of the circle which starts with P on the straight line and ends with P in this same position creates what is called one *arch* of the cycloid. The length s of one arch is given by the formula

$$s = 2r \int_0^{2\pi} \sin\frac{\theta}{2}\, d\theta,$$

where r is the radius of the circle and θ is the angle through which the circle has turned. Find the length of one arch of a cycloid generated by a circle having a radius of 4 in.

60. The area bounded by one arch of the cycloid described in the previous exercise and the straight line is given by the formula

$$A = r^2 \int_0^{2\pi} (1 - \cos\theta)^2\, d\theta.$$

Evaluate this integral.

20–4. Applications of the Integral Calculus

The integral calculus has many applications in science and engineering. We have already considered the application to the problem of computing an area; we shall discuss two additional applications in this section.

There are two types of problems to which the methods of the integral calculus apply. The solution to problems of the first type is furnished by the definite integral; the solution to problems of the second type is expressed by an indefinite integral. In problems of the second type data are frequently given that determine a particular value of the constant of integration.

As an example of the first type we have considered the problem of finding an area, and in this section we shall consider the problem of

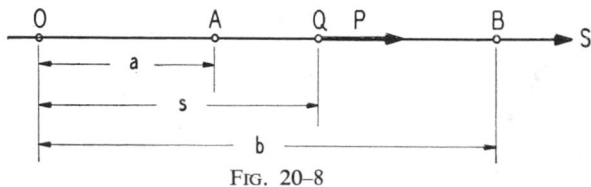

Fig. 20–8

finding the work done by a variable force. As an example of the second type we shall consider the motion of an object when the force acting on the object is known.

Let us consider an object on a number line under the influence of a force acting along the line. In many situations the force will depend on the position of the object at a point corresponding to a number s, and can therefore be thought of as a function $F(s)$. It is usual to adopt the convention of giving this function a positive sign when the force is acting in the positive direction along the line and a negative sign when the force is acting in the negative direction. For example, if one electric charge is at the origin and another is at the number $s > 0$ on the line, the second charge being attracted to the first, then it can be shown in a laboratory that the force acting on the second charge is $F(s) = -\dfrac{k^2}{s^2}$, where k is a constant, the negative sign being chosen because the force acts in the negative direction along the number line.

Consider now a positive, constant force P acting on an object that moves on the number line from A to B, as shown in Fig. 20–8. Then the **work** is defined as

$$P(b - a). \tag{1}$$

More generally, for any positive or negative constant force P acting along the line on an object that moves from any initial point at *a* to any terminal point at *b* (regardless of whether *b* is to the right or left of *a*), the work is defined by expression 1.   Thus the work is positive if *P* and $(b - a)$ are both positive or both negative.   In other words, the work is positive if the motion is in the direction of the force, negative otherwise. In the first case we say that *the force does work*, in the second case we say that *the work is done against the force*.   For example, if a weight is lifted upward the work done by the downward force of gravity is negative; i.e., the work is done against gravity.   On the other hand, the person doing the lifting exerts a force in the direction of the lift, and therefore the work done by this force is positive.

Consider now the case where the force is given by a function $F(s)$ which is not a constant, and suppose that the object is moved from *A*

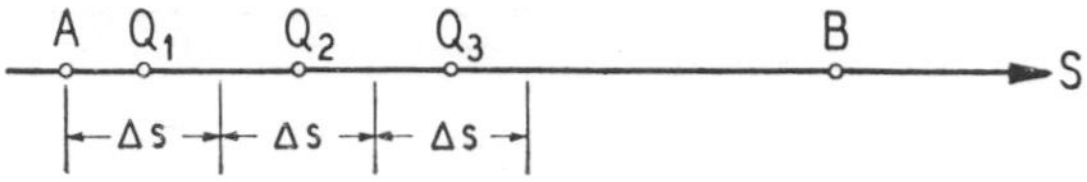

FIG. 20–9

to *B* in Fig. 20–8.   The method used to compute the work in this case will be patterned after the method used in Sec. 20–1 to compute areas. First divide the distance *AB* into a number of equal segments of length Δs (Fig. 20–9).   If the length of these segments is small, it may be assumed that the force is nearly constant along each segment.   If Q_1 is an arbitrary point on the first segment whose distance from *A* is s_1, then the force at all the other points of this segment has approximately the same value $P_1 = F(s_1)$ as at Q_1, and therefore the work along this segment is roughly

$$F(s_1) \cdot \Delta s.$$

In the same way it may be inferred that the work along each of the following segments is

$$F(s_2) \cdot \Delta s, \qquad F(s_3) \cdot \Delta s, \qquad \cdots$$

and that therefore the total work done along the distance *AB* is approximately

$$F(s_1) \cdot \Delta s + F(s_2) \cdot \Delta s + F(s_3) \cdot \Delta s + \cdots.$$

The work can be computed to any desired degree of precision by dividing the distance *AB* into an increasing number of parts.

This discussion becomes identical with that of Sec. 20–1 if the graph of the function $P = F(s)$ is plotted (Fig. 20–10).   Hence we may state that the work done if the object moves from A to B is given by the shaded area and has, therefore, the value

$$W = \int_a^b F(s)\, ds.$$

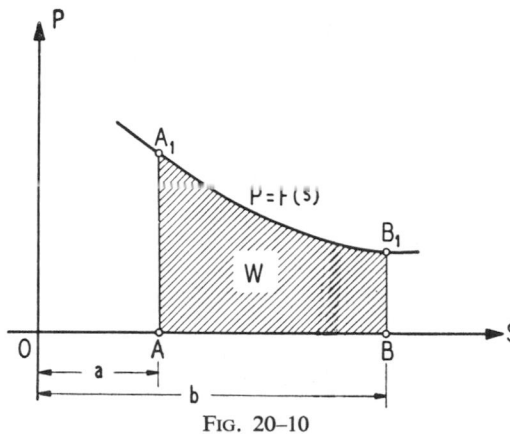

FIG. 20–10

Thus the following rule may be stated.

In order to find the work done by a variable force on a point moving in a straight line, choose on this line a point of reference O from which the distance s of the variable point is measured. For each point of the line find as a function of s the force P acting at the point, obtaining

$$P = F(s).$$

Then the work done during the motion from the point $s = a$ to the point $s = b$ is given by the definite integral

$$W = \int_a^b F(s)\, ds.$$

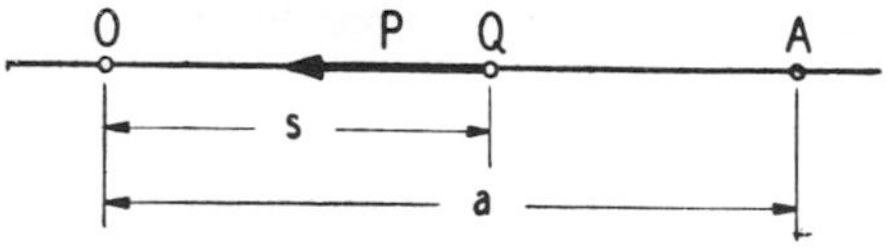

FIG. 20–11

Example 1. An object is moving on a line under the influence of a force which is proportional to the distance s between the object and the point O of reference and directed to the point O. What is the work required to move the object from O to a point A so that $OA = a$?

The force P is directed toward O as shown by the vector in Fig. 20–11, whence P is negative. Since P is proportional to s and s is positive,

$$P = F(s) = -k^2 s$$

where k^2 is a positive constant. The force which must be applied from O to the right is therefore a positive one equal in absolute value to P or k^2s. Hence the work done in moving the object from O to A is

$$\int_0^a k^2s \, ds = \left[k^2 \frac{s^2}{2} \right]_0^a = \frac{1}{2} k^2 a^2.$$

As an example of the second type of problem mentioned in this section, we shall consider the motion of an object when the force producing the motion is a known function of the time. Let us assume that a small object Q of mass m is moving along a straight line (Fig. 20–12) and that its instantaneous position is defined by its distance s from a point of reference O. Since the object moves, s may be regarded as a function of the time t that has elapsed since the motion started. In Chapter 19

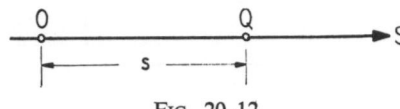

FIG. 20–12

it was shown that the velocity v of the moving object Q, being the instantaneous rate of change of s with respect to the time, is given by the derivative

$$v = \frac{ds}{dt},$$

and that the acceleration a of the object, being the instantaneous rate of change of the velocity with respect to the time, is

$$a = \frac{dv}{dt} = \frac{d^2s}{dt^2}.$$

A fundamental law of mechanics states that the force F acting on an object, provided proper units for all the quantities involved are chosen, is given by the product of mass and acceleration:

$$F = ma.$$

If F is known as a function of t, this formula can be written as

$$m \frac{dv}{dt} = F(t) \quad \text{or} \quad \frac{dv}{dt} = \frac{1}{m} F(t).$$

By this relation v is a function of t whose derivative is known, and therefore v can be found by computing an indefinite integral. Since $\frac{ds}{dt} = v$, s can be found by computing another indefinite integral after v has been found.

Since the mass m is always considered to be positive, the relation $F = ma$ implies that the acceleration has the same sign as the force. Thus the acceleration is positive if the force is acting in the positive direction and negative if the force is acting in the negative direction.

Example 2.   An object is thrown vertically upward from the surface of the earth with an initial velocity v_0.   If the object is acted on only by the force of gravity, find its distance above the earth as a function of the time t that has elapsed since the object was thrown, it being given that the acceleration produced by the force of gravity is $g = 32.2$ ft per sec per sec.

Let us first choose a coordinate system consisting of a vertical number line with positive direction upward and origin at the point from which the object is thrown.   If $s = s(t)$ denotes the position of the object t sec after being thrown, then $s(0) = 0$.   Since the force of gravity is acting downward,

$$F = ma = m\frac{d^2s}{dt^2} = -gm = -32.2m, \tag{2}$$

where m is the mass of the object.   Since m is positive, the negative sign is chosen, as F, which is acting downward, is negative.   From equation 2 we obtain

$$a = \frac{d^2s}{dt^2} = \frac{dv}{dt} = -32.2,$$

from which

$$v = -\int 32.2\, dt = -32.2t + C,$$

where C is the constant of integration.   This constant has in this problem a value that can be determined by using the information that when $t = 0$, $v = v_0$. Substituting these values in the formula for v, we obtain

$$v_0 = -32.2 \cdot 0 + C,$$

and therefore

$$v = C - 32.2t = v_0 - 32.2t.$$

In order to find s we use the formula

$$\frac{ds}{dt} = v = v_0 - 32.2t,$$

from which we find that

$$s = \int (v_0 - 32.2t)\, dt = v_0 t - 16.1t^2 + C'.$$

In order to find the value that has to be ascribed to C' for the purpose of solving the present problem, we use the fact that, for $t = 0$, the moving object was at $s = 0$.   Substituting these values in the formula for s, we obtain $C' = 0$, whence

$$s = v_0 t - 16.1t^2.$$

At what time does the object return to earth?   We may answer this question by noticing that the object is at the earth when $s = 0$, and that therefore we can find when the object is at the earth by solving the equation

$$v_0 t - 16.1t^2 = 0.$$

We find that $t = 0$ and $v_0/16.1$. The first result we identify as the time of starting, and the second as the time at which the object returns to earth.

To find the values of the constants of integration in this problem, we used the conditions that were known for the beginning of the motion. These conditions are called **initial conditions.** In particular, we used the fact that, when $t = 0$, the velocity is v_0, and the fact that, when $t = 0$, the distance s from the starting point is zero. Initial conditions of this kind can always be used to find the constants of integration in problems of this type.

It is important in the specific problems involving the applications of the type introduced in this section to note carefully the units used in measuring each quantity that is considered. If distance is measured in feet and time in seconds, then velocity, which arises as a distance divided by time, is measured in feet per second, and acceleration, which arises as a velocity divided by time, is measured in feet per second per second. Similarly, if distance is measured in feet and force in pounds, then work, which arises as the product of a distance and a force, is measured in a unit called the foot-pound, the amount of work required to lift one pound a distance of one foot. Other basic units of distance, time, and force lead to similar units for velocity, acceleration, and work.

Exercises

Find the areas with the boundaries given in exercises 1–10, plotting a graph of each.

1. The curve $y = 9x^2$, the x axis, and the lines $x = 2$ and $x = 3$.
2. The curve $y = 4x - x^2$ and the x axis.
3. The curve $y = x^3 - 3x + 2$ and the x axis.
4. The curve $y = 16 - x^4$ and the x axis.
5. The curve $y = \sin x$ and the portion of the x axis between 0 and π.
6. The curve $y = \cos x$ and the portion of the x axis between $-\pi/2$ and $\pi/2$.
7. The portion of the curve $y = \sin x$ between $x = 0$ and $x = \pi$ and the line $y = \frac{1}{2}$.
8. The curve $y = 4x - x^2$ and the line $y = x$.
9. The curve $y = x^2$ and the line $y = x + 2$.
10. The curve $y = x^3$ and the line $y = 8$.

11. The force that acts on a particle on a line is proportional to the square of its distance from the origin and is directed to the origin. Find the work required to move the particle from a to b, where $0 < a < b$. What is the work required to move the particle from c to d, where $c < d < 0$?

12. The force that acts on a particle on a line is proportional to the nth power, where $n > 0$, of its distance from the origin and is directed to the origin. Find the work required to move the particle from a to b, where $0 < a < b$. What is the work required to move the particle from c to d, where $c < d < 0$?

13. If one electric charge is at the origin of a number line and another charge is at a distance s from the first charge on the line, and if there is an attractive force between these charges (rather than a force of repulsion), then the force acting on the second charge is proportional to $1/s^2$. How much work is required to move the charge from a point a on

the line to a point b, where $0 < a < b$?   How much work is required to move the charge from a to infinity in the positive direction?

14. Solve exercise 13 under the assumption that the force acting on the second charge is proportional to $1/s^n$, where $n > 1$.

Hooke's law for spiral springs states that the force exerted by a spring when stretched or compressed is proportional to the distance the spring is stretched or compressed, respectively.   Thus, if F is the force exerted by the spring and s is the amount by which the spring is stretched or compressed, then $F = ks$, where k is the constant depending on the physical characteristics of the spring.   The constant k can be determined if a force F_0 and the corresponding stretching or compression s_0 are known.   Then $k = \dfrac{F_0}{s_0}$ and $F = \dfrac{F_0}{s_0} s$.   Hooke's law is used in exercises 15–20 below.

15. If a spring exerts a force of 5 lb when stretched 1 in., how much work must be done to stretch the spring 4 in.?

16. If a spring exerts a force of 25 lb when stretched 10 in., how much work must be done to stretch the spring 20 in.?

17. A spring whose normal length is 24 in. exerts a force of 10 lb when stretched 2 in. Find the work done in stretching the spring from a length of 28 in. to a length of 32 in.

18. A spring whose normal length is 8 in. exerts a force of 4 lb when compressed to a length of 6 in.   Find the work done in compressing the spring from a length of 6 in. to a length of 2 in.

19. A spring exerts a force of 12 lb when stretched 2 in.   How much longer than its normal length can this spring be stretched by 9 ft-lb of work?

20. If the spring of exercise 19 is already stretched 3 in., how much further can it be stretched by an additional 1.75 ft-lb of work?

In exercises 21 through 30 assume that the acceleration of gravity is 32.0 ft per sec per sec and that the objects dealt with in each exercise are acted on only by the force of gravity.

21. A baseball is thrown straight upward with an initial velocity of 96 ft per sec.   How high does it go?   How long is it in the air?

22. If a stone is thrown straight upward with an initial velocity of 144 ft per sec, how long does it take the stone to reach its highest point?   How long is the stone in the air?

23. An artillery shell is shot straight upward with an initial velocity of 1600 ft per sec. How long does it take the shell to reach an altitude of 30,000 ft?   What is the maximum altitude reached by the shell?

24. If the initial velocity of the shell in exercise 23 is doubled, by what factors are the time to reach 30,000 ft and the maximum altitude reached by the shell multiplied?

25. If a ball is dropped from a height of 144 ft above the earth, how long before it hits the ground?

26. What is the time of fall of a bomb dropped from an altitude of 30,000 ft if air resistance is neglected?   What is the terminal velocity of the bomb, i.e., the velocity of the bomb when it strikes the earth?

27. A rocket is shot straight downward from an altitude of 30,000 ft above the earth with an initial velocity of 800 ft per sec.   How long does it take the rocket to hit the earth?   What is the terminal velocity of the rocket, i.e., the velocity of the rocket when it hits the earth?

28. With what initial velocity must a baseball be thrown upward if it is desired that the maximum height it attains should be 144 ft?

29. What is the minimum muzzle velocity (i.e., initial velocity) that can be given to a shell if it is desired that the shell reach an altitude of at least 40,000 ft when fired straight upward?

30. What is the muzzle velocity (i.e., initial velocity) of a shell that has a velocity of 64 ft per sec at an altitude of 40,000 ft, if the shell is fired straight upward from the surface of the earth?

31. Find the area in the first quadrant bounded by the x axis, the y axis, and the curve $y = x^3 - 6x^2 + 8x + 3$.

32. Find the area above the x axis bounded by the x axis and the curve $y = x^3 - 6x^2 + 3x + 10$.

33. Find the area in the first quadrant bounded by the x axis, the y axis, the line $x = 4$, and the curve $y = 2x(4 + x^2)^{-2}$.

34. Find the area bounded by the curves $x^2 - 4y = 0$ and $x - 2y + 4 = 0$.

35. Find the area bounded by the curves $x - 2y + 7 = 0$ and $x^2 - 2x - 4y + 9 = 0$.

36. The weight w in pounds of a body is inversely proportional to the square of the distance x in miles of the body from the center of the earth, or $w = \dfrac{k}{x^2}$, where k is a positive constant. If the radius of the earth is 4000 miles, find the work in mile-pounds required to fire a rocket weighing 8700 lb from the surface of the earth to a point 120 miles above the surface of the earth.

37. A positive electric charge of Q electrostatic units which is concentrated in a small sphere exerts a force of repulsion of Q/r^2 dynes on a positive electrostatic unit at the distance of r cm. Compute the work done by the electric force if the unit charge moves from a distance of 5 cm to a distance of 12 cm from the charge Q. (The work of a force of 1 dyne along a distance of 1 cm is called an erg.)

38. A dredge hoists a scoop filled with sand and water through a distance of 36 ft. Find the work in foot-pounds required to hoist the load if water leaks out of the scoop at such a rate that the total weight P in pounds of the load is given by the expression $P = 2620 - \dfrac{x^2}{3}$, where x in feet is the height through which the scoop has been hoisted.

39. An object starting from rest at the origin and moving in a straight line has an acceleration in feet per second per second given by the relation $a = \dfrac{t}{6} - \dfrac{2}{3}$, t being the time in seconds after starting. Find expressions for the speed and the position of the object at the end of t_1 sec.

40. A relief parcel is released from a helicopter and falls with a constant acceleration $g = 32.2$ ft per sec per sec. How high is the helicopter if the time of descent of the parcel is 1.8 sec?

41. A particle of water in a geyser is thrown vertically upward with an initial speed of 78 ft per sec. If friction and air resistance are neglected, and if the acceleration due to gravity is 32.2 ft per sec per sec, find the time required for the water to reach the highest point in its travel and the maximum height to which the water rises.

42. A sinusoidal force acting along a straight line may be given by the formula $P = F \sin \omega t$, where F is a positive constant and t is time. If this force is acting on a small object of mass m and if the object starts from rest at $t = 0$, find a formula for its position on the line at any time t, and show that the velocity of the object oscillates between 0 and $\dfrac{2F}{m\omega}$.

Progress Report

This chapter is concerned with the connection between computing an area and finding a function whose derivative is given. A few consequences of this connection are also considered.

The chapter first considers the approximate computation of an area. For practical approximate computations the trapezoidal formula is derived. The definite integral $\int_a^b f(x)\,dx$ for $f(x) \geq 0$ is defined as the area bounded by the curve $y = f(x)$, the x axis, and the lines $x = a$ and $x = b$, this area being approximated as closely as we please by a sum of rectangles. Then the connection between the definite integral and the problem of finding a function whose derivative is given is established in this important theorem:

If $f(x)$ is given and $F(x)$ is a function such that

$$\frac{dF(x)}{dx} = f(x),$$

then

$$\int_a^b f(x)\,dx = F(b) - F(a). \tag{1}$$

Since this theorem states that the definite integral can easily be evaluated by means of equation 1 if a function $F(x)$ can be found whose derivative is $f(x)$, the next problem considered was how to find a function whose derivative is given, that is, how to find the indefinite integral of a function. A few simple techniques of finding indefinite integrals were studied.

The chapter concluded with a discussion of the application of the definite integral to problems involving work and the application of the indefinite integral to problems involving the motion of an object on a line when the force acting on the object is known.

Appendix———————————————

Review of Fundamentals
of Geometry

In the following pages we shall briefly summarize a few of the terms and facts from geometry that have been used in the discussions in this book.

In plane geometry, which deals entirely with configurations in the plane, there are certain fundamental conceptions which we must assume from experience, without definition. These include the notion of a **point,** a **straight line,** and an **area.** Understanding that a straight line extends infinitely far in both directions, we define a **ray** to be a point on a line with the part of the line on one side of the point and a **segment** to be that part of a straight line between two points on the line. It is assumed that one and only one straight line can be drawn through two points.

Experience leads us to believe that two straight lines can meet, at most, in one point, called the **intersection** of the lines. The **angle** formed by two intersecting lines has been defined in Chapter 4. The intersection of the lines is called the **vertex** of the angle. In Chapter 4 the **degree,** a unit for the measurement of angles, was also introduced. Special names are sometimes used to refer to angles classified according to their magnitude as follows:

Size of Angle	Term	
90°	**Right angle**	
180°	**Straight angle**	
Between 0° and 90°	**Acute angle**	**Oblique angles**
Between 90° and 180°	**Obtuse angle**	

If two angles have a sum of 90°, they are called **complementary** angles. If two angles have a sum of 180°, they are called **supplementary** angles.

If two intersecting lines form a right angle, the lines are said to be **perpendicular.** If two lines do not intersect, they are said to be **parallel.** There are many theorems concerning parallel and perpendicular lines. The most important are:

1. Two lines that are perpendicular to the same line are parallel.

2. Two lines that are parallel to the same line are parallel.

3. If two parallel lines are cut by a third line (Fig. A–1), the **corresponding angles** are equal (for example, $\angle 1 = \angle 5$, $\angle 2 = \angle 6$, $\angle 3 = \angle 7$, $\angle 4 = \angle 8$), the **alternate interior angles** are equal (for example, $\angle 3 = \angle 6$, $\angle 4 = \angle 5$), and the **alternate exterior angles** are equal (for example, $\angle 1 = \angle 8$, $\angle 2 = \angle 7$).

From Fig. A–1, we see that $\angle 1 = \angle 4$, $\angle 2 = \angle 3$, etc. Such pairs of angles are called **vertical angles.**

4. Vertical angles are equal.

5. Two angles having their sides respectively parallel or perpendicular are either equal or supplementary.

If three or more intersecting straight lines form a figure that completely encloses a portion of the plane, that figure is called a **polygon.** The

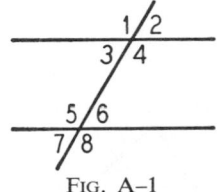

Fig. A–1

line segments that bound the enclosure are called the **sides** of the polygon, and the sum of their lengths is called the **perimeter** of the polygon. The **area of a polygon** is the area of the portion of the plane enclosed by the polygon. If two polygons have their corresponding angles and sides equal, they are said to be **congruent.** If two polygons have their corresponding angles equal, they are said to be **similar.**

A **triangle** is a polygon having three sides. The sides of a triangle taken two at a time form the **three angles of the triangle.** An exterior angle of a triangle is an angle formed by one side of the triangle and the extension of another side through their point of intersection. An **altitude** of a triangle is the perpendicular line drawn from any vertex to the opposite side. A triangle one of whose angles is a right angle is a **right triangle.** A triangle none of whose angles are right angles is an **oblique triangle.** A triangle two of whose sides are equal in length is an **isosceles triangle.** A triangle all three of whose sides are equal is an **equilateral triangle.**

If the corresponding sides and angles of two triangles are equal, the triangles are **congruent.** If the corresponding angles of two triangles are equal, the triangles are **similar.**

Some important theorems concerning triangles are:

1. The sum of the lengths of any two sides of a triangle is greater than the length of the third side.

2. If two sides of a triangle are unequal, the angle opposite the longer side is greater than the angle opposite the shorter side, and conversely.

3. The sum of the angles of a triangle is 180°.

4. The corresponding sides of similar triangles are proportional.

5. Two triangles are congruent if:

(*a*) Two sides and the included angle of one are equal, respectively, to two sides and the included angle of the other.

(*b*) Two angles and a side of one are equal, respectively, to two angles and the corresponding side of the other.

(*c*) Three sides of one are equal, respectively, to the three sides of the other.

6. Two triangles are similar if:

(*a*) Two angles of one are equal, respectively, to two angles of the other.

(*b*) An angle of one equals an angle of the other and the including sides are proportional.

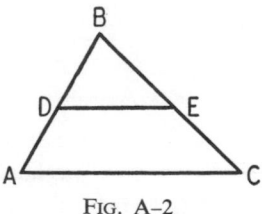

Fig. A–2

7. In an isosceles triangle, the angles opposite the equal sides are equal.

8. The angles of an equilateral triangle are each equal to 60°.

9. If *DE* is parallel to *AC* in Fig. *A*–2, then triangle *BDE* is similar to triangle *BAC*.

The side opposite the right angle in a right triangle is called the **hypotenuse.** The other two sides are the **legs** of the triangle. The following theorems about right triangles are useful:

1. If a triangle is a right triangle, then the square of the length of the hypotenuse is equal to the sum of the squares of the lengths of the other two sides, and conversely. This theorem is called the theorem of **Pythagoras.**

2. If one of the acute angles of a right triangle is 30°, then the side opposite that angle has a length equal to one-half the hypotenuse, and the legs and hypotenuse are proportional, respectively, to 1, $\sqrt{3}$, 2.

3. If one of the acute angles of a right triangle is 45°, the legs are equal, and the legs and hypotenuse are proportional, respectively, to 1, 1, $\sqrt{2}$.

4. Two right triangles are congruent if:

(*a*) Two sides of one are equal, respectively, to two sides of the other.

(*b*) An acute angle and a side of one are equal, respectively, to an acute angle and the corresponding side of the other.

5. Two right triangles are similar if:

(*a*) An acute angle of one equals an acute angle of the other.

(*b*) Two sides of one are proportional, respectively, to the corresponding sides of the other.

6. If ABC is a right triangle with hypotenuse AB, then CD, the altitude to the hypotenuse, forms two right triangles ACD, BCD such that $\triangle ACD$ is similar to $\triangle BCD$, $\triangle ACD$ is similar to $\triangle ABC$, $\triangle BCD$ is similar to $\triangle ABC$.

A polygon having four sides is a **quadrilateral.** A quadrilateral that has two and only two sides parallel is called a **trapezoid.** The parallel sides are called the **bases.** A **parallelogram** is a quadrilateral that has both pairs of opposite sides parallel. Either pair of parallel sides may be called **bases.** The perpendicular distance between two parallel sides of a parallelogram or trapezoid is called an **altitude.** A **rectangle** is a parallelogram having four right angles. A **square** is a rectangle that has its adjacent sides equal. The most important facts about quadrilaterals are:

1. If the opposite sides of a quadrilateral are equal, then the quadrilateral is a parallelogram; and, conversely, if a quadrilateral is a parallelogram, then its opposite sides are equal.

2. If the opposite angles of a quadrilateral are equal, then it is a parallelogram; and conversely.

3. If one pair of opposite sides of a quadrilateral are equal and parallel, then it is a parallelogram.

Many questions in plane geometry and in other more advanced work are simplified somewhat by use of the notion of the locus of a point. If C is a given condition and S is a set of points such that (*a*) every point that satisfies C is a point of S, and (*b*) every point that does not satisfy C is not in S, then S is called the **locus of a point satisfying C.**

The locus of a point in the plane that satisfies the condition that the point is at a fixed distance from a fixed point in the plane is called a **circle** (Fig. A–3). The fixed point is the **center** of the circle. The length of a circle is called the **circumference,** and the **area** of a circle is the area of the portion of the plane enclosed by the circle. Either of the two parts of a circle between two points on the circle is called an **arc** of the circle. A **tangent** is a line that touches a circle in one and only one point, called the **point of tangency.** A **secant** is a line that cuts a circle in two points. A **chord** is a line segment joining two points on a circle. A **diameter** is a chord that passes through the center of a circle, and a **radius** is a line segment from the center of a circle to any point on the circle.   Any arc of a circle that is cut off by the two sides of an angle is called an **arc intercepted by** or **subtended by the angle.**   An angle formed by two radii of a circle is called a **central angle.**   The figure formed by a central angle and its intercepted arc is called a **sector** of the circle.   An **inscribed angle** is an angle formed by two chords that meet at a point on the circle.

In the circle in Fig. A–3, O is the center, and the portion of the circle between A and B, or between A and F, is an arc.   The line KH is a tangent with point of tangency D; the line IJ is a secant; EC and FB

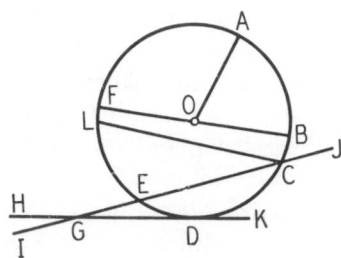

FIG. A–3

are chords; FB is also a diameter; OF, OA, and OB are radii; CD and ED are arcs intercepted by angle KGJ; AB is the arc intercepted by the central angle BOA; and ABO is a sector. Angle LCE is an inscribed angle.

Some of the more useful theorems concerning the circle are:

1. A tangent is perpendicular to the radius drawn to the point of tangency.

2. An inscribed angle that intercepts a semicircle is a right angle.

3. On the same circle or on circles with the same radius, central angles are proportional to the lengths of their intercepted arcs.

4. The length of a chord of a circle is less than or equal to the length of the diameter.

5. The ratio of the circumference to the length of a diameter of a circle is the same for all circles.   This ratio, which is denoted by π, is an irrational number having approximately the value 3.14159.

Solid geometry deals with configurations in space.   A few of its terms and facts are summarized in the following paragraphs.

Any straight line passing through two points of a plane lies entirely in the plane.   A plane can be determined by three points, by two intersecting lines, by a line and a point not on the line, or by two parallel lines. If two planes intersect, their intersection is a straight line.

A line L is perpendicular to a plane if every line in the plane that passes through the intersection of L and the plane is perpendicular to L. Lines that are perpendicular to the same plane are parallel. Planes that are perpendicular to the same or parallel lines are parallel to each other.

If a line and a plane do not intersect, they are said to be **parallel.** If two planes do not intersect, they are said to be **parallel.** Let two planes, P_1 and P_2, intersect in a line L, and let L_1 be a line in P_1 which meets L at P and is perpendicular to L; let L_2 be a line in P_2 which meets L at P and is perpendicular to L. Then the angles between L_1 and L_2 are defined to be the **angles between the planes** P_1 **and** P_2; if L_1 and L_2 are perpendicular, the planes P_1 and P_2 are said to be perpendicular.

If C is a plane curve and L is a line that passes through C but does not lie in the plane of C, the surface found by moving L about C always parallel to its original position is called a **cylindrical surface.** Each position of the moving line L is called an **element** (or **ruling**) of the surface. All the elements are parallel. When two parallel planes cut all the elements of a cylindrical surface, a **cylinder** is formed and the parallel sections are called **bases** of the cylinder. A **right circular cylinder** is a cylinder having bases that are perpendicular to the elements and that cut the surface in circles. The **axis** of a right circular cylinder is the line between the centers of the bases.

If two segments in space have a common end point, the angle between their projections on a horizontal plane is called the **horizontal angle** between the segments.

FORMULAS FROM GEOMETRY

For a **triangle** with side b and altitude to that side h,
$$\text{Area} = \tfrac{1}{2}bh.$$

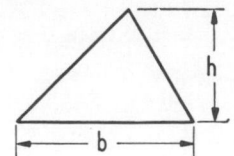

For a **parallelogram** with side b and altitude to that side h,
$$\text{Area} = bh.$$

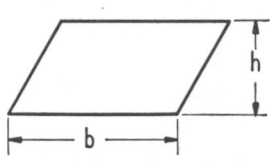

For a **rectangle** of length b and width h,
$$\text{Area} = bh.$$

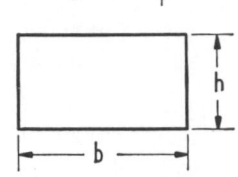

For a **square** of side b,
$$\text{Area} = b^2.$$

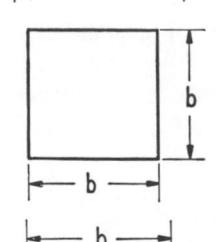

For a **trapezoid** with parallel bases b and b' and altitude h,
$$\text{Area} = \tfrac{1}{2}h(b + b').$$

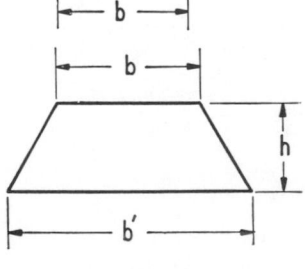

For a **circle** with radius r,
$$\text{Circumference} = 2\pi r,$$
$$\text{Area} \qquad\quad = \pi r^2.$$

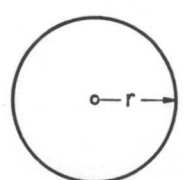

In a circle of radius r, a **sector** with central angle θ measured in radians has an
$$\text{Area} = \tfrac{1}{2}r^2\theta.$$

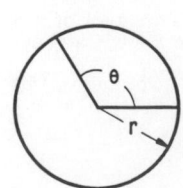

For a **rectangular solid** of length l, height h, and width w,

Total surface area = $2(lw + lh + wh)$,

Volume   = lwh.

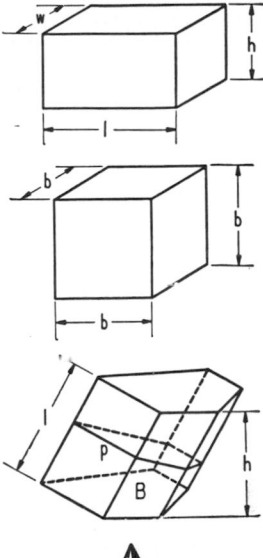

For a **cube** with the length of one edge b,

Total surface area = $6b^2$,

Volume   = b^3.

For a **prism** with lateral edge l, altitude h, perimeter of a right section p, and area of the base B,

Lateral area = pl,

Total area   = $pl + 2B$,

Volume   = Bh.

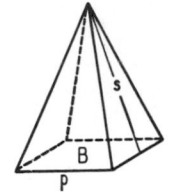

For a **regular pyramid** with slant height s, perimeter of the base p, and area of the base B,

Lateral area = $\frac{1}{2}ps$,

Total area   = $\frac{1}{2}ps + B$.

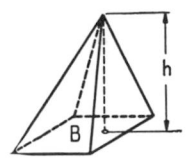

For a **pyramid** with altitude h and area of the base B.
Volume = $\frac{1}{3}Bh$.

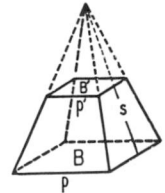

For a **frustum of a regular pyramid** with slant height s, bases of perimeter p and p', and areas B and B' respectively,

Lateral area = $\frac{1}{2}s(p + p')$,

Total area   = $\frac{1}{2}s(p + p') + B + B'$.

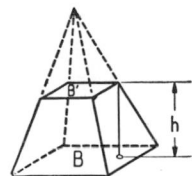

For a **frustum of a pyramid** with altitude h and areas of the bases B and B',

Volume = $\frac{1}{3}h(B + B' + \sqrt{BB'})$.

For a **cylinder** with lateral edge l, altitude h, perimeter of a right section p, and base area B,

$$\text{Lateral area} = pl,$$
$$\text{Total area} \;\; = pl + 2B,$$
$$\text{Volume} \qquad = hB.$$

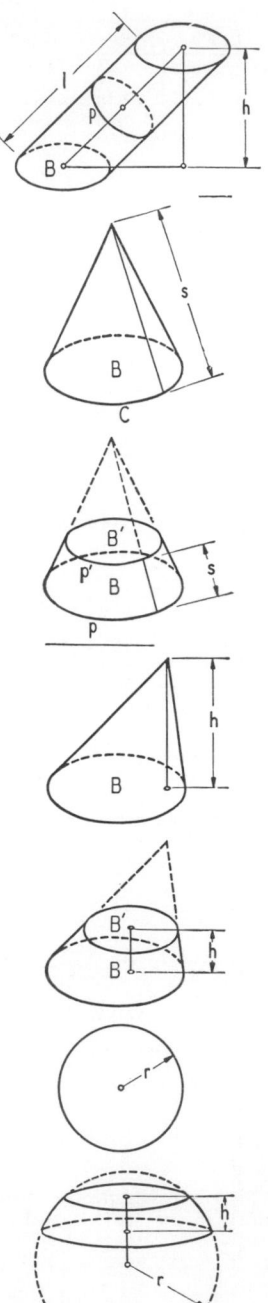

For a **right circular cone** with slant height s, circumference of the base C, and area of the base B,

$$\text{Lateral area} = \tfrac{1}{2}sC,$$
$$\text{Total area} \;\; = \tfrac{1}{2}sC + B.$$

For a **frustum of a right circular cone** with slant height s, bases of perimeter p and p' and area B and B', respectively,

$$\text{Lateral area} = \tfrac{1}{2}s(p + p'),$$
$$\text{Total area} \;\; = \tfrac{1}{2}s(p + p') + B + B'.$$

For a **cone** with altitude h and base area B,

$$\text{Volume} = \tfrac{1}{3}Bh.$$

For a **frustum of a cone** with altitude h and bases of area B and B',

$$\text{Volume} = \tfrac{1}{3}h(B + B' + \sqrt{BB'}).$$

For a **sphere** of radius r,

$$\text{Surface area} = 4\pi r^2,$$
$$\text{Volume} \qquad = \tfrac{4}{3}\pi r^3.$$

In a sphere of radius r, a **zone** with altitude h has

$$\text{Area} = 2\pi hr.$$

TABLE 1—Squares and Square Roots

N	N²	√N	√10N	N	N²	√N	√10N
1.00	1.0000	1.00000	3.16228	**1.50**	2.2500	1.22474	3.87298
1.01	1.0201	1.00499	3.17805	1.51	2.2801	1.22882	3.88587
1.02	1.0404	1.00995	3.19374	1.52	2.3104	1.23288	3.89872
1.03	1.0609	1.01489	3.20936	1.53	2.3409	1.23693	3.91152
1.04	1.0816	1.01980	3.22490	1.54	2.3716	1.24097	3.92428
1.05	1.1025	1.02470	3.24037	1.55	2.4025	1.24499	3.93700
1.06	1.1236	1.02956	3.25576	1.56	2.4336	1.24900	3.94968
1.07	1.1449	1.03441	3.27109	1.57	2.4649	1.25300	3.96232
1.08	1.1664	1.03923	3.28634	1.58	2.4964	1.25698	3.97492
1.09	1.1881	1.04403	3.30151	1.59	2.5281	1.26095	3.98748
1.10	1.2100	1.04881	3.31662	**1.60**	2.5600	1.26491	4.00000
1.11	1.2321	1.05357	3.33167	1.61	2.5921	1.26886	4.01248
1.12	1.2544	1.05830	3.34664	1.62	2.6244	1.27279	4.02492
1.13	1.2769	1.06301	3.36155	1.63	2.6569	1.27671	4.03733
1.14	1.2996	1.06771	3.37639	1.64	2.6896	1.28062	4.04969
1.15	1.3225	1.07238	3.39116	1.65	2.7225	1.28452	4.06202
1.16	1.3456	1.07703	3.40588	1.66	2.7556	1.28841	4.07431
1.17	1.3689	1.08167	3.42053	1.67	2.7889	1.29228	4.08656
1.18	1.3924	1.08628	3.43511	1.68	2.8224	1.29615	4.09878
1.19	1.4161	1.09087	3.44964	1.69	2.8561	1.30000	4.11096
1.20	1.4400	1.09545	3.46410	**1.70**	2.8900	1.30384	4.12311
1.21	1.4641	1.10000	3.47851	1.71	2.9241	1.30767	4.13521
1.22	1.4884	1.10454	3.49285	1.72	2.9584	1.31149	4.14729
1.23	1.5129	1.10905	3.50714	1.73	2.9929	1.31529	4.15933
1.24	1.5376	1.11355	3.52136	1.74	3.0276	1.31909	4.17133
1.25	1.5625	1.11803	3.53553	1.75	3.0625	1.32288	4.18330
1.26	1.5876	1.12250	3.54965	1.76	3.0976	1.32665	4.19524
1.27	1.6129	1.12694	3.56371	1.77	3.1329	1.33041	4.20714
1.28	1.6384	1.13137	3.57771	1.78	3.1684	1.33417	4.21900
1.29	1.6641	1.13578	3.59166	1.79	3.2041	1.33791	4.23084
1.30	1.6900	1.14018	3.60555	**1.80**	3.2400	1.34164	4.24264
1.31	1.7161	1.14455	3.61939	1.81	3.2761	1.34536	4.25441
1.32	1.7424	1.14891	3.63318	1.82	3.3124	1.34907	4.26615
1.33	1.7689	1.15326	3.64692	1.83	3.3489	1.35277	4.27785
1.34	1.7956	1.15758	3.66060	1.84	3.3856	1.35647	4.28952
1.35	1.8225	1.16190	3.67423	1.85	3.4225	1.36015	4.30116
1.36	1.8496	1.16619	3.68782	1.86	3.4596	1.36382	4.31277
1.37	1.8769	1.17047	3.70135	1.87	3.4969	1.36748	4.32435
1.38	1.9044	1.17473	3.71484	1.88	3.5344	1.37113	4.33590
1.39	1.9321	1.17898	3.72827	1.89	3.5721	1.37477	4.34741
1.40	1.9600	1.18322	3.74166	**1.90**	3.6100	1.37840	4.35890
1.41	1.9881	1.18743	3.75500	1.91	3.6481	1.38203	4.37035
1.42	2.0164	1.19164	3.76829	1.92	3.6864	1.38564	4.38178
1.43	2.0449	1.19583	3.78153	1.93	3.7249	1.38924	4.39318
1.44	2.0736	1.20000	3.79473	1.94	3.7636	1.39284	4.40454
1.45	2.1025	1.20416	3.80789	1.95	3.8025	1.39642	4.41588
1.46	2.1316	1.20830	3.82099	1.96	3.8416	1.40000	4.42719
1.47	2.1609	1.21244	3.83406	1.97	3.8809	1.40357	4.43847
1.48	2.1904	1.21655	3.84708	1.98	3.9204	1.40712	4.44972
1.49	2.2201	1.22066	3.86005	1.99	3.9601	1.41067	4.46094
1.50	2.2500	1.22474	3.87298	**2.00**	4.0000	1.41421	4.47214
N	N²	√N	√10N	N	N²	√N	√10N

TABLE 1—Squares and Square Roots—*Continued*

N	N²	√N	√10N	N	N²	√N	√10N
2.00	4.0000	1.41421	4.47214	**2.50**	6.2500	1.58114	5.00000
2.01	4.0401	1.41774	4.48330	2.51	6.3001	1.58430	5.00999
2.02	4.0804	1.42127	4.49444	2.52	6.3504	1.58745	5.01996
2.03	4.1209	1.42478	4.50555	2.53	6.4009	1.59060	5.02991
2.04	4.1616	1.42829	4.51664	2.54	6.4516	1.59374	5.03984
2.05	4.2025	1.43178	4.52769	2.55	6.5025	1.59687	5.04975
2.06	4.2436	1.43527	4.53872	2.56	6.5536	1.60000	5.05964
2.07	4.2849	1.43875	4.54973	2.57	6.6049	1.60312	5.06952
2.08	4.3264	1.44222	4.56070	2.58	6.6564	1.60624	5.07937
2.09	4.3681	1.44568	4.57165	2.59	6.7081	1.60935	5.08920
2.10	4.4100	1.44914	4.58258	**2.60**	6.7600	1.61245	5.09902
2.11	4.4521	1.45258	4.59347	2.61	6.8121	1.61555	5.10882
2.12	4.4944	1.45602	4.60435	2.62	6.8644	1.61864	5.11859
2.13	4.5369	1.45945	4.61519	2.63	6.9169	1.62173	5.12835
2.14	4.5796	1.46287	4.62601	2.64	6.9696	1.62481	5.13809
2.15	4.6225	1.46629	4.63681	2.65	7.0225	1.62788	5.14782
2.16	4.6656	1.46969	4.64758	2.66	7.0756	1.63095	5.15752
2.17	4.7089	1.47309	4.65833	2.67	7.1289	1.63401	5.16720
2.18	4.7524	1.47648	4.66905	2.68	7.1824	1.63707	5.17687
2.19	4.7961	1.47986	4.67974	2.69	7.2361	1.64012	5.18652
2.20	4.8400	1.48324	4.69042	**2.70**	7.2900	1.64317	5.19615
2.21	4.8841	1.48661	4.70106	2.71	7.3441	1.64621	5.20577
2.22	4.9284	1.48997	4.71169	2.72	7.3984	1.64924	5.21536
2.23	4.9729	1.49332	4.72229	2.73	7.4529	1.65227	5.22494
2.24	5.0176	1.49666	4.73286	2.74	7.5076	1.65529	5.23450
2.25	5.0625	1.50000	4.74342	2.75	7.5625	1.65831	5.24404
2.26	5.1076	1.50333	4.75395	2.76	7.6176	1.66132	5.25357
2.27	5.1529	1.50665	4.76445	2.77	7.6729	1.66433	5.26308
2.28	5.1984	1.50997	4.77493	2.78	7.7284	1.66733	5.27257
2.29	5.2441	1.51327	4.78539	2.79	7.7841	1.67033	5.28205
2.30	5.2900	1.51658	4.79583	**2.80**	7.8400	1.67332	5.29150
2.31	5.3361	1.51987	4.80625	2.81	7.8961	1.67631	5.30094
2.32	5.3824	1.52315	4.81664	2.82	7.9524	1.67929	5.31037
2.33	5.4289	1.52643	4.82701	2.83	8.0089	1.68226	5.31977
2.34	5.4756	1.52971	4.83735	2.84	8.0656	1.68523	5.32917
2.35	5.5225	1.53297	4.84768	2.85	8.1225	1.68819	5.33854
2.36	5.5696	1.53623	4.85798	2.86	8.1796	1.69115	5.34790
2.37	5.6169	1.53948	4.86826	2.87	8.2369	1.69411	5.35724
2.38	5.6644	1.54272	4.87852	2.88	8.2944	1.69706	5.36656
2.39	5.7121	1.54596	4.88876	2.89	8.3521	1.70000	5.37587
2.40	5.7600	1.54919	4.89898	**2.90**	8.4100	1.70294	5.38516
2.41	5.8081	1.55242	4.90918	2.91	8.4681	1.70587	5.39444
2.42	5.8564	1.55563	4.91935	2.92	8.5264	1.70880	5.40370
2.43	5.9049	1.55885	4.92950	2.93	8.5849	1.71172	5.41295
2.44	5.9536	1.56205	4.93964	2.94	8.6436	1.71464	5.42218
2.45	6.0025	1.56525	4.94975	2.95	8.7025	1.71756	5.43139
2.46	6.0516	1.56844	4.95984	2.96	8.7616	1.72047	5.44059
2.47	6.1009	1.57162	4.96991	2.97	8.8209	1.72337	5.44977
2.48	6.1504	1.57480	4.97996	2.98	8.8804	1.72627	5.45894
2.49	6.2001	1.57797	4.98999	2.99	8.9401	1.72916	5.46809
2.50	6.2500	1.58114	5.00000	**3.00**	9.0000	1.73205	5.47723
N	**N²**	**√N**	**√10N**	**N**	**N²**	**√N**	**√10N**

TABLE 1—Squares and Square Roots—*Continued*

N	N²	√N	√10N	N	N²	√N	√10N
3.00	9.0000	1.73205	5.47723	**3.50**	12.2500	1.87083	5.91608
3.01	9.0601	1.73494	5.48635	3.51	12.3201	1.87350	5.92453
3.02	9.1204	1.73781	5.49545	3.52	12.3904	1.87617	5.93296
3.03	9.1809	1.74069	5.50454	3.53	12.4609	1.87883	5.94138
3.04	9.2416	1.74356	5.51362	3.54	12.5316	1.88149	5.94979
3.05	9.3025	1.74642	5.52268	3.55	12.6025	1.88414	5.95819
3.06	9.3636	1.74929	5.53173	3.56	12.6736	1.88680	5.96657
3.07	9.4249	1.75214	5.54076	3.57	12.7449	1.88944	5.97495
3.08	9.4864	1.75499	5.54977	3.58	12.8164	1.89209	5.98331
3.09	9.5481	1.75784	5.55878	3.59	12.8881	1.89473	5.99166
3.10	9.6100	1.76068	5.56776	**3.60**	12.9600	1.89737	6.00000
3.11	9.6721	1.76352	5.57674	3.61	13.0321	1.90000	6.00833
3.12	9.7344	1.76635	5.58570	3.62	13.1044	1.90263	6.01664
3.13	9.7969	1.76918	5.59464	3.63	13.1769	1.90526	6.02495
3.14	9.8596	1.77200	5.60357	3.64	13.2496	1.90788	6.03324
3.15	9.9225	1.77482	5.61249	3.65	13.3225	1.91050	6.04152
3.16	9.9856	1.77764	5.62139	3.66	13.3956	1.91311	6.04979
3.17	10.0489	1.78045	5.63028	3.67	13.4689	1.91572	6.05805
3.18	10.1124	1.78326	5.63915	3.68	13.5424	1.91833	6.06630
3.19	10.1761	1.78606	5.64801	3.69	13.6161	1.92094	6.07454
3.20	10.2400	1.78885	5.65685	**3.70**	13.6900	1.92354	6.08276
3.21	10.3041	1.79165	5.66569	3.71	13.7641	1.92614	6.09098
3.22	10.3684	1.79444	5.67450	3.72	13.8384	1.92873	6.09918
3.23	10.4329	1.79722	5.68331	3.73	13.9129	1.93132	6.10737
3.24	10.4976	1.80000	5.69210	3.74	13.9876	1.93391	6.11555
3.25	10.5625	1.80278	5.70088	3.75	14.0625	1.93649	6.12372
3.26	10.6276	1.80555	5.70964	3.76	14.1376	1.93907	6.13188
3.27	10.6929	1.80831	5.71839	3.77	14.2129	1.94165	6.14003
3.28	10.7584	1.81108	5.72713	3.78	14.2884	1.94422	6.14817
3.29	10.8241	1.81384	5.73585	3.79	14.3641	1.94679	6.15630
3.30	10.8900	1.81659	5.74456	**3.80**	14.4400	1.94936	6.16441
3.31	10.9561	1.81934	5.75326	3.81	14.5161	1.95192	6.17252
3.32	11.0224	1.82209	5.76194	3.82	14.5924	1.95448	6.18061
3.33	11.0889	1.82483	5.77062	3.83	14.6689	1.95704	6.18870
3.34	11.1556	1.82757	5.77927	3.84	14.7456	1.95959	6.19677
3.35	11.2225	1.83030	5.78792	3.85	14.8225	1.96214	6.20484
3.36	11.2896	1.83303	5.79655	3.86	14.8996	1.96469	6.21289
3.37	11.3569	1.83576	5.80517	3.87	14.9769	1.96723	6.22093
3.38	11.4244	1.83848	5.81378	3.88	15.0544	1.96977	6.22896
3.39	11.4921	1.84120	5.82237	3.89	15.1321	1.97231	6.23699
3.40	11.5600	1.84391	5.83095	**3.90**	15.2100	1.97484	6.24500
3.41	11.6281	1.84662	5.83952	3.91	15.2881	1.97737	6.25300
3.42	11.6964	1.84932	5.84808	3.92	15.3664	1.97990	6.26099
3.43	11.7649	1.85203	5.85662	3.93	15.4449	1.98242	6.26897
3.44	11.8336	1.85472	5.86515	3.94	15.5236	1.98494	6.27694
3.45	11.9025	1.85742	5.87367	3.95	15.6025	1.98746	6.28490
3.46	11.9716	1.86011	5.88218	3.96	15.6816	1.98997	6.29285
3.47	12.0409	1.86279	5.89067	3.97	15.7609	1.99249	6.30079
3.48	12.1104	1.86548	5.89915	3.98	15.8404	1.99499	6.30872
3.49	12.1801	1.86815	5.90762	3.99	15.9201	1.99750	6.31664
3.50	12.2500	1.87083	5.91608	**4.00**	16.0000	2.00000	6.32456
N	N²	√N	√10N	N	N²	√N	√10N

TABLE 1—Squares and Square Roots—*Continued*

N	N²	√N̄	√10N̄	N	N²	√N̄	√10N̄
4.00	16.0000	2.00000	6.32456	**4.50**	20.2500	2.12132	6.70820
4.01	16.0801	2.00250	6.33246	4.51	20.3401	2.12368	6.71565
4.02	16.1604	2.00499	6.34035	4.52	20.4304	2.12603	6.72309
4.03	16.2409	2.00749	6.34823	4.53	20.5209	2.12838	6.73053
4.04	16.3216	2.00998	6.35610	4.54	20.6116	2.13073	6.73795
4.05	16.4025	2.01246	6.36396	4.55	20.7025	2.13307	6.74537
4.06	16.4836	2.01494	6.37181	4.56	20.7936	2.13542	6.75278
4.07	16.5649	2.01742	6.37966	4.57	20.8849	2.13776	6.76018
4.08	16.6464	2.01990	6.38749	4.58	20.9764	2.14009	6.76757
4.09	16.7281	2.02237	6.39531	4.59	21.0681	2.14243	6.77495
4.10	16.8100	2.02485	6.40312	**4.60**	21.1600	2.14476	6.78233
4.11	16.8921	2.02731	6.41093	4.61	21.2521	2.14709	6.78970
4.12	16.9744	2.02978	6.41872	4.62	21.3444	2.14942	6.79706
4.13	17.0569	2.03224	6.42651	4.63	21.4369	2.15174	6.80441
4.14	17.1396	2.03470	6.43428	4.64	21.5296	2.15407	6.81175
4.15	17.2225	2.03715	6.44205	4.65	21.6225	2.15639	6.81909
4.16	17.3056	2.03961	6.44981	4.66	21.7156	2.15870	6.82642
4.17	17.3889	2.04206	6.45755	4.67	21.8089	2.16102	6.83374
4.18	17.4724	2.04450	6.46529	4.68	21.9024	2.16333	6.84105
4.19	17.5561	2.04695	6.47302	4.69	21.9961	2.16564	6.84836
4.20	17.6400	2.04939	6.48074	**4.70**	22.0900	2.16795	6.85565
4.21	17.7241	2.05183	6.48845	4.71	22.1841	2.17025	6.86294
4.22	17.8084	2.05426	6.49615	4.72	22.2784	2.17256	6.87023
4.23	17.8929	2.05670	6.50384	4.73	22.3729	2.17486	6.87750
4.24	17.9776	2.05913	6.51153	4.74	22.4676	2.17715	6.88477
4.25	18.0625	2.06155	6.51920	4.75	22.5625	2.17945	6.89202
4.26	18.1476	2.06398	6.52687	4.76	22.6576	2.18174	6.89928
4.27	18.2329	2.06640	6.53452	4.77	22.7529	2.18403	6.90652
4.28	18.3184	2.06882	6.54217	4.78	22.8484	2.18632	6.91375
4.29	18.4041	2.07123	6.54981	4.79	22.9441	2.18861	6.92098
4.30	18.4900	2.07364	6.55744	**4.80**	23.0400	2.19089	6.92820
4.31	18.5761	2.07605	6.56506	4.81	23.1361	2.19317	6.93542
4.32	18.6624	2.07846	6.57267	4.82	23.2324	2.19545	6.94262
4.33	18.7489	2.08087	6.58027	4.83	23.3289	2.19773	6.94982
4.34	18.8356	2.08327	6.58787	4.84	23.4256	2.20000	6.95701
4.35	18.9225	2.08567	6.59545	4.85	23.5225	2.20227	6.96419
4.36	19.0096	2.08806	6.60303	4.86	23.6196	2.20454	6.97137
4.37	19.0969	2.09045	6.61060	4.87	23.7169	2.20681	6.97854
4.38	19.1844	2.09284	6.61816	4.88	23.8144	2.20907	6.98570
4.39	19.2721	2.09523	6.62571	4.89	23.9121	2.21133	6.99285
4.40	19.3600	2.09762	6.63325	**4.90**	24.0100	2.21359	7.00000
4.41	19.4481	2.10000	6.64078	4.91	24.1081	2.21585	7.00714
4.42	19.5364	2.10238	6.64831	4.92	24.2064	2.21811	7.01427
4.43	19.6249	2.10476	6.65582	4.93	24.3049	2.22036	7.02140
4.44	19.7136	2.10713	6.66333	4.94	24.4036	2.22261	7.02851
4.45	19.8025	2.10950	6.67083	4.95	24.5025	2.22486	7.03562
4.46	19.8916	2.11187	6.67832	4.96	24.6016	2.22711	7.04273
4.47	19.9809	2.11424	6.68581	4.97	24.7009	2.22935	7.04982
4.48	20.0704	2.11660	6.69328	4.98	24.8004	2.23159	7.05691
4.49	20.1601	2.11896	6.70075	4.99	24.9001	2.23383	7.06399
4.50	20.2500	2.12132	6.70820	**5.00**	25.0000	2.23607	7.07107
N	N²	√N̄	√10N̄	N	N²	√N̄	√10N̄

TABLE 1—Squares and Square Roots—*Continued*

N	N²	√N	√10N	N	N²	√N	√10N
5.00	25.0000	2.23607	7.07107	**5.50**	30.2500	2.34521	7.41620
5.01	25.1001	2.23830	7.07814	5.51	30.3601	2.34734	7.42294
5.02	25.2004	2.24054	7.08520	5.52	30.4704	2.34947	7.42967
5.03	25.3009	2.24277	7.09225	5.53	30.5809	2.35160	7.43640
5.04	25.4016	2.24499	7.09930	5.54	30.6916	2.35372	7.44312
5.05	25.5025	2.24722	7.10634	5.55	30.8025	2.35584	7.44983
5.06	25.6036	2.24944	7.11337	5.56	30.9136	2.35797	7.45654
5.07	25.7049	2.25167	7.12039	5.57	31.0249	2.36008	7.46324
5.08	25.8064	2.25389	7.12741	5.58	31.1364	2.36220	7.46994
5.09	25.9081	2.25610	7.13442	5.59	31.2481	2.36432	7.47663
5.10	26.0100	2.25832	7.14143	**5.60**	31.3600	2.36643	7.48331
5.11	26.1121	2.26053	7.14843	5.61	31.4721	2.36854	7.48999
5.12	26.2144	2.26274	7.15542	5.62	31.5844	2.37065	7.49667
5.13	26.3169	2.26495	7.16240	5.63	31.6969	2.37276	7.50333
5.14	26.4196	2.26716	7.16938	5.64	31.8096	2.37487	7.50999
5.15	26.5225	2.26936	7.17635	5.65	31.9225	2.37697	7.51665
5.16	26.6256	2.27156	7.18331	5.66	32.0356	2.37908	7.52330
5.17	26.7289	2.27376	7.19027	5.67	32.1489	2.38118	7.52994
5.18	26.8324	2.27596	7.19722	5.68	32.2624	2.38328	7.53658
5.19	26.9361	2.27816	7.20417	5.69	32.3761	2.38537	7.54321
5.20	27.0400	2.28035	7.21110	**5.70**	32.4900	2.38747	7.54983
5.21	27.1441	2.28254	7.21803	5.71	32.6041	2.38956	7.55645
5.22	27.2484	2.28473	7.22496	5.72	32.7184	2.39165	7.56307
5.23	27.3529	2.28692	7.23187	5.73	32.8329	2.39374	7.56968
5.24	27.4576	2.28910	7.23878	5.74	32.9476	2.39583	7.57628
5.25	27.5625	2.29129	7.24569	5.75	33.0625	2.39792	7.58288
5.26	27.6676	2.29347	7.25259	5.76	33.1776	2.40000	7.58947
5.27	27.7729	2.29565	7.25948	5.77	33.2929	2.40208	7.59605
5.28	27.8784	2.29783	7.26636	5.78	33.4084	2.40416	7.60263
5.29	27.9841	2.30000	7.27324	5.79	33.5241	2.40624	7.60920
5.30	28.0900	2.30217	7.28011	**5.80**	33.6400	2.40832	7.61577
5.31	28.1961	2.30434	7.28697	5.81	33.7561	2.41039	7.62234
5.32	28.3024	2.30651	7.29383	5.82	33.8724	2.41247	7.62889
5.33	28.4089	2.30868	7.30068	5.83	33.9889	2.41454	7.63544
5.34	28.5156	2.31084	7.30753	5.84	34.1056	2.41661	7.64199
5.35	28.6225	2.31301	7.31437	5.85	34.2225	2.41868	7.64853
5.36	28.7296	2.31517	7.32120	5.86	34.3396	2.42074	7.65506
5.37	28.8369	2.31733	7.32803	5.87	34.4569	2.42281	7.66159
5.38	28.9444	2.31948	7.33485	5.88	34.5744	2.42487	7.66812
5.39	29.0521	2.32164	7.34166	5.89	34.6921	2.42693	7.67463
5.40	29.1600	2.32379	7.34847	**5.90**	34.8100	2.42899	7.68115
5.41	29.2681	2.32594	7.35527	5.91	34.9281	2.43105	7.68765
5.42	29.3764	2.32809	7.36206	5.92	35.0464	2.43311	7.69415
5.43	29.4849	2.33024	7.36885	5.93	35.1649	2.43516	7.70065
5.44	29.5936	2.33238	7.37564	5.94	35.2836	2.43721	7.70714
5.45	29.7025	2.33452	7.38241	5.95	35.4025	2.43926	7.71362
5.46	29.8116	2.33666	7.38918	5.96	35.5216	2.44131	7.72010
5.47	29.9209	2.33880	7.39594	5.97	35.6409	2.44336	7.72658
5.48	30.0304	2.34094	7.40270	5.98	35.7604	2.44540	7.73305
5.49	30.1401	2.34307	7.40945	5.99	35.8801	2.44745	7.73951
5.50	30.2500	2.34521	7.41620	**6.00**	36.0000	2.44949	7.74597
N	N²	√N	√10N	N	N²	√N	√10N

TABLE 1—SQUARES AND SQUARE ROOTS—*Continued*

N	N²	√N	√10N	N	N²	√N	√10N
6.00	36.0000	2.44949	7.74597	**6.50**	42.2500	2.54951	8.06226
6.01	36.1201	2.45153	7.75242	6.51	42.3801	2.55147	8.06846
6.02	36.2404	2.45357	7.75887	6.52	42.5104	2.55343	8.07465
6.03	36.3609	2.45561	7.76531	6.53	42.6409	2.55539	8.08084
6.04	36.4816	2.45764	7.77174	6.54	42.7716	2.55734	8.08703
6.05	36.6025	2.45967	7.77817	6.55	42.9025	2.55930	8.09321
6.06	36.7236	2.46171	7.78460	6.56	43.0336	2.56125	8.09938
6.07	36.8449	2.46374	7.79102	6.57	43.1649	2.56320	8.10555
6.08	36.9664	2.46577	7.79744	6.58	43.2964	2.56515	8.11172
6.09	37.0881	2.46779	7.80385	6.59	43.4281	2.56710	8.11788
6.10	37.2100	2.46982	7.81025	**6.60**	43.5600	2.56905	8.12404
6.11	37.3321	2.47184	7.81665	6.61	43.6921	2.57099	8.13019
6.12	37.4544	2.47386	7.82304	6.62	43.8244	2.57294	8.13634
6.13	37.5769	2.47588	7.82943	6.63	43.9569	2.57488	8.14248
6.14	37.6996	2.47790	7.83582	6.64	44.0896	2.57682	8.14862
6.15	37.8225	2.47992	7.84219	6.65	44.2225	2.57876	8.15475
6.16	37.9456	2.48193	7.84857	6.66	44.3556	2.58070	8.16088
6.17	38.0689	2.48395	7.85493	6.67	44.4889	2.58263	8.16701
6.18	38.1924	2.48596	7.86130	6.68	44.6224	2.58457	8.17313
6.19	38.3161	2.48797	7.86766	6.69	44.7561	2.58650	8.17924
6.20	38.4400	2.48998	7.87401	**6.70**	44.8900	2.58844	8.18535
6.21	38.5641	2.49199	7.88036	6.71	45.0241	2.59037	8.19146
6.22	38.6884	2.49399	7.88670	6.72	45.1584	2.59230	8.19756
6.23	38.8129	2.49600	7.89303	6.73	45.2929	2.59422	8.20366
6.24	38.9376	2.49800	7.89937	6.74	45.4276	2.59615	8.20975
6.25	39.0625	2.50000	7.90569	6.75	45.5625	2.59808	8.21584
6.26	39.1876	2.50200	7.91202	6.76	45.6976	2.60000	8.22192
6.27	39.3129	2.50400	7.91833	6.77	45.8329	2.60192	8.22800
6.28	39.4384	2.50599	7.92465	6.78	45.9684	2.60384	8.23408
6.29	39.5641	2.50799	7.93095	6.79	46.1041	2.60576	8.24015
6.30	39.6900	2.50998	7.93725	**6.80**	46.2400	2.60768	8.24621
6.31	39.8161	2.51197	7.94355	6.81	46.3761	2.60960	8.25227
6.32	39.9424	2.51396	7.94984	6.82	46.5124	2.61151	8.25833
6.33	40.0689	2.51595	7.95613	6.83	46.6489	2.61343	8.26438
6.34	40.1956	2.51794	7.96241	6.84	46.7856	2.61534	8.27043
6.35	40.3225	2.51992	7.96869	6.85	46.9225	2.61725	8.27647
6.36	40.4496	2.52190	7.97496	6.86	47.0596	2.61916	8.28251
6.37	40.5769	2.52389	7.98123	6.87	47.1969	2.62107	8.28855
6.38	40.7044	2.52587	7.98749	6.88	47.3344	2.62298	8.29458
6.39	40.8321	2.52784	7.99375	6.89	47.4721	2.62488	8.30060
6.40	40.9600	2.52982	8.00000	**6.90**	47.6100	2.62679	8.30662
6.41	41.0881	2.53180	8.00625	6.91	47.7481	2.62869	8.31264
6.42	41.2164	2.53377	8.01249	6.92	47.8864	2.63059	8.31865
6.43	41.3449	2.53574	8.01873	6.93	48.0249	2.63249	8.32466
6.44	41.4736	2.53772	8.02496	6.94	48.1636	2.63439	8.33067
6.45	41.6025	2.53969	8.03119	6.95	48.3025	2.63629	8.33667
6.46	41.7316	2.54165	8.03741	6.96	48.4416	2.63818	8.34266
6.47	41.8609	2.54362	8.04363	6.97	48.5809	2.64008	8.34865
6.48	41.9904	2.54558	8.04984	6.98	48.7204	2.64197	8.35464
6.49	42.1201	2.54755	8.05605	6.99	48.8601	2.64386	8.36062
6.50	42.2500	2.54951	8.06226	**7.00**	49.0000	2.64575	8.36660
N	**N²**	**√N**	**√10N**	**N**	**N²**	**√N**	**√10N**

TABLE 1—Squares and Square Roots—*Continued*

N	N²	√N	√10N	N	N²	√N	√10N
7.00	49.0000	2.64575	8.36660	**7.50**	56.2500	2.73861	8.66025
7.01	49.1401	2.64764	8.37257	7.51	56.4001	2.74044	8.66603
7.02	49.2804	2.64953	8.37854	7.52	56.5504	2.74226	8.67179
7.03	49.4209	2.65141	8.38451	7.53	56.7009	2.74408	8.67756
7.04	49.5616	2.65330	8.39047	7.54	56.8516	2.74591	8.68332
7.05	49.7025	2.65518	8.39643	7.55	57.0025	2.74773	8.68907
7.06	49.8436	2.65707	8.40238	7.56	57.1536	2.74955	8.69483
7.07	49.9849	2.65895	8.40833	7.57	57.3049	2.75136	8.70057
7.08	50.1264	2.66083	8.41427	7.58	57.4564	2.75318	8.70632
7.09	50.2681	2.66271	8.42021	7.59	57.6081	2.75500	8.71206
7.10	50.4100	2.66458	8.42615	**7.60**	57.7600	2.75681	8.71780
7.11	50.5521	2.66646	8.43208	7.61	57.9121	2.75862	8.72353
7.12	50.6944	2.66833	8.43801	7.62	58.0644	2.76043	8.72926
7.13	50.8369	2.67021	8.44393	7.63	58.2169	2.76225	8.73499
7.14	50.9796	2.67208	8.44985	7.64	58.3696	2.76405	8.74071
7.15	51.1225	2.67395	8.45577	7.65	58.5225	2.76586	8.74643
7.16	51.2656	2.67582	8.46168	7.66	58.6756	2.76767	8.75214
7.17	51.4089	2.67769	8.46759	7.67	58.8289	2.76948	8.75785
7.18	51.5524	2.67955	8.47349	7.68	58.9824	2.77128	8.76356
7.19	51.6961	2.68142	8.47939	7.69	59.1361	2.77308	8.76926
7.20	51.8400	2.68328	8.48528	**7.70**	59.2900	2.77489	8.77496
7.21	51.9841	2.68514	8.49117	7.71	59.4441	2.77669	8.78066
7.22	52.1284	2.68701	8.49706	7.72	59.5984	2.77849	8.78635
7.23	52.2729	2.68887	8.50294	7.73	59.7529	2.78029	8.79204
7.24	52.4176	2.69072	8.50882	7.74	59.9076	2.78209	8.79773
7.25	52.5625	2.69258	8.51469	7.75	60.0625	2.78388	8.80341
7.26	52.7076	2.69444	8.52056	7.76	60.2176	2.78568	8.80909
7.27	52.8529	2.69629	8.52643	7.77	60.3729	2.78747	8.81476
7.28	52.9984	2.69815	8.53229	7.78	60.5284	2.78927	8.82043
7.29	53.1441	2.70000	8.53815	7.79	60.6841	2.79106	8.82610
7.30	53.2900	2.70185	8.54400	**7.80**	60.8400	2.79285	8.83176
7.31	53.4361	2.70370	8.54985	7.81	60.9961	2.79464	8.83742
7.32	53.5824	2.70555	8.55570	7.82	61.1524	2.79643	8.84308
7.33	53.7289	2.70740	8.56154	7.83	61.3089	2.79821	8.84873
7.34	53.8756	2.70924	8.56738	7.84	61.4656	2.80000	8.85438
7.35	54.0225	2.71109	8.57321	7.85	61.6225	2.80179	8.86002
7.36	54.1696	2.71293	8.57904	7.86	61.7796	2.80357	8.86566
7.37	54.3169	2.71477	8.58487	7.87	61.9369	2.80535	8.87130
7.38	54.4644	2.71662	8.59069	7.88	62.0944	2.80713	8.87694
7.39	54.6121	2.71846	8.59651	7.89	62.2521	2.80891	8.88257
7.40	54.7600	2.72029	8.60233	**7.90**	62.4100	2.81069	8.88819
7.41	54.9081	2.72213	8.60814	7.91	62.5681	2.81247	8.89382
7.42	55.0564	2.72397	8.61394	7.92	62.7264	2.81425	8.89944
7.43	55.2049	2.72580	8.61974	7.93	62.8849	2.81603	8.90505
7.44	55.3536	2.72764	8.62554	7.94	63.0436	2.81780	8.91067
7.45	55.5025	2.72947	8.63134	7.95	63.2025	2.81957	8.91628
7.46	55.6516	2.73130	8.63713	7.96	63.3616	2.82135	8.92188
7.47	55.8009	2.73313	8.64292	7.97	63.5209	2.82312	8.92749
7.48	55.9504	2.73496	8.64870	7.98	63.6804	2.82489	8.93308
7.49	56.1001	2.73679	8.65448	7.99	63.8401	2.82666	8.93868
7.50	56.2500	2.73861	8.66025	**8.00**	64.0000	2.82843	8.94427
N	N²	√N	√10N	N	N²	√N	√10N

TABLE 1—Squares and Square Roots—*Continued*

N	N²	√N	√10N	N	N²	√N	√10N
8.00	64.0000	2.82843	8.94427	**8.50**	72.2500	2.91548	9.21954
8.01	64.1601	2.83019	8.94986	8.51	72.4201	2.91719	9.22497
8.02	64.3204	2.83196	8.95545	8.52	72.5904	2.91890	9.23038
8.03	64.4809	2.83373	8.96103	8.53	72.7609	2.92062	9.23580
8.04	64.6416	2.83549	8.96660	8.54	72.9316	2.92233	9.24121
8.05	64.8025	2.83725	8.97218	8.55	73.1025	2.92404	9.24662
8.06	64.9636	2.83901	8.97775	8.56	73.2736	2.92575	9.25203
8.07	65.1249	2.84077	8.98332	8.57	73.4449	2.92746	9.25743
8.08	65.2864	2.84253	8.98888	8.58	73.6164	2.92916	9.26283
8.09	65.4481	2.84429	8.99444	8.59	73.7881	2.93087	9.26823
8.10	65.6100	2.84605	9.00000	**8.60**	73.9600	2.93258	9.27362
8.11	65.7721	2.84781	9.00555	8.61	74.1321	2.93428	9.27901
8.12	65.9344	2.84956	9.01110	8.62	74.3044	2.93598	9.28440
8.13	66.0969	2.85132	9.01665	8.63	74.4769	2.93769	9.28978
8.14	66.2596	2.85307	9.02219	8.64	74.6496	2.93939	9.29516
8.15	66.4225	2.85482	9.02774	8.65	74.8225	2.94109	9.30054
8.16	66.5856	2.85657	9.03327	8.66	74.9956	2.94279	9.30591
8.17	66.7489	2.85832	9.03881	8.67	75.1689	2.94449	9.31128
8.18	66.9124	2.86007	9.04434	8.68	75.3424	2.94618	9.31665
8.19	67.0761	2.86182	9.04986	8.69	75.5161	2.94788	9.32202
8.20	67.2400	2.86356	9.05539	**8.70**	75.6900	2.94958	9.32738
8.21	67.4041	2.86531	9.06091	8.71	75.8641	2.95127	9.33274
8.22	67.5684	2.86705	9.06642	8.72	76.0384	2.95296	9.33809
8.23	67.7329	2.86880	9.07193	8.73	76.2129	2.95466	9.34345
8.24	67.8976	2.87054	9.07744	8.74	76.3876	2.95635	9.34880
8.25	68.0625	2.87228	9.08295	8.75	76.5625	2.95804	9.35414
8.26	68.2276	2.87402	9.08845	8.76	76.7376	2.95973	9.35949
8.27	68.3929	2.87576	9.09395	8.77	76.9129	2.96142	9.36483
8.28	68.5584	2.87750	9.09945	8.78	77.0884	2.96311	9.37017
8.29	68.7241	2.87924	9.10494	8.79	77.2641	2.96479	9.37550
8.30	68.8900	2.88097	9.11043	**8.80**	77.4400	2.96648	9.38083
8.31	69.0561	2.88271	9.11592	8.81	77.6161	2.96816	9.38616
8.32	69.2224	2.88444	9.12140	8.82	77.7924	2.96985	9.39149
8.33	69.3889	2.88617	9.12688	8.83	77.9689	2.97153	9.39681
8.34	69.5556	2.88791	9.13236	8.84	78.1456	2.97321	9.40213
8.35	69.7225	2.88964	9.13783	8.85	78.3225	2.97489	9.40744
8.36	69.8896	2.89137	9.14330	8.86	78.4996	2.97658	9.41276
8.37	70.0569	2.89310	9.14877	8.87	78.6769	2.97825	9.41807
8.38	70.2244	2.89482	9.15423	8.88	78.8544	2.97993	9.42338
8.39	70.3921	2.89655	9.15969	8.89	79.0321	2.98161	9.42868
8.40	70.5600	2.89828	9.16515	**8.90**	79.2100	2.98329	9.43398
8.41	70.7281	2.90000	9.17061	8.91	79.3881	2.98496	9.43928
8.42	70.8964	2.90172	9.17606	8.92	79.5664	2.98664	9.44458
8.43	71.0649	2.90345	9.18150	8.93	79.7449	2.98831	9.44987
8.44	71.2336	2.90517	9.18695	8.94	79.9236	2.98998	9.45516
8.45	71.4025	2.90689	9.19239	8.95	80.1025	2.99166	9.46044
8.46	71.5716	2.90861	9.19783	8.96	80.2816	2.99333	9.46573
8.47	71.7409	2.91033	9.20326	8.97	80.4609	2.99500	9.47101
8.48	71.9104	2.91204	9.20869	8.98	80.6404	2.99666	9.47629
8.49	72.0801	2.91376	9.21412	8.99	80.8201	2.99833	9.48156
8.50	72.2500	2.91548	9.21954	**9.00**	81.0000	3.00000	9.48683
N	N²	√N	√10N	N	N²	√N	√10N

TABLE 1—SQUARES AND SQUARE ROOTS—*Continued*

N	N²	√N	√10N	N	N²	√N	√10N
9.00	81.0000	3.00000	9.48683	**9.50**	90.2500	3.08221	9.74679
9.01	81.1801	3.00167	9.49210	9.51	90.4401	3.08383	9.75192
9.02	81.3604	3.00333	9.49737	9.52	90.6304	3.08545	9.75705
9.03	81.5409	3.00500	9.50263	9.53	90.8209	3.08707	9.76217
9.04	81.7216	3.00666	9.50789	9.54	91.0116	3.08869	9.76729
9.05	81.9025	3.00832	9.51315	9.55	91.2025	3.09031	9.77241
9.06	82.0836	3.00998	9.51840	9.56	91.3936	3.09192	9.77753
9.07	82.2649	3.01164	9.52365	9.57	91.5849	3.09354	9.78264
9.08	82.4464	3.01330	9.52890	9.58	91.7764	3.09516	9.78775
9.09	82.6281	3.01496	9.53415	9.59	91.9681	3.09677	9.79285
9.10	82.8100	3.01662	9.53939	**9.60**	92.1600	3.09839	9.79796
9.11	82.9921	3.01828	9.54463	9.61	92.3521	3.10000	9.80306
9.12	83.1744	3.01993	9.54987	9.62	92.5444	3.10161	9.80816
9.13	83.3569	3.02159	9.55510	9.63	92.7369	3.10322	9.81326
9.14	83.5396	3.02324	9.56033	9.64	92.9296	3.10483	9.81835
9.15	83.7225	3.02490	9.56556	9.65	93.1225	3.10644	9.82344
9.16	83.9056	3.02655	9.57079	9.66	93.3156	3.10805	9.82853
9.17	84.0889	3.02820	9.57601	9.67	93.5089	3.10966	9.83362
9.18	84.2724	3.02985	9.58123	9.68	93.7024	3.11127	9.83870
9.19	84.4561	3.03150	9.58645	9.69	93.8961	3.11288	9.84378
9.20	84.6400	3.03315	9.59166	**9.70**	94.0900	3.11448	9.84886
9.21	84.8241	3.03480	9.59687	9.71	94.2841	3.11609	9.85393
9.22	85.0084	3.03645	9.60208	9.72	94.4784	3.11769	9.85901
9.23	85.1929	3.03809	9.60729	9.73	94.6729	3.11929	9.86408
9.24	85.3776	3.03974	9.61249	9.74	94.8676	3.12090	9.86914
9.25	85.5625	3.04138	9.61769	9.75	95.0625	3.12250	9.87421
9.26	85.7476	3.04302	9.62289	9.76	95.2576	3.12410	9.87927
9.27	85.9329	3.04467	9.62808	9.77	95.4529	3.12570	9.88433
9.28	86.1184	3.04631	9.63328	9.78	95.6484	3.12730	9.88939
9.29	86.3041	3.04795	9.63846	9.79	95.8441	3.12890	9.89444
9.30	86.4900	3.04959	9.64365	**9.80**	96.0400	3.13050	9.89949
9.31	86.6761	3.05123	9.64883	9.81	96.2361	3.13209	9.90454
9.32	86.8624	3.05287	9.65401	9.82	96.4324	3.13369	9.90959
9.33	87.0489	3.05450	9.65919	9.83	96.6289	3.13528	9.91464
9.34	87.2356	3.05614	9.66437	9.84	96.8256	3.13688	9.91968
9.35	87.4225	3.05778	9.66954	9.85	97.0225	3.13847	9.92472
9.36	87.6096	3.05941	9.67471	9.86	97.2196	3.14006	9.92975
9.37	87.7969	3.06105	9.67988	9.87	97.4169	3.14166	9.93479
9.38	87.9844	3.06268	9.68504	9.88	97.6144	3.14325	9.93982
9.39	88.1721	3.06431	9.69020	9.89	97.8121	3.14484	9.94485
9.40	88.3600	3.06594	9.69536	**9.90**	98.0100	3.14643	9.94987
9.41	88.5481	3.06757	9.70052	9.91	98.2081	3.14802	9.95490
9.42	88.7364	3.06920	9.70567	9.92	98.4064	3.14960	9.95992
9.43	88.9249	3.07083	9.71082	9.93	98.6049	3.15119	9.96494
9.44	89.1136	3.07246	9.71597	9.94	98.8036	3.15278	9.96995
9.45	89.3025	3.07409	9.72111	9.95	99.0025	3.15436	9.97497
9.46	89.4916	3.07571	9.72625	9.96	99.2016	3.15595	9.97998
9.47	89.6809	3.07734	9.73139	9.97	99.4009	3.15753	9.98499
9.48	89.8704	3.07896	9.73653	9.98	99.6004	3.15911	9.98999
9.49	90.0601	3.08058	9.74166	9.99	99.8001	3.16070	9.99500
9.50	90.2500	3.08221	9.74679	**10.00**	100.000	3.16228	10.0000
N	N²	√N	√10N	N	N²	√N	√10N

TABLE 2—Common Logarithms *

N	0	1	2	3	4	5	6	7	8	9
0	....	0000	3010	4771	6021	6990	7782	8451	9031	9542
1	0000	0414	0792	1139	1461	1761	2041	2304	2553	2788
2	3010	3222	3424	3617	3802	3979	4150	4314	4472	4624
3	4771	4914	5051	5185	5315	5441	5563	5682	5798	5911
4	6021	6128	6232	6335	6435	6532	6628	6721	6812	6902
5	6990	7076	7160	7243	7324	7404	7482	7559	7634	7709
6	7782	7853	7924	7993	8062	8129	8195	8261	8325	8388
7	8451	8513	8573	8633	8692	8751	8808	8865	8921	8976
8	9031	9085	9138	9191	9243	9294	9345	9395	9445	9494
9	9542	9590	9638	9685	9731	9777	9823	9868	9912	9956
10	0000	0043	0086	0128	0170	0212	0253	0294	0334	0374
11	0414	0453	0492	0531	0569	0607	0645	0682	0719	0755
12	0792	0828	0864	0899	0934	0969	1004	1038	1072	1106
13	1139	1173	1206	1239	1271	1303	1335	1367	1399	1430
14	1461	1492	1523	1553	1584	1614	1644	1673	1703	1732
15	1761	1790	1818	1847	1875	1903	1931	1959	1987	2014
16	2041	2068	2095	2122	2148	2175	2201	2227	2253	2279
17	2304	2330	2355	2380	2405	2430	2455	2480	2504	2529
18	2553	2577	2601	2625	2648	2672	2695	2718	2742	2765
19	2788	2810	2833	2856	2878	2900	2923	2945	2967	2989
20	3010	3032	3054	3075	3096	3118	3139	3160	3181	3201
21	3222	3243	3263	3284	3304	3324	3345	3365	3385	3404
22	3424	3444	3464	3483	3502	3522	3541	3560	3579	3598
23	3617	3636	3655	3674	3692	3711	3729	3747	3766	3784
24	3802	3820	3838	3856	3874	3892	3909	3927	3945	3962
25	3979	3997	4014	4031	4048	4065	4082	4099	4116	4133
26	4150	4166	4183	4200	4216	4232	4249	4265	4281	4298
27	4314	4330	4346	4362	4378	4393	4409	4425	4440	4456
28	4472	4487	4502	4518	4533	4548	4564	4579	4594	4609
29	4624	4639	4654	4669	4683	4698	4713	4728	4742	4757
30	4771	4786	4800	4814	4829	4843	4857	4871	4886	4900
31	4914	4928	4942	4955	4969	4983	4997	5011	5024	5038
32	5051	5065	5079	5092	5105	5119	5132	5145	5159	5172
33	5185	5198	5211	5224	5237	5250	5263	5276	5289	5302
34	5315	5328	5340	5353	5366	5378	5391	5403	5416	5428
35	5441	5453	5465	5478	5490	5502	5514	5527	5539	5551
36	5563	5575	5587	5599	5611	5623	5635	5647	5658	5670
37	5682	5694	5705	5717	5729	5740	5752	5763	5775	5786
38	5798	5809	5821	5832	5843	5855	5866	5877	5888	5899
39	5911	5922	5933	5944	5955	5966	5977	5988	5999	6010
40	6021	6031	6042	6053	6064	6075	6085	6096	6107	6117
41	6128	6138	6149	6160	6170	6180	6191	6201	6212	6222
42	6232	6243	6253	6263	6274	6284	6294	6304	6314	6325
43	6335	6345	6355	6365	6375	6385	6395	6405	6415	6425
44	6435	6444	6454	6464	6474	6484	6493	6503	6513	6522
45	6532	6542	6551	6561	6571	6580	6590	6599	6609	6618
46	6628	6637	6646	6656	6665	6675	6684	6693	6702	6712
47	6721	6730	6739	6749	6758	6767	6776	6785	6794	6803
48	6812	6821	6830	6839	6848	6857	6866	6875	6884	6893
49	6902	6911	6920	6928	6937	6946	6955	6964	6972	6981
50	6990	6998	7007	7016	7024	7033	7042	7050	7059	7067
N	0	1	2	3	4	5	6	7	8	9

* Ralph G. Hudson, *The Engineers' Manual*, Second Edition, New York, John Wiley & Sons, Inc., 1939, pp. 250–251. Reprinted by permission.

TABLE 2—COMMON LOGARITHMS—*Continued*

N	0	1	2	3	4	5	6	7	8	9
50	6990	6998	7007	7016	7024	7033	7042	7050	7059	7067
51	7076	7084	7093	7101	7110	7118	7126	7135	7143	7152
52	7160	7168	7177	7185	7193	7202	7210	7218	7226	7235
53	7243	7251	7259	7267	7275	7284	7292	7300	7308	7316
54	7324	7332	7340	7348	7356	7364	7372	7380	7388	7396
55	7404	7412	7419	7427	7435	7443	7451	7459	7466	7474
56	7482	7490	7497	7505	7513	7520	7528	7536	7543	7551
57	7559	7566	7574	7582	7589	7597	7604	7612	7619	7627
58	7634	7642	7649	7657	7664	7672	7679	7686	7694	7701
59	7709	7716	7723	7731	7738	7745	7752	7760	7767	7774
60	7782	7789	7796	7803	7810	7818	7825	7832	7839	7846
61	7853	7860	7868	7875	7882	7889	7896	7903	7910	7917
62	7924	7931	7938	7945	7952	7959	7966	7973	7980	7987
63	7993	8000	8007	8014	8021	8028	8035	8041	8048	8055
64	8062	8069	8075	8082	8089	8096	8102	8109	8116	8122
65	8129	8136	8142	8149	8156	8162	8169	8176	8182	8189
66	8195	8202	8209	8215	8222	8228	8235	8241	8248	8254
67	8261	8267	8274	8280	8287	8293	8299	8306	8312	8319
68	8325	8331	8338	8344	8351	8357	8363	8370	8376	8382
69	8388	8395	8401	8407	8414	8420	8426	8432	8439	8445
70	8451	8457	8463	8470	8476	8482	8488	8494	8500	8506
71	8513	8519	8525	8531	8537	8543	8549	8555	8561	8567
72	8573	8579	8585	8591	8597	8603	8609	8615	8621	8627
73	8633	8639	8645	8651	8657	8663	8669	8675	8681	8686
74	8692	8698	8704	8710	8716	8722	8727	8733	8739	8745
75	8751	8756	8762	8768	8774	8779	8785	8791	8797	8802
76	8808	8814	8820	8825	8831	8837	8842	8848	8854	8859
77	8865	8871	8876	8882	8887	8893	8899	8904	8910	8915
78	8921	8927	8932	8938	8943	8949	8954	8960	8965	8971
79	8976	8982	8987	8993	8998	9004	9009	9015	9020	9025
80	9031	9036	9042	9047	9053	9058	9063	9069	9074	9079
81	9085	9090	9096	9101	9106	9112	9117	9122	9128	9133
82	9138	9143	9149	9154	9159	9165	9170	9175	9180	9186
83	9191	9196	9201	9206	9212	9217	9222	9227	9232	9238
84	9243	9248	9253	9258	9263	9269	9274	9279	9284	9289
85	9294	9299	9304	9309	9315	9320	9325	9330	9335	9340
86	9345	9350	9355	9360	9365	9370	9375	9380	9385	9390
87	9395	9400	9405	9410	9415	9420	9425	9430	9435	9440
88	9445	9450	9455	9460	9465	9469	9474	9479	9484	9489
89	9494	9499	9504	9509	9513	9518	9523	9528	9533	9538
90	9542	9547	9552	9557	9562	9566	9571	9576	9581	9586
91	9590	9595	9600	9605	9609	9614	9619	9624	9628	9633
92	9638	9643	9647	9652	9657	9661	9666	9671	9675	9680
93	9685	9689	9694	9699	9703	9708	9713	9717	9722	9727
94	9731	9736	9741	9745	9750	9754	9759	9763	9768	9773
95	9777	9782	9786	9791	9795	9800	9805	9809	9814	9818
96	9823	9827	9832	9836	9841	9845	9850	9854	9859	9863
97	9868	9872	9877	9881	9886	9890	9894	9899	9903	9908
98	9912	9917	9921	9926	9930	9934	9939	9943	9948	9952
99	9956	9961	9965	9969	9974	9978	9983	9987	9991	9996
100	0000	0004	0009	0013	0017	0022	0026	0030	0035	0039
N	0	1	2	3	4	5	6	7	8	9

TABLE 3

Natural Trigonometric Functions for Decimal Fractions of a Degree

Deg.	Sin	Tan	Cot	Cos	Deg.	Deg.	Sin	Tan	Cot	Cos	Deg.
0.0	0.00000	0.00000	∞	1.0000	90.0	.5	.07846	.07870	12.706	.9969	.5
.1	.00175	.00175	573.0	1.0000	.9	.6	.08020	.08046	12.429	.9968	.4
.2	.00349	.00349	286.5	1.0000	.8	.7	.08194	.08221	12.163	.9966	.3
.3	.00524	.00524	191.0	1.0000	.7	.8	.08368	.08397	11.909	.9965	.2
.4	.00698	.00698	143.24	1.0000	.6	.9	.08542	.08573	11.664	.9963	.1
.5	.00873	.00873	114.59	1.0000	.5	5.0	0.08716	0.08749	11.430	0.9962	85.0
.6	.01047	.01047	95.49	0.9999	.4	.1	.08889	.08925	11.205	.9960	.9
.7	.01222	.01222	81.85	.9999	.3	.2	.09063	.09101	10.988	.9959	.8
.8	.01396	.01396	71.62	.9999	.2	.3	.09237	.09277	10.780	.9957	.7
.9	.01571	.01571	63.66	.9999	.1	.4	.09411	.09453	10.579	.9956	.6
1.0	0.01745	0.01746	57.29	0.9998	89.0	.5	.09585	.09629	10.385	.9954	.5
.1	.01920	.01920	52.08	.9998	.9	.6	.09758	.09805	10.199	.9952	.4
.2	.02094	.02095	47.74	.9998	.8	.7	.09932	.09981	10.019	.9951	.3
.3	.02269	.02269	44.07	.9997	.7	.8	.10106	.10158	9.845	.9949	.2
.4	.02443	.02444	40.92	.9997	.6	.9	.10279	.10334	9.677	.9947	.1
.5	.02618	.02619	38.19	.9997	.5	6.0	0.10453	0.10510	9.514	0.9945	84.0
.6	.02792	.02793	35.80	.9996	.4	.1	.10626	.10687	9.357	.9943	.9
.7	.02967	.02968	33.69	.9996	.3	.2	.10800	.10863	9.205	.9942	.8
.8	.03141	.03143	31.82	.9995	.2	.3	.10973	.11040	9.058	.9940	.7
.9	.03316	.03317	30.14	.9995	.1	.4	.11147	.11217	8.915	.9938	.6
2.0	0.03490	0.03492	28.64	0.9994	88.0	.5	.11320	.11394	8.777	.9936	.5
.1	.03664	.03667	27.27	.9993	.9	.6	.11494	.11570	8.643	.9934	.4
.2	.03839	.03842	26.03	.9993	.8	.7	.11667	.11747	8.513	.9932	.3
.3	.04013	.04016	24.90	.9992	.7	.8	.11840	.11924	8.386	.9930	.2
.4	.04188	.04191	23.86	.9991	.6	.9	.12014	.12101	8.264	.9928	.1
.5	.04362	.04366	22.90	.9990	.5	7.0	0.12187	0.12278	8.144	0.9925	83.0
.6	.04536	.04541	22.02	.9990	.4	.1	.12360	.12456	8.028	.9923	.9
.7	.04711	.04716	21.20	.9989	.3	.2	.12533	.12633	7.916	.9921	.8
.8	.04885	.04891	20.45	.9988	.2	.3	.12706	.12810	7.806	.9919	.7
.9	.05059	.05066	19.74	.9987	.1	.4	.12880	.12988	7.700	.9917	.6
3.0	0.05234	0.05241	19.081	0.9986	87.0	.5	.13053	.13165	7.596	.9914	.5
.1	.05408	.05416	18.464	.9985	.9	.6	.13226	.13343	7.495	.9912	.4
.2	.05582	.05591	17.886	.9984	.8	.7	.13399	.13521	7.396	.9910	.3
.3	.05756	.05766	17.343	.9983	.7	.8	.13572	.13698	7.300	.9907	.2
.4	.05931	.05941	16.832	.9982	.6	.9	.13744	.13876	7.207	.9905	.1
.5	.06105	.06116	16.350	.9981	5	8.0	0.13917	0.14054	7.115	0.9903	82.0
.6	.06279	.06291	15.895	.9980	.4	.1	.14090	.14232	7.026	.9900	.9
.7	.06453	.06467	15.464	.9979	.3	.2	.14263	.14410	6.940	.9898	.8
8	.06627	.06642	15.056	.9978	.2´	.3	.14436	.14588	6.855	.9895	.7
.9	.06802	.06817	14.669	.9977	.1	.4	.14608	.14767	6.772	.9893	.6
4.0	0.06976	0 06993	14.301	0.9976	86.0	.5	.14781	.14945	6.691	.9890	.5
.1	.07150	.07168	13.951	.9974	.9	.6	.14954	.15124	6.612	.9888	.4
.2	.07324	.07344	13.617	.9973	.8	.7	.15126	.15302	6.535	.9885	.3
.3	.07498	.07519	13.300	.9972	.7	.8	.15299	.15481	6.460	.9882	.2
.4	.07672	.07695	12.996	.9971	.6	.9	.15471	.15660	6.386	.9880	.1
Deg.	Cos	Cot	Tan	Sin	Deg.	Deg.	Cos	Cot	Tan	Sin	Deg.

TABLE 3—*Continued*

NATURAL TRIGONOMETRIC FUNCTIONS FOR DECIMAL FRACTIONS OF A DEGREE

Deg.	Sin	Tan	Cot	Cos	Deg.	Deg.	Sin	Tan	Cot	Cos	Deg.
9.0	0.15643	0.15838	6.314	0.9877	81.0	.5	.2334	.2401	4.165	.9724	.5
.1	.15816	.16017	6.243	.9874	.9	.6	.2351	.2419	4.134	.9720	.4
.2	.15988	.16196	6.174	.9871	.8	.7	.2368	.2438	4.102	.9715	.3
.3	.16160	.16376	6.107	.9869	.7	.8	.2385	.2456	4.071	.9711	.2
.4	.16333	.16555	6.041	.9866	.6	.9	.2402	.2475	4.041	.9707	.1
.5	.16505	.16734	5.976	.9863	.5	14.0	0.2419	0.2493	4.011	0.9703	76.0
.6	.16677	.16914	5.912	.9860	.4	.1	.2436	.2512	3.981	.9699	.9
.7	.16849	.17093	5.850	.9857	.3	.2	.2453	.2530	3.952	.9694	.8
.8	.17021	.17273	5.789	.9854	.2	.3	.2470	.2549	3.923	.9690	.7
.9	.17193	.17453	5.730	.9851	.1	.4	.2487	.2568	3.895	.9686	.6
10.0	0.1736	0.1763	5.671	0.9848	80.0	.5	.2504	.2586	3.867	.9681	.5
.1	.1754	.1781	5.614	.9845	.9	.6	.2521	.2605	3.839	.9677	.4
.2	.1771	.1799	5.558	.9842	.8	.7	.2538	.2623	3.812	.9673	.3
.3	.1788	.1817	5.503	.9839	.7	.8	.2554	.2642	3.785	.9668	.2
.4	.1805	.1835	5.449	.9836	.6	.9	.2571	.2661	3.758	.9664	.1
.5	.1822	.1853	5.396	.9833	.5	15.0	0.2588	0.2679	3.732	0.9659	75.0
.6	.1840	.1871	5.343	.9829	.4	.1	.2605	.2698	3.706	.9655	.9
.7	.1857	.1890	5.292	.9826	.3	.2	.2622	.2717	3.681	.9650	.8
.8	.1874	.1908	5.242	.9823	.2	.3	.2639	.2736	3.655	.9646	.7
.9	.1891	.1926	5.193	.9820	.1	.4	.2656	.2754	3.630	.9641	.6
11.0	0.1908	0.1944	5.145	0.9816	79.0	.5	.2672	.2773	3.606	.9636	.5
.1	.1925	.1962	5.097	.9813	.9	.6	.2689	.2792	3.582	.9632	.4
.2	.1942	.1980	5.050	.9810	.8	.7	.2706	.2811	3.558	.9627	.3
.3	.1959	.1998	5.005	.9806	.7	.8	.2723	.2830	3.534	.9622	.2
.4	.1977	.2016	4.959	.9803	.6	.9	.2740	.2849	3.511	.9617	.1
.5	.1994	.2035	4.915	.9799	.5	16.0	0.2756	0.2867	3.487	0.9613	74.0
.6	.2011	.2053	4.872	.9796	.4	.1	.2773	.2886	3.465	.9608	.9
.7	.2028	.2071	4.829	.9792	.3	.2	.2790	.2905	3.442	.9603	.8
.8	.2045	.2089	4.787	.9789	.2	.3	.2807	.2924	3.420	.9598	.7
.9	.2062	.2107	4.745	.9785	.1	.4	.2823	.2943	3.398	.9593	.6
12.0	0.2079	0.2126	4.705	0.9781	78.0	.5	.2840	.2962	3.376	.9588	.5
.1	.2096	.2144	4.665	.9778	.9	.6	.2857	.2981	3.354	.9583	.4
.2	.2113	.2162	4.625	.9774	.8	.7	.2874	.3000	3.333	.9578	.3
.3	.2130	.2180	4.586	.9770	.7	.8	.2890	.3019	3.312	.9573	.2
.4	.2147	.2199	4.548	.9767	.6	.9	.2907	.3038	3.291	.9568	.1
.5	.2164	.2217	4.511	.9763	.5	17.0	0.2924	0.3057	3.271	0.9563	73.0
.6	.2181	.2235	4.474	.9759	.4	.1	.2940	.3076	3.251	.9558	.9
.7	.2198	.2254	4.437	.9755	.3	.2	.2957	.3096	3.230	.9553	.8
.8	.2215	.2272	4.402	.9751	.2	.3	.2974	.3115	3.211	.9548	.7
.9	.2233	.2290	4.366	.9748	.1	.4	.2990	.3134	3.191	.9542	.6
13.0	0.2250	0.2309	4.331	0.9744	77.0	.5	.3007	.3153	3.172	.9537	.5
.1	.2267	.2327	4.297	.9740	.9	.6	.3024	.3172	3.152	.9532	.4
.2	.2284	.2345	4.264	.9736	.8	.7	.3040	.3191	3.133	.9527	.3
.3	.2300	.2364	4.230	.9732	.7	.8	.3057	.3211	3.115	.9521	.2
.4	.2317	.2382	4.198	.9728	.6	.9	.3074	.3230	3.096	.9516	.1
Deg.	Cos	Cot	Tan	Sin	Deg.	Deg.	Cos	Cot	Tan	Sin	Deg.

TABLE 3—*Continued*

NATURAL TRIGONOMETRIC FUNCTIONS FOR DECIMAL FRACTIONS OF A DEGREE

Deg.	Sin	Tan	Cot	Cos	Deg.	Deg.	Sin	Tan	Cot	Cos	Deg.
18.0	0.3090	0.3249	3.078	0.9511	72.0	.5	.3827	.4142	2.414	.9239	.5
.1	.3107	.3269	3.060	.9505	.9	.6	.3843	.4163	2.402	.9232	.4
.2	.3123	.3288	3.042	.9500	.8	.7	.3859	.4183	2.391	.9225	.3
.3	.3140	.3307	3.024	.9494	.7	.8	.3875	.4204	2.379	.9219	.2
.4	.3156	.3327	3.006	.9489	.6	.9	.3891	.4224	2.367	.9212	.1
.5	.3173	.3346	2.989	.9483	.5	23.0	0.3907	0.4245	2.356	0.9205	67.0
.6	.3190	.3365	2.971	.9478	.4	1	.3923	.4265	2.344	.9198	.9
.7	.3206	.3385	2.954	.9472	.3	.2	.3939	.4286	2.333	.9191	.8
.8	.3223	.3404	2.937	.9466	.2	.3	.3955	.4307	2.322	9184	.7
.9	.3239	.3424	2.921	.9461	.1	.4	.3971	.4327	2.311	.9178	.6
19.0	0.3256	0.3443	2.904	0.9455	71.0	.5	.3987	.4348	2.300	.9171	.5
.1	.3272	.3463	2.888	.9449	.9	.6	.4003	.4369	2.289	.9164	.4
.2	.3289	.3482	2.872	.9444	.8	.7	.4019	.4390	2.278	.9157	.3
.3	.3305	.3502	2.856	.9438	.7	.8	.4035	.4411	2.267	.9150	.2
.4	.3322	.3522	2.840	.9432	.6	.9	.4051	.4431	2.257	.9143	.1
.5	.3338	.3541	2.824	.9426	.5	24.0	0.4067	0.4452	2.246	0.9135	66.0
.6	.3355	.3561	2.808	.9421	.4	.1	.4083	.4473	2.236	.9128	.9
.7	.3371	.3581	2.793	.9415	.3	.2	.4099	.4494	2.225	.9121	.8
.8	.3387	.3600	2.778	.9409	.2	.3	.4115	.4515	2.215	.9114	.7
.9	.3404	.3620	2.762	.9403	.1	.4	.4131	.4536	2.204	.9107	.6
20.0	0.3420	0.3640	2.747	0.9397	70.0	.5	.4147	.4557	2.194	.9100	.5
.1	.3437	.3659	2.733	.9391	.9	.6	.4163	.4578	2.184	.9092	.4
.2	.3453	.3679	2.718	.9385	.8	.7	.4179	.4599	2.174	.9085	.3
.3	.3469	.3699	2.703	.9379	.7	.8	.4195	.4621	2.164	.9078	.2
.4	.3486	.3719	2.689	.9373	.6	.9	.4210	.4642	2.154	.9070	.1
.5	.3502	.3739	2.675	.9367	.5	25.0	0.4226	0.4663	2.145	0.9063	65.0
.6	.3518	.3759	2.660	.9361	.4	.1	.4242	.4684	2.135	.9056	.9
.7	.3535	.3779	2.646	.9354	.3	.2	.4258	.4706	2.125	.9048	.8
.8	.3551	.3799	2.633	.9348	.2	.3	.4274	.4727	2.116	.9041	.7
.9	.3567	.3819	2.619	.9342	.1	.4	.4289	.4748	2.106	.9033	.6
21.0	0.3584	0.3839	2.605	0.9336	69.0	.5	.4305	.4770	2.097	.9026	.5
.1	.3600	.3859	2.592	.9330	.9	.6	.4321	.4791	2.087	.9018	.4
.2	.3616	.3879	2.578	.9323	.8	.7	.4337	.4813	2.078	.9011	.3
.3	.3633	.3899	2.565	.9317	.7	.8	.4352	.4834	2.069	.9003	.2
.4	.3649	.3919	2.552	.9311	.6	.9	.4368	.4856	2.059	.8996	.1
.5	.3665	.3939	2.539	.9304	.5	26.0	0.4384	0.4877	2.050	0.8988	64.0
.6	.3681	.3959	2.526	.9298	.4	.1	.4399	.4899	2.041	.8980	.9
.7	.3697	.3979	2.513	.9291	.3	.2	.4415	.4921	2.032	.8973	.8
.8	.3714	.4000	2.500	.9285	.2	.3	.4431	.4942	2.023	.8965	.7
.9	.3730	.4020	2.488	.9278	.1	.4	.4446	.4964	2.014	.8957	.6
22.0	0.3746	0.4040	2.475	0.9272	68.0	.5	.4462	.4986	2.006	.8949	.5
.1	.3762	.4061	2.463	.9265	.9	.6	.4478	.5008	1.997	.8942	.4
2	.3778	.4081	2.450	.9259	.8	.7	.4493	.5029	1.988	.8934	.3
.3	.3795	.4101	2.438	.9252	.7	.8	.4509	.5051	1.980	.8926	.2
.4	.3811	.4122	2.426	.9245	.6	.9	.4524	.5073	1.971	.8918	.1
Deg.	Cos	Cot	Tan	Sin	Deg.	Deg.	Cos	Cot	Tan	Sin	Deg.

TABLE 3—*Continued*

NATURAL TRIGONOMETRIC FUNCTIONS FOR DECIMAL FRACTIONS OF A DEGREE

Deg.	Sin	Tan	Cot	Cos	Deg.	Deg.	Sin	Tan	Cot	Cos	Deg.
27.0	0.4540	0.5095	1.963	0.8910	63.0	.5	.5225	.6128	1.6319	.8526	.5
.1	.4555	.5117	1.954	.8902	.9	.6	.5240	.6152	1.6255	.8517	.4
.2	.4571	.5139	1.946	.8894	.8	.7	.5255	.6176	1.6191	.8508	.3
.3	.4586	.5161	1.937	.8886	.7	.8	.5270	.6200	1.6128	.8499	.2
.4	.4602	.5184	1.929	.8878	.6	.9	.5284	.6224	1.6066	.8490	.1
.5	.4617	.5206	1.921	.8870	.5	32.0	0.5299	0.6249	1.6003	0.8480	58.0
.6	.4633	.5228	1.013	.8862	.4	.1	.5314	.6273	1.5941	.8471	.9
.7	.4648	.5250	1.905	.8854	.3	.2	.5329	.6297	1.5880	.8462	.8
.8	.4664	.5272	1.897	.8846	.2	.3	.5344	.6322	1.5818	.8453	.7
.9	.4679	.5295	1.889	.8838	.1	.4	5358	6346	1.5757	.8443	.6
28.0	0.4695	0.5317	1.881	0.8829	62.0	.5	.5373	.6371	1.5697	.8434	.5
.1	.4710	.5340	1.873	.8821	.9	.6	.5388	.6395	1.5637	.8425	.4
.2	.4726	.5362	1.865	.8813	.8	.7	.5402	.6420	1.5577	.8415	.3
.3	.4741	.5384	1.857	.8805	.7	.8	.5417	.6445	1.5517	.8406	.2
.4	.4756	.5407	1.849	.8796	.6	.9	.5432	.6469	1.5458	.8396	.1
.5	.4772	.5430	1.842	.8788	.5	33.0	0.5446	0.6494	1.5399	0.8387	57.0
.6	.4787	.5452	1.834	.8780	.4	.1	.5461	.6519	1.5340	.8377	.9
.7	.4802	.5475	1.827	.8771	.3	.2	.5476	.6544	1.5282	.8368	.8
.8	.4818	.5498	1.819	.8763	.2	.3	.5490	.6569	1.5224	.8358	.7
.9	.4833	.5520	1.811	.8755	.1	.4	.5505	.6594	1.5166	.8348	.6
29.0	0.4848	0.5543	1.804	0.8746	61.0	.5	.5519	.6619	1.5108	.8339	.5
.1	.4863	.5566	1.797	.8738	.9	.6	.5534	.6644	1.5051	.8329	.4
.2	.4879	.5589	1.789	.8729	.8	.7	.5548	.6669	1.4994	.8320	.3
.3	.4894	.5612	1.782	.8721	.7	.8	.5563	.6694	1.4938	.8310	.2
.4	.4909	.5635	1.775	.8712	.6	.9	.5577	.6720	1.4882	.8300	.1
.5	.4924	.5658	1.767	.8704	.5	34.0	0.5592	0.6745	1.4826	0.8290	56.0
.6	.4939	.5681	1.760	.8695	.4	.1	.5606	.6771	1.4770	.8281	.9
.7	.4955	.5704	1.753	.8686	.3	.2	.5621	.6796	1.4715	.8271	.8
.8	.4970	.5727	1.746	.8678	.2	.3	.5635	.6822	1.4659	.8261	.7
.9	.4985	.5750	1.739	.8669	.1	.4	.5650	.6847	1.4605	.8251	.6
30.0	0.5000	0.5774	1.7321	0.8660	60.0	.5	.5664	.6873	1.4550	.8241	.5
.1	.5015	.5797	1.7251	.8652	.9	.6	.5678	.6899	1.4496	.8231	.4
.2	.5030	.5820	1.7182	.8643	.8	.7	.5693	.6924	1.4442	.8221	.3
.3	.5045	.5844	1.7113	.8634	.7	.8	.5707	.6950	1.4388	.8211	.2
.4	.5060	.5867	1.7045	.8625	.6	.9	.5721	.6973	1.4335	.8202	.1
.5	.5075	.5890	1.6977	.8616	.5	35.0	0.5736	0.7002	1.4281	0.8192	55.0
.6	.5090	.5914	1.6909	.8607	.4	.1	.5750	.7028	1.4229	.8181	.9
.7	.5105	.5938	1.6842	.8599	.3	.2	.5764	.7054	1.4176	.8171	.8
.8	.5120	.5961	1.6775	.8590	.2	.3	.5779	.7080	1.4124	.8161	.7
.9	.5135	.5985	1.6709	.8581	.1	.4	.5793	.7107	1.4071	.8151	.6
31.0	0.5150	0.6009	1.6643	0.8572	59.0	.5	.5807	.7133	1.4019	.8141	.5
.1	.5165	.6032	1.6577	.8563	.9	.6	.5821	.7159	1.3968	.8131	.4
.2	.5180	.6056	1.6512	.8554	.8	.7	.5835	.7186	1.3916	.8121	.3
.3	.5195	.6080	1.6447	.8545	.7	.8	.5850	.7212	1.3865	.8111	.2
.4	.5210	.6104	1.6383	8536	.6	.9	.5864	.7239	1.3814	.8100	.1
Deg.	Cos	Cot	Tan	Sin	Deg.	Deg.	Cos	Cot	Tan	Sin	Deg.

TABLE 3—*Continued*

NATURAL TRIGONOMETRIC FUNCTIONS FOR DECIMAL FRACTIONS OF A DEGREE

Deg.	Sin	Tan	Cot	Cos	Deg.	Deg.	Sin	Tan	Cot	Cos	Deg.
36.0	0.5878	0.7265	1.3764	0.8090	54.0	.5	.6494	.8541	1.1708	.7604	.5
.1	.5892	.7292	1.3713	.8080	.9	.6	.6508	.8571	1.1667	.7593	.4
.2	.5906	.7319	1.3663	.8070	.8	.7	.6521	.8601	1.1626	.7581	.3
.3	.5920	.7346	1.3613	.8059	.7	.8	.6534	.8632	1.1585	.7570	.2
.4	.5934	.7373	1.3564	.8049	.6	.9	.6547	.8662	1.1544	.7559	.1
.5	.5948	.7400	1.3514	.8039	.5	41.0	0.6561	0.8693	1.1504	0.7547	49.0
.6	.5962	.7427	1.3465	.8028	.4	.1	.6574	.8724	1.1463	.7536	.9
.7	.5976	.7454	1.3416	.8018	.3	.2	.6587	.8754	1.1423	.7524	.8
.8	.5990	.7481	1.3367	.8007	.2	.3	.6600	.8785	1.1383	.7513	.7
.9	.6004	.7508	1.3319	.7997	.1	.4	.6613	.8816	1.1343	.7501	.6
37.0	0.6018	0.7536	1.3270	0.7986	53.0	.5	.6626	.8847	1.1303	.7490	.5
.1	.6032	.7563	1.3222	.7976	.9	.6	.6639	.8878	1.1263	.7478	.4
.2	.6046	.7590	1.3175	.7965	.8	.7	.6652	.8910	1.1224	.7466	.3
.3	.6060	.7618	1.3127	.7955	.7	.8	.6665	.8941	1.1184	.7455	.2
.4	.6074	.7646	1.3079	.7944	.6	.9	.6678	.8972	1.1145	.7443	.1
.5	.6088	.7673	1.3032	.7934	.5	42.0	0.6691	0.9004	1.1106	0.7431	48.0
.6	.6101	.7701	1.2985	.7923	.4	.1	.6704	.9036	1.1067	.7420	.9
.7	.6115	.7729	1.2938	.7912	.3	.2	.6717	.9067	1.1028	.7408	.8
.8	.6129	.7757	1.2892	.7902	.2	.3	.6730	.9099	1.0990	.7396	.7
.9	.6143	.7785	1.2846	.7891	.1	.4	.6743	.9131	1.0951	.7385	.6
38.0	0.6157	0.7813	1.2799	0.7880	52.0	.5	.6756	.9163	1.0913	.7373	.5
.1	.6170	.7841	1.2753	.7869	.9	.6	.6769	.9195	1.0875	.7361	.4
.2	.6184	.7869	1.2708	.7859	.8	.7	.6782	.9228	1.0837	.7349	.3
.3	.6198	.7898	1.2662	.7848	.7	.8	.6794	.9260	1.0799	.7337	.2
.4	.6211	.7926	1.2617	.7837	.6	.9	.6807	.9293	1.0761	.7325	.1
.5	.6225	.7954	1.2572	.7826	.5	43.0	0.6820	0.9325	1.0724	0.7314	47.0
.6	.6239	.7983	1.2527	.7815	.4	.1	.6833	.9358	1.0686	.7302	.9
.7	.6252	.8012	1.2482	.7804	.3	.2	.6845	.9391	1.0649	.7290	.8
.8	.6266	.8040	1.2437	.7793	.2	.3	.6858	.9424	1.0612	.7278	.7
.9	.6280	.8069	1.2393	.7782	.1	.4	.6871	.9457	1.0575	.7266	.6
39.0	0.6293	0.8098	1.2349	0.7771	51.0	.5	.6884	.9490	1.0538	.7254	.5
.1	.6307	.8127	1.2305	.7760	.9	.6	.6896	.9523	1.0501	.7242	.4
.2	.6320	.8156	1.2261	.7749	.8	.7	.6909	.9556	1.0464	.7230	.3
.3	.6334	.8185	1.2218	.7738	.7	.8	.6921	.9590	1.0428	.7218	.2
.4	.6347	.8214	1.2174	.7727	.6	.9	.6934	.9623	1.0392	.7206	.1
.5	.6361	.8243	1.2131	.7716	.5	44.0	0.6947	0.9657	1.0355	0.7193	46.0
.6	.6374	.8273	1.2088	.7705	.4	.1	.6959	.9691	1.0319	.7181	.9
.7	.6388	.8302	1.2045	.7694	.3	.2	.6972	.9725	1.0283	.7169	.8
.8	.6401	.8332	1.2002	.7683	.2	.3	.6984	.9759	1.0247	.7157	.7
.9	.6414	.8361	1.1960	.7672	.1	.4	.6997	.9793	1.0212	.7145	.6
40.0	0.6428	0.8391	1.1918	0.7660	50.0	.5	.7009	.9827	1.0176	.7133	.5
.1	.6441	.8421	1.1875	.7649	.9	.6	.7022	.9861	1.0141	.7120	.4
.2	.6455	.8451	1.1833	.7638	.8	.7	.7034	.9896	1.0105	.7108	.3
.3	.6468	.8481	1.1792	.7627	.7	.8	.7046	.9930	1.0070	.7096	.2
.4	.6481	.8511	1.1750	.7615	.6	.9	.7059	.9965	1.0035	.7083	.1
						45.0	0.7071	1.0000	1.0000	0.7071	45.0
Deg.	Cos	Cot	Tan	Sin	Deg.	Deg.	Cos	Cot	Tan	Sin	Deg.

TABLE 4

LOGARITHMS OF TRIGONOMETRIC FUNCTIONS FOR DECIMAL FRACTIONS OF A DEGREE

Deg.	L. Sin	L. Tan	L. Cot	L. Cos	Deg.
0.0	$-\infty$	$-\infty$	∞	0.0000	90.0
.1	7.2419	7.2419	2.7581	0.0000	.9
.2	7.5429	7.5429	2.4571	0.0000	.8
.3	7.7190	7.7190	2.2810	0.0000	.7
.4	7.8439	7.8439	2.1561	0.0000	.6
.5	7.9408	7.9409	2.0591	0.0000	.5
.6	8.0200	8.0200	1.9800	0.0000	.4
.7	8.0870	8.0870	1.9130	0.0000	.3
.8	8.1450	8.1450	1.8550	0.0000	.2
.9	8.1961	8.1962	1.8038	9.9999	.1
1.0	8.2419	8.2419	1.7581	9.9999	89.0
.1	8.2832	8.2833	1.7167	9.9999	.9
.2	8.3210	8.3211	1.6789	9.9999	.8
.3	8.3558	8.3559	1.6441	9.9999	.7
.4	8.3880	8.3881	1.6119	9.9999	.6
.5	8.4179	8.4181	1.5819	9.9999	.5
.6	8.4459	8.4461	1.5539	9.9998	.4
.7	8.4723	8.4725	1.5275	9.9998	.3
.8	8.4971	8.4973	1.5027	9.9998	.2
.9	8.5206	8.5208	1.4792	9.9998	.1
2.0	8.5428	8.5431	1.4569	9.9997	88.0
.1	8.5640	8.5643	1.4357	9.9997	.9
.2	8.5842	8.5845	1.4155	9.9997	.8
.3	8.6035	8.6038	1.3962	9.9996	.7
.4	8.6220	8.6223	1.3777	9.9996	.6
.5	8.6397	8.6401	1.3599	9.9996	.5
.6	8.6567	8.6571	1.3429	9.9996	.4
.7	8.6731	8.6736	1.3264	9.9995	.3
.8	8.6889	8.6894	1.3106	9.9995	.2
.9	8.7041	8.7046	1.2954	9.9994	.1
3.0	8.7188	8.7194	1.2806	9.9994	87.0
.1	8.7330	8.7337	1.2663	9.9994	.9
.2	8.7468	8.7475	1.2525	9.9993	.8
.3	8.7602	8.7609	1.2391	9.9993	.7
.4	8.7731	8.7739	1.2261	9.9992	.6
.5	8.7857	8.7865	1.2135	9.9992	.5
.6	8.7979	8.7988	1.2012	9.9991	.4
.7	8.8098	8.8107	1.1893	9.9991	.3
.8	8.8213	8.8223	1.1777	9.9990	.2
.9	8.8326	8.8336	1.1664	9.9990	.1
4.0	8.8436	8.8446	1.1554	9.9989	86.0
.1	8.8543	8.8554	1.1446	9.9989	.9
.2	8.8647	8.8659	1.1341	9.9988	.8
.3	8.8749	8.8762	1.1238	9.9988	.7
.4	8.8849	8.8862	1.1138	9.9987	.6
Deg.	L. Cos	L. Cot	L. Tan	L. Sin	Deg.

TABLE 4—*Continued*

LOGARITHMS OF TRIGONOMETRIC FUNCTIONS FOR DECIMAL FRACTIONS OF A DEGREE

Deg.	L. Sin	L. Tan	L. Cot	L. Cos	Deg.
4.5	8.8946	8.8960	1.1040	9.9987	85.5
.6	8.9042	8.9056	1.0944	9.9986	.4
.7	8.9135	8.9150	1.0850	9.9985	.3
.8	8.9226	8.9241	1.0759	9.9985	.2
.9	8.9315	8.9331	1.0669	9.9984	.1
5.0	8.9403	8.9420	1.0580	9.9983	85.0
.1	8.9489	8.9506	1.0494	9.9983	.9
.2	8.9573	8.9591	1.0409	9.9982	.8
.3	8.9655	8.9674	1.0326	9.9981	.7
.4	8.9736	8.9756	1.0244	9.9981	.6
.5	8.9816	8.9836	1.0164	9.9980	.5
.6	8.9894	8.9915	1.0085	9.9979	.4
.7	8.9970	8.9992	1.0008	9.9978	.3
.8	9.0046	9.0068	0.9932	9.9978	.2
.9	9.0120	9.0143	0.9857	9.9977	.1
6.0	9.0192	9.0216	0.9784	9.9976	84.0
.1	9.0264	9.0289	0.9711	9.9975	.9
.2	9.0334	9.0360	0.9640	9.9975	.8
.3	9.0403	9.0430	0.9570	9.9974	.7
.4	9.0472	9.0499	0.9501	9.9973	.6
.5	9.0539	9.0567	0.9433	9.9972	.5
.6	9.0605	9.0633	0.9367	9.9971	.4
.7	9.0670	9.0699	0.9301	9.9970	.3
.8	9.0734	9.0764	0.9236	9.9969	.2
.9	9.0797	9.0828	0.9172	9.9968	.1
7.0	9.0859	9.0891	0.9109	9.9968	83.0
.1	9.0920	9.0954	0.9046	9.9967	.9
.2	9.0981	9.1015	0.8985	9.9966	.8
.3	9.1040	9.1076	0.8924	9.9965	.7
.4	9.1099	9.1135	0.8865	9.9964	.6
.5	9.1157	9.1194	0.8806	9.9963	.5
.6	9.1214	9.1252	0.8748	9.9962	.4
.7	9.1271	9.1310	0.8690	9.9961	.3
.8	9.1326	9.1367	0.8633	9.9960	.2
.9	9.1381	9.1423	0.8577	9.9959	.1
8.0	9.1436	9.1478	0.8522	9.9958	82.0
.1	9.1489	9.1533	0.8467	9.9956	.9
.2	9.1542	9.1587	0.8413	9.9955	.8
.3	9.1594	9.1640	0.8360	9.9954	.7
.4	9.1646	9.1693	0.8307	9.9953	.6
.5	9.1697	9.1745	0.8255	9.9952	.5
.6	9.1747	9.1797	0.8203	9.9951	.4
.7	9.1797	9.1848	0.8152	9.9950	.3
.8	9.1847	9.1898	0.8102	9.9949	.2
.9	9.1895	9.1948	0.8052	9.9947	.1
Deg.	L. Cos	L. Cot	L. Tan	L. Sin	Deg.

TABLE 4—*Continued*

LOGARITHMS OF TRIGONOMETRIC FUNCTIONS FOR DECIMAL FRACTIONS OF A DEGREE

Deg.	L. Sin	L. Tan	L. Cot	L. Cos	Deg.
9.0	9.1943	9.1997	0.8003	9.9946	81.0
.1	9.1991	9.2046	0.7954	9.9945	.9
.2	9.2038	9.2094	0.7906	9.9944	.8
.3	9.2085	9.2142	0.7858	9.9943	.7
.4	9.2131	9.2189	0.7811	9.9941	.6
.5	9.2176	9.2236	0.7764	9.9940	.5
.6	9.2221	9.2282	0.7718	9.9939	.4
.7	9.2266	9.2328	0.7672	9.9937	.3
.8	9.2310	9.2374	0.7626	9.9936	.2
.9	9.2353	9.2419	0.7581	9.9935	.1
10.0	9.2397	9.2463	0.7537	9.9934	80.0
.1	9.2439	9.2507	0.7493	9.9932	.9
.2	9.2482	9.2551	0.7449	9.9931	.8
.3	9.2524	9.2594	0.7406	9.9929	.7
.4	9.2565	9.2637	0.7363	9.9928	.6
.5	9.2606	9.2680	0.7320	9.9927	.5
.6	9.2647	9.2722	0.7278	9.9925	.4
.7	9.2687	9.2764	0.7236	9.9924	.3
.8	9.2727	9.2805	0.7195	9.9922	.2
.9	9.2767	9.2846	0.7154	9.9921	.1
11.0	9.2806	9.2887	0.7113	9.9919	79.0
.1	9.2845	9.2927	0.7073	9.9918	.9
.2	9.2883	9.2967	0.7033	9.9916	.8
.3	9.2921	9.3006	0.6994	9.9915	.7
.4	9.2959	9.3046	0.6954	9.9913	.6
.5	9.2997	9.3085	0.6915	9.9912	.5
.6	9.3034	9.3123	0.6877	9.9910	.4
.7	9.3070	9.3162	0.6838	9.9909	.3
.8	9.3107	9.3200	0.6800	9.9907	.2
.9	9.3143	9.3237	0.6763	9.9906	.1
12.0	9.3179	9.3275	0.6725	9.9904	78.0
.1	9.3214	9.3312	0.6688	9.9902	.9
.2	9.3250	9.3349	0.6651	9.9901	.8
.3	9.3284	9.3385	0.6615	9.9899	.7
.4	9.3319	9.3422	0.6578	9.9897	.6
.5	9.3353	9.3458	0.6542	9.9896	.5
.6	9.3387	9.3493	0.6507	9.9894	.4
.7	9.3421	9.3529	0.6471	9.9892	.3
.8	9.3455	9.3564	0.6436	9.9891	.2
.9	9.3488	9.3599	0.6401	9.9889	.1
13.0	9.3521	9.3634	0.6366	9.9887	77.0
.1	9.3554	9.3668	0.6332	9.9885	.9
.2	9.3586	9.3702	0.6298	9.9884	.8
.3	9.3618	9.3736	0.6264	9.9882	.7
.4	9.3650	9.3770	0.6230	9.9880	.6
Deg.	L. Cos	L. Cot	L. Tan	L. Sin	Deg.

TABLE 4—*Continued*

LOGARITHMS OF TRIGONOMETRIC FUNCTIONS FOR DECIMAL FRACTIONS OF A DEGREE

Deg.	L. Sin	L. Tan	L. Cot	L. Cos	Deg.
13.5	9.3682	9.3804	0.6196	9.9878	76.5
.6.	9.3713	9.3837	0.6163	9.9876	.4
.7	9.3745	9.3870	0.6130	9.9875	.3
.8	9.3775	9.3903	0.6097	9.9873	.2
.9	9.3806	9.3935	0.6065	9.9871	.1
14.0	9.3837	9.3968	0.6032	9.9869	76.0
.1	9.3867	9.4000	0.6000	9.9867	.9
.2	9.3897	9.4032	0.5968	9.9865	.8
.3	9.3927	9.4064	0.5936	9.9863	.7
.4	9.3957	9.4095	0.5905	9.9861	.6
.5	9.3986	9.4127	0.5873	9.9859	.5
.6	9.4015	9.4158	0.5842	9.9857	.4
.7	9.4044	9.4189	0.5811	9.9855	.3
.8	9.4073	9.4220	0.5780	9.9853	.2
.9	9.4102	9.4250	0.5750	9.9851	.1
15.0	9.4130	9.4281	0.5719	9.9849	75.0
.1	9.4158	9.4311	0.5689	9.9847	.9
.2	9.4186	9.4341	0.5659	9.9845	.8
.3	9.4214	9.4371	0.5629	9.9843	.7
.4	9.4242	9.4400	0.5600	9.9841	.6
.5	9.4269	9.4430	0.5570	9.9839	.5
.6	9.4296	9.4459	0.5541	9.9837	.4
.7	9.4323	9.4488	0.5512	9.9835	.3
.8	9.4350	9.4517	0.5483	9.9833	.2
.9	9.4377	9.4546	0.5454	9.9831	.1
16.0	9.4403	9.4575	0.5425	9.9828	74.0
.1	9.4430	9.4603	0.5397	9.9826	.9
.2	9.4456	9.4632	0.5368	9.9824	.8
.3	9.4482	9.4660	0.5340	9.9822	.7
.4	9.4508	9.4688	0.5312	9.9820	.6
.5	9.4533	9.4716	0.5284	9.9817	.5
.6	9.4559	9.4744	0.5256	9.9815	.4
.7	9.4584	9.4771	0.5229	9.9813	.3
.8	9.4609	9.4799	0.5201	9.9811	.2
.9	9.4634	9.4826	0.5174	9.9808	.1
17.0	9.4659	9.4853	0.5147	9.9806	73.0
.1	9.4684	9.4880	0.5120	9.9804	.9
.2	9.4709	9.4907	0.5093	9.9801	.8
.3	9.4733	9.4934	0.5066	9.9799	.7
.4	9.4757	9.4961	0.5039	9.9797	.6
.5	9.4781	9.4987	0.5013	9.9794	.5
.6	9.4805	9.5014	0.4986	9.9792	.4
.7	9.4829	9.5040	0.4960	9.9789	.3
.8	9.4853	9.5066	0.4934	9.9787	.2
.9	9.4876	9.5092	0.4908	9.9785	.1
Deg.	L. Cos	L. Cot	L. Tan	L. Sin	Deg.

TABLE 4—*Continued*

LOGARITHMS OF TRIGONOMETRIC FUNCTIONS FOR DECIMAL FRACTIONS OF A DEGREE

Deg.	L. Sin	L. Tan	L. Cot	L. Cos	Deg.
18.0	9.4900	9.5118	0.4882	9.9782	72.0
.1	9.4923	9.5143	0.4857	9.9780	.9
.2	9.4946	9.5169	0.4831	9.9777	.8
.3	9.4969	9.5195	0.4805	9.9775	.7
.4	9.4992	9.5220	0.4780	9.9772	.6
.5	9.5015	9.5245	0.4755	9.9770	.5
.6	9.5037	9.5270	0.4730	9.9767	.4
.7	9.5060	9.5295	0.4705	9.9764	.3
.8	9.5082	9.5320	0.4680	9.9762	.2
.9	9.5104	9.5345	0.4655	9.9759	.1
19.0	9.5126	9.5370	0.4630	9.9757	71.0
.1	9.5148	9.5394	0.4606	9.9754	.9
.2	9.5170	9.5419	0.4581	9.9751	.8
.3	9.5192	9.5443	0.4557	9.9749	.7
.4	9.5213	9.5467	0.4533	9.9746	.6
.5	9.5235	9.5491	0.4509	9.9743	.5
.6	9.5256	9.5516	0.4484	9.9741	.4
.7	9.5278	9.5539	0.4461	9.9738	.3
.8	9.5299	9.5563	0.4437	9.9735	.2
.9	9.5320	9.5587	0.4413	9.9733	.1
20.0	9.5341	9.5611	0.4389	9.9730	70.0
.1	9.5361	9.5634	0.4366	9.9727	.9
.2	9.5382	9.5658	0.4342	9.9724	.8
.3	9.5402	9.5681	0.4319	9.9722	.7
.4	9.5423	9.5704	0.4296	9.9719	.6
.5	9.5443	9.5727	0.4273	9.9716	.5
.6	9.5463	9.5750	0.4250	9.9713	.4
.7	9.5484	9.5773	0.4227	9.9710	.3
.8	9.5504	9.5796	0.4204	9.9707	.2
.9	9.5523	9.5819	0.4181	9.9704	.1
21.0	9.5543	9.5842	0.4158	9.9702	69.0
.1	9.5563	9.5864	0.4136	9.9699	.9
.2	9.5583	9.5887	0.4113	9.9696	.8
.3	9.5602	9.5909	0.4091	9.9693	.7
.4	9.5621	9.5932	0.4068	9.9690	.6
.5	9.5641	9.5954	0.4046	9.9687	.5
.6	9.5660	9.5976	0.4024	9.9684	.4
.7	9.5679	9.5998	0.4002	9.9681	.3
.8	9.5698	9.6020	0.3980	9.9678	.2
.9	9.5717	9.6042	0.3958	9.9675	.1
22.0	9.5736	9.6064	0.3936	9.9672	68.0
.1	9.5754	9.6086	0.3914	9.9669	.9
.2	9.5773	9.6108	0.3892	9.9666	.8
.3	9.5792	9.6129	0.3871	9.9662	.7
.4	9.5810	9.6151	0.3849	9.9659	.6
Deg.	L. Cos	L. Cot	L. Tan	L. Sin	Deg.

TABLE 4—*Continued*

LOGARITHMS OF TRIGONOMETRIC FUNCTIONS FOR DECIMAL FRACTIONS OF A DEGREE

Deg.	L. Sin	L. Tan	L. Cot	L. Cos	Deg.
22.5	9.5828	9.6172	0.3828	9.9656	67.5
.6	9.5847	9.6194	0.3806	9.9653	.4
.7	9.5865	9.6215	0.3785	9.9650	.3
.8	9.5883	9.6236	0.3764	9.9647	.2
.9	9.5901	9.6257	0.3743	9.9643	.1
23.0	9.5919	9.6279	0.3721	9.9640	67.0
.1	9.5937	9.6300	0.3700	9.9637	.9
.2	9.5954	9.6321	0.3679	9.9634	.8
.3	9.5972	9.6341	0.3659	9.9631	.7
.4	9.5990	9.6362	0.3638	9.9627	.6
.5	9.6007	9.6383	0.3617	9.9624	.5
.6	9.6024	9.6404	0.3596	9.9621	.4
.7	9.6042	9.6424	0.3576	9.9617	.3
.8	9.6059	9.6445	0.3555	9.9614	.2
.9	9.6076	9.6465	0.3535	9.9611	.1
24.0	9.6093	9.6486	0.3514	9.9607	66.0
.1	9.6110	9.6506	0.3494	9.9604	.9
.2	9.6127	9.6527	0.3473	9.9601	.8
.3	9.6144	9.6547	0.3453	9.9597	.7
.4	9.6161	9.6567	0.3433	9.9594	.6
.5	9.6177	9.6587	0.3413	9.9590	.5
.6	9.6194	9.6607	0.3393	9.9587	.4
.7	9.6210	9.6627	0.3373	9.9583	.3
.8	9.6227	9.6647	0.3353	9.9580	.2
.9	9.6243	9.6667	0.3333	9.9576	.1
25.0	9.6259	9.6687	0.3313	9.9573	65.0
.1	9.6276	9.6706	0.3294	9.9569	.9
.2	9.6292	9.6726	0.3274	9.9566	.8
.3	9.6308	9.6746	0.3254	9.9562	.7
.4	9.6324	9.6765	0.3235	9.9558	.6
.5	9.6340	9.6785	0.3215	9.9555	.5
.6	9.6356	9.6804	0.3196	9.9551	.4
.7	9.6371	9.6824	0.3176	9.9548	.3
.8	9.6387	9.6843	0.3157	9.9544	.2
.9	9.6403	9.6863	0.3137	9.9540	.1
26.0	9.6418	9.6882	0.3118	9.9537	64.0
.1	9.6434	9.6901	0.3099	9.9533	.9
.2	9.6449	9.6920	0.3080	9.9529	.8
.3	9.6465	9.6939	0.3061	9.9525	.7
.4	9.6480	9.6958	0.3042	9.9522	.6
.5	9.6495	9.6977	0.3023	9.9518	.5
.6	9.6510	9.6996	0.3004	9.9514	.4
.7	9.6526	9.7015	0.2985	9.9510	.3
.8	9.6541	9.7034	0.2966	9.9506	.2
.9	9.6556	9.7053	0.2947	9.9503	.1
Deg.	L. Cos	L. Cot	L. Tan	L. Sin	Deg.

TABLE 4—*Continued*

LOGARITHMS OF TRIGONOMETRIC FUNCTIONS FOR DECIMAL FRACTIONS OF A DEGREE

Deg.	L. Sin	L. Tan	L. Cot	L. Cos	Deg.
27.0	9.6570	9.7072	0.2928	9.9499	63.0
.1	9.6585	9.7090	0.2910	9.9495	.9
.2	9.6600	9.7109	0.2891	9.9491	.8
.3	9.6615	9.7128	0.2872	9.9487	.7
.4	9.6629	9.7146	0.2854	9.9483	.6
.5	9.6644	9.7165	0.2835	9.9479	.5
.6	9.6659	9.7183	0.2817	9.9475	.4
.7	9.6673	9.7202	0.2798	9.9471	.3
.8	9.6687	9.7220	0.2780	9.9467	.2
.9	9.6702	9.7238	0.2762	9.9463	.1
28.0	9.6716	9.7257	0.2743	9.9459	62.0
.1	9.6730	9.7275	0.2725	9.9455	.9
.2	9.6744	9.7293	0.2707	9.9451	.8
.3	9.6759	9.7311	0.2689	9.9447	.7
.4	9.6773	9.7330	0.2670	9.9443	.6
.5	9.6787	9.7348	0.2652	9.9439	.5
.6	9.6801	9.7366	0.2634	9.9435	.4
.7	9.6814	9.7384	0.2616	9.9431	.3
.8	9.6828	9.7402	0.2598	9.9427	.2
.9	9.6842	9.7420	0.2580	9.9422	.1
29.0	9.6856	9.7438	0.2562	9.9418	61.0
.1	9.6869	9.7455	0.2545	9.9414	.9
.2	9.6883	9.7473	0.2527	9.9410	.8
.3	9.6896	9.7491	0.2509	9.9406	.7
.4	9.6910	9.7509	0.2491	9.9401	.6
.5	9.6923	9.7526	0.2474	9.9397	.5
.6	9.6937	9.7544	0.2456	9.9393	.4
.7	9.6950	9.7562	0.2438	9.9388	.3
.8	9.6963	9.7579	0.2421	9.9384	.2
.9	9.6977	9.7597	0.2403	9.9380	.1
30.0	9.6990	9.7614	0.2386	9.9375	60.0
.1	9.7003	9.7632	0.2368	9.9371	.9
.2	9.7016	9.7649	0.2351	9.9367	.8
.3	9.7029	9.7667	0.2333	9.9362	.7
.4	9.7042	9.7684	0.2316	9.9358	.6
.5	9.7055	9.7701	0.2299	9.9353	.5
.6	9.7068	9.7719	0.2281	9.9349	.4
.7	9.7080	9.7736	0.2264	9.9344	.3
.8	9.7093	9.7753	0.2247	9.9340	.2
.9	9.7106	9.7771	0.2229	9.9335	.1
31.0	9.7118	9.7788	0.2212	9.9331	59.0
.1	9.7131	9.7805	0.2195	9.9326	.9
.2	9.7144	9.7822	0.2178	9.9322	.8
.3	9.7156	9.7839	0.2161	9.9317	.7
.4	9.7168	9.7856	0.2144	9.9312	.6
Deg.	L. Cos	L. Cot	L. Tan	L. Sin	Deg.

TABLE 4—*Continued*

LOGARITHMS OF TRIGONOMETRIC FUNCTIONS FOR DECIMAL FRACTIONS OF A DEGREE

Deg.	L. Sin	L. Tan	L. Cot	L. Cos	Deg.
31.5	9.7181	9.7873	0.2127	9.9308	58.5
.6	9.7193	9.7890	0.2110	9.9303	.4
.7	9.7205	9.7907	0.2093	9.9298	.3
.8	9.7218	9.7924	0.2076	9.9294	.2
.9	9.7230	9.7941	0.2059	9.9289	.1
32.0	9.7242	9.7958	0.2042	9.9284	58.0
.1	9.7254	9.7975	0.2025	9.9279	.9
.2	9.7266	9.7992	0.2008	9.9275	.8
.3	9.7278	9.8008	0.1992	9.9270	.7
.4	9.7290	9.8025	0.1975	9.9265	.6
.5	9.7302	9.8042	0.1958	9.9260	.5
.6	9.7314	9.8059	0.1941	9.9255	.4
.7	9.7326	9.8075	0.1925	9.9251	.3
.8	9.7338	9.8092	0.1908	9.9246	.2
.9	9.7349	9.8109	0.1891	9.9241	.1
33.0	9.7361	9.8125	0.1875	9.9236	57.0
.1	9.7373	9.8142	0.1858	9.9231	.9
.2	9.7384	9.8158	0.1842	9.9226	.8
.3	9.7396	9.8175	0.1825	9.9221	.7
.4	9.7407	9.8191	0.1809	9.9216	.6
.5	9.7419	9.8208	0.1792	9.9211	.5
.6	9.7430	9.8224	0.1776	9.9206	.4
.7	9.7442	9.8241	0.1759	9.9201	.3
.8	9.7453	9.8257	0.1743	9.9196	.2
.9	9.7464	9.8274	0.1726	9.9191	.1
34.0	9.7476	9.8290	0.1710	9.9186	56.0
.1	9.7487	9.8306	0.1694	9.9181	.9
.2	9.7498	9.8323	0.1677	9.9175	.8
.3	9.7509	9.8339	0.1661	9.9170	.7
4	9.7520	9.8355	0.1645	9.9165	.6
.5	9.7531	9.8371	0.1629	9.9160	.5
.6	9.7542	9.8388	0.1612	9.9155	.4
.7	9.7553	9.8404	0.1596	9.9149	.3
.8	9.7564	9.8420	0.1580	9.9144	.2
.9	9.7575	9.8436	0.1564	9.9139	.1
35.0	9.7586	9.8452	0.1548	9.9134	55.0
.1	9.7597	9.8468	0.1532	9.9128	.9
.2	9.7607	9.8484	0.1516	9.9123	.8
.3	9.7618	9.8501	0.1499	9.9118	.7
.4	9.7629	9.8517	0.1483	9.9112	.6
.5	9.7640	9.8533	0.1467	9.9107	.5
.6	9.7650	9.8549	0.1451	9.9101	.4
.7	9.7661	9.8565	0.1435	9.9096	.3
.8	9.7671	9.8581	0.1419	9.9091	.2
.9	9.7682	9.8597	0.1403	9.9085	.1
Deg.	L. Cos	L. Cot	L. Tan	L. Sin	Deg.

TABLE 4—*Continued*

LOGARITHMS OF TRIGONOMETRIC FUNCTIONS FOR DECIMAL FRACTIONS OF A DEGREE

Deg.	L. Sin	L. Tan	L. Cot	L. Cos	Deg.
36.0	9.7692	9.8613	0.1387	9.9080	54.0
.1	9.7703	9.8629	0.1371	9.9074	.9
.2	9.7713	9.8644	0.1356	9.9069	.8
.3	9.7723	9.8660	0.1340	9.9063	.7
.4	9.7734	9.8676	0.1324	9.9057	.6
.5	9.7744	9.8692	0.1308	9.9052	.5
.6	9.7754	9.8708	0.1292	9.9046	.4
.7	9.7764	9.8724	0.1276	9.9041	.3
.8	9.7774	9.8740	0.1260	9.9035	.2
.9	9.7785	9.8755	0.1245	9.9029	.1
37.0	9.7795	9.8771	0.1229	9.9023	53.0
.1	9.7805	9.8787	0.1213	9.9018	.9
.2	9.7815	9.8803	0.1197	9.9012	.8
.3	9.7825	9.8818	0.1182	9.9006	.7
.4	9.7835	9.8834	0.1166	9.9000	.6
.5	9.7844	9.8850	0.1150	9.8995	.5
.6	9.7854	9.8865	0.1135	9.8989	.4
.7	9.7864	9.8881	0.1119	9.8983	.3
.8	9.7874	9.8897	0.1103	9.8977	.2
.9	9.7884	9.8912	0.1088	9.8971	.1
38.0	9.7893	9.8928	0.1072	9.8965	52.0
.1	9.7903	9.8944	0.1056	9.8959	.9
.2	9.7913	9.8959	0.1041	9.8953	.8
.3	9.7922	9.8975	0.1025	9.8947	.7
.4	9.7932	9.8990	0.1010	9.8941	.6
.5	9.7941	9.9006	0.0994	9.8935	.5
.6	9.7951	9.9022	0.0978	9.8929	.4
.7	9.7960	9.9037	0.0963	9.8923	.3
.8	9.7970	9.9053	0.0947	9.8917	.2
.9	9.7979	9.9068	0.0932	9.8911	.1
39.0	9.7989	9.9084	0.0916	9.8905	51.0
.1	9.7998	9.9099	0.0901	9.8899	.9
.2	9.8007	9.9115	0.0885	9.8893	.8
.3	9.8017	9.9130	0.0870	9.8887	.7
.4	9.8026	9.9146	0.0854	9.8880	.6
.5	9.8035	9.9161	0.0839	9.8874	.5
.6	9.8044	9.9176	0.0824	9.8868	.4
.7	9.8053	9.9192	0.0808	9.8862	.3
.8	9.8063	9.9207	0.0793	9.8855	.2
.9	9.8072	9.9223	0.0777	9.8849	.1
40.0	9.8081	9.9238	0.0762	9.8843	50.0
.1	9.8090	9.9254	0.0746	9.8836	.9
.2	9.8099	9.9269	0.0731	9.8830	.8
.3	9.8108	9.9284	0.0716	9.8823	.7
.4	9.8117	9.9300	0.0700	9.8817	.6
Deg.	L. Cos	L. Cot	L. Tan	L. Sin	Deg.

TABLE 4—*Continued*

LOGARITHMS OF TRIGONOMETRIC FUNCTIONS FOR DECIMAL FRACTIONS OF A DEGREE

Deg.	L. Sin	L. Tan	L. Cot	L. Cos	Deg.
40.5	9.8125	9.9315	0.0685	9.8810	49.5
.6	9.8134	9.9330	0.0670	9.8804	.4
.7	9.8143	9.9346	0.0654	9.8797	.3
.8	9.8152	9.9361	0.0639	9.8791	.2
.9	9.8161	9.9376	0.0624	9.8784	.1
41.0	9.8169	9.9392	0.0608	9.8778	49.0
.1	9.8178	9.9407	0.0593	9.8771	.9
.2	9.8187	9.9422	0.0578	9.8765	.8
.3	9.8195	9.9438	0.0562	9.8758	.7
.4	9.8204	9.9453	0.0547	9.8751	.6
.5	9.8213	9.9468	0.0532	9.8745	.5
.6	9.8221	9.9483	0.0517	9.8738	.4
.7	9.8230	9.9499	0.0501	9.8731	.3
.8	9.8238	9.9514	0.0486	9.8724	.2
.9	9.8247	9.9529	0.0471	9.8718	.1
42.0	9.8255	9.9544	0.0456	9.8711	48.0
.1	9.8264	9.9560	0.0440	9.8704	.9
.2	9.8272	9.9575	0.0425	9.8697	.8
.3	9.8280	9.9590	0.0410	9.8690	.7
.4	9.8289	9.9605	0.0395	9.8683	.6
.5	9.8297	9.9621	0.0379	9.8676	.5
.6	9.8305	9.9636	0.0364	9.8669	.4
.7	9.8313	9.9651	0.0349	9.8662	.3
.8	9.8322	9.9666	0.0334	9.8655	.2
.9	9.8330	9.9681	0.0319	9.8648	.1
43.0	9.8338	9.9697	0.0303	9.8641	47.0
.1	9.8346	9.9712	0.0288	9.8634	.9
.2	9.8354	9.9727	0.0273	9.8627	.8
.3	9.8362	9.9742	0.0258	9.8620	.7
.4	9.8370	9.9757	0.0243	9.8613	.6
.5	9.8378	9.9772	0.0228	9.8606	.5
.6	9.8386	9.9788	0.0212	9.8598	.4
.7	9.8394	9.9803	0.0197	9.8591	.3
.8	9.8402	9.9818	0.0182	9.8584	.2
.9	9.8410	9.9833	0.0167	9.8577	.1
44.0	9.8418	9.9848	0.0152	9.8569	46.0
.1	9.8426	9.9864	0.0136	9.8562	.9
.2	9.8433	9.9879	0.0121	9.8555	.8
.3	9.8441	9.9894	0.0106	9.8547	.7
.4	9.8449	9.9909	0.0091	9.8540	.6
.5	9.8457	9.9924	0.0076	9.8532	.5
.6	9.8464	9.9939	0.0061	9.8525	.4
.7	9.8472	9.9955	0.0045	9.8517	.3
.8	9.8480	9.9970	0.0030	9.8510	.2
.9	9.8487	9.9985	0.0015	9.8502	.1
45.0	9.8495	0.0000	0.0000	9.8495	45.0
Deg.	L. Cos	L. Cot	L. Tan	L. Sin	Deg.

TABLE 5
Natural Trigonometric Functions for Angles in Radians

Rad.	Sin	Tan	Cot	Cos
.00	.00000	.00000	∞	1.00000
.01	.01000	.01000	99.997	0.99995
.02	.02000	.02000	49.993	.99980
.03	.03000	.03001	33.323	.99955
.04	.03999	.04002	24.987	.99920
.05	.04998	.05004	19.983	.99875
.06	.05996	.06007	16.647	.99820
.07	.06994	.07011	14.262	.99755
.08	.07991	.08017	12.473	.99680
.09	.08988	.09024	11.081	.99505
.10	.09983	.10033	9.9666	.99500
.11	.10978	.11045	9.0542	.99396
.12	.11971	.12058	8.2933	.99281
.13	.12963	.13074	7.6489	.99156
.14	.13954	.14092	7.0961	.99022
.15	.14944	.15114	6.6166	.98877
.16	.15932	.16138	6.1966	.98723
.17	.16918	.17166	5.8256	.98558
.18	.17903	.18197	5.4954	.98384
.19	.18886	.19232	5.1997	.98200
.20	.19867	.20271	4.9332	.98007
.21	.20846	.21314	4.6917	.97803
.22	.21823	.22362	4.4719	.97590
.23	.22798	.23414	4.2709	.97367
.24	.23770	.24472	4.0864	.97134
.25	.24740	.25534	3.9163	.96891
.26	.25708	.26602	3.7591	.96639
.27	.26673	.27676	3.6133	.96377
.28	.27636	.28755	3.4776	.96106
.29	.28595	.29841	3.3511	.95824
.30	.29552	.30934	3.2327	.95534
.31	.30506	.32033	3.1218	.95233
.32	.31457	.33139	3.0176	.94924
.33	.32404	.34252	2.9195	.94604
.34	.33349	.35374	2.8270	.94275
.35	.34290	.36503	2.7395	.93937
.36	.35227	.37640	2.6567	.93590
.37	.36162	.38786	2.5782	.93233
.38	.37092	.39941	2.5037	.92866
.39	.38019	.41105	2.4328	.92491
Rad.	Sin	Tan	Cot	Cos

TABLE 5—*Continued*

NATURAL TRIGONOMETRIC FUNCTIONS FOR ANGLES IN RADIANS

Rad.	Sin	Tan	Cot	Cos
.40	.38942	.42279	2.3652	.92106
.41	.39861	.43463	2.3008	.91712
.42	.40776	.44657	2.2393	.91309
.43	.41687	.45862	2.1804	.90897
.44	.42594	.47078	2.1241	.90475
.45	.43497	.48306	2.0702	.90045
.46	.44395	.49545	2.0184	.89605
.47	.45289	.50797	1.9686	.89157
.48	.46178	.52061	1.9208	.88699
.49	.47063	.53339	1.8748	.88233
.50	.47943	.54630	1.8305	.87758
.51	.48818	.55936	1.7878	.87274
.52	.49688	.57256	1.7465	.86782
.53	.50553	.58592	1.7067	.86281
.54	.51414	.59943	1.6683	.85771
.55	.52269	.61311	1.6310	.85252
.56	.53119	.62695	1.5950	.84726
.57	.53963	.64097	1.5601	.84190
.58	.54802	.65517	1.5263	.83646
.59	.55636	.66956	1.4935	.83094
.60	.56464	.68414	1.4617	.82534
.61	.57287	.69892	1.4308	.81965
.62	.58104	.71391	1.4007	.81388
.63	.58914	.72911	1.3715	.80803
.64	.59720	.74454	1.3431	.80210
.65	.60519	.76020	1.3154	.79608
.66	.61312	.77610	1.2885	.78999
.67	.62099	.79225	1.2622	.78382
.68	.62879	.80866	1.2366	.77757
.69	.63654	.82534	1.2116	.77125
.70	.64422	.84229	1.1872	.76484
.71	.65183	.85953	1.1634	.75836
.72	.65938	.87707	1.1402	.75181
.73	.66687	.89492	1.1174	.74517
.74	.67429	.91309	1.0952	.73847
.75	.68164	.93160	1.0734	.73169
.76	.68892	.95045	1.0521	.72484
.77	.69614	.96967	1.0313	.71791
.78	.70328	.98926	1.0109	.71091
.79	.71035	1.0092	.99084	.70385
Rad.	Sin	Tan	Cot	Cos

TABLE 5—*Continued*

NATURAL TRIGONOMETRIC FUNCTIONS FOR ANGLES IN RADIANS

Rad.	Sin	Tan	Cot	Cos
.80	.71736	1.0296	.97121	.69671
.81	.72429	1.0505	.95197	.68950
.82	.73115	1.0717	.93309	.68222
.83	.73793	1.0934	.91455	.67488
.84	.74464	1.1156	.89635	.66746
.85	.75128	1.1383	.87848	.65998
.86	.75784	1.1616	.86091	.65244
.87	.76433	1.1853	.84365	.64483
.88	.77074	1.2097	.82668	.63715
.89	.77707	1.2346	.80998	.62941
.90	.78333	1.2602	.79355	.62161
.91	.78950	1.2864	.77738	.61375
.92	.79560	1.3133	.76146	.60582
.93	.80162	1.3409	.74578	.59783
.94	.80756	1.3692	.73034	.58979
.95	.81342	1.3984	.71511	.58168
.96	.81919	1.4284	.70010	.57352
.97	.82489	1.4592	.68531	.56530
.98	.83050	1.4910	.67071	.55702
.99	.83603	1.5237	.65631	.54869
1.00	.84147	1.5574	.64209	.54030
1.01	.84683	1.5922	.62806	.53186
1.02	.85211	1.6281	.61420	.52337
1.03	.85730	1.6652	.60051	.51482
1.04	.86240	1.7036	.58699	.50622
1.05	.86742	1.7433	.57362	.49757
1.06	.87236	1.7844	.56040	.48887
1.07	.87720	1.8270	.54734	.48012
1.08	.88196	1.8712	.53441	.47133
1.09	.88663	1.9171	.52162	.46249
1.10	.80121	1.9648	.50897	.45360
1.11	.89570	2.0143	.49644	.44466
1.12	.90010	2.0660	.48404	.43568
1.13	.90441	2.1198	.47175	.42666
1.14	.90863	2.1759	.45959	.41759
1.15	.91276	2.2345	.44753	.40849
1.16	.91680	2.2958	.43558	.39934
1.17	.92075	2.3600	.42373	.39015
1.18	.92461	2.4273	.41199	.38092
1.19	.92837	2.4979	.40034	.37166
Rad.	Sin	Tan	Cot	Cos

TABLE 5—*Continued*

NATURAL TRIGONOMETRIC FUNCTIONS FOR ANGLES IN RADIANS

Rad.	Sin	Tan	Cot	Cos
1.20	.93204	2.5722	.38878	.36236
1.21	.93562	2.6503	.37731	.35302
1.22	.93910	2.7328	.36593	.34365
1.23	.94249	2.8198	.35463	.33424
1.24	.94578	2.9119	.34341	.32480
1.25	.94898	3.0096	.33227	.31532
1.26	.95209	3.1133	.32121	.30582
1.27	.95510	3.2236	.31021	.29628
1.28	.95802	3.3413	.29928	.28672
1.29	.96084	3.4672	.28842	.27712
1.30	.96356	3.6021	.27762	.26750
1.31	.96618	3.7471	.26687	.25785
1.32	.96872	3.9033	.25619	.24818
1.33	.97115	4.0723	.24556	.23848
1.34	.97348	4.2556	.23498	.22875
1.35	.97572	4.4552	.22446	.21901
1.36	.97786	4.6734	.21398	.20924
1.37	.97991	4.9131	.20354	.19945
1.38	.98185	5.1774	.19315	.18964
1.39	.98370	5.4707	.18279	.17981
1.40	.98545	5.7979	.17248	.16997
1.41	.98710	6.1654	.16220	.16010
1.42	.98865	6.5811	.15195	.15023
1.43	.99010	7.0555	.14173	.14033
1.44	.99146	7.6018	.13155	.13042
1.45	.99271	8.2381	.12139	.12050
1.46	.99387	8.9886	.11125	.11057
1.47	.99492	9.8874	.10114	.10063
1.48	.99588	10.983	.09105	.09067
1.49	.99674	12.350	.08097	.08071
1.50	.99749	14.101	.07091	.07074
1.51	.99815	16.428	.06087	.06076
1.52	.99871	19.670	.05084	.05077
1.53	.99917	24.498	.04082	.04079
1.54	.99953	32.461	.03081	.03079
1.55	.99978	48.078	.02080	.02079
1.56	.99994	92.621	.01080	.01080
1.57	1.00000	1255.8	.00080	.00080
1.58	.99996	−108.65	−.00920	−.00920
1.59	.99982	−52.067	−.01921	−.01920
Rad.	Sin	Tan	Cot	Cos

TABLE 5—*Continued*

NATURAL TRIGONOMETRIC FUNCTIONS FOR ANGLES IN RADIANS

Rad.	Sin	Tan	Cot	Cos
1.60	.99957	−34.233	−.02921	−.02920
1.61	.99923	−25.495	−.03922	−.03919
1.62	.99879	−20.307	−.04924	−.04918
1.63	.99825	−16.871	−.05927	−.05917
1.64	.99761	−14.427	−.06931	−.06915
1.65	.99687	−12.599	−.07937	−.07912
1.66	.99602	−11.181	−.08944	−.08909
1.67	.99508	−10.047	−.09953	−.09904
1.68	.99404	−9.1208	−.10964	−.10899
1.69	.99290	−8.9492	−.11977	−.11892
1.70	.99166	−7.6966	−.12993	−.12884
1.71	.99033	−7.1373	−.14011	−.13875
1.72	.98889	−6.6524	−.15032	−.14865
1.73	.98735	−6.2281	−.16056	−.15853
1.74	.98572	−5.8535	−.17084	−.16840
1.75	.98399	−5.5204	−.18115	−.17825
1.76	.98215	−5.2221	−.19149	−.18808
1.77	.98022	−4.9534	−.20188	−.19789
1.78	.97820	−4.7101	−.21231	−.20768
1.79	.97607	−4.4887	−.22278	−.21745
1.80	.97385	−4.2863	−.23330	−.22720
1.81	.97153	−4.1005	−.24387	−.23693
1.82	.96911	−3.9294	−.25449	−.24663
1.83	.96659	−3.7712	−.26517	−.25631
1.84	.96398	−3.6245	−.27590	−.26596
1.85	.96128	−3.4881	−.28669	−.27559
1.86	.95847	−3.3608	−.29755	−.28519
1.87	.95557	−3.2419	−.30846	−.29476
1.88	.95258	−3.1304	−.31945	−.30430
1.89	.94949	−3.0257	−.33051	−.31381
1.90	.94630	−2.9271	−.34164	−.32329
1.91	.94302	−2.8341	−.35284	−.33274
1.92	.93965	−2.7463	−.36413	−.34215
1.93	.93618	−2.6632	−.37549	−.35153
1.94	.93262	−2.5843	−.38695	−.36087
1.95	.92896	−2.5095	−.39849	−.37018
1.96	.92521	−2.4383	−.41012	−.37945
1.97	.92137	−2.3705	−.42185	−.38868
1.98	.91744	−2.3058	−.43368	−.39788
1.99	.91341	−2.2441	−.44562	−.40703
2.00	.90930	−2.1850	−.45766	−.41615
Rad.	Sin	Tan	Cot	Cos

TABLE 6—Degrees to Radians *

Degs.	0.0	0.1	0.2	0.3	0.4	0.5	0.6	0.7	0.8	0.9
0	0.0000	0.0017	0.0035	0.0052	0.0070	0.0087	0.0105	0.0122	0.0140	0.0157
1	0.0175	0.0192	0.0209	0.0227	0.0244	0.0262	0.0279	0.0297	0.0314	0.0332
2	0.0349	0.0367	0.0384	0.0401	0.0419	0.0436	0.0454	0.0471	0.0489	0.0506
3	0.0524	0.0541	0.0559	0.0576	0.0593	0.0611	0.0628	0.0646	0.0663	0.0681
4	0.0698	0.0716	0.0733	0.0750	0.0768	0.0785	0.0803	0.0820	0.0838	0.0855
5	0.0873	0.0890	0.0908	0.0925	0.0942	0.0960	0.0977	0.0995	0.1012	0.1030
6	0.1047	0.1065	0.1082	0.1100	0.1117	0.1134	0.1152	0.1169	0.1187	0.1204
7	0.1222	0.1239	0.1257	0.1274	0.1292	0.1309	0.1326	0.1344	0.1361	0.1379
8	0.1396	0.1414	0.1431	0.1449	0.1466	0.1484	0.1501	0.1518	0.1536	0.1553
9	0.1571	0.1588	0.1606	0.1623	0.1641	0.1658	0.1676	0.1693	0.1710	0.1728
10	0.1745	0.1763	0.1780	0.1798	0.1815	0.1833	0.1850	0.1868	0.1885	0.1902
11	0.1920	0.1937	0.1955	0.1972	0.1990	0.2007	0.2025	0.2042	0.2059	0.2077
12	0.2094	0.2112	0.2129	0.2147	0.2164	0.2182	0.2199	0.2217	0.2234	0.2251
13	0.2269	0.2286	0.2304	0.2321	0.2339	0.2356	0.2374	0.2391	0.2409	0.2426
14	0.2443	0.2461	0.2478	0.2496	0.2513	0.2531	0.2548	0.2566	0.2583	0.2601
15	0.2618	0.2635	0.2653	0.2670	0.2688	0.2705	0.2723	0.2740	0.2758	0.2775
16	0.2793	0.2810	0.2827	0.2845	0.2862	0.2880	0.2897	0.2915	0.2932	0.2950
17	0.2967	0.2985	0.3002	0.3019	0.3037	0.3054	0.3072	0.3089	0.3107	0.3124
18	0.3142	0.3159	0.3176	0.3194	0.3211	0.3229	0.3246	0.3264	0.3281	0.3299
19	0.3316	0.3334	0.3351	0.3368	0.3386	0.3403	0.3421	0.3438	0.3456	0.3473
20	0.3491	0.3508	0.3526	0.3543	0.3560	0.3578	0.3595	0.3613	0.3630	0.3648
21	0.3665	0.3683	0.3700	0.3718	0.3735	0.3752	0.3770	0.3787	0.3805	0.3822
22	0.3840	0.3857	0.3875	0.3892	0.3910	0.3927	0.3944	0.3962	0.3979	0.3997
23	0.4014	0.4032	0.4049	0.4067	0.4084	0.4102	0.4119	0.4136	0.4154	0.4171
24	0.4189	0.4206	0.4224	0.4241	0.4259	0.4276	0.4294	0.4311	0.4328	0.4346
25	0.4363	0.4381	0.4398	0.4416	0.4433	0.4451	0.4468	0.4485	0.4503	0.4520
26	0.4538	0.4555	0.4573	0.4590	0.4608	0.4625	0.4643	0.4660	0.4677	0.4695
27	0.4712	0.4730	0.4747	0.4765	0.4782	0.4800	0.4817	0.4835	0.4852	0.4869
28	0.4887	0.4904	0.4922	0.4939	0.4957	0.4974	0.4992	0.5009	0.5027	0.5044
29	0.5061	0.5079	0.5096	0.5114	0.5131	0.5149	0.5166	0.5184	0.5201	0.5219
30	0.5236	0.5253	0.5271	0.5288	0.5306	0.5323	0.5341	0.5358	0.5376	0.5393
31	0.5411	0.5428	0.5445	0.5463	0.5480	0.5498	0.5515	0.5533	0.5550	0.5568
32	0.5585	0.5603	0.5620	0.5637	0.5655	0.5672	0.5690	0.5707	0.5725	0.5742
33	0.5760	0.5777	0.5794	0.5812	0.5829	0.5847	0.5864	0.5882	0.5899	0.5917
34	0.5934	0.5952	0.5969	0.5986	0.6004	0.6021	0.6039	0.6056	0.6074	0.6091
35	0.6109	0.6126	0.6144	0.6161	0.6178	0.6196	0.6213	0.6231	0.6248	0.6266
36	0.6283	0.6301	0.6318	0.6336	0.6353	0.6370	0.6388	0.6405	0.6423	0.6440
37	0.6458	0.6475	0.6493	0.6510	0.6528	0.6545	0.6562	0.6580	0.6597	0.6615
38	0.6632	0.6650	0.6667	0.6685	0.6702	0.6720	0.6737	0.6754	0.6772	0.6789
39	0.6807	0.6824	0.6842	0.6859	0.6877	0.6894	0.6912	0.6929	0.6946	0.6964
40	0.6981	0.6999	0.7016	0.7034	0.7051	0.7069	0.7086	0.7103	0.7121	0.7138
41	0.7156	0.7173	0.7191	0.7208	0.7226	0.7243	0.7261	0.7278	0.7295	0.7313
42	0.7330	0.7348	0.7365	0.7383	0.7400	0.7418	0.7435	0.7453	0.7470	0.7487
43	0.7505	0.7522	0.7540	0.7557	0.7575	0.7592	0.7610	0.7627	0.7645	0.7662
44	0.7679	0.7697	0.7714	0.7732	0.7749	0.7767	0.7784	0.7802	0.7819	0.7837
45	0.7854	0.7871	0.7889	0.7906	0.7924	0.7941	0.7959	0.7976	0.7994	0.8011
	0'	6'	12'	18'	24'	30'	36'	42'	48'	54'

$90° = 1.5708$ radians $\qquad 30° = \frac{\pi}{6}, \quad 45° = \frac{\pi}{4}, \quad 60° = \frac{\pi}{3}, \quad 90° = \frac{\pi}{2}$ radians

$180° = 3.1416$ radians $\qquad 120° = \frac{2\pi}{3}, \quad 135° = \frac{3\pi}{4}, \quad 150° = \frac{5\pi}{6}, \quad 180° = \pi$ radians

$270° = 4.7124$ radians $\qquad 210° = \frac{7\pi}{6}, \quad 225° = \frac{5\pi}{4}, \quad 240° = \frac{4\pi}{3}, \quad 270° = \frac{3\pi}{2}$ radians

$360° = 6.2832$ radians $\qquad 300° = \frac{5\pi}{3}, \quad 315° = \frac{7\pi}{4}, \quad 330° = \frac{11\pi}{6}, \quad 360° = 2\pi$ radians

* Ralph G. Hudson, *The Engineers' Manual*, Second Edition, New York. John Wiley & Sons, Inc., 1939, pp. 276–277. Reprinted by permission.

TABLE 6—Degrees to Radians—*Continued*

Degs.	0.0	0.1	0.2	0.3	0.4	0.5	0.6	0.7	0.8	0.9
45	0.7854	0.7871	0.7889	0.7906	0.7924	0.7941	0.7959	0.7976	0.7994	0.8011
46	0.8029	0.8046	0.8063	0.8081	0.8098	0.8116	0.8133	0.8151	0.8168	0.8186
47	0.8203	0.8221	0.8238	0.8255	0.8273	0.8290	0.8308	0.8325	0.8343	0.8360
48	0.8378	0.8395	0.8412	0.8430	0.8447	0.8465	0.8482	0.8500	0.8517	0.8535
49	0.8552	0.8570	0.8587	0.8604	0.8622	0.8639	0.8657	0.8674	0.8692	0.8709
50	0.8727	0.8744	0.8762	0.8779	0.8796	0.8814	0.8831	0.8849	0.8866	0.8884
51	0.8901	0.8919	0.8936	0.8954	0.8971	0.8988	0.9006	0.9023	0.9041	0.9058
52	0.9076	0.9093	0.9111	0.9128	0.9146	0.9163	0.9180	0.9198	0.9215	0.9233
53	0.9250	0.9268	0.9285	0.9303	0.9320	0.9338	0.9355	0.9372	0.9390	0.9407
54	0.9425	0.9442	0.9460	0.9477	0.9495	0.9512	0.9529	0.9547	0.9564	0.9582
55	0.9599	0.9617	0.9634	0.9652	0.9669	0.9687	0.9704	0.9721	0.9739	0.9756
56	0.9774	0.9791	0.9809	0.9826	0.9844	0.9861	0.9879	0.9896	0.9913	0.9931
57	0.9948	0.9966	0.9983	1.0001	1.0018	1.0036	1.0053	1.0071	1.0088	1.0105
58	1.0123	1.0140	1.0158	1.0175	1.0193	1.0210	1.0228	1.0245	1.0263	1.0280
59	1.0297	1.0315	1.0332	1.0350	1.0367	1.0385	1.0402	1.0420	1.0437	1.0455
60	1.0472	1.0489	1.0507	1.0524	1.0542	1.0559	1.0577	1.0594	1.0612	1.0629
61	1.0647	1.0664	1.0681	1.0699	1.0716	1.0734	1.0751	1.0769	1.0786	1.0804
62	1.0821	1.0838	1.0856	1.0873	1.0891	1.0908	1.0926	1.0943	1.0961	1.0978
63	1.0996	1.1013	1.1030	1.1048	1.1065	1.1083	1.1100	1.1118	1.1135	1.1153
64	1.1170	1.1188	1.1205	1.1222	1.1240	1.1257	1.1275	1.1292	1.1310	1.1327
65	1.1345	1.1362	1.1380	1.1397	1.1414	1.1432	1.1449	1.1467	1.1484	1.1502
66	1.1519	1.1537	1.1554	1.1572	1.1589	1.1606	1.1624	1.1641	1.1659	1.1676
67	1.1694	1.1711	1.1729	1.1746	1.1764	1.1781	1.1798	1.1816	1.1833	1.1851
68	1.1868	1.1886	1.1903	1.1921	1.1938	1.1956	1.1973	1.1990	1.2008	1.2025
69	1.2043	1.2060	1.2078	1.2095	1.2113	1.2130	1.2147	1.2165	1.2182	1.2200
70	1.2217	1.2235	1.2252	1.2270	1.2287	1.2305	1.2322	1.2339	1.2357	1.2374
71	1.2392	1.2409	1.2427	1.2444	1.2462	1.2479	1.2497	1.2514	1.2531	1.2549
72	1.2566	1.2584	1.2601	1.2619	1.2636	1.2654	1.2671	1.2689	1.2706	1.2723
73	1.2741	1.2758	1.2776	1.2793	1.2811	1.2828	1.2846	1.2863	1.2881	1.2898
74	1.2915	1.2933	1.2950	1.2968	1.2985	1.3003	1.3020	1.3038	1.3055	1.3073
75	1.3090	1.3107	1.3125	1.3142	1.3160	1.3177	1.3195	1.3212	1.3230	1.3247
76	1.3265	1.3282	1.3299	1.3317	1.3334	1.3352	1.3369	1.3387	1.3404	1.3422
77	1.3439	1.3456	1.3474	1.3491	1.3509	1.3526	1.3544	1.3561	1.3579	1.3596
78	1.3614	1.3631	1.3648	1.3666	1.3683	1.3701	1.3718	1.3736	1.3753	1.3771
79	1.3788	1.3806	1.3823	1.3840	1.3858	1.3875	1.3893	1.3910	1.3928	1.3945
80	1.3963	1.3980	1.3998	1.4015	1.4032	1.4050	1.4067	1.4085	1.4102	1.4120
81	1.4137	1.4155	1.4172	1.4190	1.4207	1.4224	1.4242	1.4259	1.4277	1.4294
82	1.4312	1.4329	1.4347	1.4364	1.4382	1.4399	1.4416	1.4434	1.4451	1.4469
83	1.4486	1.4504	1.4521	1.4539	1.4556	1.4573	1.4591	1.4608	1.4626	1.4643
84	1.4661	1.4678	1.4696	1.4713	1.4731	1.4748	1.4765	1.4783	1.4800	1.4818
85	1.4835	1.4853	1.4870	1.4888	1.4905	1.4923	1.4940	1.4957	1.4975	1.4992
86	1.5010	1.5027	1.5045	1.5062	1.5080	1.5097	1.5115	1.5132	1.5149	1.5167
87	1.5184	1.5202	1.5219	1.5237	1.5254	1.5272	1.5289	1.5307	1.5324	1.5341
88.	1.5359	1.5376	1.5394	1.5411	1.5429	1.5446	1.5464	1.5481	1.5499	1.5516
89	1.5533	1.5551	1.5568	1.5586	1.5603	1.5621	1.5638	1.5656	1.5673	1.5691
90	1.5708	1.5725	1.5743	1.5760	1.5778	1.5795	1.5813	1.5830	1.5848	1.5865
	0′	6′	12′	18′	24′	30′	36′	42′	48′	54′

$90° = 1.5708$ radians	$30° = \dfrac{\pi}{6}$, $\quad 45° = \dfrac{\pi}{4}$, $\quad 60° = \dfrac{\pi}{3}$, $\quad 90° = \dfrac{\pi}{2}$ radians
$180° = 3.1416$ radians	$120° = \dfrac{2\pi}{3}$, $135° = \dfrac{3\pi}{4}$, $150° = \dfrac{5\pi}{6}$, $\quad 180° = \pi$ radians
$270° = 4.7124$ radians	$210° = \dfrac{7\pi}{6}$, $225° = \dfrac{5\pi}{4}$, $240° = \dfrac{4\pi}{3}$, $\quad 270° = \dfrac{3\pi}{2}$ radians
$360° = 6.2832$ radians	$300° = \dfrac{5\pi}{3}$, $315° = \dfrac{7\pi}{4}$, $330° = \dfrac{11\pi}{6}$, $360° = 2\pi$ radians

TABLE 7—Constants with Their Common Logarithms

	Number	Logarithm
Base of Naperian logarithms	$e = 2.71828183$	0.4342945
Modulus of common logs., $\log_{10}e =$	$u = 0.43429448$	9.6377843–10
Reciprocal of modulus	$\dfrac{1}{u} = 2.30258509$	.3622157
Circumference of a circle in degrees . .	$= 360$	2.5563025
Circumference of a circle in minutes . .	$= 21600$	4.3344538
Circumference of a circle in seconds . .	$= 1296000$	6.1126050
Radian expressed in degrees	$= 57.29578$	1.7581226
Radian expressed in minutes	$= 3437.7468$	3.5362739
Radian expressed in seconds	$= 206264.806$	5.3144251
Ratio of a circumference to diameter . .	$\pi = 3.14159265$	0.4971499
$\pi = 3.14159\,26535\,89793\,23846\,26433\,8328$	$g = 981$	2.9916690

Number	Logarithm	Number	Logarithm
$2\pi = 6.28318531$	0.7981799	$\pi^2 = 9.86960440$	0.9942997
$4\pi = 12.56637061$	1.0992099	$\dfrac{1}{\pi^2} = 0.10132118$	9.0057003–10
$\dfrac{\pi}{2} = 1.57079633$	0.1961199	$\sqrt{\pi} = 1.77245385$	0.2485749
$\dfrac{\pi}{3} = 1.04719755$	0.0200286	$\dfrac{1}{\sqrt{\pi}} = 0.56418958$	9.7514251–10
$\dfrac{4\pi}{3} = 4.18879020$	0.6220886	$\sqrt{\dfrac{3}{\pi}} = 0.97720502$	9.9899857–10
$\dfrac{\pi}{4} = 0.78539816$	9.8950899–10	$\sqrt{\dfrac{4}{\pi}} = 1.12837917$	0.0524551
$\dfrac{\pi}{6} = 0.52359878$	9.7189986–10	$\sqrt[3]{\pi} = 1.46459189$	0.1657166
$\dfrac{1}{\pi} = 0.31830989$	9.5028501–10	$\dfrac{1}{\sqrt[3]{\pi}} = 0.68278406$	9.8342834–10
$\dfrac{1}{2\pi} = 0.15915494$	9.2018201–10	$\sqrt[3]{\pi^2} = 2.14502940$	0.3314332
$\dfrac{3}{\pi} = 0.95492966$	9.9799714–10	$\sqrt{\dfrac{3}{4\pi}} = 0.62035049$	9.7926371–10
$\dfrac{4}{\pi} = 1.27323954$	0.1049101	$\sqrt[3]{\dfrac{\pi}{6}} = 0.80599598$	9.9063329–10

Number	Logarithm
If the radius $r = 1$, the length of the arc is	
for 1 degree $= \dfrac{\pi}{180} = 0.01745329$	8.2418774–10
for 1 minute $= \dfrac{\pi}{10800} = 0.00029089$	6.4637261–10
for 1 second $= \dfrac{\pi}{648000} = 0.00000485$	4.6855749–10
$\sin 1'' = 0.00000485$	4.6855749–10

772

TABLE 8

Greek Alphabet

A, α	Alpha	I, ι	Iota	P, ρ	Rho
B, β	Beta	K, κ	Kappa	Σ, σ, s	Sigma
Γ, γ	Gamma	Λ, λ	Lambda	T, τ	Tau
Δ, δ	Delta	M, μ	Mu	Υ, υ	Upsilon
E, ϵ	Epsilon	N, ν	Nu	Φ, ϕ, φ	Phi
Z, ζ	Zeta	Ξ, ξ	Xi	X, χ	Chi
H, η	Eta	O, o	Omicron	Ψ, ψ	Psi
Θ, θ, ϑ	Theta	Π, π	Pi	Ω, ω	Omega

Prefixes Used with Units

Prefixes	Abbreviation	Meaning	Scientific or Engineering Notation
micromicro	$\mu\mu$	0.000 000 000 001	10^{-12}
millimicro	$m\mu$	0.000 000 001	10^{-9}
micro	μ	0.000 001	10^{-6}
milli	m	0.001	10^{-3}
centi	c	0.01	10^{-2}
deci	d	0.1	10^{-1}
deka	dk	10	10
hekto	h	100	10^2
kilo	k	1000	10^3
mega	M	1000 000	10^6

Table 8—*Continued*

Quantities and Units Used in the Exercises and Examples

Quantity	Symbol	Basic Unit	Common Multiples
Length	l	meter	millimeter, centimeter, decimeter, kilometer
		foot	inch, yard, mile
Mass	m	gram	milligram, centigram, kilogram
		slug	
Time	t	second	minute, hour
Force	F	dyne	
		pound	ton
Velocity	v	meters per second	centimeters per second
		feet per second	miles per hour
Acceleration	a	meters per second per second	centimeters per second per second
		feet per second per second	miles per hour per hour
Temperature	T	degree	
Charge	Q	coulomb	
Potential	E or V	volt	microvolt, millivolt, kilovolt
Current	I	ampere	microampere, milliampere
Frequency	f	cycle per second	kilocycle per second, megacycle per second
Resistance	R	ohm	microhm, megohm
Conductance	G	mho	micromho, millimho
Inductance	L	henry	microhenry, millihenry
Capacitance	C	farad	micromicrofarad, microfarad
Energy	W	joule	
Power	P	watt	microwatt, milliwatt, kilowatt
Magnetic flux density	B	gauss	kilogauss
Magnetic field intensity	H	oersted	

Answers to the Odd-Numbered Exercises

Chapter 1

Page 5. Sec. 1–2

1. $4 > 2$. **3.** $-4 < 0$. **5.** $-4 < -2$. **7.** $5 > -5$.
9. $6 > -2$. **11.** $1 > -4$. **13.** $-3 > -10$. **15.** $-5 < 6$.

Page 7. Sec. 1–3

1. $2 < 5$. **3.** $-3 < 4$. **5.** $-8 < -6$.
7. $\frac{3}{2} > -\frac{5}{2}$. **9.** $-6.2 < -3.1$. **11.** $-4\frac{1}{2} < -4$.
13. $-3 < 5$. **15.** $-100 < 50$. **17.** $-6, -2, 3, \pi$.
19. $-3.5, 0, 1.2, 6.1$. **21.** $-16, -10, 0, 8$. **23.** $-10, -4.5, 0.5, 1$.
25. $0, 2, 3, 5$. **27.** $\pi, \pi, \frac{3}{2}, \frac{3}{2}$. **29.** $6, 10, 12, 8$.

Pages 10–11. Sec. 1–8

1. (a) -6, (b) -16, (c) 10, (d) -3, (e) 17.
3. (a) -11, (b) 2, (c) 17, (d) 0, (e) 1.
5. (a) -15, (b) 49, (c) -8, (d) -24, (e) -20.
7. (a) 4, (b) -29, (c) -59, (d) -49, (e) -46.
9. (a) -24, (b) -84, (c) -437, (d) 252, (e) -84.
11. (a) -192, (b) 900, (c) 1152, (d) 2208, (e) 0.
13. (a) 0, (b) -4, (c) 1, (d) -2, (e) -2.
15. (a) $\frac{1}{2}$, (b) $-\frac{1}{2}$, (c) not defined, (d) 0, (e) $-\frac{1}{3}$.
17. (a) -12, (b) -18, (c) 0, (d) 198, (e) 180.

Page 12. Sec. 1–9

1. 2. **3.** 8. **5.** 15. **7.** -20. **9.** 37.
11. -22. **13.** -76. **15.** 148. **17.** -27. **19.** 29.

Pages 16–18. Sec. 1–10

1. (a) $\frac{2}{9}$, (b) $\frac{4}{3}$, (c) $\frac{17}{2}$. **3.** (a) $\frac{64}{125}$, (b) $\frac{100}{91}$, (c) $-\frac{2}{3}$. **5.** 100.
7. 60. **9.** 288. **11.** 3528.
13. 8. **15.** 25. **17.** 22.
19. 9. **21.** $\frac{3}{8}$. **23.** $\frac{17}{20}$.
25. 0. **27.** $\frac{11}{294}$. **29.** $\frac{3}{7}$.
31. $-\frac{3}{25}$. **33.** $\frac{6}{49}$. **35.** $-\frac{9}{5}$.
37. $\frac{3}{2}$. **39.** $\frac{4}{3}$. **41.** -1.
43. $\frac{7}{25}$. **45.** $\frac{3}{10}$. **47.** $\frac{81}{16}$.
49. $\frac{14}{9}$ tons. **51.** 26 lb. **53.** $\frac{1}{8}$.
55. $258\frac{3}{4}$ board-ft. **57.** $10{,}240$ lb per sq in. **59.** 1 gram.

775

Page 22. Sec. 1–11

1. $10^8, 10^4, 10^2, 10^0, 10^{-1}, 10^{-3}, 10^{-6}$.   **3.** 2^8.

5. 5^3.   **7.** $\dfrac{3^5}{2}$.   **9.** $\dfrac{2}{5 \cdot 3}$.

11. 1.   **13.** -1.   **15.** $2 \cdot 5^2$.

17. $\dfrac{5 \cdot 2^2}{3}$.   **19.** $\dfrac{1}{2^5 \cdot 3}$.   **21.** $-\dfrac{7}{2^2 \cdot 5}$.

23. $-\dfrac{1}{2^{15}}$.   **25.** 5^4.   **27.** 3^5.

Page 24. Sec. 1–12

1. 3.3×10^4 ft-lb per min.   **3.** 3.085×10^6 times the mass.

5. 7.91×10^3 mi.   **7.** 5.87×10^{12} mi.

9. 1.6617×10^{-24} gram.   **11.** 1.16×10^{-5} cycles per sec.

13. 30,000,000,000,000,000,000 molecules.   **15.** 0.000,031 in.

17. 2,000,000,000 light-years.   **19.** 0.000,000,000,000,000,000,159,1 coulomb.

21. 4000 to 8000 A.   **23.** 22,000 ft per min.

Page 27. Sec. 1–13

1. 2.   **3.** 4.   **5.** 3.   **7.** 3.

9. 5.   **11.** 3.   **13.** 1.   **15.** 2.

17. 1.   **19.** 4.   **21.** 1; 1.   **23.** 2.

Pages 30–31. Sec. 1–16

1. 7.33×10^2.   **3.** -2.51×10^2.   **5.** 9.408×10^{-1}.

7. 1.6.   **9.** 3.8×10^3.   **11.** 2.1×10^2.

13. 5.57×10^2.   **15.** 2.27×10^{-3}.   **17.** 5.4×10^3.

19. 5.62×10^2.   **21.** 3.4.   **23.** 8.5.

25. 1.1×10.   **27.** 4.6×10^{-1}.   **29.** 7.352×10^{-2}.

31. 3.93×10^3.   **33.** 7.9×10^{-2}.   **35.** -2.50×10^4.

37. 1.50×10^2.   **39.** 3.03×10^{-10}.   **41.** 3.946×10^{-2}.

43. 3.4.   **45.** 1.97×10^2.   **47.** 4.136×10^{-2}.

49. 3.22.

Page 37. Sec. 1–20

1. 6.80×10.   **3.** 8.46×10.   **5.** 7.97×10^2.

7. 7.05×10.   **9.** 1.678×10^2.   **11.** 9.17×10^{-3}.

13. 1.765×10^2.   **15.** 2.327.   **17.** 1.959×10^{-4}.

19. 1.299×10^2.   **21.** 2.362×10^{-1}.   **23.** 2.752×10^3.

Page 38. Sec. 1–21

1. 1.300×10.   **3.** 6.25×10^{-1}.   **5.** 5.16.

7. 8.48×10^{-5}.   **9.** 1.518×10^3.   **11.** 1.226×10^3.

13. 8.48×10^{-3}.   **15.** 8.82×10^{-2}.   **17.** 7.43×10^3.

19. 5.17×10.   **21.** 1.190×10^{-4}.   **23.** 1.465×10^6.

Page 39. Sec. 1–22

1. 1.690×10^2.
7. 6.23×10^{-1}.
13. 1.011×10^{-5}.
19. 4.52×10^{-7}.

3. 5.78×10^5.
9. 3.21×10^7.
15. 1.918×10^5.
21. 6.79×10.

5. 1.812×10^2.
11. 7.81×10.
17. 3.33×10^6.
23. 2.10×10^3.

Page 40. Sec. 1–23

1. 6.00.
7. 5.06.
13. 2.166×10^{-1}.
19. 2.43×10.

3. 1.897×10.
9. 1.972×10.
15. 9.81×10^{-3}.
21. 2.54×10^{-2}.

5. 1.600×10.
11. 1.300×10^2.
17. 8.35×10^2.
23. 2.54×10^5.

Pages 42–43. Sec. 1–25

1. 4.15×10.
7. 2.60×10^{-3}.
1? 2.147.
19. 3.44×10.

3. 1.054×10^8.
9. 4.07.
15. 7.26×10^{-1}.

5. 3.99.
11. 1.235.
17. 4.11.

Chapter 2

Pages 47–48. Sec. 2–2

1. $6a$.
7. $-V_1 + 8V_3$.
13. $6R_1 + R_2 - 3R_3$.
19. $4x + 8y - 6z + 13w$.
25. $-2E + 7IX$.
31. $4I_1 + 3I_2 - 11I_3$.
37. $2x$.
43. $E - 4I_1R_1 + 4I_2R_2$.
49. $2E - RI - IZ$.

3. $14xy$.
9. $6x - 4y$.
15. $-10E + 2RI - 5ZI$.
21. $7V$.
27. $20ax - 13by - 9cz$.
33. $Z_1 + 5Z_2 + 11$.
39. $3ac - by$.
45. $-6x + 5y + z - 23$.

5. $8E_1 - 3E_2 - 10E_3$.
11. $6ab + 4cd$.
17. $2R_1 + 2R_2 + 2R_3 + 14R_4$.
23. $a + 2b$.
29. $3x^2 + 8xy - 2y^2$.
35. $2R_1 + 10R_2 - 6R_3 + 3R_4$.
41. 0.
47. $8y - 6x$.

Pages 48–49. Sec. 2–3

1. $40x^3y^2$.
7. $56a^3b^4c^3d^3$.
13. $-30R_1R_2R_3I^5$.
19. $120I^6r^3t^3$.
25. $30x^{11}y^8$.

3. $30a^9$.
9. $30I^3R^3E$.
15. $90i^3t^4r^2s^3$.
21. $-105x^5y^5z^4$.
27. $80a^7b^9$.

5. $12IR^2$.
11. $-6W^4X^3Y^3$.
17. $-60w^2x^2y^3z^3$.
23. $40a^5b^4c^5d^4$.
29. $R_a^5 R_b^5 R_c^5$.

Pages 49–50. Sec. 2–4

1. $x^2 + 3x$.
7. $x^2 - y^2$.
13. $5a^4 - 5b^4$.
17. $36R^2I_1^2 + 12R^2I_1I_2 - 12RI_1^2E_0 + 12RI_1E_0$.
21. $2I^3R + 2IR^2 - 4I - 3I^2R - 3R^2 + 6$.
25. $2x^4 - 5x^3 + 6x^2 - 13x + 10$.
27. $8m^2n^2 - 2mn^2p + 8m^2np - 15n^2p^2 + 43mnp^2 - 30m^2p^2$.
29. $15x^3 - 6x^3y + 8x^2y^2 + 18xy^2 + 10x^2y + 8xy^3 + 12y^3$.

3. $2I^2RX + 2I^2RZ$.
9. $12I^2 - 5I\quad 2$.
15. $12a^2bc - 20a^2cd + 8abc^2 + 28abcd$.

5. $2R_1R_2 - R_1 + 8R_2 - 4$.
11. $2r^4 + r^2z - 3z^2$.
19. $2P^3 + 5P^2 + P - 3$.
23. $x^2 - 2xy + y^2 - z^2$.

Page 50. Sec. 2–5

1. $4a^5$.

3. $\dfrac{a}{2Z}$.

5. b.

7. $\dfrac{4a^3b^2}{c^2}$.

9. $(a+b)bt$.

11. $\frac{1}{2}M^2(x+y)$.

13. $6a^2b^3cd(a+1)^2$.

15. $(L^2-2Z+5)^2T^{n-1}$.

17. N^2MQ^{3-2r}.

19. $S^{m-1}R^mV^3$.

Page 51. Sec. 2–6

1. $xy+3xz$.

3. $R_1+IR_1-I^2$.

5. $\dfrac{L^2}{R}-1$.

7. $3b^n+5a^{2n-1}-2a^{n-2}b$.

9. $\dfrac{(a+b)^2}{2}+\dfrac{a+b}{k}$.

11. $\dfrac{B^n}{C}+\dfrac{2C^{n-1}}{A}-\dfrac{AC}{3B^2}$.

13. $y^5-\dfrac{3py^2}{4}+\dfrac{p^2y}{2}$.

15. $\dfrac{6y^2Z(p+1)^m}{x^3}-\dfrac{3Z^2}{xy}+\dfrac{x^{m-3}(p+1)}{3y^2Z}$.

17. $y^3x^2Z^2(p+1)^{2m-1}-\dfrac{x^4Z^3(p+1)^{m-1}}{2}+\dfrac{x^{m+2}(p+1)^m}{18y}$.

19. $6R^2Z+\dfrac{8}{R^3Z^2L^4}-\dfrac{4}{RZL}$.

Pages 52–53. Sec. 2–7

1. $2x+10$.

3. e^3+e^2-3e.

5. $\lambda^2f^2+2\lambda fu+u^2$.

7. $y^2b^2-6ybx+9x^2$.

9. $R^2-\dfrac{4R}{3}+\dfrac{4}{9}$.

11. I^2R^2-1.

13. $r^2-9r-36$.

15. $a^2-7a-60$.

17. y^2-2y+1.

19. $\mu^2+16\mu+64$.

21. L^4-x^2.

23. $R^2-\dfrac{4R}{5}+\dfrac{4}{25}$.

25. $vu+2au-vb-2ab$.

27. $E_1^2+10E_1+25$.

29. $3a^2x^2+6a^2xz+3a^2z^2$.

31. $\dfrac{1}{R_1^2}-\dfrac{1}{R_2^2}$.

33. $\theta^2-\frac{1}{4}$.

35. $9a^2+18ab+9b^2$.

37. $196a^2-196ab+49b^2$.

39. $\dfrac{a^2}{b^2}-\dfrac{p^2}{q^2}$.

41. $\dfrac{L^2}{R^2}-\dfrac{4}{i^2}$.

43. $x^2+4x-21$.

45. $4-I^2$.

47. $I^4R^2-L^2M^2$.

49. $-R_1^2R_2^2+2R_1R_2-1$.

Pages 55–56. Sec. 2–8

1. $a(y^2+x^2+v^2)$.

3. $17x^2(1-17x)$.

5. $4L^2(L-2R)$.

7. $2ab(b+2a-3)$.

9. $a^2l^2(a+3l-7al)$.

11. $W\left(\dfrac{L}{R}+C+W^2P\right)$.

13. $x^2y^2Z^2(xZ-y+1)$.

15. $2a^2(a-2b^2+4a^2b)$.

17. $7R_1R_2(R_1-2R_2+1)$.

19. $(x-y)^2(a^2-3)$.

21. $(x+1)^2$.

23. $(x+a)(3-c)$.

25. $p(3+q)(1+rs)$.

27. $x^2y(x+y)(xy+1)$.

29. $(L-M)(L+R)$.

31. $(\alpha_1^2+\alpha_2^2)(\theta^2+\beta)$.

33. $(3C-I_p)(E_p-I_p)$.

35. $(a+b)(a+x-1)$.

37. $(x+3)^2$.

39. $(M-R)^2$.

41. $(4R^2+1)(2R-1)(2R+1)$.

43. $\frac{1}{4}(C_1+C_2)(C_1-C_2)$.

45. $(i^2t+rs^2)(i^2t-rs^2)$.

47. $(3f+1)^2$.

49. $(3e-7)^2$.

51. $Z^2(y^2+1)(y+1)(y-1)$.

53. $(3R_1-4R_2)^2$.

55. $(x-6)(x-2)$.

57. $(y+6)(y-1)$.

59. $(q+14)(q+2)$.

61. $(K+4)(K-17)$.

63. $\pi(\alpha+7)(\alpha-2)$.

65. $p^2(q-8)(q-1)$.

67. $3(y-7)(y+4)$.

69. $2(R_1+R_2)(R_1+2R_2)$.

71. $k(m+\pi n)^2$.

73. $(ax^2+7m)(ax^2+5m)$.

75. $(C_1^2C_2+3)(C_1^2C_2+1)$.

77. $(S+2V^2L^2)(S+V^2L^2)$.

79. $cu(i_p+3e)(i_p+2e)$.

Page 57. Sec. 2–9

1. $35a^2b^4$.

3. $180R^3L^2$.

5. $ax(x + a)(x - a)$.

7. $(x - 2)(x + 3)(x + 1)$.

9. $y^{11}(y + 1)$.

11. $s(r + s)(r - s)$.

13. $6(e - 3)^2(e + 3)$.

15. $(x - b)(x + b)(x - a)$. **17.** $(L + 3)(L + 1)(L - 1)$.

19. $(x + y)^2(x - y)$.

Pages 59–60. Sec. 2–10

1. $\dfrac{x^2 + y^2}{x^2y^2}$.

3. $\dfrac{2(t + 1)}{t}$.

5. $\dfrac{3}{4 - x}$.

7. $-\dfrac{2r}{(r + 1)(r - 1)}$.

9. 0.

11. $-\dfrac{bc + ad}{b(b + d)}$.

13. $\dfrac{3}{y^n}$.

15. $\dfrac{y}{3x + y}$,

17. $-\dfrac{3(x - 4)}{(2x - 1)(2x + 1)}$.

19. $\dfrac{135y^{n+1} - 50b^{n+2}c^2 + 28bc^n}{210b^6c^4y}$.

21. $\dfrac{2a^{n+1}}{(a^n + 1)(a^n - 1)}$.

23. $\dfrac{4p + 1}{(1 - p)(1 + p)}$.

25. $\dfrac{4x^my^m}{(x^m - y^m)(x^m + y^m)}$.

27. $\dfrac{4Z^2 + 8Z + 5}{(2Z + 1)^2}$.

29. $\dfrac{x^2 + 5x + 2}{x(x + 1)^2}$.

31. 0.

33. $\dfrac{-3x^2 - 3x + 2}{(x - 1)^2(x + 1)}$.

35. $\dfrac{11x^2 - 2x - 7}{(3x + 1)(3x - 1)}$.

37. $\dfrac{2z^2 - 3z + 4}{(2z - 1)^2}$.

39. $-\dfrac{2x + 9}{(x + 3)(x - 3)(x - 2)(x + 2)}$.

41. $\dfrac{e_1 - e_4}{(e_1 - e_2)(e_2 - e_3)(e_3 - e_4)}$.

43. $-\dfrac{2(x - 13)}{x - 7}$.

45. $\dfrac{x(\alpha^2\beta - \alpha^2\gamma - \alpha\beta^2 + \beta^2\gamma + \alpha\gamma^2 - \beta\gamma^2)}{(\alpha - \beta)(\alpha - \gamma)(\beta - \gamma)}$.

47. $\dfrac{6L^2 + 29L + 3}{L(L + 5)}$.

49. $\dfrac{2c^2 + 8cd - d^2}{(c - d)^2(c + 2d)}$.

Pages 62–63. Sec. 2–11

1. $\dfrac{7b}{2ax}$.

3. $\dfrac{g^2n^2}{t^2m^2}$.

5. $\dfrac{e_2e_3^3e_4}{e_1e_5e_6^3}$.

7. 1.

9. 5.

11. $\dfrac{3E_1}{E_1 + E_2}$.

13. $(H - K)^2$.

15. $\dfrac{1}{x}$.

17. 1.

19. $\dfrac{(x + 7)(x + 1)}{(x - 3)(x - 2)}$.

21. $\left(\dfrac{i_1i_2}{e_1e_2}\right)^2$.

23. $\dfrac{X_c^2X_l^2}{12\pi r}$.

25. $\dfrac{E_2(E_1 + E_2)}{E_1}$.

27. $\dfrac{(u^2 + v^2)^2}{(u + v)^2}$.

29. $M - \dfrac{1}{M}$.

31. $\dfrac{p^2 + q^2}{q}$.

33. 1.

35. $\dfrac{5}{3y}$.

37. $\dfrac{ad}{2bc}$.

39. $-(g + 1)$.

41. $\dfrac{r_2 + r_1}{r_2 - r_1}$.

43. $\dfrac{(x - y)(x + y)}{y^2}$.

45. $\dfrac{(x + 3)(x - 1)}{(x - 3)(x + 2)}$.

47. $\left(\dfrac{a + b}{2a + b}\right)^3$.

49. x.

Pages 67–68. Sec. 2–12

1. 5.   **3.** 13.   **5.** $\frac{9}{2}$.   **7.** 10.   **9.** $a - 1$.
11. 8.   **13.** 7.   **15.** 18.   **17.** $-\frac{65}{3}$.   **19.** 4.
21. -4.   **23.** 8.   **25.** $\frac{18}{25}$.   **27.** $-\frac{27}{4}$.   **29.** $a^2 - 1$.

31. 5.3.   **33.** $b - 2a$.   **35.** 2.00.   **37.** $\dfrac{(a-2)(a+2)}{a-1}$.   **39.** 21.

41. $\dfrac{a(c-b)}{2a+b+c}$.   **43.** 2.   **45.** 1.   **47.** -1.   **49.** $2a$.

Page 70. Sec. 2–13

1. 3, -3.   **3.** $\frac{9}{2}$, $-\frac{9}{2}$.   **5.** 3, 3.   **7.** $-\frac{3}{2}$, $-\frac{3}{2}$.
9. $\frac{5}{2}$, $\frac{5}{2}$.   **11.** 3, -1.   **13.** 6, -3.   **15.** 4, 2.
17. 5, -5.   **19.** $\frac{3}{35}$, $-\frac{3}{35}$.   **21.** 0, 5.   **23.** 0, $-\frac{2}{3}$.
25. 2, -5.   **27.** 6, -1.   **29.** 4, -4.   **31.** -4, -2.

Pages 71–73. Sec. 2–14

1. $R_3 = R_T - R_1 - R_2$.   **3.** $L_1 = L_T - L_2 + 2M$.

5. $I_{xo} = I_x - Ad^2, A = \dfrac{I_x - I_{xo}}{d^2}$.   **7.** $E = IR$.

9. $R = \dfrac{P}{I^2}$.   **11.** $d = \dfrac{Wh}{F}, W = \dfrac{Fd}{h}$.

13. $R = \dfrac{ApV}{t}, t = \dfrac{ApV}{R}$.   **15.** $g = \dfrac{v^2 - v_0^2}{2h}, v_0^2 = v^2 - 2gh$.

17. $a = \dfrac{Fg}{w}, w = \dfrac{Fg}{a}$.   **19.** $t = \dfrac{d}{v}$.

21. $m = \dfrac{2fd}{v^2}, d = \dfrac{mv^2}{2f}$.   **23.** $l = \dfrac{RA}{k}$.

25. $T = \dfrac{Qd}{K(t_2 - t_1)a}, a = \dfrac{Qd}{K(t_2 - t_1)T}$.   **27.** $\alpha = \dfrac{l_t - l_0 - l_0\beta t^2}{l_0 t}, \beta = \dfrac{l_t - l_0 - l_0\alpha t}{l_0 t^2}$.

29. $\phi = BA$.   **31.** $a^2 = \dfrac{9V\eta}{2g(d_1 - d_2)}$.

33. $r_1 = \dfrac{r_2 r_3}{r_4}$.   **35.** $l = \dfrac{10^8 F}{22.5BI}, B = \dfrac{10^8 F}{22.5lI}$.

37. $V_j = \dfrac{P_n g J}{V_a} + V_a$.   **39.** $N_2 = -\dfrac{10^8 lM}{1.26N_1 A\mu}$.

41. $V_a = \dfrac{4W}{\pi\rho D^2(e^2 - h^2)}$.   **43.** $A = \dfrac{10^8 Cd}{8.84K}$.

45. $f = \dfrac{X_L}{2\pi L}, L = \dfrac{X_L}{2\pi f}$.   **47.** $d^2 = \dfrac{L}{0.0251n^2 l}, l = \dfrac{L}{0.0251 d^2 n^2}$.

49. $e_2 = e_1 + \dfrac{i_{avg}(t_2 - t_1)}{C}, C = i_{avg}\dfrac{t_2 - t_1}{e_2 - e_1}$.

Pages 75–77. Sec. 2–15

1. 5860 watts.   **3.** 170 watts.   **5.** 11,000 ft.
7. 240 biquadratic in.   **9.** 56.68 cu in.   **11.** 2.4 dynes.
13. 4.7×10^{12} dynes.   **15.** 5.2 ft per sec per sec.   **17.** 1100 cm per sec.
19. 21 mph.   **21.** 1.1×10^5 ohms.

Pages 80–81. Sec. 2–16

1. 2, 3.   **3.** 1, 3.   **5.** 5, 3.   **7.** -5, 8.

9. 10, 1.   **11.** -3, 2.   **13.** 3, 2.   **15.** 3, 4.

17. 1, 2.   **19.** -2, 3.   **21.** 3, 1.   **23.** 3, 2.

25. $\frac{11}{5}$, $-\frac{2}{5}$.   **27.** $\frac{40}{19}$, $-\frac{87}{19}$.   **29.** 1, $\frac{1}{2}$.   **31.** $\frac{1}{2}$, $\frac{1}{3}$.

33. 5, 4.   **35.** 20, 20.   **37.** -2, 3.   **39.** $\frac{1}{3}$, $-\frac{1}{3}$.

Pages 85–86. Sec. 2–17

1. 18, 14.   **3.** 64, 32.   **5.** 9.

7. $64°$, $32°$, $84°$.   **9.** 10, 11, 12.

11. A fires 15 kg at 5 kg per min, D fires 30 kg at 10 kg per min, C fires 6 kg at 2 kg per min.

13. 11, 13; -11, -13.   **15.** \$30.

17. 100 ma in circuit 1, 80 ma in circuit 2, 60 ma in circuit 3.

19. 12.   **21.** 84.   **23.** 82.

25. 44.   **27.** $\frac{2}{7}$ days.   **29.** $\frac{100}{9}$ days.

31. $\frac{24}{35}$ days.   **33.** $43\frac{31}{83}$ min.   **35.** $1\frac{1}{2}$ gal.

Pages 90–91. Sec. 2–18

1. 8 hr; 40 mi; 32 mi.   **3.** 40 mph.   **5.** 50 mph, $33\frac{1}{3}$ mph.

7. $26\frac{2}{3}$ hr.   **9.** 8 volts.   **11.** $36\frac{2}{3}$ amp.

13. 15 ohms.   **15.** $176\frac{1}{3}$ ohms.   **17.** 2000 amp.

19. 450 lb.   **21.** 8 ft from the 400-lb weight.   **23.** 3300 lb, 3900 lb.

Chapter 3

Pages 96–99. Sec. 3–4

1. $A = f(h) = 2h.$ **3.** $A = f(e) = 6e^2.$ **5.** $A = f(r) = \pi r^2.$

7. $A = f(r) = 4\pi r^2.$ **9.** $I = f(t) = 2t.$ **11.** $A = f(W, L) = WL.$

13. $V = f(h, A) = \frac{1}{3}hA.$ **15.** $A = f(b, h) = \frac{1}{2}bh.$ **17.** $V = f(h, r) = \pi r^2 h.$

19. $d = f(t, v) = vt.$ **21.** $9, -3, -7.$ **23.** $-\frac{1}{4}, \frac{2}{7}, 6.$

25. $0, 4\frac{7}{8}, 3.$ **27.** $0, 0, -abc, (d - a)(d - b)(d - c).$

29. $0, 0, 0, a^3(a - 1)^2(a + 1).$

31. $a + \dfrac{1}{a}, \quad \left(a^2 + \dfrac{1}{a^2}\right)\left(a + \dfrac{1}{a}\right).$

33. $0, \dfrac{1 - t^2}{1 + t^2}, \quad \left(\dfrac{1 - t}{1 + t}\right)^2, \quad \dfrac{t - 1}{t + 1}, \quad \dfrac{1 + t}{1 - t}, \quad t.$

35. $b + a.$ **37.** $2x + h - 4.$ **39.** $\dfrac{1}{\sqrt{x + \alpha} + \sqrt{x}}.$

41. 5, 2.   **43.** 4, 10.   **45.** 2, 870.

47. $4, 0, 0, -1, 12.$ **49.** $0, -5, -2, -1.$ **51.** $0, -2$, not defined, $\frac{3}{4}$.

53. 6, 9, 8, 0.   **55.** 2, 3, −6, 6.   **57.** 0, not defined, −$\frac{9}{2}$, 1.

59. 2, $\frac{1}{2}$, $\frac{4}{9}$, $\frac{2}{5}$.   **61.** $\frac{x}{2}\sqrt{36 - x^2}$.   **63.** $\frac{h^2}{4}$.

65. $x^2 + \frac{144}{x}$.   **67.** $h = \frac{36}{\pi r^2}$, $S = 2\pi r^2 + \frac{72}{r}$.

69. (a) $A = 50x$, (b) $A = 45x + 200$.

71. $A = \frac{x^2}{16} + \frac{\sqrt{3}(36 - x)^2}{36}$.

73. $y = 2\sqrt{64 - x^2}$, $V = 2\pi x^2\sqrt{64 - x^2}$, $S = 2\pi x^2 + 4\pi x\sqrt{64 - x^2}$.

75. $d = 5t$.

Page 100. Sec. 3–5

1. $x = \frac{y}{4} - 3$.   **3.** $x = \pm\sqrt{y - 4}$.   **5.** $q = \sqrt[3]{p - 1}$.

7. $r = \pm\sqrt{x + a^2}$.   **9.** $d = \frac{C}{\pi}$.   **11.** $d = 2\sqrt{\frac{A}{\pi}}$.

13. $r = \sqrt[3]{\frac{3V}{4\pi}}$.   **15.** $e = \sqrt[3]{V}$.   **17.** $A = 6\sqrt[3]{V^2}$.

19. $A = \sqrt[3]{36\pi V^2}$.

Page 117. Sec. 3–9

43. 3.0, −2.0.   **45.** 2.2, −3.2.   **47.** 3.4, −2.4.
49. 1.9, −1.5, −0.3.   **51.** 1.2, −1.8, −3.4.

Pages 120–121. Sec. 3–10

1. $y = 4x + 3$.   **3.** $y = x - 1$.   **5.** $y = 16(x^2 - 2x + 1)$.

7. $y = \frac{x^2}{4} - 4$.   **9.** $y = \pm x\sqrt{x}$.   **19.** $t = 1, 5$; (0, 0), (8, 8).

21. 4 sec; 6 sec; $t = 0, 5$.   **23.** 8 sec; $t = \pm4$.   **25.** $t = 3, -1, -5$.
27. 44.1 ft; $3\frac{1}{8}$ sec; $\frac{1}{8}$ sec; 3 sec.
29. $t = 2$ sec, 2000 ft from the foot of the cliff.

Pages 124–126. Sec. 3–11

1. 2.25.   **3.** $g = 5.12 \times 10^8/d^2$, 8.00 ft per sec per sec.
5. 15.4 in.   **7.** 21.3 in.   **9.** 85,800 lb.
11. 285 lb.   **13.** 2.8×10^{-3} coulombs.   **15.** 17.4 in.
17. 1.95 joules.   **19.** 3 times.   **21.** 8 times.
23. 42 ft.   **25.** 2.50 cents; 56.9 kw-hr.

Pages 128–129. Sec. 3–12

1. 3, 3.   **3.** 4, −6.   **5.** 5, $-\frac{15}{4}$.
7. $\frac{5}{2}$, $-\frac{15}{2}$.   **9.** $\frac{7}{2}$, no y intercept.   **11.** 7, −6.
13. 0, 0.   **15.** 8, −8.   **17.** 8, −16.
19. 7.00, −3.50.   **21.** 333, −555.   **23.** 32, $-\frac{160}{9}$.

Pages 131–132. Sec. 3–13

1. Intersect at $(3, -2)$.   **3.** Parallel.   **5.** Intersect at $(2, 0)$.

7. Intersect at $(4, -3)$.   **9.** Parallel.   **11.** Coincide.

13. Intersect at $(35, -15)$.   **15.** Parallel.   **17.** Parallel.

19. Intersect at $(\tfrac{12}{7}, -1)$.   **21.** Yes, 36.   **23.** No.

Chapter 4

Page 137. Sec. 4–2

1. $17.60°$.   **3.** $161.5°$.   **5.** $141.58°$.   **7.** $0.70°$.

9. $18.3231°$.   **11.** $212.008°$.   **13.** $0.0217°$.   **15.** $-43.4597°$.

17. $34.5°$.   **19.** $21.0°$.   **21.** $112.2228°$.   **23.** $56.8014°$.

25. $27° 6'$.   **27.** $63° 52'$.   **29.** $612° 23' 0''$.   **31.** $7° 30' 3''$.

33. $-38° 41' 8''$.   **35.** $26° 12' 18.2''$.   **37.** $20° 17'$.   **39.** $-0° 1'$.

Pages 141–144. Sec. 4–3

1. $\dfrac{\pi}{12}, \dfrac{\pi}{6}$.   **3.** $\dfrac{\pi}{2}, \pi, \dfrac{3\pi}{2}$.   **5.** $\dfrac{3\pi}{4}, \dfrac{7\pi}{6}$.

7. $\dfrac{7\pi}{18}, \dfrac{5\pi}{6}$.   **9.** $\dfrac{10\pi}{3}$.   **11.** $-\dfrac{5\pi}{6}$.

13. $-\dfrac{31\pi}{18}$.   **15.** $\dfrac{11\pi}{10}, -\dfrac{11\pi}{10}$.   **17.** $45.00°, 135.00°$.

19. $360.00°, 720.00°$.   **21.** $120.00°, 420.00°$.   **23.** $20.00°$.

25. $585.00°$.   **27.** $-100.00°$.   **29.** $-540.00°, 540.00°$.

31. 0.541.   **33.** 4.472.   **35.** 0.2700.

37. 4.4968.   **39.** 0.22007.   **41.** 3.41420.

43. 1.04052.   **45.** 1.66708.   **47.** $69.84°$.

49. $138.4°$.   **51.** $265.8°$.   **53.** $1.83°$.

55. $324.99°$.   **57.** $177.99°$.   **59.** $325.035°$.

61. 0.5451.   **63.** 1.718.   **65.** 4.477.

67. 0.867.   **69.** 1.9004.   **71.** 3.84768.

73. 3.466723.   **75.** 6.463176.   **77.** 6.0 rad, $340°$.

79. 6.4 in.   **81.** 2814 mi.   **83.** 9109 mi.

85. 1036 mph, 1520 ft per sec, 0.2618 rad per hr.

87. 1.061 ft.

89. $R = \dfrac{17.6s}{\pi d}$, $A = \dfrac{35.2s}{d}$, 14 revolutions per sec, 86 rad per sec.

91. 25.1 sq in., 40.2 sq in., 81.5 sq in., 17.8 sq in.

93. 61.1 sq in.   **95.** 3.016 rad, $172.8°$.   **97.** 188.6 ft per sec.

Pages 151–152. Sec. 4–7

[For the answers to Exercises 1–55 the trigonometric functions are listed in this order: sine, cosine, tangent, cotangent, secant, and cosecant.]

1. $\dfrac{4}{5}, \dfrac{3}{5}, \dfrac{4}{3}, \dfrac{3}{4}, \dfrac{5}{3}, \dfrac{5}{4}.$

3. $\dfrac{3\sqrt{10}}{10}, -\dfrac{\sqrt{10}}{10}, -3, -\dfrac{1}{3}, -\sqrt{10}, \dfrac{\sqrt{10}}{3}.$

5. $\dfrac{4\sqrt{97}}{97}, \dfrac{9\sqrt{97}}{97}, \dfrac{4}{9}, \dfrac{9}{4}, \dfrac{\sqrt{97}}{9}, \dfrac{\sqrt{97}}{4}.$

7. $\dfrac{\sqrt{17}}{17}, \dfrac{4\sqrt{17}}{17}, \dfrac{1}{4}, 4, \dfrac{\sqrt{17}}{4}, \sqrt{17}.$

9. $-\dfrac{\sqrt{2}}{2}, \dfrac{\sqrt{2}}{2}, -1, -1, \sqrt{2}, -\sqrt{2}.$

11. $0, 1, 0,$ undefined, $1,$ undefined.

13. $-1, 0,$ undefined, $0,$ undefined, $-1.$

15. $\dfrac{5\sqrt{106}}{106}, -\dfrac{9\sqrt{106}}{106}, -\dfrac{5}{9}, -\dfrac{9}{5}, -\dfrac{\sqrt{106}}{9}, \dfrac{\sqrt{106}}{5}.$

17. $-\dfrac{\sqrt{15}}{4}, -\dfrac{1}{4}, \sqrt{15}, \dfrac{\sqrt{15}}{15}, -4, -\dfrac{4\sqrt{15}}{15}.$

19. $\dfrac{3\sqrt{130}}{130}, \dfrac{11\sqrt{130}}{130}, \dfrac{3}{11}, \dfrac{11}{3}, \dfrac{\sqrt{130}}{11}, \dfrac{\sqrt{130}}{3}.$

21. $\dfrac{3\sqrt{13}}{13}, \dfrac{2\sqrt{13}}{13}, \dfrac{3}{2}, \dfrac{2}{3}, \dfrac{\sqrt{13}}{2}, \dfrac{\sqrt{13}}{3}.$

23. $\dfrac{7}{12}, -\dfrac{\sqrt{95}}{12}, -\dfrac{7\sqrt{95}}{95}, -\dfrac{\sqrt{95}}{7}, -\dfrac{12\sqrt{95}}{95}, \dfrac{12}{7}.$

25. $\dfrac{5}{91}, -\dfrac{8\sqrt{129}}{91}, -\dfrac{5\sqrt{129}}{1032}, -\dfrac{8\sqrt{129}}{5}, -\dfrac{91\sqrt{129}}{1032}, \dfrac{91}{5}.$

27. $\dfrac{4}{5}, -\dfrac{3}{5}, -\dfrac{4}{3}, -\dfrac{3}{4}, -\dfrac{5}{3}, \dfrac{5}{4}.$

[For the answers to Exercises 29–55 the quadrant in which the angle terminates is given by Roman numerals, I, II, III, IV.]

29. I: $\dfrac{3}{7}, \dfrac{2\sqrt{10}}{7}, \dfrac{3\sqrt{10}}{20}, \dfrac{2\sqrt{10}}{3}, \dfrac{7\sqrt{10}}{20}, \dfrac{7}{3};$

    II: $\dfrac{3}{7}, -\dfrac{2\sqrt{10}}{7}, -\dfrac{3\sqrt{10}}{20}, -\dfrac{2\sqrt{10}}{3}, -\dfrac{7\sqrt{10}}{20}, \dfrac{7}{3}.$

31. II: $\dfrac{\sqrt{65}}{9}, -\dfrac{4}{9}, -\dfrac{\sqrt{65}}{4}, -\dfrac{4\sqrt{65}}{65}, -\dfrac{9}{4}, \dfrac{9\sqrt{65}}{65};$

    III: $-\dfrac{\sqrt{65}}{9}, -\dfrac{4}{9}, \dfrac{\sqrt{65}}{4}, \dfrac{4\sqrt{65}}{65}, -\dfrac{9}{4}, -\dfrac{9\sqrt{65}}{65}.$

33. I: $\dfrac{\sqrt{3}}{2}, \dfrac{1}{2}, \sqrt{3}, \dfrac{\sqrt{3}}{3}, 2, \dfrac{2\sqrt{3}}{3};$

    II: $\dfrac{\sqrt{3}}{2}, -\dfrac{1}{2}, -\sqrt{3}, -\dfrac{\sqrt{3}}{3}, -2, \dfrac{2\sqrt{3}}{3};$

III: $-\dfrac{\sqrt{3}}{2}, -\dfrac{1}{2}, \sqrt{3}, \dfrac{\sqrt{3}}{3}, -2, -\dfrac{2\sqrt{3}}{3}$;

IV: $-\dfrac{\sqrt{3}}{2}, \dfrac{1}{2}, -\sqrt{3}, -\dfrac{\sqrt{3}}{3}, 2, -\dfrac{2\sqrt{3}}{3}$.

35. I: $\dfrac{7\sqrt{170}}{170}, \dfrac{11\sqrt{170}}{170}, \dfrac{7}{11}, \dfrac{11}{7}, \dfrac{\sqrt{170}}{11}, \dfrac{\sqrt{170}}{7}$;

II: $\dfrac{7\sqrt{170}}{170}, -\dfrac{11\sqrt{170}}{170}, -\dfrac{7}{11}, -\dfrac{11}{7}, -\dfrac{\sqrt{170}}{11}, \dfrac{\sqrt{170}}{7}$;

III: $-\dfrac{7\sqrt{170}}{170}, -\dfrac{11\sqrt{170}}{170}, \dfrac{7}{11}, \dfrac{11}{7}, -\dfrac{\sqrt{170}}{11}, -\dfrac{\sqrt{170}}{7}$;

IV: $-\dfrac{7\sqrt{170}}{170}, \dfrac{11\sqrt{170}}{170}, -\dfrac{7}{11}, -\dfrac{11}{7}, \dfrac{\sqrt{170}}{11}, -\dfrac{\sqrt{170}}{7}$.

37. I: $\dfrac{7\sqrt{149}}{149}, \dfrac{10\sqrt{149}}{149}, \dfrac{7}{10}, \dfrac{10}{7}, \dfrac{\sqrt{149}}{10}, \dfrac{\sqrt{149}}{7}$;

III: $-\dfrac{7\sqrt{149}}{149}, -\dfrac{10\sqrt{149}}{149}, \dfrac{7}{10}, \dfrac{10}{7}, -\dfrac{\sqrt{149}}{10}, -\dfrac{\sqrt{149}}{7}$.

39. III: $-\dfrac{2}{5}, -\dfrac{\sqrt{21}}{5}, \dfrac{2\sqrt{21}}{21}, \dfrac{\sqrt{21}}{2}, -\dfrac{5\sqrt{21}}{21}, -\dfrac{5}{2}$;

IV: $-\dfrac{2}{5}, \dfrac{\sqrt{21}}{5}, -\dfrac{2\sqrt{21}}{21}, -\dfrac{\sqrt{21}}{2}, \dfrac{5\sqrt{21}}{21}, -\dfrac{5}{2}$.

41. II: $\dfrac{2\sqrt{5}}{5}, -\dfrac{\sqrt{5}}{5}, -2, -\dfrac{1}{2}, -\sqrt{5}, \dfrac{\sqrt{5}}{2}$;

IV: $-\dfrac{2\sqrt{5}}{5}, \dfrac{\sqrt{5}}{5}, -2, -\dfrac{1}{2}, \sqrt{5}, -\dfrac{\sqrt{5}}{2}$.

43. $-1, 0,$ undefined, $0,$ undefined, $-1.$

45. I: $\dfrac{\sqrt{2}}{2}, \dfrac{\sqrt{2}}{2}, 1, 1, \sqrt{2}, \sqrt{2}$;

IV: $-\dfrac{\sqrt{2}}{2}, \dfrac{\sqrt{2}}{2}, -1, -1, \sqrt{2}, -\sqrt{2}.$

47. I when $m > 0$ and IV when $m < 0$: $\dfrac{m}{\sqrt{1+m^2}}, \dfrac{1}{\sqrt{1+m^2}}, m, \dfrac{1}{m}, \sqrt{1+m^2}, \dfrac{\sqrt{1+m^2}}{m}$;

III when $m > 0$ and II when $m < 0$:

$$-\dfrac{m}{\sqrt{1+m^2}}, -\dfrac{1}{\sqrt{1+m^2}}, m, \dfrac{1}{m}, -\sqrt{1+m^2}, -\dfrac{\sqrt{1+m^2}}{m}.$$

49. II when $a>0$ and I when $a<0$:

$$\frac{\sqrt{9-16a^2}}{3}, \; -\frac{4a}{3}, \; -\frac{\sqrt{9-16a^2}}{4a}, \; -\frac{4a}{\sqrt{9-16a^2}}, \; -\frac{3}{4a}, \; \frac{3}{\sqrt{9-16a^2}};$$

III when $a > 0$ and IV when $a < 0$:

$$-\frac{\sqrt{9-16a^2}}{3}, \; -\frac{4a}{3}, \; \frac{\sqrt{9-16a^2}}{4a}, \; \frac{4a}{\sqrt{9-16a^2}}, \; -\frac{3}{4a}, \; -\frac{3}{\sqrt{9-16a^2}}.$$

51. III: $-\dfrac{1}{7}, \; -\dfrac{4\sqrt{3}}{7}, \; \dfrac{\sqrt{3}}{12}, \; 4\sqrt{3}, \; -\dfrac{7\sqrt{3}}{12}, \; -7;$

IV: $-\dfrac{1}{7}, \; \dfrac{4\sqrt{3}}{7}, \; -\dfrac{\sqrt{3}}{12}, \; -4\sqrt{3}, \; \dfrac{7\sqrt{3}}{12}, \; -7.$

53. II: $\dfrac{4\sqrt{2}}{9}, \; -\dfrac{7}{9}, \; -\dfrac{4\sqrt{2}}{7}, \; -\dfrac{7\sqrt{2}}{8}, \; -\dfrac{9}{7}, \; \dfrac{9\sqrt{2}}{8};$

III: $-\dfrac{4\sqrt{2}}{9}, \; -\dfrac{7}{9}, \; \dfrac{4\sqrt{2}}{7}, \; \dfrac{7\sqrt{2}}{8}, \; -\dfrac{9}{7}, \; -\dfrac{9\sqrt{2}}{8}.$

55. III: $-0.1875, \; -0.9823, \; 0.1909, \; 5.239, \; -1.018, \; -5.333;$

IV: $-0.1875, \; 0.9823, \; -0.1909, \; -5.239, \; 1.018, \; -5.333.$

57. $\dfrac{25}{81}.$ **59.** $\dfrac{448}{4225}.$ **61.** $\dfrac{216 + 283\sqrt{55}}{36(8 + \sqrt{55})}.$

63. $\dfrac{5}{18}.$ **65.** $-\dfrac{25 + 6\sqrt{41}}{41}.$

67. θ in III: $\left(\sqrt{65} - \dfrac{8}{63}\right)F;$ θ in IV: $-\left(\sqrt{65} + \dfrac{8}{63}\right)F.$

69. $-24\sqrt{2}.$ **71.** $E.$

Page 156. Sec. 4–8

[For the answers to Exercises 1–31 the trigonometric functions are listed in this order: sine, cosine, tangent, cotangent, secant, and cosecant.]

1. $\dfrac{\sqrt{3}}{2}, \; -\dfrac{1}{2}, \; -\sqrt{3}, \; -\dfrac{\sqrt{3}}{3}, \; -2, \; \dfrac{2\sqrt{3}}{3}.$ **3.** $-\dfrac{\sqrt{3}}{2}, \dfrac{1}{2}, \; -\sqrt{3}, \; -\dfrac{\sqrt{3}}{3}, 2, \; -\dfrac{2\sqrt{3}}{3}.$

5. $\dfrac{1}{2}, \dfrac{\sqrt{3}}{2}, \dfrac{\sqrt{3}}{3}, \; \sqrt{3}, \dfrac{2\sqrt{3}}{3}, \; 2.$ **7.** $-\dfrac{\sqrt{3}}{2}, \; -\dfrac{1}{2}, \; \sqrt{3}, \dfrac{\sqrt{3}}{3}, \; -2, \; -\dfrac{2\sqrt{3}}{3}.$

9. $\dfrac{\sqrt{2}}{2}, \; -\dfrac{\sqrt{2}}{2}, \; -1, \; -1, \; -\sqrt{2}, \; \sqrt{2}.$ **11.** $\dfrac{\sqrt{2}}{2}, \dfrac{\sqrt{2}}{2}, \; 1, 1, \; \sqrt{2}, \; \sqrt{2}.$

13. $0, \; -1, \; 0, \;$ undefined, $-1,$ undefined.   **15.** $0, \; 1, \; 0, \;$ undefined, $1,$ undefined.

17. $0, \; -1, \; 0, \;$ undefined, $-1,$ undefined.   **19.** $-\dfrac{\sqrt{3}}{2}, \dfrac{1}{2}, \; -\sqrt{3}, \; -\dfrac{\sqrt{3}}{3}, 2, \; -\dfrac{2\sqrt{3}}{3}.$

21. $-\dfrac{\sqrt{3}}{2}, \dfrac{1}{2}, -\sqrt{3}, -\dfrac{\sqrt{3}}{3}, 2, -\dfrac{2\sqrt{3}}{3}$.   **23.** 1, 0, undefined, 0, undefined, 1.

25. 1, 0, undefined, 0, undefined, 1.   **27.** $\dfrac{1}{2}, \dfrac{\sqrt{3}}{2}, \dfrac{\sqrt{3}}{3}, \sqrt{3}, \dfrac{2\sqrt{3}}{3}, 2$.

29. 0, -1, 0, undefined, -1, undefined.   **31.** $-\dfrac{1}{2}, \dfrac{\sqrt{3}}{2}, -\dfrac{\sqrt{3}}{3}, -\sqrt{3}, \dfrac{2\sqrt{3}}{3}, -2$.

35. E.   **37.** 1.   **39.** Undefined.

41. $\dfrac{13}{6}$.   **43.** Undefined.   **45.** $-\dfrac{\sqrt{2}}{2} Z_1 + \dfrac{7}{2} Z_2$.

Pages 158–159. Sec. 4–9

1. $\cos 73°$.   **3.** $\sec 65°$.   **5.** $\csc 21°$.   **7.** $\sec 62.47°$.

9. $\sin 77.32°$.   **11.** $\sec 89.953°$.   **13.** $\cos 23°$.   **15.** $\csc 88.57°$.

17. $30°$.   **19.** $30°$.   **21.** $16.25°$.   **23.** $6°$.

25. $0°$.   **27.** 5.   **29.** 1.   **31.** $15°$.

33. $26°$.   **35.** No value of θ satisfies.

Pages 161–162. Sec. 4–10

1. 0.12014.   **3.** 0.9806.   **5.** 0.8660.   **7.** 0.7912.   **9.** 2.500.

11. 0.8829.   **13.** 1.779.   **15.** 0.6600.   **17.** 0.9549.   **19.** 0.8884.

21. 3.516.   **23.** 0.5347.   **25.** 0.5360.   **27.** 0.8987.   **29.** 1.045.

31. 0.510.   **33.** 12.50.   **35.** 0.523.   **37.** 0.148.   **39.** 84.1.

41. 1440.   **43.** 1.91.   **45.** 335.

Page 163. Sec. 4–10

1. $5.70°$.   **3.** $27.10°$.   **5.** $51.80°$.   **7.** $69.40°$.   **9.** $84.10°$.

11. $46.30°$.   **13.** $70.40°$.   **15.** $37.74°$.   **17.** $52.22°$.   **19.** $31.57°$.

21. $22.13°$.   **23.** $62.28°$.   **25.** $71.46°$.   **27.** $59.10°$.   **29.** $48.69°$.

31. No value of θ satisfies.   **33.** $89.75°$.   **35.** $88.1°$.

Page 166. Sec. 4–11

1. 0.6018.   **3.** -0.7813.   **5.** 0.5543.   **7.** -0.9003.   **9.** -0.9245.

11. 2.924.   **13.** 0.09237.   **15.** 0.8934.   **17.** -0.9995.   **19.** 0.6101.

21. -0.6626.   **23.** 0.3034.   **25.** -0.2371.   **27.** 0.3153.   **29.** 0.06662.

31. 1.147.   **33.** -1.5.   **35.** -2151.   **37.** $-0.473 E_1 + 0.881 E_2$. **39.** -24.6.

41. -14.74.

Pages 168–169. Sec. 4–12

1. 33.80°, 146.20°.   **3.** 70.30°, 250.30°.   **5.** 135.30°, 224.70°.

7. 107.20°, 287.20°.   **9.** 30.00°, 330.00°.   **11.** 90.00°, 270.00°.

13. 32.30°, 147.70°, 212.30°, 327.70°. **15.** 61.45°, 118.55°.   **17.** 158.55°, 338.55°.

19. 10.56°, 169.44°, 190.56°, 349.44°.   **21.** No value of θ satisfies.

23. No value of θ satisfies.

25. 31.30°, 148.70°, 391.30°, 508.70°, −328.70°, −211.30°, etc.

27. 42.40°, 222.40°, 402.40°, 582.40°, −317.60°, −137.60°, etc.

29. 110.65°, 290.65°, 470.65°, 650.65°, −249.35°, −69.35°, etc.

33. 46.84°, 226.84°.   **35.** No value of θ satisfies.

37. 60.58°, 240.58°.   **39.** 88.3°, 271.7°.

41. 47.1°, 312.9°.

Page 171. Sec. 4–13

1. −0.4407.   **3.** −0.4910.   **5.** 1.024.   **7.** −0.02443.

9. −1.0000.   **11.** −0.8129.   **13.** −1.0000.   **15.** 0.0000.

Pages 172–173. Sec. 4–14

[For the answers to Exercises 1–13 the trigonometric functions are listed in this order: sine, cosine, tangent, cotangent, secant, cosecant.]

1. $\alpha: \dfrac{4}{5}, \dfrac{3}{5}, \dfrac{4}{3}, \dfrac{3}{4}, \dfrac{5}{3}, \dfrac{5}{4};$ $\beta: \dfrac{3}{5}, \dfrac{4}{5}, \dfrac{3}{4}, \dfrac{4}{3}, \dfrac{5}{4}, \dfrac{5}{3}.$ **3.** $\alpha: \dfrac{3}{5}, \dfrac{4}{5}, \dfrac{3}{4}, \dfrac{4}{3}, \dfrac{5}{4}, \dfrac{5}{3};$ $\beta: \dfrac{4}{5}, \dfrac{3}{5}, \dfrac{4}{3}, \dfrac{3}{4}, \dfrac{5}{3}, \dfrac{5}{4}.$

5. $\alpha: \dfrac{9}{41}, \dfrac{40}{41}, \dfrac{9}{40}, \dfrac{40}{9}, \dfrac{41}{40}, \dfrac{41}{9};$ $\beta: \dfrac{40}{41}, \dfrac{9}{41}, \dfrac{40}{9}, \dfrac{9}{40}, \dfrac{41}{9}, \dfrac{41}{40}.$

7. $\alpha: \dfrac{4\sqrt{65}}{65}, \dfrac{7\sqrt{65}}{65}, \dfrac{4}{7}, \dfrac{7}{4}, \dfrac{\sqrt{65}}{7}, \dfrac{\sqrt{65}}{4};$

$\beta: \dfrac{7\sqrt{65}}{65}, \dfrac{4\sqrt{65}}{65}, \dfrac{7}{4}, \dfrac{4}{7}, \dfrac{\sqrt{65}}{4}, \dfrac{\sqrt{65}}{7}.$

9. $\alpha: \dfrac{5}{11}, \dfrac{4\sqrt{6}}{11}, \dfrac{5\sqrt{6}}{24}, \dfrac{4\sqrt{6}}{5}, \dfrac{11\sqrt{6}}{24}, \dfrac{11}{5};$

$\beta: \dfrac{4\sqrt{6}}{11}, \dfrac{5}{11}, \dfrac{4\sqrt{6}}{5}, \dfrac{5\sqrt{6}}{24}, \dfrac{11}{5}, \dfrac{11\sqrt{6}}{24}.$

11. $\alpha: \dfrac{\sqrt{3}}{3}, \dfrac{\sqrt{6}}{3}, \dfrac{\sqrt{2}}{2}, \sqrt{2}, \dfrac{\sqrt{6}}{2}, \sqrt{3};$

$\beta: \dfrac{\sqrt{6}}{3}, \dfrac{\sqrt{3}}{3}, \sqrt{2}, \dfrac{\sqrt{2}}{2}, \sqrt{3}, \dfrac{\sqrt{6}}{2}.$

13. α: 0, 1, 0, not defined, 1, not defined;

    β: 1, 0, not defined, 0, not defined, 1.

15. $\alpha = 53.13°,\ \beta = 36.87°.$ **17.** $\alpha = \beta = 45.00°.$

19. $\alpha = 62.97°,\ \beta = 27.03°.$ **21.** $\alpha = 59.03°,\ \beta = 30.97°.$

23. $\alpha = 33.56°,\ \beta = 56.44°.$ **25.** $\alpha = 55.15°,\ \beta = 34.85°.$

27. $\alpha = 48.19°,\ \beta = 41.81°.$ **29.** $\alpha = 2.26°,\ \beta = 87.74°.$

Pages 177–180. Sec. 4–15

1. $\alpha = 31.6°$, $\beta = 58.4°$, $b = 15.0$.   **3.** $\alpha = 55.94°$, $\beta = 34.06°$, $c = 382.9$.

5. $\beta = 81.40°$, $b = 147.1$, $c = 148.7$.   **7.** $\alpha = 12.25°$, $\beta = 77.75°$, $c = 217.3$.

9. $\alpha = 54.8°$, $\beta = 35.2°$, $b = 26.9$.   **11.** $\alpha = 17°$, $\beta = 73°$, $c = 1.40$.

13. $\alpha = 46.05°$, $\beta = 43.95°$, $a = 8,945$.   **15.** $\alpha = 75.68°$, $a = 149.5$, $c = 154.3$.

17. These two given angles do not determine a unique triangle.

19. No triangle possible.   **21.** 41 ft.

23. 55.1 ft.   **25.** 197.5 ft.

27. 6400 ft.   **29.** Yes, 328-ft guy wires will hold safely.

31. 5.61 in.   **33.** 2.68 ft.   **35.** 38°.

37. 17.3°, 54.6 ft.   **39.** 76°.   **41.** 1.3 min.

43. 0.650 in.   **45.** 0.14°, 0.18°.   **47.** 11,500 ft, 251 mph.

49. 18.0 mi, 337.5°.

Chapter 5

Page 199. Sec. 5–9

1. 4.   **3.** 4.   **5.** 6.   **7.** $\frac{1}{2}$.

9. 12.   **11.** 5.   **13.** None.   **15.** None.

17. None.   **19.** None.   **21.** None.   **23.** None.

Page 203. Sec. 5–11

1. π, 1.   **3.** $\frac{\pi}{2}$, none.   **5.** π, none.   **7.** $\frac{2\pi}{3}$, 2.   **9.** $\frac{\pi}{2}$, 3.

11. 4π, $\frac{1}{2}$.   **13.** $\frac{\pi}{3}$, 3.   **15.** 2, 1.   **17.** 1, none.   **19.** 2, none.

21. 2, $\frac{4}{3}$.   **23.** $\frac{2}{3}$, 2.   **25.** $\frac{\pi}{2}$, none.   **27.** $\frac{\pi}{2}$, 5.   **29.** π, none.

Pages 207–208. Sec. 5–13

1. 3, π, π, $\frac{\pi}{2}$.   **3.** 2, $\frac{2\pi}{3}$, π, $\frac{\pi}{3}$.   **5.** 2, π, $\frac{\pi}{2}$, $\frac{\pi}{4}$.

7. 3, $\frac{2\pi}{3}$, $-\frac{\pi}{4}$, $-\frac{\pi}{12}$.   **9.** 4, $\frac{\pi}{2}$, $-\pi$, $-\frac{\pi}{4}$.   **11.** None, π, $-\pi$, $-\pi$.

13. None, $\frac{\pi}{3}$, $-\frac{\pi}{4}$, $-\frac{\pi}{12}$.   **15.** None, $\frac{\pi}{4}$, $-\frac{\pi}{2}$, $-\frac{\pi}{8}$.   **17.** None, $\frac{2\pi}{3}$, π, $\frac{\pi}{3}$.

19. None, π, $\frac{\pi}{6}$, $\frac{\pi}{12}$.   **21.** None, $\frac{2\pi}{3}$, $-\frac{\pi}{3}$, $-\frac{\pi}{9}$.   **23.** 2, 8π, $\frac{5\pi}{12}$, $\frac{5\pi}{3}$.

25. 5, 2, $-\frac{\pi}{3}$, $-\frac{1}{3}$.   **27.** 4, 1, $\frac{\pi}{6}$, $\frac{1}{12}$.   **29.** $\frac{3}{2}$, $\frac{1}{3}$, 2π, $\frac{1}{3}$.

31. None, $\frac{1}{3}$, $\frac{3\pi}{2}$, $\frac{1}{2}$.   **33.** None, 1, $-\frac{\pi}{6}$, $-\frac{1}{12}$.   **35.** 4, 2, $-\frac{\pi}{9}$, $-\frac{1}{9}$.

1. $T = 1.96$ sec, $f = 0.51$ cycles per sec.   **3.** 6.2 cm.

5. 377 rad per sec.

7. 6.28×10^4 rad per sec.

9. 60 cycles per sec.

11. 10 kc per sec.

13. 0.0167 sec.

15. 1.39×10^{-6} sec.

17. 12.0 sec.

19. 0.0167 sec.

21. 22.4 sec.

Chapter 6

Pages 227–228. Sec. 6–3

1. 0.76, 157°.   **3.** 1.15, 113°.   **5.** 1.85, 68°.

7. 2.12, 318°.   **9.** 6.48, 251°.   **11.** 2.31, 84°.

13. 5.3, 167°.   **15.** 1.03, 137°.   **17.** 0.71, 273°.

19. 1.54, 96°.   **21.** 44 mi, 35°.   **23.** 2.9 mi.

25. 135 mi, 20°.   **27.** 178 mph, 11°.   **29.** 148 lb, 66°.

Pages 229–230. Sec. 6–4

29. 53 lb, 37 lb.   **31.** 175 lb, 51 lb, 168 lb.

Pages 235–236. Sec. 6–5

1. $(-6, 5)$.   **3.** $(-3, 1)$.   **5.** $(13, 25)$.   **7.** $(-1, -4)$.

9. $(5, -11)$.   **11.** $(-22, -8)$.   **13.** $(-11, 15)$.   **15.** $(-17, -9)$.

17. $(-9, 23)$.   **19.** $(0, -30)$.   **21.** 6.40, 51.3°.   **23.** 13.0, 292.6°.

25. 7.62, 203.2°.   **26.** 7.00, 0.0°.   **29.** 5.83, 239.0°.   **31.** 44.0, 287.2°.

33. 45.0, 36.9°.   **35.** 22.8, 217.9°.   **37.** 32.0, 270.0°.   **39.** $(13.7, 14.6)$.

41. $(-9.07, 11.4)$.   **43.** $(0.502, 4.34)$.   **45.** $(-124, 28.4)$.   **47.** $(253, -464)$.

49. $(2940, -1370)$.   **51.** $(0.257, -0.629)$.   **53.** $(-40.5, 13.3)$.   **55.** $(4.01, -4.04)$.

57. 45.2, 42.9°.   **59.** 528, 203.1°.   **61.** 0.0740, 247.8°.   **63.** 0.00690, 2.0°.

65. 648, 56.5°.   **67.** 18.4, 146.9°.   **69.** 0.502, 50.3°.   **71.** 3.28, 79.1°.

Pages 241–244. Sec. 6–7

1. 143 mi.   **3.** 27 mi, 102°.   **5.** 28 knots, 306°.   **7.** 37 mph, 319°.

9. 78°, 250 mph.   **11.** 134 lb.   **13.** 84 lb, no, no.   **15.** 216 lb, 48°.

17. 77°.   **19.** 9.4 lb.   **21.** 41 lb.   **23.** 286 lb.

25. 230 lb.   **27.** 53 mph, 16°.   **29.** 72 mph, 202°.

Chapter 7

Pages 247–248. Sec. 7–1

1. $49P^2 - 25$.   **3.** $25k^2 - 10k + 1$.   **5.** $3a^2 - 8ax - 35x^2$.

7. $x^4 - 6x^2a + 9a^2$.   **9.** $xe_1 + ye_1 - xe_2 - ye_2$.   **11.** $a^2 - 2a - 35$.

3. $15x^4 + x^2 - 2$.   **15.** $2 + 3R^2 - 14R^4$.

7. $S_1^2 + S_2^2 + S_3^2 + 2S_1S_2 + 2S_1S_3 + 2S_2S_3$.

9. $a^2 + b^2 + c^2 - 2ab - 2ac + 2bc$.   **21.** $p^2 + q^2 + 2pq - 4p - 4q + 4$.

23. $9Z_1^2 + Z_2^2 - 6Z_1Z_2 + 12Z_1 - 4Z_2 + 4$.   **25.** $w^2 + 4v^2 + 9z^2 + 4wv + 6wz + 12vz$.

27. $L^4 + 4L^3 + 10L^2 + 12L + 9$.   **29.** $8a^3 + 27$.

31. $a^6 - b^6$.   **33.** $r_1^3 + r_2^3$.

35. $\frac{1}{8} - x^3y^3$.   **37.** $27t^3 + 108t^2 + 144t + 64$.

39. $m^8 - 4m^6d + 6m^4d^2 - 4m^2d^3 + d^4$.

41. $E_1^5 - 5E_1^4E_2 + 10E_1^3E_2^2 - 10E_1^2E_2^3 + 5E_1E_2^4 - E_2^5$.

43. $x^7 + 7x^6y + 21x^5y^2 + 35x^4y^3 + 35x^3y^4 + 21x^2y^5 + 7xy^6 + y^7$.

45. $\dfrac{x^8}{16} + \dfrac{x^6y}{6} + \dfrac{x^4y^2}{6} + \dfrac{2x^2y^3}{27} + \dfrac{y^4}{81}$.   **47.** $x^{10} - 5x^7 + 10x^4 - 10x + \dfrac{5}{x^2} - \dfrac{1}{x^5}$.

49. $\dfrac{x^4}{y^4} + \dfrac{4x^2}{y^2} + 6 + \dfrac{4y^2}{x^2} + \dfrac{y^4}{x^4}$.   **51.** $4x^{2m} + 12x^my^n + 9y^{2n}$.

53. $36x^{2n} + 15x^ny^m - 75y^{2m}$.   **55.** $a^{6m} + 1$.

61. 1.1041.   **63.** 0.81707.

65. 33.632.   **67.** 0.66483.

69. 1.2202.   **71.** 1029.1.

Pages 250–251. Sec. 7–2

1. $(x - 15)(x + 4)$.   **3.** $(10a - 3)(a + 2)$.   **5.** $(3e + 4)(2e - 5)$.

7. $(4a + 1)(3 - 2a)$.   **9.** $(1 - y)(y^2 + y + 1)$.   **11.** $(r + s)(r^2 - rs + s^2)$.

13. $3(2y + 1)(y - 1)$.   **15.** $(7x + 6)(x - 1)$.   **17.** $(a^2 + 14)(a^2 - 3)$.

19. $(2x - 5y)(4x^2 + 10xy + 25y^2)$.   **21.** $(2m + 3n)(4m^2 - 6mn + 9n^2)$.

23. $(a + b + c)(a^2 - ab - ac + b^2 + 2bc + c^2)$.   **25.** $(x - y)(x + y)(x^2 + xy + y^2)(x^2 - xy + y^2)$.

27. $2(3m - 2np)^2$.   **29.** $(x + y + z - 6)(x + y + z + 6)$.

31. $(x + y + z)^2$.   **33.** $4(3X - 1)(X + 2)$.

35. $2(a^2 + 1)^2$.   **37.** $(3P - 2Q)(3P + 2Q)(9P^2 + 4Q^2)$.

39. $5(x - y)(4x + 3y)$.   **41.** $(5Z_1 + 2Z_2)(2Z_1 - 3Z_2)$.

43. $(3 - x - y - z)(x^2 + y^2 + z^2 + 2xy + 2xz + 2yz + 3x + 3y + 3z + 9)$.

45. $(3\alpha^2 - 2)(\alpha^2 - 5)$.   **47.** $(6p - 5)(2p + 3)$.

49. $(3x^2 - 2y)(x^2 - 3y)$.   **51.** $(x^4 + 13y)(x^4 - 7y)$.

53. $2s(3r^2 + s^2)$.

55. $(x_1 + x_2 - x_3 + x_4)(x_1^2 + 2x_1x_2 + x_2^2 + x_1x_3 + x_2x_3 - x_1x_4 - x_2x_4 + x_3^2 - 2x_3x_4 + x_4^2)$.

57. $(4\alpha + 3\beta)(16\alpha^2 - 12\alpha\beta + 9\beta^2)$.

59. $(x + y + z + w)(x^2 + 2xy + y^2 - xz - yz - xw - yw + z^2 + 2zw + w^2)$.

61. $(3x + 3y + 1)(x + y + 2)$.   **63.** $(s_1 + s_2 + 1)(2s_1 + 2s_2 + 5)$.

65. $(5x^n + 3)(x^n - 2)$.   **67.** $(2x^ny^m + 3)(x^ny^m - 1)$.

69. $\left(\dfrac{2}{L} - \dfrac{T}{3}\right)\left(\dfrac{4}{L^2} + \dfrac{2T}{3L} + \dfrac{T^2}{9}\right)$.   **71.** $(r - s - t)(r - s + t)$.

73. $(Z^2 + 1)(Z - 1)$.   **75.** $(1 - r_2 - r_1)(1 + r_2 + r_1)$.

77. $(x - y + z)(x + y - z)$.   **79.** $(R + S)(R - S + 1)$.

81. $(m + 2n)^3$.   **83.** $(p + 2r + t)^2$.

85. $(3x - 4y - 2z)^2$.   **87.** $(3a + b - 2c)^2$.

89. $(3x_1 - 4x_1x_2 + 5)^2$.

Pages 252–254. Sec. 7–3

1. $\dfrac{13x - 22y}{60}$.

3. $\dfrac{3x^2 + 2xy + 5y^2}{x^3 y^3}$.

5. $\dfrac{3\mu + 2}{\mu^3 + 27}$.

7. $\dfrac{10z^2}{3x^3 y}$.

9. $-\dfrac{a}{b}$.

11. $\dfrac{I_1^{\,2} + I_1 I_2 + I_2^{\,2}}{I_1 + I_2}$.

13. $\dfrac{(\mu + \beta)(\mu^2 + \mu\beta + \beta^2)}{\beta}$.

15. $\dfrac{(E_g + E_p)^2}{E_g^{\,2} + E_g E_p + E_p^{\,2}}$.

17. 0.

19. $\dfrac{v(v - u + uv)}{u^3 - v^3}$.

21. $\dfrac{2P^2}{P^3 - 1}$.

23. $\dfrac{2a^3}{a^4 + a^2 r^2 + r^4}$.

25. $-\dfrac{s^3(r + s)}{r^2(s^2 + sr + r^2)}$.

27. $\dfrac{z + 2}{z + 3}$.

29. $\dfrac{x + y}{x^2 + xy + y^2}$.

31. $\dfrac{x^3}{x - 1}$.

33. $\dfrac{I_1^{\,2} - I_1 I_2 + I_2^{\,2}}{I_1^{\,2} + I_2^{\,2}}$.

35. $\dfrac{2(a^2 + T^2)}{a^3 + T^3}$.

37. $\dfrac{e_1^{\,2} + e_2^{\,2}}{e_1^{\,4} + e_1^{\,2}e_2^{\,2} + e_2^{\,4}}$.

39. $\dfrac{n + 5}{n^4 - 1}$.

41. $\dfrac{a^2 + b^2}{(a + b)^2}$.

43. $\dfrac{Y - 2X}{Y}$.

45. $E^2 - 4$.

47. $\dfrac{2(I_1 + I_2)}{I_1^{\,3} - I_2^{\,3}}$.

49. $\dfrac{K}{K + 1}$.

Pages 256–257. Sec. 7–4

1. $\frac{7}{8}$.

3. $R + 1$.

5. $x + y$.

7. $\dfrac{abc}{6}$.

9. $\dfrac{ir + er + ei}{re^2 + ei^2 + ir^2}$.

11. $\dfrac{pq^2}{(p + q)(p - q)^2}$.

13. $-\dfrac{2sr}{s^2 + r^2}$.

15. $-\dfrac{(C_1^{\,2} + C_2^{\,2})^2}{4C_1^{\,2}C_2^{\,2}}$.

17. $\dfrac{1 + E - E^2}{2E^2 - 1}$.

19. $\dfrac{C^2 + 1}{C^3}$.

21. 1.

23. $\dfrac{3x - 1}{2x - 1}$.

25. $\dfrac{E_2^{\,3}(E_1 - E_2)}{E_1^{\,3}(E_1 + E_2)}$.

27. $\dfrac{n(n^3 + n + 1)}{n^2 + 1}$.

29. $\dfrac{C_1^{\,2} - C_1 + 1}{2C_1 - 1}$.

Pages 261–262. Sec. 7–5

1. 6.

3. 3, 4.

5. $-1, 1$.

7. $\frac{2}{3}$.

9. $-\frac{2}{7}$.

11. -3.

13. No solution.

15. No solution.

17. No solution.

19. -6.

21. -2.

23. 1, $3\frac{1}{4}$.

25. $\frac{3}{4}$.

27. 0, $-\frac{2}{3}$.

29. No solution.

31. -4.

33. $-1\frac{31}{47}$.

35. 1.

37. $-\frac{3}{4}$.

Pages 263–264. Sec. 7–6

1. $\frac{11}{31}$. **3.** $\frac{34}{71}$. **5.** 7. **7.** 26.

9. 58. **11.** $\frac{3}{8}$. **13.** $\frac{aN}{a+b}, \frac{bN}{a+b}$. **15.** $2\frac{6}{7}$ hr.

17. 20 ft. **19.** 6 mph. **21.** 4 mph. **23.** 40 mph, 50 mph.

Pages 266–268. Sec. 7–7

1. $V = \dfrac{RT}{P}, \; T = \dfrac{PV}{R}$. **3.** $f = \dfrac{300 \cdot 10^6}{\lambda}$. **5.** $f_s = \dfrac{4r_s}{\pi d^2}, \; d^2 = \dfrac{4r_s}{\pi f_s}$.

7. $d = \dfrac{8.84 KA}{10^8 C}$. **9.** $W = \dfrac{8EIy}{l^3}, \; E = \dfrac{Wl^3}{8Iy}$. **11.** $f = \dfrac{3T \sin \alpha}{2rW}, \; r = \dfrac{3T \sin \alpha}{2fW}$.

13. $Q_1 = \dfrac{Q_2}{1 - \epsilon_c}, \; Q_2 = (1 - \epsilon_c)Q_1$. **15.** $k = \dfrac{nf_c}{f_s + nf_c}$.

17. $J = \dfrac{W}{E_2 - E_1 - Q}, \; E_1 = \dfrac{J(E_2 - Q) - W}{J}$. **19.** $\omega_1 = \dfrac{tT + J_m \omega_2}{J_m}$.

21. $R_L = \dfrac{Ar_p}{1 - A}, \; r_p = \dfrac{R_L(1 - A)}{A}$. **23.** $w = \dfrac{R_L^2}{2(M - aR_L)}$.

25. $l = \dfrac{0.8r^2N^2 - 6rL - 10tL}{9L}, \; t = \dfrac{0.8r^2N^2 - 6rL - 9lL}{10L}$.

27. $f = \dfrac{rr'}{(\mu - 1)(r' - r)}, \; r = \dfrac{fr'(\mu - 1)'}{r' + f(\mu - 1)}$.

29. $J = \dfrac{U_2^2 - U_1^2}{64.34(H_1 - H_2)}, \; H_1 = \dfrac{U_2^2 - U_1^2}{64.34J} + H_2$.

31. $C_1 = \dfrac{C_2 C_3 C_T}{C_2 C_3 - C_2 C_T - C_3 C_T}, \; C_2 = \dfrac{C_1 C_3 C_T}{C_1 C_3 - C_1 C_T - C_3 C_T}, \; C_3 = \dfrac{C_1 C_2 C_T}{C_1 C_2 - C_1 C_T - C_2 C_T}$.

33. $L_1 = \dfrac{K_1 A_1}{K_2 A_2}(K_2 A_2 R - L_2), \; A_2 = \dfrac{K_1 A_1 L_2}{K_2(K_1 A_1 R - L_1)}$.

35. $t_1 = t_2 - \dfrac{L}{e_{avg}}(i_2 - i_1), \; t_2 = t_1 + \dfrac{L}{e_{avg}}(i_2 - i_1)$.

37. $t = \dfrac{h - h_0}{mh_0 - lh}, \; l = \dfrac{mth_0 - h + h_0}{ht}$.

39. $W_A = \dfrac{AM_A(W_B M_C + M_B W_C)}{M_B M_C(1 - A)}, \; M_A = \dfrac{W_A M_B M_C(1 - A)}{A(W_B M_C + M_B W_C)}$.

Chapter 8

Pages 270–271. Sec. 8–1

1. L^8. **3.** $x^3 y^4$. **5.** $81 s^4 r^4$. **7.** $Q_1^8 Q_2^{12}$. **9.** $-27 L_1^3 L_2^6$.

11. Z^8. **13.** $\dfrac{1}{a^2 b^3}$. **15.** F_1. **17.** $27 r_1^6 r_2^5$. **19.** $a^7 b^7$.

21. $8z_1{}^3z_2{}^6z_3{}^9$.   **23.** x^5.   **25.** a^8.   **27.** $\dfrac{1}{a^5b^{10}c^5}$.   **29.** $\dfrac{f_1}{4f_2{}^2}$.

31. x^m.   **33.** $32b^{4m}$.   **35.** $\dfrac{1}{v^{2m}}$.   **37.** y^{6q}.   **39.** p^{4n}.

41. $-\dfrac{b^2}{a^7c^8}$.   **43.** $\dfrac{1}{x^{3n}y}$.   **45.** $r_x{}^m r_y{}^m$.   **47.** $\dfrac{1}{R_1{}^2R_2{}^2}$.   **49.** $\cos\theta$.

51. $\tan^3 B$.   **53.** 2^{6q+10}.   **55.** $\frac{1}{9}$.   **57.** 2^{2n^2-3}.   **59.** $\dfrac{s^8-1}{s^4}$.

61. $\dfrac{\cos^{12}\beta}{\sin^9\alpha}$.   **63.** $z^{4m(m-n)}$.   **65.** $\dfrac{1}{r^2 s^{2n}}$.

Pages 273–274. Sec. 8–2

1. 3.   **3.** 7.   **5.** 12.   **7.** 30.   **9.** 25.

11. 2.   **13.** -1.   **15.** 5.   **17.** 10.   **19.** $\frac{2}{3}$.

21. $\frac{12}{13}$.   **23.** $14\,rs$.   **25.** $-\frac{2}{5}$.   **27.** $-\dfrac{a}{3}$.   **29.** $3u$.

31. $-3M_x$.   **33.** $2a$.   **35.** $-\frac{1}{2}L^2$.   **37.** $-2\sin\theta$.   **39.** $25\tan A$.

41. $-7(\sin\beta)^2$.   **43.** $-L_x{}^2L_y$.   **45.** $-E_1{}^2E_2E_3$.   **47.** 3.   **49.** 8.

51. 12.   **53.** 20.   **55.** 7.   **57.** 14.   **59.** 0.0100.

61. $\frac{1}{2}$.   **63.** $\frac{1}{4}$.   **65.** 0.1300.   **67.** $\frac{3}{5}$.   **69.** $\frac{4}{11}$.

71. -2.   **73.** -4.   **75.** -6.   **77.** -0.100.   **79.** $-\frac{1}{3}$.

81. $-\frac{3}{4}$.   **83.** ± 8.   **85.** ± 9.   **87.** ± 16.   **89.** ± 0.40.

91. $\pm\frac{4}{7}$.   **93.** $\pm\frac{3}{20}$.

Pages 276–278. Sec. 8–4

1. 6.   **3.** 2.   **5.** -4.   **7.** 2.

9. 1.   **11.** 0.0700.   **13.** $\frac{1}{5}$.   **15.** $\dfrac{1}{9.61}$.

17. $\frac{1}{2}$.   **19.** $\frac{1}{2}$.   **21.** -11.   **23.** $\frac{10}{9}$.

25. 6.   **27.** 5.   **29.** $\dfrac{1}{25y^3}$.   **31.** $\dfrac{4}{\sin^3\theta}$.

33. $\dfrac{1}{6y^4}$.   **35.** s^4.   **37.** $\dfrac{x^2y^3\tan\theta}{3}$.   **39.** $\dfrac{e_3}{e_1{}^2e_2{}^3}$.

41. $8E_1E_2{}^3E_3{}^5$.   **43.** $\dfrac{xy}{y-x}$.   **45.** $\dfrac{a^2-ab+b^2}{ab}$.   **47.** $5x^{-3}$.

49. nN^{-2}.   **51.** $3L^{-4}$.   **53.** $E_1{}^{-1}E_2{}^{-1}E_3{}^{-1}$.   **55.** $ax^{-3}y^{-3}z^{-3}$.

57. $7ap^2q^{-3}$.   **59.** $(m+n)(p-q)^{-1}$.   **61.** $\sqrt[3]{x}$.   **63.** $\sqrt[4]{L^3}$.

65. $a\sqrt[3]{z}$.   **67.** $\sqrt[3]{2V}$.   **69.** $\sqrt{3M_1{}^3}$.   **71.** $\sqrt[4]{9e_1e_2e_3}$.

73. $\sqrt{e_1+e_2}$.   **75.** $\sqrt{(L_x+L_y)^3}$.   **77.** $\sqrt{a^2+b^2}$.   **79.** $a^{\frac{1}{2}}$.

81. $L_1{}^{\frac{1}{3}}$.   **83.** $d^{\frac{5}{3}}$.   **85.** $b^{\frac{7}{8}}$.   **87.** $(\cos\beta)^{\frac{1}{3}}$.

89. $r^{\frac{3}{2}}$.   **91.** $(R_1+R_2)^{\frac{1}{2}}$.   **93.** $(L_1+L_2+L_3)^{\frac{1}{3}}$.   **95.** $(\sin A+\cos A)^{\frac{1}{2}}$.

Pages 280–281. Sec. 8–5

1. 1.

3. c^2.

5. $\dfrac{3}{M}$.

7. $\dfrac{r}{s}$.

9. $L^{\frac{1}{4}}$.

11. $\dfrac{32e_1{}^2}{e_2{}^3}$.

13. $\dfrac{2}{K}$.

15. $9u^4v^2$.

17. $\dfrac{1}{n^{2x}}$.

19. $\dfrac{x^{3n}y^{2n^2}z^{n^2}}{z^n}$.

21. x^{m^2}.

23. $\dfrac{1}{x^{\frac{1}{2}}}$.

25. $x + \dfrac{1}{x} + 2$.

27. $u^{\frac{1}{2}} - v^{\frac{1}{2}}$.

29. $\dfrac{x^{4n} - 1}{x^{2n}}$.

31. $r + \dfrac{6}{r} - 5$.

33. $a - b$.

35. $x^3 + y^{\frac{3}{2}}$.

37. 3.

39. $-\frac{1}{8}$,

41. 8.

43. 5.00

45. 9.

47. $\frac{1}{32}$.

49. 1.

51. $(a^{\frac{1}{3}} + 2b^{\frac{1}{3}})(a^{\frac{1}{3}} - 2b^{\frac{1}{3}})$.

53. $(R + L^{-1})(R - L^{-1})$.

55. $(e_1 - e_2{}^{-1})^2$.

57. $(p^4 - q^{\frac{1}{3}})^2$.

59. $(z^{\frac{1}{3}} - 1)(z^{\frac{1}{3}} - 3)$.

61. $(2s_1{}^{\frac{1}{3}} + 5s_2{}^{\frac{1}{3}})^2$.

63. $(3h^{-1} - 2k)(h^{-1} + k)$.

Pages 284–286. Sec. 8–6

1. $3\sqrt{10}$.

3. $2\sqrt[3]{3}$.

5. $2\sqrt[4]{3}$.

7. $-2\sqrt[3]{5}$.

9. $a^2\sqrt[3]{a^2}$.

11. $2E^3\sqrt{5E}$.

13. $3\sqrt[3]{2}\sin\alpha$.

15. $2\tan A\sqrt{7\tan A}$.

17. $0.200xy^2\sqrt[3]{x^2}$.

19. $3a^3b^2\sin\theta\sqrt{2\sin\theta}$.

21. $2\sqrt{1 - M}$.

23. $(x^2 + y^2)\sqrt{x^2 + y^2}$.

25. $\sqrt{18}$.

27. $\sqrt[4]{48}$.

29. $\sqrt{50R}$.

31. $\sqrt[4]{162R_1{}^{11}}$.

33. $\sqrt{(a + b)^3}$.

35. $\sqrt{60rs}$.

37. $\sqrt[3]{\sin\theta\cos\theta}$.

39. $\sqrt[4]{\dfrac{p}{(p - q)^3}}$.

41. $\frac{1}{3}\sqrt{3}$.

43. $\frac{1}{2}\sqrt[3]{2}$.

45. $\frac{1}{2}\sqrt[4]{14}$.

47. $\dfrac{1}{L_2}\sqrt{L_1L_2}$.

49. $\dfrac{1}{5y}\sqrt{15xy}$.

51. $\dfrac{mn}{3}\sqrt[3]{9}$.

53. $\dfrac{1}{\sin A}\sqrt{a\sin A}$.

55. $\frac{1}{2}\sqrt[4]{8\tan\theta}$.

57. $-\dfrac{2E}{3R}\sqrt[3]{6E^2R}$.

59. $\dfrac{1}{4b^2}\sqrt{2b(3a + 1)}$.

61. $x\sqrt{x}$.

63. $2\sqrt{2}$.

65. $\sqrt[3]{3}$.

67. $b\sqrt{2b}$.

69. $RL\sqrt{6R}$.

71. $\sqrt{a}\tan A$.

73. $\dfrac{1}{7z}\sqrt{42uvz}$.

75. $\dfrac{a}{b\sin B}\sqrt{\sin A\sin B}$.

77. $\sqrt{5(E_1 - E_2)}$.

79. $\frac{1}{3}\sqrt{3}$.

81. $\dfrac{ab}{3}\sqrt{ab}$.

83. $\frac{3}{2}\sqrt{5}$.

85. $\dfrac{2}{uv}\sqrt{u}.$ **87.** $\tfrac{2}{5}\sqrt[3]{25E}.$ **89.** $\dfrac{2x}{3ab}\sqrt{xyb}.$

91. $\dfrac{2L^2}{5}\sqrt{5}.$ **93.** $\dfrac{2}{Z^2}\sqrt{4Z^4}.$ **95.** $\dfrac{t(r-s)}{4}\sqrt[3]{4}.$

97. $\dfrac{1}{R}\sqrt[3]{R^3+R^2}.$ **99.** $\dfrac{1}{2y}\sqrt[3]{xy^3+24}.$ **101.** $(a+b)\sqrt{a-2b}.$

103. $\dfrac{v^2-u}{v^2}\sqrt{5}.$

Pages 286–287. Sec. 8–7

1. $6\sqrt{2}.$ **3.** $46\sqrt{3}.$ **5.** $17\sqrt[3]{3}.$

7. $4\sqrt{L}.$ **9.** $(2+9a)\sqrt[3]{5x}.$ **11.** $11\sqrt{6}.$

13. $-\tfrac{6}{7}\sqrt{7}.$ **15.** $(2\sin A - 3a\cos A)\sqrt{\sin A}.$

17. $-\dfrac{2E_2}{E_1{}^2-E_2{}^2}\sqrt{E_1{}^2-E_2{}^2}.$ **19.** $\tfrac{13}{6}\sqrt[3]{36}.$ **21.** $\dfrac{3(2u-1)}{4v}\sqrt{2uv}.$

23. $a(7a-4)\sqrt[3]{3ab^2}.$ **25.** $(5b-4y^2+by)\sqrt{2b^2-3y}.$

Pages 288–289. Sec. 8–8

1. $42\sqrt{2}.$ **3.** $48\sqrt{6}.$ **5.** $80a\sqrt{3}.$ **7.** $36\sqrt[3]{4}.$

9. $x^2\sqrt[3]{x}.$ **11.** $18L.$ **13.** $\tfrac{108}{5}.$ **15.** $\tfrac{1}{6}\sqrt[3]{84}.$

17. $2\sqrt{3}.$ **19.** $s\sqrt[12]{9s^7}.$ **21.** $ab\sqrt[20]{a^2b^{11}c^{13}}.$ **23.** $\tfrac{3}{2}\sqrt[12]{72}.$

25. $x^2\sqrt[3]{2}.$ **27.** $3\sin\theta\sqrt{\sin\theta}.$ **29.** $x+y-2\sqrt{xy}.$ **31.** $2.$

33. $-18-21\sqrt{2}.$ **35.** $\sqrt{e_1{}^2-e_2{}^2}.$ **37.** $3\sqrt{5}.$

39. $\sin\theta\sqrt[12]{\sin^5\theta\cos^6\theta}.$ **41.** $20.$ **43.** $a^2b\sqrt[12]{a^3b^8}.$

45. $2(M-\sqrt{M^2-1}).$ **47.** $x_1+x_2+x_3+2\sqrt{x_1x_2}+2\sqrt{x_2x_3}+2\sqrt{x_1x_3}.$

49. $(E_1+E_2)\sqrt[12]{(E_1+E_2)^5(E_1-E_2)^{10}}.$

51. $2a+3b-49c.$ **53.** $40+12\sqrt{15}.$

Pages 291–292. Sec. 8–9

1. $2.$ **3.** $3.$ **5.** $3\sqrt[3]{\sin\theta}.$ **7.** $2\sqrt{\cos A}.$

9. $\tfrac{2}{3}\sqrt{6}.$ **11.** $\tfrac{1}{3}\sqrt[6]{108}.$ **13.** $\sqrt[6]{250}.$ **15.** $\sqrt[6]{rs}.$

17. $\sqrt{2}-\sqrt{5}.$ **19.** $-\dfrac{2+\sqrt{6}}{2}.$ **21.** $11+6\sqrt{3}.$ **23.** $\dfrac{\sqrt{R_1}+\sqrt{R_2}}{R_1-R_2}.$

25. $\dfrac{e_1+e_2+2\sqrt{e_1e_2}}{e_1-e_2}.$ **27.** $\dfrac{i_1+i_2-2\sqrt{i_1i_2}}{i_1-i_2}.$ **29.** $\dfrac{1}{m}\sqrt[12]{m^{11}n^2}.$ **31.** $\dfrac{111+13\sqrt{21}}{172}.$

33. $\dfrac{\sqrt{I_p}(\sqrt{I_p+1}-1)}{I_p}.$ **35.** $\sqrt{15}.$

37. $\dfrac{\sin A+\cos A-2\sqrt{\sin A\cos A}}{\sin A-\cos A}.$ **39.** $\dfrac{x^2-5y^2+\sqrt{x^4-y^4}}{5x^2-13y^2}.$

Pages 294–295. Sec. 8–10

1. 22.   **3.** No solution.   **5.** $\frac{7}{2}$.   **7.** 4.   **9.** 5.

11. 5.   **13.** 49.   **15.** 9.   **17.** 4.   **19.** 12.   **21.** -1.

23. 2.   **25.** 2.   **27.** $\frac{81}{49}$.   **29.** 0.   **31.** 7.   **33.** 5.

Page 296. Sec. 8–11

1. 1.69.   **3.** 2.12.   **5.** 3.58.   **7.** 3.76.   **9.** 0.632.   **11.** -0.462.

13. 0.925.   **15.** 8.32.   **17.** 3.54.   **19.** 7.61.   **21.** 0.521.

Pages 297–299. Sec. 8–12

1. $r = \sqrt{\dfrac{A}{\pi}}$.   **3.** $E = \sqrt{PR}$.   **5.** $r = \sqrt{\dfrac{V}{\pi h}}$.

7. $h = \dfrac{D^2}{1.063^2}$.   **9.** $t = \sqrt{\dfrac{2s}{g}}$.   **11.** $a = \sqrt[3]{V}$.

13. $r = \sqrt[3]{\dfrac{3V}{4\pi}}$.   **15.** $W = I^2R,\ R = \dfrac{W}{I^2}$.   **17.** $L = \dfrac{\lambda^2}{1884^2 C},\ C = \dfrac{\lambda^2}{1884^2 L}$.

19. $L_1 = \dfrac{M^2}{K^2 L_2},\ L_2 = \dfrac{M^2}{K^2 L_1}$.   **21.** $\lambda = \dfrac{2\pi V^2}{g}$.   **23.** $R = \dfrac{V^2}{M^2 g \gamma t},\ t = \dfrac{V^2}{M^2 R g \gamma}$.

25. $D = \sqrt{\dfrac{3.095 Lwz}{Pg}}$.   **27.** $r = \dfrac{e \sin \theta}{2c}\sqrt{\dfrac{a\beta}{\pi S_p}},\ \sin \theta = \dfrac{2rc}{e}\sqrt{\dfrac{\pi S_p}{a\beta}}$.

29. $V_1 = V_2\sqrt[\alpha]{\dfrac{Q_1}{Q_2}},\ V_2 = V_1\sqrt[\alpha]{\dfrac{Q_2}{Q_1}}$.   **31.** $N = \dfrac{1}{a}\sqrt{\dfrac{L(6a + 9b + 10c)}{0.315\mu h}}$.

33. $s_1 = \sqrt[4]{12 I_g + s_2{}^4},\ s_2 = \sqrt[4]{s_1{}^4 - 12 I_g}$.   **35.** $T = 273\left(\dfrac{V_T{}^2}{V_0{}^2} - 1\right)$.

37. $s_1 = \sqrt{12 k_g{}^2 - s_2{}^2},\ s_2 = \sqrt{12 k_g{}^2 - s_1{}^2}$.   **39.** $l = x\sqrt{\dfrac{3W}{W - 3V_x}},\ x = l\sqrt{\dfrac{W - 3V_x}{3W}}$.

41. $\dfrac{v}{c} = \dfrac{\sqrt{m^2 - m_0{}^2}}{m}$.   **43.** $E_p + E_g = \dfrac{(a+b)[a + b(\mu+1)]^3 \cdot 10^{12}}{r_p{}^2 A_1{}^2}$.

Chapter 9

Page 310. Sec. 9–1

1. 2.8×10.   **3.** 1.1×10^3.   **5.** 3.7×10^2.   **7.** 5.0×10^4.

9. 3.2.   **11.** 6.0×10^{-2}.   **13.** 1.1×10^{-1}.   **15.** 5.9×10.

Pages 312–313. Sec. 9–2

1. $\log_2 8 = 3$.   **3.** $\log_4 64 = 3$.   **5.** $\log_{12} \frac{1}{144} = -2$.   **7.** $\log_7 \frac{1}{343} = -3$.

9. $\log_{10} 0.01 = -2$.   **11.** $\log_{10} 10{,}000 = 4$.   **13.** $\log_6 \frac{1}{1296} = -4$.   **15.** $\log_{13} 2197 = 3$.

17. $\log_5 125 = 3$.   **19.** $\log_2 2.3 = 1.20$.   **21.** $\log_2 10 = 3.32$.   **23.** $4^2 = 16$.

25. $4^4 = 256.$ **27.** $10^2 = 100.$ **29.** $7^4 = 2401.$ **31.** $10^3 = 1000.$

33. $3^{-4} = \frac{1}{81}.$ **35.** $6^3 = 216.$ **37.** $6^{-1} = \frac{1}{6}.$ **39.** $1.3^2 = 1.69.$

41. $10^{\frac{1}{2}} = \sqrt{10}.$ **43.** 9.   **45.** 3.   **47.** $-4.$

49. $\frac{1}{2}.$ **51.** $-5.$ **53.** 2.   **55.** $\frac{1}{2}.$

57. $\frac{2}{3}.$ **59.** 0.   **61.** 16.   **63.** 49.

65. $\frac{1}{6}.$ **67.** 625.   **69.** 1.44.   **71.** 7.

73. 1.   **75.** $\frac{1}{243}.$ **77.** 144.

Page 314. Sec. 9-3

1. $-9.$ **3.** $-3.$ **5.** 2.   **7.** $-3.$ **9.** 4.   **11.** $2x + 1.$

13. $-2.$ **15.** 1.   **17.** $2(x + 1).$ **19.** $-\frac{1}{3}.$ **21.** 2.   **23.** 15.

25. 4.   **27.** 5.   **29.** $x^3 - 1.$ **31.** $x^2 - y^2.$ **33.** $e^2.$ **35.** 1.

Pages 318-319. Sec. 9-5

1. $\log_B L + \log_B M + \log_B N.$ **3.** $\log_B L + \log_B M - \log_B N.$

5. $2 \log_B M + 4 \log_B N.$ **7.** $-n \log_B T.$

9. $\frac{1}{3}(5 \log_3 C + 7 \log_3 D).$ **11.** $n \log_{10} L + \frac{1}{2} \log_{10} S - m \log_{10} T.$

13. $\frac{1}{5}(3 \log_5 E - \log_5 C - 2 \log_5 D).$ **15.** $\dfrac{1}{m} \log_e x - n \log_e y.$

17. $\log_{10} LMN.$ **19.** $\log_{10} x^2 y^3.$ **21.** $\log_e QRS.$ **23.** $\log_e \dfrac{3y\sqrt{2x}}{\sqrt[3]{2z}}.$

25. $\log_a \dfrac{13\sqrt{6}}{768}.$ **27.** $\log_B x^2 z^3 \sqrt{y}.$ **29.** $\log_e \dfrac{x^m y}{\sqrt[4]{z^n}}.$ **31.** 0.7781.

33. 0.3495.   **35.** 1.8060.   **37.** 0.7386.   **39.** 1.5506.

41. 1.4858.   **43.** 4.1209.   **45.** $-1.6121.$ **47.** $\frac{5}{8}.$

49. $-\frac{1}{9}.$ **51.** $\frac{16}{5}.$ **53.** 2.   **55.** 4.

57. $\frac{7}{4}.$ **59.** $\frac{1}{3}.$

Page 323. Sec. 9-9

1. 0.   **3.** 2.   **5.** -2 or $8 - 10.$ **7.** 1.

9. -1 or $9 - 10.$ **11.** 0.   **13.** 0.   **15.** 0.

17. 3.   **19.** -4 or $6 - 10.$ **21.** -1 or $9 - 10.$ **23.** -7 or $3 - 10.$

25. 6.   **27.** -12 or $8 - 20.$ **29.** -16 or $4 - 20.$ **31.** 37,410.

33. 374,100.   **35.** 3.741.   **37.** 0.3741.   **39.** 0.000,000,037,41.

41. 0.000,003,741.

43. 3,741,000,000,000.

45. 0.000,000,000,000,000,000,000,000,000,037,41.

Page 325. Sec. 9–10

[In the answers to exercises 1–21, the characteristic is given first and the matissa second.]

1. 6, 0.4134.   **3.** 2, 0.4816.   **5.** 0, 0.37216.

7. $9 - 10$ or -1, 0.1736.   **9.** $7 - 10$ or -3, 0.3619.   **11.** -10, 0.1026.

13. 6, 0.5563.   **15.** 5, 0.7984.   **17.** $9 - 10$ or -1, 0.2367.

19. $2 - 10$ or -8, 0.1735.   **21.** 11, 0.4287.   **23.** 2.6830.

25. 0.9004.   **27.** $9.4997 - 10$.   **29.** $7.9004 - 10$.

31. $9.9004 - 10$.   **33.** 3.6830.   **35.** 4.7300.

37. 6.9004.   **39.** 7.7300.   **41.** 10.4997.

Page 327. Sec. 9–11

1. 0.6609.   **3.** 1.4393.   **5.** 2.9605.   **7.** $9.2405 - 10$.   **9.** 4.9528.

11. 5.3404.   **13.** 3.5416.   **15.** 6.7672.   **17.** 827.   **19.** 1.70.

21. 3.56×10^8.   **23.** 3980.   **25.** 2.10×10^5.   **27.** 63.2.   **29.** 6.31.

Page 330. Sec. 9–12

1. 0.5665.   **3.** 2.0157.   **5.** 5.3743.   **7.** $9.6747 - 10$.

9. $8.9276 - 10$.   **11.** $7.4874 - 10$.   **13.** 1.2425.   **15.** 2.8749.

17. $9.6974 - 10$.   **19.** $7.8934 - 10$.   **21.** 1.8032.   **23.** 3.6827.

25. $6.8584 - 10$.   **27.** 5.7953.   **29.** $8.9673 - 10$.   **31.** $2.4602 - 10$.

33. $7.9257 - 10$.   **35.** 468.6.   **37.** 4.184×10^6.   **39.** 1.762.

41. 0.000,224,7.   **43.** 0.003,468.   **45.** 3.757×10^{-6}.   **47.** 53.65.

49. 1.021.   **51.** 4.827.   **53.** 0.009,344.   **55.** 0.039,59.

57. 7.552×10^{-21}.

Pages 334–335. Sec. 9–13

1. 14,470.   **3.** 1.686.   **5.** 6.021.   **7.** 0.000,473,6.   **9.** 0.065,37.

11. 2.859.   **13.** 0.2511.   **15.** 0.1339.   **17.** 2,712,000.   **19.** 0.002,481.

21. 3209.   **23.** 0.003,510.   **25.** 0.4047.   **27.** 0.7245.   **29.** 94.34.

31. 3,196,000.   **33.** 4.701.   **35.** 0.9708.   **37.** 9.338.   **39.** 41.83.

41. 2.524.   **43.** 0.5849.   **45.** 5.034.   **47.** 4.821.   **49.** 24.20.

51. 11.06.   **53.** 1.892.   **55.** 2.317.   **57.** 547.0.   **59.** 10.97.

61. 6.542.   **63.** 0.4186.   **65.** 1.401.   **67.** 0.3230.   **69.** 0.3912.

71. 140.1.   **73.** 0.096,78.   **75.** 0.023,45.   **77.** 0.8220.   **79.** 10.54.

Page 337. Sec. 9–14

1. 157,500.   **3.** 53.8.   **5.** 0.154.   **7.** 2.361.

9. 1.77.   **11.** 0.335.   **13.** 0.064,30.   **15.** 0.928.

17. 9.32.   **19.** 0.0670.   **21.** 10.5.   **23.** 12.1 horsepower.

25. 35.60 sec.

Pages 340–341, Sec. 9–15

1. 0.033,41.   **3.** 0.001,450.   **5.** 2.835.   **7.** 0.7805.

9. 0.3839.   **11.** 1.024.   **13.** 0.885.   **15.** 0.0431.

17. 0.130.   **19.** 0.003,62.   **21.** −201.0.   **23.** 1.132.

25. −0.000,000,251,7.   **27.** −7.725.   **29.** −1.303.   **31.** 4.548.

33. 3.091.   **35.** -4.23×10^8.   **37.** 78.52.   **39.** 2.448×10^8.

41. 348.   **43.** −21,660.   **45.** −0.000,075,5.   **47.** 0.037,84.

49. 0.004,549.   **51.** −1760.   **53.** 8.04.   **55.** 44.1.

57. −76.3.   **59.** 14.5.   **61.** 627.   **63.** 5,360.

65. 9.64.

Pages 343–345. Sec. 9–16

1. $\beta = 34°$, $b = 16$, $c = 29$.   **3.** $\beta = 58.2°$, $a = 356$, $b = 575$.

5. $\alpha = 35.74°$, $a = 39.94$, $b = 55.50$.   **7.** $\beta = 65.84°$, $b = 0.7745$, $c = 0.8490$.

9. $\alpha = 56.36°$, $a = 4.122$, $c = 4.951$.   **11.** $\alpha = 37.72°$, $\beta = 52.28°$, $c = 10,120$.

13. $\alpha = 53.4°$, $\beta = 36.6°$, $a = 4.26 \times 10^8$.   **15.** $x_A = 4.3$, $y_A = 2.1$.

17. $x_A = -68$, $y_A = 53$.   **19.** $x_A = -7180$, $y_A = -3280$.

21. $x_A = 0.0414$, $y_A = -0.0112$.   **23.** $x_A = -416.4$, $y_A = 380.0$.

25 $x_A = 70.99$, $y_A = 65.70$.   **27.** $A = 18.6$, $\alpha = 11.3°$.

29. $A = 9.32$, $\alpha = 120.4°$.   **31.** $A = 10.51$, $\alpha = 215.09°$.

33. $A = 789.4$, $\alpha = 336.90°$.   **35.** 375, 55.0°.

37. 1,230, 223.2°.   **39.** 994.2, 80.32°.

41. 307, 158.2°.   **43.** 13.85, 37.79°.

45. 629 ft.   **47.** 13 sec.

Page 346. Sec. 9–17

1. 4.426.   **3.** 6.564.   **5.** 7.923,

7. 1.956.   **9.** −0.4624.   **11.** 553.

13. 1.75.   **15.** 0.1002.   **17.** 0.00234.

Page 347. Sec. 9–18

1. 1.6021.   **3.** 2.084.   **5.** 1.794.   **7.** 2.5.

9. −3.33.   **11.** −7.44.   **13.** 4.   **15.** 4.127.

Pages 352–353. Sec. 9–20

1. 2.2 db.   **3.** 251.2.   **5.** 1.9 db.   **7.** 4.89×10^3 maxwells per sq cm.   **9.** 202 ohms.

Chapter 10

Pages 357–358. Sec. 10–1

13. $\cos \theta$.   **15.** $\tan \theta$.   **17.** $1 + \sin^2 \theta$.   **19.** 1.

21. $\sec \theta \csc \theta$.   **23.** $- \cos^2 \alpha$.   **25.** $E_m \sec A$.   **27.** $\cot \theta$.

29. $\sin A$.   **31.** $\cot \theta$.   **33.** $\tan \alpha + 1$.   **35.** $2 \sec^2 A$.

37. $\sin \theta \cos \theta$.   **39.** $\cos \theta$.

47. $\cos \theta = \sqrt{1 - \sin^2 \theta}$, $\tan \theta = \dfrac{\sin \theta}{\sqrt{1 - \sin^2 \theta}}$, $\cot \theta = \dfrac{\sqrt{1 - \sin^2 \theta}}{\sin \theta}$,

$\sec \theta = \dfrac{1}{\sqrt{1 - \sin^2 \theta}}$, $\csc \theta = \dfrac{1}{\sin \theta}$.

49. $\sin \theta = \dfrac{-\tan \theta}{\sqrt{1 + \tan^2 \theta}}$, $\cos \theta = \dfrac{-1}{\sqrt{1 + \tan^2 \theta}}$, $\cot \theta = \dfrac{1}{\tan \theta}$,

$\sec \theta = -\sqrt{1 + \tan^2 \theta}$, $\csc \theta = \dfrac{\sqrt{1 + \tan^2 \theta}}{-\tan \theta}$.

51. $\sin \theta = - \dfrac{\sqrt{\sec^2 \theta - 1}}{\sec \theta}$, $\cos \theta = \dfrac{1}{\sec \theta}$, $\tan \theta = -\sqrt{\sec^2 \theta - 1}$,

$\cot \theta = - \dfrac{1}{\sqrt{\sec^2 \theta - 1}}$, $\csc \theta = - \dfrac{\sec \theta}{\sqrt{\sec^2 \theta - 1}}$.

53. $\cos \theta = -\sqrt{1 - \sin^2 \theta}$, $\tan \theta = \dfrac{-\sin \theta}{\sqrt{1 - \sin^2 \theta}}$, $\cot \theta = \dfrac{\sqrt{1 - \sin^2 \theta}}{-\sin \theta}$,

$\sec \theta = - \dfrac{1}{\sqrt{1 - \sin^2 \theta}}$, $\csc \theta = \dfrac{1}{\sin \theta}$.

55. $\sin \theta = - \dfrac{\tan \theta}{\sqrt{1 + \tan^2 \theta}}$, $\cos \theta = - \dfrac{1}{\sqrt{1 + \tan^2 \theta}}$, $\cot \theta = \dfrac{1}{\tan \theta}$,

$\sec \theta = - \sqrt{1 + \tan^2 \theta}$, $\csc \theta = - \dfrac{\sqrt{1 + \tan^2 \theta}}{\tan \theta}$.

57. $\sin \theta - \dfrac{\sqrt{\sec^2 \theta - 1}}{-\sec \theta}$, $\cos \theta = \dfrac{1}{\sec \theta}$, $\tan \theta = -\sqrt{\sec^2 \theta - 1}$,

$\cot \theta = - \dfrac{1}{\sqrt{\sec^2 \theta - 1}}$, $\csc \theta = \dfrac{-\sec \theta}{\sqrt{\sec^2 \theta - 1}}$.

Pages 361–364. Sec. 10–2

17. Values of A and B for which $\cot A + \cot B = 0$.

19. Either A or $B = 0°$, $90°$, $180°$, $270°$, etc., and values of A and B such that $A \pm B = 90°$, $270°$, etc.

21. $A = 45°$, $90°$, $135°$, $225°$, $270°$, $315°$, etc.

23. $A = 0°$, $90°$, $180°$, $270°$, etc.   **25.** $A = 90°$, $270°$, etc.

27. $A = 90°, 270°$, etc.

29. $A = 90°, 270°$, etc.

31. $\dfrac{\sqrt{3}}{2}, \dfrac{1}{2}, \sqrt{3}$.

33. $\dfrac{\sqrt{6}+\sqrt{2}}{4}, \dfrac{\sqrt{6}-\sqrt{2}}{4}, 2+\sqrt{3}$.

35. 1, 0, not defined.

37. $\dfrac{1}{2}, \dfrac{\sqrt{3}}{2}, \dfrac{\sqrt{3}}{3}$.

39. $-\dfrac{1}{2}, \dfrac{\sqrt{3}}{2}, -\dfrac{\sqrt{3}}{3}$.

41. $-\sin \alpha$.

43. $\dfrac{\sqrt{2}}{2}(\sin\theta - \cos\theta)$.

45. $\tfrac{1}{2}(\sin\phi + \sqrt{3}\cos\phi)$.

47. $-\sin A$.

49. $-\tan A$.

51. $\dfrac{2\sqrt{2}+\sqrt{3}}{6}, \dfrac{2\sqrt{6}-1}{6}, \dfrac{2\sqrt{2}-\sqrt{3}}{6}, \dfrac{2\sqrt{6}+1}{6}$.

53. $\dfrac{22\sqrt{37}}{185}, \dfrac{21\sqrt{37}}{185}, \dfrac{14\sqrt{37}}{185}, \dfrac{27\sqrt{37}}{185}$.

55. $\dfrac{8+3\sqrt{21}}{25}, \dfrac{2(2\sqrt{21}-3)}{25}, \dfrac{8-3\sqrt{21}}{25}, \dfrac{2(2\sqrt{21}+3)}{25}$.

57. $\dfrac{24-5\sqrt{21}}{65}, \dfrac{2(5+6\sqrt{21})}{65}, \dfrac{24+5\sqrt{21}}{65}, \dfrac{2(5-6\sqrt{21})}{65}$.

59. $\dfrac{3\sqrt{10}-\sqrt{30}}{20}, \dfrac{3\sqrt{30}+\sqrt{10}}{20}, -\dfrac{3\sqrt{10}+\sqrt{30}}{20}, \dfrac{\sqrt{10}-3\sqrt{30}}{20}$.

63. $\sin 220° = 2\sin 110° \cos 110°$, $\cos 220° = \cos^2 110° - \sin^2 110°$,

$\sin 220° = -\sqrt{\dfrac{1-\cos 440°}{2}}$, $\cos 220° = -\sqrt{\dfrac{1+\cos 440°}{2}}$.

65. $\sin 3\theta = 2\sin\dfrac{3\theta}{2}\cos\dfrac{3\theta}{2}$, $\cos 3\theta = \cos^2\dfrac{3\theta}{2} - \sin^2\dfrac{3\theta}{2}$,

$\sin 3\theta = \pm\sqrt{\dfrac{1-\cos 6\theta}{2}}$, $\cos 3\theta = \pm\sqrt{\dfrac{1+\cos 6\theta}{2}}$.

67. $\sin 100° = 2\sin 50° \cos 50°$, $\cos 100° = \cos^2 50° - \sin^2 50°$,

$\sin 100° = \sqrt{\dfrac{1-\cos 200°}{2}}$, $\cos 100° = -\sqrt{\dfrac{1+\cos 200°}{2}}$.

69. $\sin\dfrac{\theta}{5} = 2\sin\dfrac{\theta}{10}\cos\dfrac{\theta}{10}$, $\cos\dfrac{\theta}{5} = \cos^2\dfrac{\theta}{10} - \sin^2\dfrac{\theta}{10}$,

$\sin\dfrac{\theta}{5} = \pm\sqrt{\dfrac{1-\cos\dfrac{2\theta}{5}}{2}}$, $\cos\dfrac{\theta}{5} = \pm\sqrt{\dfrac{1+\cos\dfrac{2\theta}{5}}{2}}$.

71. $\sin\tfrac{2}{3}\omega t = 2\sin\tfrac{1}{3}\omega t \cos\tfrac{1}{3}\omega t$, $\cos\tfrac{2}{3}\omega t = \cos^2\tfrac{1}{3}\omega t - \sin^2\tfrac{1}{3}\omega t$,

$\sin\dfrac{2}{3}\omega t = \pm\sqrt{\dfrac{1-\cos\tfrac{4}{3}\omega t}{2}}$, $\cos\dfrac{2}{3}\omega t = \pm\sqrt{\dfrac{1+\cos\tfrac{4}{3}\omega t}{2}}$.

73. $\dfrac{\sqrt{3}}{2}, \dfrac{1}{2}$.

75. $0, -1$.

77. $0, 1$.

79. $\dfrac{\sqrt{2 - \sqrt{3}}}{2}, \dfrac{\sqrt{2 + \sqrt{3}}}{2}$.

81. $1, 0$.

83. $0, 1$.

85. $\dfrac{3\sqrt{7}}{8}, \dfrac{1}{8}, -\dfrac{\sqrt{2}}{4}, -\dfrac{\sqrt{14}}{4}$.

87. $-\dfrac{120}{169}, \dfrac{119}{169}, -\dfrac{\sqrt{26}}{26}, \dfrac{5\sqrt{26}}{26}$.

89. $-\dfrac{120}{169}, -\dfrac{119}{169}, \dfrac{3\sqrt{13}}{13}, \dfrac{2\sqrt{13}}{13}$.

91. $0, 1, 1, 0$.

Pages 366–368. Sec. 10–3

9. $\frac{1}{2}[\cos 12° + \cos 84°]$.

11. $\frac{1}{2}[\sin 3\omega t + \sin 7\omega t]$.

13. $\frac{1}{2}\left[\cos\dfrac{2\pi}{9} - \cos\dfrac{4\pi}{9}\right]$.

15. $\frac{1}{2}[\cos 2x + 1]$.

17. $\frac{1}{2}[\sin 2\beta + \sin 2(\alpha + \theta)]$.

19. $\frac{1}{2}\sin\left(2\alpha - \dfrac{2\pi}{3}\right)$.

21. $2 \sin 2\omega t \sin \omega t$.

23. $-2\cos\dfrac{7\pi}{8}\sin\dfrac{5\pi}{8}$.

25. $2\cos\theta\cos 30°$.

27. $2\sin\dfrac{7\theta}{24}\cos\left(\omega t + \dfrac{\theta}{24}\right)$.

29. $-2\sin\left(4\theta + \dfrac{\alpha}{2}\right)\sin\left(\theta - \dfrac{\alpha}{2}\right)$.

31. $-2\sin\left(\dfrac{3\omega t}{4} + \dfrac{7\pi}{12}\right)\cos\left(\dfrac{\omega t}{4} + \dfrac{\pi}{2}\right)$.

Pages 374–375. Sec. 10–4

1. $4\sin x$.

3. $\sqrt{10}\sin(\omega t - 18.43°)$.

5. $\sqrt{19}\sin(x - 6.59°)$.

7. $10.2\sin(x - 48°)$.

9. $115\sin(\theta - 86°)$.

11. $3.0\sin(\omega t - 65°)$.

13. 0.0.

15. $4.21\sin(x - 12.0°)$.

17. $13.9\sin(\omega t + 54°)$.

19. $3.61\sin(\omega t - 13.9°)$.

Pages 378–379. Sec. 10–5

1. $37.00°, 143.00°$.

3. $235.10°, 304.90°$.

5. $143.57°, 216.43°$.

7. $54.99°, 305.01°$.

9. No value of β satisfies the equation.

11. $12.44°, 77.56°, 192.44°, 257.56°$.

13. $231.88°$.

15. No value of α that satisfies the equation lies between $0°$ and $360°$.

17. $23.04°, 66.96°, 203.04°, 246.96°$.

19. $246.4°$.

21. $60°, 120°, 240°, 300°$.

23. $90°, 270°$.

25. $41.81°, 125.27°, 234.73°, 318.19°$.

27. No value of α satisfies the equation.

29. $0°, 180°, 225°, 315°$.

31. $199.47°, 340.53°$.

33. $30°, 150°$.

35. $48.59°, 131.41°, 203.58°, 336.42°$.

37. $60°, 300°$.

39. $0.00°, 4.87°, 40.13°, 90.00°, 94.87°, 130.13°, 180.00°, 184.87°, 220.13°, 270.00°, 274.87°, 310.13°$.

41. 15.45°, 35.89°, 195.45°, 215.89°.   **43.** All of values of θ.

45. 120°, 240°.   **47.** 30°, 150°, 270°.

49. 60°, 180°, 300°.   **51.** 120°, 240°.

53. 0°, 90°, 180°, 270°.   **55.** 0°, 60°, 90°, 120°, 180°, 240°, 300°.

57. 15°, 45°, 75°, 195°, 225°, 255°.

Pages 382–384. Sec. 10–6

1. $y = \arccos x$.   **3.** $y = \arctan x$.   **5.** $y = \tfrac{1}{2} \arctan 2z$.

7. $y = -\pi + \arcsin z$.   **9.** $y = \arctan \dfrac{x+2}{3}$.   **11.** $y = \tfrac{1}{3} \arcsin \dfrac{m-n}{x}$.

13. $y = \tfrac{1}{4} \arctan 4x$.   **15.** $y = \arcsin \dfrac{B}{A}$.   **17.** $y = 3 \arccos \dfrac{3x-2}{2}$.

19. $y = 1 + \arccos 2(M + 2N)$.   **21.** $y = 1 + \tfrac{3}{7} \operatorname{arcsec} \dfrac{2a-b}{c}$.

23. $x = \cos y$.   **25.** $x = \dfrac{3}{7a} \cos \theta$.   **27.** $x = \dfrac{1}{2m} \sec \dfrac{\theta}{3}$.

29. $x = 4 \tan \dfrac{9M-2}{N}$.   **31.** $x = \dfrac{B}{A} \sin \alpha$.   **33.** $x = y + \tan \dfrac{\theta-1}{2}$.

35. $x = \dfrac{\sqrt{2}\, E_m}{2} \sin (377t - 30°)$.   **37.** $x = -A \tan \theta$.

39. $x = 1 - N \cos (\phi - y)$.   **41.** $x = \tfrac{3}{5}\left(\dfrac{\pi}{2} + \sin \theta \right)$.   **43.** 45°.

45. −390°, −150°, −30°, 210°, 330°, 570°.   **47.** 30°.

49. 45°.   **51.** −450°, −270°, −90°, 90°, 270°, 450°.

53. −510°, −210°, −150°, 150°, 210°, 510°.   **55.** 0°.

57. −450°, −270°, −90°, 90°, 270°, 450°.   **59.** 2π.

61. $\dfrac{\pi}{3}$.   **63.** $\dfrac{5\pi}{2}$.   **65.** 19.1°.

67. −322.05°, −217.95°, 37.95°, 142.05°, 397.95°, 502.05°.

69. 7.66°.   **71.** −101.28°, −78.72°, 18.72°, 41.28°, 138.72°, 161.28°.

73. 4.12°.   **75.** −314.75°, −225.25°, 45.25°, 134.75°, 405.25°, 494.75°.

77. 88.64°.   **79.** 26°.   **81.** −75°.

83. 0.3371.   **85.** −5.005.   **87.** 1.5.

89. $\tfrac{1}{2}$.   **91.** 52.   **93.** 231.90°.

Chapter 11

Page 391. Sec. 11–4

1. $b = 44$, $c = 43$, $\gamma = 72°$.   **3.** $a = 40.7$, $c = 31.2$, $\alpha = 68.5°$.

5. $a = 9.32$, $c = 21.7$, $\gamma = 165.4°$.   **7.** $b = 54.0$, $c = 30.0$, $\beta = 96.5°$.

9. $b = 19.35$, $c = 15.67$, $\alpha = 42.33°$.   **11.** $\beta = 69°$.

13. $c = 9.34$.   **15.** $b = 2.36$.

17. $a = 9.30$.   **19.** $b = 302.4$.

1. $a = 11, \beta = 80°, \gamma = 80°$.

3. $a = 5.6, \beta = 60°, \gamma = 45°$.

5. $c = 48, \alpha = 38°, \beta = 119°$.

7. $b = 9.58, \alpha = 13.4°, \gamma = 77.2°$.

9. $a = 1764, \beta = 31.47°, \gamma = 46.84°$.

11. $a = 2.6$.

13. $a = 154$.

15. $c = 0.922$.

17. $\beta = 39.9°$.

19. $b = 0.4319$.

1. No solution.

3. No solution.

5. $c = 40.8, \alpha = 45.4°, \gamma = 68.6°$

7. $a_1 = 60.6, \alpha_1 = 37.3°, \beta_1 = 77.2°; \; a_2 = 20.3, \alpha_2 = 11.7°, \beta_2 = 102.8°$.

9. $b = 3.03, \alpha = 90.0°, \beta = 36.2°$.

11. $\alpha_1 = 83.0°, \alpha_2 = 32.4°$.

13. $\alpha = 62.7°$.

15. $\beta_1 = 23.28°, \beta_2 = 156.72°$.

17. $\gamma = 90.0°$.

19. $\alpha = 14.10°$.

1. No solution.

3. $\alpha = 33°, \beta = 130°, \gamma = 17°$.

5. $\alpha = 42.4°, \beta = 83.6°, \gamma = 53.9°$.

7. $\alpha = 25.3°, \beta = 121.5°, \gamma = 33.2°$.

9. $\alpha = 157.04°, \beta = 13.22°, \gamma = 9.73°$.

11. $\alpha = 73°$.

13. No solution.

15. $\alpha = 19.2°$.

17. $\gamma = 66.3°$.

1. C: $C = 30, \gamma = 57°$.

3. C: $C = 32, \gamma = 183°$.

5. C: $C = 960, \gamma = 272°$.

7. C: $C = 4.24, \gamma = 241.0°$.

9. C: $C = 55, \gamma = 192.5°$.

11. C: $C = 1.369, \gamma = 164.9°$.

13. C: $C = 2090, \gamma = 75.1°$.

15. C: $C = 169, \gamma = 256.2°$.

17. (a) **B**: $B = 18, \; \beta = 10°$;   **C**: $C = 10, \; \gamma = 112°$.   (b) **B**: $B = 18, \; \beta = 60°$;   **C**: $C = 5.9, \gamma = 330°$.   (c) **B**: $B = 12, \beta = 270°$;   **C**: $C = 28, \gamma = 60°$.

19. (a) **B**: $B = 120, \; \beta = 10°$;   **C**: $C = 220, \; \gamma = 112°$.   (b) **B**: $B = 200, \beta = 60°$;   **C**: $C = 65, \gamma = 150°$.   (c) **B**: $B = 130, \beta = 90°$;   **C**: $C = 87, \gamma = 60°$.

21. (a) **B**: $B = 38, \; \beta = 190°$;   **C**: $C = 17, \; \gamma = 292°$.   (b) **B**: $B = 35, \; \beta = 240°$;   **C**: $C = 16, \gamma = 150°$.   (c) **B**: $B = 31, \beta = 90°$;   **C**: $C = 62, \gamma = 240°$.

23. (a) **B**: $B = 160, \; \beta = 10°$;   **C**: $C = 570, \; \gamma = 112°$.   (b) **B**: $B = 450, \beta = 60°$;   **C**: $C = 330, \gamma = 150°$.   (c) **B**: $B = 660, \beta = 90°$;   **C**: $C = 120, \beta = 240°$.

25. (a) **B**: $B = 5.4, \; \beta = 190°$;   **C**: $C = 6.0, \; \gamma = 292°$.   (b) **B**: $B = 7.15, \beta = 240°$;   **C**: $C = 0.63, \; \gamma = 330°$.   (c) **B**: $B = 1.3, \; \beta = 270°$;   **C**: $C = 6.1, \gamma = 240°$.

27. (a) **B**: $B = 590, \; \beta = 190°$;   **C**: $C = 1180, \gamma = 112°$.   (b) **B**: $B = 340, \beta = 60°$;   **C**: $C = 1380, \; \gamma = 150°$.   (c) **B**: $B = 2760, \beta = 90°$;   **C**: $C = 2040, \; \gamma = 240°$.

29. (*a*) **B**: $B = 6.25, \beta = 190°$; **C**: $C = 52.4, \gamma = 112°$. (*b*) **B**: $B = 28.2, \beta = 60°$; **C**: $C = 46.1, \gamma = 150°$. (*c*) **B**: $B = 92.1, \beta = 90°$; **C**: $C = 51.6, \gamma = 240°$.

31. (*a*) **B**: $B = 640, \beta = 10°$; **C**: $C = 243, \gamma = 112°$. (*b*) **B**: $B = 555, \beta = 60°$; **C**: $C = 291, \gamma = 330°$. (*c*) **B**: $B = 582, \beta = 270°$; **C**: $C = 1059, \gamma = 60°$.

33. 2220 ft, 1650 ft.   **35.** 118°, 16°, 46°.   **37.** 1280 ft.

39. 1764 ft.   **41.** 128 mi.   **43.** 35 mi.

45. 252,000 sq ft.   **47.** 111 ft.   **49.** 53°.

51. 145 mi.   **53.** 7400 ft.   **55.** 19.35 mi bearing 57.7°.

57. 14.5 mph bearing 207°.   **59.** 15.6 ft.   **61.** 365.5 ft.

63. 5.5 lb, 74°, 38°.   **65.** 430 lb, 9°, 12°.   **67.** 322 lb, 247 lb.

69. 165 lb at 34°, 155 lb at 37°.   **71.** 150 lb.

73. 710 lb.   **75.** 33 ft.   **77.** 2900 lb.

79. 2°, 2.5 hr.   **81.** 10°, 4.7 mi, 0.39 hr.

Chapter 12

Page 412. Sec. 12–1

1. $-j$.   **3.** -1.   **5.** 1.   **7.** 1.   **9.** 1.   **11.** j.   **13.** -1.

15. j.   **17.** $j9$.   **19.** $j8$.   **21.** $-j7$.   **23.** $j0.1$.   **25.** $j3\sqrt{2}$.   **27.** $j2\sqrt{10}$.

29. $j\frac{5}{11}$.   **31.** $j\frac{4}{7}\sqrt{3}$.   **33.** $j4$.   **35.** $j\frac{4}{5}$.   **37.** $j2.45$.   **39.** $j2.24$.   **41.** $j3.32$.

43. $-j4.47$. **45.** $-j2.08$. **47.** $-j9.73$. **49.** $j4$.   **51.** -26.   **53.** 14.

Pages 414–415. Sec. 12–3

1 $3 - j2$.   **3.** $-1 - j$.   **5.** $-7 + j$.

7. $-j$.   **9.** j.   **11.** -15.

13. $x + jy$.   **15.** $3 - j4$.   **17.** $x = 4, y = 3$.

19. $x = 3, y = -2$.   **21.** $x = 2, y = 1$.

Pages 416–418. Sec. 12–4

1. $4 - j$.   **3.** $1 - j13$.   **5.** $18 + j$.

7. $0.5 + j2.2$.   **9.** 0.70.   **11.** $j10$.

13. 13.   **15.** -225.   **17.** -4.

19. $\frac{5}{6} + j\frac{1}{6}$.   **21.** $23 + j7$.   **23.** $-7 + j24$.

25. $-j6$.   **27.** $48 + j11$.   **29.** $-4 + j2\sqrt{5}$.

31. $31 + j42\sqrt{2}$.   **33.** $37 + j10\sqrt{5}$.   **35.** $-2 + j2$.

37. $-\sqrt{2} - j5$.   **39.** $-9 - j40$.   **41.** j.

43. $\frac{1}{2} + j\frac{1}{2}$.   **45.** $-\frac{3}{5} - j\frac{1}{5}$.   **47.** $-j$.

49. $\frac{23}{85} + j\frac{44}{85}$.   **51.** $-\frac{5}{26} - j\frac{25}{26}$.   **53.** $\frac{1}{83} + j\frac{9\sqrt{2}}{166}$.

Pages 425–426. Sec. 12–6

1. $1.41(\cos 45° + j \sin 45°) = 1.41\underline{/45°}$.

3. $1(\cos 0° + j \sin 0°) = 1\underline{/0°}$.

5. $3(\cos 0° + j \sin 0°) = 3\underline{/0°}$.

7. $1(\cos 90° + j \sin 90°) = 1\underline{/90°}$.

9. $3(\cos 90° + j \sin 90°) = 3\underline{/90°}$.

11. $9.220(\cos 102.53° + j \sin 102.53°) = 9.220\underline{/102.53°}$.

13. $2(\cos 60° + j \sin 60°) = 2\underline{/60°}$.

15. $3.225(\cos 60.26° + j \sin 60.26°) = 3.225\underline{/60.26°}$.

17. $0.6009(\cos 326.31° + j \sin 326.31°) = 0.6009\underline{/326.31°}$.

19. $j6$.   **21.** -2.

23. $\frac{1}{2}$.   **25.** $4.330 - j2.500$.

27. $7.128 + j3.632$.   **29.** $0.77 - j2.27$.

31. $-3.38 + j5.10$.   **33.** $0.222 - j1.695$.

Page 428. Sec. 12–7

1. $e^{j \cdot 0}$.   **3.** $e^{j(\pi/2)}$.   **5.** $\sqrt{2}\, e^{j(\pi/4)}$.   **7.** $\sqrt{2}\, e^{j(3\pi/4)}$.

9. $2e^{j(\pi/6)}$.   **11.** $\sqrt{73}\, e^{j5.071}$.   **13.** $6.5\, e^{j4.11}$.   **15.** $11.2\, e^{-j1.31}$.

17. $3e^{j0.7156}$.   **19.** j.   **21.** -1.   **23.** $-j$.

25. -1.   **27.** $0.0707 + j0.9975$.   **29.** $-0.9041 - j0.4274$.

31. j.

Pages 432–433. Sec. 12–10

1. $20\underline{/53°}$, $12.04 + j15.97$.   **3.** $42\underline{/75°}$, $10.87 + j40.57$.

5. $14.87\underline{/70.35°}$, $5 + j14$.   **7.** $13.04\underline{/175.60°}$, $-13 + j$.

9. $3.63\underline{/21.4°}$, $3.38 + j1.32$.   **11.** $3.808\underline{/113.20°}$, $-1.500 + j3.500$.

13. $125.0\underline{/159.36°}$, $-117.0 + j44.06$.   **15.** $186,000\underline{/130.5°}$, $-121,000 + j141,000$.

17. $3.4\underline{/303°}$, $1.9 - j2.9$.   **19.** $0.088\underline{/210°}$, $-0.076 - j0.044$.

21. $-3 + j5$.   **23.** $4\underline{/120°}$.

25. $-1 + j5$.   **27.** $-1 + j2$.

29. $-1 - j2$.   **31.** $\pm(0.7071 - j0.7071)$.

33. $\pm(1.8 - j1.0)$.

35. $-j$, $0.8660 + j0.5000$, $-0.8660 + j0.5000$.

37. $\pm(0.9239 + j0.3827)$, $\pm(0.3827 - j0.9239)$.

Pages 434–435. Sec. 12–11

1. $7.7 - j0.29$, $7.7\underline{/-2°}$.   **3.** $2.048 + j4.512$, $4.955\underline{/65.59°}$.

5. $8.6 - j1.0$, $8.7\underline{/-7°}$.   **7.** $9.0 + j17.4$, $19.6\underline{/63°}$.

9. $14.1 - j4.2$, $14.7\underline{/-17°}$.   **11.** $-178 + j1921$, $1929\underline{/95.29°}$.

13. $-17.2 - j27.5,\ 32\underline{/-122^\circ}.$ **15.** $5.12 - j22.8,\ 23\underline{/-77^\circ}.$

17. $-0.042 + j0.044,\ 0.061\underline{/133^\circ}.$ **19.** $0.11 - j2.0,\ 2.0\underline{/-87^\circ}.$

21. $2^{20},\ 2^{20}\underline{/0^\circ}.$ **23.** $1900 + j300,\ 1920\underline{/9^\circ}.$

25. $-320 + j180,\ 360\underline{/151^\circ}.$ **27.** $-3.2 + j1.1,\ 3.4\underline{/161^\circ}.$

29. $j,\ 1\underline{/90^\circ};\ -0.8660 - j0.5000,\ 1\underline{/210^\circ};\ 0.8660 - j0.5000,\ 1\underline{/330^\circ}.$

31. $2.87 + j0.91,\ 3.0\underline{/18^\circ};\ -2.87 - j0.91,\ 3.0\underline{/198^\circ}.$

33. $1.3 + j2.6,\ 2.9\underline{/64^\circ};\ -1.3 - j2.6,\ 2.9\underline{/244^\circ}.$

35. $0.3827 + j0.9239,\ 1\underline{/67.5^\circ};\ -0.9239 + j0.3827,\ 1\underline{/157.5^\circ};\ -0.3827 - j0.9239,$
$1\underline{/247.5^\circ};\ 0.9239 - j0.3827,\ 1\underline{/337.5^\circ}.$

37. $-1.0260 + j2.8191,\ 3\underline{/110^\circ};\ -1.9284 - j2.2980,\ 3\underline{/230^\circ};\ 2.9544 - j0.5208,\ 3\underline{/350^\circ}.$

Page 439. Sec. 12–12

1. 20 ohms, 5.8 amp.

3. 65 volts.

5. 4.2 amp, 133 volts.

7. 21 ohms, 5.3 amp, 96-volt resistance voltage drop, 53-volt reactance voltage drop.

9. 33 ohms, 103 ohms.

Chapter 13

Page 444. Sec. 13–1

1. $x = 2,\ y = 3,\ z = 6.$ **3.** $p = 3,\ q = -2,\ r = 1.$

5. $u = 3,\ v = 1,\ w = -1.$ **7.** $R_1 = 1,\ R_2 = 2,\ R_3 = 3.$

9. $A = 22,\ B = 25,\ C = 27.$ **11.** $E_1 = 1,\ E_2 = 2,\ E_3 = -2.$

13. $x = \frac{1}{6},\ y = -\frac{2}{3},\ z = \frac{7}{12}.$ **15.** $R_1 = 1,\ R_2 = \frac{3}{2},\ R_3 = \frac{5}{2}.$

17. $x_1 = \frac{3}{5},\ x_2 = \frac{2}{3},\ x_3 = -3.$ **19.** $t = -1,\ x = 1,\ y = -2,\ z = 0.$

Page 449. Sec. 13–2

1. $x = -0.851,\ y = 1.986.$ **3.** $r = 13.8,\ s = 4.96.$

5. $R_1 = -1.20,\ R_2 = 3.30,\ R_3 = -5.57.$ **7.** $x = 0.475,\ y = 0.734,\ z = 0.981.$

Pages 456–457. Sec. 13–4

1. $x = 1,\ y = 2.$ **3.** $x = 5,\ y = -3.$ **5.** $x = 0,\ y = 0.$

7. $z_1 = 1,\ z_2 = 1.$ **9.** $a = 0.1,\ b = 0.2.$ **11.** $P_1 = \frac{22}{35},\ P_2 = \frac{177}{35}.$

13. $z_1 = \frac{11}{9},\ z_2 = -\frac{5}{3}.$ **15.** $r = 0,\ s = 0.$ **17.** $p = \frac{64}{5},\ q = \frac{133}{5}.$

19. $x = 1,\ y = 1.$

1. $x = \frac{49}{11}, y = -\frac{3}{11}$. **3.** Inconsistent. **5.** Dependent.

7. Dependent. **9.** Inconsistent. **11.** Dependent.

13. Inconsistent. **15.** Dependent.

17. $k = 5$, dependent; $k = -5$, inconsistent; $k \neq \pm 5$, independent.

19. $k = 0$, inconsistent; $k = 26$, inconsistent; $k \neq 0$ or 26, independent.

21. $k = 16$, dependent; $k \neq 16$, independent.

23. $k = 0$, dependent; $k \neq 0$, independent.

Pages 463–464. Sec. 13–6

1. No. . **3.** Yes. **5.** No. **7.** No. **9.** No.

11. $k = \frac{1}{3}$. **13.** $k = \pm 1$. **15.** $k = 8, 14$. **17.** $k = -\frac{243}{32}$.

Pages 473–474. Sec. 13–7

1. 0. **3.** 40. **5.** 8. **7.** 51. **9.** -24.

11. 1. **13.** -9. **15.** 0. **17.** -65. **19.** 0.

21. $x^3 - 3a^2x + 2a^3$. **23.** -168. **25.** 10,560. **27.** 0.

29. 285,567. **31.** $13x - 15$.

33. $a_1(x_2 - x_3) + a_2(x_3 - x_1) + a_3(x_1 - x_2)$. **35.** $(a_1 - a_2)(a_2 - a_3)(a_3 - a_1)$.

37. $x(y_1 - y_2) + x_1(y_2 - y) + x_2(y - y_1)$. **39.** $abc(a - b)(b - c)(c - a)$.

Pages 480–482. Sec. 13–8

1. Independent. **3.** Not independent. **5.** Not independent.

7. Independent. **9.** $x = -t, y = t, z = t$. **11.** $r_1 = 0, r_2 = 0, r_3 = 0$.

13. $M = -4t, N = t, P = t$. **15.** $E_1 = t, E_2 = t, E_3 = t$.

17. $x = 2t, y = t, z = 0$. **19.** The equations are equivalent.

21. $x = 2t, y = t, z = 5t$. **23.** $x - 6t, y = -3t, z = -5t$.

25. $x = 3t, y = t, z = t$. **27.** The equations are equivalent.

29. $x = -t, y = t, z = 0$.

31. $A = 4t, B = t, C = -5t$; $A = 4, B = 1, C = -5$.

33. $A = t(1 - m), \quad B = t(1 + m^2), \quad C = -tm(1 + m); \quad A - 1 - m, \quad B = 1 + m^2,$
$C = -m(1 + m)$.

35. $k = \frac{3}{2}$. **37.** $k = 0, 2$. **39.** No real value of k.

Page 484. Sec. 13–9

1. 12.9 amp.

3. $I_1 - I_2 - I_3 = 0, I_1R_1 + I_3R_3 = E_1, -I_2R_2 + I_3R_3 = E_2$.

Chapter 14

Pages 489–490. Sec. 14–3

5. -4.5, 1.5.   **7.** No real roots.   **9.** 4.4, 1.6.

11. (a) No real values; (b) no real values; (c) no real values; (d) 0, 2; (e) 1, 1; (f) no real values; (g) 2.2, -0.2; (h) no real values.

Pages 491–492. Sec. 14–5

1. ± 2.24.   **3.** ± 2.04.   **5.** ± 2.24.   **7.** $\pm j1.14$.

9. -0.885, 0.000.   **11.** 7, -1.   **13.** $\frac{1}{2}$, -1.   **15.** ± 4.69.

17. 8, -4.   **19.** 3.5, -1.   **21.** ± 4.90.

23. 8.42×10^8 cm per sec.   **25.** 100 volts.   **27.** 7.44 in.

29. 295 rpm.

Pages 496–498. Sec. 14–6

1. 5.73, 2.27.   **3.** -0.146, -6.85.   **5.** 2.56, -1.56.

7. 0.477, -10.5.   **9.** $\frac{1}{2}(5 \pm j\sqrt{7})$.   **11.** $-3 \pm j$.

13. $\frac{1}{4}(-3 \pm j\sqrt{15})$.   **15.** 3.16, -0.158.   **17.** $\frac{1}{3}(-4 \pm j\sqrt{5})$.

19. $\frac{1}{2}(-\sqrt{3} \pm j)$.   **21.** $\frac{1}{3}(4 \pm j\sqrt{5})$.   **23.** 0.783, -0.383.

25. $\frac{1}{3}(1 \pm j\sqrt{14})$.   **27.** 1.41, -0.0787.   **29.** $\frac{1}{4}\left(\sqrt{3} \pm j\sqrt{8\sqrt{2}-3}\right)$.

31. $0.601 \pm j0.553$.   **33.** 3, -2.   **35.** 1.69, -1.59.

37. 6, $\frac{4}{3}$.   **39.** 0.768, -0.434.

41. 6.01×10^{-5}, -9.43×10^{-5}.   **43.** Unequal real roots.

45. Double real root.   **47.** Double real root.   **49.** Unequal complex roots.

51. Unequal complex roots.   **53.** 3, -4.   **55.** 8.

57. 4.   **59.** 3, $-\frac{34}{9}$.

61. Unequal real roots if $k < 1$; double real root if $k = 1$; unequal complex roots if $k > 1$.

63. Unequal real roots if $p < 4$; double real root if $p = 4$; unequal complex roots if $p > 4$.

65. Unequal real roots if $m > -\frac{1}{4}$; double real root if $m = -\frac{1}{4}$; unequal complex roots if $m < -\frac{1}{4}$.

67. Unequal real roots if $|k| > 2$; double real root if $|k| = 2$; unequal complex roots if $|k| < 2$.

69. 16.   **71.** No roots.   **73.** 9.

75. 3.   **77.** -2.   **79.** 6.05.

81. 3.   **83.** 1, 5.   **85.** 16, 17; -17, -16.

87. 14, 16; -16, -14.   **89.** 5, 6.   **91.** 40.65 in. by 31.95 in.

93. 3.81 in. or 12.94 in.   **95.** 9 mph.   **97.** 194 rpm.

99. 36.5 amp, 274 amp.

Pages 500–501. Sec. 14–7

1. $-5, -7$.

3. $-\frac{2}{5}, -\frac{4}{5}$.

5. $-\frac{1}{2}, -4$.

7. $-\frac{3}{5}, -\frac{7}{5}$.

9. $0, 10$.

11. $x^2 - 5x + 6 = 0$.

13. $x^2 - 2x - 3 = 0$.

15. $x^2 - 7x = 0$.

17. $20x^2 - x - 1 = 0$.

19. $x^2 - 3 = 0$.

21. $x^2 + 1 = 0$.

23. $x^2 - 2x - 2 = 0$.

25. $x^2 - 4x + 13 = 0$.

27. $(y - 2)(y + 3)$.

29. $(z - 1.19)(z + 4.19)$.

31. $(7x - y)(3x + 2y)$.

33. $(m - 2.62n)(m - 0.382n)$.

35. $(R_1 + 0.472R_2)(R_1 + 3.53R_2)$.

37. $(z - 2 - j3)(z - 2 + j3)$.

Page 503. Sec. 14–8

1. $\pm 1.41, \pm j$.

3. $\pm 1, \pm j1.41$.

5. $\pm 1.73, \pm j2.65$.

7. $\pm j1.41, \pm j2.24$.

9. 16.

11. $1, -8$.

13. $0.457, 0.549, -1.22, -1.46$.

15. $1, 1, -0.382, -2.62$.

17. $-2, -4, -3 \pm j2$.

19. $\frac{1}{4}, -\frac{1}{3}$.

21. $1, -3, -1 \pm j$.

23. $2, 0$.

25. $6, -1, 2, -3$.

27. $30°, 150°, 270°$, etc.

29. $71.57°, 251.57°, 135°, 315°$, etc.

31. $22.5°, 112.5°, 202.5°, 292.5°$, etc.

33. $x = \ln (u + \sqrt{u^2 + 1})$.   **35.** $x = \ln \sqrt{\dfrac{1 + v}{1 - v}}$.

Page 507. Sec. 14–10

1. $(4, 5), (5, 4)$.

3. $(2.44, 3.39), (3.39, 2.44)$.

5. $(-2, 4), (8, -1)$.

7. $(0, 1), (1, 0)$.

9. $(2 + j2.83, -1 + j2.83), (2 - j2.83, -1 - j2.83)$.

11. $(1.04, 0.854), (-1.04, -0.854), (1.04, -0.854), (-1.04, 0.854)$.

13. $(2, 1), (-2, -1), (2, -1), (-2, 1)$.   **15.** $(2 + j3, 1 + j2), (2 - j3, 1 - j2)$.

17. $(4, 1), (-4, -1), (j, -j4), (-j, j4)$.   **19.** $(1.58, 0.00), (-1.58, 0.00)$.

21. $(\frac{1}{2}(-2 + j\sqrt{6}), -1 - j\sqrt{6}), (\frac{1}{2}(-2 - j\sqrt{6}), -1 + j\sqrt{6})$.

23. $(k, k), (-k, -k)$.

25. $-\dfrac{mr}{\sqrt{1 + m^2}}, \dfrac{r}{\sqrt{1 + m^2}}$.

27. $(\frac{1}{2}(1 + \sqrt{1 - 4k}), \frac{1}{2}(1 - \sqrt{1 - 4k})), (\frac{1}{2}(1 - \sqrt{1 - 4k}), \frac{1}{2}(1 + \sqrt{1 - 4k})); k = \frac{1}{4}$.

29. $(\frac{1}{2}(-k + \sqrt{18 - k^2}), \frac{1}{2}(k + \sqrt{18 - k^2})), (\frac{1}{2}(-k - \sqrt{18 - k^2}), \frac{1}{2}(k - \sqrt{18 - k^2}));$
$k = \pm\sqrt{18}$.

Pages 509–511.   Sec. 14–11

1. $r = \pm 2\sqrt{\dfrac{5I_d}{3m} - h^2}, \; h = \pm\dfrac{1}{2}\sqrt{\dfrac{20I_d}{3m} - r^2}.$

3. $b = \pm\sqrt{12k_0^2 - h^2}, \; h = \pm\sqrt{12k_0^2 - b^2}.$

5. $v_1 = \pm\sqrt{v_2^2 + \dfrac{2Fs}{m}}, \; v_2 = \pm\sqrt{v_1^2 - \dfrac{2Fs}{m}}.$

7. $U_1 = \pm\sqrt{U_2^2 - 64.34J(H_1 - H_2)}, \; U_2 = \pm\sqrt{U_1^2 + 64.34J(H_1 - H_2)}.$

9. $I = \dfrac{E_g \pm\sqrt{E_g^2 - 4r_gP_0}}{2r_g}.$

11. $t = \dfrac{-v_0 \pm\sqrt{v_0^2 + 2as}}{a}.$

13. $x_0 = \dfrac{x_cA \pm\sqrt{x_c^2A^2 - 4AJ_0}}{2A}.$

15. $m = \dfrac{f \pm\sqrt{f^2 - 16\pi^2v_f^2R^2}}{8\pi^2v_f^2}.$

17. $x = \dfrac{-Kv \pm\sqrt{K^2v^2 + 4Kv}}{2}.$

19. $P = W\left(1 \pm\sqrt{1 + \dfrac{2h}{L}}\right), \; W = \dfrac{PL}{2h}\left(-1 \pm\sqrt{1 + \dfrac{2h}{L}}\right).$

21. $p_2 = \dfrac{1}{2}\left(p_1 \pm\sqrt{p_1^2 - 3.56\dfrac{G^2T_1}{a_t^2}}\right).$

23. $f = \dfrac{1}{2\pi\sqrt{LC}}, \; 2.05 \times 10^5$ cycles per sec.

25. 13.5 in.

27. 5 amp.

29. $t = \dfrac{1}{g}(kv_0 \pm\sqrt{k^2v_0^2 - 2gh}).$

31. $r_g = \dfrac{1}{2}\left(r_p \pm\sqrt{r_p^2 - \dfrac{4\mu r_p}{K}}\right).$

33. $2.49 \times 10^6.$

35. 7.1 in., 16.9 in.

Page 515.   Sec. 14–12

1. $3x^4 + 3x^3 - 17x^2 - 6x + 7.$

3. $-7y^5 - 5y^4 + 3y^3 - 15y^2 - 9y + 11.$

5. $2a^4 - a^3b - 5a^2b^2 + 7ab^3 - 3b^4.$

7. $x^3 + 3x^2 - 3x - 8, \; -1.$

9. $r^4 - 5r^3 + 14r^2 - 19r + 33, \; -68.$

11. $y^3 - y^2 + y - 4, \; -4y + 2.$

13. $2E^2 + E - 2, \; 0.$

15. $Z^2 + 4, \; 0.$

17. $x^2 + x + 1, \; 0.$

19. $y^2 + y + 1, \; 2.$

21. $a^3 + a^2 + a + 1, \; 0.$

23. $b^3 - b^2 + b - 1, \; 2.$

25. $x^2 + xy + y^2, \; 0.$

27. $9x^2 - 6xy + 4y^2, \; 0.$

Pages 517–518. Sec. 14–13

1. $x^2 + 3 + \dfrac{5}{x-2}$.

3. $R^3 - 2R^2 + 2R - 1$.

5. $Z^3 + 2Z^2 + 2Z + 7 + \dfrac{9}{Z-2}$.

7. $x^4 - 4x^3 + 4x^2 - 3x + 3 + \dfrac{-10}{x+1}$.

9. $y^4 - y^3 + 2y^2 + 3y - 2 + \dfrac{-20}{y-1}$.

11. -22.

13. -35.
15. 464.
17. 8.
19. 765.

21. No.
23. Yes.
25. No.
27. Yes.

29. Yes.
31. No.
33. No.
35. Yes.

37. No.
39. 19.
41. $-1, \tfrac{3}{2}$.

Page 521. Sec. 14–16

1. 3, 2.
3. $-3, 2$.
5. 3, 2.
7. $4.19, -1.19$.

9. $-1.75 \pm j1.20$.
11. $0.414, -2.41$.
13. $2, \tfrac{2}{3}$.
15. $\tfrac{1}{2}, -1$.

17. $-0.333 \pm j1.11$.
19. $1.00 \pm j1.41$.
21. 2, 1.
23. $\pm j2$.

25. $6, -4$.

Page 524. Sec. 14–17

1. $1, 2, -2$.
3. -1.
5. $-1, 2, -3$.

7. $-\tfrac{1}{2}, 2, 3$.
9. $\tfrac{1}{2}$.
11. $-1. 2$.

13. $-1, -1, -2, 3$.
15. No rational roots.
17. $\tan \theta = -1, 2, -3, 4$.

19. $\sin \alpha = 1, -1, -1, \tfrac{1}{2}$.
21. $-2, 4.30, 0.697$.
23. $-3, 0.500 \pm j0.866$.

25. $-2, 1 \pm j$.
27. $\tfrac{1}{2}, 0.414, -2.41$.
29. $-\tfrac{1}{2}, 1.28, -0.781$.

31. $1, 2, -3 \pm j$.
33. $-1, -1, -0.500 \pm j0.866$.

35. $\tan \theta = 2, 2, -3, -4$.
37. $1, 2, -2, 3, 4$.

Page 528. Sec. 14–18

1. -2.5.
3. $-1.7, -1.4, 3.0$.
5. -3.1.

7. 1.7.
9. $-0.3, -1.5, 1.9$.
11. 1.5.

13. $2.9, -3.4$.
15. No real roots.
17. $1.0 \times 10^{-9}, 4.8$.

19. $1.0 \times 10^{-8}, 4.3$.
21. 0.0, 5.0.
23. $-0.8, 2.0$.

25. 0.4, 2.2.
27. 0.4.
29. 2.6.

31. 0.5.

Pages 535–536. Sec. 14–19

1. 1.26.
3. -1.32.
5. $-0.879, 1.35, 2.53$.

7. 0.121.
9. $-1.31, 1.14$.
11. $1.00, -1.60$.

13. 3.42.
15. ± 1.57.
17. 0.804 ft.

19. 0.68 in.
21. $2.12 \text{ ft} \times 2.12 \text{ ft} \times 1.12 \text{ ft}$. **23.** 3.85 in.

Chapter 15

Pages 538–539. Sec. 15–1

1. $\sqrt{34}$.   **3.** $4\sqrt{2}$.   **5.** 20.   **7.** $\sqrt{5}$.

9. $4\sqrt{13}$.   **29.** $(-1, -2)$, 5.   **31.** $(-1, 0)$.

Page 540. Sec. 15–2

1. $(4, 3)$.   **3.** $(-6, \frac{1}{2})$.   **5.** $(1, 2)$.

7. $(\frac{1}{2}, -1)$, $(2, -\frac{5}{2})$, $(\frac{11}{2}, \frac{1}{2})$.   **9.** $(-7, -2)$.

Page 543. Sec. 15–3

1. 1, 45.0°.   **3.** -4, 104.0°.   **5.** $\frac{4}{5}$, 38.7°.   **7.** $\frac{4}{3}$, 53.1°.

17. 7.   **19.** 2.   **21.** -1.   **23.** -6.

Pages 544–545. Sec. 15–4

1. $\frac{1}{2}$, 3, -2.   **3.** -1, 1, $\frac{1}{3}$.   **5.** $\frac{1}{8}$, 2, $-\frac{1}{2}$.

15. Rectangle.   **17.** Not a rectangle.   **19.** Rectangle.

Pages 547–548. Sec. 15–5

1. 31.0°.   **3.** 49.4°.   **5.** 84.8°.

7. 74.7°.   **9.** 81.8°.   **11.** 36.9°, 40.6°, 102.5°.

13. 63.4°, 40.6°, 76.0°.   **15.** 102.1°, 17.6°, 60.3°.   **17.** 41.1°, 69.4°, 69.4°.

19. $\frac{1}{2}$.   **21.** 1.   **23.** $-8-5\sqrt{3} = -16.66$.

Pages 550–551. Sec. 15–6

1. 10.   **3.** 1.   **5.** 23.   **7.** 13.

9. 0.   **11.** 16.   **13.** 46.   **15.** 15.

17. 18.   **19.** 36.   **21.** 48.   **23.** 12.

25. 50.

Pages 554–555. Sec. 15–11

1. $x - 3y + 5 = 0$.   **3.** $x + 4y - 26 = 0$.   **5.** $5x + y + 26 = 0$.

7. $3x - y - 12 = 0$.   **9.** $y + 5 = 0$.   **11.** $y = \frac{1}{3}x + 2$.

13. $y = -3x + 5$.   **15.** $y = -\frac{2}{5}x - 1$.   **17.** $y = 8x$.

19. $y = -4x - 2$.   **21.** $x - y = 0$.   **23.** $x - \sqrt{3}y + 1 + 4\sqrt{3} = 0$.

25. $\sqrt{3}x + y - 2 = 0$.   **27.** $y = x + 2$.   **29.** $y = \dfrac{\sqrt{3}}{3}x - 3$.

31. $y = -\dfrac{\sqrt{3}}{3}x$.   **33.** -1, 4.   **35.** $-\frac{5}{4}$, 5.

37. $\frac{1}{2}$, 4.   **39.** 4, 4.   **41.** $\frac{3}{5}$, $-\frac{1}{5}$.

43. $1, -\frac{14}{5}$.

45. $x - y - 2 = 0, \ x + y - 4 = 0$.

47. $2x + y - 6 = 0, \ x - 2y - 13 = 0$.

49. $3x - 2y + 2 = 0, \ 2x + 3y + 10 = 0$.

51. $x - 2y + 10 = 0, \ 2x + y + 5 = 0$.

53. $x = 0$.

55. $x - 7 = 0$.

57. $y + 5 = 0$.

59. $18.4°$.

61. $45°$.

63. $90°$.

Page 556. Sec. 15–12

1. $4x - 3y - 0$.

3. $x + y - 1 = 0$.

5. $2x + 3y - 6 = 0$.

7. $2x + y + 1 = 0$.

9. $x - y + 4 = 0$.

11. $x + y = 0$.

13. $x + 8y + 15 - 0$.

15. $2x - y - 5 = 0$.

17. $3x - 4y - 11 = 0, \ 9x + y - 33 = 0, \ 6x + 5y - 22 = 0$.

19. $3x + 2y - 8 = 0, \ 2x + 3y - 7 = 0, \ 4x + y - 9 = 0$.

21. $7x - 13y - 30 = 0, \ 2x - 11y - 28 = 0, \ 5x - 2y - 2 = 0$.

23. $2x + y + 1 = 0$.

25. $x + 3y + 13 = 0$.

Page 558. Sec. 15–13

1. $3x + 5y - 15 = 0$.

3. $4x - y + 8 = 0$.

5. $5x + 3y + 15 = 0$.

7. $x + 4y - 4 = 0$.

9. $3x - 8y - 24 = 0$.

11. $5, -3$.

13. $5, -2$.

15. $6, 8$.

17. $\frac{9}{2}, -\frac{3}{2}$.

19. $-20, 60$.

21. $9, \frac{9}{7}$.

23. $x + y - 8 = 0$.

25. $8x + 9y - 60 = 0, \ 2x + y - 10 = 0$.

27. $x + y - 20 = 0$.

Pages 559–560. Sec. 15–14

1. 4.47.

3. 3.79.

5. 1.94.

7. $3.53; 4.24; 4.02$.

9. 2.69.

11. 3.54.

13. (a) $\frac{9}{2}$; (b) 19; (c) $\frac{25}{2}$; (d) $\frac{97}{2}$.

Pages 560–562. Sec. 15–15

1. $6x - y - 25 = 0$.

3. $(7, 5), (0, 6)$.

5. $\frac{1}{2}\sqrt{122} = 5.52$.

9. $x + 6y - 4 = 0$.

11. $(-1, 1)$.

13. 90.

15. $3x + 5y - 30 = 0$.

17. $(1, 3)$.

19. $10x - y = 0$.

21. $A = \frac{1}{2}\begin{vmatrix} x_2 & y_2 \\ x_3 & y_3 \end{vmatrix}$.

23. (a) $2x - y - 8 = 0, 2x - 5y = 0, y + 2 = 0$; (b) $2x - 5y - 16 = 0, y - 2 = 0$, $2x - y + 8 = 0$; (c) $x + 2y - 4 = 0, \quad 5x + 2y = 0, \quad x + 1 = 0$; (d) $5x + 2y - 11 = 0, \quad x - 5 = 0, \quad x + 2y + 9 = 0$; (e) 16; (f) 4, $\frac{16}{10}\sqrt{29}, \frac{16}{5}\sqrt{5}$.

25. $x + y + 1 = 0$.

27. (a) $\frac{3}{5}$, (b) $\frac{6}{5}$.

29. $d = 0.0010w + 0.30; \ 0.30$.

31. 17.2.

Chapter 16

Pages 565–566. Sec. 16–1

1. $x^2 + y^2 - 8x - 6y - 11 = 0.$ **3.** $x^2 + y^2 - 49 = 0.$

5. $x^2 + y^2 - 6y - 55 = 0.$ **7.** $x^2 + y^2 - 4x + 10y + 18 = 0.$

9. $x^2 + y^2 - 2ax - 2ay + a^2 = 0.$ **11.** $x^2 + y^2 - 2ax + 2ay + a^2 = 0.$

13. $x^2 + y^2 - 2ay - 3a^2 = 0.$ **15.** $x^2 + y^2 - 6x = 0.$

17. $x^2 + y^2 - 2ax - 2by = 0.$ **19.** $x^2 + y^2 + 4x - 8y + 16 = 0.$

21. $x^2 + y^2 - 2ax - 2ay + a^2 = 0.$ **23.** $5x^2 + 5y^2 - 10x - 20y - 11 = 0.$

25. $5x^2 + 5y^2 + 10x - 30y + 1 = 0.$ **27.** $x^2 + y^2 - 4x + 10y + 3 = 0.$

29. $x^2 + y^2 - 6x - 8y + 20 = 0.$ **31.** $x^2 + y^2 - 2ax - 2ay = 0.$

33. $x^2 + y^2 - 4x = 0,\ x^2 + y^2 + 4x = 0.$

35. $x^2 + y^2 - 4x - 4y + 4 = 0,\ x^2 + y^2 + 16x - 16y + 64 = 0.$

37. $x^2 + y^2 - 4x + 2y - 36 = 0.$ **39.** $x^2 + y^2 - 11x + 14 = 0.$

41. $x^2 + y^2 - 4x - 6y + 9 = 0.$

Page 568. Sec. 16–2

1. Point circle at $(-2, 1)$.   **3.** $(0, -8)$, 6.   **5.** No circle.

7. Point circle at $(-1, -2)$.   **9.** $(\frac{3}{2}, \frac{5}{2})$, 3.   **11.** No circle.

13. $(-3, 4)$, 4.   **15.** $(-\frac{5}{6}, -2)$, $\frac{13}{6}$.   **17.** $(1, -2)$, $\frac{5}{2}$.

19. -6.

Pages 570–571. Sec. 16–3

1. $x^2 + y^2 - x + y = 0.$ **3.** $x^2 + y^2 - 16x - 2y = 0.$

5. $4x^2 + 4y^2 - 24x - 21y + 20 = 0.$ **7.** $x^2 + y^2 - 4y - 21 = 0.$

9. $x^2 + y^2 + 2x + 4y - 20 = 0.$ **11.** $x^2 + y^2 + 4x - 46 = 0.$

13. $5x^2 + 5y^2 - 12x + 64y - 560 = 0.$

15. $x^2 + y^2 + 6x + 2y + 6 = 0,\ 13x^2 + 13y^2 + 30x - 6y - 34 = 0.$

17. $x^2 + y^2 - 4x - 8y + 4 = 0,\ x^2 + y^2 - 20x - 40y + 100 = 0.$

19. $x^2 + y^2 + 2x - 2y + 1 = 0,\ x^2 + y^2 + 10x - 10y + 25 = 0.$

21. $x^2 + y^2 - 14x + 39 = 0,\ x^2 + y^2 - 6x - 8y + 15 = 0.$

23. $x^2 + y^2 - 4x - 4y + 4 = 0,\ 25x^2 + 25y^2 - 20x + 20y + 4 = 0.$

25. $x^2 + y^2 - 3x - 2y = 0.$

27. $3x^2 + 3y^2 + 17x - 16y + 25 = 0.$

29. $x^2 + y^2 + 6x + 8y - 56 = 0.$

Pages 575–576. Sec. 16–4

1. $x^2 + 4xy - 12y - 9 = 0.$ **3.** $5x^2 - x^2y - 14xy + 3xy^2 - 5x = 0.$

5. $x^4 - y^4 - 4x^2 + 4y^2 = 0.$

7. $2x^3 - 3x^2 - x^2y + x + 2xy^2 - y + 3xy - 2y^2 - y^3 = 0.$

9. $x = 0$, $x - 3 = 0$.

11. $x + 3 = 0$, $x - 3 = 0$.

13. $3x + 2y = 0$, $3x - 2y = 0$.

15. $x = 0$, $y = 0$, $x - y = 0$, $x + y = 0$.

17. $x - y = 0$, $x + y + 2 = 0$.

19. $x^2 + y^2 - 1 = 0$, $x - y = 0$.

21. $24x - 25y - 23 = 0$.

23. $x - 4y - 2 = 0$.

25. $5x - y - 18 = 0$.

27. $x + 2y - 5 = 0$.

29. $5x - 10y - 27 = 0$.

31. $2x + y - 4 = 0$.

35. $4x - 3y - 18 = 0$.

37. $8x + 4y - 1 = 0$.

39. $x + 1 = 0$.

41. No common chord, no points of intersection.

43. $3x + 4y - 9 = 0$, $2x - 3y + 7 = 0$, $x + 7y - 16 = 0$.

Pages 579–580. Sec. 16–6

1. $x - 4 = 0$, straight line.

3. $y - 5 = 0$, straight line.

5. $x^2 + y^2 - 25 = 0$, circle.

7. $x^2 + y^2 + 4x + 6y - 51 = 0$, circle.

9. $x - 5 = 0$, straight line.

11. $x - 3y + 4 = 0$, straight line.

13. $x^2 + y^2 - 8x + 12 = 0$, circle.

15. $x^2 - 3y^2 + 32y - 64 = 0$, a curve not previously studied.

17. $3x^2 + 4y^2 + 6x - 9 = 0$, a curve not previously studied.

19. $8x + 3 = 0$, $8x - 3 = 0$, straight lines.

Pages 590–591. Sec. 16–10

1. $x^2 + y^2 - 16 = 0$.

3. $y = 0$.

5. $y - 6 = 0$.

7. $y - x \tan\sqrt{x^2 + y^2} = 0$.

9. $x - 8 - 0$.

11. $x^2 + y^2 - 4x^2y^2 = 0$.

13. $y^2 - 2x - 1 = 0$.

15. $x^2 + y^2 - x + y = 0$.

17. $x^2 + y^2 - 2x - 3y = 0$.

19. $x^4 + x^2y^2 - y^2 = 0$.

21. $r \sin \theta = 4$.

23. $r = 1$.

25. $r = 4 \cos \theta$.

27. $\sin \theta + \cos \theta = 0$.

29. $r \cos \theta - r \sin \theta = 3$.

31. $r = 8 \cos \theta$.

33. $r^2 \cos 2\theta = 4$.

35. $r = \tan \theta \sec \theta$.

37. $r = 4a \cot \theta \csc \theta$.

39. $r = \cos \theta - \sin \theta$.

Chapter 17

Pages 596–597. Sec. 17–2

1. $(0, 0)$, $(1, 0)$, $x + 1 = 0$, 4.

3. $(0, 0)$, $(0, 1)$, $y + 1 = 0$, 4.

5. $(0, 0)$, $(4, 0)$, $x + 4 = 0$, 16.

7. $(0, 0)$, $(-2, 0)$, $x - 2 = 0$, 8.

9. $(0, 0)$, $(0, 2)$, $y + 2 = 0$, 8.

11. $(0, 0)$, $(0, \frac{3}{2})$, $y + \frac{3}{2} = 0$, 6.

13. $(0, 0)$, $(8, 0)$, $x + 8 = 0$, 32.

15. $(0, 0)$, $(0, -3)$, $y - 3 = 0$, 12.

17. $(0, 0)$, $(\frac{1}{4}, 0)$, $4x + 1 = 0$, 1.

19. $(0, 0)$, $(0, \frac{4}{5})$, $5y + 4 = 0$, $\frac{16}{5}$.

21. $(0, 0)$, $(-\frac{5}{6}, 0)$, $6x - 5 = 0$, $\frac{10}{3}$.   **23.** $y^2 = 8x$.

25. $x^2 = 16y$.   **27.** $x^2 = -32y$.   **29.** $x^2 = 20y$.   **31.** $y^2 = -16x$.

33. $3x^2 = -32y$.   **35.** $y^2 = 16x$.   **37.** $x^2 = 16y$.   **39.** $y^2 = 8x$.

41. $5x^2 = 9y$.   **43.** $4x^2 = -y$.   **45.** $x^2 = 32y - 256$.

Pages 601–602. Sec. 17–3

1. $(\pm 5, 0)$, $(\pm \sqrt{21}, 0)$, 5, 2.   **3.** $(\pm 10, 0)$, $(\pm 8, 0)$, 10, 6.

5. $(\pm 3, 0)$, $(\pm \sqrt{5}, 0)$, 3, 2.   **7.** $(0, \pm 4)$, $(0, \pm 2\sqrt{3})$, 4, 2.

9. $(0, \pm 6)$, $(0, \pm \sqrt{35})$, 6, 1.   **11.** $(\pm 14, 0)$, $(\pm 8\sqrt{3}, 0)$, 14, 2.

13. $(0, \pm 4)$, $(0, \pm \frac{1}{3}\sqrt{44})$, 4, $\frac{10}{3}$.   **15.** $(\pm \sqrt{3}, 0)$, $(\pm \frac{1}{2}\sqrt{3}, 0)$, $\sqrt{3}$, $\frac{3}{2}$.

17. $(0, \pm 5)$, $(0, \pm \sqrt{21})$, 5, 2.   **19.** $21x^2 + 25y^2 = 2100$.

21. $16x^2 + 25y^2 = 400$.   **23.** $16x^2 + 25y^2 = 400$.

25. $49x^2 + 4y^2 = 196$.   **27.** $x^2 + 9y^2 = 225$.

29. $25x^2 + y^2 = 25$.   **31.** $x^2 + 5y^2 = 9$.

33. $9x^2 + 4y^2 = 144$.   **35.** $16x^2 + 25y^2 = 400$.

37. $x^2 + 4y^2 = 12$.   **39.** $9x^2 + 25y^2 = 900$.

41. $16x^2 + 25y^2 + 32x - 384 = 0$.

Pages 606–607. Sec. 17–4

1. $(\pm 1, 0)$, $(\pm \sqrt{2}, 0)$, 1, 1, $y = \pm x$.   **3.** $(\pm 2, 0)$, $(\pm 2\sqrt{2}, 0)$, 2, 2, $y = \pm x$.

5. $(0, \pm 3)$, $(0, \pm 3\sqrt{2})$, 3, 3, $y = \pm x$.   **7.** $(0, \pm 6)$, $(0, \pm 3\sqrt{5})$, 6, 3, $y = \pm 2x$.

9. $(\pm 6, 0)$, $(\pm 3\sqrt{5}, 0)$, 6, 3, $y = \pm \frac{1}{2}x$.   **11.** $(0, \pm 5)$, $(0, \pm \sqrt{34})$, 5, 3, $y = \pm \frac{5}{3}x$.

13. $(\pm 8, 0)$, $(\pm 8\sqrt{2}, 0)$, 8, 8, $y = \pm x$.   **15.** $(\pm 10, 0)$, $(\pm 2\sqrt{26}, 0)$, 10, 2, $y = \pm \frac{1}{5}x$.

17. $(\pm 4, 0)$, $(\pm 2\sqrt{13}, 0)$, 4, 6, $y = \pm \frac{3}{2}x$.   **19.** $(0, \pm \frac{5}{2})$, $(0, \pm \frac{5}{2}\sqrt{5})$, $\frac{5}{2}$, 5, $y = \pm \frac{1}{2}x$.

21. $(0, \pm \frac{5}{2})$, $(0, \pm \frac{5}{6}\sqrt{13})$, $\frac{5}{2}$, $\frac{5}{3}$, $y = \pm \frac{3}{2}x$.

23. $(\pm 5, 0)$, $(\pm 3\sqrt{5}, 0)$, 5, $2\sqrt{5}$, $y = \pm \frac{2}{5}\sqrt{5}x$,

25. $16x^2 - 9y^2 = 144$.   **27.** $144y^2 - 25x^2 = 3600$.

29. $225y^2 - 64x^2 = 14,400$.   **31.** $64x^2 - 225y^2 = 14,400$.

33. $9x^2 - 16y^2 = 144$.   **35.** $3x^2 - 11y^2 = 64$.

37. $y^2 - 2x^2 = 14$.   **39.** $16x^2 - 9y^2 = 256$.

41. $5y^2 - 20x^2 = 144$.

Pages 609–610. Sec. 17–6

1. $(0, -3), (-2, -2), (1, 1), (-3, -2)$.   **3.** $(1, -2), (3, 2), (5, 0), (4, -7)$.

5. $(2, 6), (1, 5), (0, 8), (3, 2)$.

7. (a) $(0, -9)$; (b) $(6, -6)$; (c) $(-1, 1)$; (d) $(7, -1)$; (e) $(4, 4)$.

9. $x' - 2y' = 0$.   **11.** $3y' - x' - 6 = 0$.

13. $7x' + y' - 4 = 0$.   **15.** $x'^2 + y'^2 - 16 = 0$.

17. $36x'^2 + 36y'^2 - 169 = 0$.   **19.** $x'^2 - 8y' = 0$.

21. $4x'^2 + 9y'^2 - 16 = 0$.   **23.** $x'^2 - y'^2 - 16 = 0$.

25. $x'^2 + y'^2 = r^2$.

Pages 612–613. Sec. 17–7

1. $(3, 2)$; $(5, 2)$; $(5, -2), (5, 6)$; 8.

3. $(-1, -3)$; $(-1, 0)$; $(-7, 0), (5, 0)$; 12.

5. $(-8, 20)$; $(-8, \frac{39}{2})$; $(-9, \frac{39}{2}), (-7, \frac{39}{2})$; 2.

7. $(3, -3)$; $(5, -3)$; $(5, -7), (5, 1)$; 8.

9. $(\frac{1}{4}, 0)$; $(\frac{1}{4}, \frac{1}{8})$; $(0, \frac{1}{8}), (\frac{1}{2}, \frac{1}{8})$; $\frac{1}{2}$.

11. $(8, -10)$; $(8, -\frac{21}{2})$; $(7, -\frac{21}{2}), (9, -\frac{21}{2})$; 2.

13. $(2, -2)$; $(2 \pm 2\sqrt{3}, -2)$; $(6, -2), (-2, -2)$; $(2, 0), (2, -4)$.

15. $(2, -2)$; $(2, -2 \pm 2\sqrt{3})$; $(2, 2), (2, -6)$; $(4, -2), (0, -2)$.

17. $(-1, 0)$; $(3, 0), (-5, 0)$; $(4, 0), (-6, 0)$; $(-1, 3), (-1, -3)$.

19. $(-\frac{3}{2}, 1)$; $(-\frac{3}{2} \pm \sqrt{2}, 1)$; $(-3, 1), (0, 1)$; $(-\frac{3}{2}, \frac{1}{2}), (-\frac{3}{2}, \frac{3}{2})$.

21. $(-2, 0)$; $(-2 \pm \sqrt{19}, 0)$; $(-12, 0), (8, 0)$; $(-2, \pm 9)$.

23. $(a, 0)$; $(a, \pm a)$; $(a, \pm a\sqrt{2})$; $(0, 0), (2a, 0)$.

25. $(1, -2)$; $(5, -2), (-3, -2)$; $(6, -2), (-4, -2)$; $(1, 1), (1, -5)$; $3x - 4y - 11 = 0, 3x + 4y + 5 = 0$.

27. $(3, 1)$; $(6, 1), (0, 1)$; $(3 \pm \sqrt{13}, 1)$; $(3, 3), (3, -1)$; $2x - 3y - 3 = 0, 2x + 3y - 9 = 0$.

29. $(-2, 3)$; $(10, 3), (-14, 3)$; $(11, 3), (-15, 3)$; $(-2, 8), (-2, -2)$; $5x - 12y + 46 = 0, 5x + 12y - 26 = 0$.

31. $(0, -1)$; $(0, 0), (0, -2)$; $(0, -1 \pm \sqrt{17})$; $(4, -1)$; $(-4, -1)$; $x - 4y - 4 = 0, x + 4y + 4 = 0$.

33. $(-4, 0)$; $(-4, 6), (-4, -6)$; $(-4, \pm 3\sqrt{5})$; $(-7, 0), (-1, 0)$; $2x - y + 8 = 0, 2x + y + 8 = 0$.

35. $(3, -7)$; $(3, -3), (3, -11)$; $(3, -7 \pm 4\sqrt{2})$; $(7, -7), (-1, -7)$; $x - y - 10 = 0, x + y + 4 = 0$.

37. $x^2 - 4x + 32y - 60 = 0$.

39. $16x^2 + 25y^2 + 100y - 300 = 0$.

41. $24x^2 + 25y^2 - 96x - 100y - 404 = 0$.

43. $169x^2 + 144y^2 + 1014x + 648y + 729 = 0$.

45. $9y^2 - 16x^2 - 72y - 32x - 16 = 0$.

47. $9x^2 - 16y^2 - 36x - 64y - 172 = 0$.

Page 615. Sec. 17–8

1. $y^2 = -8(x+1)$.

3. $(x - \frac{1}{2})^2 = y + \frac{13}{4}$.

5. $(y-1)^2 = -16(x-6)$.

7. $\dfrac{(x - \frac{1}{2})^2}{\frac{1}{4}} + \dfrac{y^2}{1} = 1$.

9. $\dfrac{x^2}{4} + \dfrac{(y-3)^2}{9} = 1$.

11. $\dfrac{(x-1)^2}{4} + \dfrac{(y-1)^2}{2} = 1$.

13. $\dfrac{(y-1)^2}{1} - \dfrac{(x+2)^2}{2} = 1$.

15. $\dfrac{(x+10)^2}{100} - \dfrac{y^2}{100} = 1$.

17. $\dfrac{(x+1)^2}{8} - \dfrac{(y-1)^2}{2} = 1$.

19. $\dfrac{(x-8)^2}{4} + \dfrac{(y+2)^2}{1} = 1$.

21. $(x-2)^2 = 2(y-3)$.

23. $\dfrac{(x-1)^2}{9} + \dfrac{(y-2)^2}{4} = 1$.

25. $\dfrac{(x-1)^2}{9} - \dfrac{(y+2)^2}{16} = 1$.

27. $\dfrac{(x-1)^2}{16} + \dfrac{(y-3)^2}{4} = 1$.

29. $\dfrac{(x+3)^2}{\frac{9}{2}} - \dfrac{(y-2)^2}{\frac{9}{4}} = 1$.

Page 618. Sec. 17–9

1. $x'^2 - y'^2 = 4$.　　　**3.** $x'^2 - y'^2 = 2k$.　　　**5.** $2x'y' = -k$.

7. (a) $(3.60, -0.23)$;　(b) $(1.63, -4.83)$;　(c) $(-2.00, 3.46)$;　(d) $(-1.27, 6.20)$;
　(e) $(-4.46, -0.27)$.

9. $3x'^2 - y'^2 = 10$.　　　**11.** $4x'^2 + y'^2 = 4$.　　　**13.** $x''^2 - y''^2 = 4$.

Pages 621–622. Sec. 17–10

1. $4x'^2 - y'^2 = 4$.　　　**3.** $x'^2 + 9y'^2 = 0$.　　　**5.** $6x'^2 + y'^2 = 10$.

7. $7x'^2 - 6y'^2 = 42$.　　　**9.** $y'^2 - x'^2 = 1$.　　　**11.** $y'^2 - 9x'^2 = 81$.

13. $25(x' - \frac{17}{5})^2 + 100(y' - \frac{6}{5})^2 = 1600$.

15. $x'^2 = 10(y' + 4)$.

17. $3(y' - \frac{9}{5})^2 - 2(x' + \frac{8}{5})^2 = 6$.

19. $7(x' - 10\sqrt{2})^2 - 3y'^2 = 1400$.

Chapter 18

Pages 628–629. Sec. 18–3

1. $3\sqrt{2}$.　　　**3.** $\sqrt{30}$.　　　**5.** $10\sqrt{2}$.　　　**7.** $\sqrt{62}$.　　　**9.** $\frac{1}{2}\sqrt{861}$.

15. Sphere with radius 3 and center at the origin.

17. Sphere with radius 7 and center at $(-1, 5, -3)$.

19. $4x - 3y + 5z = 4$, a plane.

Pages 633–634. Sec. 18–5

1. $\cos \alpha = 0.5455$, $\cos \beta = -0.8182$, $\cos \gamma = 0.1818$; $\alpha = 56.9°$, $\beta = 144.9°$, $\gamma = 79.5°$.

3. $\cos \alpha = 0.0000$, $\cos \beta = 0.0000$, $\cos \gamma = -1.0000$; $\alpha = 90.0°$, $\beta = 90.0°$, $\gamma = 180.0°$.

5. $\cos \alpha = 0.7071$, $\cos \beta = 0.0000$, $\cos \gamma = -0.7071$; $\alpha = 45.0°$, $\beta = 90.0°$, $\gamma = 135.0°$.

7. $\cos \alpha = 0.8571$, $\cos \beta = -0.4286$, $\cos \gamma = 0.2857$; $\alpha = 31.0°$, $\beta = 115.4°$, $\gamma = 73.4°$.

9. $\cos \alpha = 0.4575$, $\cos \beta = 0.4575$, $\cos \gamma = 0.7625$; $\alpha = 62.8°$, $\beta = 62.8°$, $\gamma = 40.3°$.

11. $\cos \alpha = 0.2540$, $\cos \beta = -0.8890$, $\cos \gamma = 0.3810$; $\alpha = 75.3°$, $\beta = 152.8°$, $\gamma = 67.6°$.

13. $\cos \beta = \pm \frac{1}{2}\sqrt{3}$; $\beta = 30°$, $150°$.　　　15. $\cos \gamma = \pm \frac{1}{2}\sqrt{2}$; $\gamma = 45°$, $135°$.

17. $\cos \alpha = \pm \frac{1}{2}\sqrt{2}$; $\alpha = 45°$, $135°$.　　　19. $\cos \gamma = 0$, $\gamma = 90°$.

21. $\alpha = 0°$, $\beta = 90°$, $\gamma = 90°$; $\cos \alpha = 1$, $\cos \beta = 0$, $\cos \gamma = 0$.

23. $\alpha = 90°$, $\beta = 90°$, $\gamma = 0°$; $\cos \alpha = 0$, $\cos \beta = 0$, $\cos \gamma = 1$.

Pages 637–638. Sec. 18–7

1. -9, -10, 1.　　　3. 4, 1, -8.　　　5. -8, 5, -2.

7. $\cos \alpha = 1.0000$, $\cos \beta = 0.0000$, $\cos \gamma = 0.0000$; $\alpha = 0.0°$, $\beta = 90.0°$, $\gamma = 90.0°$.

9. $\cos \alpha = \cos \beta = \cos \gamma = 0.5774$; $\alpha = \beta = \gamma = 54.7°$.

11. $\cos \alpha = 0.3333$, $\cos \beta = -0.6667$, $\cos \gamma = 0.6667$; $\alpha = 70.5°$, $\beta = 131.8°$, $\gamma = 48.2°$.

13. $\cos \alpha = 0.1111$, $\cos \beta = -0.4444$, $\cos \gamma = 0.8889$; $\alpha = 83.6°$, $\beta = 116.4°$, $\gamma = 27.3°$.

15. $\cos \alpha = 0.2673$, $\cos \beta = -0.5345$, $\cos \gamma = 0.8018$; $\alpha = 74.5°$, $\beta = 122.3°$, $\gamma = 36.7°$.

17. $90.0°$.　　　19. $39.5°$.　　　21. $19.1°$.

23. L_1, L_3, L_6 are parallel; L_2, L_4, L_5 are parallel.

25. L_2, L_6 are parallel; L_3, L_4, L_5 are parallel.

27. Groups of mutually perpendicular lines: L_1, L_2, L_3; L_1, L_3, L_4; L_2, L_3, L_5; L_3, L_4, L_5.

35. $74.2°$, $105.8°$.　　　39. $5\sqrt{2}$, $\sqrt{146}$.

Pages 639–640. Sec. 18–8

1. $y^2 - z^2 = 0$, the two planes containing the x axis that make angles of $45°$ with the xy plane.

3. $x^2 - z^2 = 0$, the two planes containing the y axis that make angles of $45°$ with the xy plane.

5. $9x^2 - z^2 = 0$, the two planes containing the y axis that make angles whose tangents are 3 with the xy plane.

7. $x^2 = 9$, the two planes perpendicular to the x axis at $x = 3$ and $x = -3$.

9. $x^2 + y^2 + z^2 = 64$, the sphere with the center at the origin and radius 8.

11. $x^2 + y^2 + (z - 1)^2 = 36$, the sphere with center at $(0, 0, 1)$ and radius 6.

13. $(x + 1)^2 + (y - 2)^2 + (z - 3)^2 = 16$, the sphere with center at $(-1, 2, 3)$ and radius 4.

15. $z = 3$, the plane perpendicular to the z axis at $z = 3$.

17. $2x - 10y - 10z = 11$, the plane which is the perpendicular bisector of the segment joining the two given points.

19. $y^2 + z^2 = 25$, the right circular cylinder of radius 5 whose axis is the x axis.

21. $x^2 + y^2 = 36$, the right circular cylinder of radius 6 whose axis is the z axis.

23. The xz plane.

25. The plane perpendicular to the x axis at $x = 5$.

27. The plane perpendicular to the z axis at $z = 1$.

29. The plane containing the y axis that bisects the right angle between the positive x axis and the positive z axis.

31. The plane containing the z axis and the line $y = 3x$ in the xy plane.

33. The planes perpendicular to the x axis at $x = 1$ and $x = -1$.

35. The right circular cylinder of radius 1 whose axis is the z axis.

37. The sphere with radius 4 and center at the origin.

39. The sphere with radius 3 and center at the point $(1, 2, 4)$.

41. $x^2 + y^2 + z^2 - 4x + 4y - 6z - 13 = 0$.

43. $3x^2 + 4y^2 + 4z^2 - 48 = 0$.

Pages 643–644. Sec. 18–11

1. $x + 2y + z = 0$.

3. $y - 2 = 0$.

5. $2x + y - 3z - 16 = 0$.

7. $12x - 6y - 3z - 22 = 0$.

9. 6, 6, 2; 1, 1, 3.

11. $-4, -6, -3$; 3, 2, 4.

13. $2, -10, \frac{5}{2}$; $5, -1, 4$.

15. 2, 3, no z intercept; 3, 2, 0.

17. No x intercept, 3, -12; 0, 4, -1.

19. Parallel.

21. Parallel.

23. Neither.

25. Parallel.

27. $27x - 14y - 17z = 0$.

29. $2x + z - 10 = 0$.

31. $bcx + acy + abz = abc$.

33. $x + 3y - z + 3 = 0$.

35. $x - 2y + 2z - 15 = 0$.

37. $x + 2y + 3z + 13 = 0$.

39. $x - y - z = 0$.

41. $5x + 7y + z - 22 = 0$.

45. $60°$.

47. $54.7°$.

1. $\dfrac{x-4}{-3} = \dfrac{y-2}{1} = \dfrac{z-3}{2}$.

3. $\dfrac{x-1}{-3} = \dfrac{y}{6} = \dfrac{z+5}{5}$.

5. $\dfrac{x-3}{5} = \dfrac{y-4}{-1} = \dfrac{z-5}{6}$.

7. $\dfrac{x}{-1} = \dfrac{y}{2} = \dfrac{z+3}{3}$.

9. $\dfrac{x-2}{\sqrt{2}} = \dfrac{y-3}{1} = \dfrac{z+4}{1}$.

11. $\dfrac{x-2}{-1} = \dfrac{y}{3} = \dfrac{z-1}{8}$.

13. $\dfrac{x}{2} = \dfrac{y}{4} = \dfrac{z}{7}$.

15. $\dfrac{y-4}{4} = \dfrac{z-3}{3}$, $x = 1$.

17. $(\frac{3}{4}, -\frac{1}{6}, 0)$, $(\frac{7}{6}, 0, \frac{1}{4})$, $(0, -\frac{7}{2}, \frac{9}{4})$.

19. $(3, -2, 1)$.

23. $(-9, 6, -4)$.

25. $\dfrac{x+3}{-1} = \dfrac{z-5}{2}$, $y = 4$.

27. $x = 1$, $z = 3$.

29. $\dfrac{x-\frac{31}{5}}{1} = \dfrac{y+\frac{22}{5}}{1} = \dfrac{z-0}{1}$.

31. $x = 3 + 3t$, $y = -2 - t$, $z = 4 + 2t$.

Chapter 19

1. (a) $\Delta t = 1$ min, $\Delta T = 380°$; (b) $\Delta t = 1$ min, $\Delta T = 180°$;

(c) $\Delta t = 1$ min, $\Delta T = -100°$; (d) $\Delta t = 1$ min, $\Delta T = -120°$.

3. 0.0969, 0.0511, 0.0107, 0.0054, 0.0010. 5. 0.1513, 0.0822, 0.0177, 0.0090, 0.0018.

7. 0.0134, 0.0068, 0.0014, 0.0007, 0.0001. 9. 0.03464, 0.01734, 0.00867, 0.00173.

11. 0.0314, 0.0158, 0.0079, 0.0016. 13. 0.0153, 0.0078, 0.0039, 0.0008.

15. 0.03529, 0.01761, 0.00880, 0.00176. 17. 0.0698, 0.0343, 0.0170, 0.0034.

19. 0.599, 0.279, 0.135, 0.026.

21. -0.0032, -0.0019, -0.0006, -0.0003.

23. -0.0104, -0.0062, -0.0020, -0.0010.

25. -0.0154, -0.0092, -0.0031, -0.0015.

27. 125, 21, 2.01; 225, 41, 4.01; 425, 81, 8.01; 825, 161, 16.01.

29. 45.2202, 7.7717, 0.7489; 77.6393, 14.2189, 1.3926; 110.5276, 20.7860, 2.0489.

31. -3, 3, 1.5, 0.3.

33. 7, 3.25, 0.61.

35. -0.00119, -0.00061, 0.00032.

37. 1.75, -0.6364, -0.3333.

39. -1.265, -0.440, -0.134.

41. 8.5000π, 4.0000π, 0.7600π.

43. 228, 54.75, 10.83.

45. -38.28.

47. 7937.5π.

49. $11.6°$.

51. 46 ft, 14 ft, -18 ft, -34 ft.

Pages 663–664. Sec. 19–2

1. 16.162° per cm of mercury.

3. 10,000 gausses per oersted, 2500 gausses per oersted, 800 gausses per oersted.

5. 0.0365, 0.0125, 0.0007.   **7.** 0.0167, 0.0127, 0.0084.

9. −28.65, −38.20, −52.1.   **11.** 11, 10.1, 10.01

13. 16, 128, 257.6,   **15.** 0.5073, 0.4807, 0.5086.

17. 3681, 2400, 3681.   **19.** 499 cu ft per ft.

21. 5300 lb per sq in.

Pages 667–668. Sec. 19–3

1. 0.145, 0.045, 1.05.   **3.** 0.289, 0.144, 0.192.   **5.** 10.00, 10.20, 10.02.

7. 0.921, 0.622, 1.000.   **9.** −0.218, −0.718, −0.922.

11. −0.001, −0.004, −0.004.   **13.** 1.40, 0.52, 0.17.

15. 450 lb per sq in. per percentage of zinc, 650 lb per sq in. per percentage of zinc, 1150 lb per sq in. per percentage of zinc.

17. 7.5 thousandths of an inch per degree, 6.25 thousandths of an inch per degree, −6.25 thousandths of an inch per degree.

19. 0.0425 bhp per rpm, 0.0400 bhp per rpm, 0.0225 bhp per rpm.

21. 0.01 ft per degree, 0.01 ft per degree, 0.01 ft per degree.

23. −0.718 cu in. per lb per sq in., −0.225 cu in. per lb per sq in., −0.577 cu in. per lb per sq in.

Pages 671–672. Sec. 19–4

1. 6.   **3.** 0.001.   **5.** 8.   **7.** 2.   **9.** $\frac{4}{5}$.   **11.** −2.

13. $-\frac{1}{3}$.   **15.** 0.   **17.** 30.   **19.** 60.   **21.** $5I$.   **23.** 0.

25. 10.   **27.** 0.   **29.** 1.0000.   **31.** 1.0000.   **33.** $\dfrac{C_1 C_3}{C_1 + C_3}$.

Pages 677–678. Sec. 19–6

1. 40, −20, 0.   **3.** 12, 27, 3.   **5.** 40, 4, −4.   **7.** 24, 108, 18.

9. 5, 74, 14.   **11.** $48I$.   **13.** $2\pi r$.   **15.** 9.

17. $32t$.   **19.** $18 + 7.2t$.   **21.** $6t^2 - 2t$.   **23.** $-\dfrac{1}{200C^2}$.

25. $-\dfrac{10}{(1 + R)^2}$.   **27.** $4w - 3 - \dfrac{1}{w^2}$.   **29.** $-400\pi^2$ dynes per sec.

Pages 683–684. Sec. 19–9.

1. $4x^3$.   **3.** $9x^8$.   **5.** $\frac{6}{5}\sqrt[5]{x}$.   **7.** $\frac{5}{2}x^{\frac{3}{2}}$.

9. $-\dfrac{6}{x^7}$.   **11.** $-\dfrac{1}{2x^{\frac{3}{2}}}$.   **13.** $27x^2$.   **15.** $\frac{2}{3}x$.

17. $-\dfrac{14}{5t^3}$.   **19.** $\dfrac{2}{\sqrt[3]{t^2}}$.   **21.** 5.   **23.** $9x^2$.

25. $24x$.   **27.** $8t-1$.   **29.** $21t^2+10t-1$.   **31.** $\dfrac{2}{x^2}+\dfrac{2}{x^3}$.

33. $\dfrac{6}{t^2}-\dfrac{48}{t^3}$.   **35.** $\dfrac{3}{2\sqrt{x}}-\dfrac{5}{3\sqrt[3]{x^2}}$.   **37.** $\cos x-\sin x$.   **39.** $4\cos x+\sin x$.

41. $-6\sin\theta+5$.   **43.** $12(1+x)^3$.   **45.** $\dfrac{t}{\sqrt{t^2+5}}$.   **47.** $-\dfrac{2}{(4+t)^2}$.

49. $-1.5\sin 3\theta$.   **51.** $15{,}080\cos 377t$.   **53.** $3400\cos(680t+28)$.

55. $4\sin 2t$.   **57.** $\cos 2x$.   **59.** $3x^2(\cos 3x-x\sin 3x)$.

61. $\dfrac{(1+t)^3(1+9t)}{2\sqrt{t}}$.   **63.** $10(1+v)(13+17v)(5+7v)^2$.

65. $-192x^3(3-x^2)^2(3+x^2)^2$.   **67.** $-\dfrac{31}{(5t-2)^2}$.

69. $\dfrac{\theta\cos\theta-2\sin\theta}{\theta^3}$.   **71.** $\dfrac{\cos x+(1+x)\sin x}{\cos^2 x}$.

73. $\sec\theta\tan\theta$.   **75.** $\dfrac{\cos x\,(1+2\sin^2 x)}{\cos^2 2x}$.

77. $-\dfrac{4t}{(1+t^2)^2}$.   **79.** $\dfrac{2v(v^4+2v^2-1)}{(v^2+1)^2}$.

81. $\dfrac{(4x^2+3)^2(20x^2-3)}{2x^2}$.   **83.** $-\dfrac{1}{2(1+v)\sqrt{1+v}}$.

85. $\dfrac{-1}{30\omega^3\sqrt{4+\dfrac{1}{30\omega^2}}}$.   **87.** $-\dfrac{bm+an}{(m+nt)^2}$.

89. $-\dfrac{2kU\sin t\cos t}{\sqrt{m+U\cos^2 t}}$.   **91.** $M\omega\cos\omega t-N\omega\sin\omega t$.

93. $E_1\omega\cos(\omega t+\alpha_1)+E_2\omega\cos(\omega t+\alpha_2)$.

Page 687. Sec. 19–10

1. (a) $4x-y-2=0$, (b) $8x+y+8=0$, (c) $y=0$.

3. (a) $3x-y-6=0$, (b) $2y-3=0$, (c) $3x+y=0$.

5. (a) $3x+4y-25=0$, (b) $4x+3y-25=0$, (c) $y-5=0$.

7. (a) $x-y=0$, (b) $4x\sqrt{2}-8y+4\sqrt{2}-\pi\sqrt{2}=0$, (c) $y-1=0$.

9. (a) $x-y+3=0$, (b) $18x-6x\sqrt{3}+12y-6-3\pi-18\sqrt{3}+\pi\sqrt{3}$,

    (c) $6x+2y-2-3\pi=0$.

11. $0.009t^2$, $0.018t$.   **13.** $v_0 + gt$, g.

15. $14t \sin 3t + 21t^2 \cos 3t$, $7(2 - 9t^2) \sin 3t + 84t \cos 3t$.

17. $6t^2 \cos 2t - 4t^3 \sin 2t + 3$, $12t \cos 2t - 24t^2 \sin 2t - 8t^3 \cos 2t$.

19. 160 volts.   **21.** $(4t + 60.06)$ volts.

23. $(100t^2 - 37t + 19.4)$ volts.

25. $200 \sin (377t + 0.1) + 1131 \cos (377t + 0.1)$ volts.

27. 450 lb.   **29.** $(1350t + 301.2)$ lb.

31. $(300t^2 + 517.2t - 544.8)$ lb.

33. $529.20 \cos (12.6t + 0.2) - 101.28 \sin (12.6t + 0.2)$ lb.

Pages 693–694. Sec. 19–11

1. Min: $(2, -24)$.   **3.** Max: $(\frac{1}{4}, \frac{1}{8})$.

5. Min: $(-\frac{1}{3}, -\frac{4}{3})$.   **7.** Min: $(\frac{5}{2}, -\frac{13}{4})$.

9. Max: $(\frac{3}{2}, 56)$.   **11.** Max: $(-1, 3)$; min: $(1, -1)$.

13. Max: $(-\frac{1}{9}\sqrt{6}, \frac{2}{27}\sqrt{6})$; min: $(\frac{1}{9}\sqrt{6}, -\frac{2}{27}\sqrt{6})$.

15. Max: $(-1, -2)$; min: $(1, 2)$.   **17.** Max: $(0, 1)$.

19. Max: $(-3, 333)$; min: $(\frac{5}{3}, \frac{1}{4})$.   **21.** Max: $\left(\dfrac{\pi}{2}, 2\right)$; min: $\left(-\dfrac{\pi}{2}, -2\right)$.

23. Max: $(0, 5)$; min: $(\pi, -1)$.

25. Max: $(0.936, 1.760)$; min: $(2.574, -0.369)$.

27. $3.83°$ C.   **29.** Squares 3 in. on a side.

31. Diameter $= \dfrac{8}{\sqrt[3]{\pi}}$ in., height $= \dfrac{8}{\sqrt[3]{\pi}}$ in.   **33.** Side of base $= 12$ yd. height $= 6$ yd.

35. $45°$.

37. Radius $= 2\sqrt{6}$ cm, volume $= 16\pi\sqrt{3}$ cu cm.

Pages 697–699. Sec. 19–12

1. $dy = 10\, dx$.   **3.** $ds = (80 - 32t)\, dt$.   **5.** $du = (15x^2 - 4x + 1)dx$.

7. $dy = -\dfrac{1}{x^2}\, dx$.   **9.** $du = -\dfrac{2v}{(v^2 + 1)^2}\, dv$.   **11.** $dy = \dfrac{x}{(1-x^2)\sqrt{1-x^2}}\, dx$.

13. $du = -\dfrac{4v}{(1 + v^2)^2}\, dv$.   **15.** $dv = \dfrac{u^2 - 4u - 1}{(u - 2)^2}\, du$.   **17.** $dr = 3(\sin\theta + \theta \cos\theta)\, d\theta$.

19. 814.3 cu in.   **21.** 137.2 cu in.   **23.** 62.83 cu ft.

25. 32.88 ft.   **27.** 1 per cent.   **29.** 3.033.

31. 3.005.   **33.** 0.7108.   **35.** 1.007.

Chapter 20

Pages 705–707. Sec. 20–1

1. 22.0. **3.** 21.5. **5.** 162. **7.** 15.7.

9. 17.5. **11.** 1.11. **13.** 0.992. **15.** 0.997.

17. 0.701. **19.** 0.304. **21.** 333. **23.** 236.

25. 1.47. **27.** 41.5. **29.** 19.0. **31.** 0.352.

33. 0.566. **35.** 1.16.

37. If $\Delta x = 1$, error $= 16$, 1.56%; if $\Delta x = \frac{1}{2}$, error $= 4$, 0.39%.

39. 3.44, 0.09. **41.** 0.637.

Pages 713–715. Sec. 20–2

17. 9. **19.** 42. **21.** 180. **23.** 12. **25.** 45.

27. $\frac{255}{8}$. **29.** $\frac{2}{3}$. **31.** $\frac{46}{3}$. **33.** 27. **35.** $\frac{8}{3}$.

37. $\frac{216}{5}$. **39.** $\frac{134}{3}$. **41.** $\frac{3}{4}$. **43.** $\sqrt{5} - 1$. **45.** $\frac{\pi}{2}$.

47. $\frac{3\pi}{4}$. **49.** $\frac{1}{8}(\pi - 2)$. **51.** $\frac{4}{3}$. **53.** $\frac{1}{12}(8 - 5\sqrt{2})$. **55.** 1.

57. $\frac{1}{4}(2 - \sqrt{2})$. **59.** $\frac{1}{4}$. **61.** $\frac{1666}{3}$.

Pages 719–720. Sec. 20–3

1. $\frac{1}{5}x^5 + C$. **3.** $r^4 + C$. **5.** $-\frac{1}{x} + C$.

7. $\frac{3}{5}t\sqrt[3]{t^2} + C$. **9.** $\frac{1}{3}x^3 - x^2 + x + C$. **11.** $a^2t + \frac{1}{2}bt^2 - \frac{1}{3}t^3 + C$.

13. $-\frac{a}{5x^5} - \frac{b}{3x^3} - \frac{c}{x} + C$. **15.** $\frac{1}{4}(x - 1)^4 + C$. **17.** $\frac{2}{9}(3x + 2)^{\frac{3}{2}} + C$.

19. $-\frac{1}{15(5x+8)^3} + C$. **21.** $-3\cos x - 4\sin x + C$. **23.** $10\theta - \frac{1}{6}\cos\theta + C$.

25. $\frac{1}{12}(3x^4 + 8x^3 + 6x^2) + C$. **27.** $\sqrt{x^2 + 2} + C$. **29.** $\frac{1}{3}(x^2 - 5)^{\frac{3}{2}} + C$.

31. $-\frac{1}{(2r^2 + 1)^2} + C$. **33.** $-\frac{1}{3}\cos 3x + C$. **35.** $\frac{3}{2}\sin 2\theta - \cos 2\theta + C$.

37. $\frac{3}{2}\sin(2t + 4) + \frac{5}{4}\sin(4t - 6) + C$.

39. $\frac{1}{15}\sqrt{4x - 2}(6x^2 - x - 1) + C$.

41. $\frac{1}{2}\sqrt{t^4 + 4t} + C$. **43.** $\frac{625}{4}$. **45.** $\frac{45}{2}$.

47. $\frac{65}{6}$. **49.** 2. **51.** $\frac{11}{4}m^4$.

53. $\frac{1935}{4}$. **55.** $\frac{2}{\omega}$. **57.** $\frac{4}{15}$.

59. 32 in.

Pages 726–728. Sec. 20–4

1. 57.

3. $\frac{27}{4}$.

5. 2.

7. $\sqrt{3} - \dfrac{\pi}{3}$.

9. $\frac{3}{2}$.

11. $\frac{1}{3}k^2(b^3 - a^3)$, $\frac{1}{3}k^2(c^3 - d^3)$.

13. $k^2\left(\dfrac{1}{a} - \dfrac{1}{b}\right)$, $\dfrac{k^2}{a}$.

15. 40 in.-lb.

17. 10 ft-lb.

19. 6 in.

21. 144 ft, 6 sec.

23. 25 sec, 40,000 ft.

25. 3 sec.

27. 25 sec, 1600 ft per sec.

29. 1600 ft per sec.

31. $\frac{4.5}{4}$.

33. $\frac{1}{5}$.

35. 9.

37. $\frac{7}{60}$ Q erg.

39. $\frac{1}{12}(t_1{}^2 - 8t_1)$, $\frac{1}{36}(t_1{}^3 - 12t_1{}^2)$.

41. 2.42 sec, 94.5 ft.

Index